Pediatric Exercise Medicine

From Physiologic Principles to Health Care Application

Oded Bar-Or, MD, FACSM
Director, Children's Exercise and Nutrition Centre,
McMaster University

Thomas W. Rowland, MD
Baystate Medical Center

Human Kinetics

Library of Congress Cataloging-in-Publication Data

Bar-Or, Oded.
  Pediatric exercise medicine : from physiologic principles to
health care application / Oded Bar-Or, Thomas W. Rowland.
      p. ; cm.
Includes bibliographical references and index.
  ISBN 0-88011-597-1 (Hard cover)
  1. Exercise therapy for children. 2. Exercise for children.
  [DNLM: 1. Exercise--physiology--Adolescent. 2.
Exercise--physiology--Child. 3. Exercise Therapy--Adolescent. 4.
Exercise Therapy--Child. 5. Exertion--Adolescent. 6. Exertion--Child.
WB 541 B223p 2004] I. Rowland, Thomas W. II. Title.
RJ53.E95B37 2004
615.8'2'083--dc22   2003015757

ISBN: 0-88011-597-1

Copyright © 2004 by Oded Bar-Or and Thomas W. Rowland

Permission notices for material reprinted in this book from other sources can be found on pages xiii-xviii.

Acquisitions Editor: Loarn D. Robertson, PhD
Developmental Editors: Rebecca Crist and Renee Thomas Pyrtel
Assistant Editors: Sandria Washington, Ann M. Augspurger, and Kim Thoren
Copyeditor: Joyce Sexton
Proofreader: Erin Cler
Indexer: Betty Frizzéll
Permission Manager: Dalene Reeder
Graphic Designer: Andrew Tietz
Graphic Artists: Angela K. Snyder and Kathleen Boudreau-Fuoss
Photo Manager: Kareema McLendon
Cover Designer: Robert Reuther
Art Managers: Kelly Hendren and Kareema McLendon
Illustrator: Boguslaw Wilk
Printer: Sheridan Books

Printed in the United States of America

10 9 8 7 6 5 4 3 2 1

Human Kinetics
Web site: www.HumanKinetics.com

United States: Human Kinetics
P.O. Box 5076
Champaign, IL 61825-5076
800-747-4457
e-mail: humank@hkusa.com

Canada: Human Kinetics
475 Devonshire Road Unit 100
Windsor, ON N8Y 2L5
800-465-7301 (in Canada only)
e-mail: orders@hkcanada.com

Europe: Human Kinetics
107 Bradford Road
Stanningley
Leeds LS28 6AT, United Kingdom
+44 (0) 113 255 5665
e-mail: hk@hkeurope.com

Australia: Human Kinetics
57A Price Avenue
Lower Mitcham, South Australia 5062
08 8277 1555
e-mail: liaw@hkaustralia.com

New Zealand: Human Kinetics
Division of Sports Distributors NZ Ltd.
P.O. Box 300 226 Albany
North Shore City
Auckland
0064 9 448 1207
e-mail: blairc@hknewz.com

To Marilyn and Margot

CONTENTS

Chapter 3   Climate, Body Fluids, and the Exercising Child   69

PART II   CLINICAL PERSPECTIVES OF CHILDREN AND EXERCISE

Chapter 4   Children and Exercise in a Clinical Context—an Overview   105

Chapter 5   Physical Activity and Preventive Health Care in Children and Adolescents   117

PART III   EXERCISE AND PEDIATRIC DISEASES

Chapter 6   Pulmonary Diseases   139

Chapter 7   Cardiovascular Diseases   177

Chapter 8   Endocrine Diseases   219

Chapter 9   Nutritional Diseases   237

Chapter 10   Neuromuscular and Musculoskeletal Diseases   269

Chapter 11   Hematologic, Oncologic, and Renal Diseases   305

Chapter 12   Emotional and Mental Disorders   323

PREFACE

The preface to *Pediatric Sports Medicine for the Practitioner*, which was the first incarnation of this book, stated: "Much knowledge has been generated in recent years by scientists investigating the triad: *child-exercise-health*. Yet little of this information is available in pediatric textbooks for *application* by the clinician." Today, 21 years later, an abundance of scientific knowledge has accumulated on the relationship between children's health and physical activity/inactivity. Nevertheless, relatively little information has been incorporated into pediatric textbooks. The main objective of this second edition is, therefore, to further bridge the gap between the exercise scientist and the health practitioner.

Because of methodologic and ethical constraints, research in pediatric exercise physiology and its health implications has lagged behind that conducted with adults. However, since the mid-1970s interest in the effects of exercise on children has grown considerably. For example, a Medline search for the combination *children* and *exercise* identified 84 peer-reviewed articles a year during 1970 to 1975, 221 articles a year during 1985-1989, and 494 articles a year in 2000-2002. Such a major surge of interest has also been shown for the relationship between exercise and pediatric disease. For example, a Medline search for *children* and *exercise* and *obesity* yielded four peer-reviewed articles a year for the period 1970-1974. This has risen to 20 articles a year in 1985-1989 and has shot up to 104 articles a year in 2000-2002—a staggering 26-fold increase in three decades.

One reason for the growing interest in pediatric exercise medicine is the increasing involvement of children and youth in high-level sports. Today's prepubertal athlete is often exposed to training regimens that a decade or two ago were considered too demanding even for adult athletes. In some sports, such as female gymnastics, children and adolescents excel and reach world standards. In other sports, such as tennis or diving, athletes usually do not reach their peak before the third decade of life, but their specialized training might start as early as the first decade.

Another reason for this information explosion is the increasing interest in the role played by an inactive lifestyle in the current obesity epidemic and its comorbidities, and the possible use of enhanced physical activity in the treatment of obesity and other chronic pediatric diseases. A growing number of pediatric medical centers are developing exercise laboratories and counseling services to explore and explain the beneficial and detrimental effects of exercise on children's health.

In 1983, when *Pediatric Sports Medicine for the Practitioner* appeared, the bulk of information regarding physiologic and clinical aspects of pediatric exercise was summarized in seven proceedings of the European Group of Pediatric Work Physiology. This group has since published eleven more proceedings. Another boost of interest in pediatric exercise research occurred in 1986, with the establishment of the North American Society for Pediatric Exercise Medicine. In addition to its biennial scientific meetings, the society initiated the *Pediatric Exercise Science* journal that has now reached its 15th year.

This book is divided into three parts. *Part I—Exercise Physiology of the Healthy Child* (chapters 1-3) addresses the physiologic responses to exercise, as well as habitual physical activity, in healthy children. It focuses on the effects of growth and development, comparing responses to exercise and activity patterns of children, adolescents, and adults. *Part II—Clinical Perspectives of Children and Exercise* (chapters 4-5) provides an overview of the relevance of exercise to children's current and future health. It highlights the effects of physical activity/inactivity for child health, the effects of illness on physical activity and fitness, and the relationship between physical activity in early years and the development of risk factors for adulthood chronic diseases. *Part III—Exercise and Pediatric Diseases* (chapters 6-12) represents the main part of this book. It provides an in-depth analysis of the relationships between specific diseases, physical activity, and physical fitness. Individual chapters cover pulmonary, cardiovascular, endocrine, nutritional, neuromuscular, musculoskeletal, hematologic, oncologic, and renal diseases, as well as emotional and mental disorders. The book concludes with six *appendixes* that provide normative data for fitness performances, methodologic aspects of exercise testing and of the assessment of physical activity and energy expenditure, the energy equivalents of various activities and sports, the principles of scaling for size, and a glossary of terms often used in exercise physiology.

The book is intended for a wide variety of health professionals whose young clientele requires counseling regarding physical activity and sports. These include pediatricians, family physicians, general practitioners, physiatrists, physical therapists, nutritionists and kinesiologists, as well as school medical personnel. In addition, the extensively referenced chapters will be useful to scientists and graduate and undergraduate students in exercise physiology.

The preface to the first edition stated that "to many clinicians, exercise, as a clinical entity, is a 'black box' yet to be opened." It is our hope that this second edition, like its predecessor, will stimulate health professionals to further integrate physical activity and exercise into their overall strategy of child care.

ACKNOWLEDGMENTS

A large number of people inspired and helped us in the production of this book. We are particularly grateful to several of them, who made it really happen. First, we thank Rainer Martins, who several years ago came up with the idea of publishing a second edition of *Pediatric Sports Medicine for the Practitioner*. Rainer has shown patience and perseverance in making sure that we bring this idea to fruition. Marilyn Bar-Or has been most instrumental in going over the text and providing linguistic advice and constructive criticism. Boguslaw ("Bogdan") Wilk has worked beyond the call of duty, and, with much dedication, to produce the graphs. Moreover, Bogdan's work provided much of the information regarding dehydration–rehydration. Shirley Lampan has shown great devotion over the years in assisting at the various stages of the manuscript preparation. Last but not least, our gratitude to the graduate students, post-doctoral Fellows and visiting scientists at the Children's Exercise and Nutrition Centre, whose original work and ideas have provided a wealth of material for this book. These include Cameron ("Joe") Blimkie, Randy Calvert, Pascale Duché, Bareket Falk, Gail Frost, John Hay, Helge Hebestreit, Heidi Keller, Susi Kriemler, Desiree Maltais, Marilyn McNiven, Flavia Meyer, Dawn Parker, Michael Riddell, Eric Small, Kensaku Suei, Brian Timmons, Vish Unnithan, Edgar Van Mil, Bea Volpe-Ayub, Dianne Ward, and Brian Wilson.

CREDITS

Figure 1.3 Reprinted, by permission, from H.C.G. Kemper, R. Verschuur, and L. de Mey, 1989, "Longitudinal changes of aerobic fitness in youth ages 12 to 13," *Pediatric Exercise Science* 1: 257-270.

Figure 1.6 Reprinted, by permission, from S. Zanconato, D.M. Cooper, and Y. Armon, 1991, "Oxygen cost and oxygen uptake dynamics and recovery with 1 min of exercise in children and adults," *Journal of Applied Physiology* 71: 993-998.

Figure 1.9 Reprinted, by permission, from G. Frost, J. Dowling, K. Dyson, and O. Bar-Or, 1997, "Cocontraction in three age groups of children during treadmill locomotion," *Journal of Electromyology and Kinesiology* 7: 179-186.

Figure 1.10 Reprinted, by permission, from J.M. Hausdorff, L. Zemany, C.-K. Peng, and A.L. Goldberger, 1999, "Maturation of gait dynamics: Stride-to-stride variability and its temporal organization in children," *Journal of Applied Physiology* 86: 1040-1047.

Figure 1.11 Adapted, by permission, from O. Bar-Or, 1982, "Physiologische gesetzmassigkeiten sportlicher aktivitat beim kind." In *Kinder im leistungssport*, edited by H. Howald and E. Han (Basel: Birkhauser), 18-30.

Figure 1.13 Adapted, by permission, from O. Inbar and O. Bar-Or, 1986, "Anaerobic characteristics in male children and adolescents," *Medicine in Science and Sports Exercise* 18: 264-226.

Figure 1.14a and b Reprinted, by permission, from E. Dore et al., 2000, "Dimensional changes cannot account for all differences in short-term cycling power during growth." *International Journal of Sports Medicine* 21: 360-365.

Figure 1.16 Adapted, by permission, from J.C. Martin and R.M. Malina, 1998, "Developmental variations in anaerobic performance associated with age and sex." In *Pediatric anaerobic performance*, edited by E. Van Praagh (Champaign, IL: Human Kinetics), 45-64.

Figure 1.20 Adapted, by permission, from S. Zanconato et al., 1993, "31P Magnetic resonance spectroscopy of leg muscle metabolism during exercise in children and adults," *Journal of Applied Physiology* 74: 2214-2218.

Figure 1.22 Adapted, by permission, from Y. Armon et al., 1991, "Oxygen uptake dynamics during high-intensity exercise in children and adults," *Journal of Applied Physiology* 70: 841-848.

Figure 1.23a and b Reprinted, by permission, from O. Bar-Or, 1982, "Physiologische gesetzmassigkeiten sportlicher aktivitat beim kind." In *Kinder im leistungssport*, edited by H. Howald and E. Han (Basel: Birkhauser), 18-30.

Figure 1.24a and b Reprinted, by permission, from H. Hebestreit, K. Mimura, and O. Bar-Or, 1993, "Recovery of muscle power after high-intensity short-term exercise: Comparison between boys and men," *Journal of Applied Physiology* 74: 2875-2880.

Figure 1.29a Reprinted, by permission, from K.R. Turley and J.H. Wilmore, 1997, "Cardiovascular responses to treadmill and cycle ergometer exercise in children and adults," *Journal of Applied Physiology* 83: 948-957.

Figure 1.30 Reprinted, by permission, from Wirth et al., 1978, "Cardiopulmonary adjustment and metabolic response to maximal and submaximal physical exercise of boys and girls at different stages of maturity," *European Journal of Applied Physiology* 29: 229-240.

Figure 1.34 Reprinted, by permission, from H. Hebestreit et al., 1995, "Climate-related corrections for improved estimation of energy expenditure from heart rate in children," *Journal of Applied Physiology* 79: 47-54.

Figure 1.38 Reprinted, by permission, from K.R. Turley and J.H. Wilmore, 1997, "Cardiovascular responses to treadmill and cycle ergometer exercise in children and adults," *Journal of Applied Physiology* 83: 948-957.

Figure 1.40a-d Reprinted, by permission, from T. Reybrouck et al., 1985, "Ventilatory anaerobic threshold in healthy children: Age and sex differences," *European Journal of Applied Physiology* 54: 278-284.

Figure 1.41 Reprinted, by permission, from P.O. Åstrand, 1952, *Experimental studies of physical working capacity in relation to sex and age* (Copenhagen: Munksgaard), 86.

Figure 1.44 Reprinted, by permission, from K. Froberg and O. Lammert, 1996, "Development of muscle strength during childhood." In *The child and adolescent athlete*, edited by O. Bar-Or (London: Blackwell Scientific), 29.

Figure 1.45a Reprinted, by permission, from R.M. Malina, C. Bouchard, and O. Bar-Or, 2003, *Growth, maturation and physical activity* (Champaign, IL: Human Kinetics), 2nd ed., 356.

Figure 1.46 Adapted, by permission, from L.R. Martinez and E.M. Haymes, 1992, "Substrate utilization during treadmill running in prepubertal girls and women," *Medicine in Science and Sports Exercise* 24: 975-983.

Figure 1.47 Adapted, by permission, from B.W. Timmons, O. Bar-Or, and M.C. Riddell, 2003, "Oxidation rate of exogenous carbohydrate during exercise is higher in boys than in men," *Journal of Applied Physiology* 94: 278-284.

Figure 1.48 Reprinted, by permission, from O. Inbar and O. Bar-Or, 1975, "The effects of intermittent warm-up on 7-9 year-old boys," *European Journal of Applied Physiology* 34: 81-89.

Figure 1.49 Reprinted, by permission, from O. Inbar and O. Bar-Or, 1975, "The effects of intermittent warm-up on 7-9 year-old boys," *European Journal of Applied Physiology* 34: 81-89.

Figure 1.52 Reprinted, by permission, from B.W. Timmons and O. Bar-Or, 2003, "RPE during prolonged exercise with and without carbohydrate ingestion in boys and men," *Medicine in Science and Sports Exercise* 35: 1901-1907.

Figure 1.54 Reprinted, by permission, from R.R. Pate and D.S. Ward, 1996, "Endurance trainability of children and youths." In *The child and adolescent athlete*, edited by O. Bar-Or (Oxford: Blackwell Scientific), 130-152.

Figure 1.56 Reprinted, by permission, from C.J.R. Blimkie and O. Bar-Or, 1996, "Trainability of muscle strength, power and endurance during childhood." In *The child and adolescent athlete*, edited by O. Bar-Or (Oxford: Blackwell Scientific), 115.

Figure 1.57 Reprinted, by permission, from T. Rowland et al., 1999, "Physiological determinants of maximal aerobic power in healthy 12-year-old boys," *Pediatric Exercise Science* 11: 317-326.

Table 1.5 Reprinted, by permission, from H. Hebestreit, K. Mimura, and O. Bar-Or, 1993, "Recovery of muscle power after high-intensity short-term exercise: Comparison between boys and men," *Journal of Applied Physiology* 74: 2875-2880.

Table 1.10 Reprinted, by permission, from R.M. Malina, C. Bouchard, and O. Bar-Or, 2003, *Growth, maturation and physical activity* (Champaign, IL: Human Kinetics).

Figure 3.2 Reprinted, by permission, from B. Falk, O. Bar-Or, and J.D. MacDougall, 1992, "Thermoregulatory responses of pre-, mid-, and late-pubertal boys to exercise in dry heat," *Medicine in Science and Sports Exercise* 24: 688-694.

Figure 3.5 Reprinted, by permission, from O. Bar-Or, 1989, "Temperature regulation during exercise in children and adolescents." In *Perspectives in exercise science and sports medicine, Vol. 2 Youth, exercise, and sport*, edited by C.V. Gisolfi and D.R. Lamb (Indianapolis: Benchmark Press), 335-367.

Figure 3.6a Reprinted, by permission, from O. Bar-Or, 1989, "Temperature regulation during exercise in children and adolescents." In *Perspectives in exercise science and sports medicine, Vol. 2 Youth, exercise, and sport*, edited by C.V. Gisolfi and D.R. Lamb (Indianapolis: Benchmark Press), 335-367.

Figure 3.7a and b From T. Araki et al., 1979, "Age differences during sweating during muscular exercise," *Journal of Fitness and Sports Medicine* 28: 239-248. Reprinted, by permission, from O. Bar-Or, 1989, "Temperature regulation during exercise in children and adolescents." In *Perspectives in exercise science and sports medicine, Vol. 2. Youth, exercise, and sport*, edited by C.V. Gisolfi and D.R. Lamb (Indianapolis: Benchmark Press), 335-367. By permission from O. Bar-Or.

Figure 3.8 Reprinted, by permission, from B. Falk et al., 1991, "Sweat lactate in exercise in children and adolescents of varying physical maturity," *Journal of Applied Physiology* 71: 1735-1740.

Figure 3.9 From O. Bar-Or, 1989, "Temperature regulation during exercise in children and adolescents." In *Perspectives in exercise science and sports medicine, Vol. 2 Youth, exercise, and sport*, edited by C.V. Gisolfi and D.R. Lamb (Indianapolis: Benchmark Press), 335-367. By permission from O. Bar-Or.

Figure 3.12 Reprinted, by permission, from O. Bar-Or, 1980, "Climate and the exercising child — A review," *International Journal of Sports Medicine* 1: 53-65.

Figure 10.20a Reprinted, by permission, from B. Falk et al., 1997, "Birth weight and physical ability," *Medicine in Science and Sports Exercise* 29: 1124-1130.

Figure 10.20b Reprinted, by permission, from H. Keller et al., 1998, "Neuromotor ability," *Developmental Medicine and Child Neurology* 40: 661-666.

Figure 10.22 Adapted, by permission, from P.N. Malleson et al., 1996, "Physical fitness and its relationship to other indices of health status in children with chronic arthritis," *Journal of Rheumatology* 23: 1059-1065.

Figure 10.25 Adapted, by permission, from J.C. Agre et al., 1987, "Physical activity capacity in children with myelomeningocele," *Archives of Physical Medicine and Rehabilitation* 68: 372-377.

Figure 10.31 Reprinted, by permission, from O. Bar-Or, 1993, "Noncardiopulmonary pediatric exercise tests." In *Pediatric laboratory exercise testing: Clinical guidelines*, edited by T.W. Rowland (Champaign, IL: Human Kinetics), 165-185.

Table 10.1 Adapted, with permission, from D.F. Parker, 1993, "Muscle performance and gross motor function of children with spastic cerebral palsy," *Developmental Medicine and Child Neurology* 35: 17-23.

Figure 11.1 Reprinted, by permission, from O. Bar-Or, 1983, *Pediatric sports medicine for the practitioner* (New York: Springer-Verlag), 252.

Figure 11.2 Reproduced with permission by the American Journal of Clinical Nutrition 30: 910-917. © 1977 AM J Clin Nutr. American Society for Clincial Nutrition.

Figure 11.3a and b Reprinted from Journal of Pediatrics, Vol. 94, P.R. Dallman and M.A.Siimes, "Percentile curves for hemoglobin and red cell volume in infancy and childhood," 26-3, Copyright (1979), with permission of Elsevier.

Table II.12 Reproduced, with permission, from G. Borg, 1998, *Borg's perceived exertion pain scale* (Champaign, IL: Human Kinetics), 47.

Figure II.4 Reprinted, by permission, from R. Robertson, 2004, Perceived exertion for practitioners: Rating effort with the OMNI picture system (Champaign, IL: Human Kinetics), 145.

Exercise Physiology of the Healthy Child

Physiologic and Perceptual Responses to Exercise in the Healthy Child

*T*o understand the sick child's response to exercise, one must first be familiar with the "normal" physiologic and perceptual responses to exercise. Whether adapting to a single bout of exercise or to repeated exercise stimuli, the child—like the adult—undergoes physiologic changes. The basic premise of this chapter is that, although such changes take place at all ages, there are growth- and maturation-related differences in response to exertion. We do not expect a 6-year-old child to run as fast or as far as a teenager, who in turn is slower and weaker than a young adult. Nor will a child have the same muscle strength as the more mature individual. On the other hand, children often perceive exercise intensities to be lower, and they recover faster from the strain of exercise, compared with adults.

Physiologic capacities have long been recognized as dependent on body and system dimensions. Without morphologic growth of the myocardium, for example, its contractile force cannot be high enough to pump sufficient blood to the growing periphery. Similarly, when the bone scaffolding of a teenager is growing, body strength or mechanical power will not follow suit unless muscle mass also develops, and this in itself may depend on hormonal and other pubertal changes.

It is beyond the scope of this chapter to methodically analyze the physiologic changes

that occur with exercise in each and every system. These have been discussed thoroughly in general texts of exercise physiology. Nor does this chapter recite the fundamentals of growth and development; the interested reader may resort to standard texts on this topic (291). The following sections present general concepts of exercise physiology and elaborate on the differences in physiologic and perceptual responses to exercise between children and older age groups. We also indicate how such differences may affect the physical capability of the child and thereby set a limit on his or her performance. The chapter concludes with a review of the effects of training on children's physiologic functions and physical performance.

Our understanding of children's physiologic responses to exercise is still deficient. More so than with adults, we are limited by ethical considerations and by methodologic constraints. There are very few investigators who would, for example, puncture the artery of a child, take a needle biopsy of a child's muscle, or insert a thermistor into the esophagus merely to satisfy curiosity. To study environmental effects, one cannot readily expose children to "hostile environments" where extreme cold, high heat and humidity, or hypoxic conditions prevail. Nor can adequate animal models be set up to study age-related differences in trainability, motor learning, or thermoregulation. Pediatric exercise physiologists are still seeking instruments and protocols appropriate for body size, body proportions, level of motivation, emotional state, and attention span of the young child, especially the preschooler. Many studies have borrowed concepts, methods, and instruments that are suitable for adults but not for children. In spite of recent technological breakthroughs, there still is a need to develop child-friendly methodologies in pediatric exercise sciences.

Owing to these constraints, our state of the art is still limited to some knowledge of the cardiovascular, pulmonary, and locomotor systems; general concepts of energy transfer and metabolism; and some phenomena of thermoregulation and body fluid shifts. Fewer data are available on hormonal, immunological, muscle cell, subcellular, or molecular phenomena. Nor have we begun to understand processes within the central or peripheral nervous system of the exercising child.

This chapter therefore focuses on the metabolic, cardiovascular, pulmonary, and musculoskeletal systems. A detailed discussion of thermoregulation is presented in chapter 3.

Size-Dependent and Size-Independent Differences

Growth and development are accompanied by major anatomical and physiologic changes. With the increase in size of body tissues and organs, there is a change in physiologic function. Understanding the relationships between function and size is extremely important in pediatric research and clinical practice. It would otherwise be hard to interpret differences among children, to decide whether a certain value is within normal, or to assess changes that occur over time in the same individual. For example, which child has a higher local muscle endurance: a girl 120 cm tall whose total work in the Wingate anaerobic test is 2.4 kilojoules (kJ), or a boy 145 cm tall with total work of 6 kJ? Or let us take a boy with muscle hypotonia who is prescribed a muscle-strengthening program. At the start of the program, when his body weight is 40 kg, the peak torque of his knee extensors is 100 newton-meters (N·m). One year

later, at a body weight of 48 kg, his peak torque is 140 N·m. Can we claim that the program was efficacious, or does the increase in strength merely reflect larger muscles?

There are three basic patterns to the function–size relationship in the growing individual. Some functional changes are proportional to changes in size (e.g., muscle strength increases as a function of its cross-sectional area, and lung diffusion capacity is related to the surface area of the alveoli walls). Others are related to qualitative changes as well as size (e.g., anaerobic muscle power, which depends on activity of rate-limiting enzymes as well as on muscle volume; or economy of locomotion, which is related to the degree of co-contraction of antagonist muscles as well as to leg length). Then there are physiologic and chemical characteristics completely independent of size (e.g., serum electrolyte level, or O_2 content in the arterial blood). These three patterns are summarized in figure 1.1. Table 1.1 lists examples of exercise-related physiologic functions in each of these patterns.

One way to determine the effect of body size on physiologic functions is through scaling. A detailed discussion of the various approaches to scaling is available in appendix V, "Scaling for Size Differences."

Metabolic Responses to Exercise in Children

For mechanical energy to be released at the myofibrillar level and effect muscle contraction, splitting of adenosine triphosphate (ATP) must take place. This high-energy compound is available in small quantities (about 4-5 mmol · kg^{-1} wet weight) in the resting muscle. However, once contractions start, there is an immediate need for reinforcement of ATP. This can be supplied from (1) limited stores of creatine phosphate (CP), (2) anaerobic glycolysis, or (3) the tricarboxylic acid cycle. The former two sources do not require addition of O_2 and are therefore called anaerobic. The latter requires O_2 and is termed aerobic. Muscle contractions that result from anaerobic reactions cannot be sustained longer than 40 to 50 sec. In contrast, muscle contractions utilizing aerobic energy turnover can last many minutes or even hours, albeit at a lower intensity than with

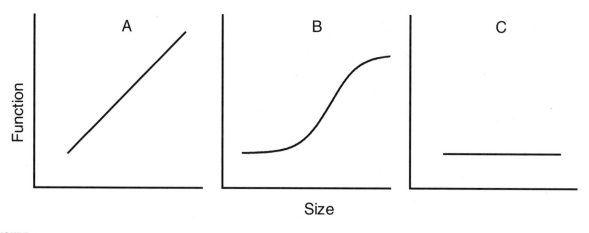

FIGURE 1.1   Relationship between function and body size. Three patterns are summarized schematically: Function is related to size (a); function is related to qualitative differences as well as size (b); and function is totally independent of size (c). Note: Graph (a) suggests a linear relationship with size, but the relationship can also be nonlinear.

TABLE 1.1   Exercise-Related Physiologic Functions and Their Relationship to Size

Predominant relationship to size	Relationship to qualitative differences and size	No relationship to size
Muscle strength	O_2 cost of locomotion	Arterial O_2 content
Maximal O_2 uptake	Anaerobic performance	Serum electrolyte level
Maximal minute ventilation	Sweating rate	Core body temperature
Lung diffusion capacity	Resistance to cold-induced hypothermia	Heat content of tissues
Rate of body heat loss	Resting metabolic rate	Rate of enzymatic action

anaerobic exercise. Even though most activities utilize both aerobic and anaerobic pathways, in the terminology of sport scientists physical tasks are subdivided into "aerobic"-type and "anaerobic"-type activities. The former include distance running, swimming, cycling, cross-country skiing, and other endurance-requiring tasks. The latter include sprinting, jumping, throwing, and other sports in which the required power intensity is high and the duration short. A question commonly asked is whether, compared with adults, children are characteristically aerobic or anaerobic performers. We shall try to offer an answer in the following sections.

Maximal Aerobic Power

The most commonly used index of maximal aerobic power has been maximal O_2 uptake, which is the highest volume of oxygen that can be consumed by the body per time unit. This value reflects the highest metabolic rate made available by aerobic energy turnover. Figure 1.2 presents cross-sectional data of the relationship between maximal O_2 uptake and chronologic age in 3,910 girls and boys, 6 to 18 years old. It is evident that, with the growth of the child, there is a concomitant increase in his or her maximal O_2 uptake. Until age 12, values grow at the same rate in both sexes, even though boys have higher values as early as age 5 (482). Maximal O_2 uptake of boys keeps increasing until the age of about 17 to 18 but increases hardly at all beyond age 14 in girls. An example of this difference is depicted in figure 1.3, which summarizes longitudinal changes in maximal O_2 uptake between ages 13 and 21 years (240).

While maximal O_2 uptake depends on respiratory and hemodynamic factors (see related sections later in this chapter and figure 4.5), it is also related to the oxidative enzymatic activity in the exercising muscles and to the size of these muscles (100; 116; 120). Because muscle size is hard to measure, most workers in the field have used an estimate of fat-free mass instead (21; 84; 147; 240; 336).

Earlier studies suggested that gender-related differences in maximal aerobic power can be explained by differences in muscle size. Indeed,

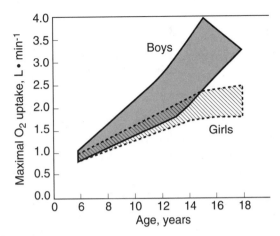

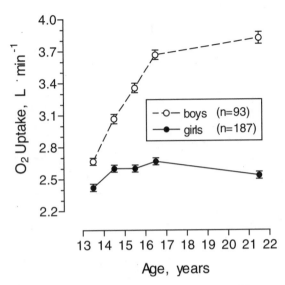

FIGURE 1.2   Maximal aerobic power and age. Absolute values of maximal O_2 uptake in girls (n = 1,730) and boys (n = 2,180) 6 to 18 years old. Each dot represents a mean of a group. The shaded areas were constructed by eyeball technique to indicate a general trend. Data by Andersen and Magel 1970 (8); Andersen et al. 1974 (11); Åstrand 1952 (25); Bar-Or et al. 1971 (45); Bar-Or and Zwiren 1973 (47); Chatterjee et al. 1979 (92); Ekblom 1969 (133); Gaisl and Buchberger 1977 (177); Hermansen and Oseid 1971 (208); Ikai et al. 1970 (217); Kobayashi et al. 1978 (245); MacDougall et al. 1979 (272); Máček et al. 1979 (275); Mocellin 1975 (311); Nagle et al. 1977 (321); Robinson1938 (360); Seliger 1970 (407); Shephard et al. 1969 (410); and Thoren 1967 (431).

FIGURE 1.3   Longitudinal changes in maximal O_2 uptake in 107 girls and 93 boys who took part in the Amsterdam Growth and Health of Teenagers Study. Vertical lines denote SEM.

Adapted, with permission, from Kemper et al. 1989 (240).

once maximal O_2 uptake was related to lean leg volume, there were no differences between boys and girls, as shown in figure 1.4 (116). More recent work, using allometric scaling (for details, see appendix V, p. 381) and magnetic resonance estimates of muscle volume, suggests that muscle volume, in itself, may not explain the higher aerobic performance of boys (18; 464).

When the maximal O_2 uptake of adolescents of different ages but the same body weight or body height is compared, it is positively related to age (422). Longitudinal studies, based on scaling by regression (19) or by multilevel modeling (21), suggest that maximal O_2 uptake increases with the pubertal stage of the child. The implication is that maximal aerobic power depends also on maturity and not only on body dimensions.

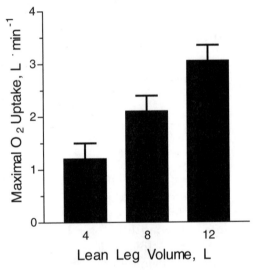

FIGURE 1.4   Maximal aerobic power and lean leg volume. Maximal O_2 uptake, determined by a progressive upright cycle ergometry test, in relationship to lean leg volume as assessed by length and circumference measurements and corrected for skinfold thickness. Subjects were 92 girls and boys.

Adapted from Davies et al. 1972 (116).

Maximal O_2 Uptake per Kilogram Body Mass

On the basis of the preceding discussion, one might conclude that maximal aerobic power is less developed in children than at older ages. For tasks that involve moving the whole body from one place to another, however, the child, whose body mass is smaller, may not need as high an absolute maximal O_2 uptake as the heavier adolescent or adult.

Although on theoretical grounds, ratio scaling is not the ideal way to compare the maximal aerobic power of people who differ in body size (see appendix V), this approach has been used in most studies. The most convenient, and traditionally accepted, way to express maximal O_2 uptake has been relative to body mass, height, surface area, or fat-free mass. A majority of studies still express maximal O_2 uptake per kilogram body mass. Such a comparison is shown in figure 1.5. While there is hardly any age-related change in maximal O_2 uptake in the boys, this function continuously declines among the girls. Such a decline may reflect an increase in body adiposity (and hence a relative decrease in fat-free mass) of girls during adolescence.

A person's maximal O_2 uptake depends on the ability of the pulmonary and cardiovascular systems to transport oxygen from the ambient air to the body cells. A discussion of the response of these systems to exercise is provided later in this chapter.

Mechanical Efficiency and Economy of Movement

When muscle contraction results in movement, external mechanical work is produced. The energy equivalent of this work is only some 20% to 25% of the chemical energy utilized during the contraction. The other 75% to 80% is converted into heat.

Mechanical efficiency (ME) is the ratio between external mechanical work (W) produced by the muscle and the chemical energy (E) utilized during the contraction. Because some energy (e) is required by the muscle in its resting state, the net energy utilized during contraction is E minus e. Converting the numerator and the denominator to the same power units (e.g., Watts, or kilocalories per minute), one obtains a nondimensional ratio for ME. When multiplied by 100, ME can then be presented in a percentage:

$$ME(\%) = W \times 100 / (E - e)$$

When work is done during a known period of time, power units rather than work units can be used in the numerator and denominator.

To assess the mechanical efficiency of the body as a whole, one must calculate the mechanical power produced by the body and the chemical power needed for that activity. The latter is done conveniently by measuring O_2 uptake and assigning 5 kcal (21 kJ) for each liter of O_2. (This value varies somewhat depending on the fuel source. It is 4.70 kcal $\cdot$ L^{-1} for fats and 5.05 kcal $\cdot$ L^{-1} for carbohydrates.) O_2 uptake at rest must be subtracted from that measured during exercise to obtain net O_2 uptake, as follows:

$$ME = \text{mechanical power} / (\text{exercise metabolic rate} - \text{resting metabolic rate})$$

One can also determine mechanical efficiency by calculating the increment in O_2 uptake, which accompanies a known increment in mechanical power. This approach obviates the need to measure resting metabolic rate (e.g., [102]).

The measurement of mechanical power poses methodologic problems. While one can determine it with good accuracy during cycle ergometry (assuming a well-calibrated ergometer and negligible power loss between the pedals and the flywheel), its assessment in other activities is more difficult. Although equations are available to calculate power output during walking or running, based on speed, slope, and body mass, these equations disregard differences in gait. Such differences exist, for example, in the extent of vertical displacement of the body (or its segments), energy transfer within and between body segments, and degree of lateral pelvic tilt, which are not included in most calculations of power output.

Improved models for determining total body mechanical power during walking or running were developed in recent years on the basis of sophisticated kinematic analysis (170). However, there is an ongoing debate among biomechanists

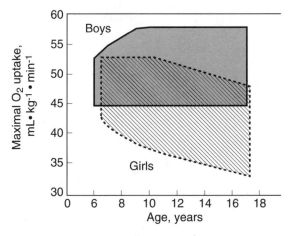

FIGURE 1.5 Maximal aerobic power per kilogram body mass and age. Data are for girls (n = 1,730) and boys (n = 2,180). Subjects, symbols, and sources are the same as in figure 1.2.

regarding the validity of these models. This lack of uncertainty detracts from our ability to use a valid value in the numerator of the mechanical efficiency equation.

Thus, interindividual differences in O_2 uptake during walking or running do not necessarily denote differences in mechanical efficiency at the cellular level, but rather a difference in the economy of locomotion (233; 316; 400). The same concept applies in other activities, such as swimming, in which O_2 uptake varies markedly among individuals due to different levels of economy of motion.

In this book we use the term "mechanical efficiency" only when both the numerator and the denominator in the equation are measurable. We use "economy of motion" instead for data in which the numerator has not been accurately determined. The concept of economy of motion is important for understanding the differences in exercise performance between children and other age groups, as well as between the genders and in children with various disabilities.

Age-Related Differences in the Metabolic Cost of Exercise

Studies in the 1950s and 1960s suggested that mechanical efficiency of cycling is similar in children, adolescents, and adults, ranging between 18% and 30% (average 25%) (21; 53; 183; 407; 410; 430; 474). More recent work, however, provides evidence that at high exercise intensities, the O_2 cost of cycling is higher in children (204; 484). For example, when O_2 uptake of cycling above the ventilatory threshold and the O_2 uptake during recovery were combined, the cumulative cost for a given mechanical output was higher in children than in adults (figure 1.6) (484). The cause for this pattern is not clear.

In contrast, there is unanimity among researchers that the O_2 cost during walking and running is higher in children when expressed per kilogram body mass (14; 25; 132; 171; 183; 254; 273; 293; 300; 315; 319; 360; 378; 383; 418; 442). This pattern is shown in figure 1.7, which presents submaximal O_2 uptake of 5- to 18-year-old girls and boys who ran on a treadmill at various speeds (25). At 10 km · hr^{-1}, for example, there was an 8 mL · kg^{-1} · min^{-1} (20%) greater metabolic cost for the 5-year-old child than for the 17-year-old adolescent. A similar difference is shown in figure 1.8 for walking at various uphill inclines.

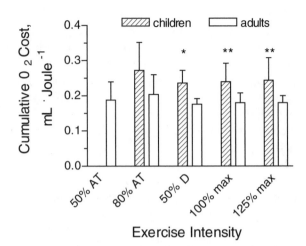

FIGURE 1.6 Cumulative O_2 cost for given mechanical work at various cycling intensities. The combined cost during and following various exercise intensities is presented in relationship to the respective mechanical work at each intensity. Subjects were ten 7- to 11-year-old children and thirteen 26- to 42-year-old adults. AT = anaerobic threshold; 50%Δ = midway between the anaerobic threshold and maximal O_2 uptake.

Reproduced, with permission, from Zanconato et al. 1991 (484).

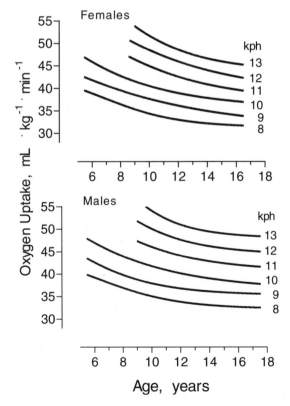

FIGURE 1.7 Submaximal O_2 uptake and age. Sixty-seven girls and 72 boys, 4 to 18 years old, ran on a treadmill at various speeds. Based on Åstrand 1952 (25).

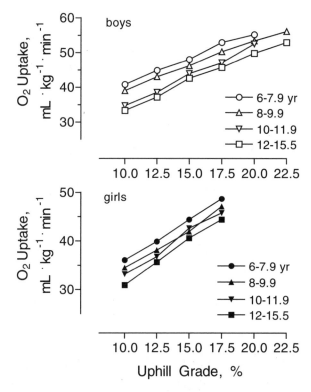

FIGURE 1.8 Oxygen cost of walking at various slopes. Mean values for 6- to 15-year-old girls (n = 64) and boys (n = 83) subdivided into four age groups. Subjects walked at 5 to 6 km · hr⁻¹ on a treadmill. Based on Skinner et al. 1971 (418).

Sallis et al. compiled data from various studies to quantify the excessive metabolic cost of locomotion in children and adolescents compared with young adults (393). As summarized in table 1.2, 5-year-old children need, on the average, 37% more oxygen (calculated per body mass unit) than do adults who perform the same walking or running tasks. This excess decreases with age.

Ariens et al. provide longitudinal observations about the economy of running. Ninety-eight females and 84 males performed identical treadmill runs (8 km · hr⁻¹ at various slopes) at ages 13, 14, 15, 16, 21, and 27 years. Their O_2 uptake per kilogram body mass decreased consistently as they grew, suggesting an increase in economy. Economy in the females was higher than in the males. These differences persisted even when allometric analysis was used to describe the possible effect of body size (14). A higher running economy for females was also found in 6-year-old children who ran on a treadmill at 8 km · hr⁻¹ (317).

A high energy cost of locomotion is also apparent in patients with neuromuscular (e.g., cerebral palsy, muscular dystrophy) or musculoskeletal (e.g., advanced kyphoscoliosis) diseases, as described in chapter 10.

Possible Explanations for the Low Economy of Locomotion in Children

As outlined in the list below, several mechanisms have been suggested to explain the higher metabolic cost of locomotion in children. Among these, the most likely are a higher resting metabolic rate, a higher stride frequency, mechanically "wasteful" locomotion style, and excessive co-contraction of antagonist muscles.

Possible causes for the high metabolic energy cost of walking and running in children follow:

Higher resting metabolic rate

Higher ventilatory cost

Higher stride frequency

Lower storage of elastic recoil forces

Greater moments of inertia due to a more distal mass distribution

Mass-contraction speed imbalance

Mechanically "wasteful" locomotion style

High co-contraction of antagonist muscles

High Resting Metabolic Rate

Can high metabolic cost during exercise be explained by a high resting metabolic rate, or is the net metabolic cost in children high irrespective

TABLE 1.2	Excess O_2 Cost of Locomotion per Kilogram Body Mass in Children of Various Ages Compared With Young Adults

Age, years	Excess cost, %
5	37
7	26
9	19
11	13
13	9
15	5
17	3

Adapted from Sallis et al. 1991 (393).

of their resting metabolic rate? When calculated per kilogram body mass, resting metabolic rate is lower in children than in adolescents, and even more so than in adults (25; 83; 99; 273; 360). Reported differences in resting metabolic rate between children and adults range from 1 to 2 ml · kg^{-1} · min^{-1} (273) to 3 to 3.5 ml · kg^{-1} · min^{-1} (360). In contrast, differences in exercise metabolic rates reach as much as 7 to 8 ml · kg^{-1} · min^{-1} (e.g., [25]). As a result, the net O$_2$ cost during walking or running is higher in children than in adolescents or adults, and a high resting O$_2$ cost cannot explain children's lower economy of locomotion.

Higher Stride Frequency

Because of their shorter legs, children use shorter strides and a higher stride frequency when compared with adolescents or adults (132; 273; 379; 433; 442; 462). Because the metabolic cost per stride seems to be similar in children, adolescents, and adults (381; 442; 462), it is likely that a shorter stride can explain, in part, the higher metabolic cost of locomotion in children. However, based on multiple regression analysis that included kinematic and electromyographic variables, stride rate was only a minor predictor of the metabolic cost of walking and running in 7- to 16-year-old girls and boys (168; 171).

Mechanically "Wasteful" Locomotion Style

It is reasonable to assume that a high metabolic cost in younger children reflects a higher mechanical cost of locomotion. This hypothesis was tested by Frost et al., who determined total body mechanical work in 7- to 16-year-old girls and boys during treadmill runs and walks at six speeds (168; 170). Based on kinematic analysis, mechanical work was calculated taking into account the work done by body segments and allowing for energy transfer between and within adjacent limb segments. In spite of considerable age-related differences in metabolic cost, total body mechanical work was similar in the different age groups. As a result, using multiple regression analysis, mechanical work was only a minor predictor of the variability in metabolic work in five of the six speeds.

High Co-Contraction of Antagonist Muscles

While stride frequency and mechanically "wasteful" gait style theories assume that a high metabolic cost reflects a high mechanical output by the body, a recent tridisciplinary study, which included metabolic, kinematic, and electromyographic analysis, suggests a different, possibly more important, mechanism (168; 171). Electromyography was used to determine the electrical activation of the knee and ankle flexors and extensors of pre- (7-8 years old), mid- (10-12 years), and late-pubertal (15-16 years) children during treadmill runs and walks at six speeds. In a normal gait cycle, when one muscle group (e.g., ankle extensors) contracts, the antagonist group (ankle flexors) relaxes most of the time. There is, however, a short period during which the two groups contract simultaneously. This is the co-contraction period. While co-contraction is important for joint stability and for the control of movement smoothness, excessive co-contraction may cause excessive metabolic cost, which cannot be discerned by kinematic or kinetic analysis. In the study by Frost et al. (171), the 7- to 8-year-old children had a distinctly higher co-contraction than did the two more mature groups (figure 1.9). When data were pooled from all three groups, co-contraction of the antagonist muscles was the second best predictor (after age) of the metabolic cost of walking and running (168). The implication is that a 7- to 8-year-old child may not yet have the required neuromotor control to optimally synchronize the action between muscle groups. It is not clear whether physical training at a young age would speed up the growth-related decline in co-contraction.

In line with these observations is a study by Hausdorff et al., who analyzed variability in stride time as an index of maturity of gait (200). As seen in figure 1.10, stride-to-stride variability in 3- to 4-year-old children was considerably higher than in the 6- to 7-year-old group, whose variability was greater than in the 11- to 14-year-old group. The variability in the latter group approached that of adults. The authors concluded that mature stride dynamics may not be completely achieved even by age 7 years.

In conclusion, data available so far suggest that the high metabolic cost of locomotion in children reflects several age-related differences. The most important among these are excessive co-contraction of antagonist muscle groups and a high stride frequency. Inconsistent stride-to-stride walking pattern, which may reflect immature neuromotor control in the first decade of life, may be another important cause. More research is needed, combining several testing disciplines,

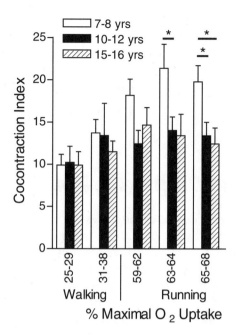

FIGURE 1.9 Co-contraction of antagonist calf muscles in pre-, mid-, and late-pubertal children who walked and ran at various intensities (expressed as % $\dot{V}O_2$ max).

Reproduced, with permission, from Frost et al. 1997 (171).

to provide further understanding of the reasons children are less economical than adolescents and adults during walking and running.

Metabolic Reserve

Whatever its underlying mechanism, the high metabolic cost of locomotion makes small children less effective "aerobic machines" than might be expected from their high maximal O_2 uptake. If one takes the difference between maximal O_2 uptake and the O_2 uptake needed for a given task to represent metabolic reserve, one can see that children are at a disadvantage. This is displayed schematically in figure 1.11. The 8-year-old boy who runs at 180 m · min^{-1} is operating at 90% of maximal aerobic power, while at the same running speed the 16-year-old requires only 75% of maximum. Thus, the "reserve" range of running speeds becomes higher with age, which explains why children are less capable than adolescents and adults of competing over long distances even though they can maintain a slow speed for long periods.

Effect of Training on the Metabolic Cost of Locomotion

It is pertinent to ask whether training can reduce the metabolic cost of locomotion in children and adolescents. Very few studies have addressed this question. In a controlled 11-week intervention, Petray and Krahenbuhl assigned 10-year-old

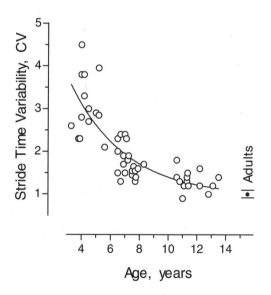

FIGURE 1.10 Stride time variability during walking as a function of age. Fifty girls and boys, ages 3 to 14 years, were monitored while walking for 8 min at their self-determined speed around a 400-m track. Stride time variability is presented as coefficient of variation (CV; standard deviation [SD] divided by mean stride time). Data for adults are derived from Hausdorff et al. 1996 (199).

Reproduced, with permission, from Hausdorff et al. 1999 (200).

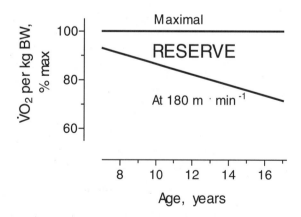

Figure 1.11 Aerobic reserve and age. Maximal O_2 uptake and O_2 uptake during a treadmill run at 180 m · min^{-1} in 134 girls and boys 7 to 16 years old. Based on data by MacDougall et al. 1979 (272).

Adapted, with permission, from Bar-Or 1982 (36).

boys to several groups that included run training, instructions regarding running style, or a combination thereof. None of these interventions were accompanied by changes in the metabolic cost of running or in kinematic variables (344). A similar lack of effect was reported following a 12-week run training intervention (269).

These findings contrast with observations regarding changes that occur in young athletes over several years. As seen in figure 1.12, middle and long distance runners who were tested repeatedly from age 10 to 12 had a sharp drop in O_2 uptake per kilogram. This decrease was considerably faster than that observed in cross-sectional comparisons of nonathletes (113; 114). A similar pattern was shown in another report (233). In contrast, the rate of drop of O_2 uptake was similar in teenage runners and nonathletes who were followed over 8 years. At each stage, however, the O_2 in the runners was lower. This difference may have resulted from a lack of random allocation to the two groups (417).

A recent experiment with trained adult distance runners was conducted to determine whether relaxation biofeedback techniques would affect the O_2 cost of treadmill running. Indeed, the economy of running increased progressively during the 6-week program (88). It is tempting to assume that the enhanced relaxation

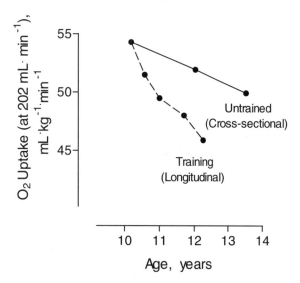

FIGURE 1.12 Age-related decline in the O_2 cost of treadmill running in male middle and long distance runners who were followed longitudinally while training. Comparison is made with cross-sectional observations of nonathlete males. Based on Daniels and Oldridge 1971 (113) and on Daniels et al. 1978 (114).

was accompanied by a decrease in co-contraction of antagonist muscles.

The possibility that economy of gait may improve with proper interventions is important to athletes, but it is also relevant to children with a neuromuscular disease, whose walking proficiency may be limited by a high metabolic cost (see chapter 10).

Anaerobic Exercise

Energy needs during exercise cannot always be met fully by oxidative energy turnover in the muscle, which requires an adequate supply of O_2. An activity that depends predominantly on nonoxidative energy turnover is considered anaerobic. Such activity typically can be sustained for 1 min or less and is of a very high intensity—that is, an intensity higher than that which elicits maximal aerobic power (hence the term supramaximal intensity). Examples are short and long sprints and jumping events. However, anaerobic energy turnover may be called upon also during an endurance event, such as an all-out effort at the final stages of a distance run, swim, or cross-country skiing race. As described in chapter 2, children often resort spontaneously to short-term bursts of activity, interspersed with a rest period (31).

Energy sources for such activities are high-energy phosphate (CP or creatine phosphate) that is stored in the muscle, or ATP that is produced through anaerobic glycolysis. The latter pathway is accompanied by enhanced production of lactate in the muscle and a subsequent increase of blood lactate levels. In contrast, the use of available ATP and CP stores is not accompanied by lactate production and, therefore, has been termed "alactic." Traditionally, it has been assumed that when a highly intense activity lasts up to 10 sec, the energy turnover is alactic, and that glycolytic energy turnover takes longer to emerge. However, with the advent of muscle biochemistry, it is now accepted that even in the first 10 sec of anaerobic exercise, glycolytic processes make a major contribution (225).

While aerobic fitness refers to an individual as a whole, anaerobic fitness is a local characteristic of a muscle or a muscle group. It is possible, for example, that a sprinter or a high jumper will display a very high anaerobic fitness of the knee extensors, but only moderate anaerobic fitness of the elbow flexors. The opposite may occur in a child with Duchenne muscular dystrophy; in such a child, anaerobic performance of the upper limbs is better preserved than the performance of the leg muscles.

The mechanical power that is generated during anaerobic exercise is considerably higher than during aerobic exercise. In children it can reach 200% to 400% of the mechanical power generated during a maximal aerobic task (69; 129; 156). With the advent of research in the field, terminology regarding anaerobic performance has become somewhat confusing. Terms such as peak power, anaerobic power, mean power, anaerobic capacity, anaerobic work capacity, local muscle endurance, and total mechanical work are sometimes used interchangeably. In this book, peak power denotes the highest mechanical power that is generated during exercise of 10 sec or less. Mean power denotes the average mechanical power that is generated during a 30-sec task such as the Wingate test.

This section focuses on the changes that occur during growth and maturation in anaerobic per-formance and the possible reasons for the low anaerobic performance of children compared with adolescents and, especially, with adults. Anaerobic performance of children with a chronic disease is addressed in subsequent chapters. Appendix II outlines methods used to assess anaerobic performance. More detailed information about anaerobic exercise is available in textbooks by Inbar et al. (222) and Van Praagh (450).

Anaerobic Performance of Children and Adolescents

The ability of children to perform anaerobic-type activities is distinctly lower than that of adolescents, whose performance is lower than in adults (20; 51; 70; 116; 119; 154; 156; 221; 224; 259; 397). Figure 1.13 summarizes the age-related pattern of anaerobic muscle power and muscle endurance of the legs and the arms of males who performed the Wingate anaerobic test. As expected, performance expressed in absolute power units is positively

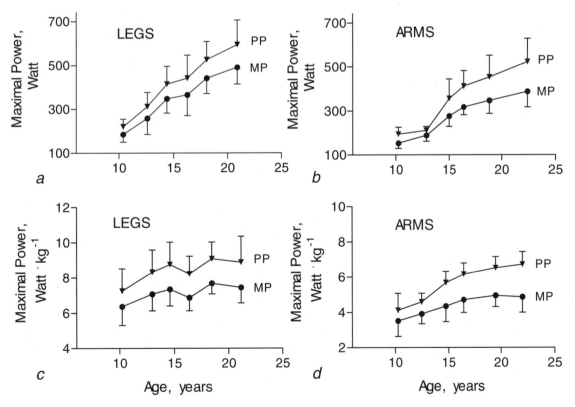

FIGURE 1.13   Anaerobic performance of male children, adolescents, and young adults. Graphs (a) and (c) summarize the mean power (MP) and peak power (PP) of 156 subjects who performed the Wingate leg cycling test. Graphs (b) and (d) summarize the MP and PP of 95 subjects who performed the Wingate arm cranking test. Vertical lines denote 1 SD.

Adapted, with permission, from Inbar and Bar-Or 1986 (221).

related to age; but, even when normalized for body mass, the power produced by an 8-year-old boy is still only 70% to 80% of that generated by a young adult (221). A similar pattern exists even when performance is scaled through allometry (20; 124).

Similar age- and gender-related changes in peak anaerobic power are obtained with the force-velocity test (figure 1.14 Van Praagh, 2000 [450a]) and with the Margaria step-running test (116; 119; 259). Irrespective of the test, the performance of the girls is lower than that of the boys. The difference becomes more apparent at age 14 years and beyond.

To sum up, whether or not children have inferior maximal aerobic power is still debatable; our conclusion will depend on the basis we choose for comparison. This is not the case with anaerobic performance, which is lower in the child than in the adult in absolute and relative terms alike, whether scaled to body mass, height squared, fat-free mass, or allometrically based exponents of mass. A graphic comparison of the growth-related difference between aerobic and anaerobic performance is shown in figure 1.15. To use a common scale, performance is shown as a percentage, taking the value at 18 years as 100%. While maximal aerobic power does not change (in boys) or even decreases (in girls) with age, there is a progressive growth-related increase in anaerobic performance. The same relative pattern in the growth of the two fitness components will be obtained, whichever mode one uses to scale the data.

The patterns just described are based on cross-sectional comparisons of various age or maturational groups. Very few studies have looked at longitudinal changes over time in anaerobic performance. Falk et al. used a mixed cross-sectional design to follow nonathlete Canadian boys ages 10.9 to 16.2 years, the youngest of whom were prepubertal and the oldest mid- to late-pubertal. While maximal aerobic power per kilogram body

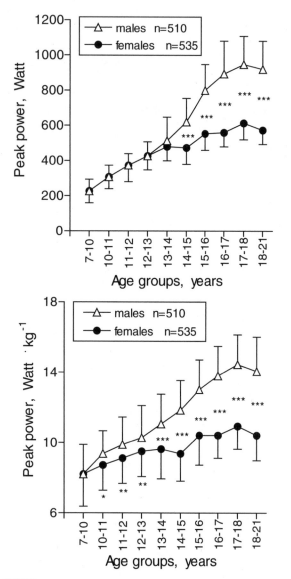

FIGURE 1.14 Peak anaerobic power, determined by the force-velocity test, in 1,045 females and males, ages 7 to 21 years. Values are presented in Watts (top figure) and Watt per kg (bottom figure).

Reproduced, with permission, from Van Praagh 2000 (450a).

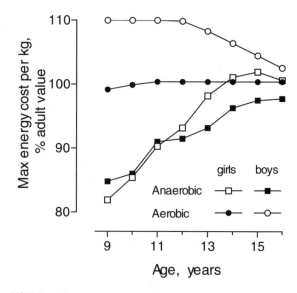

FIGURE 1.15 Development of aerobic and anaerobic characteristics. Maximal O_2 uptake and maximal performance in the Margaria step-running test in 9- to 16-year-old girls and boys. Mean values are percentages, taking the value at 18 years as 100%. Based on Kurowski 1977 (259) and Bar-Or 1982 (36).

mass did not change over time, peak anaerobic power increased by approximately 4 Watt · kg^{-1} (156). A longitudinal study by Duché et al. (129) revealed a similar pattern in French boys.

Possible Reasons for Low Anaerobic Performance in Children

In spite of growing research in this field, it is not entirely clear why children have a deficient anaerobic performance when compared with adolescents and adults. Three factors seem to be the main culprits:

- Smaller muscle mass per body mass
- Lower glycolytic capability
- Deficient neuromuscular coordination

Smaller Muscle Mass

The ability of a muscle to generate force depends on, among other factors, its cross-sectional area. The shortening velocity during muscle contraction depends on its length, among other factors. Because mechanical power is the product of force and velocity, power depends on the volume or mass of the muscle (each of which is a function of the product of cross-sectional area and length). Muscle volume in children, relative to their body mass, is smaller than in adults, and it gradually increases during childhood and adolescence (297). This may be one reason for children's lower ability to generate power. As seen in figure 1.16, the rate of increase in muscle volume during childhood and adolescence is similar to the rate of increase in peak and mean anaerobic power, when all are normalized for body mass. Yet, even when scaled for fat-free mass, muscle volume, or muscle mass, anaerobic performance is still lower in children (123). One can conclude, therefore, that the relatively small muscle mass in children is only one reason for their lower anaerobic performance.

Lower Glycolytic Capability

The markedly lower anaerobic performance of children

reflects their lower ability for anaerobic energy turnover. Several findings support this notion: Table 1.3 summarizes characteristics of biochemical substrates within the muscle that are utilized for muscle contraction. The main age-related difference is in glycolytic capability. The resting concentration of glycogen, and especially the rate of its anaerobic utilization, is lower in the child, who is therefore at a functional disadvantage when performing strenuous activities that last 10 to 60 sec.

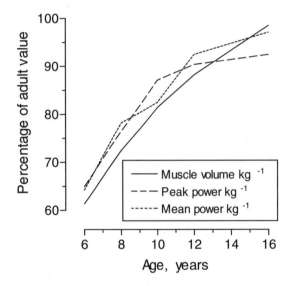

FIGURE 1.16 Estimated thigh muscle volume and peak and mean anaerobic power, all calculated per unit body mass for boys ages 6 to 16 years. Values are expressed as a percentage of values for 18-year-old men. Muscle volume was estimated based on X-ray data by Maresh 1970 (297). Data for muscle power were derived from studies of boys who performed the Wingate test (120; 156; 220; 322).

Adapted, with permission, from Martin and Malina 1998 (298).

TABLE 1.3	Substrate Availability and Utilization in Muscles of Preadolescent Boys[*]		
	RESTING VALUES		
Substrate	**Concentration in muscle (mmol · kg^{-1} wet weight)**	**Compared with older individuals**	**Utilization rate during exercise**
ATP	3.5-5.0	No change with age	Same as in adults
CP	12-22	Lower in children	Same as or less than in adults
Glycogen	45-75	Lower in children	Much less than in adults

ATP = adenosine triphosphate; CP = creatine phosphate

[*]Based on Eriksson and Saltin 1974 (146), Eriksson 1980 (139), and Karlsson 1971 (234).

One way of assessing anaerobic glycogen utilization is by measuring maximal lactate concentration in the muscle. With all its methodologic shortcomings, maximal blood lactate is often used as a surrogate measurement because, for ethical reasons, muscle biopsies are seldom performed on healthy children. The great majority of studies show that maximal blood lactate is lower in children than in adolescents or adults (25; 65; 205; 302; 351; 360; 472).

Studies by Eriksson in the 1970s showed that both submaximal and maximal lactate levels in the quadriceps muscle are lower in boys than in men (143). Figure 1.17 shows that at each level of cycling exercise the boys had a lower muscle lactate concentration, maximal level being about 35% lower than in the adults (143). Figure 1.18 summarizes maximal muscle lactate levels as a function of age, suggesting that, in boys, there is a continuous increase during the second decade of life. Studies in rats have suggested that lactate production is related to the level of circulating testosterone (256). It has therefore been suggested, but not confirmed, that the ability of boys to produce lactate (143) or to generate peak anaerobic power (162) depends on circulating testosterone and other hormonal changes during puberty, such as increase in growth hormone and insulin-like growth factors. The lower anaerobic performance of the mature female when compared with the male and the lesser age-related difference among females are in line with this notion. However, there is insufficient evidence to allow one to state that the difference between the rate of glycolysis in boys and men is fully or partially explained by differences in male hormone activity.

The rate of glycolysis is limited by the activity of such enzymes as phosphorylase, pyruvate dehydrogenase, and phosphofructokinase. The latter enzyme has been found to be less active in the muscle cells of 11- to 13-year-old boys (141; 142) or 16- to 17-year-old boys (164) than in young adults. In contrast, the phosphofructokinase level in 13- to 15-year-old girls was not different from that of young adult women. It is noteworthy that muscle phosphofructokinase in young rats is markedly lower than in mature rats (131).

An additional indicator of anaerobic capability is the degree of acidosis at which a muscle can still contract. Some adult athletes can push themselves to exercise at arterial blood pH as low as 6.80 (244), which is equivalent to a pH = 6.60 or less in the muscle cell. Untrained individuals,

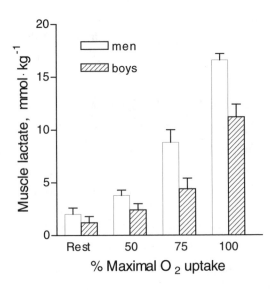

FIGURE 1.17 Lactate concentration in the quadriceps muscle of 13.5- to 14.8-year-old boys and young men, as a function of exercise intensity. Values are per wet muscle tissue. Vertical lines denote 1 SEM.

Adapted, with permission, from Eriksson et al. 1971 (143).

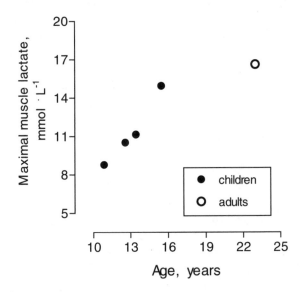

FIGURE 1.18 Muscle lactate levels at the end of a maximal O_2 uptake test in 11- to 16-year-old boys and young adults. Based on data by Eriksson and Saltin 1974 (146).

in contrast, can seldom sustain exercise when their arterial blood pH reaches 7.20. Children do not reach as high levels of acidosis as can adolescents or adults (177; 244; 258; 302; 351; 455; 483) (figure 1.19). For example, on the basis of ^{31}P magnetic resonance spectroscopy, Zanconato et al. showed that, during intense contraction of the calf muscles, children did not reach as low muscle pH levels as adults (483) (figure 1.20).

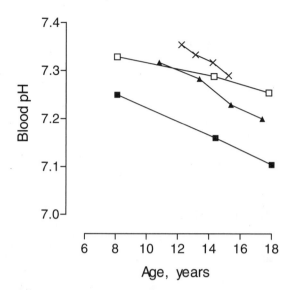

FIGURE 1.19 Acid–base balance and age. Mean values of blood pH following all-out cycle ergometer exercise (□, ▲ X) or 300-m run (■). Based on data by Kindermann et al. 1975 (243); Matejková et al. 1980 (302); and Von Ditter et al. 1977 (455).

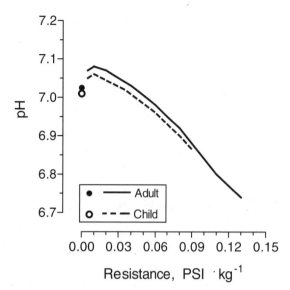

FIGURE 1.20 Muscle pH at rest and during incremental exercise to exhaustion in a 9-year-old girl and a 33-year-old man. Data were taken during ^{31}P magnetic resonance spectroscopy while subjects were performing plantar flexion against a piston that provided increasing resistance.

Adapted from Zanconato et al. 1993 (483).

Using the same technology, Kuno et al. reported a lower acidosis in the thigh muscles of adolescents compared with adults (258). Based on longitudinal observations (302), the increase in acidosis is approximately 0.01 to 0.02 pH unit each year.

Deficient Neuromuscular Coordination

Performance of anaerobic tasks, such as sprinting, cycling at maximal speed, skipping, or jumping, requires coordination among muscle groups. Moreover, success in a sprint race depends also on the reaction time and movement time in response to the starter's gun. This is in contrast to a single muscle contraction (static, isometric, or isotonic), in which the coordination among muscle groups or the speed of reaction to a stimulus is less important. Skills that require short bursts of high-intensity activities develop gradually during childhood (298). For example, vertical jumping performance improves with age, but its fastest increase occurs after the year of peak height velocity in boys (56). Although no studies have directly correlated anaerobic performance with neuromuscular coordination, some data suggest that such a relationship exists. A case in point is children who were born prematurely and had an extremely low birth weight. Compared with children of normal birth weight, they performed poorly on the Wingate test (237) and also had a longer reaction time, slower maximal pedaling

rate against "zero" resistance, deficient coordination during a vertical jump, and poor whole-body coordination at ages 5 to 7 years (158; 202; 236). For more details see the section on extremely low birth weight in chapter 10.

Relationship Between Anaerobic and Aerobic Performance

To characterize a person's exercise performance, it may be worthwhile to describe the performance using a combined index of aerobic and anaerobic fitness. Put differently, it may be useful to characterize a person's anaerobic performance relative to his or her aerobic performance. This may help, for example, a physical therapist or a physiatrist who wishes to construct a rehabilitation program, or a coach who is planning a training program for a middle distance runner.

To facilitate such an approach, scientists at the Children's Exercise & Nutrition Centre have proposed using the power ratio, which is the ratio between a person's peak anaerobic power and peak aerobic power achieved during a maximal aerobic test (69; 156). To provide a valid value, the two tests should be conducted with the same ergometer. Based on cross-sectional (69) and longitudinal (156) data, the ratio for cycling increases from about 2 at age 8 years to about 3 at age 13 to 14 in girls and age 14 to 15 in boys. Subsequently, it rises at a more moderate pace throughout adolescence. The ratio for girls is somewhat higher than for boys (figure 1.21). Based on mathematical considerations, the use of a ratio implies that the Y intercept is zero, which is not the case for the power ratio. Still, the power ratio may prove useful as a guide for planning a training or a rehabilitation program.

Heredity and Anaerobic Performance

Anaerobic characteristics depend to a certain extent on a person's genotype. Using a twins study, Komi and colleagues (252) were the first to show a greater variance in peak power (using the Margaria step-running test) among dizygotic twins than among monozygotic twins, which suggested the role of heredity in such performance. This has since been confirmed by a large-scale sibling study (416). Furthermore, heredity seems to affect fat-free mass (79), muscle size (210), proportion of muscle fiber types (414), the ratio between glycolytic and oxidative muscle enzymes (80), and the trainability of high-intensity muscle performance (416). The estimated role of heredity in anaerobic muscle characteristics is summarized in table 1.4. For a detailed review of this topic, see Simoneau and Bouchard 1998 (415).

TABLE 1.4	The Genetic Variance in Anaerobic Performance and Related Variables	
Variable		**Genetic variance**
Anaerobic performance		50%
Fat-free mass		30%
Fiber type distribution		45%
Glycolytic-to-oxidative enzyme activity		50%

Based on Simoneau and Bouchard 1998 (415).

Anaerobic Threshold

During progressively increasing exercise, a point is reached at which lactate production exceeds its elimination from the blood. At this point, which has been termed "anaerobic threshold," blood lactate starts accumulating appreciably. Related to such accumulation, there is an increase in ventilation that is disproportionate to the increase in metabolic demands. Physiologists often use the exercise level at which ventilation accelerates, rather than the point of increase in blood lactate, as an index of anaerobic threshold. For further discussion of anaerobic threshold see the section "Ventilation and Ventilatory Threshold" later in this chapter.

O_2 Uptake On-Transients

In line with the notion that children are less capable than adults of exerting anaerobically is the pattern of their O_2 uptake on-transients. Any individual, on transition from rest to exercise or from a certain exercise level to a higher one, increases his or her metabolic rate. At first, the aerobic supply of additional energy lags behind the actual demands for energy, and an O_2 deficit is contracted. The balance of chemical energy at these initial stages is facilitated by anaerobic pathways. During submaximal exercise, the aerobic energy supply gradually catches up with the demand and a new metabolic steady state ensues within 2 to 5 min.

In 1938, Robinson suggested that children have shorter O_2 uptake transients than adults (360). This was confirmed in the early 1980s by several studies (165; 278; 389). Subsequent work by Armon et al. (15) suggested that the shorter O_2 uptake transients in children occur when exercise intensity is high (above the anaerobic threshold),

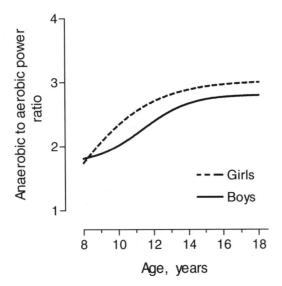

FIGURE 1.21   Changes with age in the anaerobic-to-aerobic power ratio of children and adolescents.

Adapted from Blimkie et al. 1986 (69).

whereas at lower intensities the pattern of increase in O_2 uptake is independent of age or body size (figure 1.22). It has subsequently been suggested that the reported faster O_2 uptake response in children may not occur if one excludes the very few seconds of exercise (also called phase I of the O_2 uptake transients) during which the increase in O_2 uptake results from the enhanced return of venous blood from the contracting skeletal muscles (204). Nor were there differences in O_2 uptake on-transients when intense exercise was performed for 60 sec only (484).

An interesting question is whether, due to their shorter O_2 transients, children do not need to resort as much to anaerobic pathways (hence the smaller O_2 deficit and lactate production), or whether these shorter transients are compensatory for their low glycolytic capacity. This question is still unresolved (323). Another possibility, not yet tested, is that the shorter transients in children are a reflection of a smaller body and the resulting shorter circulation time (110). Yet another explanation is that the shorter on-transients in children may reflect a higher contribution of slow-twitch muscle fibers than in adults (100). While this is an interesting hypothesis, it requires further confirmation. Autopsy data (332) suggest that 1-year-old children have a higher proportion of slow-twitch fibers than do adults. However, the adult pattern of fiber type distribution seems to be established sometime around age 6 (52).

An important functional implication of the children's shorter on-transients during intense exercise is that, compared with adults, they incur a smaller O_2 deficit at the start of such exercise. Indeed, as shown by Máček and Vávra, 10- to 11-year-old children had a much lower O_2 deficit during the first few minutes of intense exercise than 20- to 22-year-old adults (278) (figure 1.23). Likewise, recent studies show that, during high-intensity exercise, the accumulated O_2 deficit is lower in children compared with adults (89). A lower O_2 deficit is in line with the smaller and faster O_2 debt repayment that occurs in children during recovery from exercise (see discussion of recovery in the next section).

Recovery Following Exercise

People with experience in pediatric exercise testing often note that children recover quickly following strenuous exercise. Adults who complete, for example, an all-out aerobic test are usually exhausted and quite reluctant to pursue further activities for several hours. This is not the case with children, who often amaze us with a request to repeat this strenuous test 15 to 30 min after its completion, because they think they could improve on their previous performance!

Several studies have shown that, indeed, performance and physiologic functions of children recover faster than in adults following submaximal (48; 333; 484), maximal (48; 333; 484), and supramaximal (48; 206; 484) tasks. Physiologic functions that recover faster in children include heart rate (48; 333), O_2 uptake (206; 484), CO_2

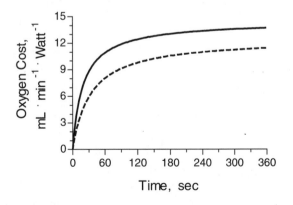
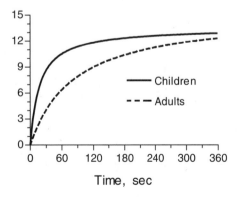

FIGURE 1.22 On-transient of O_2 uptake in children and adults. Subjects performed 6-min cycling below (left graph) and above (right graph) their anaerobic threshold.

Adapted, with permission, from Armon et al. 1991 (15).

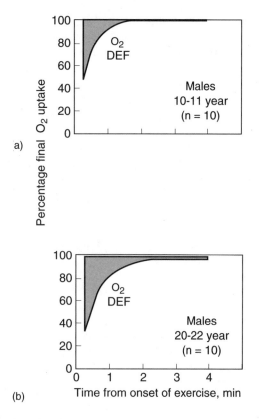

FIGURE 1.23 Oxygen deficit of children and adults. O_2 uptake transients of 10- to 11-year-old boys (top) and 20- to 22-year-old men (bottom) who cycled at 90% to 100% of their predetermined maximal aerobic power. Adapted from Máček and Vávra 1980 (278).

Reproduced, with permission, from Bar-Or 1982 (36).

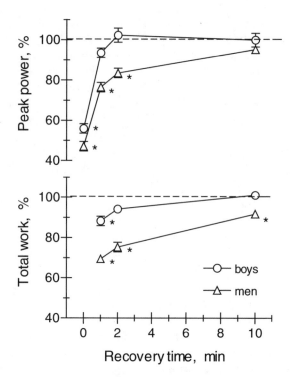

FIGURE 1.24 Recovery of peak power (top) and total mechanical work (bottom) in boys and men following intense exercise. Each day, eight 8- to 12-year-old boys and eight 18- to 23-year-old men performed pairs of Wingate tests, with various rest intervals in between. Zero time denotes the end of the first test.

Reproduced, with permission, from Hebestreit et al. 1993 (206).

output, and ventilation (206). Likewise, plasma volume, lactate, and pH return to their pre-exercise level faster in children than in adults (205). Figure 1.24 displays the recovery pattern of peak power and total mechanical work in boys and men following the Wingate anaerobic test. While the boys reached full recovery within 2 min, the men did not recover fully even after 10 min of rest (206). Figure 1.25 summarizes recovery of O_2 uptake and heart rate in the same boys and men during 10 min after the Wingate anaerobic test. Clearly, the boys recovered faster and thus had a faster repayment of O_2 debt.

To quantify the rate of recovery one can calculate a time constant, which is the time that a physiologic function takes to recover 63.2% from its level at peak exercise to its level at baseline (48; 484). Another criterion is half-time constant, the time for a 50% recovery from peak to baseline (206). As seen in table 1.5, the half-time constants for various physiologic functions are considerably

shorter in children than in adults during recovery from supramaximal exercise. The faster recovery of children, boys and girls alike, is particularly apparent when the exercise task is of high intensity, and less so at milder intensities (48; 484).

It is not entirely clear why children recover faster than adults. The mechanism for this pattern may be different for different physiologic functions. It is possible that a faster recovery of heart rate in children is due to a higher parasympathetic tone (333). Other suggested mechanisms include a smaller exercise-induced increase in catecholamine activity and a faster clearance of hydrogen ions (48). A faster recovery of O_2 uptake has been attributed to a shorter circulation time, or a shorter diffusion distance between muscle fibers and their adjacent capillaries, both of which may enhance the clearance of metabolites (206). Another possible reason is the smaller O_2 deficit incurred by children at the start of exercise, which calls for a lesser O_2 debt repayment (see the earlier section "O_2 Uptake On-Transients"). A faster

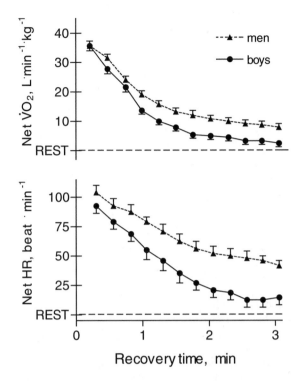

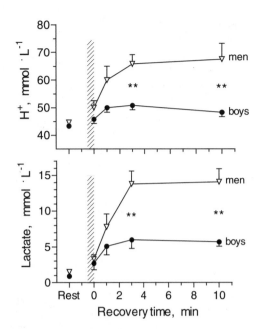

FIGURE 1.25 Recovery of O_2 uptake (top) and heart rate (bottom) in boys and men during the first 3 min following the Wingate anaerobic test. "Net" denotes the difference between the observed and the resting values. Subjects and source are the same as in figure 1.24 (206).

FIGURE 1.26 Recovery of plasma hydrogen ion (top) and lactate (bottom) in boys and men following the Wingate anaerobic test. Blood was drawn through an indwelling antecubital venous catheter. Subjects were five 8- to 11-year-old boys and five 19- to 29-year-old men.

Adapted from Hebestreit et al., 1996 (205).

TABLE 1.5	Recovery Rate of Boys and Young Men Following the Wingate Anaerobic Test	
Half-time constant, sec	**Boys**	**Men**
Heart rate	64.0 (21.1)	132.8 (68.0)
Ventilation	61.2 (23.2)	99.9 (68.0)
CO_2 output	64.4 (10.5)	93.4 (17.5)
O_2 uptake	40.9 (3.6)	52.5 (8.6)

Half-time constants were calculated for heart rate, ventilation, CO_2 output, and O_2 uptake.

Means (SD)

Reproduced, with permission, from Hebestreit et al. 1993 (206).

recovery of mechanical power is likely a result of a lesser use of anaerobic energy pathways during high-intensity exercise. This is reflected by lower plasma lactate and H+ ions at the end of exercise and their faster clearance during recovery, as shown by Hebestreit et al. (205) (figure 1.26). Although plasma acid–base status is not identical to that in the muscle cells, it is likely that the intramuscular acidosis in that study was greater in the men than in the boys during recovery.

Another explanation is a smaller exercise-induced decrease in the plasma volume of children compared with adults (205).

In adults, physiologic functions return to their pre-exercise levels faster if the person keeps exercising at a low intensity during the cool-down period (active recovery) than if the person performs no exercise during this period (passive recovery). The same pattern has been shown for children. For example, Dotan et al. found that recovery of blood lactate in 9- to 11-year-old girls and boys following bouts of short-term, high-intensity exercise (150% peak O_2 uptake) was faster when the children kept exercising at submaximal levels, compared with a passive recovery (125). This beneficial effect of active recovery occurred when the exercise was done at intensities ranging from 40% to 60% maximal O_2 uptake.

Morphologic and Functional "Specialization"

Morphologic and functional characteristics of physically active adults, especially high-level athletes, are often highly specialized. Some of these characteristics are acquired, such as increased muscle mass and strength following resistance training. Others are predominantly inherent and constitute one's "talent," such as tall stature, high maximal aerobic power, or short reaction time. Specificity of characteristics in adult athletes is obvious between sports (e.g., weightlifters vs. distance runners or gymnasts) but is also apparent within sports (105). Among runners, for example, sprinters are mesomorphic, with a well-developed musculature, but they have only average maximal aerobic power. In contrast, distance runners are usually thin, with a high maximal aerobic power and low explosive strength. Sprinters have about equal distribution of fast- and slow-twitch muscle fibers. Marathoners, on the other hand, have as many as 80% to 90% slow-twitch fibers with a highly oxidative biochemical profile.

Does such specialization exist already during childhood? In *Pediatric Sports Medicine for the Practitioner,* it was suggested that children, prepubescents in particular, are metabolic nonspecialists. This statement was based on observations related to morphologic characteristics and physical performance profiles of children. The somatotype, for example, of children who are successful athletes seldom reaches the extremes found among adult athletes. Functionally, a child who is the sprinting "star" of his or her class is often also above average in distance running and successful in a variety of team sports. Several laboratory-based studies from the 1970s showed that children who possess a high maximal O_2 uptake also perform above average anaerobically (42; 219; 303). An example is a study in which aerobic and anaerobic characteristics were assessed in 8- to 11-year-old boys, six of whom were elite U.S. distance age-group runners. As seen in figure 1.27, those boys who scored well in the Wingate anaerobic test also had a high maximal O_2 uptake per kilogram body weight, which suggests a lack of specialization.

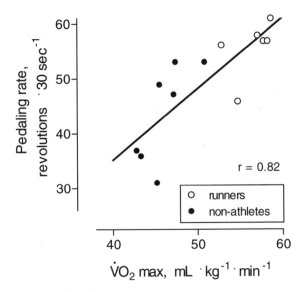

FIGURE 1.27 The child as a physiologic "nonspecialist." Individual performance (pedaling rate) in the Wingate anaerobic test plotted against maximal aerobic power. Data are for 13 boys, 8 to 11 years old, 6 of whom were elite cross country runners, the others nonathletes. Based on Mayers and Gutin 1979 (303).

The notion of nonspecialization was reinforced in subsequent studies. For example, Falgairette et al. tested the aerobic and anaerobic performance of 53 boys, 11 years old, who were swimmers and nonswimmers. They found that the two performance characteristics were moderately correlated, irrespective of whether the boys were engaged in sport or not. The authors concluded that there was neither aerobic nor anaerobic specialization during prepubertal development, and regular sporting activity induces no great changes in the bioenergetic characteristics of prepubertal boys (155). In another study, a moderate correlation between mean anaerobic power (but not peak power) and maximal O_2 uptake was reported for girls and boys ages 11.1 and 10.8 years (120).

The issue of children as metabolic nonspecialists was recently given a fresh look by T. Rowland (368). In analyzing the evidence, the author suggests that the lower specialization in child athletes may reflect child-adult differences in trainability (see related section, "Trainability of Maximal Aerobic Power"). Another possibility is that children are somatotype nonspecialists (i.e., their development into a distinct somatotype is not complete), rather than metabolic nonspecialists.

In conclusion, based on somewhat fragmentary information, it appears that compared with adult athletes, prepubertal and early-pubertal child athletes are less specialized as "anaerobic" or "aerobic" performers. The nature of this difference, however, is unclear.

Cardiovascular Response to Exercise

The main role of the cardiovascular system during exercise is to transport additional O_2 to the exercising muscles and clear CO_2 from the muscles. Other roles include transport of nutrients, metabolites, and hormones; retention of osmotic and acid–base balance; and convection of heat from body core to the periphery. This section focuses on hemodynamic responses to exercise in healthy children and how these responses change with growth and development. For a discussion of the responses of children with a heart disease, see chapter 7.

Changes that facilitate a greater O_2 supply to the muscle can best be described through the Fick principle:

$$\dot{V}O_2 = \dot{Q}\,(CaO_2 - C\bar{v}O_2)$$

$\dot{V}O_2$ is O_2 uptake, $\dot{Q}$ is cardiac output, and $CaO_2 - C\bar{v}O_2$ is the difference in O_2 content of arterial and mixed-venous blood. $\dot{Q}$ is the product of heart rate (HR) and stroke volume (SV). Therefore:

$$\dot{V}O_2 = HR \times SV\,(CaO_2 - C\bar{v}O_2)$$

For O_2 uptake to increase, either cardiac output or arterial-venous O_2 difference must rise. In fact, both rise during exercise—cardiac output through an increase in HR and in stroke volume, and arterial-venous O_2 difference through an increase in muscle blood flow. The latter results in a decreased mixed-venous O_2 content.

Table 1.6 summarizes differences in the hemodynamic response to exercise between children and adults.

Cardiac Output and Stroke Volume

As in adults, the cardiac output of children rises at the beginning of exercise, or upon transition to a higher level of exercise. A new steady state cardiac output is established within a few minutes. With a gradual increase in the intensity of upright exercise there is an initial rise in stroke volume, which reaches its maximum (approximately 35-40% above resting level) at mild-to-moderate exercise intensities and remains fairly constant at higher intensities. When exercise is performed in a supine position, stroke volume remains near constant at a wide range of intensities. In contrast to stroke volume, HR increases from rest to exercise whether the child is upright or supine, and it keeps increasing until near-maximal exercise intensity. Maximal cardiac output of healthy children reaches three to four times its resting value. Most of this increase is due to HR; only 20% to 25% is due to stroke volume. A relatively greater contribution of stroke volume has been reported in child athletes (327; 370). Altogether, stroke volume is considered a major limiting factor in the O_2 transport chain and an important determinant of maximal O_2 uptake (371; 372). As evidenced by echocardiography (334; 373) and radionuclide technique (118), the increase in stroke volume is achieved mostly by a reduction in the end-systolic left ventricular volume with little change in the end-diastolic volume. The relatively stable end-diastolic volume suggests that the enhanced systemic venous return during exercise (mostly from the exercising muscles) is matched by an increase in HR, which causes a reduction in ventricular filling time (376).

Children have a markedly lower stroke volume than adults, at all levels of exercise. This

TABLE 1.6	Central and Peripheral Hemodynamic Response to Exercise—a Comparison Between Children and Adults
Function	**Children's response (compared with adults')**
Heart rate—submax	Higher, especially at first decade
Heart rate—max	Higher
Stroke volume—submax, max	Lower
Cardiac output—submax	Somewhat lower
Arterial-mixed venous O_2 difference—submax	Somewhat higher
Blood flow to active muscle	Higher
Systolic and diastolic blood pressure—submax, max	Lower
Total peripheral resistance	Higher

is compensated for, but only in part, by a higher HR. The end result is a somewhat lower cardiac output at each metabolic level (figure 1.28) (44; 94; 127; 137; 144; 313; 362; 439). However, when expressed per body surface area (i.e., as cardiac index), cardiac output does not seem to increase with growth (309; 376).

It is not clear whether the somewhat lower cardiac output of children is of any functional significance. Quite possibly the concomitant higher arterial-mixed venous O_2 difference is sufficient to compensate the O_2 transport system during submaximal exercise (44; 94; 439). A potential deficiency due to low cardiac output may exist, however, during near-maximal and maximal exercise when peripheral O_2 extraction can no longer rise (144), or when the child is exposed to combined stresses of exercise and extreme heat. In the latter case, the circulation is called upon to simultaneously support the increased metabolic needs and the need for heat convection to the periphery. Such demands may not be met in full, and exercise in the heat cannot be sustained (127) (for details, see chapter 3).

While adult women have a higher HR and arterial-venous difference, lower stroke volume, and

a somewhat lower cardiac output than men, there are only minor gender-related differences in the hemodynamic responses of children to exercise. Heart rate is somewhat lower and stroke volume somewhat higher in boys, but there are no gender differences in cardiac output or in arterial-venous O_2 difference (44; 107; 186; 362; 369; 437). A higher stroke volume in boys reflects their larger left ventricular mass (437). Figure 1.29 summarizes hemodynamic responses to submaximal exercise in girls, boys, young women, and young men.

The degree of the child-adult differences apparently depends on the age, or the developmental stage, of the child (182; 244; 313; 362). Younger children (e.g., 8-10 years old) have a lower stroke volume at a given exercise level than do older children (e.g., 11-13 years of age), which is compensated for by a higher HR. As a result, cardiac output of young children is only marginally lower at any given metabolic rate. A related finding is presented in figure 1.30. Children who performed maximal exercise were subdivided into prepubertal, pubertal, and postpubertal groups. At a given maximal O_2 uptake, heart volume (an important factor in the ability to raise stroke volume during exercise [212]) of the prepubertal children was the smallest (476). It is possible that the prepubescents also had a lower maximal cardiac output and, true to the Fick principle, had to resort to higher peripheral O_2 extraction to achieve a certain level of maximal O_2 uptake.

Heart Rate and Exercise

Thanks to its relative ease of monitoring, HR has been the most commonly analyzed variable in exercise physiology. Its measurement has proven valuable for monitoring the cardiovascular response to exercise. In addition, its close relationship to metabolic level has made HR a useful indirect indicator of energy expenditure, as well as a means for prediction of maximal O_2 uptake. Moreover, due to its marked sensitivity to any increase or decrease in training, HR has become a valuable gauge for determining fitness and compliance to intervention programs.

This section focuses on the relationship between HR and other physiologic and psychologic variables before, during, and following exercise. Special emphasis is given to factors that modify the HR response to exercise in children, as summarized in table 1.7. A practitioner who

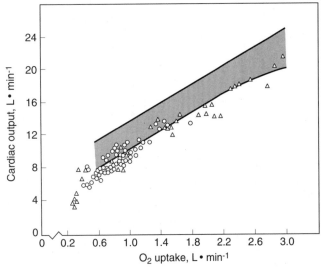

FIGURE 1.28 Cardiac output of the exercising child. Individual values for boys performing upright submaximal and maximal cycle ergometer exercise. O = Data from Godfrey et al. (186) on boys with body height of 110 to 154 cm, tested by CO_2 rebreathing. $\triangle$ = Data by Eriksson (137) on boys 13 to 14 years old, tested by dye dilution. The shaded area represents young adults performing upright exercise. Compiled from the literature by Bar-Or et al. 1971 (44).

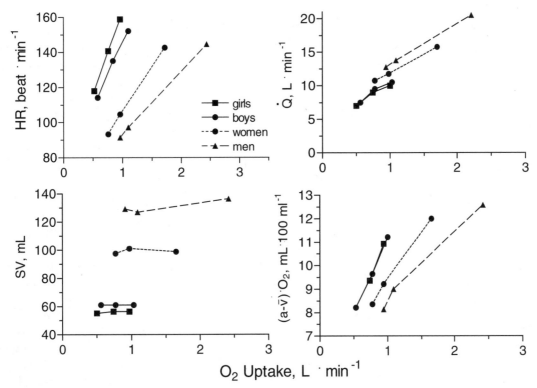

FIGURE 1.29 Hemodynamic responses to submaximal cycling exercise in girls, boys, women, and men. Twelve girls and 12 boys (7-9 years old) and 12 women and 12 men (18-26 years old) performed three submaximal bouts. Cardiac output (Q) was determined by CO_2 rebreathing. HR = heart rate; SV = stroke volume; $(a-\bar{v})O_2$ = arterio-mixed venous O_2 difference.

Reproduced, with permission, from Turley and Wilmore 1997 (439).

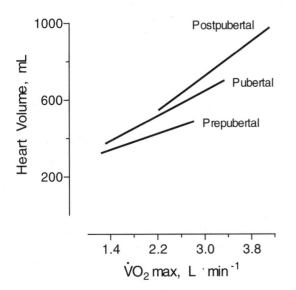

FIGURE 1.30 Heart volume, maximal O_2 uptake, and maturational stage. Fifty-one male and female swimmers were subdivided into prepubertal (10.6 ± 0.4 years), pubertal (12.5 ± 0.3), and postpubertal (16.4 ± 0.7) groups. Subjects performed an all-out cycle ergometer test. Heart volume was assessed at rest, using chest X-rays.

Reproduced, with permission, from Wirth et al. 1978 (476).

wishes to include the measurement of HR within a diagnostic repertoire should be thoroughly familiar with such factors.

Age

Submaximal HR in children declines with age (9; 25; 78; 186; 360; 386; 441; 476; 478). As shown in figure 1.31, HR can be as much as 30 to 40 beat · min⁻¹ higher in an 8-year-old child than in an 18-year-old performing the same absolute task. Such a difference is partially due to the greater relative exercise intensity performed by the younger children (figure 1.31), but it is also found at equal relative metabolic loads (476). The higher HR among young children is biologically sound, as it compensates for a lower stroke volume.

Maximal HR of children and adolescents ranges between 195 and 210 beat · min⁻¹ (9; 17; 25; 41; 360). It remains quite stable during childhood (30; 45; 375) and starts declining with age during the late teens. Such a decline is independent of gender, level of training, climate, or other environmental conditions. It is equivalent to 0.7 to 0.8 beat · min⁻¹ a year (41). If one takes the difference between

TABLE 1.7	Factors Known to Affect Heart Rate (HR) Response to Exercise Among Children and Adolescents	
Factor	Submaximal HR	Maximal HR
Age	Young > old	No effect
Sex	Females > males	No effect
Adiposity	Obese > lean	No effect
Climatic stress	↑	No effect
Emotional stress	↑	No effect
Active muscle mass	Small > large	Large > small
Body position	Upright > supine	Upright > supine
Training	↓	No effect or slight ↓
Detraining	↑	No effect
Heat acclimatization	↓	No effect
Habituation	↓	No effect
Diseases		
Anemia	↑	No effect
Anorexia nervosa	↑ (↓ if severe)	↓
A-V block	↓	↓
Chronic fatigue	↑	↓
Cyanotic heart defects	↑	↓
Dysrhythmias	↑ ↓	Various
Fever	↑	No effect
Muscle dystrophy, atrophy, and paralysis	↑	↓
Medications		
Beta-blockers	↓	↓
Methylphenidate	↑	No effect
Beta 2 sympathomimetics	↑	No effect
Thyroid hormone	↑	No effect

submaximal and maximal HR to reflect a certain heart rate reserve, a 16-year-old adolescent has a distinctly greater reserve than a 6-year-old child. This is shown graphically in figure 1.32 and is in line with the smaller metabolic reserve of children, as discussed earlier in the section "Mechanical Efficiency and Economy of Movement."

Gender

As a rule, women have a higher HR than men at any given exercise level. Traditionally, this difference has been attributed to the low blood hemoglobin concentration of the mature female (25). However, a higher HR is also found among preadolescent girls, whose hemoglobin level is not different from that of boys (44; 186; 276; 386; 438; 461), and even among children as young as 6 years of age (12; 186; 343). The degree of relative tachycardia in females ranges between 10 and 20 beat · min[-1]. Figure 1.33 summarizes an experiment in which 6- to 7-year-old girls and boys carried various loads on their backs while walking at a constant speed. Irrespective of the load, HR of the girls was some 20 beat · min[-1] higher than among the boys. The cause of high exercise HR in the young female is not clear. It could result from a lower stroke volume (44; 438) or from lower levels of habitual activity (181). One cannot rule out gender-related differences in autonomic cardiac regulation, but there are no data to substantiate such a notion. Young boys have a faster postexercise decline of HR than do girls (343), which may reflect quantitative differences in autonomic regulatory mechanisms.

Adiposity

Obese children have a higher submaximal HR than do lean ones (28; 312), which reduces their HR reserve (see the section "Cardiorespiratory Responses to Submaximal Exercise" in chapter 9).

Climatic Stress

During exercise in hot or humid climates, HR is higher than in a thermoneutral environment. An elevation above comfortable ambient conditions (23-24° C, 50-60% relative humidity) of 5° to 7° C or of 15% to 20% relative humidity may cause an increase of 10 or more beat · min[-1] and render false information. In a climatic chamber-based study, Hebestreit et al. have shown that at ambient temperature of 35° C, HR may be as much as 15 to 20 beat · min[-1] higher than at 22° C (201; 203) (figure 1.34). The average increase of HR was 1.05

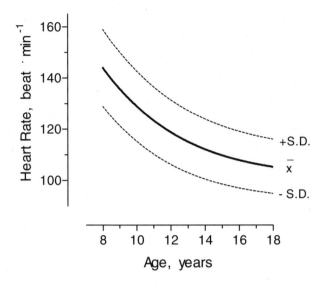

FIGURE 1.31 Submaximal heart rate and age. Subjects were 237 boys, 8 to 18 years old, who participated in a growth and maturity study in West Germany. They all performed a cycle ergometer task at 29.4 Watt.

Adapted from Bouchard et al. 1977 (78).

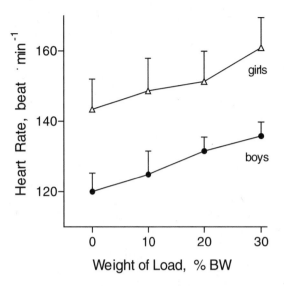

FIGURE 1.33 Exercise heart rate and gender. Nine girls and 6 boys, 6 to 7 years old, walked on a treadmill at 4 km · hr⁻¹, without carrying any load and while carrying on their back a schoolbag weighing 10%, 20%, or 30% of their body mass. Heart rate was monitored at the end of a 4-min walk in each condition. Vertical lines denote 1 SEM. Unpublished data from the Wingate Institute, Israel.

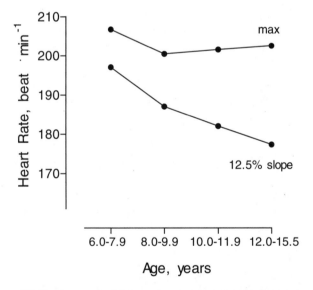

FIGURE 1.32 Heart rate reserve and age. Sixty-one girls and 83 boys, 6 to 15.5 years old, walked on a treadmill at 5.6 km · hr⁻¹, performing a progressive all-out protocol. Their mean heart rate is shown at a treadmill slope of 12.5% and at the highest attainable slope (max). Based on data by Bar-Or et al. 1971 (45) and Skinner et al. 1971 (418).

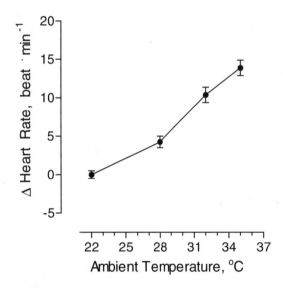

FIGURE 1.34 Effect of ambient temperature on submaximal heart rate. Twenty children (8-11 years old) performed several submaximal 5-min cycling bouts in different climatic heat stress conditions. Vertical lines denote 1 SEM.

Reproduced, with permission, from Hebestreit et al. 1995 (203).

beat · min[-1] for each 1° C above 22° C. In general, this climatic effect was similar irrespective of whether the exercise was continuous or intermittent. However, at high exercise intensities, the climatic effect was somewhat greater when children performed intermittent exercise (201).

The practical implication for a clinician who wishes to test children by exercise is to use an air-conditioned room in which temperature and humidity are fairly standardized. Another implication is for long-term HR monitoring as a means for assessing energy expenditure. To increase the validity of such monitoring when the child is exposed to climatic heat stress, one can use a correction nomogram, as discussed in appendix III (figure III.4) (203). A recent study by Kriemler et al. has shown that correction of HR for climatic heat stress in the summer reduced the estimated energy expenditure during outdoor activities by 9% in a group of 9- to 14-year-old obese boys (255).

Acclimatization to Heat

The difference in HR between the acclimatized and non-acclimatized state can be as much as 15 to 20 beat · min[-1] (38). Such a difference can easily mask any other effects on HR (see also discussion of acclimatization in chapter 3).

Emotional Stress and Habituation

As in adults, the HR of children increases with excitement or fear. It can rise by as much as 20 to 40 beat · min[-1] above the "real" value. Such an increase is apparent mostly at rest and during mild levels of exercise. For example, in a group of 4- to 10-year-old patients, mean HR was 21 beat · min[-1] above baseline just prior to the administration of anesthesia (268).

To a child, the first encounter with an exercise laboratory, its instruments, and the unfamiliar personnel and tasks can induce apprehension. In research, a session of habituation is highly recommended, during which the child should undergo all procedures and be familiarized with the equipment. Ideally, data from that session should be discarded. In a clinical setup, one should reassure the child and then take measurements, but discard HR data from low exercise levels if they look out of line with those obtained at higher intensities. One way to habituate a young child is to let him or her play prior to testing with such equipment items as the mouthpiece, nose clip, or stethoscope.

Even though this is a prudent approach, there are scant data on the actual effect of habituation in pediatric exercise testing. In a study by Frost et al., 7- to 11-year-old children repeatedly walked or ran on a treadmill. Each child was monitored during six 6-min trials. The same protocol was repeated on another day (169). While steady state HR did not differ among the six walks or the six runs on either day, HR of the very first trial was higher (by 5-10 beat · min[-1]) in the running group on the first day compared with the second. There were no interday differences in the HR of the walking group. In contrast, Unnithan et al. administered, on two separate days, three submaximal treadmill runs to boys (10.7 ± 0.7 years) and found no interday differences in their HR (443).

Active Muscle Mass

When a given mechanical power is produced by small muscles, HR will rise more than when the same power is produced by a larger muscle mass. A case in point is arm ergometry, in which HR can be 20 to 30 beat · min[-1] higher than that obtained at identical loads during leg ergometry (34; 184). Thus, when children with diseases such as cerebral palsy, poliomyelitis, muscular dystrophy, or spina bifida are tested by arm cranking, one must make allowance for the small muscle mass involved. Prediction equations and nomograms designed for leg ergometry should not be used with such patients. Unfortunately, normative data for arm ergometry in children are not available.

Training and Detraining

The HR response to exercise is a highly sensitive gauge of changes in aerobic fitness (see "Training of the Cardiovascular System" later in this chapter).

Diseases

Some diseases are accompanied by specific changes in the submaximal or maximal HR (table 1.7). For details, see the respective chapters.

Medications

Pharmacologic agents that affect the autonomic nervous system or the metabolic rate induce changes in HR. At present, specific information is scant. Propranolol, a commonly used beta-blocker, was found to reduce both the submaximal and maximal HR of healthy children by some 15 to 30 beat · min[-1] (431) (figure 1.35).

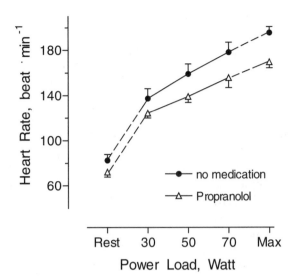

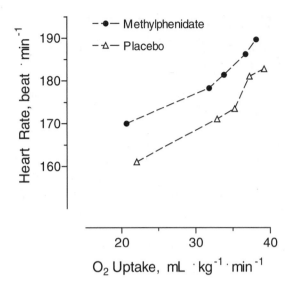

FIGURE 1.35 Beta-adrenergic blockers and exercise heart rate. Changes in heart rate with and without orally administered 10 mg propranolol. The subjects were six 11-year-old healthy boys who performed a progressive supine cycle test. Vertical lines denote 1 SEM. Data by Thorén 1967 (431).

FIGURE 1.36 Methylphenidate and exercise heart rate. Comparison between the effect of methylphenidate and placebo pills on twenty 6- to 12-year-old girls and boys with attention deficit.

Adapted from Boileau et al. 1976 (72).

Methylphenidate, which is often prescribed for children with attention deficit, induced an increase of some 10 beat · min⁻¹ in submaximal HR, but very little in maximal HR (72) (figure 1.36). Although not documented, it is likely that other drugs (e.g., thyroid hormone) would also affect children's hemodynamic response to exercise. Beta 2 sympathomimetics, often prescribed in asthma, are known to increase the resting HR. However, there seems to be little or no effect on HR during exercise (214).

Body Position

Supine exercise testing often has to be resorted to, especially in a clinical setting. In this body position, which is more favorable for venous return, resting submaximal and maximal HRs are lower than during upright exercise (108). For example, among 8- to 15-year-old healthy boys, maximal HR during upright cycling was higher by 9 beat · min⁻¹ than during supine exercise (107).

Distribution of Blood Flow During Exercise

In addition to an increase in cardiac output, there is a major redistribution of regional blood flow during exercise. In adults, an increase takes place in blood flow to the exercising muscles (skeletal

and respiratory), to the myocardium, and, at some power intensities, to the skin. Concurrently, there is a marked compensatory blood flow decline to the kidneys, splanchnic area, and nonexercising muscle. Data on peripheral blood flow of children are rather limited. Compared with young adults, 12-year-old boys had a higher muscle blood flow after exercise (246). This difference diminished when the same boys were retested 1 year later (247) and again 4 years later (248).

If indeed children have a higher muscle blood flow compared with adults, this may represent a more favorable peripheral distribution of blood during exercise. Such distribution facilitates a greater O_2 transport to the exercising muscle (in spite of a lower cardiac output) and a greater decline in the O_2 content of mixed-venous blood. As in adults, muscle blood flow of boys seems to reach a new plateau 30 to 40 sec after the onset of a new exercise level. Such adjustment is faster than that of O_2 uptake or cardiac output. It is possible, then, that muscle blood flow is not a limiting factor in a child's O_2 transport capacity during exercise.

There is some evidence that when children exercise in a hot climate, their blood flow to the skin is relatively greater than in adolescents and adults (128; 157). For details, see the section "Low Cardiac Output and Enhanced Changes in Peripheral Blood Flow" in chapter 3.

Arterial Blood Pressure and Total Peripheral Resistance in Rhythmic Exercise

The higher contractile forces of the myocardium during exercise cause an increase in intraventricular systolic pressure (figure 1.37). This is one mechanism by which more blood is made available to the periphery. Such an increase is manifested by, among other functions, a rise in arterial systolic blood pressure. In contrast, diastolic blood pressure, which depends primarily on the peripheral vascular resistance, changes little (or even decreases slightly) with exercise, and its direction cannot be predicted. When rhythmic exercise is performed, the rise in systolic blood pressure is proportional to exercise intensity and to the overall metabolic level. Such a pattern is operative in all healthy individuals,

irrespective of age. There are, however, age- or size-related quantitative differences: For a given level of exercise, a small child responds with a lower increase in systolic and diastolic pressures than does an adolescent (227; 257; 355; 424).

With the increase in mean blood pressure, total peripheral resistance (mean arterial blood pressure divided by cardiac output) decreases progressively. In spite of the lower arterial pressure in children, their peripheral resistance is higher than in adults (figure 1.38) (94; 439). There seem to be no gender differences in total peripheral resistance (192; 439).

A lower exercise blood pressure in the young child is in line with the lower cardiac output and stroke volume. There is no reason to assume that such an age-related pressure difference is either materially beneficial or detrimental to the working capacity of the young child.

Within a given age group, boys have a higher peak systolic blood pressure than girls (355), probably because of higher maximal stroke volume in the boys. For reasons unknown, African American children respond to exercise with higher arterial blood pressure than do Caucasian children (3; 424).

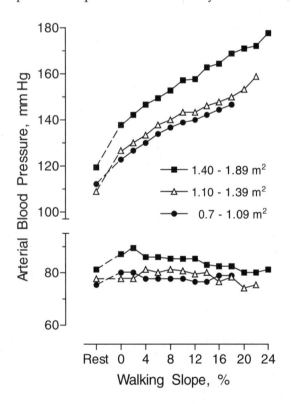

FIGURE 1.37 Systolic (top) and diastolic (bottom) blood pressure during exercise in children and adolescents. Subjects were 274 girls and boys who had been referred to an outpatient pediatric clinic for reasons other than organic cardiovascular disease. Walking was done on a treadmill at 3.5 km · hr⁻¹. The slope was raised each minute by 2%. The participants are subdivided into three groups according to body surface area.

Adapted from Riopel et al. 1979 (355).

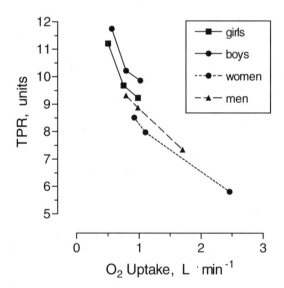

FIGURE 1.38 Total peripheral resistance (TPR) in children and adults of both genders at several levels of treadmill exercise. Subjects and methods are the same as in figure 1.29.

Reproduced, with permission, from Turley and Wilmore 1997 (439).

Arterial Blood Pressure and Total Peripheral Resistance in Static Exercise

The blood pressure responses to static contraction are different from those that occur during rhythmic contractions. Most striking is the dissociation between metabolic demands and blood pressure: Both systolic and diastolic pressures increase above and beyond those expected merely from an increase in metabolism or in cardiac output. Whenever the static effort (e.g., during resistance training) exceeds some 20% of the maximal voluntary contraction of the respective muscle group, steady state cannot be reached. Fatigue keeps increasing and exhaustion takes place within minutes, even though HR may remain as low as 110 to 120 beat · min^{-1} (41; 266). Apparently, a pressor response is activated during static contraction such that within 1 min (50% of maximal voluntary contraction), systolic pressure of adults may reach near-maximal values, with a concomitant rise in diastolic blood pressure.

In adults, the rise during static contraction depends mostly on the intensity of contraction of a muscle group, relative to the maximal voluntary contraction of that muscle group, and less on the muscle mass that performs the static contraction or the actual force generated. The Valsalva maneuver further increases the blood pressure (271). Another important finding is that static exercise can induce a tremendous rise in blood pressure. This was shown by MacDougall et al., who took direct measurements of pressure inside a brachial artery. Values as high as 350 mmHg systolic pressure and over 250 mmHg diastolic pressure were found when weightlifters maintained a two-leg contraction to exhaustion at 90% maximal voluntary contraction (274).

Although it has been assumed that children and adolescents respond like adults to static exercise, specific data on children's blood pressure response are scant and relate mostly to handgrip (163; 401; 424; 436; 454). Peak systolic blood pressure during rhythmic cycling is usually higher in children than that achieved by exhausting handgrip efforts of various combinations of intensity and duration. Turley et al. have shown that, at an equal percentage of maximal handgrip contraction, the rise in blood pressure of adults is greater than in children. For example, at 30% maximal handgrip contraction held over 3 min, the mean increase in systolic blood pressure of 7- to 9-year-old boys was 17.4 mmHg, compared with 36.2 mmHg in 18- to 26-year-old men. The respective increases for diastolic pressure were 20.2 and 29.3 mmHg. No studies are available in which the metabolic levels of rhythmic and static efforts were equated to obtain a valid comparison of their effect on blood pressure.

There is a definite need for more data on the blood pressure response of children to static activities. Such information is relevant, for example, to children with hypertension, those with cerebral a-v shunt, and patients who, due to locomotor disability, resort to static activities.

Pulmonary Response to Exercise

In addition to its role in enhancing O_2 and CO_2 transport during exercise, the respiratory system affects acid–base balance by controlling body stores of CO_2. To facilitate the increase in O_2 and CO_2 exchange, ventilation rises, bringing about a rise in alveolar ventilation. In addition, blood flow through the pulmonary capillaries increases proportionately, as does the rate of pulmonary diffusion for oxygen. The increase in ventilation is achieved by a faster respiratory rate and a larger tidal volume. This section reviews age-related differences in respiratory responses to exercise, as summarized in table 1.8.

Ventilation and Ventilatory Threshold

In absolute terms, ventilation increases with age. This is true for rest and for exercise at all levels. However, when calculated per body mass unit, ventilation at rest and during submaximal exercise is higher in children, and it decreases with age (319; 360; 382). For example, Robinson's cross-sectional data show that at age 6, submaximal ventilation per kilogram is 50% higher than at 17 years (360). In a longitudinal study, Rowland and Cunningham found a decrease of 26% in submaximal ventilation per kilogram from age 9 to 14 years in girls and boys who walked on the treadmill (382).

Whereas a 6-year-old child may reach a maximal ventilation of 30 to 40 L · min^{-1}, a young adult can reach 100 to 120 L · min^{-1} and more (25; 360).

TABLE 1.8	Respiratory Function During Exercise—a Comparison Between Children and Adults	
Function	**Children's response (compared with adults')**
Ventilation in liters per minute—submax, max	Lower
Ventilation per kilogram body weight—max	Same or higher
Ventilation per kilogram body weight—submax	Higher
Ventilatory threshold, % max O_2 uptake	Same or higher
Respiratory rate—max, submax	Higher
Tidal volume/vital capacity—max	Lower
Tidal volume/vital capacity—submax	Same or lower
Ventilatory equivalent—max, submax	Higher
Dead space/tidal volume	Same
Partial pressure of arterial CO_2	Somewhat lower

Some studies suggest that when expressed per kilogram body mass, maximal ventilation is about the same in children, adolescents, and adults (305; 319; 360). However, a 5-year longitudinal study showed that this is true between ages 9 and 14 for boys but not for girls, whose ventilation declined steadily with age (382).

Putting together submaximal and maximal data, one can infer that the ventilatory reserve (i.e., the difference between maximal and submaximal ventilation) increases with age, at least for males. This is in line with the age-related increase in metabolic reserve and HR reserve, as discussed earlier.

During exercise at incremental intensities, ventilation increases linearly with the metabolic rate until the child reaches the ventilatory threshold (also called ventilatory anaerobic threshold—VAT) at approximately 60% to 70% of maximal O_2 uptake. Beyond this level, ventilation rises in an accelerated manner relative to the rise in O_2 uptake. Such accelerated ventilation is considered a response to metabolic acidosis and an increase in CO_2 production. This increase, which is in addition to the CO_2 produced through the oxidation of energy sources in the contracting muscles, results from the need to buffer (through bicarbonate) the increasing levels of lactate and hydrogen ions. The ventilatory threshold is related to, and often coincides with, the "anaerobic threshold" (see the section "Anaerobic Threshold" earlier in this chapter).

As shown in figure 1.39, ventilation (L · min⁻¹) at any given O_2 uptake (L · min⁻¹) is higher the younger the child, and the ventilatory thresh-

old seems to appear earlier than in adolescents or adults. However, when ventilation is plotted against percent maximal O_2 uptake, the ventilatory threshold is often (102; 231; 352), but not always (286; 387), higher in young children. This pattern occurs in both genders (102; 352; 453) (figure 1.40).

Respiratory Frequency and Tidal Volume

Because ventilation is the product of respiratory frequency and tidal volume, it is important to understand the changes that occur in these two variables during growth. Exercising children have a markedly higher respiratory frequency than older individuals who perform the same task, maximal or submaximal (25; 78; 185; 305; 319; 360; 382; 387; 404). For example, during a walk at 5.6 km · hr⁻¹, 8.6% incline, a

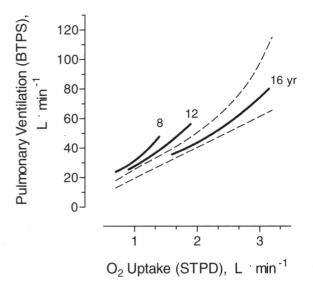

FIGURE 1.39 Ventilation during exercise at different ages. Subjects were 83 Norwegian boys, subdivided into three age groups, who cycled at two submaximal intensities and one maximal intensity. BTPS = body temperature, ambient pressure, saturated with water vapor; STPD = standard temperature and pressure, dry air. The area between the broken lines represents data for young adults (24). Data for the children are by Andersen et al. 1974 (10).

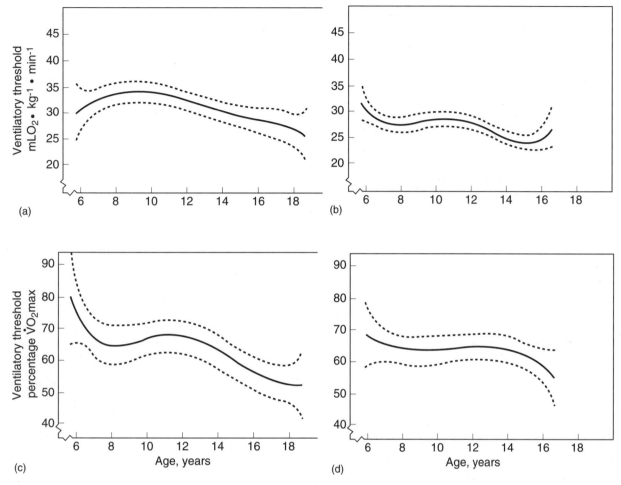

FIGURE 1.40   Ventilatory threshold (mean and 95% confidence interval) as a function of age. Cross-sectional data of 117 girls and 140 boys who exercised on a treadmill. The threshold is expressed as milliliters of O_2 per kg per min and as percentage maximal O_2 uptake.

Reproduced, with permission, from Reybrouck et al. 1985 (352).

6-year-old child breathed at a rate of some 50 cycles · min^{-1}, while a 25-year-old man needed only 25 cycles · min^{-1} (360). During a maximal running test, respiratory frequency was 70 cycles · min^{-1} in 5-year-old children compared with 50 per minute at age 17 (25). Age- or maturation-related differences in respiratory frequency during cycling may be less distinct than during running or walking, but are still apparent (78; 185; 387; 404). There seem to be no gender differences in respiratory frequency (figure 1.41) (17; 25; 382).

The age-related decline in respiratory frequency is compensated for by an even greater increase in tidal volume so that absolute ventilation increases with age, as discussed previously. The increase in tidal volume is proportional to the increase in body mass (305; 382).

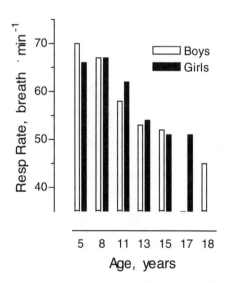

FIGURE 1.41   Respiratory rate declines with age in boys and girls. Subjects and methods are as in figure 1.7.

Reproduced, with permission, from Åstrand 1952 (25).

Does their relative tachypnea reflect shallow breathing in children? One way to express the depth of ventilation is by the ratio tidal volume/vital capacity. During submaximal tasks this ratio is marginally lower in children than in adults. During maximal exercise, however, it is only 0.42 to 0.48 in children, compared with 0.48 to 0.54 in adolescents and 0.56 to 0.59 in adults (25; 185; 360; 387). One can therefore conclude that, compared with adolescents and adults, children respond to exercise with relative tachypnea and with shallow breathing.

Ventilatory Equivalent

The ratio known as ventilatory equivalent, the number of liters of air needed to ventilate the lungs in order to supply 1 L of oxygen, is a numerical expression of ventilatory efficiency. (An alternative way to display ventilatory equivalent is ventilation per 1 L of exhaled CO_2.) A high ratio reflects low efficiency.

Ventilatory equivalent typically drops with age (figure 1.42) (10; 25; 245; 361; 377). Some studies show this pattern for submaximal and maximal levels, while others show it only for submaximal levels. This age-related trend suggests a less efficient ventilation among younger versus older children. Girls have a higher submaximal ventilatory equivalent than age-matched boys (17; 25; 382). This was not confirmed by a longitudinal

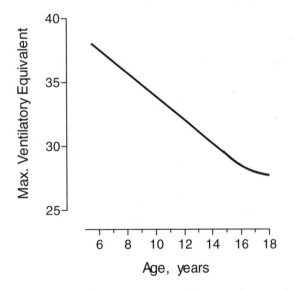

FIGURE 1.42   Ventilatory equivalent during maximal exercise, as related to age. A schematic presentation based on data by Andersen et al. 1974 (10), Åstrand 1952 (25), and Kobayashi et al. 1978 (245).

study in which some 120 girls and boys were followed up for 5 to 7 years (387).

For children, the major implication of less efficient ventilation is a greater O_2 cost of breathing. This may contribute to their relatively high metabolic demands during submaximal tasks (see the section "Mechanical Efficiency and Economy of Movement" earlier in this chapter).

Alveolar Ventilation and Gas Exchange

It is the alveolar rather than the pulmonary ventilation that determines gas exchange in the alveoli. In spite of their shallow breathing, children's alveolar ventilation is adequate for gas exchange (16; 101; 176; 186; 411). In fact, children have mild alveolar hyperventilation, as judged from their arterial CO_2 pressure (PCO_2). During various levels of treadmill (12; 360) or cycling (185; 186) exercise, children's arterial PCO_2 has been lower (33-36 mmHg) than among adults (40 mmHg). The relative alveolar hyperventilation in children may be explained by two mechanisms: a lower CO_2 threshold for driving the respiration (187) and a greater relative dead space due to the fast, shallow respirations.

There are no systematic data to indicate age-related differences in the rate of pulmonary diffusion during exercise. The scant findings range from lower (249) to higher (13; 409) values in children, compared with adults.

Vital Capacity and Exercise Performance

Vital capacity is strongly related to body size, particularly height, during growth (17; 347; 348; 405). Historically, vital capacity was considered an index of fitness, based on high values among some groups of athletes. However, when body size is partialled out, correlation between athletic performance of healthy children and their vital capacity is low (109). Vital capacity of girl swimmers (26; 136) and boy swimmers (140) was higher than expected from their body size, but no correlation was shown between swimming performance of these children and their vital capacity. With the exception of elite endurance athletes, only in advanced lung or chest wall disease will vital capacity become a limiting factor in exercise performance (see chapters 6 and 10).

Effects of Growth and Maturation on Muscle Strength

In this section we will discuss muscle strength in normal children by focusing on three subtopics. First, we will consider what strength is and how it is measured. Then we will consider the development of strength during growth and maturation. Lastly, we will look at factors that determine strength during growth.

Strength and Its Measurement

Muscle strength is defined as the highest mechanical force or torque that can be generated by a muscle or a muscle group. Strength is an important component of fitness in health and disease, because weak muscles may markedly limit a person's physical fitness and daily physical abilities. There are three types of muscle action: concentric (in which the muscle shortens), eccentric (the muscle lengthens, because external forces that act in an opposite direction to its action are greater than the force or torque generated by the muscle itself), and isometric (no change in muscle length, because of opposing forces that equal the force generated by the muscle). The maximal muscle performance in each of these three actions can be used to describe muscle strength.

The physical fitness literature often includes another category of muscle strength, explosive strength (the maximal force or torque produced by the muscle in the shortest possible time period, as in vertical jump) (e.g., [58]). Strictly speaking, however, this performance characteristic is not strength, because it is measured by mechanical power and not by force or torque. For further discussion of high-intensity muscle power, see the section titled "Anaerobic Exercise."

Unlike aerobic fitness, which refers to the body as a whole, strength is a local characteristic of each muscle or muscle group. As a result, a person's strength varies from one muscle group to the other, and the correlation between muscle groups is low to moderate (23; 167). It is possible, for example, for a child to have strong lower limbs and weak arms or to show major differences in strength of the right and left sides of the body.

To characterize a person's strength, one should therefore take measurements of several muscle groups.

Clinicians often assess a patient's strength, as part of a physical examination, by asking the patient to flex or extend a muscle group against resistance provided by the clinician. This approach, at best, can yield a rough, qualitative assessment of strength. While useful as a qualitative clinical tool, this procedure is ineffective when one wants to compare a child's strength with normative data, or monitor small changes in strength during growth or as a result of treatment or natural progression of a disease. To achieve a more accurate measurement of muscle strength, one should use a dynamometer, which can provide objective, quantifiable, and reproducible information. A handgrip dynamometer, for example, provides information about isometric strength of the finger flexors in each hand. More sophisticated dynamometers are available in exercise laboratories, and in some physical therapy and rehabilitation clinics, to determine isometric, concentric, and eccentric strength of larger muscle groups (e.g., knee extensors, knee flexors, elbow flexors), as well as small ones (e.g., finger flexors) (253).

Development of Strength During Growth and Maturation

Abundant cross-sectional studies (e.g., [23; 95; 115; 172; 207; 211; 213; 230; 232; 235; 292; 326; 337; 349; 366; 367; 425; 427; 457; 468]) and several longitudinal studies (e.g., [90; 241; 264]) have documented changes in muscle strength during childhood and adolescence. These have been summarized over the years in various reviews (e.g., [23; 58; 61; 71; 167; 228; 291; 428]). In general, strength increases as the child grows and the muscle mass increases. However, the relationships between strength, size, and muscle mass differ among muscle groups and among the strength characteristics (e.g., static vs. dynamic vs. explosive strength).

During the prepubertal years, the pace of increase in strength is similar in girls and boys. Muscle strength in the two groups is similar, or is slightly higher in boys (figure 1.43). Thereafter, strength becomes progressively greater in boys. It starts accelerating in boys at approximately age 13 to 14 years, whereas the increase in girls' strength shows little or no acceleration. The gender difference in muscle size and strength becomes apparent more in the upper body than in the legs. While the greater strength of maturing boys may reflect their higher levels of circulating testosterone (366) and faster somatic growth, it may also result from their greater involvement in sport and other physical activities that strengthen the muscles.

The velocity of strength increase during adolescence varies among individuals. Longi-

tudinal studies show that the fastest increase occurs usually during the year of peak height velocity, or during the subsequent year, which coincides with the year of peak weight velocity. Figure 1.44 summarizes the velocity of increase in static arm pull of several European groups of boys and girls who were tested periodically before, during, and following the time of peak height velocity.

Because strength depends on various morphological, neurological, biomechanical, and hormonal factors, the development of strength during childhood and adolescence is not merely a function of chronological age. It is often more related to maturational age, typically expressed as skeletal age (291). For example, early maturers are stronger than late or average maturers of the same chronological age (98), even when differences in body mass are corrected for (90). The correlations between strength and skeletal age are greater in boys than in girls (58). In boys, the advantage of early maturation is apparent mostly at ages 13 to 16 years (167) and diminishes later, due to catch-up by the late maturers (57; 97) (figure 1.45). In girls, the maturation-related strength differences are apparent mostly at ages 11 to 15 years, and they subsequently diminish.

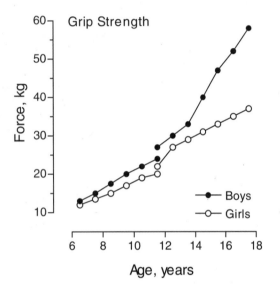

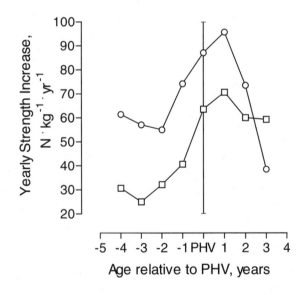

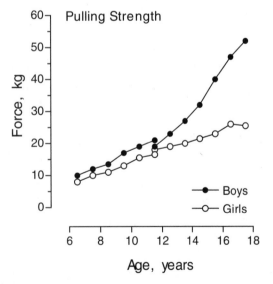

FIGURE 1.43   Changes in strength with chronological age. Cross-sectional data for elbow flexors and knee extensors of girls (open symbols) and boys (filled symbols), 5 to 17 years of age. Values are presented in newtons (N).

Reprinted from Malina et al., 2004 (291).

FIGURE 1.44   Yearly increase in upper (□) and lower (○) body isometric strength of boys. Values are aligned according to peak height velocity (PHV).

Reproduced, with permission, from Froberg and Lammert 1996 (167).

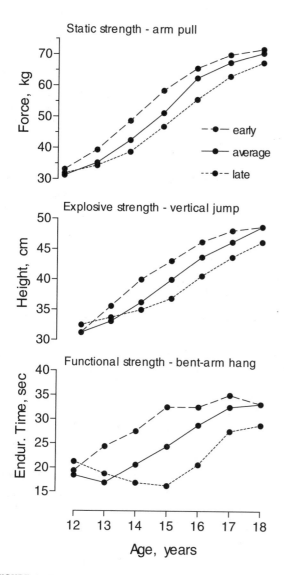

FIGURE 1.45 Longitudinal changes in strength of early-, average-, and late-maturing Belgian boys. Division into the maturational groups was determined by the difference between skeletal and chronological age of each subject. Based on data from the Leuven Growth Study (57).

Reproduced, with permission, from Malina et al. 2004 (291).

Factors That Determine Strength During Growth

Strength is determined mostly by two factors: the cross-sectional area of the contracting muscle and the number of motor units that are activated at any given time. The cross-sectional area reflects the size, and to a lesser extent, the number, of muscle fibers (229). While isometric strength is not affected by the length of the muscle, its contraction velocity during dynamic exercise (and the resulting mechanical power) is related

to muscle length, which keeps increasing as the child grows (228).

It is still unclear whether or not the increase in strength during growth reflects changes in muscle contractile characteristics. For example, 16-year-old boys were found to activate a higher percentage of their available motor units in the quadriceps muscle than 10-year-old boys. However, there was no age difference in the percentage of motor unit activation in the elbow flexors (61; 64).

The factors just referred to determine the strength of a single muscle or a muscle group, as measured in a laboratory setting. In a real-life situation (e.g., sport or daily activities), the coordination between and among muscle groups is also important. For example, success in discus throwing depends not only on the strength of each muscle but also on the precise timing at which each muscle group contracts. Likewise, getting up from a chair requires the coordination among several muscle groups of the lower limbs, the upper limbs, and the trunk.

Effects of Growth and Maturation on Bone

The normal development of bone mass during childhood and adolescence is considered important in the prevention of bone thinning (osteopenia) and fractures in older years, particularly in females. Bone mineral density normally rises during the growing years, peaks by early adulthood, and then declines in persons who are elderly. Promoting accretion of bone mass in the pediatric years through proper nutrition and exercise, then, may ameliorate the degree of bone loss as one ages.

During the growing years, bone increases in size, mineral content, and mineral density. Measures of cortical (outer compact) and trabecular (inner spongy) bone density can be made by both absorptiometry and computed tomography techniques. The former, which is used more commonly, can provide only a two-dimensional, or areal, estimate of mineral content relative to bone size (grams per square centimeter), while the depth or thickness of the bone tissue is ignored. This is not a true volume density as measured by tomography (grams per cubic centimeter). This accounts for some of the variability in the use of absorptiometry for estimation of bone mineralization. It is important to recognize, as well, that bone density varies markedly from one bone to another, between sites in the same bone, and according to factors such as gender and race (66; 238; 239).

Both bone mineral content and areal bone density gradually increase during the prepubertal years, and values for boys and girls are similar. Osteoblastic activity and bone development at this time are triggered by pituitary growth hormone secretion, probably mediated via insulin-like growth factors (IGF). A significant correlation has been observed between serum IGF levels and areal bone density in both prepubertal boys and girls.

At puberty, growth of bone density accelerates in response to the anabolic actions of the sex hormones. Gains in bone density between the ages of 10 and 15 years (typically 15-25%) are greater than at any other time in life. Most evidence suggests that estrogen is most critical to bone development during adolescence in both males and females, although testosterone contributes to cortical bone size. Increases in bone mass during puberty, however, are more evident in males than females. Greater values for men become evident by late adolescence, and gender differences become more obvious during early adulthood.

The age of peak bone mineral density for women is between 20 and 25 years of age (the data are not clear in males), and almost all women have maximized total bone mineral content by age 26 years. Rate of biological maturation appears to have no influence on age of peak bone density (239).

Three factors appear critical for normal bone development: (1) adequate hormonal stimulation, (2) sufficient dietary calcium intake, and (3) mechanical stresses of gravity and physical activity. Estrogen deficiency at menopause may play an important role in the development of osteoporosis in persons who are elderly. In young females, hypoestrogenemia associated with oligoamenorrhea that accompanies intense sport training may have the same effect. Calcium is essential for bone mineralization, and bone mass has been positively associated with dietary intake in premenopausal females. Limited data in men have not so far revealed this association.

The importance of weight-bearing exercise on bone development is illustrated by the distinct bone atrophy that occurs when the skeleton is immobilized by bed rest, plaster casting, or denervation. Typical loss of bone mass when an extremity is casted for a sport injury, for instance, is about 15 to 20%. Immobilization causes not only bone resorption but also increased calcium excretion in the urine, a phenomenon also observed in astronauts after a period of weightlessness (223).

Forces created by normal muscle contraction and gravitational load create miniscule deformations in bone structure (106). These changes trigger biochemical responses that stimulate cell proliferation, differentiation, and bone formation. The intensity of forces is probably a more effective stimulus for this effect than the number of stress repetitions is (296).

In adults, higher bone mineral density is observed in active versus inactive women, female athletes versus nonathletes, and women involved in weight-bearing sports as compared to non-weight-bearing activities (238). In children and adolescents, cross-sectional studies have also demonstrated a positive association between bone mineral density and weight-bearing activity (419). In a 6-year longitudinal study, Bailey et al. found that recreationally active boys and girls gained 9% and 17% greater bone mineral content, respectively, than those who were inactive (29). (The effects of athletic training on bone health in children and adolescents are addressed later in this chapter.)

In the Amsterdam Growth and Health Longitudinal Study, both 27-year-old males and females who reported higher levels of habitual physical activity in the previous 15 years demonstrated a higher peak bone mineral density than those who were less active (241). In that study, physical activity measured as peak strain created a more osteogenic effect than physical activity measured as energy expenditure.

Prolonged Exercise

Prolonged activities are defined here as those that last 30 min or more and are continuous in nature. Although such activities are contrary to a child's pattern of spontaneous exercise (31), many children have been pursuing these types of activities since distance events (e.g., running, swimming, cross-country skiing) became popular within children's sport. Prolonged activities are also practiced by other young athletes and dancers as part of their training regimen.

The first published studies on children's response to prolonged exercise, which came from Czechoslovakia (277; 280; 283) and West Germany (244), suggested that, in general, the responses of children and adolescents are similar to those of adults. When exercise intensity is 60% to 70% of maximal O_2 uptake and lasts about 30 to 60 min, HR reaches a new plateau within the first few minutes; but it then gradually drifts upward and may be some 10% to 15% higher after 40 to 60 min than at 10 min of exercise (93; 94; 277; 280; 434). Ventilation may rise during that time by some 2 L · min^{-1}, and O_2 uptake by 1 to 2 mL · kg^{-1} · min^{-1}. Whereas the rise in HR and ventilation can be explained by an increase in core temperature and a mild level of dehydration (which is accompanied by a reduction in plasma volume), the drift in O_2 uptake may be due to a shift from carbohydrate to fat utilization by the active muscles. This shift is reflected by a concomitant reduction in the respiratory exchange ratio, RER (i.e., CO_2 production/O_2 uptake) (22; 279; 300; 354; 385; 435).

The main age-related differences in the response to prolonged exercise are a lower accumulation of blood lactate (244; 280; 300), a milder increase in serum potassium (54), and a milder decrease in plasma volume (94; 281) in children. Heart rate is higher in children throughout the prolonged activity, but there is no age-related difference in the rate of HR increase over time (94; 435). Stroke volume is higher in adults and remains so throughout a prolonged bout of exercise (94).

Whether the lower accumulation of blood lactate signifies less production in the muscle or a faster removal from the blood is not clear. It may, however, result from a lesser use of carbohydrates and a greater reliance on fat as energy source. This idea, based on age-related differences in respiratory exchange ratio, was documented in some (300; 354; 435), but not all (22; 385), studies. An example of children's greater reliance on fat is shown in figure 1.46. It is not clear why children have a preferential use of fat as an energy source. One possibility is that they have greater intramuscular triglyceride deposits (52). Another hypothesis is that a higher use of fats is an outcome of a less-developed glycogenolytic or glycolytic system (435).

Using stable isotope 13C, Riddell (354) and Timmons (435) have shown that, even though children use less endogenous carbohydrate and more fat, they utilize exogenous carbohydrate (provided as a beverage) faster than do adults during prolonged exercise (figure 1.47). The cause for such faster utilization rate in children is not clear.

In conclusion, apart from low economy of locomotion (see earlier section, "Mechanical Efficiency and Economy of Movement"), there do not seem to be any underlying physiologic factors that would make children less suitable than adults for prolonged continuous exercise. A child's preference for activities of shorter duration must be explained by psychologic and cultural factors, such as shorter attention span, the need for recreational stimuli, and a lower socially induced motivation for long-term exercise.

Warm-Up Effect

Warming up is a common practice among athletes and dancers. Warm-up should be practiced by anyone who exercises. Stated dividends include both improvement of performance and prevention of injuries. There are a number of possible physiologic mechanisms by which warm-up

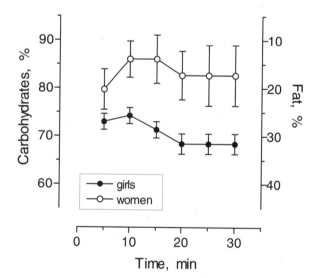

FIGURE 1.46 Relative utilization of fat and carbohydrate energy sources in women and in girls. Ten 8- to 10-year-old prepubertal girls and ten 20- to 32-year-old women ran for 30 min at 70% of their maximal O_2 uptake. Substrate utilization was calculated from the respiratory exchange ratio. Vertical lines denote 1 SEM.

Adapted, with permission, from Martinez and Haymes 1992 (300).

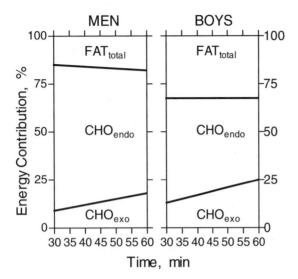

FIGURE 1.47 Effect of the ingestion of 6% carbohydrate (4% sucrose, 2% glucose) on substrate utilization in men and boys during prolonged exercise. Twelve boys (9.8 ± 0.1 years old) and 10 men (22.1 ± 0.5 years) cycled at 70% maximal O_2 uptake for 60 min. Total fat utilization and endogenous carbohydrate utilization (CHOendo) were calculated from the respiratory exchange ratio. Utilization rate of the exogenous carbohydrate (CHOexo) was determined using 13C isotope, which was added to the beverage and collected in the expired air (354).

Adapted, with permission, from Timmons et al. 2003 (435).

exerts its benefits: a faster impulse propagation along neurons; faster muscle contraction (396); a "shift to the right" in the oxyhemoglobin dissociation curve (thus greater O_2 availability to tissues); decrease in O_2 deficit at the beginning of exercise; improved coronary blood flow and myocardial oxygenation (preventing myocardial ischemia at the start of intense exercise); reduction in the risk of exercise-induced asthma (117) (see the section "Refractory Period" in chapter 6); increased mechanical efficiency of the muscles and reduction of viscosity within the muscle, between tendons and their sheath, and inside the joints. While warming up is practiced by athletes at all levels, its beneficial effect in preventing running-related injuries is controversial (449).

Proponents of a psychologic effect of warm-up claim that much of the improved performance is due to the attitude of the athlete who will not perform "all-out" without warming up. To circumvent such a psychologic mechanism, Inbar and Bar-Or selected young children who had been found to be ignorant of the concept of warm-up (218). These 7- to 9-year-old nonathletic boys performed a 4-

min cycling task at a supramaximal intensity and a 30-sec anaerobic task (the Wingate anaerobic test) with and without warming up. Warm-up consisted of 15-min intermittent treadmill runs raising rectal temperature by 0.52° C and elevating HR to 150 ± 10 beat · min⁻¹. The time interval between this prior exercise and the criterion task was 4 min. The results are shown in figures 1.48 and 1.49. Warming up helped the children utilize greater aerobic resources and reach steady state faster than without prior exercise. It also improved the score in the Wingate test, increasing the total mechanical work by 7%, but the increase in peak power (4%) was not significant. Intermittent warm-up in these children was found to be more effective than continuous warm-up, even when the two induced identical increases in metabolic level.

Passive warming of the muscle through immersion in hot water can also improve anaerobic performance. For example, following immersion in a water bath at 44° C, peak anaerobic power during cycling increased in adults by 20% (396). Conversely, cooling down (water temperature of 12° C) caused a 25% decrease in their peak power. One practical implication is that warming up should be extended if the child is exposed to a cold environment.

Various protocols can be used for warming up, depending on the task that follows, on climatic conditions, and on individual habits. One should, however, attempt to include the following three components in any warm-up: nonspecific activities that raise the core temperature, stretching exercises, and specific activities (e.g., a pitcher in baseball should throw the ball, using the actual pitching motion). A warm-up routine that is too intense may induce a reduction in performance due to fatigue (398). Based on practical experience, the optimal duration of warming up at the start of a nonspecific activity session is 8 to 12 min. The time interval between the end of the warm-up and the start of competition varies among athletes and sporting events. From a physiologic point of view it should not exceed 4 to 5 min.

Perception of Exercise Intensity

While this book focuses on physiologic and pathophysiologic aspects of pediatric exercise,

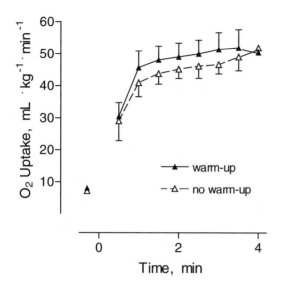

FIGURE 1.48 Warm-up effect on O$_2$ uptake in strenuous exercise. Twelve 7- to 9-year-old boys performed a 4-min one-stage cycling task with and without prior warm-up. Vertical bars denote 1 SD. See text for procedures.

Reproduced, with permission, from Inbar and Bar-Or 1975 (218).

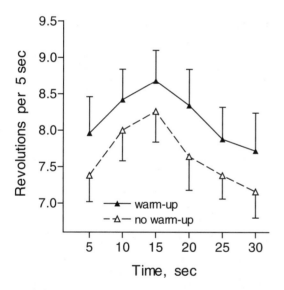

FIGURE 1.49 Warm-up effect on anaerobic performance. Twelve 7- to 9-year-old boys performed the 30-sec cycling Wingate anaerobic test with and without prior warm-up. See text for procedures. Vertical bars denote 1 SD.

Reproduced, by permission, from Inbar and Bar-Or 1975 (218).

it is important also to consider the perceptual responses of children to exercise. Specifically, one should understand how children perceive (i.e., detect and interpret) sensory cues from their working muscles, as well as other effort-related sensations. The pioneer in applying psychophysical principles to the analysis of effort perception is

Gunnar Borg, whose notion of rating of perceived exertion (RPE) was first introduced in 1962 (73). It was further developed over the years in numerous laboratories.

Several RPE scales have been constructed to quantify the way people perceive their physical effort. Appendix II offers a description of scales that can be used with children. Other chapters in this book address RPE data as they pertain to a specific disease. This section focuses on the way children perceive their exercise intensity and on age-related differences in RPE. While most research addresses the overall perception of effort, some investigators have focused on specific body parts—for example, determining the perception of effort in the legs and in the chest (287). This approach is clinically useful when one wishes to quantify, for example, the dyspnea in a patient with a pulmonary disease (242), or the perception of leg fatigue in a patient with muscular dystrophy or atrophy.

Most studies on RPE address the estimation of effort intensity by the child (i.e., how the child perceives a certain effort intensity while exercising). Fewer data are available regarding the way RPE is used for production of exercise intensity (i.e., children's ability to select an exercise intensity that is prescribed to them as a number on an RPE scale). This section addresses both aspects. The interested reader is also referred to review articles (40; 46; 76; 148; 263; 284; 359) and monographs (75; 325).

Age-Related Differences in Rating of Perceived Exertion

As discussed in chapter 2, children are habitually more active than adults. There are several environmental, physiologic, and psychosocial causes for this difference. One explanation is that adults, including those who are elderly, opt for a sedentary lifestyle because they perceive exercise as more fatiguing than children do.

Do children indeed perceive exercise to be easier? To address this possibility, Bar-Or compiled data from six studies performed in the mid-1970s at the Wingate Institute on some 1,300 males, 7 to 68 years old (589 at ages 7-17). All subjects performed an identical multistage cycle ergometer test (35). Rating of perceived exertion was determined at each stage, using the Borg 6 to 20 category scale (74). A methodological dilemma was how to compare across ages the exercise intensity at which RPE was determined. One cannot simply express RPE for an absolute mechanical power, because a 50-Watt load, for example, represents a much more intense task for a 10-year-old boy than for a young adult. One way of overcoming this is by calculating the RPE at a given physiologic strain. If submaximal HR is taken as an index of cardiovascular strain, the ratio RPE/HR can then represent the subjective strain at a given level of objective strain. Figure 1.50 summarizes this ratio for various age groups at different exercise intensities. Excluding the youngest group of 7- to 9-year-old gymnasts, there was an age-related pattern in which children had a lower RPE/HR than did the adolescents, who in turn gave a lower rating than the adults. Because submaximal HR at any given exercise level is higher in young children, RPE was also calculated at a certain percentage of maximal HR. This is a more prudent way to compare cardiovascular strain in people who vary in age. The results are shown schematically in figure 1.51, in which the ratio RPE/% maximal HR is lower in children than in adolescents, and even more so than in adults. The author (35) concluded that, indeed, exercising at a certain physiologic strain is perceived to be easier by children than by older individuals.

Subsequent studies, on a smaller scale, have confirmed that RPE at a given % maximal HR is lower in children than in adults (251; 448). However, more recently, Mahon et al. reported that RPE of nine boys (age 10.5 ± 0.7 years) and nine young men was similar when exercise was performed at the ventilatory threshold (286). Other comparisons by the same group showed that children's RPE was higher at ventilatory threshold (287; 288).

It is hard to reconcile the differences among these studies, but the bulk of the evidence appears to suggest that, during short bouts of exercise (i.e., up to 10 min), children's RPE is lower for a given physiologic strain than in adults. The reason for this pattern is not clear. The lower rating by children might be explained by their lower blood lactate, blood acidity (356), and O_2 deficit. However, lower blood lactate levels in children are not necessarily accompanied by a lower RPE (286; 388). As discussed earlier (in the section "Recovery Following Exercise"), children recover faster than adults following a strenuous

physical task. The lower RPE in children is in line with their faster recovery. It is possible that children's eagerness to repeat a strenuous task is, in part, related to their low rating of perceived effort.

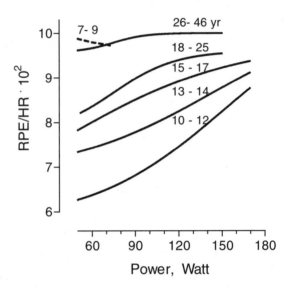

FIGURE 1.50 Development of exercise perception with age. The ratio of rating of perceived exertion (RPE) to heart rate (HR) at various power loads. Data on 904 males, 7 to 46 years old, who exercised on the cycle ergometer. Subjects are subdivided into six age groups. Based on Bar-Or 1977 (35).

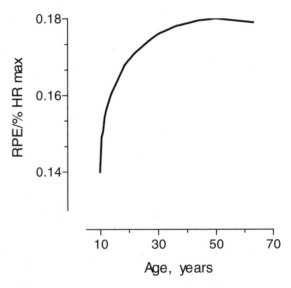

FIGURE 1.51 Exercise perception and age. Rating of perceived exertion (RPE) at a given percentage of maximal heart rate (RPE/%HRmax) in 1,307 males, 9 to 68 years old, who exercised at 100 W. Schematic presentation, based on Bar-Or 1977 (35).

From Bar-Or 1982 (36).

In contrast to RPE response to a short bout of exercise, children seem to have a faster rise in RPE during prolonged exercise of 15 min or more. As seen in figure 1.52, RPE of 10- to 11-year-old boys increased significantly faster than in young men, when fairly intense exercise (cycling at 70% maximal O_2 uptake) was sustained over 60 min (434). The faster increase in the children's RPE was maintained even when RPE was expressed as the ratio RPE/HR. A similarly faster increase in RPE was reported for 10- to 13-year-old boys, compared with young men, during a 40-min cycling bout at ventilatory threshold (94) and for children 10.7 ± 0.8 years old during 16-min cycling at ventilatory threshold (288). The faster increase in children's RPE is hard to explain by differences in central or peripheral physiologic cues. For example, during the 60-min cycling experiment, the rate of increase in blood lactate levels was in fact slower in the boys than in the men. Likewise, even though respiratory rate and ventilatory equivalent (ventilation divided by O_2 uptake) were higher in the boys, there was no correlation between RPE and respiratory rate or ventilatory equivalent in either group (434). Similarly, in the 40-min exercise study, the drop in plasma volume was smaller in the boys (94).

It is tempting to speculate that children's preference for short-term exercise results, in part, from their fast increase in perceived effort during prolonged exercise. Confirming or refuting this hypothesis requires further research.

The process of perceiving and interpreting a given effort is complex. It involves a feedback mechanism by which peripheral signals (e.g., from the exercising muscles and possibly joints and tendons) and central signals (e.g., tissue acidity, muscle blood flow, skin temperature) are sent to the sensory cortex. There also may be a feed-forward mechanism by which the sensory cortex receives signals from the motor cortex about an intended motor activity (43; 86; 359). The information derived from these processes then needs to be interpreted and converted to a number on an RPE scale, which is presented to the child. The latter stage requires a certain cognitive and semantic ability. A question then remains as to whether there is a critical age under which children cannot rate their perceived exertion. In the large-scale study just described (35), the youngest group of 7- to 9-year-old gymnasts had a much higher rating than any other child

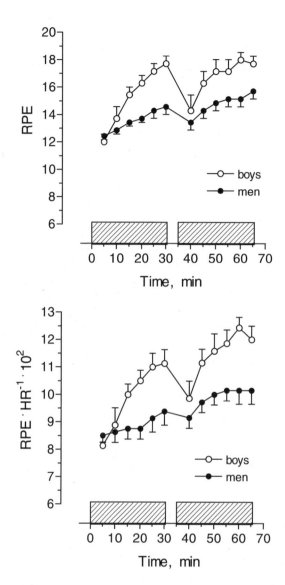

FIGURE 1.52 Rating of perceived exertion (RPE) during prolonged exercise in boys and men. Ten 9- to 10-year-old boys and ten 20- to 25-year-old men cycled at 70% of their maximal O_2 uptake for 30 × 2-min periods, with a 5-min rest period in between. The top graph depicts RPE changes over time. The bottom graph depicts the ratio between RPE and heart rate (HR). Data are mean and 1 SEM.

Reproduced, with permission, from Timmons et al. 2003 (434).

or adolescent group. The author suggested that this might have reflected their inability to provide valid ratings at such a young age. Indeed, the RPE-to-HR correlation (considered an index of validity) in that group was lower than in the more mature groups. More recently it was shown that children as young as 8 (460) can provide valid estimations of RPE using the Borg 6 to 20 scale. Likewise, 8-year-old Japanese children could use this scale, but younger children could not understand it (310). It thus seems that the critical age for using the Borg category scale is around 8 years, but more studies are needed that are designed to address this question.

In an attempt to expand the age range at which RPE can be used, several alternative scales have been developed in recent years. In addition to numbers, these scales include a pictorial display of different degrees of effort. Examples include a stick diagram of a person performing various activities next to the Borg 6 to 20 category scale (328), a heart riding a bicycle (267), a person carrying different loads while riding a bicycle (149), a child riding a bicycle up a slope (358), and a child running up a slope (444). While some of these scales are intended for estimation of effort, others (e.g., [357; 469]) are used also for production of effort. One of these scales (the OMNI scale) has been found valid for effort estimation by children as young as 6 years (444). Another scale (CERT), when used with 8- to 11-year-old children, yielded higher correlations with HR than did the Borg 6 to 20 category scale (261). For more information about RPE scales see appendix II.

Rating of Perceived Exertion As a Means for Production and Self-Regulation of Effort

As discussed in the section on training, a prescription of physical activity includes four elements: the type of activity, its intensity, the frequency of the sessions, and the total duration of the program. The most challenging element of prescription, especially for children, is the intensity. There are various ways of prescribing exercise intensity; one is through the production mode of RPE. This is usually implemented in two sessions. The child is first introduced to the concept of RPE while performing a multistage exercise (see appendix II for details). As in the estimation mode, the child is asked to rate several intensities. Emphasis is given to the lightest (or near lightest, e.g., 7 on the 6 to 20 category scale) and the hardest (or near hardest, e.g., 19) intensities. This process, called anchoring, helps the child to experience the sensations that occur at a range of intensities that may subsequently be prescribed. The actual prescription occurs in the second session. The child is given instructions such as "Jog for 5 min at an intensity of 8 (on the scale), then run for another 10 min at an intensity of 13, and for 2 min at number 16 . . ." and so on.

Several studies have examined the suitability of the RPE production mode for healthy children (150; 151; 262; 469-471) and those with a disease (357; 458-460). Altogether, this approach, using various scales, seems feasible for children as young as 8 years (357; 460; 469). Younger children may not be cognitively ready for production or self-regulation of RPE (469). An attempt was made to apply the production mode (using the CERT, Children's Effort Rating Table) to a field setting of a physical education class for children ages 6 to 11 years. Only a minority of the pupils could consistently self-regulate their activity using this method (104).

Immune Response to Exercise

Exercise often triggers response by the immune system. The magnitude and direction of this response depend, among others, on the intensity and duration of the exercise and on the person's nutritional and health status. While most data in this area relate to adults, recent years have seen an increasing interest in children's immune response to exercise (71 a-c; 133a; 323a-b; 334a; 343a; 399a; 412b; 434a). This section overviews such changes in response to an acute bout of exercise and the differences between individuals of various training and activity levels. For further information related to adult responses, see reviews by Nieman (324a) and by Pedersen and Hoffman-Goetz (342a).

Response to a Single Exercise Bout

The typical and easily discernable immune response to high-intensity exercise is an increase in the concentration of immune cells (neutrophils, monocytes, and lymphocytes and their subsets) in peripheral blood during exertion and a brief transient decline in their numbers during recovery. The latter stage, called "open window," has been considered to increase a person's susceptibility to infection (324a). Among these white blood cells, natural killer (NK) cells show the greatest cellular response to exercise (342a). In addition to cellular immune changes, exercise also triggers changes in cytokines, which, among other functions, help to regulate signaling of innate and specific immune responses. Two cytokines that have been studied extensively as to their response to exercise are interleukin 6 (IL-6) and tumor necrosis factor α (TNF-α).

The type of exercise that triggers these responses varies among studies. Exercise duration in studies with children ranges from 30 sec (71c) to 90 minutes (323a; 399a). Exercise intensity ranges from 70% of maximal aerobic power (434a) to a supramaximal anaerobic task (Wingate anaerobic test). In some studies, immune response was triggered by real-life sports situations such as a soccer (343a; 399a), wrestling (323a), and water polo (323b) practice.

Do children have the same immune response to exercise as do adults? Several authors have suggested that children and adults respond in the same manner (133a; 343a; 412a). However, their observations were not based on a direct comparison. In a recent study Timmons and colleagues (434a) exposed prepubertal and early-pubertal boys (mean age 9.8 yr) and young adult men (22.1 yr) to 60 min cycling at 70% of predetermined maximal O_2 uptake. Immune cells and cytokines were measured before, immediately after, and 60 min after exercise. There were several intergroup differences: The boys had a lower increase in total leukocyte, NK cells, and NK T cell counts, as well as IL-6 immediately after exercise. After 60 min of recovery, lymphocytes and T cells were suppressed in the men but not in the boys. IL-6 remained elevated in the men and close to resting values in the boys. The authors concluded that both the increase in immune response immediately after exercise and the suppression of response during recovery are milder in boys than in men. It has yet to be determined whether these differences depend on age or maturational level and whether they also occur in females.

Activity, Training, and the Immune Response

While a single bout of high-intensity exercise may be followed by a transitory suppression of immune activity, a habitually active lifestyle in adults has been suggested to increase a person's protection against infections. Data are inconsis-

tent as to whether this protective effect exists also in children and youth. A yearlong follow-up study of Finnish 12-year-old girls and boys compared the incidence and severity of viral and bacterial respiratory infections in those who took part in organized training (ice hockey, swimming, or track and field) compared with those who did not train but participated in extracurricular sport activities (334a). There was no intergroup difference in the occurrence of respiratory infections or in the use of antibiotics. Likewise, there was no difference in the immune response (leukocytes and natural killer cells) to a high-intensity 30-sec cycling task in trained and untrained 9- to 17-year-old boys. In contrast, a study of randomly selected eighth graders (244a) revealed a lower occurrence of viral upper respiratory tract infections in highly active eighth graders, compared with moderately active controls from the same schools (figure 1.53). Subjects were followed up for one month during the Canadian winter.

With such scarcity of data it is difficult to reconcile the differences of outcome in these studies.

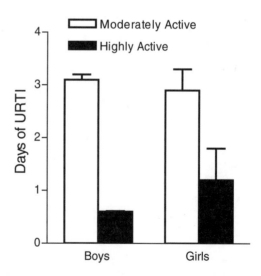

FIGURE 1.53 Activity level and susceptibility to upper respiratory tract infections (URTI). Subjects were eighth graders (126 boys and 130) from the Niagara region in Canada who recorded the occurrence (number of times per month) and duration of cold and flu symptoms during the month of January. All children were subdivided, based on questionnaires, into "moderately" and "highly" active groups. The graph depicts the number of days during that month that the subjects reported URTI. Vertical lines denote SEM. The intergroup difference in the boys was significant ($p < 005$). Based on Klentrou et al. 2003 (244a).

More research is needed on this important area, which will address the effect of physical activity on the susceptibility to infections other than in the respiratory tract. It is also possible that the effect of habitual physical activity will be different in younger children and in those affected by diseases that suppress the immune system.

Training

Physical training is the process by which exercise, repeated during weeks and months, induces morphologic and functional changes in body tissues and systems. Mostly affected are the skeletal muscles, myocardium, blood vessels, adipose tissue, bones, ligaments, tendons, and the central nervous and endocrine systems. Understanding the principles and effects of training is important to scientist, clinician, and educator alike. This is particularly so in view of our growing knowledge regarding the perils of sedentarism and the health benefits of an active lifestyle. It is also important to obtain better insight into training principles in light of the increasing demands placed on young athletes at all levels.

This section focuses on principles of training, as well as the specific effects of training on various body systems (cardiovascular, pulmonary, musculoskeletal) and fitness components (aerobic, anaerobic, and strength) in healthy children and adolescents. Trainability of motor skills in the growing child is discussed in textbooks of motor development and is not dealt with here. Subsequent chapters present discussion of the relevance of training to patients with a specific disease or illness.

Terminology

In the first edition of this book, a distinction was made between conditioning and training, which are often used interchangeably. "Conditioning" was used for intervention programs in which emphasis is on a general increase of the metabolic level and overall fitness (e.g., a mixture of running and calisthenics). "Training," on the other hand, referred to programs that emphasize a specific fitness component or a specific part of the body (e.g., sprint training or strength training of the elbow flexors). For simplicity's sake, this text uses "training" for both types of programs.

Trainability in this book denotes the degree of responsiveness of a body system, organ, tissue, or a fitness component to a training program.

Methodologic Constraints

There are three main approaches to the study of training. One is to use cross-sectional comparison, in which a group of highly fit children is compared to a less fit group. Another approach is correlational analysis, in which a certain function or morphologic characteristic (e.g., maximal cardiac output or muscle size) is correlated with the level of physical activity. The third approach is to use an interventional, longitudinal design, in which functions or morphologic characteristics are measured before, during, and after the intervention. Correlational studies and cross-sectional designs can elucidate differences among subjects of various fitness levels and generate hypotheses, but they cannot yield any cause-and-effect inference. This can be achieved only through an interventional approach.

Research on the training of children has inherent methodologic obstacles and pitfalls. In adults, changes in function between pre- and post-intervention can be attributed with fair certainty to the training program. Not so with children or adolescents. Here, changes due to growth, development, and maturation often outweigh and mask those induced by the intervention. It is intriguing that many of the physiologic changes that result from training also take place in the natural process of growth and maturation. Such is the case, for example, with the reduction of submaximal HR and ventilatory equivalent, or the increase of stroke volume, maximal blood lactate, muscle strength, or economy of running.

Partialling out growth and developmental effects from the changes seen following a training program has been a major challenge. One approach is to use the scaling principles (see appendix V) by which size-related changes can be predicted and accounted for in a training study (417; 456). This approach, while useful in the absence of a control group, is not an adequate substitute for a controlled design, as it does not allow one to tease out such effects as development, maturation, habituation, or learning (190).

A control group is part and parcel of a well-designed intervention study. Ideally, allocation to an intervention group or control group must be random, or subjects in the two groups should be carefully matched for physical characteristics and pre-intervention fitness level. While many studies with healthy children have followed this approach, interventions with pediatric patients often violate this principle. Such studies all too often include an intervention group of patients without controls, or with a "convenience sample" of controls (e.g., those who refuse to train or who live far away from the training site). The main reasons for the lack of suitable controls in clinically oriented studies are an insufficient number of subjects and, in particular, ethical constraints that preclude the inclusion of a "nontreatment" group. To overcome geographic and other logistic barriers in recruiting patients, some clinical studies have used a home-based program. An important advantage of this approach is that it mimics "real-life" situations for many patients who otherwise would not be able to enhance their activity level. The main challenge in home-based programs is the researcher's ability to quantify the intervention and confirm compliance.

Another challenge in clinical studies is that the patients often need medication or other therapies. Because these cannot be stopped, one cannot exclude the possibility of an interaction between the training intervention and the other treatments. Finally, children who volunteer for a training study (healthy ones in particular) are likely to be more physically active than the general child population. As a result, one cannot assume that the controls will be sedentary during the study, or that the intervention group was not already trained prior to the study.

In view of these constraints, our state-of-the-art knowledge on training of children, particularly pediatric patients, is still fragmentary. One must look upon some of the data as reflecting changes that occur with training, but do not necessarily result from it.

Principles of Physical Training

The clinically oriented reader who may wish to incorporate exercise into a therapeutic regimen is not expected to master the techniques of training. This role should be left to experts in adapted physical education (a subspecialty of physical education, dealing with the clinically disabled child), kinesiologists, or physical therapists with a special interest in fitness and sport. It is important, however, that a clinician who considers prescribing an activity be conversant with the rationale, principles, and terminology of training. These are presented in the sections that follow. Discussion of a more technical nature can be found in textbooks of exercise physiology (e.g., [91; 473]).

Specificity of Training

The changes that take place in body tissues as a result of chronic exercise are stimulus specific: A particular activity may induce a change in one tissue and not in another. Furthermore, a tissue may respond in one way to a certain exercise stimulus and in another way to a different one (64; 134; 270). Distance running, for example, will induce myocardial hypertrophy, an increase in mitochondrial volume, improvement of the oxidative capacity within muscle fibers, and preferential hypertrophy of slow-twitch fibers. In contrast, knee extension against high resistance will barely affect the myocardium, but it will induce hypertrophy (and possibly hyperplasia) of the quadriceps—mostly the fast-twitch fibers—and enhanced recruitment of motor units. An example of specificity of training is seen in a study by Fournier et al. of 16- to 17-year-old boys. A 5-month sprint training regimen increased the activity of muscle phosphofructokinase (an "anaerobic" enzyme) by 20.6%, with no change in succinyl-dehydrogenase (an oxidative enzyme). In contrast, endurance training induced a 42% increase in succinyl-dehydrogenase activity and only 6% in phosphofructokinase activity (164).

Specificity of the stimulus can be nicely illustrated in strength training (391). Each of the following factors will modify the response to a strength training regimen:

1. The involved muscle(s)
2. Type of contraction (concentric, eccentric, isometric)
3. Intensity of contraction and number of repetitions and sets
4. Velocity of contractions
5. Joint angle at which contraction is performed
6. Movement pattern

The fine nuances of specific responses to training are especially important to athletes, who strive to attain perfection. They are less crucial in the clinical context to those interested in the general improvement of physical working capacity. Nevertheless, an intelligent approach to prescribing exercise to a disabled or detrained child should tailor the type of activity to his or her current functional deficiencies and future needs. While some children, especially those with musculoskeletal disabilities, may require prescription of activities "custom made" to their residual ability, others can benefit from sports practiced by the general population.

Different sports develop different components of physical fitness. Table 1.9 summarizes the major benefits to fitness that various sports can induce. The fitness components are presented in this table in popular rather than physiologic terms. General stamina, for example, stands for maximal aerobic

TABLE 1.9   Typical Effects of Various Sports on Fitness

Type of sport	Stamina	Local muscle endurance	Muscle strength	Speed	Agility	Flexibility	Body weight control
Individual sports							
Boxing	+++	+++	++	+++	+++		+++
Cycling (long and middle distance)	+++	+++	++	++	−	−	+++
Figure skating	++	+++	++	+	+++	+++	++
Golf	−	−	+	+	−	+	++
Gymnastics	+	+++	+++	++	+++	+++	++
Horseback riding	−	++	+	−	−	−	−
Jumping (track and field)	+	+	+++	+++	+	+++	+
Rowing	+++	+++	+++	+	−	+	+++
Running							
Sprint	+	+	+++	+++	++	+	+
Middle distance	+++	+++	++	++	+	+	+++
Distance	+++	+	+	+	−	−	+++
Sailing	+	+++	++	+	++	−	+
Skiing: downhill and slalom	++	+++	++	++	+++	+	+
Cross country	+++	++	++	+	+	−	+++
Swimming	+++	+++	++	++	+	++	+++
Tennis (squash)	++	++	++	++	+++	+	++
Throwing (discus, etc.)	+	++	+++	+++	++	++	+
Walking	++	+	+	−	−	−	++
Weightlifting	−	++	+++	+++	+	+	+
Wrestling, judo	++	+++	+++	++	+++	+++	++
Team sports							
Baseball	+	++	+	++	+++	+	+
Basketball, soccer	++	++	+	++	+++	+	++
Football (American)	++	+++	+++	+++	++	+	++
Ice hockey	++	+++	++	+++	+++	+	++
Volleyball	+	++	++	++	+++	++	+
Water polo	+++	+++	++	++	++	++	++

*− = Hardly any effect; + = some effect; ++ = much; +++ = very much.

power, and local muscular endurance for mean anaerobic power. The importance assigned to each sport in improving a certain fitness component is the most typical effect, but there are many variations. For example, volleyball is not the sport of choice for weight control. It can, however, be practiced in such a way that calorie expenditure is high and body weight is controlled. Similarly, soccer is not the most suitable sport for developing strength of the trunk muscles. However, by incorporating certain elements (calisthenics, resistance training), a soccer player can greatly develop these muscles.

Dosage of Training

In analogy to other forms of prescription, training can be characterized by its intensity, its frequency, the duration of each session, and the overall duration of the program.

Intensity

The intensity of an activity is determined either by the metabolic demands (e.g., O_2 uptake), by the strain on the cardiovascular system (e.g., HR), or in the case of strength training by the weight to be lifted, pushed, or pulled. Although often described in absolute terms, intensity should also be described in relationship to the current fitness level of the individual. For example, to a disabled (or a small) child with maximal O_2 uptake of 1.0 L $\cdot$ min^{-1}, an activity that requires 0.75 L $\cdot$ min^{-1} of O_2 is more intense than it is to a healthy or a larger child with maximal O_2 uptake of 1.5 L $\cdot$ min^{-1}. Similarly, the lifting of 20 kg is less intense for a muscle with a maximal voluntary contraction of 40 kg than it is for a muscle with a maximum of only 30 kg. It is therefore advisable to describe the intensity of a training task as a percentage of the individual's current capability in this task.

It has long been established that trainability is positively related to the intensity of the activities performed. This is especially true for maximal aerobic power and strength but also applies to other fitness components.

An important concept to recognize is that of the intensity threshold. This is the intensity of exercise below which little or no training effect can be discerned. For young adults, an intensity threshold of maximal aerobic power is 60% to 70% of maximal O_2 uptake (which corresponds to the anaerobic threshold), or 70% to 80% of maximal HR. There are fewer data about training thresholds for children (285), but experience shows that their aerobic intensity threshold is at least as high as that for adults. The threshold for developing strength is about 60% to 65% of maximal voluntary contraction.

The dosage of a training program must be progressive in its demands. Once an individual has improved a certain fitness component, the intensity previously sufficient to induce changes may now be below the threshold and may no longer be sufficient. An example is a muscle with maximal voluntary contraction of 30 kg trained by lifting 20 kg. As a result of the training, the muscle becomes stronger and its maximal voluntary contraction is now 35 kg. At this level, to achieve a further training effect, the load may have to be increased to 23 or 25 kg.

Frequency

The number of exercise sessions a week may determine the success of a program. Because of interaction with the effects of intensity and duration of each session, there is no single optimal frequency that is suitable for all programs. While some training effect can be achieved by one weekly session, it is generally advisable for patients to practice at least two or three times a week. Some athletes, children included, train as often as 12 to 15 times a week! Such a regimen is obviously not recommended for a child who is undergoing a therapeutic program or who merely wishes to improve his or her fitness. The frequency of sessions for such children also depends on logistic limitations (e.g., ability of parents to drive the child to the sport facility).

During the last decade, several guidelines were published regarding the desirable frequency of activity sessions for the general child and youth population. These range from three sessions a week (394) to daily sessions (Canada Activity Guide for Children and Youth 2002, www.paguide.com). As discussed in chapter 5, such guidelines are not based on valid evidence that the recommended frequency is indeed optimal for the prevention of disease or for maintenance of fitness.

Duration of a Session

As with frequency, there is no single optimal duration. One should allow approximately 5 to

10 min for warming up at the beginning and 5 to 7 min for cooling down at the end of each session. The main portion of the session, during which the intensity threshold is exceeded, should last 15 to 30 min. Thus, the overall duration will be about 35 to 45 min. If the main part of the session is too long, the child may become overly fatigued. The duration should also be tailored to the attention span of the participants. A preschooler or a child who is mentally retarded may lose interest within 5 to 10 min. A more mature or better-motivated child can maintain interest for much longer.

Duration of the Program

As with any therapeutic program, to yield any effect there is a minimal period of time for which intervention should be sustained. While some training effects can be attained within 1 to 2 weeks, most clinically oriented programs last at least 6 to 8 weeks. As a rule, the longer the program the more effective it is (provided that the principle of progression is adhered to). Overall duration should also be determined by the specific goals. A weight control regimen, for example, may require 6 to 8 months to produce noticeable effects. In contrast, the strengthening of a specific muscle group may require as little as 1 month.

An important aim of exercise programs for healthy and disabled children alike is to form habits, to acquire skills, and to enjoy sport. Such educational goals, and not only the physiologic ones, should also dictate the duration and content of the program.

Window of Opportunity for Trainability

A question confronting pediatric exercise scientists is whether there are specific developmental stages at which the growing individual is most or least responsive to training stimuli. This question is also important to the coach who wishes to plan a young athlete's career, and to the physical therapist, physiatrist, or pediatrician who wishes to include physical training in his or her clinical management repertoire. In a broader biological sense such a topic relates to the basic question: Are there phases in the formative years when certain characteristics are most (or least) sensitive to environmental stimuli? The following sections address the scant data about optimal developmental stages in trainability.

Factors Affecting Trainability

Factors that affect trainability may differ among the various body systems and fitness components. There are, however, some common factors to be considered, as summarized in the following list.

- Age
- Gender
- Time elapsed from peak height velocity
- Maturity (skeletal, sexual)
- Body size and composition
- Motor proficiency
- Pretraining fitness level
- Genotype
- Interaction between genotype and the environment

Trainability of Maximal Aerobic Power

Among adults who undergo an aerobic training program, there is an age-related trend: The younger the individual, the more trainable he or she is (395). By extrapolating from such a trend, one might expect children to be even more trainable than young adults, but this is not the case.

Numerous studies have addressed the aerobic trainability of children, and their outcome is inconsistent. Some show that adolescents, children, or both responded in a "predictable" manner to specific training regimens (33; 121; 126; 141; 190; 191; 269; 289; 290; 295; 301; 329; 412; 421; 463; 467; 480). There is, however, much evidence to suggest that aerobic trainability in prepubescents, particularly in those less than 10 years old, is often lower than expected (7; 37; 47; 59; 111; 113; 114; 180; 198; 215; 245; 304; 314; 340; 364; 374; 384; 403; 412; 420; 423; 445; 465; 479).

The reason for such inconsistency is not clear, but it seems that much of the difference in outcome results from differences in the intensity and dosage of training, as well as the mode of training (e.g., short interval runs vs. longer distance runs). Pate and Ward (339) identified 11 studies between 1969 and 1986 that fulfilled the following strict scientific criteria: (1) The study included random or pair-matched controls; (2) the training protocol was clearly described; (3) training took place outside physical education classes (where the dosage of activities may not adhere to scientific standards); (4) training focused exclusively on endurance; (5) physiologic measures (maximal O_2 uptake, anaerobic threshold) were used as performance criteria; (6) proper statistical methods were performed; (7) the study was published in a peer-reviewed journal. The age range of subjects in these strictly selected studies was 5 to 16 years. All but one of the studies showed a greater increase in maximal O_2 uptake in the intervention group than in the control. The increase ranged between 1% and 16%, with an average increase of 9.7% (figure 1.54). Anaerobic threshold, measured in two of the studies, showed an even greater increase.

Notwithstanding, even when the training dosage is adequate, trainability of young children (10 years or younger) seems lower than in older age groups. This pattern is summarized in table 1.10, which divides subjects from 28 studies into three age groups: less than 10 years, 10 to 13, and 14 or older. Out of the 13 studies with the youngest age group, only one (8%) showed an increase of more than 5% in maximal O_2 uptake. This compares with 75% in the middle age group and 100% in the oldest age group (291). A similar pattern was obtained through a meta-analysis of 23 studies of subjects 13 years old or younger, in which the mean increase in maximal O_2 uptake was 4.8% (341). On the basis of several additional reviews, one may conclude that prepubescents do improve their aerobic performance, but the extent of improvement seems somewhat lower than in more mature adolescents (39; 285; 339; 375).

Some studies show little or no increase in maximal O_2 uptake, even though running performance improves with training (47; 113; 269; 314; 479). An example is a study in which 91 girls and boys, 9 to 10 years old, performed interval running two, three, or four times a week for 9 weeks. Although the participants markedly improved their running performance, there was no discernible improvement in maximal O_2 uptake (47). Similar results were obtained for 5-year-old children who ran 750 to 1,500 m five times a week during 14 months (479) or for other prepubescents who practiced middle distance and distance running (314; 423).

How can one reconcile an improvement in distance running with little or no concomitant increase in maximal O_2 uptake? One possibility is that the child improves running economy (114), thereby delaying fatigue toward the end of a race. Another explanation is an improvement in anaerobic performance, even though the training program is aimed at increasing endurance (135; 190; 399). It is also possible that maximal aerobic power of the prepubescent does improve but that maximal O_2 uptake is not the most sensitive indicator of this (111; 303; 364; 402). The latter scenario has been shown in some (50; 289) but not all (365) studies, in which the percent increase in anaerobic threshold was greater than in maximal O_2 uptake. A fourth possible reason for the apparent lack of training effects in prepubescents could be the high habitual activity of the "controls." It is quite possible that even though they did not participate in the regimented program, these children were so active in their free time that intergroup differences were quite negligible (112; 198; 479; 481).

At what developmental stage does aerobic trainability reach its maximum? Reports in

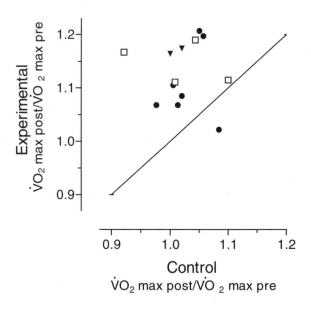

FIGURE 1.54 Changes between pre- and posttraining maximal O_2 uptake in a training group versus its control group in 11 studies. Each point represents a study mean (one study is represented by three points, one for each of three subgroups). ● = prepubescents; □ = pubescents or mixed; ▼ = postpubescents.

Reproduced, with permission, from Pate and Ward 1996 (339).

TABLE 1.10	Relative Changes in Maximal O_2 Uptake (ml · kg⁻¹ · min⁻¹) Associated With Training in Children and Adolescents					
Age, years	N	≤ 0%	1-5%	6-10%	11-15%	>15%
≤ 10	13	4	8			1
10-13	12	1	2	3	2	4
14+	3			1		2

Based on reviews by Mocellin 1975 (311), Rowland 1985 (374), and Pate and Ward 1990 (338). N refers to the number of training studies in the indicated age range.

Reproduced, with permission, from Malina et al. 2004 (291).

this regard are conflicting. One might attempt to reconcile such findings by aligning the subjects according to the age at which they reach their peak height velocity. On the basis of this approach, Kobayashi et al. (245) suggested that the effectiveness of aerobic training was greatest at, or around, the year of peak height velocity. However, another study using a similar technique could not identify such a critical stage in aerobic trainability (7).

Mechanisms for Improved Maximal Aerobic Power

Aerobic performance depends on the adequacy of the O_2 transport chain, in which ambient oxygen is delivered to the muscle mitochondria. This chain includes the respiratory and cardiovascular systems, as well as the muscle itself. Further details about training-related changes in the cardiovascular and respiratory systems are given in the following sections. Because of methodological constraints, only scant data are available regarding changes that occur in the children's muscles. These are summarized in table 1.11.

The main training-related change in the organelles of muscle fibers of adults is an increase in the number and volume of mitochondria. While no such information is available for children, children's capability for aerobic energy turnover is enhanced following endurance training as shown by increased glycogen storage and oxidative enzyme activity (142; 143; 164). Skeletal muscle hypertrophy, and possibly hyperplasia, occurs in the chronically active muscle of adults. Hypertrophy, but not hyperplasia, has been found also among adolescents undergoing endurance training (164; 224), but not following sprint train-

ing (164). The cross-sectional area of slow-twitch muscle fibers, as well as some of the fast-twitch fibers, of 16- to 17-year-old boys increased 10% to 30% following a 3-month endurance program. There is no evidence for a change in fiber type distribution in children as a result of training (142; 224). In adults, training induces an increase in muscle capillarization and the creation of collateral circulation to the skeletal muscle. This aspect has not been investigated in children.

Trainability of Anaerobic Power

Even though research on anaerobic trainability of children and adolescents is sparse, compared with research on aerobic trainability, there have been several publications since the 1980s that show a training-related increase in anaerobic performance (85; 96; 190; 191; 304; 320; 330; 365). There are far fewer studies that document the effect of anaerobic training on histochemical and biochemical muscle characteristics (85; 139; 164). Improvement in anaerobic performance is discerned within a few weeks of training, and in this respect, children are not different from adults. Performance improvement has been documented irrespective of the criterion test. Using the Wingate test, both peak power and mean power increase (190; 191; 304; 365). Likewise, there is an increase in peak power during the force-velocity test (330), time to exhaustion during treadmill sprints (320), sprinting velocity (85), and total numbers of pull-ups and push-ups (96).

Figure 1.55 is an example of trainability of muscle endurance in response to a high-intensity swimming program of 9- to 11-year-old girls and boys. While a control group showed no improvement in performance, the performance of the swimmers improved by approximately 100%.

With the small number of controlled studies, it is impossible to tell whether anaerobic trainability is age related, although there is some suggestion, based on an uncontrolled study, that age 11 to 13 years is the most sensitive stage for training of dynamic endurance of the elbow flexors in boys (216). Nor is it possible to determine

Variable	Change	Information available on children	Information on adults only
Mitochondria—number	Increase		X
Mitochondria—volume	Increase		X
Glycogen stores	Increase	X	
Triglycerides—stores	Increase		X
Triglycerides—utilization	Increase		X
Myoglobin content	Increase		X
Oxidative enzyme activity (e.g., succinyl dehydrogenase, cytochrome oxidase, palmityl coenzyme A synthetase)	Increase	X	

TABLE 1.11 Aerobic Training-Induced Changes in Muscle Metabolism of Children and Adults

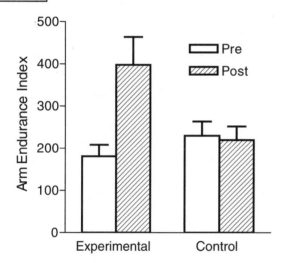

FIGURE 1.55 Effect of training on anaerobic performance of children. Fifteen girls (n = 2) and boys (n = 13) took part in a high-intensity 7-month swim program. There were 15 controls, with the same gender composition. Muscle endurance (mean ± SEM) is presented as a combined index of push-ups, pull-ups, and body mass.

Adapted from Clarke and Vaccaro 1979 (96).

whether trainability depends on the intensity of the training stimulus and on the pretraining fitness level.

Mechanisms for Anaerobic Trainability

On the basis of our knowledge of factors that determine anaerobic performance in children, as well as adult studies regarding the mechanism of muscle trainability, one can assume that the improvement in children's anaerobic performance is achieved through neural and biochemical changes. Neural changes may include enhanced activation of motor units, reduction in antagonist muscle co-contraction, and improved intermuscular coordination. Unfortunately, there are no studies that were designed to decipher the possible neurological mechanisms in children.

There is better evidence regarding biochemical changes that accompany anaerobic training. As shown by Eriksson (138; 141), Cadefu (85), and Fournier (164), based on muscle biopsy data, there is an increase in the activity of anaerobic enzymes such as phosphofructokinase and other glycolytic enzymes. There is also an increase in resting muscle glycogen level and its utilization rate during exercise, and in maximal muscle lactate. These data suggest that anaerobic training is accompanied by an increase in the glycogen flux. There is also evidence for a rise in muscle ATP and CP levels at rest, which may facilitate alactic performance. Biochemical changes that occur with training are summarized in table 1.12.

Training of Muscle Strength

It was stated in the first edition of this book that "the definitive study on (strength) trainability of children has yet to be performed." Indeed, until the early 1980s, the studies addressing this topic were a mere handful that could not yield a clear picture regarding the effects of resistance training on children's strength (324; 363; 457). The prevailing thought at that time was that strength cannot be improved before puberty (209).

This field has since been given much impetus, and it is quite clear today that, indeed, children's strength is trainable even in the first decade of life (64; 68; 122; 130; 152; 153; 159; 160; 174; 175; 197; 306; 307; 335; 345; 346; 350; 390; 392; 408; 413; 466; 467). When expressed as percent improvement, the trainability of prepubescents is as good as, or even better than, that in more mature groups (161;

TABLE 1.12 Training of the Anaerobic System in Children[*]		
Variable (in muscle)	**Pretraining level (vs. adults)**	**Response to training**
% Fast-twitch fibers	Similar	No change (?)
CP (rest)	Similar	Increase
ATP (rest)	Similar	Increase
ATP utilization	Low	Increase
Glycogen (rest)	Low	Increase
Glycogen utilization	Low	Increase
PFK	Low	Increase
Lactate—submax	Low	No change
Lactate—max	Low	Increase

[*]Based on muscle biopsies done on 12 boys, 11-15 years old, who underwent a combined aerobic–anaerobic training program for 6 to 16 weeks. Based on Eriksson 1972 (138).

197; 342; 345; 363; 390). However, when trainability is expressed as an absolute increase in strength (in kilograms or newtons), prepubertal children show a lesser improvement. Factors considered to affect strength trainability include chronological and skeletal age (174), pubertal stage (345), gender (71), and adiposity (160). Figure 1.56 is an example of the effect of training on several muscle groups and modes of muscle contraction.

Possible Mechanisms for Strength Increase

One of the main differences between the response of children and that of more mature groups to strength training is that children's enhanced strength is not accompanied by an increase in muscle size (335; 350; 467). With the absence of muscle hypertrophy, the alternative mechanism for greater force generation has been attributed to neural and neuromotor changes. One possible process that has been suggested is an increase in motor unit activation. Using an interpolated twitch technique, Blimkie (63) and Ramsay (350) found an approximately 10% increase in the number of activated motor units following 10 weeks of strength training in prepubescent boys. Ozmun et al. (335) reported a 16.8% increase in the integrated electromyogram of the elbow flexors in prepubescent girls and boys who trained for 8 weeks. This pattern is suggestive of an increase in neuromuscular activation. While more research is warranted in this field, it seems that the increase in muscle strength,

particularly at the early stages (e.g., first 6-8 weeks) of training, is facilitated in part by neural and neuromotor changes.

Safety Considerations

Potentially, resistance training has risks, mostly to joints, ligaments, and muscles. However, accumulating experience suggests that with appropriate supervision and correct techniques (see the following guidelines), the risk is minimal (4; 55; 62; 193; 353; 429). It is important, however, to distinguish between resistance training and weightlifting. The latter refers to competitive powerlifts or Olympic weightlifting. Such routines, when practiced by children, may prove riskier than resistance training. Until further information is available, these routines should be discouraged, as stated by the American Academy of Pediatrics (55) and the American Orthopedic Society for Sports Medicine (87).

Guidelines for Resistance Training

The following guidelines, as suggested by Blimkie and Bar-Or (64), were composed for nonexperts who wish to optimize safety and effectiveness in childhood resistance training:

1. Check for physical and medical contraindication.

2. Ensure experienced supervision, preferably by an adult, when using free weights or training machines.

3. Ascertain proper technique.

4. Warm up with calisthenics and stretches.

5. Begin a program with exercise that uses body weight as resistance before progressing to free weights or weight training machines.

6. Individualize training loads whether using free weights or machines.

7. Train all major muscle groups and both flexors and extensors.

8. Exercise the muscles through their entire range of motion.

9. Alternate days of training with rest days, limiting participation to no more than three times a week.

10. When using free weights or machines, progress gradually from light loads, high repetitions

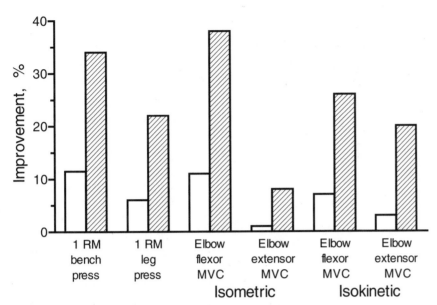

FIGURE 1.56 Improvement in muscle strength as a result of a 20-min, three-per-week training program that emphasized the elbow flexors and knee extensors. Subjects were 9- to 11-year-old healthy boys. Values are presented as percent increase in strength of the training groups (cross-hatched bars) and the controls. 1RM = repetition maximum (the maximal weight that can be pressed in a single press); MVC = maximal voluntary contraction. Based on data by Ramsay et al. 1990 (350).

Reproduced, with permission, from Blimkie and Bar-Or 1996 (64).

(>15), and few sets (two to three) to heavier loads, fewer repetitions (6-8), and three to four sets.

11. Cool down after training, using stretch exercises for major joints and muscle groups.

12. When selecting equipment, check for durability, stability, sturdiness, and safety.

13. Consider sharp or persistent pain as a warning, and seek medical advice.

Training of the Cardiovascular System

Morphologic and functional changes in the cardiovascular system that take place with training are summarized in table 1.13.

Morphologic Changes

Myocardial mass (depicted by echocardiography) and heart volume (by X-rays and by echocardiography) are higher in child endurance athletes than in nonathletes, and they increase as a result of endurance training (2; 133; 145; 179; 265; 331; 452). Cardiac hypertrophy results from an increased pressure load and, to a lesser

extent, from volume load. Training-induced collateral coronary circulation has been shown in animal studies, but there are no data available for children.

Although stroke volume is related to heart volume, the myocardial mass of children does not seem to affect their athletic capability (250). Total blood volume and total hemoglobin are higher in trained boys than in untrained ones (250), and these values increase as a result of training (144). A higher blood volume facilitates a higher venous return as well as more effective heat convection from body core to periphery. The increase in total hemoglobin raises the O_2-carrying capacity of the blood. Hemoglobin concentration, on the other hand, does not increase with training. In fact, some endurance athletes have a sport-induced iron deficiency.

Physiologic Changes

Based on cross-sectional data, stroke volume at rest and during all levels of exercise is higher in children with high aerobic fitness than in less fit children (327; 370; 371; 432) (figure 1.56). Intervention studies show that stroke volume, left ventricular filling, or both increase with training in pubertal children (144; 178; 198; 331). The increase in diastolic filling occurs in the early (rapid) phase of the diastole (331). An increased stroke volume may reflect higher blood volume and improved venous return, but it may also reflect improved myocardial contractility (475).

Figure 1.57 depicts differences in index (stroke volume per body surface area) between "high fit" and "low fit" prepubescent and early-pubescent boys 12.2 ± 0.5 years old. While stroke index increased progressively in both groups, it was higher (by approximately 10-12 mL · m⁻²) in the high fit group. However, the ratio of maximal stroke index to stroke index at rest was similar in the two groups (371). The authors suggested that

TABLE 1.13   Training-Induced Cardiovascular Changes

Variable	Change
Morphologic Heart volume	Increase
Left ventricular dimension	Increase
Myocardial	Concentric hypertrophy
Blood volume	Increase
Total hemoglobin	Slight increase
Functional Stroke volume—submax, max	Increase
Ventricular filling	Enhanced
Heart rate—submax	Decrease
Heart rate—max	No change or decrease
Cardiac output—submax	No change or decrease
Cardiac output—max	Increase
Myocardial O_2 requirements	Decrease
$(a-\bar{v})O_2$ difference—submax, max	No change
Muscle blood flow—submax, max	No change
Systolic blood pressure—submax	No change
Diastolic blood pressure—submax, max	No change or increase
Total peripheral resistance—submax, max	No change

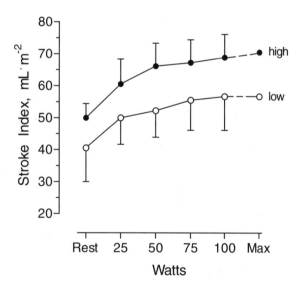

FIGURE 1.57 Stroke index in eight fit and eight less fit boys (12.2 ± 0.5 years) at rest and at submaximal and maximal cycling exercise. The two groups represented the highest and lowest quartiles, respectively, of performers in a 1-mi run within school physical education class. Stroke volume was assessed by Doppler echocardiography. Values are mean and 1 SD.

Reproduced, with permission, from Rowland et al. 1999 (371).

factors that augment left ventricular preload (e.g., a higher plasma volume, larger intrinsic left ventricular chamber size, or resting bradycardia) are responsible for differences in maximal O_2 uptake at that age.

The HR of trained athletes is lower at rest and during all levels of exercise, and it shows a faster post-exertional recovery than in nonathletes. A reduction of resting and submaximal HR is a most sensitive response to training (47; 82; 111; 145; 198; 406) and often precedes any changes in maximal aerobic power (111). The reduction in maximal HR is milder, seldom exceeding 5 to 7 beats · min⁻¹.

The mechanism of training-induced bradycardia is not clear. Bradycardia is likely secondary to an increase in stroke volume. A suggestion has been made that the bradycardia may reflect a stronger parasympathetic and a weaker sympathetic drive (145). However, as evidenced by HR variability, a 13-week training program of 10- to 11-year-old children did not induce changes in the sympathetic-to-parasympathetic relationship (294). Whatever its mechanism, a slower HR is accompanied by reduced myocardial work and O_2 uptake. The sensitivity of HR to training and detraining has been the basis of fitness tests in which maximal aerobic power is indirectly assessed from submaximal HR (1; 27; 53; 77; 84) (see appendix II, section "Determination of Maximal Aerobic Power").

With the opposing changes in stroke volume and HR, submaximal cardiac output changes little at a given metabolic level (178). Maximal cardiac output increases with training in children (145) and in adults, in proportion to the increase in maximal O_2 uptake.

In adults, the maximal arterial-venous O_2 difference increases with training, reflecting an increased peripheral O_2 extraction. In children, however, neither the submaximal (178) nor the maximal (145) arterial-venous O_2 difference changed with training. This discrepancy could result from the fact that the arterial-venous O_2 gradient is already wider in the untrained child than in the untrained adult.

Neither submaximal nor maximal muscle blood flow changed following a 40-week training program in 12-year-old boys (247). In adults, blood supply to the active muscles diminishes during submaximal exercise and increases during maximal exercise as a result of training.

Systolic arterial blood pressure at rest decreases in adults following training, by about 5-10 mmHg. A similar response has been shown in adolescents with mild hypertension (195; 196) (see also section "Systemic Hypertension," chapter 7). Blood pressure during submaximal exercise does not change with training in normotensive prepubertal boys (266), nor are there training-related changes in total peripheral resistance (145).

Training of the Pulmonary System

Even though adequate ventilation and alveolar-to-capillary diffusion are essential components of the O_2 transport system during exercise, there is little information regarding the effects of training on the pulmonary system of healthy children and adolescents. Most of the recent research in this field has focused on patients with cystic fibrosis, asthma, or a neuromuscular disease, as discussed in chapters 6 and 10. Much of our knowledge regarding the possible effects of training is inferred from cross-sectional comparisons between young athletes and nonathletes, and less so from properly designed intervention studies. The reported training effects as summarized in table 1.14, therefore, need further confirmation.

Pulmonary Functions at Rest

In a 2-year controlled follow-up of 13- to 16-year-old athletes from India, Lakhera et al. found that the athletes had higher lung volumes and capacities at rest than the controls. However, there were no training-related effects on vital capacity, other volumes and capacities within the vital capacity, forced expiratory volume in 1 sec, or maximal voluntary ventilation (260). The authors concluded that the development of the lung during

TABLE 1.14	Training-Induced Pulmonary Changes in Children
Function	**Change**
Vital capacity	No change (may increase in swimming)
Ventilation—submax	Decrease
Ventilation—max	Increase
Respiratory rate—submax	Decrease
Tidal volume—max	Increase
Ventilatory equivalent—submax, max	Decrease or no change
Respiratory muscle endurance	Increase
Pulmonary diffusing capacity	No change

adolescence under proper nutritional and health conditions is governed by the process of growth with no or negligible additional effects of physical activity. Similar findings and conclusions were reported by Baxter-Jones and Helms in a mixed longitudinal, cross-sectional 3-year follow-up of young British athletes (49) and in other studies (47; 446; 447). In contrast, as shown by Courteix et al. (103), a 1-year intense swimming program (12 hr a week) induced significantly greater changes in five 9-year-old girl swimmers, compared with controls. The changes included an increase in vital capacity, total lung capacity, and functional residual capacity, as well as several flow measurements at rest. The mechanism proposed by the authors was that, during prepuberty, intense swimming promotes isotropic lung growth by harmonizing the development of the airways and of alveolar lung spaces. Training-induced changes were reported also by others (133; 136; 198). It has been suggested that the trainability of lung functions may depend on the duration of the program; the longer the program, the greater is the likelihood of changes (375). This observation, however, is based mostly on cross-sectional data (426).

Ventilatory Functions During Exercise

Cross-sectional data show that trained children have a lower respiratory frequency and ventilatory equivalent for O_2 uptake, during submaximal exercise, than seen in the untrained (188). This suggests a more economical ventilatory pattern in the athletes. At maximal exercise, trained individuals have somewhat higher ventilation, but the difference is smaller than that in maximal O_2 uptake (426).

With aerobic training, ventilation and respiratory rate during a standard task are reduced, and O_2 extraction from the inspired air is greater. This important effect of training may reflect a lower reliance on anaerobic metabolic pathways, a lower O_2 cost of exercise, or a reduction in chemoreceptor sensitivity (423). In contrast, maximal ventilation increases with training (6; 421), in proportion to the increase in maximal O_2 uptake. Some studies show that the increase is due to an increase in the maximal tidal volume without changes in respiratory rate (380), while others report also an increase in respiratory rate (6; 421).

Specific training of the respiratory muscles, such as isometric expiratory effort, is accompanied by an increase in endurance of these muscles (i.e., their ability to sustain a high level of ventilation for prolonged periods). Such an effect has been shown among healthy adults and among children with cystic fibrosis (see chapter 6). While athletes have a high pulmonary diffusing capacity, this apparently does not increase in adolescents as a result of training above and beyond the increase in their pulmonary blood flow (249).

Training and the Bone

The rapid increase in bone density associated with the hormonal influences of puberty suggests that adolescence might represent a critical period for bone response to weight-bearing exercise. This idea was supported by the findings in a cross-sectional study of 91 female racket sport athletes ages 7 to 17 (194). Significantly greater values of areal bone mineral density (BMD) of the proximal humerus, humeral shaft, and distal radius were observed in Tanner stages 3, 4, and 5 compared to the values in nonathletic controls. However, there were no differences between groups in prepubertal subjects. Moreover, training time did not correlate with BMD in the Tanner stage 1 and 2 players but did at higher levels of sexual maturation.

MacKelvie at al. reviewed interventional studies that examined the effects of exercise training on bone density in children and adolescents (282). In the two studies that were restricted to the prepubertal age group, the improvement in bone mineral content between exercisers and controls was 1.2% to 5.6%, depending on measurement site (81; 173). In early-pubertal girls, Morris et al. found that a 10-month program of games and weight training caused a greater rise in total body, lumbar spine, and proximal femur bone mineral than in controls (318). However, two studies in postmenarcheal adolescent girls involving weight training and plyometrics demonstrated no effects on bone mineral content compared to values in control subjects (60; 477).

MacKelvie et al. concluded from this information that the exercise-related response of bone mass and structure occurs particularly in girls who are at the start of their puberty. Clearly, additional research will be needed before a precise understanding of age-specific responses of bone to exercise training is at hand.

To sum up, data about age- and maturation-related differences in trainability of any body system are still inconclusive. Definitive studies that address this issue will have to include nonathletic children, adolescents, and young adults who, at the start of the program, will all be at the same fitness level. In addition, the training dosage will have to be sufficient and carefully equated among all groups. Until such projects are launched, our knowledge on the optimal ages of trainability will be tentative at best.

Physiologic Effect of Detraining

Performance and physiologic functions deteriorate fast whenever one's activity level is appreciably reduced. This process is the reverse of training. It often occurs as a result of bed rest, immobilization by cast, or during an off-season period when an athlete reduces his or her training load.

An understanding of the processes that underlie detraining is not only important to athletes and coaches. It is as important to the clinician, who often must decide whether to prescribe bed rest, cast, cessation of training, or other means of immobilization.

Studies with healthy adults have shown that bed rest can induce, within several days, a dramatic deterioration in function and performance (166). For example, when young adult volunteers underwent bed rest for 7 days, complying with the routine of a clinical ward, there was a 67% reduction in maximal O_2 uptake; peak aerobic power; total hemoglobin; and plasma, blood, and cell volumes. There was also a drop in the resting urinary excretion of norepinephrine, but not in muscle strength (166). When bed rest is superimposed on a surgical trauma, deterioration is even faster, possibly because of the catabolic sequelae of surgery. In young adult endurance athletes, 2 months of inactivity resulted in a reduction in stroke volume during upright exercise. The authors concluded: "Despite many years of intense training, inactivity for only a few weeks results in loss of this adaptation (an increase in stroke volume) in conjunction with regression of left ventricular hypertrophy" (299).

The degree of detraining, according to some studies, is in direct relationship to the initial level of fitness; those who are more fit experience a greater drop in performance (189).

While it can be assumed that children's response to detraining is the same as that of adults, direct information pertaining to children is rather scanty. In one study, high school girls were followed during 23 weeks after cessation of track training (308). By the seventh week their submaximal HR rose by about 10 beat $\cdot$ min^{-1}, and there was a slowing of HR recovery after exercise. Although the mechanical economy of running did not change, there was a gradual decrease in ventilatory efficiency, as shown by a rise in the ventilatory equivalent (the ratio ventilation/O_2 uptake). Aerobic and anaerobic muscle enzymes in adolescent boys decreased

6 months after the termination of an endurance and a sprint program, respectively (5; 164). For example, adolescent soccer players were tested at the end of the competition season and again 4 to 8 weeks into the off-season period when they were not training. Biopsies of the vastus lateralis revealed a decrease in the cross-sectional area of type I and type II fibers and a reduction in the activity of creatine kinase, citrate synthase, phosphofructokinase, lactate dehydrogenase, and aspartate aminotransferase enzymes (5).

Another function that deteriorates with cessation of training is muscle strength (67; 408). In one study, prepubertal boys were tested at the end of a resistance training program and several weeks later. By that time, the strength of several muscle groups converged toward values found in a nontraining control group (67).

Detraining of children is most important in the clinical context. For example, 12-year-old boys who were immobilized by cast or by traction in an orthopedic ward were compared to boys who were hospitalized, but not immobilized. Arterial blood pressure of the former was higher, and some children in that group had enhanced urinary excretion of calcium (440). Detrimental effects of detraining have also been shown for children with muscle dystrophy. At one stage of their progressive disease, these children can still walk, but the slightest further diminution in activity may render them immobile for the rest of their life. For such children unjustified bed rest may be catastrophic, as discussed in chapter 10 (figure 10.31). Similar patterns have been reported for children with juvenile idiopathic arthritis (226).

Habitual Activity and Energy Expenditure in the Healthy Child

*U*nderstanding the importance of physical activity (PA) and energy expenditure (EE) to a child's health, and of the factors that affect PA and EE during childhood and adolescence, is pivotal for this book. This chapter is intended to introduce the reader to the components of PA and EE, describe the changes that occur in the PA and EE of healthy children and adolescents during the years of growth, and analyze the forces that determine these changes. Subsequent chapters discuss the relevance of activity patterns to specific diseases. Appendix III describes the methods used for measuring PA and EE.

Definitions

The terminology used in research on physical activity, energy expenditure, and their components is not always clear. This has caused confusion in describing activity patterns, in choosing methods to measure or assess activity and EE, and in interpreting the findings. The purpose of this section is to define terms and concepts regarding PA and EE as used in this book.

• *Physical activity.* Physical activity has been defined in various ways. A commonly accepted definition is "any bodily movement produced by skeletal muscles and resulting in energy expenditure" (p. 6) (7). Physical activity can be looked at from three points of view: *mechanical, physiologic,*

and *behavioral.* To a biomechanist, PA comprises elements such as force, velocity, acceleration, angles, inertia, mechanical power, or mechanical work. A physiologist analyzes PA in terms of metabolism, using measures such as O_2 uptake, metabolic energy (e.g., in kilocalories or kilojoules), metabolic power (kcal · min^{-1} or kJ · min^{-1}), or MET (see following list). A behaviorist is interested in the type of an activity (e.g., running vs. calisthenics vs. baseball); in the environment in which the child functions (e.g., playground, school); in the use of toys or apparatus; in interactions with others (e.g., friends, family members); and in who initiated the activity (the child, a parent, a friend).

• *Energy expenditure (EE).* The biochemical energy utilized by the body in order to perform its functions. Energy expenditure can be expressed as metabolic work (e.g., megajoules) or power (e.g., kilojoules per minute).

• *Diet-induced energy expenditure (DEE).* Also called diet-induced thermogenesis, diet-induced energy expenditure reflects the energy expended during and after a meal to digest, absorb, and transport the food.

• *Exercise.* A subset of PA used within the context of a training program or a laboratory test. This is distinguished from a spontaneous, nonregimented PA.

• *Exercise energy expenditure (EEE).* The energy expended for the performance of a PA.

• *Metabolic equivalent (MET)*. The ratio between exercise energy expenditure during an activity and the resting energy expenditure (EEE/REE). It is expressed as a nondimensional number. Metabolic equivalent is used in an attempt to cancel out the effect of body mass; MET values for activities performed by children are listed in table III.2 (appendix III).

• *Moderate-to-vigorous activity*. Activity that requires at least as much effort as brisk or fast walking.

• *Oxygen uptake ($\dot{V}O_2$)*. The volume of oxygen consumed by the body in a given period. Often used as a surrogate measure of EE.

• *Physical activity level (PAL)*. The ratio between 24-hr total energy expenditure and resting energy expenditure (TEE/REE). Like MET, it is used as an index of EE in which the effect of body size is canceled.

• *Power*. Work divided by time (also, force × velocity). Expressed usually in Watts, but also in liters per minute of O_2 uptake. Clinicians sometimes use "power" erroneously, to denote *"strength"* (maximal force that a muscle or a muscle group can generate).

• *Reactivity*. Changes in activity behavior of a person resulting from the knowledge that he or she is being observed.

• *Resting energy expenditure (REE)*. Energy expenditure while the subject is fully rested. Usually taken at the supine position, first thing in the morning prior to breakfast. Also called *resting metabolic rate*.

• *Total energy expenditure (TEE)*. Energy expended over 24 hr. It comprises resting energy expenditure, exercise energy expenditure, and diet-induced energy expenditure.

• *Adaptive energy expenditure*. This is another minor component of total energy expenditure, considered infrequently. It reflects increases in metabolic rate due to ambient conditions, such as climatic heat and cold stresses. Thus,

TEE = REE + EEE + DEE + Adaptive EE

• *Work*. Energy used by the body during a physical task. Usually expressed in Joules. Can be mechanical energy or metabolic energy.

Physical Activity and Physical Fitness

Intuitively, one would assume that physical activity and fitness are strongly related (active people being more fit than their less-active peers). This, however, does not seem to be the case. In a comprehensive review, Morrow and Freedson (23) reported the results of 25 cross-sectional studies that investigated the relationship between aerobic fitness and habitual PA in children and youth (males mostly). Out of these, 14 studies showed no significant relationship, while the other showed only a modest correlation (mostly r < 0.20). None of these or subsequent cross-sectional studies have shown a dose-response relationship between aerobic fitness and PA.

In a longitudinal observation of females and males between the ages of 13 and 27 years, Kemper and colleagues report that both maximal O_2 uptake and the highest running slope on the treadmill did correlate with the daily PA as assessed by an interview (14a). However, an increase in PA over the 15 years was accompanied by only 2 to 5% in aerobic fitness. The reason for such poor relationship is not clear. It may reflect the error that is inherent to estimates of PA and other confounding variables such as adiposity, age, maturation, and cultural and socioeconomic background.

Age and Maturational Changes in Physical Activity and Energy Expenditure

Children differ from adolescents and adults in the type of activities that they pursue, as well as in the amount and intensity of these activities. Likewise, adolescents differ from adults in their activity behaviors. This section first describes activities that are favored by children and adolescents, then presents an analysis of age and maturational changes in the amount of physical activity and energy expenditure.

Favorite Activities of Children and Adolescents

In the first decade of life, PA is usually spontaneous and nonorganized and typically consists of intermittent brief bouts. In contrast, older children and adolescents resort often to organized activities of a more prolonged nature. For example, 6- to 10-year-old girls and boys were found through direct observations to perform predominantly intermittent activities. The median duration for light-to-moderate activity bouts was 6 sec. Most of the high-intensity bouts did not exceed 3 sec, and 95% of these lasted less than 15 sec (figure 2.1) (4).

In a survey of 433 Canadian day care centers (34), the most frequently reported spontaneous outdoor activities included climbing, running races, jumping, cycling, swinging, and sliding. Structured activities included walks, games, field trips, ball games, and cycling. In a nationwide U.S. survey, approximately 84% of first through fourth graders took part in activities available in community organizations such as parks and recreation organizations, sport leagues, churches, YMCAs, YWCAs, scouts, and farm clubs (32). As reported by parents, the activities practiced most commonly included swimming, racing/sprinting, baseball/softball, cycling, and soccer (33).

The pattern just described undergoes changes in the *second decade of life*. An example of the choice of recreational activities of young people 10 years or older has been provided in a nationwide survey in Canada (42). Participants were asked to check activities "that are not related to work." The two most popular activities for girls and for boys ages 10 to 14 years were cycling and swimming. In the 15- to 19-year-age range, the girls preferred walking and swimming, while the boys preferred cycling and swimming. This pattern reflects activities pursued in the summer months. Preference changes markedly in other seasons. For example, during winter the most popular activity among Canadian boys is ice hockey.

Activity preferences also depend on geographic location and cultural traditions. In the United States, as evidenced in a national survey (29), basketball was the most popular sport among 16- to 18-year-old boys, and swimming was the preferred activity among the girls. Other popular activities among these U.S. youth include tackle football and baseball/softball for the boys and disco/popular dance, baseball/softball, and jogging/fast walking among the girls.

Effect of Age on the Amount of Physical Activity and Energy Expenditure

While the types of preferred activities vary among geographic and climatic regions, cultural traditions, and socioeconomic status, there is a universal pattern in the effect of age on the *amount of time* spent on physical activities and sport. Irrespective of their gender or their geographic, climatic, ethnic, or cultural background, children and adolescents become less active as they grow older (figure 2.2). This decline in activity is reflected in the total daily energy expended, as well as the energy expended on physical activities, both of which decrease with age. While some studies suggest that the decline in activity starts in the second decade of life (31; 48), others report a decline as early as age 6 years (38) or even earlier (45). Figure 2.3 summarizes this pattern using cross-sectional data from various countries. When expressed per kilogram body weight, total EE declines already after age 1 year. These data are of particular importance because they are all based on the doubly labeled water method, which is the "gold standard" for the measurement of total EE (45) (appendix III).

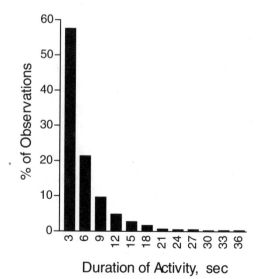

FIGURE 2.1   Distribution of children's high-intensity activities by the duration of the activity.

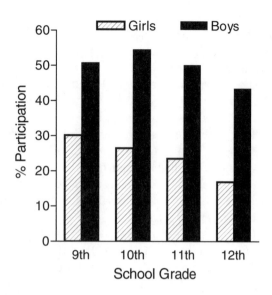

FIGURE 2.2 Age- and gender-related decline in physical activity. Percentage of U.S. 9th through 12th graders who participated in vigorous physical activity three or more times per week. Based on 11,631 students sampled for the United States Youth Risk Behavior Survey 1992 (10).

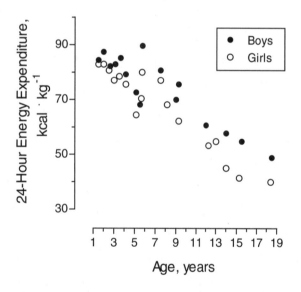

FIGURE 2.3 Decline in daily energy expenditure per kilogram body weight in healthy children and adolescents. Data are based on several studies from various countries, in which energy expenditure was measured by doubly labeled water. Adapted from Torún et al. 1996 (45).

A similar pattern was found in longitudinal measurements of Dutch girls and boys, whose EE was assessed periodically through heart rate monitoring. Both gender groups showed a continuous decline throughout the observation period, starting at age 6 (38; 48).

The decline is not limited to spontaneous, recreational activities. It is also apparent in participation in physical education classes. As shown in the first U.S. National Children and Youth Fitness Survey, conducted in the early 1980s (31), 97% of first through sixth graders attended physical education classes. In grades 11 and 12, attendance dropped to 49%. Attendance declined further during subsequent years: Daily participation in physical education among high school students was 42% in 1991 and only 25% in 1995 (12). According to the 1991 Youth Risk Behavior Survey (11), 71% of grade 9 students enrolled in physical education classes. This rate declined to 59% in grade 10, to 42% in grade 11, and to 38% in grade 12.

An age-related decline in physical education attendance occurs also in other countries such as the United Kingdom (9). In several Canadian school systems, participation in physical education classes is compulsory throughout elementary and middle school, but for one term only during the entire last 3 years of high school. As a result, the great majority of pupils opt out of physical education during these years.

Gender Differences in Physical Activity and Energy Expenditure

In general, reduced PA among girls starts earlier, and the reduction is faster, than in males. For example, based on the 1990 U.S. Youth Risk Behavior Survey (10) and as shown in figure 2.2, the percentage of 9th through 12th graders who participated in vigorous PA 3 or more days per week was markedly lower in the girls than in the boys. Likewise, in the Amsterdam Growth Study, the activity level of Dutch adolescents, as assessed by an interview (47), was considerably lower in girls than in boys at age 13. This difference was apparent mostly in high-intensity activities. In an earlier observation from that longitudinal project (48), the decline in EE assessed by heart rate monitoring was apparent at age 11 in the females and only at age 14 in the males. A review of nine studies performed in various countries on 6- to 18-year-old girls and boys concluded that the males were 14% more active as evidenced by questionnaires (36). The difference increased to

TABLE 2.1	Percentage of U.S. High School Students (Grades 9-12) Who Participated in Vigorous Physical Activity Three or More Days per Week by Gender and by Race/Ethnicity		
Category	Females	Males	Total
White	27.5	51.4	39.3
African American	17.4	42.7	29.2
Hispanic	20.9	49.9	34.5
Total	24.8	49.6	37.0

Based on 11,631 students sampled for the United States Youth Risk Behavior Survey 1990 (10).

23% in favor of the boys when objective measurements such as heart rate monitoring were used. The average decline in activity each year was 2.6% to 7.4% in the females, compared with only 1.8% to 2.7% in the males. The U.S. National Children and Youth Fitness Survey I (25; 30) sampled 8,800 students in grades 5 through 12 and assessed their activity through a questionnaire. The girls were consistently less active than the boys, with the difference ranging from 5% to 15%. As seen in table 2.1, the gender difference is pervasive in U.S. youth, irrespective of their ethnic background. While African American and Hispanic girls are less active than white girls, all three ethnic groups are less active than the respective groups of boys.

The reasons for the lower activity levels among girls are not clear. There are societal and biological factors that may limit PA among girls, particularly during and after puberty. At this stage of their life, girls become interested in social pursuits other than sport. To some extent the reason may be that success in sport has the connotation of "masculinity." A reduction in activity may also be secondary to biological changes that affect a girl's success in sport. These include an increase in body fatness, widening of the pelvis, discomfort before and during the menstrual period, and a reduction in blood hemoglobin levels.

Tracking of Habitual Physical Activity

Tracking is defined as a measure of the stability of a certain variable over time, or the maintenance over time of a relative position within a population. Body height, for example, tracks well during the years of growth, because a child whose height is at a certain percentile is likely to remain close to that percentile throughout childhood and adolescence. To determine tracking of a certain variable, one must measure the variable at least twice, usually over a span of several years. It is reported as the correlation between the two measurements (calculated usually as rank correlation).

How well does PA behavior track from childhood, or adolescence, into the adult years? This is an important question if one assumes that physically active children, compared with sedentary children, have a greater potential to become active adults. A number of studies, mostly from Europe and the United States, have addressed tracking of PA during growth (1-3; 15; 16; 20; 24; 28; 35; 38; 43; 47). These studies report mostly on tracking from early to mid-childhood or from adolescence to young adulthood. There is a great variability among reported data, which reflects the variety of methods used to assess PA as well as cultural and societal differences (1; 3; 15; 16; 20; 24; 28; 35; 37; 43; 47). As might be expected, short-term tracking of PA (i.e., over 2-3 years) is stronger than long-term tracking. Physical activity tracks moderately during childhood and during the transition into adolescence, as it does over a span of 2 to 3 years during adolescence. However, the correlation coefficients are lower over spans of 5 and 6 years, and they further diminish when one compares activity between early adolescence and young adulthood. In the Amsterdam Growth Study, for example (table 2.2), correlation coefficients for boys were 0.44 from age 13 to 16, 0.20 from age 13 to 21, and 0.05 from age 13 to 27. The same general pattern was apparent for the females (47). A similar trend has been found in Finland, based on the Cardiovascular Risk in Young Finns Study (28) and in a retrospective U.S. study (43). Moderate-to-low correlation coefficients have been found also among Danish females and males whose activity behavior was assessed at ages 15 to 19 years and again 8 years later (1). Altogether, tracking seems to be stronger in groups that are at the extremes of the PA spectrum, that is, the most and the least active (20).

TABLE 2.2	Tracking Correlation Coefficients in Total Weekly Habitual Physical Activity of Males and Females[*]		
Age span, years	Males (n = 84)	Females (n = 98)
13-16	0.44	0.58
13-21	0.20	0.18
13-27	0.05	0.17

[*]Activity was assessed by interviews.

Adapted from Van Mechelen and Kemper 1995 (47).

Factors That Affect Physical Activity in Children and Adolescents

As summarized in table 2.3, various biological, familial, psychological, societal, and cultural factors may influence activity behavior and EE of children and adolescents. Physical activity and EE may also be modified by variations in the physical environment, such as climate, weather, and seasonal changes.

Studies that have addressed these relationships are usually based on correlational analysis, which reflects associations, but not cause and effect. The following is a brief discussion of some biological factors, as well as factors related to the physical environment, that may modify PA and EE during growth and maturation.

Biological Factors

Heredity appears to have some influence on habitual PA patterns. On the basis of their observations from the 1981 Canada Fitness Survey, Perusse et al. (26) concluded that environmental influences on PA are stronger than hereditary ones. In a subsequent project, the same group (27) studied 375 families of French descent. Using a self-recorded activity log, they assessed the "level of habitual physical activity" (all daily activities, ranging from those that require a very low EE to those that require a high metabolic level) and "exercise participation" (activities that require at least 5 MET). Associations were computed for pairs that included "biological relatives" (parent vs. natural child and biological siblings, including dizygotic and monozygotic twins) and "nonbiological relatives" (e.g., spouses, uncle

vs. nephew or niece, and siblings by adoption). The authors concluded that the level of habitual PA is significantly influenced by heredity (29% heritability). In contrast, there was no genetic influence on exercise participation.

Two earlier studies of twin children in the United States (39; 49) also showed heritability of the overall activity pattern, less for the type of activity chosen by the children. One of these studies (39) showed that even when parents erroneously labeled their monozygotic twins as dizygotic and vice versa, the concordance in PA was stronger in the real monozygotes than in the real dizygotes.

One can therefore conclude that there is a significant genetic effect—the strength of which varies among studies—on a child's general activity pattern. However, the intensity of children's activities, and the specific types of sports they are engaged in, is strongly influenced by environmental factors. It is the latter relationship that is of public health importance. More research is needed to determine whether the phenotypical expressions just described can be generalized to all age groups.

Both *undernutrition* and *obesity* are associated with physical hypoactivity of children and youth. Protein-calorie undernutrition often results in reduced resting metabolic rate and 24-hr EE (45), which may reflect the body's attempt to conserve energy. For details, see chapter 9.

Health status is a major determinant of PA. As a group, children with a chronic disease or a physical or mental disability are less active than their healthy peers (18; 19; 46). This aspect is discussed in detail in other chapters of this book.

Pubertal changes, mostly in females, are often accompanied by a reduction in PA. This has partially been explained by psychosocial factors (e.g., notion that maturing girls become interested in pursuits other than sport, or that "success in sports requires masculinity") (8). However, reduction in PA may result also from biological changes as described in the earlier section on gender-related differences.

Motor skills may determine PA patterns (21). Although athletic success reflects training, it also depends on innate skill. Children and adolescents pursue activities in which they are skilled and successful.

TABLE 2.3	**Factors That Have Been Shown or Suggested to Affect Activity Behavior and Energy Expenditure of Children and Adolescents**

Biological
    Heredity
	Psychological
	    Self-efficacy

Adiposity and nutrition	Self-schema for activity
Health status	Perception of barriers to activity
Sexual maturation	Attitudes about activity
Motor skills	Beliefs about activity
Physical fitness	

Social and cultural
    Parent attitudes and behaviors
	The physical environment
	    Availability of activity facilities

Peer attitudes and behaviors	Seasonal variation
Socioeconomic status	Climatic changes
Cultural and ethnic values	Day of the week and holidays
Time spent on TV viewing	Safety considerations
Time spent on computer games	Entering the work force versus school and studies

The Physical Environment

Availability of activity facilities may affect children's participation in exercise. Children's activity depends on the proximity of their home to playgrounds, parks, and recreation centers. In some communities, though, because of safety, financial, and logistical considerations, living close to such facilities does not guarantee that the child will use them.

Seasonal and climatic variation may affect activity behavior (6; 41). In temperate and cold climatic regions, the level of PA often increases in the summer months. Indeed, according to a U.S. nationwide survey (29), boys in 5th through 12th grades spent twice as many hours per week on afternoon activities during the summer compared with the winter. Among girls, the ratio was 2.3:1 in favor of summertime activities. This seasonal trend was not age dependent. In the same survey, 90% of the students reported an appropriate PA pattern during the summer, compared with only 68% for the winter. Rates for fall and spring were 74% and 79%, respectively. The same pattern was obtained for activities with a carryover value for the future and for activities that induce "sweating and hard breathing." This pattern is similar for girls and boys.

In some countries, such as Canada, climatic and seasonal variations may be a significant factor due to the sharp interseasonal climatic variations. The effect of fewer daylight hours during the winter deserves particular attention, as this is a further limitation on children's ability to play outdoors. The relevance of the seasons to a child's activity was addressed in the Trois-Rivières regional experiment (Quebec), in which 546 girls and boys were divided into a physical education enrichment group (additional 5 hr per week) and a control group (40 min of required PE) (14). In the 1981 Canada Fitness Survey, there was an almost 2:1 ratio in the number of hours devoted to participation in the 10 most popular activities during the summer compared with winter (40) (pp. 117-118). A similar trend was reported for children from Finland, another country with sharp winter-to-summer climatic differences. In that study (44), parents reported twice as much outdoor activity in the summer than in the winter for 3- and 6-year-old girls (6.7 vs. 3.3 hr per day) and boys (7.0 vs. 3.7 hr per day). The seasonal differences are not so clear when comparison is made between spring and fall (41).

Seasonal variations may occur also in relation to *summer vacation and prolonged holidays.* While most children are more active during these periods, others (whose main activity is school based) show a reduction in activity. The greater activity level during the summer probably reflects the finding that children are more active when outdoors, as shown in U.S. (17) and Japanese (22) studies.

Spontaneous activity during *weekends* seems to be greater than during *weekdays.* For example, in the Trois-Rivières study, vigorous activities during the spring were performed less during weekdays than during weekends (41). Evaluation of children's activity pattern and EE should therefore cover weekends and weekdays alike.

Safety concerns, particularly in urban areas, have become dominant in the decisions of parents concerning whether a child is allowed to spend time outdoors or to walk to and from school (5; 12).

School versus work site is another environmental factor to be considered. Ilmarinen and Rutenfranz from Germany conducted a longitudinal follow-up of 25 girls and 26 boys between ages 14 and 17 (13). They analyzed the possibility that PA declines once a person leaves school and starts to work. At age 16, 73% of the boys and 52% of the girls quit school and joined the labor force. When observed 1 year later (recall question-naires), the boys who started working did not differ in their PA from those staying at school. However, among the girls, the occupational group was half as active as the group of girls remaining at school. The main difference was the extent of participation in sport. The authors speculate that once a person joins the workforce, there is less time available for leisure activities and a greater fatigue at the end of the working day, particularly among females.

In conclusion, it is evident that activity behaviors in children and youth are markedly affected by personal factors and numerous environmental factors.

3

Climate, Body Fluids, and the Exercising Child

In the preceding chapters, physical exertion was discussed in isolation from other stressors. For the sake of simplicity, we analyzed children's responses to exercise in a "neutral" climate: neither too cold nor too warm, neither very humid nor very dry. Yet such conditions are quite hypothetical. In many geographic regions, the climate is not neutral and may impose on the child an added environmental stress, especially during outdoor exercise.

Climate can play a major, even crucial, role in performance, subjective comfort, and health. Abundant data are available on the combined effects of exercise and climate on adults, with relevance to industry, sport, the military, and public health. Even though children are habitually more active than adults and perform many of their recreational activities out-of-doors, their response to exercise in hostile climates had not been thoroughly studied until recent years. One reason for the paucity of information on children is that military and occupational issues are irrelevant to children in most societies. Another reason is ethical: Experiments under "hostile" environments may entail major discomfort and risk to the health and well-being of children, which is incompatible with ethical guidelines for human experimentation.

In this chapter we first introduce the basic physical and physiologic concepts of thermoregulation and acquaint the reader with the nomenclature used by the environmental physiologist. Theoretically, according to geometric and functional principles, children are less efficient thermoregulators than are adults, especially in extreme climatic conditions. An analysis of this topic follows, with data on heat acclimatization and heat and cold tolerance. Involuntary dehydration often accompanies prolonged exercise, particularly in climates that are hot, humid, or both. The resulting disruption of water and electrolyte balance may be deleterious to performance and to health. We give special attention to the means by which involuntary dehydration can be prevented.

Certain children are at potentially high risk of undergoing heat-related illnesses such as heatstroke or heat exhaustion. Children who have certain diseases are at especially high risk. Problems encountered by these high-risk groups are discussed, as well as the precautions to be taken for their prevention. The chapter concludes with recommendations for the conduct of sports events in a hot and a cold climate.

Heat Stress and Heat Strain

Climatic heat stress denotes a combination of environmental conditions that stress the thermoregulatory system. Heat strain, on the other hand, is the physiologic and mental response to heat stress. Two individuals exposed to an identical heat stress may respond with a different heat strain.

Ambient temperature is just one component of climatic heat stress, and not necessarily the most important. Others are air humidity, air movement (wind), and radiant heat. The main source of the latter is solar radiation, but such heated objects as artificial turf can also generate marked radiant heat.

Various heat stress indices have been constructed that include one or more of these components. A popular index, originally designed for the military but also utilized in industry and sport, is the wet bulb globe temperature (WBGT). This incorporates air temperature, humidity, and radiation as measured by three thermometers: a dry bulb (DB), a wet bulb (WB), and a black globe (G), respectively:

$$WBGT = 0.7\ WB + 0.2\ G + 0.1\ DB$$

In this index, air temperature accounts for only 10% of heat stress, while radiation is taken as 20% and humidity as 70%! The implication is that a highly humid and mildly warm day can be more stressful than a very hot but dry one. For indoor use, when radiant heat is less important, the index is called WBT, taken as 0.7 WB + 0.3 DB. Psychrometers that monitor WB, DB, and G can be operated in the field by the coach, teacher, or team physician.

Another index, the "effective temperature," combines humidity, ambient temperature, and air velocity and is especially suitable for indoor use. Other indices take into account the type of clothing (which may interfere with sweat evaporation and heat dissipation) and the type and intensity of physical activity. A detailed description of these indices is available (93). Heat strain components include sweating rate, rectal and skin temperatures, and heart rate. Other functions such as skin blood flow, cardiac output, or ventilation play a role in thermoregulation but are less often measured. Mental functions considered to reflect heat strain are mental acuity, thermal comfort, and perception of heat intensity.

Heat Production and Heat Exchange

Living cells continuously generate heat. A major component of this metabolic heat (M) is generated by muscle, in proportion to the intensity and duration of muscular activity. Some 75% to 80% of the chemical energy used for muscle contraction is converted into heat. Dissipation of this heat (which may exceed 10 times the resting metabolic rate) is a major challenge to the thermoregulatory system during and following exercise.

Heat can penetrate the body from the environment through conduction (CD), convection (CV), and radiation (R). It can also be dissipated from the body by these three avenues. The direction and intensity of such heat transfer depend on the temperature gradient between skin temperature and ambient temperature (for CD and CV) and on the gradient between skin temperature and the surrounding objects (for R). Another avenue for heat dissipation is evaporation (E) of sweat, or of water from the epidermis and from the respiratory mucosa. Evaporation of 1 L of water requires 2.43 mjoule (580 kcal) at 33° C. It is especially important during intense exertion or whenever ambient temperature is high and dissipation by CD, CV, or R becomes ineffective. The efficiency of evaporative cooling depends largely on sweating rate, but also on water vapor pressure in the air (high humidity attenuates evaporation), wind velocity, and air temperature (stronger wind and warmer air both enhance evaporation). Because of the insulative quality of air, heat transfer by CD between air and skin is negligible. During water immersion, however, CD is a major avenue for heat exchange, because thermal conductivity of water is some 25 to 30 times that of air.

These components of heat generation and transfer can be put into a "heat balance" equation, in which S denotes heat storage within the body:

$$M \pm CD \pm CV \pm R - E = S$$

When S is zero, combined heat generation and penetration equal heat dissipation, and the body is considered to be in a thermal balance. A negative S indicates heat loss. Ordinarily, this does not occur during exercise (even on a cold day) unless the exercise is performed in water. Note that conduction, convection, and radiation appear in the equation with both a plus and a minus sign, to denote heat gain or loss, respectively. Evaporation, on the other hand, appears only with a minus sign, because it induces a cooling effect only.

Physiologic and Behavioral Means of Thermoregulation

There is a narrow range of climatic conditions, termed "neutral zone," in which metabolic heat is passively dissipated to the surroundings. The body is then kept in thermal balance without active participation of the thermoregulatory apparatus. This zone is not constant. It varies among and within individuals depending on their activity level, clothing, body surface area, and amount of subcutaneous insulating fat. In women it may vary at different stages of the menstrual cycle. In a resting naked individual of average adiposity, the neutral zone is 25° to 27° C, 50% to 60% relative humidity. Any deviation from this neutral zone will induce, via hypothalamic centers, a physiologic response. When heat stress rises mildly, vasodilation takes place in the skin, thus increasing convection from body core to the periphery. Skin temperature then rises, facilitating better heat dissipation by convection, conduction, and radiation. When heat stress further increases, sweat is produced by the eccrine glands and spreads over the skin so that heat is also dissipated by evaporation. Evaporation of sweat in humans is the single most important means for heat dissipation during exercise.

When the climate is mildly cool, peripheral vasoconstriction takes place, decreasing convection from core to skin and reducing heat dissipation. In addition, the resting metabolic rate rises. The latter process has been termed cold-induced thermogenesis. Further environmental cooling will induce shivering (i.e., rhythmic, involuntary, and uncoordinated muscle contractions), which generates heat and compensates for heat loss. Intense shivering can produce heat in excess of five times resting metabolic rate.

Thermoregulation is achieved not only by physiologic processes but also by behavior. Migration of birds, licking of fur, panting, and resorting to daytime shelter in the desert are just a few examples from the animal kingdom. Humans resort extensively to behavioral means when confronted by heat or cold stress. These include the use of clothes to increase insulation, the seeking of shade or the wearing of a hat to reduce solar radiation, the use of fans to increase evaporation, or curling up in bed to reduce the effective surface area of the body. Behavioral means of thermoregulation allow us to widen the range of climates in which we can function, and they are instrumental in the proper conduct of athletic activities and for prevention of heat- or cold-related illness.

Geometric and Physiologic Characteristics of Children Relevant to Thermoregulation

Children have several geometric and physiologic characteristics that affect their thermoregulatory capability, as summarized in table 3.1. Figure 3.1 compares heat production, penetration, and dissipation in a 1-year-old infant, an 8-year-old child, and a young adult.

TABLE 3.1	Geometric and Physiologic Characteristics of Children As Related to Their Thermoregulatory Ability	
Variable	Compared with adults	Implications for thermoregulation
Surface area per unit body mass	Larger	Greater heat gain in hot climate, less in cold climate
Metabolic heat production per unit body mass during walking or running	Larger	Greater strain on the thermodissipatory systems
Sweating rate	Much lower (until mid-puberty)	Lower capacity for evaporative cooling
Cardiac output at a given metabolic level	Somewhat lower	Lower capacity for heat convection to the periphery; potentially limited O_2 transport during intense exercise in the heat

Reproduced, with permission, from Bar-Or 1989 (16).

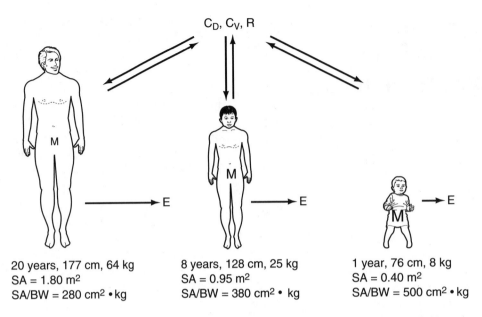

C_D, C_V, R

20 years, 177 cm, 64 kg
SA = 1.80 m²
SA/BW = 280 cm² • kg

8 years, 128 cm, 25 kg
SA = 0.95 m²
SA/BW = 380 cm² • kg

1 year, 76 cm, 8 kg
SA = 0.40 m²
SA/BW = 500 cm² • kg

FIGURE 3.1 Schematic presentation of the rate of heat production and transfer in an infant, a child, and an adult. CD = conduction; CV = convection; R = radiation; E = evaporation; M = metabolic heat production; SA = surface area of body; BW = body weight. The length of arrows represents rate of heat transfer per unit mass.

Large Surface Area-to-Mass Ratio

Although the child shown in figure 3.1 has a smaller absolute surface area than the adult, his surface area per unit mass is some 36% greater than in the adult. The respective difference between the 1-year-old infant and the adult is 78%. Because the heat flux between two objects depends on their area of contact (or, in the case of radiation, the effective surface area), heat transfer to and from the body is greater in the infant and the child than in the adult for a kilogram of body mass. The difference is depicted in the diagram by the length of the arrows for conduction, convection, and radiation. The higher the temperature gradient between air and skin, the greater the difference in the heat flux. In practical terms, the geometric difference is an asset to the child in mildly warm environments (when ambient temperature is still lower than skin temperature) (43; 48). It is also an advantage during intense exercise performed in mildly cool weather, when greater heat dissipation is advantageous. However, it becomes a liability in extremes of heat and cold alike, when heat transfer should be minimized. The smaller the child, the greater is this liability. One should realize, however, that there is a certain degree of overlap in the surface area-to-mass ratio between large children and small adults (8; 9). Thus, surface area per mass ratio per se does not predict age-related differences in thermoregulatory effectiveness.

High Metabolic Heat Production

Children expend more chemical energy per unit mass than do adults while performing similar walking or running tasks (see figures 1.7 and 1.8 and the section on energy cost of locomotion in chapter 1). As a result they produce greater metabolic heat per kilogram, which, during intense or prolonged exercise, subjects their thermoregulatory system to a greater strain. A 6-year-old child, for example, would generate 15% to 20% more heat per kilogram body mass than a 16-year-old adolescent if they both ran at the same speed.

Low Cardiac Output and Enhanced Changes in Peripheral Blood Flow

Ideally, both cardiac output and skin blood flow should increase during exercise in the heat in order to facilitate enhanced convection of heat from body core to the periphery (135). Studies performed in a thermoneutral environment (see chapter 1, "Cardiac Output and Stroke Volume" section) and a single study in the heat (51) suggest that exercising children, compared with adults, have a somewhat lower cardiac output at a given O_2 uptake. Notwithstanding, children have a higher peripheral blood flow as measured immediately after (51; 58; 85) or during (145) exercise in the heat. In one study, skin blood flow was measured in several sites. On the back and chest it was higher in 10- to 11-year-old boys than in young men, but the boys had a lower blood flow in the forearm (146). Maximal skin blood flow to the forearm is higher in children than in adults (103). Figure 3.2 shows that prepubertal boys had a higher forearm blood flow following 20-min bouts of exercise at 42° C compared with mid- and late-pubescent boys.

A potential disadvantage of the pattern just described is that the higher peripheral blood flow, when combined with a greater surface-to-mass ratio and a somewhat lower cardiac output, may result in compromised blood supply to internal organs, including the brain. This may be one factor for the lower heat tolerance of children compared with adults (51). Indirect evidence that blood supply to vital organs may be compromised in young children was obtained in a study in which Finnish infants (2 years old),

children, and adolescents were exposed for 10 min to sauna-like conditions (70° C, 20% relative humidity). The youngest age group (age 2-5 years) was the only one that did not increase its cardiac output during the 10-min exposure (79). Soon after the exposure, two of the young children had a vasovagal collapse (80).

The mechanism for a higher peripheral blood flow in children is not clear. It may reflect structural differences in skin blood vessels (103) or a higher sensitivity to vasoactive peptides (168).

Although much more research is needed on the hemodynamic responses of children to exercise in the heat, it seems that compared with adults, children rely less on evaporation of sweat and more on convection of blood flow to the periphery. It is of interest that during exposures to a cold climate, children seem to have a more effective peripheral vasoconstriction compared with adults (148). For details, see section "Physical and Physiologic Responses to Cold Climate."

Sweating Pattern

There is more to sweating than sweating rate. Sweating pattern refers also to sweating threshold, sensitivity of the sweat glands to changes in body temperatures, and the population density of heat-activated sweat glands. Most available information within the pediatric context is related to sweating rate and the population density of heat-activated sweat glands.

Sweating Rate

The sweating apparatus is apparently fully developed by the third year of life (89), but even so, children perspire less than adults (5; 43; 51; 71; 75; 143; 145; 150; 160; 163). The lower sweating rate of children is apparent not only in absolute terms but also when corrected for unit surface area (figure 3.3). While prepubescents seldom produce more than 400 to 500 mL sweat per square meter per hour, adults exposed to identical conditions produce as much as 700 to 800 mL per square meter per hour. Young women usually perspire more than girls (50; 51; 108), but this age-related difference is smaller than among males. Sweating rate in boys is higher than in girls (82) at all maturational stages (100).

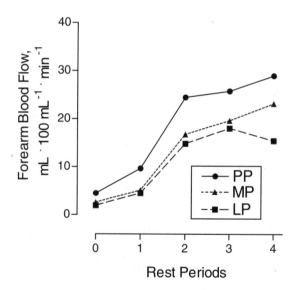

FIGURE 3.2 Forearm blood flow before entering a climatic chamber (42° C, 20% relative humidity) and following each of three bouts of exercise (20 min at 50% $\dot{V}O_2$max) in boys. PP = prepubescents; MP = midpubescents; LP = late pubescents. * = p<0.05. Values are mean ± SEM.

Reproduced, with permission, from Falk et al. 1992 (58).

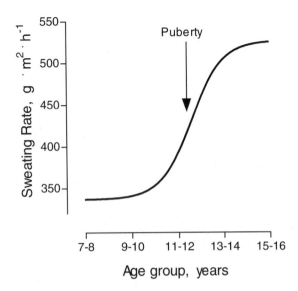

FIGURE 3.3 Development of sweating rate. Forty boys, 7 to 16 years old, exercised at a moderate intensity (heart rate 160-170 beat · min^{-1}) on the cycle ergometer, at 29° C, 60% relative humidity. Exercise time was 15 to 35 min. The arrow indicates the age at which pubertal changes were first noted. Data by Araki et al. 1979 (5).

The transition from the childhood pattern to the adulthood pattern of sweating takes place during puberty, at least among boys (figure 3.3) (5). Indeed, cross-sectional (56) and longitudinal (101) studies have shown that the sweating rate of pre- and early-pubescent boys is lower than in mid or late puberty (56; 59). Figure 3.4 summarizes the cross-sectional data for boys. Corresponding differences among females of various maturational stages are less pronounced.

Population Density of Heat-Activated Sweat Glands

As evidenced by a small database, the total number of eccrine sweat glands in humans seems to become fixed at age 2, at about 2 to 2.3 million (82; 89). Their recruitment during exercise occurs mostly at the first few minutes, and thereafter the number of active glands is quite constant (87), with some 1 to 1.7 million glands being activated at any given time (22).

Do children perspire less because they activate fewer glands or because of lower sweat production by each gland? As seen in figure 3.5, the population density of active glands is greater in children than in adolescents or adults. This pattern holds for females and males alike. It is,

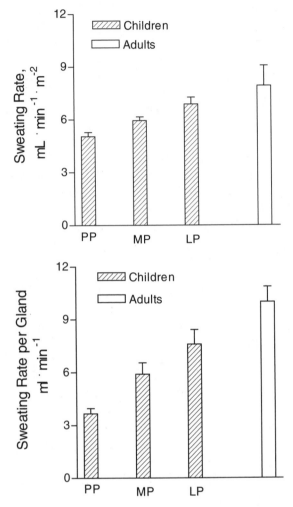

FIGURE 3.4 Sweating pattern in relationship to maturational stage. The top graph displays sweating rate per square meter of skin. The bottom graph displays sweating rate per single gland. Cross-sectional data of 36 boys and sixteen 20- to 23-year-old men who cycled intermittently (50% maximal O$_2$ uptake) at 42° C, 20% relative humidity. Tanner stage was determined from pubic hair development. PP = prepubertal; MP = midpubertal; LP = late pubertal. Vertical lines denote SEM. Based on Bar-Or 1989 (15) and Falk et al. 1992 (55).

therefore, the output per gland rather than the number of glands that limits the sweating rate of children (see next section on sweat production by a single gland).

Sweat Production by a Single Gland

Children's lower sweating rate and the greater population density of their active glands suggest that the sweat production per single gland

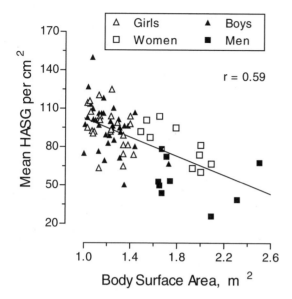

FIGURE 3.5 Relationship between population density of heat-activated sweat glands (HASG) and body surface area. Each data point represents a mean value, calculated from several sites in one subject, who was resting in a hot, dry chamber. Based on data by Bar-Or 1976 (11) and Bar-Or et al. 1968 (22).

Reproduced, with permission, from Bar-Or 1989 (15).

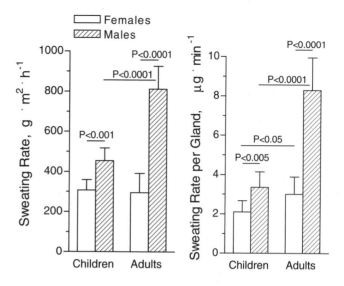

FIGURE 3.6 Total sweating rate and sweat production per gland in 9-year-old girls (n = 7) and boys (n = 7) and in young women (n = 7) and men (n = 8) sitting at 34° C, 40% relative humidity. Based on data by Kawahata 1960 (82).

Reproduced, with permission, from Bar-Or 1989 (16).

is lower in children than in adults. Indeed, whether at rest (82), during exercise in the heat (75; 145), or in response to pilocarpine iontophoresis (74), sweat production per gland is 2.5 times higher in men than in boys. This difference seems greater in the trunk than on the limbs (145). Girls also have a lower production per gland, compared with women, but the difference is much smaller than among males. Figure 3.6 is an example of such a pattern.

Sweating Threshold and Sensitivity of Sweat Glands to Thermal Stimuli

One index for the response of sweat glands to heat stress is the core temperature at which sweating starts, the sweating threshold. Another is the increase in sweating rate for a given increase in core temperature, the sweating sensitivity. As seen in figure 3.7 (top portion), a prepubescent boy started sweating only when his core temperature increased by 0.7° C, compared with a young man who started sweating when the increase in his core temperature was as little as 0.2° C (5). The authors of that study did not provide absolute core temperature values, but the pattern

does suggest that boys have a higher sweating threshold. Furthermore, the slope of the boy's increase in sweating rate was lower than for the man, suggesting a lower sensitivity. A higher sweating threshold (163) and a lower sweating sensitivity (75) among boys were also suggested in other studies. There are no data, however, on age-related differences in sweating threshold and sensitivity among females.

Mechanisms for Low Sweat Production

It is not entirely clear why sweat output per gland is smaller in children. In part, this may result from a smaller duct length and cross-sectional area of the coil (90), but neural and hormonal causes cannot be excluded. While no studies have directly correlated sweating rate with circulating androgens during puberty, this relationship may exist, since increases in sweat production among boys occur only after puberty has started. Correspondingly, sweat output in females is less dependent on maturation than in males (figure 3.6). Among adults, some studies indirectly suggest that testosterone may enhance sweating, while other experiments could not confirm this effect (128).

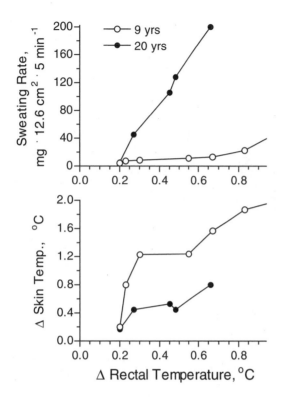

FIGURE 3.7 Sweating rate and increase in mean skin temperature in relationship to increase in rectal temperature. Comparison between a 9-year-old boy and a 20-year-old man who performed a continuous exercise task on the cycle ergometer at 29° C, 60% relative humidity.

Adapted from Araki et al. 1979 (5). Reproduced, with permission, from Bar-Or 1989 (13).

Shibasaki and colleagues (144) suggest that the lower sweating response in children is due to "underdeveloped peripheral mechanisms" rather than a low central drive. One possible peripheral mechanism is a low anaerobic energy turnover in the sweat glands. This is in analogy to children's lesser reliance on anaerobic pathways during muscle contraction (see chapter 1, section on anaerobic performance). Indeed, a relationship was found (60) between sweat production per gland and the amount of sweat lactate per gland in circumpubertal boys (figure 3.8).

Functional Implications of Low Sweat Production

Is the lower sweating rate in children an advantage or a handicap? Children's greater reliance on heat dissipation via convection, conduction, and radiation, along with their lesser dependence on

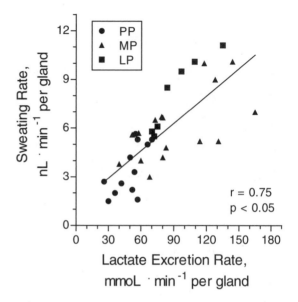

FIGURE 3.8 Relationship between sweating rate per gland and sweat lactate excretion per gland in 36 circumpubertal boys who exercised (50% $\dot{V}O_2$max) intermittently at 42° C, 18% relative humidity. PP = prepubertal; MP = midpubertal; LP = late pubertal.

Reproduced, with permission, from Falk et al. 1991 (60).

evaporation, is an economical thermoregulatory pattern that minimizes water loss. On the other hand, their lower sweating rate can be looked upon as a functional disadvantage because of a lower evaporative capacity. Insufficient evaporation results in high skin temperature and a less favorable temperature gradient for convection of heat from body core to periphery, as seen in the lower part of figure 3.7. While such a disadvantage may reduce thermoregulatory effectiveness when the child is exposed to a very high climatic heat stress, it does not seem to affect thermoregulation at thermoneutral conditions or at a mild to moderate heat stress (see upcoming section on effectiveness of thermoregulation during exercise).

It should be emphasized that sweating is not synonymous with evaporation. For example, sweat that drips from the skin, or that is trapped under the clothing, may not evaporate and consequently will not cool the skin. Furthermore, it is possible that the typical sweating pattern of children (many small drops) is more conducive to evaporative cooling than is the pattern in adults (fewer, but larger, drops). This possibility needs experimental examination.

The Apoeccrine Glands

An additional puberty-related change in sweating pattern is the appearance of apoeccrine sweat glands, which combine the morphological characteristics of eccrine and apocrine glands (138). These glands have been found through biopsies of axillary skin. Their relevance to thermoregulation is not yet clear. Figure 3.9 displays cross-sectional data obtained in male children and adolescents. It is not known whether the apoeccrine glands appear also in females during puberty, but they have been found in adult females.

Effectiveness of Thermoregulation and Heat Tolerance During Exercise

Thermoregulatory effectiveness in the context of this section is the ability to maintain thermal homeostasis during exposure to heat stress. A commonly used index for thermoregulatory effectiveness is the increase in core temperature. Heat tolerance is the ability to sustain a certain task during exposure to heat stress. Various criteria have been developed for heat tolerance. These

relate to industry, the military, or space-oriented research (93). One criterion, also used in research with children, is the climatic heat stress beyond which a prescribed task cannot be completed. Incompletion is determined either by the appearance of dizziness, aggressiveness, apathy, disorientation, nausea, exhaustion, marked headache, or abdominal cramps, or by heat strain indicators (e.g., rectal temperature higher than 39.4° C, heart rate exceeding 90% of maximum).

On the basis of their geometric and physiologic characteristics (see earlier section on these characteristics), children might be expected to be less effective thermoregulators than adults and less tolerant of climatic heat, particularly when exposed to climatic extremes. However, available literature on this topic has yielded seemingly conflicting results. While some authors have suggested that, compared with adults, children thermoregulate less effectively (47; 51; 70; 80; 147; 150; 160; 163), others could not confirm ineffective thermoregulation among children (25; 43; 48; 58; 67; 122; 148). The confusion results mostly from an inconsistency of climatic conditions in the various studies. Indeed, the relative ability of children to withstand climatic heat and cold stresses depends, to a great extent, on the degree of these stresses, as represented by the air-to-skin temperature gradient (figure 3.10). The following several sections therefore address separately the responses to thermoneutral, warm, very hot, and cold environments.

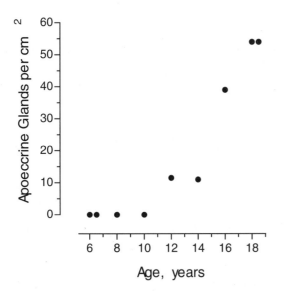

FIGURE 3.9 Population density of axillary apoeccrine sweat glands in 6- to 18-year-old males. Data by Sato et al. 1987 (138).

Reproduced, with permission, from Bar-Or 1989 (16).

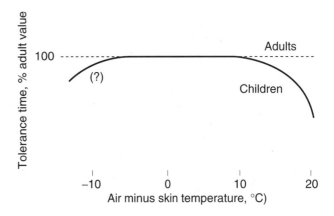

FIGURE 3.10 Ability of children to tolerate climatic heat and cold stress, compared with adults. A schematic presentation of the relationship between relative tolerance time of children as a function of the air-to-skin temperature gradient, taking adult values as 100%.

Thermoneutral Environment

Thermoneutral denotes a range of climatic conditions in which the need to dissipate or preserve heat is minimal (e.g., when a dressed person rests in an air-conditioned room). Figure 3.11 summarizes an experiment in which children and young men exercised at 21° C, 67% relative humidity, for 60 min. Although the children's evaporative heat loss was lower than in the men, their convective and radiative heat loss was greater. The end result was a very similar heat storage. Likewise, the increase in rectal temperature was identical (1.2° C) in the two groups, suggesting a similar thermoregulatory effectiveness (43). A similar pattern was reported (51) when 12-year-old girls and young women were exposed to 28° C, 45% relative humidity. In another study (67), 11-year-old boys performed several 60-min cycling tasks at 22° ± 4° C, 35% to 55% relative humidity, and reached thermal balance on each occasion. In conclusion, even though children rely less on evaporative heat loss, they compensate by a higher convective plus radiative loss. Using this strategy they are effective thermoregulators when exposed to a thermoneutral environment.

Warm Environment

Warm denotes air temperature of up to 8° to 10° C above skin temperature (e.g., 30-40° C). With few exceptions, children dissipate heat effectively under these conditions (48; 51; 58; 70; 71; 75; 98; 122; 145). For example, out of twenty-three 10- to 13-year-old boys who walked (6 km · hr⁻¹, 0% grade) for 60 min at 30° C, 80% relative humidity, only one could not complete the prescribed task. Core temperature approached plateau within 30 min (48). When divided according to somatotype, the boys with endomorphy thermoregulated somewhat less effectively than the ectomorphs or mesomorphs. This observation is in line with earlier studies (70; 71) in which obese or overweight 9- to 11-year-old girls and boys thermoregulated less effectively than their leaner counterparts. The girls, as a group, completed the prescribed task (3 × 20-min walks at 4.8 km · h⁻¹, 0% grade) at ambient temperatures as high as 42° C, and their core temperature and heat storage were similar to those found among 22-year-old women who performed an identical task (21).

When pre-, mid-, and late-pubertal boys exercised (3 × 20-min cycling at 50% $\dot{V}O_2$max) at 42° C, 18% relative humidity, there were no intergroup differences in heart rate or rectal and skin temperatures (58). In point of fact, these heat strain indices tended to be higher in the late-pubertal group, whose heat storage per kilogram body mass was significantly higher than in the midpubertal group. A major key to the thermoregulatory effectiveness of the prepubertal boys was a higher forearm blood flow (see figure 3.2), which compensated for their lower sweating rate.

These studies suggest an adequate thermoregulatory effectiveness in children who exercise in warm climates. There are, however, exceptions. In one study (51), for example, only 40% of 12-year-old girls exposed to 35° C, 65% relative humidity,

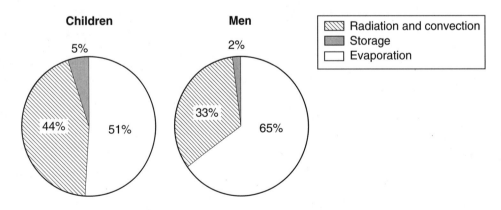

FIGURE 3.11　Heat dissipation and storage, as a percentage of metabolic heat, in children and adults who exercised in a thermoneutral environment. Subjects were 13 girls (13.8 ± 0.7 years) and boys (12.9 ± 0.8 years) and eight adults (36.1 ± 6.7 years) who ran (68% VO_2max) for 60 min at 21° C, 67% relative humidity.

Based on Davies 1981 (43).

completed two 1-hr walks (30% $\dot{V}O_2max$). This compared with 100% completion among young women.

Little information is available on tolerance levels to heat when exercise is intense and short. Performance of a 30-sec supramaximal cycling task (the Wingate anaerobic test) was not adversely affected by warm-humid (30° C, 90% relative humidity) or hot-dry (38-39° C, 25% relative humidity) climate in 10- to 12-year-old girls and boys (49). In fact, based on theoretical considerations (10; 137), anaerobic performance should improve with an increase in muscle temperature.

Very Hot Environment

Very hot denotes air temperatures as high as 40° to 50° C. For ethical reasons, very few studies have been performed under these conditions. These suggest that children are less capable of enduring a prescribed activity in very hot climates, or that they respond to such conditions with a higher heat strain compared with adults (51; 70; 80; 163). An example is presented in figure 3.12, which shows that 9- to 11-year-old girls could not complete an intermittent walking task at 50° C, 15% relative humidity, whereas young women did complete this task.

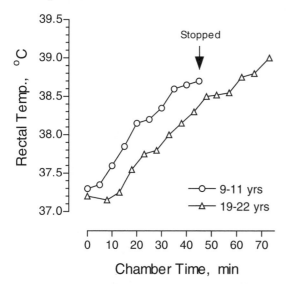

FIGURE 3.12 Heat tolerance of girls (n = 12) and women (n = 29). Changes in rectal temperature during an exposure to extremely hot climate (50° C, 15% relative humidity). Subjects walked intermittently at 4.8 km · h⁻¹. Based on data by Haymes et al. 1974 (70) and Bar-Or et al. 1969 (21).

Reproduced, with permission, from Bar-Or 1980 (12).

It has been our experience and that of others (51; 97; 122) that children who feel the need to terminate an exposure to a warm or hot climate often do so before the objective indices of their physiological strain (e.g., core temperature and heart rate) reach high values. They sometimes complain of dizziness, but most often they suddenly demand to be taken out of the chamber with no accompanying complaint. The mechanism for such a self-imposed termination is not clear, but it may reflect insufficient blood supply to certain body regions.

Physical and Physiologic Responses to Cold Climate

When a person is exposed to the cold, heat loss from the skin is enhanced. The rate of heat loss depends on the skin-to-air (or water) temperature gradient and on thermal conductivity of the medium to which the person is exposed. Thermal conductivity of water is 25 to 30 times greater than that of air, and indeed, heat loss through conduction and convection is much faster in water than in air. This is the main reason that wetting of the clothes in a cold environment may lead to a rapid, sometimes devastating, heat loss. Another factor that affects heat loss is the surface area of the skin (or, more precisely, the skin area that is in contact with the environment). As discussed at the start of this chapter (see also figure 3.1), the smaller the person, the larger the surface area-to-mass ratio. As a result, assuming no physiologic compensatory mechanisms, the smaller the person, the faster the heat loss.

Another important factor that, through convection, increases the rate of heat loss is the wind. In this context, the windchill factor is a commonly used index that quantifies the effect of cold on the skin, as well as on overall heat loss. Windchill is greatly dependent on wind velocity. For example, during exposure to 0° C, a wind of 24 km · hr⁻¹ will cause heat loss as though the actual air temperature were minus 10.5° C. The equivalent windchill factor at 40 km · hr⁻¹ wind will be minus 16° C. A wind effect is created not only by the blowing wind but also by the moving individual. Thus, in calm air of minus 10° C, the windchill factor is minus 31° C for a skater or a skier who advances at 40 km · hr⁻¹. A young cross-country skier should be advised to fully cover the head, including the face, when the windchill factor drops to minus 20° to 23° C.

One seldom realizes that even during cold exposure, exertion may be accompanied by sweating. A cross-country skier, for example, can lose several liters of sweat during a 50-km race. If the sweat penetrates beyond the clothing, its evaporation will further add to body cooling. A similar effect will occur when the clothing gets wet for any other reason. Another important avenue for evaporative cooling is through the respiratory tract. Respiratory heat loss is a combination of convective cooling (due to a temperature gradient between the respiratory tract mucosa and the inspired air) and evaporative cooling (due to humidification of the inspired air). A child can lose as much as several hundreds of kilocalories through respiration during 1 hr of exercise on a cold, dry day. For details, see the section on asthma in chapter 6.

As shown in the heat balance equation at the start of this chapter, it is the balance between heat loss through convection, conduction, radiation, and evaporation, on the one hand, and metabolic heat production on the other that determines whether the heat content of the body decreases or increases. There are two physiologic lines of defense against excessive body cooling. The first is peripheral vasoconstriction, to curtail heat loss. Through a decrease in blood flow to the skin, heat convection from body core to the periphery is reduced. This results in a lower skin temperature and a reduction in the skin-to-air temperature gradient. The final outcome is a lower heat loss to the environment through conduction, convection, and radiation.

The second line of defense is the enhancement of metabolic heat production. In the resting person, this is first induced by hormonal and neural mechanisms, with no increase in muscle action (nonshivering thermogenesis). If the resulting increase in heat production is insufficient, shivering (i.e., a rhythmic, involuntary muscle contraction) starts. Shivering is a most potent response that can increase metabolic rate as much as fivefold!

Enhanced metabolic heat production can also be achieved by physical activity. When activity is intense enough, core temperature will not decline. It may often rise, even if exercise is performed on a cold day. In such strenuous prolonged races as

marathon runs, core temperature may reach 40° to 42° C, even when the effective temperature is as low as 5° C (153). Another example is cross-country skiing, even when clothing is light and the ambient temperature is below freezing (149). In intermittent activities such as ice hockey, there is an overall high level of metabolic heat production and, even while off the ice, the child is in no danger of body cooling (96; 121).

Temperature Regulation During Swimming

An exception to the patterns just described occurs when exercise is performed in water. Because of the high thermal conductivity of water (and its specific heat, which is approximately 1,000 times that of air), large heat losses occur even when the skin-to-water temperature gradient is as low as 3° to 4° C. The rate of heat loss may be as much as 30 times higher during swimming than it is, for example, during cycling (113). It is not surprising therefore that the first study ever to document effects of cold on the exercising child was done with swimmers. Figure 3.13 summarizes changes in core temperature (measured as oral temperature) of 8- to 19-year-old club swimmers of both sexes, who swam in 20.3° C water at a speed of 30 m · min⁻¹. This corresponded to a

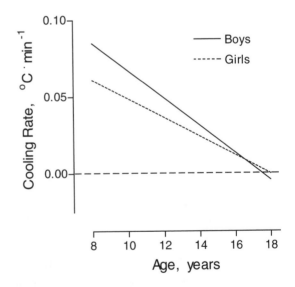

FIGURE 3.13 Rate of cooling of body core during swimming (20.3° C water) in relationship to age. Subjects were sixteen 8- to 19-year-old female and male trained swimmers.

Adapted, with permission, from Sloan and Keatinge 1973 (147).

metabolic level of four to five MET. While most of the older participants managed to maintain their oral temperature at preswim levels, some of the younger swimmers had a 2° to 3° C drop in temperature. Furthermore, the older swimmers stayed in the water for about 30 min, whereas the youngest had to leave after 18 to 20 min due to a marked subjective distress (147). The inability of the young swimmers to sustain their activity in cool water was related to their higher surface area per heat-producing unit mass and especially the lesser thickness of their subcutaneous fat. Adipose tissue has the highest insulative capacity of any body tissue. The thicker the fat layer, the better insulation it provides to the swimmer (24; 84), especially when blood flow to the skin is reduced.

Water temperature in swimming pools is usually kept at above 25° C, but there are still many children who swim outdoors at below 20° C. The question of excessive cooling in young swimmers is therefore not merely academic. It is possible that extreme cold-induced distress forces swimmers out of the water before hypothermia becomes dangerous. Still, a potential risk exists for a lean, overzealous, small-sized swimmer who may be reluctant to leave the water in spite of impending danger. Coaches and health practitioners should be aware of such a possibility. Early warning signs to look for are euphoria and disorientation.

Temperature Regulation During Rest in a Cool Climate

Several studies (6; 104; 162) have been conducted on children who rested, minimally dressed, in a "cool" room (16-20° C). The subjects sustained their initial core temperature throughout the session, and their overall physiologic responses to the cold were at least as effective as in adults. In a study that did not provide comparison with adults (54), boys dressed in sweat suits sat for 110 min (and also performed 10-min mild exercise in the middle) at 7°, 13°, or 22° C. Their rectal temperature decreased fastest at the last 50 min of the coolest condition and kept decreasing even 30 min after return to a thermoneutral environment. Cognitive functions were not affected by climate. It seems from these studies that under a mild cold stress, children have adequate physiologic responses to compensate for their large surface area-to-mass ratio. However, when the

exposure lasts more than 1 hr (54), core temperature may drop.

Temperature Regulation During Exercise in Cold Air

Only one study is available that documented physiologic responses of pre- and early-pubertal children who exercised when exposed to cold air (148). Eight 11- to 12-year-old boys and 11 young men (19-34 years old) rested for 20 min and then cycled (30% $\dot{V}O_2$max) for 40 min at 5° C, 40% relative humidity, while wearing only shorts, socks, and sneakers. A previous exposure was conducted, for comparison, at 21° C. The boys, in spite of their larger surface area-to-mass ratio, had at least as effective a thermoregulation as did the men (figure 3.14). This resulted from a higher thermogenesis during rest and exercise (figure 3.15), as well as a greater reduction of skin temperature (suggesting a greater peripheral vasoconstriction) in the limbs, compared with the adults. Based on further analysis by these authors (148), it seems that the more effective compensatory response among the boys is not merely attributable to their larger surface area-to-mass ratio, but is associated also with their early maturational stage.

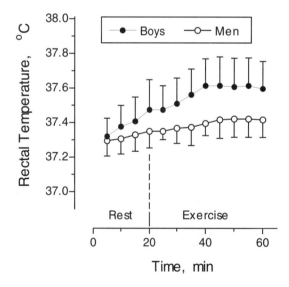

FIGURE 3.14 Rectal temperature in pre- and early-pubertal boys (closed circles) and in young men (open circles) during 20-min rest and 40-min exercise (30% $\dot{V}O_2$max) at 5° C. Vertical lines denote SEM.

Reproduced, with permission, from Smolander et al. 1992 (148).

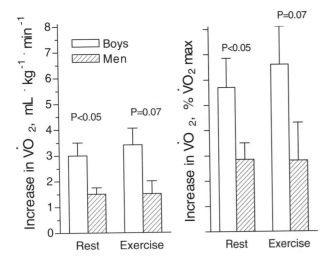

FIGURE 3.15 Effect of ambient cold on thermogenesis during rest and exercise in boys and young men. Values reflect excess oxygen uptake at 5° C compared with 21° C. Vertical lines denote SEM.

Reproduced, with permission, from Smolander et al. 1992 (148).

Implications of Cold Climate for Health

Exposure to the cold may result in local skin cooling and in generalized heat loss. Local cooling with an eventual cold-related injury (such as frostbite) may occur both at rest and during exercise. Skin sites commonly affected are the chin, cheeks, ears, fingertips, and toes. In contrast, generalized body cooling is seldom a problem during exercise, because metabolic heat production, if intense enough, more than compensates for heat loss from the skin. Generalized cooling becomes a problem whenever activities are not accompanied by a high metabolic rate, as in mountain hikes or during snowshoeing (see earlier discussion for aquatic activities).

Acclimatization and Acclimation to Exercise in the Heat

Acclimatization to heat comprises adaptive changes over time (usually 1-2 weeks) that result from continuous or repeated exposures to naturally occurring climatic heat stress. It is manifested by improved thermoregulation,

physical and mental performance, and thermal comfort. Acclimation is a similar process, but the exposures are performed under artificially controlled conditions, usually in a climatic chamber. Acclimation and acclimatization are particularly important to individuals who are abruptly confronted with a warmer environment, whether due to a heat wave or during travel to a warmer geographic region. Because most studies on this topic were performed under controlled climatic chamber conditions, in this section we usually refer to acclimation. During the unacclimated state, physical and mental tasks are performed at the cost of a greater physiologic strain. Attempts to perform intense prolonged activities at the unacclimated state increase the risk for heat-related illness (see also section on insufficient acclimatization).

The major physiologic changes that occur progressively during acclimation are as follows:

- A decrease in heart rate and rectal and skin temperatures at a given metabolic level
- An increase in sweating rate and in the sensitivity of the sweating apparatus to increments in core temperature
- A drop in electrolyte (especially Na+ and Cl–) concentration in the sweat
- A reduction in the rating of perceived exercise intensity
- An increase in thermal comfort

Acclimation—Children Versus Adults

In adults, a reasonable degree of acclimation can be achieved following four to seven exposures to the combined stresses of heat and exercise. Effective exposures should last 1 to 3 hr each, at a rate of three to seven per week. The intensity of exercise should increase gradually, so that by the end of acclimation one can perform at par with one's performance in the cooler climate. The acclimated state can be retained for up to 7 to 10 days without further exposure. It is then gradually lost.

Young teenagers acclimate to exercise in the heat, but to a lesser degree than do older adolescents or young adults (163). The ability of boys 8 to 10 years old to acclimate to exercise in dry heat, compared with that of 20- to 23-year-old men, has been studied extensively at the Wingate Institute in Israel (20; 75-77). Body temperatures and heart rate declined and sweating rate increased to the same degree in both age groups during a 2-week acclimation program (figure 3.16). The main age-related difference was in the rate of acclimation. While the adults reached a certain level of acclimation within two sessions, the children needed four to five sessions for a similar result. This relatively sluggish response is depicted in figure 3.17. A similar "delay" in response was found for the sensitivity of the sweating apparatus to changes in core temperatures. Whereas adults acclimate through a major increase in sweating rate, children show only moderate increases in sweating rate (75; 120; 163; 164). It does seem, though, that naturally acclimatized children have a considerably higher sweating rate (129) than non-acclimatized children. Children can acclimate to the heat when they exercise in neutral environments (76), and when they rest in a hot climate (77). In adults such protocols are only partially effective. It seems, therefore, that children lag behind adults in the rate of physiologic acclimation and therefore require a longer and more gradual program. On the other hand, they acclimate, albeit slowly, under less stringent protocols than are commonly recommended for adults.

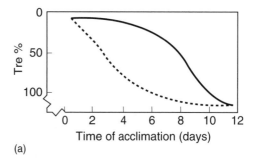

(a)

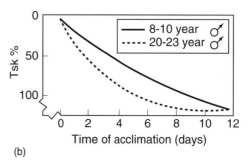

(b)

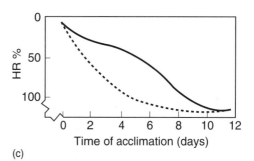

(c)

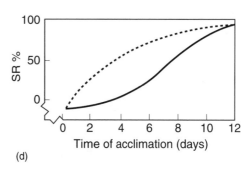

(d)

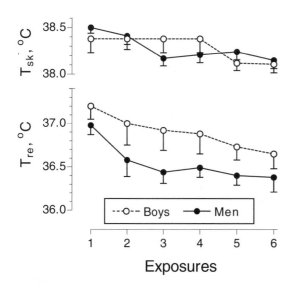

FIGURE 3.16 Heat acclimation of boys (n = 9) and young men (n = 9). Changes in rectal (Tre) and mean skin (Tsk) temperatures during a 2-week acclimation program. Values are as obtained at the end of each of six 80-min exposures to 43° C, 21% relative humidity. Subjects cycled intermittently at 40% to 50% of their maximal O₂ uptake. Vertical lines denote 1 SEM.

Reproduced, with permission, from Inbar 1978 (75).

FIGURE 3.17 Rate of heat acclimation in boys (n = 9) and young men (n = 9). Schematic presentation of changes in rectal temperature (Tre), mean skin temperature (Tsk), heart rate (HR), and sweating rate (SR) during a 2-week acclimation program. Values are presented as percent of final acclimation, baseline being 0%. Conditions and protocols as in figure 3.16.

Reproduced, with permission, from Bar-Or 1980 (12).

Perceptual Changes With Acclimation

There is a subjective component to acclimation. Concurrently with a decrease in physiologic strain, lassitude is diminished and there is an improvement in general well-being. To gauge such subjective improvement, children in the project just described were asked to rate the intensity of exercise that they were performing (20) using the rating of perceived exertion (RPE) category scale of Borg (see chapter 1, section "Age-Related Differences in Rating of Perceived Exertion"). Even though work rate and environmental conditions were identical in all exposures, the rating markedly declined from one session to another, indicating that the same task gradually seemed easier. Figure 3.18 is a comparison of such changes between boys and men. The ratio RPE/HR is taken to represent subjective difficulty at a given physiologic strain. The rate of decline in RPE/HR during the 2-week program is faster in the children, reaching lower final levels than in the adults. This phenomenon implies that even though children's physiologic acclimation is slow, their subjective improvement is faster than

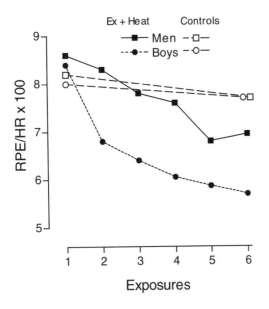

FIGURE 3.18 Changes in the ratio of rating of perceived exertion/heart rate (RPE/HR) during a 2-week acclimation program in boys and men. The "ex + heat" groups are as shown in figure 3.16. The controls were seven boys and seven men who were tested on the first and last days of the program but not exposed to exercise and heat in between. Based on data by Bar-Or and Inbar 1977 (20).

Reproduced, with permission, from Bar-Or 1980 (12).

in adults. Although subjective well-being may be looked upon as an advantage, it may also signify a potential hazard. While the insufficiently acclimated adult is usually reluctant to exert in the heat, a child who is not yet acclimated may be more daring, in spite of a marked objective strain.

Effect of Training on Thermoregulation

Aerobic training in adults is accompanied by better heat tolerance and thermoregulatory effectiveness (66; 112). Trained people have an enhanced sweating rate and a lower threshold for sweating. In 8- to 10-year-old boys, a 2-week aerobic conditioning program (six sessions at an intensity of 85% maximal heart rate) induced a reduction in core temperature, but there was no increase in sweating rate and no decrease in sweating threshold (76). The extent to which aerobic fitness improves thermoregulatory effectiveness in children seems to be minor (46; 48; 104).

Fluid and Electrolyte Balance

Maintaining adequate levels of water and electrolytes in the body is an important component of homeostasis. Exercise, with the accompanying sweat loss, the shifts of water among fluid compartments, and the movement of electrolytes between the intracellular and extracellular compartments, presents a major challenge to these homeostatic elements. Water and electrolyte losses in the sweat are especially high when climatic heat stress is superimposed on exercise. This section will describe the fluid shifts during exercise, the loss of electrolytes and water in the exercising child, the resulting "involuntary dehydration," and the means of preventing dehydration in children who exercise in the heat.

Water Shifts During Exercise

At the start of strenuous exercise, plasma volume drops by about 10% to 15%. Some decline is evident as early as 10 sec following the onset of highly intense exercise (134) and is probably due to an increase in intracapillary hydrostatic pressure. Another mechanism that follows and causes further escape of water from the intravascular compartment is hyperosmolarity of the interstitial compartment, secondary to efflux of K+ and metabolites from the contracting muscle fibers. Exercise of prolonged duration (30 min or more), if not accompanied by sufficient fluid intake, results in a further slow decline in plasma volume (39). This decline reflects total body dehydration, mostly due to sweat loss. It may also result from further osmotic drive. In dry climates, or during exposure to high altitude, large amounts of water evaporate from the upper and lower airways. If dehydration is prevented by adequate fluid replenishment, there is no further drop in plasma volume. When fluid is fully replenished during 90- to 120-min exercise, plasma volume rises gradually and eventually reaches its pre-exercise level (40).

Urinary output declines during exercise, partially compensating for sweat loss. Such a decline results from a reduced renal plasma flow and glomerular filtration rate. Urinary output is further reduced with dehydration through an increase in antidiuretic hormone activity. In contrast, sweating rate is not reduced during progressive dehydration as long as the fluid deficit does not exceed 5% to 6% of initial body weight. It seems that to support heat dissipation and thermal balance, the body surrenders its fluid balance.

Electrolyte Loss During Exercise—Maturational and Gender Effects

Human sweat contains more than 99% water. Its electrolyte concentration is lower than in the extracellular fluid. Sweat osmolality of adults seldom exceeds 180 mmol · L^{-1} (40) as compared with about 300 mmol · L^{-1} in adults' body fluids. Na+ and Cl–, but not K+, content rises with the increase in sweating rate. In contrast, concentration of Mg++ and Ca++ decreases at a high sweating rate (39). Conditioning and acclimatization to heat both induce a decrease in ionic concentration of sweat, while the sweating rate rises. As a general rule, sweating without fluid replenishment is accompanied by an increase in body fluid osmolarity. However, even though sweat is hypotonic, prolonged and repeated exercise in hot weather can induce marked salt loss and hyponatremia.

In children and adolescents, sweat is more hypotonic than in adults (5; 47; 108). Figure 3.19 depicts the effect of age on sweat chloride concentration, which is often twice as high in young adults as in prepubescents. A similar pattern occurs for Na+. In contrast, sweat K+ content is higher in pre- and midpubescents than in adults (108). Among adults, females have a lower sweat NaCl content than males. This difference, however, does not appear in children or pubescent girls and boys (108). Because of the low NaCl content in children's sweat, combined with a low sweating rate, their overall salt loss is lower than in adults, even when calculated per kilogram body weight. This pattern occurs in females and males alike.

Hypohydration and Its Effects

The marked fluid shifts that occur with exercise, especially when the sweating rate is increased, may result in a fluid deficit. Such a state of fluid deficit is referred to here as hypohydration. The actual process during which negative fluid balance takes place is referred to here as dehydration. In this section we first describe two types of dehydration that are common in physically active children and then discuss the implications of hypohydration for the performance and well-being of the child. Hypohydration accompanying several pediatric diseases is discussed later in the text.

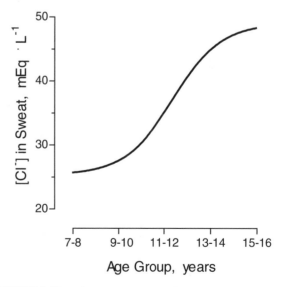

FIGURE 3.19 Changes in sweat chloride concentration during growth. Subjects and conditions are as in figure 3.3. Sweat was collected from the pectoralis area. Data by Araki et al. 1979 (5).

Involuntary Dehydration

It has long been recognized (123) that people who exercise for prolonged periods, particularly in hot climates, do not drink sufficient amounts to replenish fluid loss, even when allowed to drink ad libitum. The term "voluntary dehydration" was coined for this phenomenon (133). The term has been changed in recent years to involuntary dehydration. In adults, fluid deficits due to involuntary dehydration have been found to range between 1.5% and 7% of initial body weight, depending on the climate, duration and intensity of the activity, and the type of fluids used for replenishment. Some striking examples are available from marathon races: Among 63 adult runners who completed a race, mean sweating rate was 0.96 L · hr^{-1} and fluid intake only 0.13 L · hr^{-1}. The result was a fluid loss of 5.2% of initial body weight. The winner of the race had a 6.9% fluid loss (124).

Involuntary dehydration has been also documented in children who exercised in the heat without being forced to drink (17; 18; 88; 129; 131; 165). Figure 3.20 presents a comparison of the fluid loss of boys when forced to drink water periodically and when allowed to drink water ad libitum during intermittent exercise in the heat. Although sweating rate was virtually identical in the two exposures (figure 3.21), water intake

in the voluntary drinking exposure was only 66% of the intake needed to replenish sweat and urinary losses. Urinary output in that session was only 68% of the respective output during the forced-drinking session, but this decrease in urine volume was not enough to prevent hypohydration. One can assume that the fluid deficit would have been greater had the session lasted longer or had the exercise intensity been greater. Indeed, we recently tested 8- to 17-year-old athletes who ran a triathlon held at approximately 40° C on the Pacific coast of Costa Rica. Over 35% of the participants reached hypohydration levels greater than those described in laboratory experiments. One boy lost 4.5% of his initial body weight in spite of drinking periodically throughout the race.

Deliberate Dehydration in Sport

Another type of dehydration occurs when individuals deliberately induce a negative fluid balance, either by depriving themselves of fluid intake or by inducing excessive fluid loss. Deliberate dehydration is common among athletes who wish to "make weight" just prior to competition (e.g., wrestlers, weightlifters, boxers, judo competitors, and jockeys). The most common procedure is to induce sweating, but some individuals reduce fluid and food intake, take diuretics or

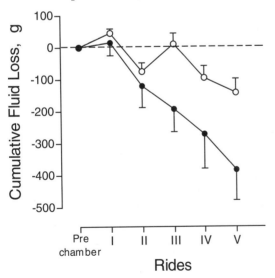

FIGURE 3.20 Involuntary dehydration in boys cycling intermittently in the heat (39° C, 45% relative humidity). Cumulative fluid loss during 3.5 hr in 11 boys, 10 to 12 years old. Vertical line is 1 SD.

Reproduced, with permission, from Bar-Or et al. 1980 (18).

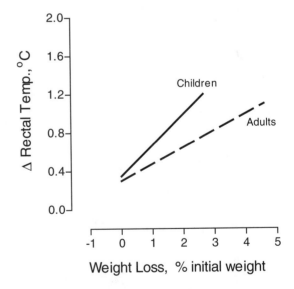

FIGURE 3.21 Relationship between the rise in rectal temperature and the level of hypohydration in children and adults exposed to exercise in dry heat.

Adapted, with permission, from Bar-Or et al. 1980 (18).

laxatives, and even induce vomiting. In North America, deliberate dehydration is often found among school-age wrestlers (72; 116; 152). In a survey among 10% of all interscholastic high school wrestlers in Iowa (158; 159), body weight was measured repeatedly until "weighing-in." Most wrestlers lost 5% to 7% of their initial body weight during the last 10 days prior to competition; 8% of them lost 10% or more. It is safe to assume that most of the weight loss represented a fluid rather than an energy deficit. Especially disheartening was the marked loss among the very lightweight categories (47 kg or less), which in some youngsters reached 15% of body weight! Most of these athletes consulted a friend or a coach, but not a physician, as to the best way of reaching their target body weight.

Attempts were made in the 1980s and 1990s to curtail deliberate weight loss among high school wrestlers. One example is the Wisconsin Wrestling Minimum Weight Project (117). The project was based on a three-pronged approach: (1) estimate of % body fat (skinfold thickness) to determine the lowest allowable competitive weight, (2) limiting the weekly weight loss, and (3) nutrition education. This approach has been adopted by other states. A survey of schools was conducted in Wisconsin to determine the effectiveness of the project 2 years after its launch (118). The results were most encouraging: decreases in the highest weight loss and the weight lost to certify prior to a match, reductions in the weekly weight cycle and the duration of the longest fast, and a decrease in bulimic behaviors of the wrestlers.

Implications for Performance and Health

Hypohydration results in physiologic dysfunction (1; 18; 37; 40; 139; 169) and may be detrimental to performance and health. There is a reduction in plasma volume, stroke volume, cardiac output, renal blood flow, glomerular filtration rate, and liver glycogen content. With exercise, water depletion is proportionately greater in the extracellular than in the intracellular compartment. Heart rate at rest and in submaximal exercise is elevated. Rehydration within 1 hr following 4% to 5% hypohydration can induce a return to normal in the hemodynamic function of high school wrestlers (1).

Hypohydration is accompanied by an electrolyte deficit, especially Na+ and Cl–, but also K+, Ca++, and Mg++. Such losses notwithstanding, plasma concentration of these ions may be elevated due to hemoconcentration. Body fluid osmolality is at first high, but with intake of water may eventually become normal or low. Drinking water in large amounts to replenish sweat losses may induce hyponatremia (7; 65). When extreme, this can lead to seizures and other neurological manifestations (65).

Thermoregulation may become inefficient when the triad heat stress, exercise, and hypohydration is in effect. Convection by blood of heat from body core to skin is particularly disrupted. This results in a rise of core temperature, which is proportional to the fluid deficit. Figure 3.21 is a comparison of changes in rectal temperature between children and adults who progressively dehydrated during exercise in the heat. For each 1% weight loss, the adults had a 0.15° C rise in temperature, compared with 0.28° C in the children (18). It is not clear whether this greater increase in body heating has clinical significance. At mild to moderate hypohydration (up to 3-4%), sweating rate remains fairly constant (19; 36); but at higher levels of hypohydration, or when plasma osmolality is markedly increased, there may be a reduction in sweat output. This further impedes heat dissipation.

Decrements in performance that accompany hypohydration include a reduction in muscular strength (26), in the time that strenuous activity can be sustained (36; 136), and in mental alertness (92). Maximal O_2 uptake is usually not reduced (22; 136) unless hypohydration is extreme. Reaction time to visual cues is not prolonged (92); nor is anaerobic capacity, as measured by the 30-sec Wingate anaerobic test, affected by hypohydration of up to 5% (78).

To the clinician, hypohydration should signal impending danger, because it further compromises the child's thermoregulation (see later section, "Health Hazards in Hot Climates"). There are no data on the effect on growth of repeated hypohydration, but such a possibility cannot be ignored in youngsters, such as wrestlers, who repeatedly undergo negative fluid and mineral balance. In addition, chronic loss of K+ in sweat and urine may contribute to muscle fatigue and cramps.

Means of Enhancing Thirst

To minimize hypohydration and electrolyte losses, fluid replacement must become part and parcel of the conduct of any prolonged physical activity. This axiom has been applied in industry, in the military, and in sport. Any inhabitant of warm climates recognizes the necessity for proper drinking habits (42; 129; 132; 157), but adequate drinking is also important when people engage in prolonged activity in temperate and even cold climates.

To obtain sufficient fluid replenishment and minimize the likelihood of dehydration, one must consider the following:

- Fluids must be selected that do not quench thirst but instead stimulate further drinking.

- Gastric emptying should be rapid in order to avoid fluid stasis and gastric distension.

- Absorption at the small intestine should not be delayed.

- Fluid quantity must be sufficient to replenish any previous dehydration, as well as the estimated losses during the activity.

- Mineral content in the fluid is important but should not be exaggerated. It must be coordinated with mineral intake during meals.

To understand how to lead an active child to consume sufficient fluids, one must consider perceptual and physiologic aspects.

Perceptual Considerations

As shown in studies with adults, fluid flavor, color, and temperature affect the palatability of a drink. This, in turn, affects voluntary drinking. Although water is an adequate beverage for immediate replenishment, it quenches thirst and does not stimulate further drinking. The increase in palatability of a beverage, in itself, can stimulate thirst (73; 156; 165). In one study (165), the addition of grape flavoring to water induced enhancement of voluntary drinking by 45% in 9- to 12-year-old boys who exercised intermittently in the heat for 3 hr (figure 3.22, top). With this increased consumption, the children were able to remain euhydrated (i.e., neither hypo- nor hyperhydrated) for 90 min, but they then started dehydrating (figure 3.22, bottom).

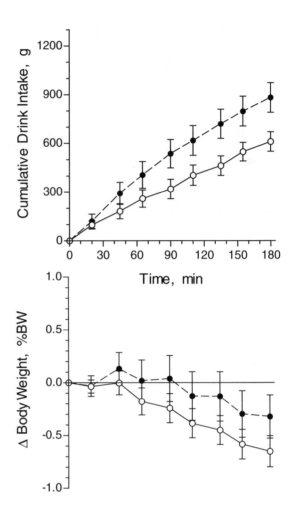

FIGURE 3.22 Effect of drink flavoring on voluntary drink intake and on body hydration. Cumulative drinking volume (top) and body weight changes (bottom) were monitored during two 3-hr intermittent exercise (50% $\dot{V}O_2$max) sessions at 35° C, 45% to 50% relative humidity. In two sessions, twelve 9- to 12-year-old boys were presented with a beverage that they were allowed to drink ad libitum. In another session, the drink was unflavored water (O) and in the other, grape-flavored water (●). Both beverages were chilled to 8° to 10° C. Subjects were untrained and not acclimatized to the heat.

Adapted, with permission, from Wilk and Bar-Or 1996 (165).

Is there a specific flavor that is most liked by children? When Canadian boys and girls were asked to rate various drink flavors, they repeatedly preferred grape to apple, orange, or water (109). This pattern was consistent whether the children were resting in a thermoneutral environment, undergoing mild progressive dehydration, or recovering from dehydration or from maximal effort. It is conceivable, however, that children from a geographic region other than southern Ontario, or of a different cultural or socioeconomic background, would have other flavor preferences. Indeed, children in Puerto Rico, who habitually trained in a hot-humid climate, varied markedly in their favorite flavor (129). As evidenced by one study with 9- to 12-year-old girls and boys (110), the color of a beverage does not seem to affect its palatability rating.

A common fallacy is that cold water should not be taken when an individual is tired and sweaty. There are no documented reports on any detrimental effects of cold water before, during, or after exercise. As shown for adults, cold drinks have the advantage over tepid or warm drinks because they empty faster from the stomach (41) and are more palatable (27). The direct cooling effect of a cold fluid is minimal when its dilution in the overall water pool is calculated. Even so, such cooling may sometimes give the edge on a very hot day.

Physiologic Mechanisms

The two physiologic triggers for thirst are a reduction in plasma volume and an increase in body fluid osmolality (61; 66; 111; 114; 115). In a healthy active person, changes in osmolality are more important than changes in plasma volume. As shown in animal studies, an increase in osmolality stimulates hypothalamic and gastrointestinal osmoreceptors.

Based on this information, the addition of salt to a beverage should increase voluntary consumption. Indeed, 9- to 12-year-old boys increased their voluntary drinking by 46% during 3 hr of intermittent exercise in the heat when NaCl (18 mmol · L^{-1}) and carbohydrate (2% glucose, 4% sucrose) were added to grape-flavored water (figure 3.23, top). This was sufficient to prevent dehydration throughout the entire 3-hr exposure (figure 3.23, bottom). The benefit derived from such a drink is consistent, and it takes place even when children are exposed repeatedly, over 2 weeks, to exercise in the heat (166). Girls also increase their voluntary drinking in response to flavoring and to the addition of salt plus carbohydrate, even though the increase is less pronounced than in boys (Bar-Or and Wilk, unpublished data).

Participants in these studies were unacclimatized to heat, and their sweating rate did not exceed 250 mL · hr^{-1}. A question remains as to

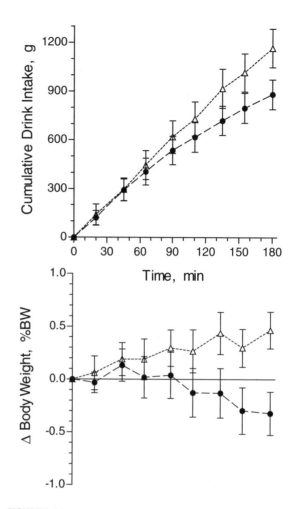

FIGURE 3.23 Effect of adding salt plus carbohydrate to a beverage. The graphs summarize the voluntary drinking volume (top) and hydration status (bottom) of boys who were presented with flavored water (•) in one session and flavored water with 6% carbohydrate plus 18 mmol L^{-1} (Δ) in the other. Subjects, conditions, and protocols were identical to those described for figure 3.25.

Adapted, with permission, from Wilk and Bar-Or 1996 (165).

whether involuntary dehydration can still be prevented when sweating is more profuse. A recent study, indeed, showed that even when sweating rate was twice as high as just described, voluntary drinking was enhanced and dehydration was prevented by the addition of a flavored 18 mmol · L^{-1} NaCl-6% carbohydrate solution (129). The subjects were highly trained and heat-acclimatized 11- to 14-year-old boys, residents of Puerto Rico, who attended a sport school.

Practical Considerations

The following are questions often asked by parents, coaches, physicians, and young athletes themselves.

How Much and How Often Should One Drink?

There is no single answer regarding the optimal amount of drink, because sweat losses depend on the specific climatic conditions, the intensity and duration of the exercise, and the child's state of acclimatization and fitness. Although a large volume of fluid, taken all at once, may enhance its emptying from the stomach, this drinking pattern is accompanied often by abdominal cramps and "heaviness." Another disadvantage of drinking large volumes all at once is that much of the water is rapidly eliminated from the body as urine. It is preferable, therefore, to drink smaller volumes of fluid periodically (e.g., every 15-20 min). An 11-year-old child, for example, who loses $400 \, mL \cdot hr^{-1}$ should drink 100 ml every 15 min rather than 200 mL every 30 min. Once dehydration starts, its reversal during exercise is almost impossible. One should therefore ascertain adequate hydration prior to the start of exercise and not wait more than 15 to 20 min for the first drink. For children who participate in road races, the official drinking stations are often too far apart and the first one is located too far into the race. Special drinking arrangements may have to be made for these young runners.

How Can One Determine Sufficient Hydration?

The simplest, most effective way for coaches, parents, and the athletes themselves to determine sufficient hydration is through weighing before and after a practice session or competition. Almost 100% of weight loss can be attributed to fluid loss. Ideally, weighing should be done without clothes and after the child has voided. A calibrated digital bathroom scale is adequate for such weighing.

If the child has lost weight, drinking volume during the practice session or competition was insufficient. This can be corrected in subsequent practice sessions, as well as on the day of competition. Another method suggested for determining adequate hydration is to measure the specific gravity of urine. This is often impossible, however, because many athletes cannot produce sufficient urine following a strenuous physical activity.

It is important to ascertain that the child is sufficiently hydrated at the start of an activity. To attain such a level the child should drink 300 to 400 mL fluid 30 to 45 min prior to the start of warm-up. This will leave enough time for voiding any extra fluid before the actual weighing.

What Should Beverage Consist Of?

The osmolarity of ingested fluid may determine gastric motility and emptying. A solution containing 18 to 20 $mmol \cdot L^{-1}$ NaCl, for example, will be emptied faster than water. Fluids with a higher salt content may retard gastric emptying, as may carbohydrate solutions that exceed 6% to 7%. Commercially prepared sport drinks usually contain minerals and carbohydrates in the right amounts. On the basis of our experience and that of others (e.g., [129]), such sport drinks are also suitable for children. Carbohydrates in a solution (e.g., glucose and sucrose) are known to enhance physical performance in adults. Whether the same occurs in children is not yet known. Juices and carbonated beverages often contain excessive amounts of sugars and a low concentration of NaCl. More economical alternatives are homemade drinks prepared according to the recommendations just cited. These can be flavored to individual taste. For children who exercise sporadically, plain water replacement is sufficient provided that the child consumes a balanced diet. It is essential, however, to remember that plain water will not stimulate enough voluntary drinking and that the child should therefore be reminded to drink above and beyond thirst. Furthermore, the consumption of plain water during a prolonged physical task may result in hyponatremia (e.g., marathon or triathlon).

Should One Use Salt Tablets?

Athletes sometimes use salt tablets or add liberal amounts of salt during mealtime. The benefits of this approach to children are unknown. On theoretical grounds, however, it may be detrimental, for the following reasons:

• Plasma aldosterone activity is markedly increased during acute exercise, such that Na+ is conserved by the kidney (39; 57). In heat-acclimatized people, it is also conserved by the sweat glands. Thus, habitual exercisers who are heat acclimatized often have a positive Na+ balance, even when not adding salt to their regular balanced diets.

• For occasional exercisers, salt content in most diets is high enough to balance sweat and urinary losses.

• Sweat, almost invariably, is hypotonic. Ingestion of salt solutions that are too concentrated will increase the hypertonicity of the interstitial space and induce further depletion of the intracellular compartment.

• Children's sweat is more dilute than that of adults.

• Studies of K+ balance, tallying intake, sweat, and urine losses, have shown that K+ can be fully preserved by a regular balanced diet, even in people who exercise often. There is therefore no need for special KCl additives.

One should therefore refrain from supplementing one's daily salt intake. Nevertheless, some athletes may benefit from additional salt. These are individuals who have an especially high NaCl concentration in their sweat and who, without salt supplementation, complain of severe exercise-induced muscle cramps. While this phenomenon has been described for adults, there are no data regarding the need for this approach for healthy child athletes.

In conclusion, thirst is an inadequate gauge for sufficient drinking during prolonged activities or in hot climates. It is a prime educational responsibility of the physician to impress this fact upon parents, teachers, coaches, and children. To enhance thirst, one should flavor the drink, choosing flavors most liked by the child. In addition, one should add NaCl (approximately 18-20 mmol · L^{-1}) and simple carbohydrates (approximately 6-7%).

Health Hazards in Hot Climates

Previous sections in this chapter have dealt with response to exercise in the heat from a physiologist's vantage point. In this section we discuss the health implications of such a response.

While this book focuses on the exercising child, it includes some information on the health hazards of hot climates to resting children. We include such information on the assumption that health hazards to the inactive child can only be magnified when a high metabolic load is superimposed on climatic heat stress. The concept of heat-related illness is introduced first, followed by epidemiologic data on the vulnerability to hot climates of infants and children. The section ends with a closer look at those groups of children who are at an especially high risk for heat-related illness.

Heat-Related Illness

Heat-related illnesses are pathologic conditions that result from exposure to heat, either at rest or during exercise. Other terms commonly used are heat disorder and heat injuries. With the increase in popularity of jogging, races such as "fun runs," and marathon races with thousands of competitors, health practitioners are being called on to treat increasing numbers of victims of heat-related illness (154). While a detailed discussion of therapy belongs in medical texts, we cannot overemphasize one common pattern shared by these conditions: They are all preventable. The health practitioner should therefore become involved before a race starts or a team commences its preseason summer training.

Table 3.2 outlines the etiology, manifestations, and principles of prevention of the major heat-related illnesses. A more detailed classification can be found elsewhere (93). Conditions range in severity from mild heat cramps and heat syncope to the often-fatal heatstroke. There is an overlap among the various conditions, and their presentation is not always clear-cut. For example, heat exhaustion due to salt depletion is often accompanied by water depletion, which in itself may present first as syncope rather than exhaustion. Heatstroke can be manifested in ways other than the classical triad of hyperpyrexia, dry skin, and neurologic deficit. One notable variation is that the person still perspires even though his or her thermoregulatory control is disrupted. Such a person may mistakenly be diagnosed with heat exhaustion rather than heatstroke. A sound therapeutic approach is to assume the worst whenever in doubt. Heat-related illnesses can occur even on mildly warm or climatically neutral days. A prolonged, intense activity such as a marathon race, bicycle road race, or soccer match, especially when accompanied by hypohydration, can induce any of the heat-related illnesses irrespective of climate (155). The two most important preventive measures are prior acclimatization (or artificial acclimation) and adequate hydration prior to and throughout the activity, as discussed in previous sections of this chapter. See also the section later in this chapter on guidelines for conduct of athletic events in the heat.

Hypothermia denotes a reduction of core temperature to 35° C or lower. There are several

TABLE 3.2   Exertional Heat-Related Illnesses. Recognition, Etiology and Prevention

Illness	Main Symptomatology	Contributing Factors	Prevention
Dehydration	Dry mouth, thirst, irritability, apathy, disorientation, dizziness, weakness, headache	Insufficient drinking before & during exercise; excessive fluid & electrolyte losses (e.g., diarrhea)	Arrive fully hydrated to playing field, drink periodically during exercise even when not thirsty, use beverages that stimulate thirst
Heat Cramps	Intense limb or whole-body pain unrelated to muscle strain, repeated or constant muscle contractions	Dehydration, major electrolyte (mostly NaCl)) loss, muscle fatigue, prolonged exertion	Enforce dietary electrolyte & fluid replacement, conditioning, and heat acclimatization
Heat Exhaustion	Inability to continue exercise, lack of (or transient) CNS signs, no severe hyperthermia	Insufficient cardiac output and blood flow to organs, prolonged high-intensity exercise in the heat	Institute intermittent rest periods in the shade; enforce acclimatization, hydration & conditioning
Exertional Hyponatremia	Plasma N^+ < 130 mmol L^{-1}, weight gain during activity, swelling of hands & feet, CNS signs (e.g., confusion, convulsions, coma	Excessive drinking (especially water) during prolonged exercise, insufficient replenishment of Na^-	Fluid intake not to exceed fluid loss, beverage and meals to include Na^+ in sufficient amounts to replenish losses
Exertional Heat Stroke	Severe hyperthermia (rectal temperature > 40^0C, 104^0F), CNS dysfunction (e.g., altered consciousness, coma, confusion, disorientation, irrational behavior)	Strenuous & prolonged exercise (mostly, but not only in hot/humid climates), major dehydration, insufficient conditioning & acclimatization, prior heat-related illness	Enforce heat acclimatization, conditioning and adequate hydration; obey guidelines for modifying activity according to climate (e.g., longer rest periods in the shade)

stages of hypothermia, the most severe of which is defined as 30° C or lower. Because of their larger surface area-to-mass ratio, small children are, theoretically, more prone to this condition. There are, however, no epidemiologic data with which to compare age- or body size-related incidence of hypothermia during exercise in the cold. At particular risk are individuals with small amounts of subcutaneous fat, as in anorexia nervosa, chronic undernutrition, and cystic fibrosis. Because of the high thermal conductivity of water, hypothermia is of particular concern in aquatic activities (see earlier section on responses to a cold climate).

Frostbite represents local crystallization of the skin or subcutaneous fat, which is likely to occur mostly in exposed areas such as the cheeks, chin, nose, and ears, but also in covered sites such as the fingertips, toes, nipples, and male genitalia. While reversible (through gradual heating), frostbite often results in permanent damage. Because of their greater peripheral vasoconstriction, children's skin temperature drops faster than in adults during exposure to the cold (148). This entails an enhanced risk for frostbite.

Bronchoconstriction due to cold inhaled air occurs often in children with asthma (see section on asthma in chapter 6), but it has been described also among healthy endurance athletes who perform high-intensity exercise in the cold. This is related to a high parasympathetic tone found among such athletes (86a; 91).

Epidemiologic Aspects of Heat-Related Illness

There are no prospective epidemiologic studies of the relationship between climatic heat stress and children's performance or health. All available reports are retrospective, usually following severe climatic heat waves. In spite of methodological constraints, these reports do convey a clear message: Infants, young children, and persons who are elderly are more affected by climatic heat waves than are adolescents or young adults.

Most susceptible are children who arrive at a hospital with hypohydration or who become hypohydrated during admission secondary to diarrhea, vomiting, and so on. Mothers of these patients often are ignorant of the need for added fluid intake on hot days (28; 34; 42; 53; 86; 141; 157). Another important situation that has led to heatstroke and death among infants is that in which they are left in a car parked in the sun. In recent years, the incidence of heatstroke-induced death among such infants (mostly 2 years and under) has exceeded 40 per year in the United States alone. Air temperature inside a car parked in the sun on a warm day can reach 60° C within 40 min. Cars with a black or other dark-hued exterior are particularly affected, because they absorb most of the solar radiation. Indeed, the radiative temperature inside the car can exceed 70° C within 40 min (Bar-Or and Wilk 2000, unpublished report).

Cold-Related Illness

Because of the enhanced heat production during exercise, cold-related illness is more likely to occur at rest, or during mild-intensity activities, than during high-intensity exercise. Still, one should be aware of the three major detrimental effects of exercise in a cold environment: hypothermia, frostbite, and bronchoconstriction.

High-Risk Pediatric Groups

The thermoregulatory response to exercise in the heat is highly variable among individuals. Variability is to some extent hereditary (97), but it depends largely on acquired differences. This

TABLE 3.3 Conditions and Diseases That Predispose the Exercising Child to Thermoregulatory Insufficiency

Condition or heat disease	Possible mechanism				
	Reduced convection to periphery	Insufficient sweating	Excessive sweating	Potential hypohydration	Other
Anorexia nervosa	X			X	Reduced subcutaneous insulation
Congenital heart disease	X		X	X	
Cystic fibrosis	X		X	X	
Diabetes (mellitus, insipidus)	X			X	
Diarrhea and vomiting	X			X	
Excessive eagerness				X	High heat production
Fever	X		X	X	Regulatory insufficiency
Hypohydration	X	X (if extreme)			
Insufficient acclimatization	X	X			
Insufficient conditioning	X	X			
Malnutrition					Reduced subcutaneous insulation
Mental deficiency					Insufficient drinking
Obesity					High heat production, low specific heat and surface area
Prior heat-related illness					Various (depending on illness)
Sickle cell anemia					Sickling, mostly with dehydration
Sweating insufficiency syndromes		X			

section highlights those groups of children who respond with a high physiologic heat strain to a given heat stress. Such a response may impede their exercise performance and may also be detrimental to their health. The early identification of children who are at high risk of developing heat-related illness is of obvious relevance for the physician. Table 3.3 outlines the conditions and diseases that may put a child at a particularly high risk. A possible mechanism is offered for the thermoregulatory deficiency.

Anorexia Nervosa

The thermoregulatory capability of patients with anorexia nervosa (AN) is often deficient (44; 107; 126). At rest they complain of cold, especially in their extremities, which is accompanied by cyanosis (acrocyanosis) (94; 95). These changes reflect a low skin temperature secondary to peripheral vasoconstriction (94). Such vasoconstriction may represent compensation for deficient insulation due to the paucity of subcutaneous fat. Core temperature is lower than normal in resting AN patients. Values lower than 36° C are not uncommon. This, most probably, results from a low resting metabolic rate and low subcutaneous thermal insulation, but it also may reflect hypothalamic

dysfunction. Even though patients with AN have a sluggish vasodilatory response to heat, they dissipate more heat by convection and radiation than by evaporation. The reverse is true for healthy individuals exposed to a similar environment.

During prolonged exercise in a thermoneutral climate, the rise of core temperature to a new plateau is slow in persons with AN compared with that in healthy controls. This could reflect a greater heat capacity (45), the specific heat of lean tissues being higher than that of adipose tissue (21). Figure 3.24 depicts the effect of a cold environment on the core temperature of a young adult female with AN compared with healthy controls. Data for the patient were taken at air temperature of 5° C and at thermoneutral conditions (23° C). Two patterns should be noted: (1) a very low rectal temperature throughout the session in the thermoneutral environment and (2) a steep decline in rectal temperature during and following exposure to 5° C. The temperature did not increase until the patient entered a warm climatic chamber, where she cycled for nearly 30 min.

In conclusion, it seems that the environmental zone in which patients with AN can maintain a constant core temperature is narrower than normal (44). Whether their deficient thermoregulatory

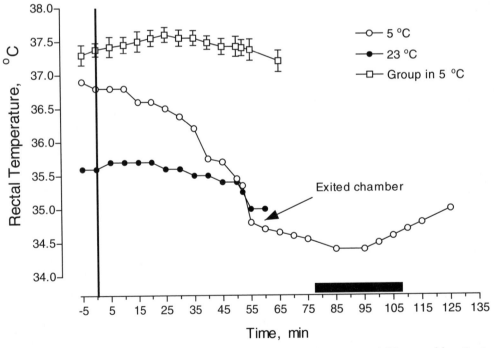

FIGURE 3.24 Effect of cold ambient air on rectal temperature in anorexia nervosa. A 21-year-old patient rested for 60 min in a climatic chamber set at 5° C and, in another session, at 23° C. Her rectal temperature is compared to that of seven healthy controls. Black bar denotes cycling after chamber reentry. Vertical lines denote SEM. Unpublished data from the author's (O. B.-O.) laboratory.

capability is due to low resting metabolic rate, insufficient insulation, or suboptimal hypothalamic control, these patients may be at risk for heat- or cold-related illness. Furthermore, some patients practice vomiting, which may induce hypohydration and electrolyte disturbances. Vomiting and its effects may occur also in patients with bulimia, who often adopt a binge-purge eating pattern.

Congenital Heart Disease

Clinical experience has shown that some infants and children with congenital heart disease (CHD) sweat excessively. This phenomenon, albeit extremely rare, has been reported in controlled indoor observations (2; 105; 125). Especially affected are those who have congestive heart failure or a right-to-left shunt. Sweat glands of some CHD patients more than 6 months old have longer ducts than normal (90). The physiologic implication of such an anatomical difference is not clear.

There are no reports regarding whether hot and humid climates actually interfere with the physical ability and well-being of the ambulatory child with a congenital cardiac problem. If one extrapolates from information on adults (31), however, there is enough evidence to recommend that the outdoor activity level of the child with severe heart defects be reduced whenever climatic heat stress is high.

Cystic Fibrosis

Infants and children with cystic fibrosis (CF) often experience marked heat prostration during climatic heat waves (42; 86; 167). It has been estimated that 15% of all New York City children with CF were hospitalized for heat prostration during the 1948 heat wave (86). One of the initial signs was profuse sweating. Within 2 to 3 days the children developed hypohydration, hyperpyrexia, and circulatory insufficiency. Serum concentration of Na+ and Cl– was low, reaching 125 and 80 milliequivalents per liter, respectively. Rehydration with an adequate salt supplement was effective. The increased sweating rate of these children may have been related to a high density of active sweat glands, as shown by pilocarpine iontophoresis (74).

Until recent years, the mechanism for the low heat tolerance of patients with CF was unknown. A 1.5-hr exposure in a climatic chamber did not reveal abnormalities in the thermoregulatory

pattern of such children (119). Nor were there aberrations in their ability to acclimate to the heat (120). A major characteristic, however, of patients with CF is the very high loss of NaCl in their sweat. Levels of Na+ and of Cl– can exceed 125 mmol · L^{-1} (88), compared with 20 to 30 mmol · L^{-1} in healthy children (108). As a result, when children with CF exercise in the heat there is a marked reduction in their serum Na+, Cl, and osmolality levels (119).

As stated earlier in the discussion of the means of enhancing thirst, an increase in extracellular osmolality is a major trigger for thirst. The abnormally low serum osmolality found in the study just mentioned (119) raised the hypothesis that CF patients who exercise in the heat may be deprived of this trigger and as a result do not drink enough to replenish their sweat losses. The final outcome would be progressive dehydration and heat intolerance. Indeed, when voluntary drinking patterns were analyzed in patients with CF who exercised intermittently for 3 hr in 31° to 33° C, 44% to 46% relative humidity, their rate of drinking unflavored water was only half that of healthy controls (figure 3.25, top). The resulting hypohydration was almost threefold in the patients compared with the controls (figure 3.25, bottom) (17). For safety considerations, the chamber protocol was such that hypohydration would not exceed 2% of initial body weight. At this mild dehydration, the only indication that heat strain was higher in the CF patients was a slower recovery of heart rate following each exercise bout. It is likely, though, that a greater degree of hypohydration would have elicited a greater deficiency in the patients' thermoregulatory effectiveness.

Can voluntary drinking of patients with CF be enhanced? Unlike what occurs in healthy children (165), the addition of flavor to water did not increase drinking volume of patients with CF during a 3-hr exposure to intermittent exercise in the heat, nor did the further addition of 30 mmol · L NaCl and 6% carbohydrates. However, when NaCl concentration was further increased to 50 mmol · L^{-1}, voluntary drinking did rise sufficiently to prevent dehydration throughout the session (figure 3.26) (88). This pattern contrasts with the response of healthy children, whose thirst can be triggered by NaCl concentration as low as 18 mmol · L^{-1} (165); and it reflects the considerably greater salt loss in the sweat of patients with CF.

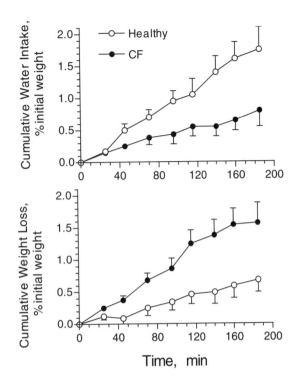

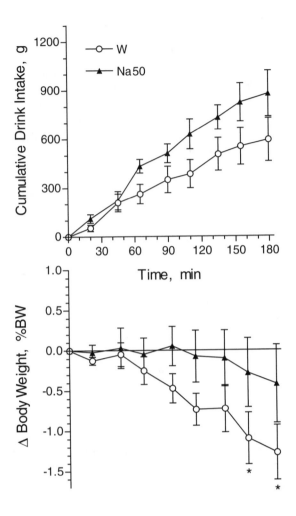

FIGURE 3.25 Cumulative voluntary drinking (top) and the degree of involuntary hypohydration (bottom) in patients with cystic fibrosis (CF) and in healthy controls. Eight patients (ages 9.5-14.1 years) and eight healthy controls performed intermittent exercise (43-47% maximal O_2 uptake) for 180 min at 31° to 33° C, 44% to 46% relative humidity. Unflavored water was available ad libitum.

Reproduced, with permission, from Bar-Or et al. 1992 (17).

In conclusion, children and adolescents with CF should be encouraged to pursue an active lifestyle. However, they must be taught to drink periodically, above and beyond thirst, particularly when exposed to climatic heat and humidity. Addition of NaCl (approximately 50 mmol · L^{-1}) to the flavored beverage will trigger thirst and stimulate further drinking. In preparing such drinks, one should try to mask the salty flavor of the NaCl solution in order to increase its palatability.

Diabetes Mellitus and Diabetes Insipidus

Regular physical activity should become integral to the management of the child with diabetes mellitus. On hot or humid days, however, caution should be exercised and special attention paid to adequate hydration. This is especially important in patients with polyuria (large urine volume), who may dehydrate and experience excessive electrolyte loss. A similar, or even greater, risk

FIGURE 3.26 The effect of NaCl solution on voluntary drinking (top) and hydration status (bottom) in patients with cystic fibrosis (CF). Six CF patients (ages 10-19 years) exercised intermittently (heart rate 140-160 beats · min^{-1}) while exposed for 3 hr to 35° C, 50% relative humidity. In one session they were given water (W) ad libitum, and in the other they were given a 50 mmol · L^{-1} solution (Na50).

Reproduced, with permission, from Kriemler et al. 1999 (88).

confronts the child with diabetes insipidus. In this rare disorder, an inability to concentrate the urine (due to a lack of antidiuretic hormone, or suppressed renal response to this hormone) results in marked polyuria and a danger of dehydration.

In adults with type 1 diabetes mellitus, exposure to a hot climate may enhance the hypoglycemic effect of exercise. This may reflect heat-induced peripheral vasodilation (62), which in turn causes an above-normal release rate of insulin from the injection site. It is not known whether this pattern applies also to children or adolescents.

Diarrhea and Vomiting

Any condition accompanied by diarrhea and vomiting, irrespective of etiology, can easily induce hypohydration and electrolyte imbalance in a child. Sweating will aggravate such disturbances, predisposing the child to heat exhaustion and heatstroke. Special precautions should be taken to include salt in the drink, for partial replacement of electrolytes.

Excessive Eagerness

The discomfort experienced by people who are exposed to heat or cold stress is protective. As a rule, an individual will try to terminate such an experience by, for example, seeking shade, stopping exercise, or getting out of cold water. Children's perception of exercise intensity while exposed to heat stress underestimates the physiologic strain that they are undergoing (12; 20). It would seem, therefore, that children do not experience the same protective discomfort that would force an adult to reduce the climatic stress. A young, ambitious, overzealous athlete may decide to conduct a practice session regardless of the hazards of the prevailing climate. Such a child or adolescent should be advised about the harmful consequences and be monitored by an adult.

Fever

Irrespective of its etiology, fever indicates some disruption of thermoregulatory control. It may therefore predispose to further hyperpyrexia in a child who is generating high metabolic heat, especially when exposed to climatic heat. Intense activities, as a rule, should be curtailed in a febrile child even if the cause is "obvious"—such as following vaccination.

Hypohydration

Hypohydration is the link between most "high-risk" conditions and thermoregulatory insufficiency. As already discussed, hypohydration can also occur in healthy children who do not replenish fluid losses before, during, or after activity.

Hypohydration is accompanied by reduced plasma volume. This impedes the convection of heat from body core to skin and reduces the central blood volume so that circulatory insufficiency may ensue. Hypohydration is almost invariably present in heat exhaustion (water depletion type) and in heatstroke. It can also lead to heat syncope.

"Thirst fever" is a syndrome seen in febrile young children with no apparent infectious etiology. The level of fever is proportional to the fluid deficit and disappears once the child is fully hydrated (35; 141).

Insufficient Acclimatization

For a full discussion of the physiologic and perceptual phenomena of children's acclimatization to exercise in the heat, see the earlier section on that topic. Insufficient acclimatization is one of the most important causes of heat-related illness. The clinician must bear in mind the sluggish rate of acclimatization in children on the one hand and their fast improving subjective sense of well-being on the other. This is especially important during climatic heat waves, whenever a practice season starts in the summer (as in American football), or when athletes travel to hot geographic regions. Retrospective studies of heatstroke strongly suggest a relationship to insufficient acclimatization. Among eight heatstroke fatalities in high school American football, six happened following the first or second session of preseason practice, before the players had a chance to acclimatize (23; 64). Another young football player was lucky to survive a heatstroke that occurred following the first practice session in the summer (127). Similar occurrences have been reported for unacclimatized college football players (151). Rural infants and preschoolers in Australia were found to have less heat-related illness than urban ones, presumably because the former were habitually more exposed to heat and therefore better acclimatized (42).

One cannot overemphasize the need to achieve acclimatization in children and adolescents who are newly exposed, or likely to be exposed, to a hot climate. Guidelines are available in other sections of this chapter.

Insufficient Conditioning

As discussed previously, conditioning per se is a means of heat acclimatization in children. One can therefore assume (even with the lack of experimental evidence) that the insufficiently conditioned child is also less acclimatized and may encounter difficulties when abruptly confronted with hot weather.

Another potential disadvantage of the less fit individual is his or her inexperience in performing strenuous tasks. There is some evidence that novice adult distance runners are more prone to heatstroke because they select a running pace incompatible with their exercise capacity (69). This issue of inexperience and heat-related illness has not been studied in children.

Mental Deficiency

The mentally deficient child does not necessarily have any thermoregulatory deficiency. It has been suggested (42), however, that such children are more prone to heat-related illness because they can neither understand nor verbalize their need for extra drinking on hot days.

Obesity

Since subcutaneous fat provides insulation to the body core, increased adiposity is an asset in cold climates (32; 83). In contrast, obesity is a distinct liability in hot climates. In response to a questionnaire, overweight women reported a lesser tolerance to heat and a greater tolerance to cold than did normal-weight women. Men had a similar tendency (68). The tolerance time of obese adults is indeed short when they exercise in a heated climatic chamber (21; 106). They respond to exercise in the heat with higher heart rate, cardiac output, and body temperatures; and their postexercise cardiovascular adaptation to orthostatic stress is less efficient than that of lean controls (21; 32; 106). Among children, those who are obese respond to a given heat stress with a higher strain than do lean children. However, in a study among 9- to 12-year-old boys, there was no difference in the ability of the lean and the obese children to complete 60 min of exercise in 46° to 48° C, 22% relative humidity (70). Nine- to 12-year-old boys who were mildly obese (31.2% fat) had a faster and greater increase in heart rate and rectal temperature than did their lean (15.5% fat) controls while walking on a treadmill in dry heat (figure 3.27) (71).

In any given individual, an inverse relationship exists between the population density of heat-activated sweat glands and skinfold thickness. As a group, persons who are obese have a lower density of active glands, especially in the trunk (22), even though their sweating rate, corrected for surface area, is the same as or even higher than that of lean people (14; 32). Sweat output per gland is therefore higher in persons who are obese. They can acclimatize well to exercise in the heat (33), but during progressive dehydration

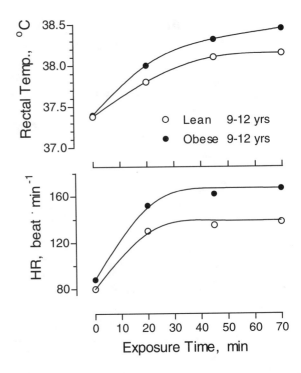

FIGURE 3.27 Obesity and heat strain. Five obese and seven lean 9- to 12-year-old boys walked intermittently (4.8 km · hr^{-1}, 5% grade) in dry heat (40-42° C, 25% relative humidity). Values are group means.

Reproduced, with permission, from Bar-Or 1980 (12), based on data by Haymes et al. 1975 (71).

their core temperature rises excessively with any given body fluid deficit (19). Obesity can mask the clinical presentation of fluid deficit, especially in young children (52). One should therefore be alert to the possibility of fluid deficit in any obese child with recent exposure to heat.

A number of mechanisms underlie the predicament of the obese child in hot climates:

- Specific heat of fat is 0.4 kcal · g^{-1} · °C^{-1}, as compared with 0.8 kcal · kg^{-1} · °C^{-1} in the fat-free mass. Thus a given amount of heat will raise the temperature of 1 g of fat twice as much as it will raise the temperature of 1 g of fat-free tissue.

- Children who are obese often have a large body mass and a small surface area-to-mass ratio. In addition, accumulating fat deposits change the contour of the body such that this ratio is further decreased. This results in a lower rate of heat transfer to and from the body. This is a disadvantage in moderately warm climates when skin temperature is higher than air temperature. Only in extreme heat will the small surface area-to-mass ratio be an advantage in reducing heat influx from the environment. It will still be a handicap as far as sweating is concerned (14; 70).

• The rise in core temperature during exercise is proportional to the relative metabolic rate (i.e., percentage of maximal O_2 uptake). Obese children as a group have a low maximal aerobic power; therefore when performing a given task they function at a relatively high percentage of their maximal aerobic power. This results in a greater rise of core temperature as well as greater fatigability.

• Fat has less water content than do most other tissues, so individuals who are obese have a low total body water content per unit mass. As a result, a certain hypohydration level, determined as percentage of initial body weight, represents for the obese a greater water deficit relative to total body water.

• Blood flow to the periphery during exercise in the heat is lower in persons who are obese than in lean people (161). This may impede convection of heat by blood, from body core to the periphery.

These physiologic liabilities have clinical implications. A relationship has been found between overweight and the risk of heatstroke and death (140). Among 125 heatstroke victims, most of the patients were somewhat overweight and actually obese (101). High school boys who died of heatstroke following football practice were markedly overweight (23; 64). These 15.3-year-old boys weighed on the average 89 kg, which probably represents some obesity and not only developed musculature.

Children who are obese, because of their low level of fitness and a high heat strain, are likely to slow down or terminate their exercise earlier than other children. One must remember to allow these individuals special consideration on hot or humid days and not to push them to their limit.

Prior Heat-Related Illness

An important question is whether an individual with a history of heat-related illness is predisposed to an above-average risk in the future. Young adults with a history of exercise-related heatstroke who were re-exposed to exercise in dry heat responded with reduced tolerance time, high rectal temperature and heart rate, but normal sweating rate (130; 142). It is not known whether such individuals have an inherent thermoregula-

tory deficiency or if they have incurred irreversible damage from the heatstroke episode.

Sickle Cell Anemia

Dehydration, exercise, and exposure to the heat have been shown to trigger sickling of red blood cells in patients with sickle cell anemia (81). This may cause occlusion of blood vessels and ischemic damage to tissues. For more details about this disease, see section "Sickle Cell Anemia" in chapter 11.

Sweating Insufficiency Syndromes

Sweating insufficiency syndromes are extremely rare and often just one manifestation of a systemic disease (63; 99). Children with such a syndrome must rely on convection and radiation for dissipating heat. A special risk exists on days when air temperature is high and heat dissipation by these avenues is impeded.

Undernutrition

Undernutrition of the hypocaloric type is associated with low subcutaneous insulation. As in AN (see previous discussion), this may lead to excessive heat loss in cold climates (29) and to an excessive rise in core temperature in hot climates, particularly when ambient temperature exceeds skin temperature (30). Undernourished children may also have a reduced sweating rate, perhaps due to local changes in the sweat glands (30).

Although these data are based on experiments with infants, it is likely that older malnourished children will respond in a similar way. One should bear in mind that, unless severely affected, the malnourished child is physically active and may therefore be adversely affected by hot climates. Reversal of undernutrition, as studied in adults (102), is accompanied by an improvement in the thermogenic response (i.e., increase in metabolic heat production) to the cold.

Guidelines for the Conduct of Athletic Events in the Heat

The following guidelines are based on position statements developed by the American College of Sports Medicine (4; 38) and by the American Academy of Pediatrics (3), as well as our own experience.

1. Ensure acclimatization/acclimation to heat. The nonacclimatized child, when first exposed to heat, must cut down on the intensity and duration of exercise and then gradually increase intensity and duration. Exposures can be conducted three to six times per week, for a total of six to eight exposures. Exercise during the session must be interspersed with rest periods.

2. Secure full hydration before practice or competition (300-400 ml fluid, 20-30 min prior to activity for a 10- to 12-year-old child) and encourage periodic drinking (every 15-20 min) during prolonged activities. Remember that thirst is not an accurate guide. Encourage active children to drink above and beyond subjective repletion. A child 10 years of age or younger should drink until repletion and then add another 100 to 125 ml. An older child should drink until repletion and then add another 200 to 250 ml.

3. Fluids should be chilled and flavored. They should include approximately 18 mmol · L^{-1} NaCl and 6% carbohydrate (e.g., glucose and sucrose). Attempt to provide the child with his or her favorite flavor.

4. Discourage excessive "making weight" by dehydration. This is a habit strongly rooted among athletes and coaches and must be eradicated, or minimized, through education. Rubberized sweat suits, laxatives, diuretics, and emetics should be absolutely disallowed. Never restrict fluids as a disciplinary or "character-building" measure.

5. Activities must be tailored to the prevailing climate. It is the prerogative of the team or school physician to postpone, curtail, or cancel activities or to increase rest periods because of climatic stress. A suggested policy is given in table 3.4. Ideally, the coach or a team physician should monitor radiation, humidity, and air temperature on the field, using a psychrometer. Those who do not possess a psychrometer can obtain information from the local weather bureau. Rest periods in well-ventilated and shaded areas are important for dissipation of stored heat during a practice.

6. Clothing should be lightweight, limited to one layer of absorbent material, and tight to the skin to facilitate the evaporation of sweat. Excessive taping and padding should be discouraged. A hat and light-colored clothing are recommended whenever feasible to reduce solar radiation. Discourage prolonged exposure of the skin to the sun.

7. Identify and screen out, as necessary, individuals who are at high risk for heat-related illness (see preceding section). A preseason examination is of value for such screening.

8. Look for and teach others about early warning symptoms and signs that precede heat-related illness. These include unexplained headache, throbbing pressure in the head, chills, nausea, piloerection on the chest and arms, disorientation, ataxia, and dry skin. Activity of children who experience any of these should be discontinued and treatment commenced.

9. Never put success before safety. Coaches, athletes, and even parents sometimes disregard

TABLE 3.4	Climatic Heat Stress and Permissible Physical Activities	
WBGT °C* (°F)	WBT °C† (°F)	Changes in activity
<25 (<77)	<15 (<59)	All activities allowed
25-27 (77-81)	15-21 (59-70)	1. Longer breaks in the shade 2. Drinking each 15 min 3. Be alert to warning symptoms of heat-related illness
27-29 (81-84)	21-24 (70-75)	As above, plus: 1. Stop activities of all unacclimatized, unconditioned, and high-risk persons 2. Limit activities of all others (drastically cut down duration of each activity; increase rest periods; disallow long distance races)
>29 (>84)	>24 (>75)	Stop all athletic activities of all participants

*WBGT (wet bulb globe temperature) = 0.7 WB + 0.2 G + 0.1 DB, where WB is wet bulb, G is black globe, and DB is dry bulb, measuring humidity, radiation, and air temperature, respectively.

†WBT = 0.7 WB + 0.3 DB (for indoor use when radiant heat is less important).

heat-related risk for the sake of athletic success. Whether a team physician, school physician, or physician or medical professional attending a competition, do your utmost to educate them. If other means fail, assert your authority.

Guidelines for the Conduct of Athletic Events in the Cold

The following guidelines are intended to prevent or ameliorate the three detrimental effects of a cold environment: hypothermia, frostbite, and exercise-induced asthma.

1. If possible, water temperature for child swimmers should be 1° to 2° C warmer than for adults.

2. Children should be encouraged to come out of the water every 15 to 20 min.

3. In long distance swimming, mostly in lakes or in the sea, apply a 1- to 2-mm layer of lanolin or petroleum jelly over the skin.

4. Most children, because of an unpleasant cold sensation, will get out of the water before hypothermia takes place. One should be aware, however, of the highly ambitious, overzealous young swimmer who may opt to stay in the water in spite of such a sensation.

5. If in doubt as to whether the child is becoming hypothermic, adopt a conservative approach by removing the child from the cold environment. The smaller the child, the greater should be your level of suspicion.

6. Heat will be conducted away from the body faster if the clothes are wet. It is important to use several layers of dry clothing for activities performed at near- or below-freezing conditions.

7. If the windchill index is minus 15° to 20° C or lower, the face must be fully covered.

8. Exercise-induced asthma can be ameliorated, or totally prevented, by covering the mouth and nose with a face mask or a scarf during activities that require high ventilation (e.g., cross-country skiing, distance running). These can be used when air temperature drops below 10° C.

PART

II

Clinical Perspectives of Children and Exercise

© Martina Sandköhler

Children and Exercise in a Clinical Context— an Overview

Any clinician who wants to incorporate exercise into his or her diagnostic or management strategy should ask the following questions about each patient:

- Is the child sufficiently active?
- If not, what is the underlying (physical, psychological, social) cause?
- What is the physical fitness status of the child?
- Can exercise be of diagnostic value?
- Will enhanced physical activity benefit the health and well-being of the child?
- Is exercise detrimental to the child's health?

This chapter provides an overview of the relevance of these questions to clinical pediatrics. Details are presented later in the book.

Habitual Activity and Disease

In adults, a sedentary lifestyle is now recognized as a primary risk for morbidity and mortality (15). In spite of this, a sedentary lifestyle among adults is accepted by many societies as normal and, in itself, is not perceived as a sign of ill health. In contrast, inactivity in a child almost invariably reflects a deviation from normality, be it physical, mental, or emotional impairment or social maladjustment. *Hypoactivity* in this book is defined as an activity level lower than that of healthy peers of a similar age, gender, and cultural and socioeconomic background. We by no means imply that today's healthy children are *sufficiently* active.

As a group, children and adolescents with a chronic disease or a physical disability are hypoactive. For example, a questionnaire-based survey was conducted in 1990 among 6- to 20-year-old girls and boys with a chronic disease, motor disability, or sensory impairment (7). The 987 participants were a near-representative sample of some 13,000 patients who lived in Ontario, Canada. Twenty-nine percent of them were categorized as "sedentary." This contrasts with only 10% of youth who were categorized as sedentary in the 1983 nationwide Canada Fitness Survey, which used a similar questionnaire (4). As seen in figure 4.1, the activity score of the subjects with a disability or chronic disease started declining at age 12 and reached a precipitous decrease in late adolescence. This pattern is similar to that of healthy children and youth (see chapter 2). However, figure 4.2 summarizes activity levels in the disabled group as a whole, compared with the healthy population. While more than 70% of the latter were categorized as "active," only 40% of the disabled group were in this category. In contrast, five times as many in the disabled group were "sedentary." A similar pattern, based on smaller samples, has been reported for various diseases and disabilities, as discussed in chapters 6 through 12.

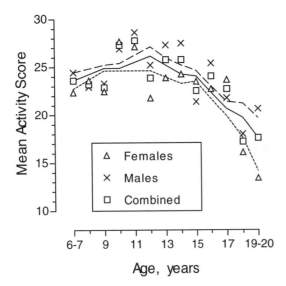

FIGURE 4.1   Habitual activity of children and youth with chronic disease, motor disability, or sensory impairment. An activity score was determined from a questionnaire similar to that used in the Canada Fitness Survey (4). Responders were 6- to 20-year-old girls (n = 472) and boys (n = 515) from Ontario, Canada.

Adapted, with permission, from Longmuir and Bar-Or 1994 (7).

Disease As a Direct and Indirect Cause of Hypoactivity

Table 4.1 summarizes pediatric conditions in which hypoactivity was documented. The list is divided into two subgroups: one in which hypoactivity is inherent to the disease, and possibly a direct result of it. The other group includes conditions in which hypoactivity is incidental to the disease. While a child with, for example, crippling arthritis or advanced muscular dystrophy is obviously limited in his or her gait and other movements, a child with asthma or diabetes can be active but often is not. The indirect restrictive effects of ill health are often imposed by others, as shown in figure 4.3. Factors such as parental overprotection, fear on the part of the child or parents, ignorance on the part of parents and teachers (and, occasionally, health practitioners), and peer-imposed social isolation all lead to hypoactivity. Impaired exercise performance also causes hypoactivity, which becomes part of a vicious circle (discussed in the section "Effects of Disease on

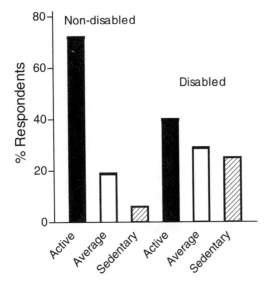

FIGURE 4.2   Comparison of activity levels in 6- to 20-year-old children and youth with a chronic disease or disability and in healthy children and youth. Based on data by Longmuir and Bar-Or 1994 (7).

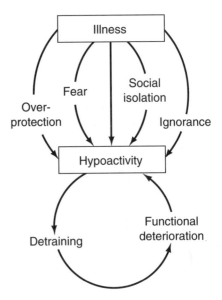

FIGURE 4.3   Direct and indirect links between hypoactivity and illness.

TABLE 4.1	Pediatric Diseases and Disabilities That Are Accompanied by Reduced Habitual Activity	
Hypoactivity inherent to disease	**Hypoactivity incidental to disease**	
Arthritis	Asthma	
Blindness	Cystic fibrosis—mild and moderate	
Brain injury	Diabetes mellitus—type 1	
Cerebral palsy	Epilepsy	
Cystic fibrosis—severe	Gynecomastia	
Cyanotic heart disease	Hemophilia	
Muscular atrophy/dystrophy	Mental retardation	
Obesity—severe	Noncyanotic heart disease	
Spina bifida	Obesity—mild and moderate	
Scoliosis—severe	"Non-disease" entities	
Undernutrition		

Physical Fitness" later in this chapter). Children and adolescents with various medical conditions sometimes perceive a given exercise intensity to be higher than do their healthy peers (1), which may be another reason why these patients tend to be hypoactive. Such a high rating of perceived exertion has been shown, for example, in patients with motor disabilities (2), obesity (17), and type 1 diabetes mellitus (11). Specific examples of other direct and indirect effects of disease on activity are listed in table 4.2.

"Non-Disease" As a Cause of Hypoactivity

Many children who have a *presumed* illness, with no organic abnormality, are denied sufficient activity. The most common example of such "non-disease" is the innocent heart murmur. In a survey of Seattle junior high schools, 93 pupils who according to the nurse's file had a "heart problem" were reevaluated by a pediatric cardiologist, and their parents were interviewed. This study was designed to ascertain how many of these children really had an organic heart disease and whether the level of activity restriction was related to the presence or absence of disease (3). As seen in figure 4.4, only one in five children had an organic heart disease. The others had either an innocent murmur or no findings at all. But the striking result was that in *both* groups, irrespective of evidence of heart disease, activity was restricted in some 40% of the participants! According to most parents, the decision to restrict their child's activity was based on the message that they had received from the physician at the time of the original diagnosis.

If these findings are typical of other school systems, hypoactive children with cardiac non-disease outnumber those with a confirmed disease. Physicians play a crucial role in the etiology of non-disease and in the resulting hypoactivity. A mere suggestion that the child should refrain from sport, or perceived uncertainty about this issue, may render the child inactive for years.

Effects of Disease on Physical Fitness

Disease can cause a decrease in physical fitness in a number of ways:

- Indirectly, through hypoactivity and deconditioning
- Lowering maximal aerobic power
- Elevating the metabolic cost of submaximal tasks
- Lowering strength, peak anaerobic power, local muscle endurance, and flexibility (range of motion)

Hypoactivity-Deconditioning-Hypoactivity: The Vicious Circle

Disease often causes hypoactivity, which leads to a deconditioning effect, a reduction in the functional ability of the child, and further hypoactivity. This vicious circle, shown in figure 4.3, can occur in any chronic disease or disability, obesity being a typical example. In some patients it follows a short period of bed rest as a result of injury, surgery, or acute exacerbation of the chronic disease. An example is a boy with muscular dystrophy who can still walk. Following some minor injury he is confined to bed for 2 to 3 weeks and as a result can no longer resume walking

TABLE 4.2	Causes of Hypoactivity in Pediatric Diseases and Disabilities As Stated by Parents or Patients
Disease	**Cause of hypoactivity**
Arthritis	Pain
Asthma	Fear of postexercise attack
Brain injury	Limited repertoire of activities
Cerebral palsy	Lack of opportunities
Diabetes mellitus	Danger of hypoglycemic crisis
Epilepsy	Fear of seizure and injury
Heart disease	Fear of "heart attack"
Hemophilia	Fear of injury and bleed
Innocent murmur	Fear of "heart attack"
Mental retardation	Isolation, social maladjustment
Obesity	Inhibition, social discrimination, low fitness

status. Deconditioning can also occur in healthy individuals who, for some reason, reduce their level of activity. For more details on the sequelae of deconditioning, see the section "Physiologic Effect of Detraining" in chapter 1.

Reduced Maximal Aerobic Power

Disease can directly affect the O_2 transport system and cause a reduction in maximal aerobic power. According to Fick's principle, maximal O_2 uptake ($\dot{V}O_2$max) equals the product of maximal cardiac output ($\dot{Q}$max) and the maximal arterial-mixed venous difference in O_2 content ($[CaO_2 - C\bar{v}O_2]$max). $\dot{Q}$max is the product of maximal stroke volume (SVmax) and maximal heart rate (HRmax). Thus:

$$\dot{V}O_2max = SVmax \times HRmax \times (CaO_2 - C\bar{v}O_2)max$$

Disease can reduce any of the three functions on the right side of the equation, thereby reducing maximal O_2 uptake. A list of such diseases and the specific physiologic function that each of them affects is presented in figure 4.5.

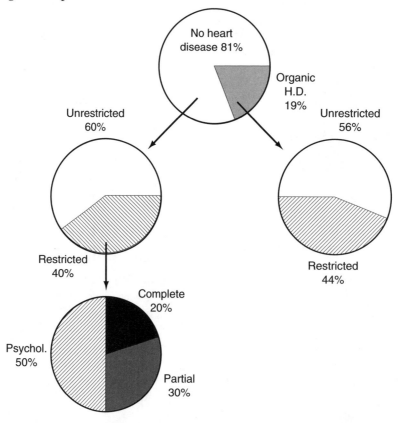

FIGURE 4.4 Cardiac "non-disease" and activity. Degree of restriction of physical activity among 93 schoolchildren who were reported in the nurse's file to have had a "heart problem" and whose condition was further diagnosed by an interview, physical examination, chest X-ray, and resting electrocardiogram. Data from Bergman and Stamm 1967 (3).

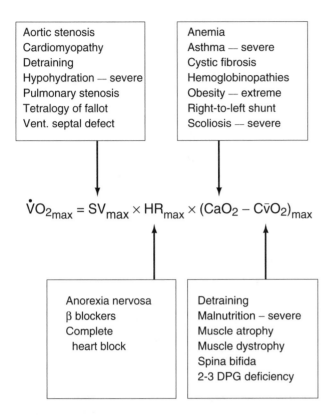

Aortic stenosis	Anemia
Cardiomyopathy	Asthma — severe
Detraining	Cystic fibrosis
Hypohydration — severe	Hemoglobinopathies
Pulmonary stenosis	Obesity — extreme
Tetralogy of fallot	Right-to-left shunt
Vent. septal defect	Scoliosis — severe

$$\dot{V}O_{2max} = SV_{max} \times HR_{max} \times (CaO_2 - C\bar{v}O_2)_{max}$$

Anorexia nervosa	Detraining
β blockers	Malnutrition – severe
Complete	Muscle atrophy
heart block	Muscle dystrophy
	Spina bifida
	2-3 DPG deficiency

FIGURE 4.5 Reduction in maximal aerobic power ($\dot{V}O_2$max) because of disease. The Fick equation and specific conditions that affect its components, thus reducing $\dot{V}O_2$max, are shown.

A subnormal *maximal stroke volume* will result from the following:

- Outflow obstruction (e.g., aortic stenosis)
- Deficient contractility (e.g., cardiomyopathy)
- Low ventricular preload (e.g., hypovolemia from dehydration)
- Deficient "forward" stroke volume (e.g., atrial septal defect)

Although *peak HR* in many diseases is lower than the age-predicted maximal HR, there is only one disease in which a low peak HR is the primary limiting factor: complete congenital heart block. Low peak HR also occurs in people whose HR is determined by a pacemaker (see chapter 7, section "Complete Heart Block and Pacemakers"). A primary low peak HR can also result from medication, notably beta-blockers. In other diseases a low peak HR may accompany, rather than cause, low maximal aerobic power. This occurs, for example, in Duchenne muscular dystrophy, where the small functional muscle mass cannot

generate sufficient metabolic and cardiac drives to induce much increase in HR or $\dot{V}O_2$. A low peak HR occurs also in advanced anorexia nervosa, possibly due to a high vagal tone.

A reduced *arterial O_2 content* can result from respiratory or cardiac disorders or from a low O_2-carrying capacity of the blood. Respiratory disorders include lung diseases such as cystic fibrosis, severe forms of bronchial asthma, or restrictive lung syndromes. They also include chest wall disorders, such as severe scoliosis, or extreme obesity, in which the alveolar ventilation is low. All these respiratory disorders may result in arterial O_2 desaturation. Cardiac abnormalities causing a reduced arterial O_2 content occur with right-to-left shunts such as tetralogy of Fallot and Eisenmenger syndrome. Anemias and hemoglobinopathies are obvious causes of low O_2-carrying capacity of the blood. Mild to moderate activities can be performed well in these disorders because of a compensatory increase in cardiac output. When activity is intense, however, cardiac output can no longer rise and performance is deficient (see chapter 11).

A high *mixed-venous* O_2 content during maximal exercise reflects low O_2 utilization by various organs, skeletal muscles in particular. This happens when muscle blood flow is low relative to blood flow to other organs (e.g., in muscle atrophy, muscle dystrophy, or severe undernutrition). Erythrocyte 2,3 diphosphoglycerate (2,3 DPG) effects a shift to the right of the O_2 dissociation curve, making more O_2 available to the tissues at a given partial O_2 pressure. A deficiency of 2,3 DPG might well reduce the availability of O_2 and result in high mixed-venous O_2 content.

High Metabolic Cost of Exercise

Even when maximal O_2 uptake is normal, a high metabolic cost of submaximal activities will leave the individual with a reduced "metabolic reserve" (figure 4.6) and impair his or her ability to sustain exercise at moderate or intense levels (see also figure 1.11 and chapter 1, section "Metabolic Responses to Exercise in Children"). Such is the case, for example, in obesity, in which the transport of excessive weight is metabolically expensive. Similarly, excessive demands are apparent during exercise in spastic or athetotic cerebral palsy (16) or in other neuromuscular and skeletal diseases accompanied by incoordination

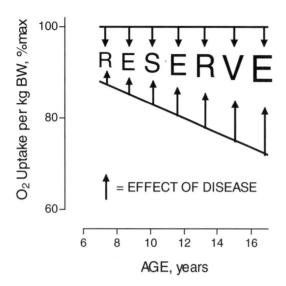

FIGURE 4.6 Effect of disease on the metabolic cost of locomotion. "Metabolic reserve" as a function of age is shown schematically. The down-pointing vertical arrows denote effect of disease on maximal O_2 uptake. The up-pointing arrows denote effect of disease on submaximal O_2 uptake.

and "wasteful" movements. O_2 uptake can also be excessive because of the high O_2 cost of breathing, as in airway obstructive syndromes, constrictive lung diseases, extreme obesity, or chest wall abnormalities.

A reduction in the *range of motion* due to a joint disease or confracture, affects the child's mobility and, thus, physical fitness. Examples are Duchenne muscular dystrophy and juvenile idiopathic arthritis.

In summary, it is apparent that some diseases affect exercise performance through more than one mechanism (e.g., obesity, asthma, cystic fibrosis, tetralogy of Fallot), whereas others cause a specific deficiency. Chapters 6 through 12 present more detailed discussion of the pathophysiology of exercise.

Exercise As a Diagnostic Tool in Pediatrics

Exercise "stress testing" has become a universally accepted tool for the assessment of coronary heart disease as well as pulmonary and other disorders in adults. Exercise in clinical assessment of children, although less publicized, is used in a grow-

ing number of pediatric clinics. In fact, there are more pediatric than adulthood disorders in which exercise testing is of clinical relevance (12). The rationale for the use of exercise in pediatric assessment is summarized in the list below. Methods of testing are outlined in appendix II.

- Assess physical fitness
- Evaluate *specific* pathophysiologic characteristics in order to provide indications for surgery, therapy, or additional tests; evaluate functional postoperative success; or diagnose a disease
- Assess adequacy of medication and other therapies
- Assess "risk" for future disease or for complications in existing disease
- Instill confidence in child and parents
- Motivate the child for further exercise or weight loss

Children are often referred to an exercise clinic to evaluate their physical fitness, which may be deficient due to a disease or to inactivity and deconditioning. For example, children with chronic renal failure, under treatment by hemodialysis, or following radiation therapy are sometimes referred to us to evaluate their ability to walk to school rather than take a taxi. We often find that the exercise level that they can comfortably sustain exceeds the energy demands of walking at a regular pace. We recommend in such a case—much to the joy of the child—that walking be permitted. Determination of fitness is also important in order to prescribe activity programs and to periodically assess progress.

Identifying a specific pathophysiologic pattern in a given disease may be of an even greater clinical relevance than the mere assessment of fitness. Various patterns can be better revealed during exercise than at rest. The rationale is based on the increase in metabolic demands during exercise that stresses, sometimes to the limit, metabolic, ventilatory, gas exchange, cardiac, vascular, neuromuscular, and thermoregulatory functions. A malfunction in a system is more likely to be discovered during stress than during rest, when functional demands are lower. Examples of such specific patterns are the appearance of ventricular dysrhythmia in complete congenital heart block or in postoperative tetralogy of Fallot, a high blood pressure in postoperative coarctation of the aorta, and post-exertional bronchial obstruction in asthma. Some findings can be used as ancillary indications for specific therapeutic interventions (e.g., need for balloon valvuloplasty with the appearance of ischemic electrocardiogram changes during exercise in a patient with aortic stenosis).

For example, a noninvasive exercise test can be used for screening to determine the necessity for an invasive test (e.g., ventricular dysrhythmias may signal a need for cardiac catheterization and electrophysiologic study). Cardiac or vascular surgery may be successful anatomically but less so functionally. This too can be revealed by stress testing (e.g., limited rise in stroke volume after Fontan operation for single ventricle).

While most children arrive at the exercise clinic with a known diagnosis, an exercise test can sometimes establish a diagnosis. This is true for growth hormone deficiency (chapter 8) and, occasionally, for asthma (through discovery of exercise-induced bronchial obstruction in a child with atypical symptoms—see chapter 6).

Another area in which exercise testing may prove beneficial is the assessment of the adequacy of drug regimens at different activity levels. While such a consideration may be obvious for insulin and for drugs that prevent exercise-induced asthma, it may also apply to other drugs such as antihypertensives, corticosteroids, or anticonvulsants. Research in this field is lacking. Another area still under investigation is the prediction, by exercise, of "risk" for future diseases such as hypertension or coronary heart disease. The usefulness of exercise for the assessment of subclinical diabetic nephropathy or neuropathy is still not clear.

Many children appearing at the exercise clinic have been inactive for prolonged periods and may have lost confidence in their ability to exercise. The successful completion of a test is, therefore, a revelation to the child, who realizes that he or she can exert strenuously. Even more impressive is the realization by some parents that their child can exercise at high intensity with no ill effect. We routinely encourage the presence of parents at the initial exercise test (and at subsequent tests if they wish). They are seated behind the child to avoid distraction, and can watch the child's reactions. We often hear an amazed parent commenting that he or she has never seen the child work so hard, develop sweat, or become red in the face. The testing situation is a learning experience in itself and can instill confidence in the patient and parent alike.

Likewise, the test and its interpretation can be used as vehicles for motivating a child to increase daily activity or, for an obese patient, to persevere in a weight-reducing program. This is especially evident when a periodic test shows improved performance.

Exercise testing, like other laboratory or clinical tests, should not be used in isolation. It must follow a thorough history-taking, including collection of data on the habitual activity of the child and other family members, attitudes at home toward physical activity, and willingness and ability of the parents to spend extra time with the child if additional activity is prescribed (e.g., driving the child to a swimming pool). Physical examination should emphasize the cardiorespiratory and musculoskeletal systems. Other laboratory tests, such as ECG, blood hemoglobin, or pulmonary functions, may often be needed. Conclusions and recommendations must integrate all these with the results of the exercise test.

Beneficial Effects of Physical Activity for the Child With a Chronic Disease

The use of physical activity for treatment is well recognized in physical therapy, occupational therapy, and adapted physical education. Training can increase muscle strength and range of joint motion; prevent contractures; and improve stamina, ambulation, and various other skills. In the context of pediatric management, conditioning and training are beneficial for additional reasons, as summarized in table 4.3.

Rarely will exercise therapy affect the pathophysiologic process itself. This may happen in obesity, where energy balance can be directly affected; in dyslipidemia, where enhanced activity may improve the lipid profile; and in type 2 diabetes mellitus, where physical activity may decrease insulin resistance. In most other diseases, the benefits of exercise are indirect and do not change the basic pathophysiologic process.

In progressive muscular dystrophy, for example, abnormal changes in the affected muscle

TABLE 4.3 Pediatric Diseases in Which Exercise Can Be Used As Therapy

Disease	Benefits
Pulmonary	
Asthma	Reduced rate and intensity of EIA*
Cystic fibrosis	Increased airway drainage, increased endurance of respiratory muscle
Cardiovascular	
Hypertension	Reduced blood pressure
Dyslipidemia	Improved lipoprotein profile
Postsurgery for a congenital defect	Hemodynamic improvement
Endocrine	
Diabetes mellitus	Better control of blood sugar levels
Nutritional	
Anorexia nervosa	Means for behavior modification
Obesity	Fat reduction, increased insulin sensitivity
Musculoskeletal	
Cerebral palsy	Ambulation, prevention of contractures, weight control
Chronic pain	Relief of pain
Idiopathic juvenile arthritis	Strengthening, increased range of motion
Muscular dystrophy	Ambulation, strengthening of residual muscle, weight control
Paralysis	Strengthening of residual muscle
Other	
Hemophilia	Mobilization, increased range of motion
Mental retardation	Increase of environmental stimuli, socialization

*EIA = exercise-induced asthma.

fibers will continue, but physical conditioning may improve the function of the residual, healthy muscle fibers. The end result is an increase in the functional level of the child and prolongation of his or her walking status (chapter 10). Another example is type 1 diabetes mellitus, in which the basic endocrine and metabolic deficiencies are not modified by conditioning, but daily diabetic control may be improved (chapter 8).

The treatment of children through enhanced physical activity is unique: By prescribing exercise we are signaling to the child that he or she can, and should, act like his or her healthy peers. We emphasize thereby abilities rather than disabilities. This is in contrast to therapy by medication, diet, or bed rest, in which the child is made to feel different from others. Furthermore, treatment by exercise is unique because it is the only therapy in which *the patient takes an active role* rather than waiting passively for others, or for the medication, to do the job. Young patients, even when they know that their disease is incurable, like this notion; and it serves as a strong motivator.

Another important characteristic of physical activity is that the more it is done the easier it becomes and the greater the sense of accomplishment. The first few sessions are the most difficult in any conditioning program: The child is unfit, lacking in skill, confidence, and motivation. Each new activity may be stressful, causing aches, pains, and frustration. This is the time when professional and parental support is most needed.

The Exercise Prescription

As in other forms of therapy, a conditioning program should be quantified regarding the *intensity, frequency,* and *duration of each session;* the *type* of activities; and the overall *duration of the program* (for details see the section "Principles of Physical Training," chapter 1).

The concept of exercise prescription has proven useful for adults in programs for prevention of, or rehabilitation after, myocardial infarction. This concept can be used, with some modifications, in pediatrics. Combining the principles outlined in chapter 1 and specific information from table 8.2 and appendix IV, one can prescribe activities, to be "filled" by a physical therapist, an exercise therapist, or a physical educator. Simple activities can be supervised by a parent. The following are examples of how an exercise prescription may be prepared.

Case 1

An 11-year-old mildly obese girl (32% body fat) weighs 43 kg (75th percentile) and is 143 cm tall (50th percentile). She is free of other diseases and takes part in physical education classes but otherwise is sedentary.

Therapeutic Goal   Reduction of fatness to 25%, without interfering with growth.

Analysis of Exercise Requirements   The 7% excess of adipose tissue is equivalent, at the present body weight, to 3.0 kg. To lose this amount the girl must achieve a negative energy balance of 21,000 kcal (88 mJoule). Assuming that half of this will be achieved by a mild reduction in energy intake, the remaining 10,500 kcal must be "burned up" by additional exercise. For 250 kcal per session, 42 sessions will be required. At a rate of three sessions per week, the overall program will last 14 weeks. The only sport that this girl likes is bowling, but she has a bicycle that she uses for errands. Since the energy equivalent of bowling is low, her program can be based on cycling. (Jogging is an alternative but is less recreational.) As seen in Appendix IV, a 43-kg child consumes about 45 kcal during a 10-min ride at 15 km · hr⁻¹ on flat terrain. This will be about 50 to 55 kcal per 10 min where the terrain is mildly sloping (as in our patient's neighborhood). Thus, a 45- to 50-min ride (11.5-12.5 km) at this comfortable pace will be sufficient to expend 250 kcal. Using the principle of progression (chapter 1, section "Principles of Physical Training"), this patient should start with a distance with which she feels comfortable and gradually increase it.

The Prescription   Type of exercise—cycling on mildly sloping (2-3%) gradient. Intensity is not important.

Frequency of sessions—three per week

Distance at each session	10 km for the first 3 weeks
	11 km for the next 3 weeks
	12 km for the next 4 weeks
	13 km for the next 4 weeks

Duration of Program   Fourteen weeks.

Medication   Medication not needed.

Furnished with this outline, a family member can easily supervise such a program.

Case 2

A 10-year-old boy with asthma withdrew 1 year ago from physical education classes because of cough, breathlessness, and wheezing, triggered by running. He is otherwise asymptomatic, but uses a steroid inhaler on a regular basis. Exercise-induced asthma was documented through an exercise provocation test. In another exercise test, peak aerobic power was 70 Watt, which is 1.5 SD below the mean for his age (see figure I.1 in appendix I).

Therapeutic Goals   (1) Resumption of normal physical activity; (2) increase of 15% in peak aerobic power.

Analysis of Therapeutic Requirements   The simplest way to enable such a boy to participate in physical education classes is to prescribe medication for the prevention of exercise-induced asthma. A β₂-adrenergic agonist (e.g., salbutamol aerosol, two puffs), taken just prior to class, is a good choice. A low exercise performance in this child most likely represents deconditioning. Resumption of activities at school will help improve his fitness to a limited extent only, so an additional conditioning program is indicated. In contrast to Case 1, in which the overall energy expenditure was a primary consideration, here the *intensity* of exercise should be emphasized. To exceed the conditioning threshold (see chapter 1, section "Principles of Physical Training"), a HR of 160 beat · min⁻¹ or more should be reached. Swimming, the least asthmogenic of sports, is the most appropriate conditioning activity for our patient (who swims well).

The Prescription   Type of exercise—swimming, any style
Intensity—HR of 155 to 160 beat · min⁻¹ during 15 to 20 min of each session
Frequency—two to three times weekly

Duration of Program   Three months.

Medication   Continue with an inhaled steroid as previously prescribed. For land-based activities that induce asthma symptoms, salbutamol, two puffs, 10 to 15 min before exercise.

Note that the prescription does not include guidelines regarding the structure of each session, its duration, or the distance to be covered. A qualified instructor is expected to plan these, according to the child's swimming proficiency, the prescribed intensity, and overall duration.

The Need for Motivation

Adults can sometimes be motivated to enhance their physical activity "because it is healthy." It is seldom that children, or even adolescents, will change their sedentary lifestyle for this reason. Furthermore, even if they do increase their habitual activity during a prescribed exercise program, it is unlikely that children will adhere to it once the program is concluded. To enhance compliance and adherence, one therefore needs to introduce into the program elements that the child considers pleasurable and "fun." Ways to do this are through games and through the provision of token rewards when the child reaches a predetermined activity goal. A similar approach can be used in order to reduce a child's sedentary pursuits, such as time spent watching television. These aspects are discussed in more detail in the context of motivational approach to treating the child with obesity (chapter 9, section "Motivational and Behavioral Aspects").

Deleterious Effects of Exercise

Under certain circumstances, exertion can be deleterious to a child's health. Detrimental effects of enhanced physical activity include sport injuries, overuse syndromes, and abnormal physiologic reactions.

Although injuries result mostly from collision and contact sports such as American football, rugby, ice hockey, and soccer, they can occur in any sport, to any participant, irrespective of competence or aspiration (13). The risk of injury can be reduced by proper conditioning; matching of opponents (by size, maturation, and level of skill); the use of such safety devices as helmets, mouth guards, and knee pads; adaptation of rules; and proper

maintenance of sport facilities and equipment. An annual preparticipation examination is recommended (6), although further evidence is needed to document the extent of the benefit of such examinations in preventing sport injuries.

Overuse syndromes may result when a movement is repeated, usually at high intensity, over months or years, causing excessive mechanical stress on bone (including stress fractures), cartilage, tendon, or muscle. These syndromes have become increasingly prevalent in recent years and are seen mostly in ambitious young athletes who train often and intensively, specializing in one or two events at the most (9).

Overuse syndromes typical for children and youth are "little league elbow" in baseball pitchers (14); low back pain in oarsmen, girl gymnasts, and horseback riders (10); and shoulder pain in swimmers (5). There are certain risk factors for overuse injuries in young athletes, as summarized in table 4.4. Discussion of sport injuries and overuse syndromes is beyond the scope of this book. The interested reader can refer to a 2000 textbook compiled by the American Academy of Orthopedic Surgeons and the American Academy of Pediatrics (13).

In this book we limit our discussion to the nontraumatic deleterious effects of acute exercise, as manifested by abnormal physiologic responses. These are found mostly in children with specific diseases but can occur also in healthy individuals, as outlined in table 4.5.

Most of the adverse reactions listed in table 4.5 are related to a single bout of exercise. Some may occur quite regularly in given patients. Examples

TABLE 4.4 Risk Factors for Overuse Injury in Young Athletes
Training error or errors in technique
Muscle-tendon imbalance
Anatomical malalignment
Unsuitable footwear
Type of playing surface
Associated disease state
Gender factors
Cultural deconditioning
Growth

From O'Neill and Micheli 1988 (8).

TABLE 4.5	Nontraumatic Deleterious Effects of Exercise in Children and Adolescents
Abnormal response	**Underlying disease or condition**
Bronchoconstriction	Asthma, atopy, history of wheezy bronchitis
Chest pain	Aortic stenosis, asthma
Delayed menarche	Healthy*
Dehydration	Healthy, cystic fibrosis
Dysrhythmia	Complete heart block, postcardiac surgery, healthy
Heatstroke, exhaustion	Dehydration, non-acclimatization, obesity, cystic fibrosis, healthy
Hematuria	Healthy, glomerulonephritis, renal calculus
Hemoglobinuria	Healthy
High blood pressure	Aortic coarctation, hypertension, obesity
Hypoglycemia	Diabetes mellitus type 1
Ischemic ST-T changes	Aortic stenosis and insufficiency, coarctation, mitral valve prolapse, familial hypercholesterolemia, sickle cell anemia
Joint and limb pain	Arthritis, chronic pain syndrome
Ketoacidosis	Diabetes mellitus (insulin deprivation)
Menstrual irregularities	Healthy*
Muscle cramps	Muscular dystrophy, mitochondrial myopathy
Proteinuria	Healthy, diabetes mellitus type 1
Seizure	Epilepsy
Sickling of red blood cells	Sickle cell anemia
Stiffness spells	Hypoparathyroidism
Sudden death	Aortic stenosis, congenital coronary anomaly, myopathy
Syncope	Aortic stenosis, complete heart block, Marfan syndrome
Ventricular tachycardia	Complete heart block

*Resulting from chronic exercise. All other responses are to acute exercise.

Adverse physiologic responses to exercise are usually preventable, or their impact can be minimized by proper precautions. Exercise-induced hypoglycemia, for example, can be averted by ensuring sufficient carbohydrate intake prior to and during activity, by reducing the insulin dosage, and by selecting an appropriate site for insulin injection (chapter 8). Heatstroke or heat exhaustion can be prevented by ensuring prior acclimatization to heat, fluid replenishment, and curtailment of activities when certain climatic conditions prevail (chapter 3). Exercise-induced asthma can be prevented, or ameliorated, by medication, by properly warming up, by changing from running to swimming, and by reducing activities on dry or very cold days (chapter 6).

In some diseases, parents or patients *assume* that exercise entails a risk that in point of fact does not exist or is minimal and is outweighed by the benefits of exercise. Such, for example, is the case with epilepsy, in which, despite popular belief, convulsions are seldom triggered by exercise (chapter 10). Parents of a boy with hemophilia sometimes keep him inactive to avoid bleeds. Current evidence suggests that such a risk is minimal. Children with hemophilia can follow normal activities providing they take the appropriate replacement therapy (chapter 11). Similar overprotection is shown by parents of children with heart disease, real or assumed, for fear of a "heart attack."

By understanding the possible deleterious effects of exercise, the clinician can prevent them or minimize their impact. A physician's role in alleviating fears of the presumed dangers of exercise is no less important.

are ischemic ST depression in aortic stenosis, high arterial blood pressure in coarctation of the aorta, hypoglycemia in type 1 diabetes mellitus, and exertional hematuria. Others appear when several unfavorable conditions interact. For example, heatstroke can occur in a healthy, but unacclimatized, child who exerts for a prolonged period of time without sufficient fluid intake. Ketoacidosis may develop in an insulin-dependent child with type 1 diabetes who exercises without having taken insulin. Chronic exercise, in addition to causing overuse syndromes, can be associated with a delay in menarche.

Physical Activity and Preventive Health Care in Children and Adolescents

*T*he idea that regular exercise and physical fitness are beneficial to health dates far back in human history. "Anyone who lives a sedentary life and does not exercise," wrote Maimonides in 1199, "even if he eats good foods and takes care of himself according to proper medical principles—all his days will be painful ones and his strength shall wane" (98). (The message is obvious, lest one be tempted to skip a day's workout.)

A Darwinian argument would hold that the association between exercise and well-being has its roots in prehistoric times, when physical fitness truly carried a survival value. The individual who was stronger, or who could run the longest or fastest, held an obvious advantage in procuring food and escaping or overpowering enemies. Genetically endowed physical fitness provided the best chances for survival, and the same could be said for the ability to improve fitness with exercise. The physiologic fitness effect—improvements in aerobic power or strength that occur with training—also raised the odds of obtaining food and surviving natural disasters, famine, and war.

It is not without some sense of irony, then, that health benefits can be observed from regular exercise in modern societies. Because now, instead of permitting one to survive "too little" in the way of resources, exercise offers salutary effects on illnesses that result from "too much"—an overabundance of food and technological progress that limits physical effort, intellectual stimula-

tion, and social interaction. Research data now convincingly confirm what the ancients knew right along—exercise is good for you. In the 21st century, regular physical activity and fitness in adults confer decreased risk for coronary artery disease, hypertension, obesity, and type 2 diabetes. There is evidence, too, that those who exercise regularly can have better emotional health and reduced chances of osteoporosis and certain forms of cancer. Physical fitness may also play an important role in reducing the infirmities of old age.

Since these diseases represent the major causes of morbidity and mortality in modern societies, physical activity and fitness may have a profound impact on the health of the population. The potential is magnified by the fact that few individuals exercise sufficiently in their daily lives to take advantage of these benefits. These observations have provided an impetus for public health measures to counter the sedentary lifestyle that characterizes most of the adult population (119).

The research information identifying positive physical health outcomes from exercise has been limited to adults. Yet increasing attention is being focused on improving physical fitness and activity in youth from a health standpoint. In this chapter the rationale for promoting exercise in children for preventive health is examined, as well as issues surrounding the promotion of fitness and habitual activity in the pediatric age

117

group. The aim of the discussion is to provide a summary of these issues. The reader is referred to several recent reviews that have comprehensively addressed this subject (6; 7; 24; 91; 100).

The Exercise-Health Link in Adults

In adults, levels of both physical activity and aerobic fitness (such as treadmill time) have been related to a reduction in all-cause mortality. When all studies are considered, sedentary individuals have a 1.2 to 2.0 times greater chance of dying during follow-up compared to those who are active (119). Increasing one's aerobic fitness also lowers the risk. Blair et al. showed that very sedentary men who improved their fitness with training had a 44% lower rate of death on 5-year follow-up than those who remained sedentary (28).

Previously, improvements in aerobic fitness were considered necessary for health benefits. This prompted recommendations for regular (three times a week) sessions of dynamic exercise (running, cycling) at relatively high exercise intensities (heart rate over 70-85% maximum). More recently, amount rather than intensity of exercise has been recognized as key to health outcomes, resulting in a shift of emphasis toward establishing habits of regular moderate activity. Currently, most recommendations call for a minimum of 30 min of moderate-intensity physical activity on most days of the week (119). In addition, endurance activities such as brisk walking should be supplemented with strength exercise.

This emphasis on regular, comfortable exercise rather than intensive training is expected to make improving activity habits more palatable to sedentary individuals. The target population is not small; only about one adult in five currently satisfies these guidelines, and 25% report no leisure time physical activity at all (119).

In examining the health benefits of regular exercise in adults, one is struck by the consistency of the supportive research data. Findings in the diseases described in this chapter have indicated a beneficial relationship with exercise in the great majority of studies, no association in a few, and a detrimental relationship only in rare circumstances. The wide range of disease processes in

which exercise can have a beneficial influence is also apparent. There are considerable differences in the pathogenesis involved, including metabolic disease influenced by vascular permeability (atherosclerosis), excess body fat (obesity), insulin resistance (type 2 diabetes mellitus), oncogenesis (colon cancer), and deposition of bone mass (osteoporosis). The multiple mechanisms by which repeated bouts of muscular activity can positively influence these diverse processes are poorly understood.

Importantly, these studies in adults have generally indicated a similar effect of physical fitness and level of habitual physical activity on health outcomes. While this might seem intuitively obvious (since regular activity can improve fitness), the two—fitness and physical activity—are in fact very different. Physical fitness describes a personal attribute, defined functionally or physiologically in terms of one's ability to perform on a motor task. A high level of fitness, which is largely inherited but can be achieved to some extent through exercise training, is characterized by a set of hemodynamic, metabolic, and anatomic features that could influence disease processes. Physical activity, on the other hand, signifies a behavior, the amount of energy expenditure in daily life, which may alter pathogenesis by different mechanisms. Understanding differences in the effects of activity and fitness on health, of course, becomes critical in formulating interventional strategies.

Coronary Artery Disease

Myocardial ischemia from atherosclerotic occlusion of the coronary arteries is the single leading cause of mortality in the United States, responsible for approximately 480,000 deaths per year (51). Some risk factors for coronary artery disease are fixed (age, gender, heredity, race), while others are modifiable (cigarette smoking, hypertension, obesity, abnormal serum lipids, sedentary lifestyle). Promoting regular exercise is thus part of an overall strategy to minimize these factors and diminish cardiac risk.

An extensive volume of research indicates that both physical activity and fitness can substantially reduce risk of fatal and nonfatal outcomes from coronary artery disease (25; 84; 119). In 36 studies assessing the role of physical activity, outcomes have typically been examined after a single baseline estimate of activity. In all but four of these reports, an inverse relationship between activity and coronary artery disease morbidity and mortality was observed. In a meta-analysis of those studies felt to be of good quality, the odds ratio, or the risk of an adverse outcome, was 1.8 times greater in sedentary compared to highly active individuals (23). Seven studies examining the relationship of physical fitness and complications of coronary artery disease have demonstrated the same beneficial effect (119).

Several different mechanisms have been postulated to account for the salutary effects of exercise on coronary artery disease. A direct effect on the atherosclerotic process is suggested by animal studies in which the development of vascular lesions has been inhibited by exercise training (67). Regular exercise may enhance myocardial perfusion by increasing coronary artery diameter, improving vascularization, altering vessel reactivity, or diminishing the risk of spasm (93). The risk of fatal arrhythmia may be reduced if fitness increases the threshold for ventricular fibrillation during ischemic events (119). Exercise training can also improve thrombolysis and diminish platelet adhesion, reducing the risk of coronary thrombosis, often the precipitating event for myocardial infarction (127).

Exercise could serve to diminish complications of coronary disease by reducing the influence of other risk factors, particularly serum lipids, obesity, and systemic hypertension (121). At least 60 studies have examined the influence of fitness and activity on serum lipids (119). Most have revealed no relationship with total or low-density lipoprotein cholesterol (LDL-C); but in approximately half, aerobic training has increased levels of high-density lipoprotein cholesterol (HDL-C), a protective factor for atherosclerosis. Studies involving moderate-intensity exercise have suggested that a favorable HDL-C response is less likely in women. In cross-sectional studies, both male and female athletes demonstrate 20% to 30% higher levels of HDL-C than nonathletes.

Importantly, multiple factors can influence lipid levels, including body fat content, cigarette use, diet, menstrual cycle, and alcohol consumption (114). These variables need to be considered in any assessment of the role of exercise in modifying serum lipids.

Systemic Hypertension

Individuals with systemic hypertension are at jeopardy for stroke, congestive heart failure, coronary artery disease, renal failure, retinopathy, aortic aneurysms, and peripheral vascular disease. The impact of hypertension on the health of the population is underscored by the fact that as many as a quarter of Americans have elevated blood pressures.

Regular physical activity and fitness in adults can both reduce the risk of developing hypertension and diminish blood pressures in individuals who are already hypertensive. Follow-up studies indicate that being active can decrease the chance of developing elevated blood pressure by approximately 30% (119). In the report by Blair et al., individuals with low fitness had a 52% greater risk of later development of hypertension compared to fit subjects (27).

A period of exercise training can reduce blood pressure levels in both hypertensive and normotensive individuals. Meta-analyses indicate that a reduction in systolic and diastolic pressure of 6 to 10 mmHg can be expected in subjects with hypertension, while more modest reductions of 3 to 4 mmHg are typical in normotensive subjects (8; 119).

The mechanisms by which regular exercise decreases resting blood pressure have not been well defined. The action of fitness in diminishing sympathetic nervous activity, with enhanced peripheral vasodilation, seems likely to be important. This effect may occur in conjunction with reduced renin-angiotensin system activity and change in baroreceptor receptivity.

Obesity

A progressive rise in body fat content is observed with increasing age, presumably reflecting a fall in energy expenditure (from reduced physical activity) that is disproportionately greater than changes in food intake. This imbalance appears to be becoming more pronounced over time, as the prevalence of obesity at all ages is rising.

Of all the influences of exercise on health, a beneficial role of physical activity in the prevention and management of obesity seems most obvious. Nonresting energy expenditure, mainly in the form of physical activity, accounts for about a third of total energy expenditure (69). Improvements in the "out" side of the energy balance equation should therefore be expected from increasing levels of daily physical activity.

Scientific documentation of a significant effect of exercise on obesity has been less than dramatic, however. Such studies are confounded by the fact that the outcome measure—body fat—by itself can inhibit physical activity. Cause-and-effect conclusions (which came first?) are therefore problematic in cross-sectional studies, many of which demonstrate lower indices of body fat in those who are more active or fit (119). Also, although the laws of thermodynamics must be obeyed, introduction of an exercise program for obese subjects could be negated if the prescribed activity results in either increased food intake or a compensatory reduction in physical activity outside of the program.

A considerable amount of physical activity is necessary to reduce body fat content. It takes about 3,500 calories of energy expenditure (equivalent to chopping wood for 10 hr or running 48 km [30 mi] to "burn off" a pound of fat. Still, such effects on body fat from such energy expenditure are accumulative, and improving physical activity habits can also increase lean body mass and potentiate dietary thermogenesis. Thus, while increasing activity alone may not be a realistic means of reducing body fat, exercise can be a valuable adjunct to dietary interventions. Obesity typically develops from very small daily disturbances in energy balance over many years. Adding regular, tolerable exercise to modest dietary restrictions may be an effective way of reversing this energy imbalance (34).

Reviews and meta-analyses of body composition changes reported with exercise training in adult obese subjects support this conclusion. These studies have indicated that small but favorable fat losses can be achieved while gains in lean body mass are promoted (13; 43). The impact of these exercise programs has been directly related to the frequency, duration, and intensity of the activity intervention.

Osteoporosis

Bone mineralization develops during the growing years, peaks in early adulthood, then declines with age. Osteoporosis, the progressive rarifaction of bone mineral density in those who are elderly, predisposes to fractures and results in significant disability and mortality, particularly in females. Potential strategies for reducing this risk include either enhancing bone density in the early years of life or lessening the decline of bone mineralization in later adulthood.

Bone mineral density is responsive to the mechanical stresses created by weight-bearing physical activity. Bone mass decreases during sustained periods of immobility and bed rest. Athletes have a greater bone mineral density than nonathletes, and cross-sectional studies in adults indicate that bone mineral density correlates with physical activity, aerobic fitness, and muscle strength. Some longitudinal studies in postmenopausal females indicate that improvements in physical activity can increase bone mass. The presence of circulating estrogen as well as dietary calcium intake appears to be important in potentiating this effect (119).

The relationship of activity and fitness to bone mineral density has been more apparent in cross-sectional than interventional studies. Active individuals typically demonstrate a 10% to 15% greater bone mineral density than inactive subjects, while exercise interventions usually show smaller gains (1-5%) in bone mass (119). The degree to which exercise can prevent the decline in bone mineral density with age is uncertain. Drinkwater concluded that "for most older women, remaining physically active does not increase bone mineral density but does prevent the bone loss that occurs inevitably if they become inactive" (p. 238) (41).

Type 2 (Non-Insulin Dependent) Diabetes Mellitus

In contrast to individuals with type 1 (juvenile) diabetes, who lack circulating insulin, those with type 2 diabetes mellitus experience hyperglycemia mainly from insulin resistance. While factors such as heredity and age are influential, a particularly close association exists between type 2 diabetes and excessive body fat. At least 170,000 persons die each year in the United States from complications of diabetes, which include coronary artery disease, stroke, and peripheral vascular disease.

Many reports have indicated a close association between physical inactivity and type 2 diabetes. In addition, prospective studies demonstrate that physical activity can prevent the development of this disease (119). Several mechanisms may contribute to this effect. (1) Physical activity provides a synergistic action with insulin in enhancing glucose uptake into muscle cells. As this effect lasts as long as 24 hr after exercise, the ability of activity to reduce hyperglycemia in these patients could be due to an overlap of acute effects rather than a response to physical training. (2) Physical activity increases sensitivity to insulin. Endurance athletes tend to have a greater insulin sensitivity than nonathletes, and insulin sensitivity is related to level of endurance fitness. (3) Physical activity may contribute to loss of body fat, an important determinant of type 2 diabetes.

The Pediatric Rationale

There are many good reasons to promote exercise in children and adolescents. Foremost, perhaps, participation in physical activities and sport adds to the enjoyment of life. Indeed, the pleasure of engaging in motor activity is probably the major reason why individuals of all ages take part in exercise. Viewed more clinically, regular exercise enhances self-esteem and confidence. Activities in group settings can have a positive effect on social development as well, promoting qualities of cooperation, discipline, and commitment. The influence of exercise on psychosocial health in youth is addressed in chapter 12. This chapter focuses solely on the means by which physical activity and fitness in children and adolescents have a bearing on physical health. Although this issue has attracted a good deal of both medical and lay attention, it is important to keep in perspective the fact that physical health represents only one of many possible reasons for encouraging exercise in children.

As Blair et al. have pointed out, physical activity, fitness, or physical activity plus fitness cannot be expected to pay direct health dividends during the growing years (26). The major causes of mortality in children are accidents (about one-half of cases), cancer, and congenital malformations. Most school days are missed because of infectious and allergic disease. In the teen years, homicide and suicide become increasingly common. Regular exercise cannot be expected to have a favorable influence on any of these conditions.

The documented exercise-health link has been established entirely in adults, and the health outcomes involve adult diseases. Children do not suffer myocardial infarction from coronary artery disease, stroke from hypertension, or femoral fractures from osteoporosis. Even obesity, increasingly common in youth, does not usually lead to medical complications during the pediatric years. By what rationale, then, can encouragement of exercise in children be supported for the promotion of physical health?

The answer lies in a cogent argument that establishing regular exercise habits in children is a prime strategy for the prevention of adult chronic disease. This concept is based on the following line of reasoning:

1. By the various mechanisms outlined earlier, physical fitness and activity are capable of significantly reducing the risk of chronic illness, particularly coronary artery disease, hypertension, obesity, osteoporosis, and type 2 diabetes.

2. Although these are diseases of adults, almost all reflect lifelong pathologic processes that begin during the pediatric years (81). Atherosclerosis, for example, starts as fatty lesions in the walls of the aorta and coronary arteries that are well established by the late teen years. Increased incidence of osteoporosis in persons who are elderly can reflect failure to maximize the development of bone density in the growing years. Eating and exercise habits in overweight youth are frequently the predecessors of adult obesity. Essential hypertension often has its onset in the late childhood and adolescent years.

3. Logically, then, encouraging exercise habits early in life as a way to ameliorate the development of these diseases should provide an effective means of preventing their clinical expression (myocardial infarction, stroke, peripheral vascular disease, bone fractures) later in the adult years (figure 5.1). Through this approach, physical activity, fitness, or both in children may pay long-term dividends in preventing complications of chronic illness in adulthood.

This rationale is not foreign to pediatric health care providers. The strategy of promoting exercise habits in children to prevent adult disease joins that of altering other behaviors in children such as diet, salt intake, and cigarette smoking, all designed to protect youth from future health problems (110). Since they are, in fact, behaviors, they are susceptible to change. On the other hand, altering human behaviors such as diet and physical activity offers serious challenges.

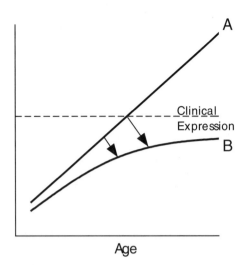

FIGURE 5.1   The pediatric rationale for promotion of exercise in children to prevent adult chronic disease. Development of risk factor (A) is reduced (B) by early and persistent exercise, preventing clinical expression (myocardial infarction, stroke, congestive heart failure).

TABLE 5.1	Mechanisms by Which Exercise in Children Might Prevent Adult Chronic Disease	
SCIENTIFIC EVIDENCE		
Decrease coronary risk factors General population		No
People with abnormal values		Yes
Prevent development of risk factors		Unknown
Primary effect on atherosclerosis		Unknown?
Improve bone mineral density		Yes
Initiate lifelong exercise behaviors		Unknown?

Is the pediatric rationale valid? The direct proof would be a straightforward one—compare long-term disease outcomes in a group of subjects who had high levels of fitness and activity as children with those of a group who were sedentary as youth. Unfortunately, no such study has been performed; and in view of the requirements of time duration, adjustment for confounding variables, consideration of change in exercise habits over time, and stability of subject cohort, it probably never will be. Consequently, direct proof that promotion of exercise in children and adolescents will lower risk of myocardial infarction, stroke, and bone fractures 50 years later is unlikely ever to be in hand.

Alternatively, however, indirect evidence can be examined (table 5.1). The pediatric rationale would be supported if recognized risk factors for adult chronic disease were diminished with exercise interventions during childhood. This is entirely feasible, given, for instance, the high frequency of risk factors for coronary disease in the pediatric population (obesity, hyperlipidemia) (7; 21). (The frequency of such factors in children and adolescents depends on the cutoff values that are chosen to define "abnormal" [7; 85]. Armstrong et al. reported that individual risk factors were evident in from 5% to 47% of British young persons ages 11 to 16 years [7].) In taking this approach, of course, one assumes (without specific proof) that reduction of risk factors is a principal mechanism by which exercise diminishes risk of chronic illness such as coronary artery disease in adults.

If exercise has a more direct effect on chronic disease processes such as atherosclerosis, indirect evidence will be more difficult to find in children. Little information is available in this age group regarding the influence of exercise on factors such as thrombolysis, coronary perfusion, and the development of atheromatous lesions.

Exercise in Children and Risk Factors for Adult Chronic Disease

The question of the relationship between disease risk factors and physical activity and fitness in young persons has drawn a large amount of research attention. Most of the data are cross-sectional, but insights are provided by interventional and long-term longitudinal studies as well. The available information has been presented in detail in two recent consensus statements (87; 100).

Serum Lipids

Current research has not provided a convincing argument that either physical activity or fitness influences lipid levels in the general pediatric population. Almost all these data are in healthy subjects with normal lipid values. Beneficial effects on lipid levels with exercise interventions have been observed in obese subjects, while the effect of exercise in patients with familial or acquired hyperlipidemias has not yet been evaluated.

Cross-Sectional Studies

Consistent with findings in adults, several early cross-sectional studies described significantly higher levels of HDL-C in athletic or highly active children (see [4] for review). Smith et al., for instance, found 32% greater HDL-C values in a group of 9- to 15-year-old trained runners compared to active control subjects (104). Subsequent studies assessing the relationship of $\dot{V}O_2$max or physical activity with serum lipids have been less impressive. Approximately equal numbers of studies have revealed positive, negative, or no relationships between aerobic fitness and HDL-C. The majority of reports have indicated no significant relationship between measures of habitual physical activity and lipid levels (see [4] for review).

Failure to consider variables such as body composition or diet may explain the findings in the highly fit subjects (although a true effect of extended athletic training on lipids cannot be dismissed). For instance, Hager et al. reported a significant inverse relationship between 1-mi (1.6-km) run time and serum HDL-C in 262 children, but when % body fat and gender were considered in the analysis, this association disappeared (57). Other authors have described relationships between fitness or activity and serum lipids that were eliminated when body fat content was taken into account (2; 22; 52; 76).

Training Studies

Eleven controlled studies performed in healthy children and adolescents to assess blood lipid responses to a period of endurance training have produced mixed results (table 5.2). None were long in duration, the longest being 20 weeks. Among these reports, no changes in HDL-C were seen in six, increases were seen in four, and a decrease was seen in one. However, Tolfrey et al. noted that many of these studies suffer from flaws in study design, particularly small sample size and questionable training duration and volume (113). They concluded that it may be premature to dismiss a possible favorable effect of long-term exercise on serum lipoprotein profile.

This argument was supported by an examination of the relationships of aerobic fitness (peak $\dot{V}O_2$), body composition, and habitual activity with serum lipoprotein levels in 71 prepubertal boys and girls (112). Bivariate correlational analysis indicated that aerobic fitness, % body fat, and habitual activity were all significantly related to lipid profile in the girls. Peak $\dot{V}O_2$ accounted for 25% of the variance of HDL-C levels. In the boys, habitual activity correlated with the ratio of total to HDL-C.

Two reports describe changes in serum lipids in response to resistance training. Weltman et al. described a 16% decrease in total cholesterol but no significant change in HDL-C after 14 weeks of

TABLE 5.2	Controlled Studies of Serum Lipid Responses to Endurance Training in Healthy Children					
					CHANGES IN	
Study	N	Age (years)	Duration (weeks)	TC	HDL-C	LDL-C
Ben-Ezra and Gallagher (20)	11	MF 8-10	10	No	No	No
Blessing et al. (29)	25	MF 13-18	16	Decrease	Increase	Decrease
Linder et al. (70)	29	M 11-17	8	No	No	No
Linder et al. (71)	103	M 7-15	4	No	No	No
Rimmer and Looney (90)	14	MF 14-17	15	Decrease	No	
Rowland et al. (97)	31	MF 10-12	13	No	No	No
Savage et al. (102)	12	M 8-9	10	No	Decrease	No
Stergioulas et al. (108)	18	M 10-12	9	Increase		
Stoedefalke et al. (109)	23	F 13-14	20	No	No	No
Tolfrey et al. (113)	28	MF 9-11	12	No	Increase	Decrease
Williford et al. (128)	12	M 11-13	15	No	Increase	Decrease

TC = total cholesterol; HDL-C = high-density lipoprotein cholesterol; LDL-C = low-density lipoprotein cholesterol.

training in 32 boys 6 to 11 years old (125). Fripp and Hodgson reported increases in HDL-C in response to a 9-week resistance program in adolescent boys (49).

Non-Interventional Longitudinal Studies

Three reports have described the relationship between changes in activity, fitness, or both and serum lipids over time. Raitakari et al. reported tracking of physical activity (by questionnaire) and coronary risk factors in a 6-year longitudinal study of Finnish adolescents (86). Activity was independently and inversely related only to serum triglycerides in the boys, but no relationship with serum lipids was seen in the females.

Janz et al. measured $\dot{V}O_2$max, physical activity, and serum lipids in 125 children for 5 years (beginning at age 10) as part of the Muscatine Study (63). With multiple regression analysis, neither activity nor aerobic fitness was a predictor for lipoprotein levels. However, in the Amsterdam Growth and Health Study, changes in physical activity significantly correlated with those in HDL-C over a period of 15 years starting in early adolescence (118).

Blood Pressure

Essential hypertension is not typically observed in young children. However, blood pressure measurements in the pediatric years offer some predictability of future adult hypertension. In the Muscatine Study, Pearson correlation coefficients between levels at ages 7 to 18 years and those at age 23 or 28 years were $r = 0.21$ to 0.39 for systolic pressures and $r = 0.11$ to 0.50 for diastolic pressures (68). Of the children who had a blood pressure over the 90th percentile at any time, 24% were above the 90th percentile for pressures as adults, 2.4 times the expected number. Given the reductions of blood pressure observed with activity and fitness in adults, it is pertinent to assess whether similar effects can be observed in youth.

Alpert and Wilmore reviewed 14 studies that examined cross-sectional relationships between either physical activity or physical fitness and resting blood pressures in healthy normotensive children (3). The findings were mixed, with many reports describing an inverse relationship between activity or fitness and blood pressure. However, as Alpert and Wilmore pointed out, "in those studies where statistical adjustments were made for body mass index (BMI), skinfold thickness, or other measures of size or fatness, the strength of these relationships generally was substantially reduced, often to levels that did not achieve statistical significance" (p. 363) (3).

Some information from short-term (3-5 years) longitudinal studies suggests that blood pressure and aerobic fitness in youth are linked over time (54; 58). However, Twisk et al. reported that the 14-year longitudinal development of physical activity (beginning at age 13 years) was not associated with changes in systolic and diastolic blood pressure in the Amsterdam Growth and Health Study (118). In addition, no relationship was observed between habitual activity in early adolescence and adult blood pressure levels.

Endurance training appears to have no influence on blood pressure in normotensive nonobese children and adolescents. Of seven such interventional studies, five revealed no effect, and in the other two, findings were minimal (3). However, Ewart et al. demonstrated significant blood pressure reductions after a semester of aerobic exercise class in a group of predominantly African American ninth graders who had blood pressures in the top third of the normal distribution (44). In this group of subjects considered to be at increased risk for adult hypertension, mean blood pressures with exercise training fell from 120/58 to 114/57. No significant changes in BMI occurred during the training period (mean 24.8 kg $\cdot$ m^{-2} before and 25.1 kg $\cdot$ m^{-2} after).

Exercise training in adolescents with mild essential hypertension has been demonstrated to be effective in lowering blood pressure. These studies are discussed in chapter 7. The role of physical activity or fitness in the management of children and adolescents with secondary hypertension has not yet been investigated.

In summary, physical fitness, activity, or both do not appear to influence blood pressure levels in normotensive youth. Associations observed between exercise and blood pressure in these children have generally been mediated by body adiposity. Limited information does indicate, however, that a period of physical training can reduce blood pressure in hypertensive adolescents, an effect similar to that observed in adults with hypertension.

Obesity

From the foregoing discussion, it is clear why obesity stands out as a particularly worrisome health risk factor in children (figure 5.2) (see chapter 9). First, obesity is closely linked to other cardiovascular risk factors—the overweight child has a greater chance of developing type 2 diabetes and is more likely to be sedentary, have elevated blood pressures, and demonstrate an unfavorable serum lipid profile compared to non-obese children. In a group of 74 children ages 7 to 13 years, Gutin et al. demonstrated a significant correlation of body fat with systolic blood pressure (r = 0.32), serum triglycerides (r = 0.14), and HDL-C (r = −0.50) (56). McMurray et al. found that among 546 obese children (BMI > 90th percentile), mean systolic and diastolic blood pressure was greater than in non-obese subjects (108/70 vs. 104/68), as was total cholesterol (4.47 vs. 4.11 mmol · L^{-1}) (77). Becque et al. reported that 97% of obese young adolescents had four or more cardiovascular risk factors (19).

Obesity has a major impact on these risk factors, as apparent relationships between physical activity or fitness and serum lipids or blood pressure often disappear when the influence of body fat is considered. The interrelations between body fatness (by skinfolds), aerobic fitness (shuttle run), blood pressure, and serum lipids were examined in 1,015 children ages 12 to 15 years in the Northern Ireland Young Hearts Project (32). Correlations between body fat and risk factors were greater than between fitness and the same factors. In multiple regression analysis, no relationship between fitness and risk factors was evident after adjustment for body fat.

Second, obese children are more likely to become obese adults, with adverse health outcomes. In a study of 11-year-old obese children, 40% were still overweight at age 26 years (106).

Four of five adolescents who are obese will maintain their obesity into the middle adult years (1). In the Harvard Growth Study, obesity in adolescence predicted a higher risk of several adverse health outcomes (particularly coronary heart disease morbidity and mortality) after 55 years of follow-up.

Third, the frequency of obesity is increasing. Dollman et al. compared BMI values in 1,463 Australian children ages 10 to 11 years in 1997 with findings in a similar study performed in 1985 (40). Values were greater in the second study, with most pronounced differences in subjects who were above the 70th percentile in the first study (figure 5.3). That is, greatest increases in body fat occurred in groups that were fatter to start with. Other reports have demonstrated similar findings (92; 115).

Issues surrounding exercise in overweight young persons are addressed in chapter 9. The following represents a summary of studies assessing the effect of physical activity on obesity in the pediatric age group.

Cross-sectional studies of the relationship between regular activity and body fat in young obese compared to non-obese subjects have revealed no consistent findings. In reviewing 13 such studies, Ward and Evans found that obese youth were significantly less active in some studies, but an equal number of reports showed no

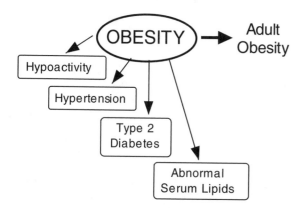

FIGURE 5.2 Cardiovascular risk factors associated with obesity in children and adolescents.

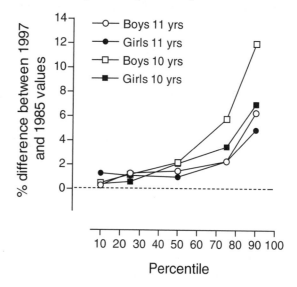

FIGURE 5.3 Comparison of mean body mass index values (kg · m^{-2}) between 1985 and 1997 surveys in 10- to 11-year-old Australian boys and girls for different percentiles (40). Greater differences are seen at higher levels of body fat content.

differences from non-obese subjects (123). Problems of cause and effect (does a sedentary lifestyle predispose to increases in body fat, or does obesity cause inactivity?), as well as confusion regarding the concept of "activity" (is it energy expenditure or body movement?), have made interpretation of these studies difficult.

In the longitudinal Amsterdam Growth and Health Study, daily activity in the period between ages 13 and 27 years was inversely related to changes in skinfold measurements (117). However, short-term (i.e., 12 week) aerobic training programs in non-obese youth generally show little or no changes in body fat.

In their review of the research literature as of 1994, Bar-Or and Baranowski concluded that numerous studies had verified that aerobic training can induce a reduction in adiposity in obese children and adolescents, but the effect is usually small unless the training is combined with dietary restriction and behavioral modification (16). These studies also suggest that intervention duration (i.e., more than 1 year) may be important, and that increasing lifestyle activities may create more persistent effects than regimented, structured exercise programs.

Ward and Bar-Or reviewed 13 studies involving aerobic exercise training in obese youth in school-based programs (122). A reduction in percent overweight after 2 to 18 months of training ranged from about 5% to 10%, but the persistence of weight loss was not assessed.

Several studies have also indicated that associated cardiovascular risk factors can be reduced in obese youth with exercise interventions. Becque et al. showed a significant rise in HDL-C (from 35.4 to 43.4 mg · dl⁻¹) and a reduction in blood pressure in obese adolescents with diet plus exercise (19). No such changes were seen in subjects with dietary intervention alone. In obese 11-year-olds, Sasaki et al. reported a 16% to 19% increase in HDL-C after 2 years of a school-based running program for obese children (101). Widhalm et al. lowered LDL-C (with no changes in HDL-C) in 124 obese children with exercise and caloric restriction at an obesity camp (126). In this study it was not possible to separate the effects of diet and exercise.

Taken as a whole, these data indicate that exercise can be an effective component of interventional strategies for children and adolescents with obesity. The most appropriate types, format,

and duration of exercise remain to be clarified. Equally important, means of sustaining fat loss from such programs over the long term need to be identified.

Osteoporosis

The role of exercise in promoting development of bone mass during childhood and adolescence as a means of preventing future osteoporosis is conceptually obvious. Bone mineral density steadily increases during the early years of life, reaching a peak at approximately 20 to 30 years. Thereafter bone density falls throughout the rest of the life span, a decline that becomes particularly exaggerated in women after menopause. Osteoporosis, the increasing rarefaction of bone in persons who are elderly, carries with it significant risk of disability and death from bone fractures.

Peak bone density in the early adult years is one of the best predictors of bone health in people who are elderly (10). It has been estimated that as much as half of the variability in bone mass at older ages can be accounted for by the extent of early bone mineralization (60). Optimizing the accrual of bone mineralization in the growing years should maximize peak bone mineral density, permitting a greater level of bone density in later life. No study has yet been performed to verify this concept. Nonetheless, there is substantial evidence to suggest, in fact, that exercise to promote bone mineralization in the pediatric years can be expected to diminish the risk of osteoporosis in later adulthood.

Bailey and Martin reviewed cross-sectional studies in the research literature as of 1994 (12). Studies in young athletes indicate that sports such as gymnastics, hockey, soccer, and volleyball can stimulate bone mineralization, but some information suggests that such an effect is less likely with swimming (as a non-weight-bearing sport) (39). In nonathletes, studies have generally indicated modest correlations between habitual physical activity and bone mineral density.

Such relationships have also been examined longitudinally. In the Amsterdam Growth and Health Study, amount of weight-bearing activity during adolescence was a significant predictor of bone mineral density in the same individuals at age 27 (124). Gunnes and Lehmann found that weight-bearing physical activity had a significant effect on change in bone mineral density in a longitudinal study of 470 healthy boys and girls (53).

Several groups have investigated the effect of a period of physical training on bone density. Morris et al. found increases of 7% to 12% in bone mineral content in 38 girls 9 to 10 years old after a 10-month exercise program conducted as a supplement to regular physical education class (80). Improvements in bone mineral content have been reported after 14 weeks of rigorous basic military training (75) and in prepubertal boys in response to an 8-month exercise program (33). Bass et al. found that 12 months of gymnastics training in 10-year-old females resulted in a 30% to 85% more rapid increase in bone density than in nonathletes (18).

However, neither Snow-Harter et al. (105) nor Blimkie et al. (30) could elicit significant changes in bone density with training of young females (8 and 6 months' duration, respectively). According to Bailey and Martin, these findings indicated that in order to enhance bone mineral density, exercise must be vigorous (12). This conclusion is consistent with studies in young animals that have demonstrated positive bone growth with exercise (47).

Preferred-limb studies have proven interesting, since factors other than exercise (genetic, metabolic) should affect bone mineralization equally in the two limbs. Consequently, limb differences in bone mineral density can be expected to reflect solely the influence of mechanical stresses from exercise (10). Greater bone density in the healthy leg by 4% to 15% has been reported in 7- to 14-year-old children who have unilateral Legg-Calvé-Perthes disease (11). The bones of the dominant arms of young baseball and tennis players as well as nonathletes have greater density than those of the nondominant arm (12).

The mechanical stress created by weight-bearing exercise promotes bone growth by stimulating osteoblastic activity. Studies indicate that short-burst, explosive activities such as skipping, stair climbing, and jumping are particularly effective in increasing bone density (65). These forms of exercise, then, are more useful in improving bone health than low-intensity sustained activities such as running that improve cardiovascular fitness (figure 5.4).

Dietary calcium and circulating estrogen potentiate the effect of exercise on promoting bone growth. Amenorrhea associated with

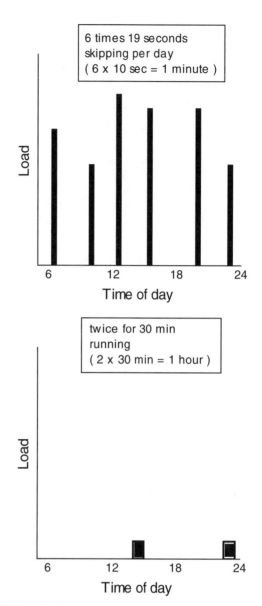

FIGURE 5.4 Comparisons of mechanical stress on bones created by short-burst exercise (top) and weight-bearing endurance exercise (bottom) (65).

high-intensity athletic training diminishes estrogen levels and can interfere with normal bone mineralization. This effect of hypoestrogenemia can increase the risk for stress fractures and, presumably in the long term, osteoporosis. Baer et al. showed that the mechanical loading on bones from running does not compensate for the negative effect of decreased estrogen levels on bone mineralization in amenorrheic adolescent runners (9). However, Moen et al. could find no difference in mean lumbar bone density in amenorrheic and eumenorrheic distance runners ages 15 to 18 years (79).

In summary, the available data support the concept that weight-bearing exercise, along with adequate calcium intake, is a valuable means for children and adolescents to maintain bone health and protect against future osteoporosis. The challenge, as emphasized by Bailey (10), is to sell this idea "to a largely disinterested teen age population who will not be at risk for another 40 or 50 years" (p. 581).

Type 2 Diabetes Mellitus

Previously considered a disease of adults, type 2 non-insulin-dependent diabetes mellitus (NIDDM) is becoming increasingly common in the pediatric age group. In a large diabetes referral program, the incidence of NIDDM in adolescents rose 10-fold (from 0.7 to 7.2 new cases per 100,000) in the years between 1982 and 1994 (83). In that center, NIDDM accounted for a third of new diagnoses of diabetes in patients between the ages of 10 and 19 years in 1994. Most were markedly obese, with an average BMI of 38 $kg \cdot m^{-2}$, and only 8% had a normal value for BMI.

The increase in NIDDM in the pediatric population has been explained by the parallel rise in incidence of childhood and adolescent obesity. Similar trends have been seen in adults. The risk of diabetes in older age groups doubles for every 20% of excess body weight, and the incidences of obesity and NIDDM are both increasing (45).

In adult studies, both obesity and physical inactivity are independent contributors to NIDDM, although excess body fat plays a more predominant role (50). Increasing activity levels may therefore act to prevent or treat NIDDM by either (1) directly affecting insulin utilization and glycemic control or (2) decreasing body fat content. The limited available information suggests that these actions of exercise in pediatric subjects simulate those described in adults.

Obese adolescents are characterized by insulin resistance, hyperinsulinism, and elevated glucose responses to feeding (58; 107). Kahle et al. investigated the effects of mild routine exercise on seven obese children (mean age 13.3 years) (64). After a 15-week exercise program, fasting glucose fell by 15%, total glucose response to a meal decreased by 15%, peak insulin response fell by 51%, and total insulin response to a meal declined by 46%. These changes occurred in the absence of significant fat loss. In a 6-year longitudinal study,

Raitakari et al. found that physically active males (but not females) had lower insulin levels than those who were sedentary (86).

Adult patients with NIDDM frequently have other coronary risk factors, including not only obesity but also systemic hypertension, hypertriglyceridemia, and low HDL-C (15). This so-called "metabolic syndrome," which creates a serious risk for cardiovascular disease, is also observed in young patients with NIDDM. Gutin et al. found that all the major risk factors for coronary artery disease and NIDDM were interrelated in a group of 7- to 13-year-old obese children (56). Gutin et al. concluded that these findings "lend support to the concept that the underlying metabolic syndrome begins quite early in life" (p. 464). Similar relationships between insulin/glucose levels and cardiovascular risk factors in children were described by Burke et al. in the Muscatine Study (35).

Steinberger and Rocchini reported that insulin resistance is directly related to lipid abnormalities in obese adolescents (107). In their study of 82 teenage patients, the degree of insulin resistance (but not body fat content) explained the largest amount of variance in levels of both LDL-C and HDL-C.

Considering the potential impact of this clustering of risk factors in young patients, early interventions to normalize body fat and treat NIDDM are important. Further research is warranted to determine the role of exercise in preventing and managing the metabolic syndrome.

Effects Independent of Traditional Risk Factors

It is possible that the exercise-health link documented in adults reflects a direct action on disease processes such as atherosclerosis or an indirect effect through variables other than the traditional risk factors. Evidence for this type of influence on physical activity or fitness in children will be difficult to come by. There are, nonetheless, some research data in young subjects suggesting that this possibility should be considered.

Fibrinolysis

Thrombosis in coronary arteries superimposed on atherosclerotic lesions poses a risk for myocardial ischemia and infarction. Evidence in adults indicates that endurance training can enhance fibrinolysis (the enzymatic breakdown of clots) and reduce platelet adhesiveness (46). Studies addressing the effect of physical activity and fitness on reducing thrombotic risk through a decrease in plasma fibrinogen levels have provided conflicting results.

Plasma fibrinogen levels in children have been associated with an abnormal lipid profile. Bao et al. reported that fibrinogen was directly related to BMI and inversely to HDL-C in 5- to 17-year-olds in the Bogalusa Heart Study (14). Isasi et al. described the cross-sectional relationship between physical fitness (as defined by heart rate during submaximal cycle testing) and plasma fibrinogen level in 193 subjects 4 to 25 years old (62). A significant inverse relationship, albeit low, was found between the two ($r = -0.24$, $p < 0.05$), which persisted after adjustment for age, sex, race/ethnicity, and BMI. No association was present between family history of early-onset coronary artery disease and fibrinogen level in the children. However, Mahon et al. could find no relationship between either VO_2max or physical activity (by recall survey) and fibrinogen level in 10- to 16-year-old children (72).

Sympathetic Activity and Heart Rate Variability

Investigations in both animals and humans suggest that enhanced sympathetic responses to stimuli are associated with an increased risk for atherosclerosis and complications of coronary artery disease (31; 74). Endurance training in dogs has raised the threshold for ventricular fibrillation by enhancing baroreflex control and vagal activity (61). These observations suggest that high sympathetic tone may serve as a risk factor for cardiovascular disease.

Cardiac autonomic function can be assessed by a determination of beat-to-beat variability in the heart rate over time. Spectral analysis has indicated that this variability can occur in a high-frequency (0.15-0.40 Hz) or low-frequency (0.04-0.15 Hz) pattern. High-frequency variability appears to reflect parasympathetic nervous activity, while low frequency is more affected by sympathetic influences. In other measures of variability using a time domain, greater variability has been associated with higher vagal (parasympathetic) tone.

In adults, most (but not all) cross-sectional studies have indicated lower heart rate variability and higher high-frequency power in highly trained individuals, reflecting greater parasympathetic tone. However, interventional training studies have revealed no consistent trend in changes in heart rate variability (78).

Little information is available in children. Kirby and Kirby reported a significant direct correlation between VO_2max and two of three markers of heart rate variability in a study of 472 healthy boys and girls (66). Gutin et al. showed a decreased ratio of low- to high-frequency power after a 4-month exercise training program in 17 obese children ages 7 to 11 years (55). These data suggest that physical fitness is associated with greater parasympathetic activity in youth. Further research will be necessary to determine the implications of these findings for their present and future cardiovascular health.

Risk Factors and Exercise in Youth: Weighing the Evidence

Is this body of research sufficiently convincing to allow us to conclude that exercise during childhood will diminish risk factors for chronic disease in adults? Although the relationship reported between physical activity, fitness, or activity plus fitness and cardiovascular risk factors has been described as "weak" (116) and "disappointing" (87), it is important to recognize what these data reveal. The research consistently indicates that optimizing habitual activity or aerobic fitness is unlikely to alter cardiovascular risk factors in healthy children who have normal blood pressure, body fat, serum lipids, and glucose tolerance. This, however, may be the wrong issue to address. The more important question: Can exercise habits or fitness slow or prevent the *development* of these risk factors as children grow into adulthood?

Currently few data exist that specifically address this issue. However, there is good reason to believe that children who are more fit and active can receive a protective effect in relation to the development of cardiovascular risk factors. This conclusion is derived from the observation that exercise interventions in children clearly act to decrease abnormal values of risk factors. This salutary effect has been observed on every risk factor that has been tested. Regular exercise is capable of lowering blood pressure, decreasing body fat, reducing insulin levels, improving glycemic control, increasing bone mineral density, and creating a more healthy serum lipid profile in youth who have abnormal values to start with (table 5.3).

Results from long-term studies such as the Amsterdam Growth and Health Study and the Harvard Growth Study provide some support for this concept. These studies are not interventional but demonstrate a longitudinal connection between exercise habits and changes in blood lipids and body fat (118). The critical question—whether or not an intervention to optimize exercise habits in children will favorably alter the progression of risk factors over time—remains to be answered.

This idea notwithstanding, the overall evidence linking exercise and cardiovascular risk factors is less impressive in children than in adults. Riddoch (87) and Twisk (116) offered several possible explanations. First, the concept that cardiovascular risk factors in children are associated with health outcomes may not be valid. Factors such as serum lipids normally change during growth and development, confusing relationships with exercise. Moreover, since most risk factors have a strong genetic component, changes resulting from augmented physical activity or fitness are likely to be small.

The concepts of activity and fitness are being forced into an adult-oriented model. In adults, the sedentary individual who sits behind a desk all day is encouraged to increase activity (regular walking, gardening) or improve aerobic fitness (by jogging three times a week). Children's activity, particularly that of younger children, is different. Children engage in play—sporadic bursts of exercise on the backdrop of a generalized higher level of motor activity. Daily energy expenditure (relative to body size) is typically high compared to that in adults, even without periods of sustained exercise. Quantifying this type of activity typical of children is difficult. Using crude techniques such as questionnaires and diaries is not likely to provide sufficiently precise data.

Children may be already optimally active. The great majority of children are both healthy and, particularly in the younger years, highly active (see chapter 2). In addition, there are no data for defining an activity or fitness threshold for "health outcomes" in youth, and arbitrary definitions of health-related fitness may be misleading.

How much scientific proof of the preventive health pediatric rationale for promotion of exercise is necessary? As Riddoch has pointed out, "no single study, or set of studies, provides definitive evidence for a meaningful health gain through being an active child. [However,] it might be said that with respect to activity/health relationships in children, absence of evidence should not be taken as evidence of absence" (p. 30) (87). It has been argued that lack of certainty of outcomes should not dissuade preventive health efforts (96). Clearly, precedence exists for major preventive health initiatives based on reasonable inference in the absence of incontrovertible scientific evidence. In the case of exercise and health outcomes in children in fact, indirect evidence provides a compelling argument for the pediatric rationale:

- Given the pediatric origins of chronic disease in adults, plus the recognized effects of exercise on health in older age groups, the rationale for early intervention is intuitively satisfying. That is, it passes the test of common sense.

- Although exercise does not appear to alter cardiovascular risk factors in healthy children, physical activity, fitness, or both clearly can ameliorate these factors in youth who have abnormal values. This suggests that regular

TABLE 5.3   Effect of Physical Activity and/or Fitness in Children on Coronary Risk Factors

	Cross-sectional	Intervention
Serum lipids General population	No	No
Subjects with elevated values		Yes
Body fat content General population	No	No
Obese subjects	Yes	
Blood pressure General population	No	No
Subjects with hypertension		Yes
Non-insulin-dependent diabetes mellitus		Yes

exercise in the growing years is likely to inhibit the increase of risk factors over time.

- Relationships of activity and fitness to cardio-vascular risk factors in youth have been observed to be either beneficial or nonexistent but never disadvantageous.

- In the case of bone health, evidence indicates that exercise interventions in childhood and adolescence are likely to optimize the development of peak bone density and diminish risk for osteoporosis.

Tracking of Physical Activity

The argument for a pediatric approach to promoting exercise for preventive health would be strengthened if a child's level of physical activity persisted, or tracked, into the adult years. That is, any beneficial effect that exercise might impose on the pathogenesis of adult disease presumably must act over an extended period of time. Little long-term effect would be expected, for instance, if a program to increase exercise habits in an obese or hypertensive adolescent resulted in only a 3-month period of increased activity.

It has been suggested, in fact, that lifestyle modification—introducing a habit of regular exercise in the child that will persist into adulthood—is a sound rationale for the promotion of exercise all by itself, independent of effects that activity might or might not have on disease processes (26; 48). Blair et al. proposed that (1) exercise in adults can be expected to have a greater impact on adult health than childhood exercise, and (2) a lifestyle of regular activity in the adult years would be more likely if such habits were started early in life (26). Rowland agreed, concluding that "the best primary strategy for improving the long-term health of children and adolescents through exercise may be creating a lifestyle pattern of regular physical activity that will carry over to the adult years rather than promoting childhood physical fitness" (p. 671) (95).

The extent to which levels of physical activity track from childhood and adolescence into the early adult years has not been impressive in the few studies that have addressed this issue. These reports were reviewed by Malina (73). As might be expected, tracking correlation coefficients in these studies decrease as the follow-up period increases. In short-term tracking studies (3-5 years) during childhood or adolescence, low to moderate correlations are typically observed (r = 0.30-0.60). Longitudinal reports of persistence of activity from adolescence to adulthood demonstrate lower correlations (r = 0.05-0.20). In Dutch adolescents, coefficients of weekly habitual physical activity (by interview questionnaire) between ages 13 and 27 years were r = 0.05 for males and 0.17 for females (120). Some evidence suggests that those who are most inactive are likely to stay that way. For example, Raitakari et al. found that 54% and 51% of males and females, respectively, who were classified as inactive at age 15 were still sedentary at age 21 (86).

Several issues surround these observations. First, the strength of longitudinal relationships of exercise habits over time may be influenced by the precision of the methodology being employed to assess physical activity. The size of the study populations has necessitated the frequent use of self-report questionnaires, which limits accuracy.

Again, the question being asked may be the wrong one. These data could be interpreted as illustrating that if nothing is done to promote physical activity in children and adolescents it is unlikely that they will become active adults. The more important issue is whether improved exercise habits introduced early in life will persist into the adult years. Information regarding long-term post-intervention tracking of exercise is virtually nonexistent. Shephard and Trudeau described follow-up findings in 30- to 35-year-old subjects who had participated in a program of enhanced physical education in first through sixth grade (103). A significantly greater percentage of women were participating as adults in at least three sessions of vigorous activity per week compared to controls (42.1% vs. 25.9%). However, no differences in activity from that of controls were observed in the men.

Tracking of physical activity does not necessarily imply that exercise habits will provide good health as an adult. As Corbin has noted, "since adults are considerably less active than children, retaining a high rank in the adult group does not assure 'adequate activity,' just 'more activity'" (p. 348) (36). He suggested a more appropriate approach of "untracking" activity in those youth who are sedentary: "We do not want inactivity in childhood to track to adulthood. Rather we must find a way to help relatively inactive children to become more active in both childhood and as adults" (p. 349).

Defining Exercise Promotion Strategies

A full discussion of approaches to encouraging exercise in youth is beyond the scope of this chapter. It is useful, however, to examine how these strategies might be shaped by insights into the mechanisms through which exercise can affect health outcomes in children and adolescents. In creating these initiatives, several key questions need addressing.

Should Efforts for Promoting Exercise in Youth Be Universal or Targeted to High-Risk Groups?

The answer to the question of whether to target depends to a large extent on how one defines a "high-risk" population. According to the general perception, children as a group are sedentary and becoming increasingly so—the victims of television, computers, and video games. If so, a generalized approach to improving exercise habits in the pediatric population would be in order.

However, Blair and his colleagues argued that "children are, by far, the most physically active and fit group in the United States" (p. 403) (26). On the basis of data quantifying the amount of daily exercise important in lowering cardiovascular risk in adults, the authors suggested that a daily energy expenditure of 12 kJ · kg (3 kcal · kg^{-1} · day^{-1}) should provide health benefits. Among students 10 to 18 years old, 88% of the girls and 94% of the boys satisfied this criterion, and three-quarters expended more than 17 kJ · kg (4 kcal · kg^{-1}) daily. This caused the authors to conclude that, at least by this standard, "an overwhelming majority of children and youth in the United States are physically active" (p. 420). By this analysis, then, efforts to encourage exercise might be more appropriately directed to the significant minority of children who are risk for persistent hypoactivity (particularly those who are obese or nonathletic or those who have chronic disease).

A large number of studies have been performed to assess the habitual daily activities of children using self-report questionnaires, diaries, and heart rate monitoring (5; 89). These have generally indicated that (1) young people spend only small amounts of time in vigorous activity, (2) boys are usually more active than girls, and (3) activity levels progressively decline with age. While these results have raised concern that youth are not adequately active, they provide no insights into the question of whether children are sufficiently active for health benefits, or whether the exercise habits of children are decreasing with time. The decline in activity with increasing age is an expected biological phenomenon (99); and by the nature of their physical activity, children can exhibit considerable daily energy expenditure without periods of intense physical activity (88).

It must be concluded that the proportion of children who are insufficiently active to derive health benefits is currently unknown. The question is a difficult one to answer. First, the desired outcome is mixed: Is the goal to improve children's activity to a threshold amount for a specific outcome (e.g., create enough mechanical stress to stimulate bone growth)? In this case, the amount of necessary activity would presumably vary (considerably) according to the benefit desired. Or, is the objective to introduce regular exercise as a lifestyle habit that will continue to adulthood? For this goal, persistence rather than a threshold amount of physical activity is more critical. For these reasons, it is difficult to decide if a general-population or at-risk group strategy is more appropriate for targeting promotion of physical activity in young persons.

What Types of Physical Activity Are Most Appropriate?

Historically, on the basis of performance on tests of motor skill, concern rose in the 1950s that American youth were insufficiently fit. The emphasis on improving physical fitness at this time was fueled by subsequent data linking aerobic fitness with health in adults. This led to the concept of "health-related fitness" and the promotion of endurance and strength performance in youth, measured by standardized school-based fitness tests. More recently, the past decade witnessed a shift in emphasis to the health benefits of moderate physical activity rather than physical fitness. The concept that habitual activity across the life span would prove most effective led to the pediatric rationale for promoting regular activity. Improving habits of physical activity then replaced fitness training as the primary goal for exercise promotion in young persons (37). That some of this activity should be weight bearing has been subsequently recognized as important in optimizing bone health.

An appreciation of the mechanisms by which activity or fitness can improve health outcomes suggests that *all* of the following forms of exercise may play a useful role (table 5.4).

Energy Expenditure

The principal role of physical activity in the prevention and treatment of obesity is "burning" calories. The greater the intensity of activity, the more calories expended and the greater the benefit in altering energy balance. Since most obese children do not tolerate vigorous exercise, however, sustained low-intensity activities are most appropriate. Endurance training programs designed to increase aerobic fitness are not necessary, since cardiovascular function is normal in most obese subjects (94).

Cardiovascular Fitness

The body's responses to regular endurance exercise can be effective in reducing risk factors for coronary artery disease. Running, cycling, or swimming performed three times a week is likely over time to reduce sympathetic tone and maintain normal blood pressure. Extended periods of such activity may play a role in favorably altering abnormal serum lipid profiles.

Weight-Bearing Activities

Applying mechanical stress on bones through weight-bearing activities stimulates the development of bone density in the growing years. While activities such as running are more effective than non-weight-bearing exercise, those activities that provide sudden bursts of stress to bones (such as jumping) are most efficacious.

Resistance Exercise

Resistance activities that improve muscle strength should be expected to improve health by preventing injuries, particularly in sport play, and diminishing the incidence and severity of chronic back disease. The research evidence remains sparse on these points, however. It is clear that resistance exercise in the proper setting can improve strength without risk in pre- as well as postpubertal subjects, but the role for emphasizing these activities in routine health promotion of children and adolescents remains to be clarified.

Stretching Exercise

Stretching exercises are important whenever the range of motion over a joint needs to be preserved or enhanced. Stretching is particularly important for patients with conditions such as spastic cerebral palsy, muscular dystrophy, and arthritis.

An optimal exercise program for preventive health, then, would include regular aerobic exercise (to affect blood pressure and HDL-C), a moderate level of daily energy expenditure (to control body fat and maintain glycemic control), resistance exercise (to promote strength), weight-bearing activities (to stimulate bone density), and stretching (to prevent contractures and improve ambulation).

In children, of course, there are many considerations besides altering disease mechanisms that dictate appropriate exercise interventions. Most importantly, choice of physical activity is highly dependent on the child's developmental stage. Age-appropriate activities, which have been outlined by several authors, emphasize several points. At young ages children are inherently

TABLE 5.4	Principal Forms of Exercise Expected to Produce Preventive Health Benefits		
	Habitual physical activity	Aerobic fitness	Resistance exercise
Abnormal serum lipids	+	+	
Hypertension		+	
Obesity	+		
Type 2 diabetes mellitus	+		
Emotional health	+		
Osteoporosis	+		+
Injury prevention			+
Chronic back disease			+

active, and the emphasis should be on providing opportunities for play (38). In later childhood, development of motor skills becomes important (17). Encouragement of lifetime activities on the adult model becomes more appropriate in the adolescent years (82).

What Are Appropriate Exercise Guidelines?

No information currently exists to allow us to confidently define dose-response relationships between exercise and health outcomes in children and adolescents. Specific recommendations for physical activity in the young have therefore been based on reasonable assumptions and data extrapolated from adult studies. Pate et al. have reviewed and provided a critique of published guidelines for both children and adolescents (82).

The threshold criterion for activity sufficient for health benefits of 3 to 4 kcal · kg^{-1} · day^{-1} for children translates into 20 to 40 min of moderate to vigorous physical activity. Corbin and Pangrazi recommended this "dose" of daily exercise as a minimum for children and suggested that 60 min per day (6-8 kcal · kg^{-1}) was even more optimal (37). In 1997, a consensus group created by the Health Education Authority in Great Britain drew on this information to recommend that all young people should participate in physical activity of at least moderate intensity for 1 hr per day (24). Examples of moderate-intensity exercise included brisk walking, steady bicycling, and playing outdoors.

The group added the recommendation that at least twice a week, some of these activities should include those that enhance muscle strength, flexibility, and bone health. Such exercise in children might involve playground activities with climbing, gymnastics, and calisthenics. Supervised formal resistance training was considered appropriate for adolescents. Pate et al. concluded that "these recommendations are consistent with pertinent body of knowledge and with the relevant official positions of government, scientific, and professional organizations. However, it is readily acknowledged that these recommendations are not supported by epidemiological or experimental research at nearly the desired level" (p. 172) (82).

An earlier (1994) Consensus Conference on Physical Activity Guidelines for Adolescents generated two recommendations: (1) All adolescents should be physically active daily, or nearly every day, as part of play, games, sport, work, transportation, recreation, physical education, or planned exercise; (2) adolescents should engage in three or more sessions per week of activities that last 20 min or more at a time and that require moderate to vigorous levels of exertion (100).

Pate et al. noted that although the first guideline is consistent with promotion of "lifestyle physical activity," it fails to identify a specific amount of daily exercise (82). The second recommendation, while specific as to amount of exercise, "espouses the traditional adult exercise prescription, which may not be ideal from a behavioral perspective with either adults or adolescents" (p. 167).

As these authors concluded, additional research is clearly needed to identify the nature and amount of physical activity necessary for health in children. Particularly important will be longitudinal investigations of childhood antecedents to habits of regular physical activity in adult life.

How do these guidelines match up to current levels of physical activity in children and adolescents? Epstein et al. reviewed 26 published studies that used heart rate monitoring as a means of assessing daily physical activity in youth ages 3 to 17 years (42). As expected, the amount of time spent in activity was inversely related to its intensity, measured as percent of heart rate reserve (the difference between maximal and resting rate). Activity declined with age and was greater in males than females. An average of 47 min a day was spent at a heart rate reserve of 40% to 50%, 29 min from 50% to 60%, and 15 min above 60% (figure 5.5). The intensity of exercise that has been advised during the recommended 30 to 60 min of daily activity for youth corresponds to approximately 40% to 50% of heart rate reserve (82). This analysis suggested, then, that most youth may already be satisfying these guidelines.

Blair et al. reviewed exercise-health studies in adults and suggested that a daily energy expenditure of 12.6 kJ · kg^{-1} · day^{-1} was an appropriate standard for positive health outcomes (26). Using questionnaire data from the National Children and Youth Fitness Study of 10- to 18-year-old subjects, they then estimated that among children and adolescents, 87% to 90% of females and 90% to 95% of males satisfied this criterion. Using a higher standard of 16.8 kJ · kg^{-1} · day^{-1}, the relative

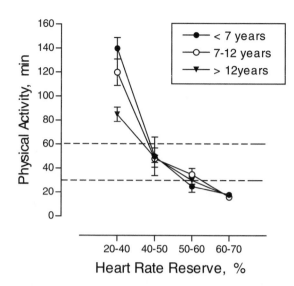

FIGURE 5.5 Minutes of daily activity time spent at different levels of heart rate reserve in youth.

Adapted, with permission, from Epstein et al. 2001 (42).

percentages were 77% to 83% and 82% to 87%. The authors concluded that while some children are sedentary, "the 'youth fitness crisis' may be exaggerated" (p. 403).

The fact that at least one adverse health outcome, obesity, is increasing in the face of these observations suggests, however, an overall inadequacy of levels of physical activity of children. This would further imply either that the levels of activity are being erroneously assessed or that the criteria recommended in these guidelines for daily activity in youth for positive health outcomes are too low.

Exercise and Pediatric Diseases

© Mary Langenfeld

Pulmonary Diseases

With the lungs being a key part of the O_2 transport chain, any disease that involves the respiratory system is likely to affect a child's physiological response to exercise, as well as physical fitness and habitual physical activity. In turn, physical exertion is likely to affect the health and well-being of patients with a pulmonary disease. This chapter reviews the interactions among physical activity, physiologic responses to exercise, physical fitness, and pediatric pulmonary diseases.

Asthma

Second to obesity, asthma is the most common pediatric chronic illness. Its prevalence in various countries ranges from 5% to 17% (120; 140; 285), and it has been on the rise in the last two decades (285; 325). Likewise, mortality from asthma among children has been rising at a fast pace (369). In adults, the prevalence of asthma seems to be higher among elite athletes than in the general population (284). The relevance of exercise to the health and well-being of the child with asthma is fourfold:

1. Acute exertion triggers bronchial obstruction and an asthmatic attack.
2. Chronic exercise is of therapeutic value.
3. Exercise is an important diagnostic tool.
4. Exercise-based research has helped to investigate the pathophysiology of asthma.

Exercise-Induced Asthma

Exercise-induced asthma (EIA) is clinically the most significant response to exercise in a child with asthma. It was originally considered to reflect the contraction of bronchial smooth muscle only (hence the term exercise-induced bronchoconstriction, EIB). Current understanding imputes to the pathology of EIA airway edema and inflammation. Clinical manifestations include any combination of cough, shortness of breath, wheezing, and chest pain/tightness. Sometimes, the only presenting symptom is exertional chest pain. As many as 70% of children and adolescents who complain of exertional chest pain may have asthma (422). Exercise-induced asthma results in increased airway resistance and lung hyperinflation, but only seldom in hypoxemia. Typically, it starts some 2 to 4 min after exercise, peaking at 5 to 10 min postexercise and disappearing spontaneously within 20 to 40 min. Sometimes it may be sustained for more than 1 hr. A *late asthmatic response* may appear some 4 to 10 hr after exercise. For details, see the section "Late Asthmatic Response" further on in this chapter.

The pulmonary functions commonly used for assessing bronchial patency and resistance to airflow are forced expiratory volume in the first second (FEV_1), peak expiratory flow rate (PEFR), and maximum midexpiratory flow rate (MMEFR, or FEF_{25-75}). While the latter apparently reflects flow in the small airways, FEV_1 and PEFR reflect flow in the larger airways. Other functions, used mostly in research and less often in a clinical context, include specific airway conductance and airway resistance. Figure 6.1 displays a typical EIA pattern, as found in a 10-year-old patient.

Infrequently, bronchial obstruction may start *during* the activity. It is not clear, though, whether this is a real phenomenon or a reflection of technical difficulties in obtaining adequate spirometric records while the patient exercises (256).

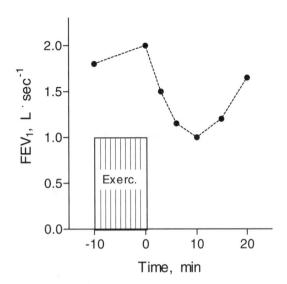

FIGURE 6.1   A typical exercise-induced asthma response. A 10-year-old girl with asthma performed 8-min exercise (constant power) in which her heart rate reached 175 to 180 beat · min^{-1}. Forced expiratory volume in the first second (FEV_1) was measured before exercise and several times postexercise.

Various criteria have been used to determine "abnormal" post-exertional bronchial obstruction. These, defined by a percentage drop (usually of FEV_1 or PEFR) from the pre-exercise value, range from 10% to 25% (8; 40; 66; 93; 113; 127; 141; 185; 205; 214; 216; 299; 311; 366; 407). Most clinicians consider a 15% drop as diagnostic. For details, see the section "The Optimal Exercise Challenge" later in this chapter.

Some authors assign importance to the combined bronchodilatory response during exercise and bronchoconstrictive response after exercise (149; 204). These combined changes presumably reflect "airway lability," which is greater in asthma patients than in healthy children. A lability index (i.e., the sum of dilation immediately after exercise and the greatest constriction following exercise) of ≥20% of the pre-exercise function (usually PEFR) is considered excessive. The clinical use of this index has declined in recent years.

Epidemiology of Exercise-Induced Asthma

When exercising in the laboratory, some 40% to 90% of patients with asthma respond with EIA. The wide range of reported response is due to the variety of protocols used by different investigators, especially the type, intensity, and duration of exercise; nonstandardized environmental temperature and humidity; the presence of allergens and pollutants; different policies regarding drug withdrawal on the day of testing; and variability in the severity of disease. Furthermore, investigators have not agreed on a universal criterion for a positive test. In spite of such variability, it is fair to state that about 70% to 75% of children with perennial asthma who are taken off medication some 6 to 8 hr before testing will respond with at least a 10% to 15% drop in FEV_1 following exercise. This rate increases when the exercise takes place in a climate that is cold, dry, or both. In some children, mostly those with mild-intermittent asthma, EIA may be the only clinical manifestation of the disease.

Some authors refer to the rates just cited as *incidence* of EIA (96; 112; 216), while others use the term *prevalence* (33; 120; 139; 256; 353). Most data, however, refer to the response to a standardized provocation test in the laboratory and do not reflect *real-life* incidence or prevalence. There is only fair correlation between the response of patients with asthma to exercise in the laboratory and their day-to-day activity-related respiratory symptoms (96; 216; 278). It is therefore more appropriate to refer to these percentages as *rate of response to a standardized exercise test*, rather than prevalence or incidence.

In the absence of definitive epidemiologic data, one can only speculate about the extent of EIA in real life. It is conceivable that any asthma patient may, at some time, experience EIA when his or her activity happens to be intense and of sufficient duration. Potentially, therefore, all patients are at risk. However, on the basis of their experience of such unpleasant episodes, many patients modify their activities in order to prevent triggering an attack. They do so spontaneously, or on the advice of others, by avoiding intense exertion or by turning to less asthmogenic sports. Thus, the rate of occurrence of EIA during a given period of time may not be high.

Although EIA occurs typically in the child with asthma, it is sometimes observed in patients with other conditions, such as allergic rhinitis (254), other atopic (nonasthmatic) disorders (66; 96; 139; 216), cystic fibrosis (98), and a history of bronchopulmonary dysplasia (23; 336). A high lability index was found in first-degree relatives of children with asthma (229); former patients with asthma (45; 204); monozygotic twin siblings of asthma patients (229); and patients with cystic fibrosis (98; 361), hay fever (204), or a past history of viral bronchiolitis (364).

Exercise-induced asthma appears also among the general child population (33; 139; 168) and teenage athletes who had no prior diagnosis of asthma (120; 331; 332). The high prevalence (10-15%) of EIA among "healthy" athletes may reflect the fact that they perceive post-exertional dyspnea as a normal response (62), but it may also result from a high vagal tone that often occurs in endurance athletes (226). Some athletes, even when they know that they have EIA, or a history of asthma, will not disclose it for fear of being excluded from their team (120).

There is evidence (1; 138) that children exposed to maternal smoking have a high rate of EIA compared with the general child population. For example, in a case-control study, the odds ratio for EIA was 2.23 in 9- to 14-year-old children with history of maternal smoking compared with children whose mother did not smoke (1). Likewise, exposure to air pollutants (212; 238; 356) and to allergens (35; 139; 275) increases the likelihood of EIA and the severity of response. Pollutants to be avoided include sulfur oxide, black smoke, and nitrogen oxide. Ozone, on the other hand, does not seem to increase the likelihood of EIA (122).

Exercise-Induced Asthma— Nature of the Exercise Provocation

The degree of post-exertional bronchoconstriction varies with the nature of the exercise. More specifically, it depends on the *type, intensity,* and *duration* of the activity.

Type of Exercise

The information commonly given by the child or parent is that activities involving running, cycling, skating, or skiing induce more discomfort than swimming. Indeed, several studies have confirmed that swimming is the least asthmogenic exercise (8; 32; 130; 156; 196; 281; 290; 320). For such a comparison to be valid, the intensity of exercise and, in particular, the ventilation must be equated among the exercise modalities. Figure 6.2 shows an example comparing effects of swimming and running in an 8-year-old boy who performed both tasks at equal O_2 uptake and ventilation and still had a considerably greater

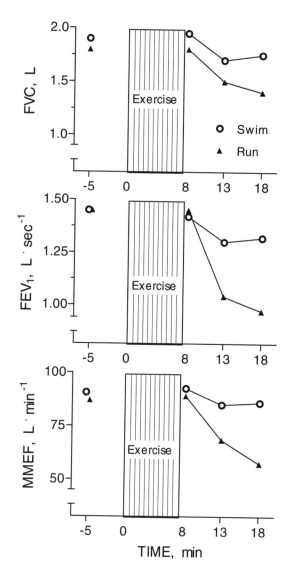

FIGURE 6.2 Swimming as a "nonprovoker" of exercise-induced asthma. Pulmonary functions of an 8-year-old boy with asthma before and after treadmill running and tethered swimming. FVC = forced vital capacity; FEV_1 = forced expiratory volume in the first second; MMEF = maximal midexpiratory flow. Both tests lasted 8 min, during which time O_2 uptake was 29 ml · kg^{-1} · min^{-1} and ventilation 34 L · min^{-1}. Ambient temperature was 27° C and relative humidity 30%. Data from the Children's Exercise & Nutrition Centre.

bronchoconstriction after running than after swimming.

What is the *most asthmogenic* type of exercise? Studies from the 1960s and early 1970s suggested that running induces greater and more consistent bronchoconstriction than walking or cycling (11; 125; 149; 201; 206; 359). It has further been suggested that free running provokes greater EIA than treadmill running (11; 352). Arm cranking was reported to be as potent a stimulus as running (8) and to cause greater EIA than leg cycling when both were done at the same absolute metabolic level (389).

The main shortcoming of the reports just mentioned is that ventilation was not equated in the comparison between modes of exercise. As described in the section "Airway Cooling As a Trigger" later in this chapter, the extent of EIA depends on the degree of airway cooling, which in turn is a function of ventilation and of climatic conditions. Indeed, subsequent studies have indicated that when ventilation, air temperature, and humidity are equal, walking, treadmill running, free running, cycling, and arm cranking induce similar degrees of bronchial obstruction (105; 222; 266).

It therefore can be concluded that although swimming is the least asthmogenic exercise, there is no single type of activity that can be considered most asthmogenic. However, an individual patient may find one specific "land activity" more asthmogenic than another.

Intensity of Exercise

The degree of EIA is related to the intensity of the exercise provocation (111; 114; 155; 215; 352; 359; 426). This relationship is presented in figure 6.3. As a rule, intense activities induce greater broncho-

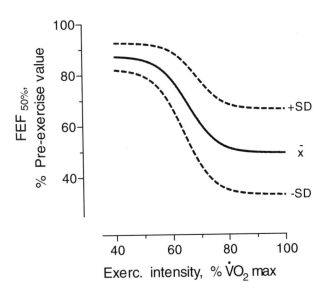

FIGURE 6.3 Exercise-induced asthma and the intensity of exercise. Nine young adults with asthma ran on a treadmill at 40%, 60%, 80%, and 100% of their maximal O_2 uptake ($\dot{V}O_2$max) for 6 min. Post-exertional forced expiratory flow at 50% of vital capacity ($FEF^{50\%}$) is expressed as a percentage of pre-exercise value. Adapted from Wilson and Evans 1981 (426).

constriction than light activities. During exercise that lasts 6 to 8 min, intensities of 70% to 85% of maximal aerobic power (heart rate = 160-180 beat · min^{-1}) have the most asthmogenic effect (114; 151; 426). This can be explained by the associated high ventilation and airway cooling. It is not clear, however, why maximal intensities (which are accompanied by the highest ventilation) do not cause the greatest EIA. One explanation is that such intense activities cannot be sustained long enough to induce sufficient airway cooling. An alternative explanation is that the increases in sympathetic tone and circulating catecholamines that accompany a highly intense activity cause bronchodilation and therefore partially inhibit the asthmatic response.

Duration of Exercise

The degree of EIA is related to the duration of exercise (206; 359). In a highly quoted study, exercise duration of 6 to 8 min was found to induce greater bronchoconstriction than shorter or longer protocols at exercise intensities equivalent to 60% to 85% of maximal aerobic power (359). More research is needed, however, to show whether a 6- to 8-minute duration is optimal at other exercise intensities: For example, the optimal duration may be less than 6 to 8 min for highly strenuous activities. A case in point is a study in which young adults with asthma had a greater drop in maximal midexpiratory flow following a supramaximal treadmill run, which exhausted them within 50 sec, than following a 7-min submaximal run (194). There are no similar published observations for children.

Prolonged activities (e.g., 20 min or more) are less asthmogenic than shorter ones. This is clinically important, because a patient who performs such prolonged activities may suffer no EIA, or may have a mild EIA *during* early parts of the activity, which will disappear by the time the exercise task is concluded. Such a phenomenon has been termed *running through* the EIA.

Refractory Period

If a patient repeats an exercise bout within 2 hr of a previous bout, the result may be little or no EIA (34; 111; 147; 262; 286; 303; 424). The time interval for such protection is known as the *refractory period* (111) (figure 6.4). Approximately 50% of all patients with EIA can benefit from this refractoriness.

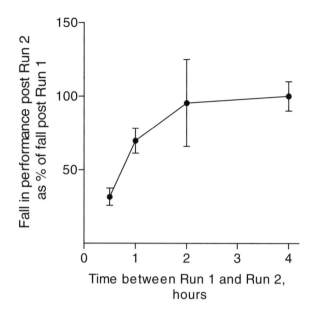

FIGURE 6.4 Duration of refractoriness following an exercise bout. Nine 9- to 14-year-old children with asthma performed pairs of 6-min treadmill runs (heart rate 185 ± 11 beat · min^{-1}). The time interval between the runs varied from 30 min to 4 hr. The degree of refractoriness is displayed as the percentage fall of peak expiratory flow rate (PEFR) in run 2 over the percentage fall of PEFR in run 1, multiplied by 100. Mean ± SEM.

Reproduced, with permission, from Edmunds et al. 1978 (111).

Although the mechanism for this phenomenon is not entirely clear, the following pattern seems to emerge:

• It is the initial exercise itself, rather than the EIA, that triggers refractoriness (34; 424; 425). In one study, however, refractoriness was not achieved when subjects were breathing warm, humid air during the first exercise bout to prevent EIA (171).

• The muscle group that performs the initial exercise bout can be different from the one performing the subsequent bout (425).

• The intensity of the initial exercise does not need to be high (259; 321). However, the threshold intensity for inducing refractoriness varies among patients.

• It is hard to predict who will respond with refractoriness and who will not.

• Several short running sprints can induce refractoriness (341), as can longer, less intense activities.

• Pretreatment for several days with indomethacin, a prostaglandin synthase inhibitor, will prevent, or attenuate, post-exertional refractoriness (286; 424). This suggests that refractoriness depends on the production of prostaglandin (most probably, inhibitory prostaglandin type E$_2$).

• Inhibitory prostaglandin type E$_2$ is released in response to the initial exercise bout, with no need for an initial EIA (424).

• During the refractory period, a person can still respond with bronchoconstriction to a challenge by histamine (172; 173) or an allergen (419).

• A second exercise bout induces less airway cooling and slower rewarming than does the initial exercise bout. It is therefore possible that repetitive exercise alters the response of bronchial microcirculation in patients with asthma (147).

Patients, parents, and coaches can use the principle of refractoriness to their advantage. Following are some recommendations:

1. Let the patient perform a moderate-intensity exercise for several minutes, or several sprints, some 45 to 60 min prior to a scheduled intense sport task. This may overlap a regular "warm-up" routine, but is not intended to replace it.

2. The intensity and duration of a suitable preliminary activity may differ among patients. The best way to select an optimal activity is by trial and error.

3. Likewise, because refractoriness is not achieved by all patients, one should use a trial and error approach to find out whether a given child will benefit from this phenomenon.

4. If one is not sure whether a certain preliminary activity on a given day will be effective in inducing refractoriness, one can still inhale a β$_2$-agonist just prior to the main event.

5. Some children may be apprehensive about performing a preliminary activity that would trigger an asthma attack on the day of competition or of an important practice session. They should be reassured about the benefits of such an approach.

Late Asthmatic Response

Although a typical EIA episode resolves spontaneously within 20 to 40 min, some patients respond with a second episode of bronchial obstruction 4 to 10 hr later (43; 53; 56; 82; 133; 227; 239; 372). The pattern of this *late asthmatic response* is displayed in figure 6.5. Based on sparse data, the prevalence of the late response is 30% to 38% (56; 373). Young asthma patients with late response to exercise have airway hyperresponsiveness to allergens such as house dust mites (227). The degree of the late response to exercise is associated with the extent of the enhanced response to an allergen. Some authors (53; 329; 434) dispute the notion of late asthmatic response, suggesting that it merely represents normal circadian variations in airway reactivity. However, based on current evidence, it seems that a late asthmatic response to exercise does exist in some patients. In obtaining medical history about EIA, one should include questions related to this phenomenon.

While the mechanism for a late asthmatic response is not known, it is possibly related to a late release of chemical mediators. In one study (239), for example, concurrently with the late response there was also an increase in neutrophil chemotactic activity. This increase did not occur in patients who did not show a late asthmatic response to exercise.

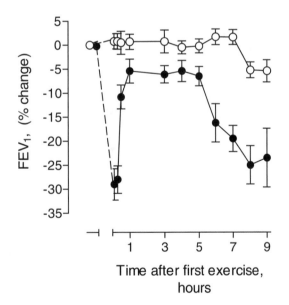

FIGURE 6.5 The late asthmatic response. Forced expiratory volume in the first second (FEV_1) was monitored on two separate days in 13 children with asthma and two adults. On one day (filled circles), the subjects exercised and then performed spirometry periodically for 9 hr. On another day (open circles), measurements were taken for 9 hr, without a preceding exercise test. Vertical lines denote 1 SEM.

Reproduced, with permission, from Orenstein 1996 (292). Based on Lee et al. 1983 (239).

Other Causes for Exercise-Induced Dyspnea

Conditions other than EIA can be accompanied by exercise-induced respiratory symptoms. The most common cause of exercise-induced dyspnea in a child is a low level of aerobic fitness. When playing or competing with other children, such a child will be exerting at a higher percentage of his or her maximal ventilation and O_2 uptake, and will therefore feel breathless before fitter peers reach this level. Typically, dyspnea in such cases appears *during* the activity and subsides spontaneously a few minutes postexercise. Exercise-induced dyspnea can occur in a healthy child, but also in children with cardiac, neuromuscular, metabolic (e.g., McArdle's syndrome), or other pulmonary (e.g., cystic fibrosis) diseases. In cystic fibrosis, exercise is very often accompanied by a vicious spell of coughing and, less often, by EIA. If a thorough history does not reveal the diagnosis, an exercise provocation test to rule out EIA is indicated. Likewise, a therapeutic trial with a β_2-agonist or cromolyn will help to exclude EIA in such conditions.

Exercise-induced dyspnea, wheezing, stridor, or a combination of these, can occur with *upper airway obstruction*, including tracheal narrowing, vocal cord dysfunction (224), and abnormal closure of the glottis (44). Typically, such conditions will not respond to β_2-agonists or to cromolyn. A flow-volume loop will show inspiratory, rather than expiratory, flattening.

Climatic Conditions and Asthma

Even when the type, intensity, and duration of exercise are standardized, EIA may not occur consistently in any given individual. The same child may respond to a certain task with marked EIA on one day and have little bronchial obstruction on another. One possible cause for such inconsistency is a change in climatic conditions.

Climate and the Resting Patient With Asthma

Patients often volunteer the information that their asthma is more severe on cold or dry days. Epidemiologic studies (107; 161; 408) and anecdotal data (68; 386) suggest a greater incidence of asthmatic attacks on cold days, especially during fall and winter. Controlled experiments have shown that in asthma patients, but not in healthy individuals, exposure to a cold shower (80) or inhalation of cold air (265; 363) induces an increase in airway resistance and a drop in FEV$_1$. By contrast, inhalation of saturated air at 37° C can prevent bronchoconstriction even while the individual is exposed to a cold shower (191).

It therefore seems that direct cooling of the airways, rather than generalized body cooling, triggers bronchoconstriction in the resting patient. On the other hand, changes in air humidity do not seem to affect the patient at rest (31; 132).

Climate and Exercise-Induced Asthma

It was first noted in the mid-1970s that air humidity and air temperature affect post-exertional asthma in the following ways:

• In neutral temperature (23-25° C), dry inspired air is more asthmogenic than humidified air (figure 6.6) (31; 420).

• Cold air is more asthmogenic than air at neutral temperatures (391). This effect is seen also among nonathletes who perform intense activities in the cold (302).

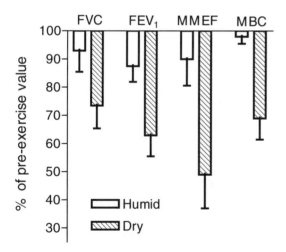

FIGURE 6.6 Air humidity and exercise-induced asthma. Pulmonary functions of ten 6- to 14-year-girls and boys with asthma, measured 10 min after each of two treadmill walks. The children were free-breathing in a climatic chamber with air temperature of 25° to 26° C and 25% ("dry") or 90% ("humid") relative humidity. FVC = forced vital capacity; FEV$_1$ = forced expiratory volume in first second; MMEF = maximal midexpiratory flow; MBC = maximal breathing capacity. Vertical lines denote 1 SEM. Based on data by Bar-Or et al. 1977 (31).

Reproduced, with permission, from Bar-Or 1980 (25).

• Warming dry air to 30° C will partially protect against EIA (79). Saturated air at 37° C renders almost complete protection against EIA (79).

This pattern has since been confirmed in numerous studies (15; 32; 103; 185; 245; 390).

The clinical implications are twofold. First, a clinician wishing to use exercise in the evaluation of EIA should attempt to monitor and, if possible, control climatic conditions during testing, especially air humidity. Second, children with asthma will be more susceptible to EIA on a cool day than during warm weather (310). Precautions should also be taken during spells of dry weather, a common phenomenon in some geographic regions—called Fohn, Sirocco, Chinook, or Hamsin. It is advisable to intensify drug therapy and, if this is of no avail, to curtail physical activity of children with asthma whenever such weather is encountered.

Mechanisms Underlying Exercise-Induced Asthma

The possible mechanisms by which exercise induces asthma can be divided into (1) those physical or chemical triggers that initiate a physiologic response and (2) the pathways by which the initial response affects the bronchial smooth muscle.

Triggering Stimuli

A variety of stimuli have been implicated as possible triggers of EIA. These include hypocapnia (125), metabolic acidosis (413), hypoxemia (21), hyperpnea (78; 188), imbalance between α and β sympathetic receptors (368), increased norepinephrine activity (162), cooling of the facial skin (208), cooling of the airways (305), and drying of the airways (170; 401). Reviews of these possible mechanisms are available elsewhere (9; 135; 151; 256; 292). The following discussion focuses on airway cooling and drying, which are currently viewed by most investigators as the prime triggers of EIA.

Airway Cooling As a Trigger

As already discussed, either cold or dry air aggravates EIA. The common denominator for these two air properties is that they both cool the respiratory mucosa—cool air by convection and dry air by evaporation. While such cooling takes place, the air becomes warmer and more humid.

An example of the effect of cooling during exercise is shown in figure 6.7. Note that the degree of cooling is similar in asthma patients and healthy controls when exercising in the same environmental conditions.

Inspired air that is at a lower temperature than the airway mucosa will cool it by *convection* according to the equation:

$$HL_C = \dot{V}_E \times HC\ (T_E - T_I)$$

in which HL_C is heat loss in kilojoules per minute; $\dot{V}_E$ = ventilation (BTPS) in liters per minute; HC = heat capacity (the product of specific heat and density) of air, which equals $0.00127\ kJ \cdot L^{-1} \cdot {}^{\circ}C^{-1}$; T_E and T_I = expiratory and inspiratory air temperature, respectively, in degrees centigrade, measured at the mouth. The factors that determine convective cooling are $\dot{V}_E$ and the temperature gradient between expired and inspired air, the former being close to body core temperature.

Inspired air that is not saturated by water vapor will cause mucosal cooling by *evaporation*, according to the equation:

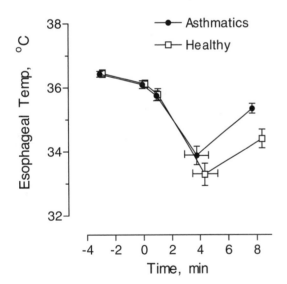

FIGURE 6.7 Changes in airway temperature due to exercise hyperpnea. Young adults with asthma and healthy controls cycled at a moderate intensity while inhaling dry air at –16° to –18° C. Esophageal temperature was measured by a thermistor at the level of the tracheal carina to estimate airway temperature at that level. Zero time denotes start of exercise, which lasted for about 4 min. Vertical and horizontal lines are 1 SEM.

Adapted from Deal et al. 1979 (104).

$$HL_e = \dot{V}_E \times HV\ (WC_E - WC_I)$$

in which HL_e is evaporative heat loss in kilojoules per minute; HV = latent heat loss of vaporization of water, which equals $2.43\ kJ \cdot g^{-1}$; WC_E and WC_I = expiratory and inspiratory water vapor content, respectively, in grams per liter. Evaporative cooling increases with the rise in gradient between the expired and inspired air and with the increase in ventilation. It also increases as the air is warmed, because relative humidity in the air drops as its temperature increases.

By combining these two equations one obtains the overall respiratory heat loss (RHL) as follows:

$$RHL = \dot{V}_E\ [HC\ (T_E - T_I) + HV\ (WC_E - WC_I)]$$

The evaporative loss is ordinarily the dominant element in the overall RHL. For example, to induce cooling of 0.006 kJoule per 1 L of air, it is sufficient to dry the inspired air (at 37° C) from 100% to 50% relative humidity. To obtain the same RHL without drying the air, one would have to cool it from 37° to minus 10° C! It should further be realized that when inspired air is warmed while progressing through the airway, its relative humidity drops, which further increases the evaporative rate from the mucosal surface. With the high ventilation of exercise, heat loss from the mucosa may occur as deep as at the 10th generation of airways (225). The greater the ventilation, the deeper is the cooling in the bronchial tree.

The increase in RHL during exercise is achieved primarily by hyperpnea (increase in ventilation) and, to a lesser extent, by an increase in T_E. Ventilation can rise as much as 10-fold in most active children, increasing RHL by that factor. Some investigators have gone as far as to suggest that the dominant, or even exclusive, role of exercise in the genesis of EIA is to increase ventilation (105; 148)—all other changes such as hypocapnia, acidosis, hormonal changes, or hypoxemia being immaterial. As a result, the term *thermally induced asthma* has been suggested to replace "exercise-induced asthma" (255).

The relationship between the degree of bronchoconstriction and RHL (106) is close to linear (figure 6.8) (79). Such a relationship holds true for a wide range of inspired air temperatures (–10 to +80° C) and levels of humidity (0-100% relative humidity) (105). There is, however, a large scatter in the degree of RHL that causes a

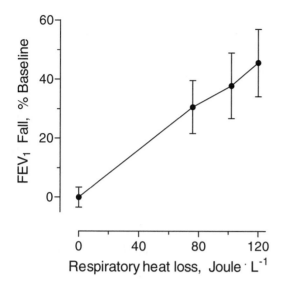

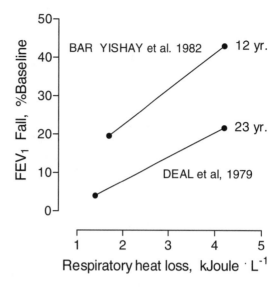

FIGURE 6.8 Exercise-induced asthma as a function of respiratory heat loss. Eight adolescents and young adults with asthma walked on a treadmill at four different combinations of air humidity and temperature. FEV_1 fall = the postexercise decrease in forced expiratory volume in the first second, expressed as a percentage of pre-exercise level. Mean ± 1 SEM. Data from Chen and Horton (1977) (79).

FIGURE 6.9 Effect of age or body size on the relationship between exercise-induced asthma and respiratory heat loss. A comparison of post-exertional fall in forced expiratory volume in the first second (FEV_1) between children and adults. Data from Bar-Yishay et al. 1982 (32) and Deal et al. 1979 (106).

given level of EIA in children with asthma (9). There are also differences in response to a given heat loss between children and adults. At a given RHL, children respond with greater EIA than do adults (figure 6.9). The reason may be that any given absolute ventilation is *relatively* greater for the child, who has a smaller surface area of respiratory mucosa.

The cooling mechanism in itself is insufficient to explain why the bronchoconstriction occurs after, and not during, exercise. One postulated mechanism is that airway cooling is followed by rapid *rewarming of the airways* after exercise, which provides the actual trigger for bronchoconstriction (146; 258). Apparently, the airways of patients with asthma rewarm faster than among healthy individuals (146). The resulting hyperemia may contribute to the narrowing of the airways (148).

Airway Drying As a Trigger

The role of airway cooling in triggering EIA is now universally accepted (256). However, airway cooling may not be the *exclusive* trigger. Several observations have raised the possibility that drying of the airways induces osmotic increases at the respiratory tract mucosa, which can trigger

EIA even without airway cooling (15; 19; 170; 324; 370; 401). Hyperosmotic solutions, when applied to the airways, induce release of histamine within 1 to 5 min, which coincides with bronchoconstriction, but there is no concomitant release of mast cell mediators (244). Interestingly, hypo-osmotic solutions can also trigger bronchoconstriction (287).

The importance of airway drying is not clear, however, as suggested by other investigators (255) who found that cooling of the airways in the lower respiratory tract causes bronchoconstriction, even in the absence of water loss.

As stated earlier, the absence of bronchoconstriction *during* exercise may be explained by the need for post-exertional rewarming of the airways. Alternatively, it may result from the bronchodilatory effect of catecholamines and the sympathetic drive, both of which are increased during exercise and subside during recovery (151). However, the asthmogenic effect of cold air is not due to suppression of catecholamine release (6). The greater sensitivity of asthma patients to airway cooling may therefore reflect their greater nonspecific airway reactivity (e.g., faster rewarming), rather than an inability to condition the inspired air.

Vagal Pathways Versus Inflammatory Mediators

While airway cooling and drying of the mucosa by exercise hyperpnea, as well as postexercise airway rewarming, are likely triggers of EIA, the exact cascade of events that follow is as yet unclear. There are two schools of thought regarding these events. One theory assumes vagal propagation from receptors located in the oropharynx, trachea, and large bronchi (261; 435). The other assumes a release from mucosal sensitized mast cells of inflammatory mediators such as histamine, bradykinin, neutrophil chemotactic factor, leukotrienes (LTC_4, LTD_4, and LTE_4), and prostaglandins (10; 111; 124; 151; 240; 247; 262). Using leukotriene antagonists, one can prevent or attenuate EIA (124; 240; 247), which is suggestive of the role played by these mediators in the genesis of EIA. However, evidence for both the vagal pathways and the mediator theories is circumstantial, based on response to drugs, in vitro findings, mechanisms found at rest, or the presence of a refractory period. A suggestion has been made that both mechanisms may be in effect: Obstruction of small airways may be a result of mediator release, whereas large airway obstruction is generated through vagal reflex pathways (257). There is no proof, though, that this distinction actually exists.

Airway Cooling and Drying and the Asthmogenicity of Swimming

The low asthmogenicity of swimming is well documented (see the section "Exercise-Induced Asthma—Nature of the Exercise Provocation"). Various theories have been put forward to explain this phenomenon: a lower pollen content over the water; high hydrostatic pressure on the chest while the body is immersed in water, which would help to reduce the expiratory muscle work; hypoventilation and CO_2 retention due to the controlled breathing pattern; peripheral vasoconstriction, which would increase central blood flow and may counteract the respiratory heat loss; a horizontal body position, which would decrease ventilation compared with an upright position; and high humidity of the inspired air, with a resulting low respiratory heat loss. Only three of these possibilities have been evaluated experimentally.

In a study by Inbar and colleagues (197), 12- to 16-year-old boys with asthma performed an 8-min shoulder extension-flexion task while standing up and, on another day, while lying prone. Ventilation, air temperature, and humidity were identical in the two trials. The two tasks induced a similar reduction of FEV_1 (20-22%), suggesting that body position per se cannot explain the low asthmogenicity of swimming. In another study by the same group (198), boys with asthma performed isocapnic hyperpnea while lying fully immersed in water and while standing upright at poolside. Again, air temperature and humidity were identical. There was no difference in the drop of FEV_1 between the two regimens, suggesting that the combination of body posture and immersion in water cannot, in itself, explain the low asthmogenicity of swimming.

It was suggested in 1977 (31) that the protective nature of swimming is due to the highly humid air that the swimmer inhales at water level. This hypothesis was tested experimentally by letting children with asthma tether-swim while inhaling dried air, thereby increasing the respiratory heat loss (32; 196). When the inspired air was dried down to 25% to 30% relative humidity, swimming did not trigger EIA (196). However, as shown in figure 6.10, drying the air further to 8% relative humidity (2 mg per L H_2O) did generate bron-

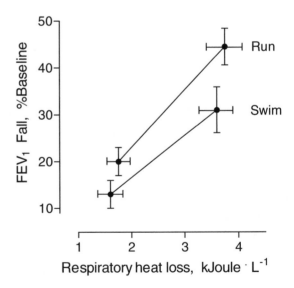

FIGURE 6.10 The protective nature of swimming against exercise-induced asthma, as a function of respiratory heat loss. Thirteen adolescents with asthma ran and swam at neutral temperature while inhaling dry or humid air. FEV_1 fall = postexercise decrease as percentage of resting FEV_1 (forced expiratory volume in the first second). Mean ± 1 SEM. Data from Bar-Yishay et al. 1982 (32).

choconstriction in the swimming patients. This confirmed the relevance of air humidity to the protective effect of swimming. Still, the degree of EIA was lower than in running, even though equal levels of heat loss were achieved (32). It thus seems that some unidentified mechanism, in addition to the protection of humid air, renders swimming less asthmogenic than other sports.

Other Physiologic Responses to Exercise

Most children with asthma respond to aerobic exercise with adequate ventilatory and hemodynamic changes. In fact, during exercise they often show greater bronchodilation than do nonasthmatics (117; 248). While ventilation is appropriate to the metabolic level of mildly and moderately affected patients (117; 121; 318), it is excessive in those who are severely affected (38; 143). Such high ventilation is associated with a high tidal volume. The work of breathing in severely affected patients is abnormally higher since pleural pressure swings during the respiratory cycle are exceedingly high and the mean pleural pressure is highly negative (377). A high tidal volume and a low respiratory frequency during exercise occur also when the asthma is mild to moderate (318). Alveolar gas exchange and lung diffusion capacity are usually normal in the exercising patient (121; 137; 143; 157). However, children who develop marked post-exertional bronchoconstriction may have a rise in alveolar-arterial O_2 pressure difference and mild O_2 desaturation (215). The integrity of pulmonary function during exercise is related to the fitness level of the patient. For example, in a study with young adult patients, expiratory airflow during exercise increased more in those who were trained compared with sedentary controls (167).

Heart volume and blood volume are normal in the child with asthma (39; 158), as are cardiac output and stroke volume during submaximal and maximal exercise (137; 158). It has been suggested (39; 416) that children with asthma resort more than healthy ones to anaerobic metabolism during submaximal exercise, as manifested by higher levels of blood lactate. Neither the mechanism nor the implications of this phenomenon have been elucidated.

In conclusion, respiratory and circulatory functions during exercise are adequate in most patients with asthma. Only severely affected patients have respiratory deficiencies that affect their O_2 transport system.

Habitual Activity of the Child With Asthma

Exercise is often equated with physical distress by the child with asthma—perhaps more so than in any other chronic disease. An attack of EIA is alarming to the child and onlookers alike. Parents tend to become overprotective (301; 339) and to limit the physical activity of their children, who are progressively perceived as "losers" by themselves and their peers. In spite of this concern, information from large-scale surveys in various countries provides equivocal evidence for hypoactivity in childhood asthma. While some suggest that the patients are less active than the general child population, others show no difference in activity or an even higher activity level in the patients.

For example, according to a survey of Swedish parents, 420 of their 10,527 children, ages 7 to 16 years, had asthma (57). During the preceding year, 40% of these children had more than 10 days of restricted physical activity due to asthma. It is noteworthy that asthma symptoms at school were usually related to sporting activities (57). In the United States, 17,110 households participated in the 1988 National Health Review Survey I (403). Almost 30% of children and adolescents with asthma had some limitation in physical activity, compared with only 5% of those without asthma. Earlier studies also documented lower activity levels among asthma patients in the United States (193; 417). In one survey, for example (417), only 52% of 262 physical education instructors stated that pupils with asthma consistently attended regular physical education classes. Forty-five percent reported that these pupils attended intermittently. In the same survey, 76% of instructors allowed the asthmatic patients to determine their own activity level in class.

The most comprehensive study to date of physical activity and childhood asthma was conducted by Nystad in Norway (283). A questionnaire was administered to 4,585 schoolchildren, with an 85% response rate. Whether determined as times per week or as hours per week, the activity level of those with asthma was not different from that of the general sample. More than 50% of all children took part in organized sport. A similar level of sport participation was found also in South African children with asthma and their healthy peers (404). Likewise, a survey among Israeli schoolchildren showed no difference in physical education attendance rate between asthma patients and their healthy peers (61).

In a Canadian study, energy expenditure of leisure time activities was calculated for 16,813 adolescents (ages 12 or older) and adults (81). On the average, energy expenditure per kilogram body weight was some 25% *higher* in males with asthma than in the general male sample (2.47 vs. 1.98 kcal $\cdot$ kg^{-1} $\cdot$ day^{-1}). Among the females, energy expenditure of those with asthma was 15% higher than in the general sample (1.77 vs. 1.54 kcal $\cdot$ kg^{-1} $\cdot$ day^{-1}). The higher energy expenditure among those with asthma was particularly evident during adolescence.

It is hard to reconcile the discrepancies among these findings. They may result from methodological differences among the surveys, as well as from cultural differences in parental and teacher attitude toward physical activity and asthma morbidity. A relatively high rate of absenteeism from physical education classes among pupils with asthma may reflect ignorance of the means for prevention of EIA, as well as fear of medico-legal consequences.

Physical Fitness of the Child With Asthma

The fitness of asthma patients has ranged from very low to Olympic championship levels. An impressive number of adults with asthma have reached world-class levels in sport, especially in swimming, in which they have often won Olympic gold medals (127), but also in more "asthmogenic" sports such as distance running, field hockey, or cycling. Asthma seems to be more prevalent among elite athletes than in the general population. A survey in Norway has shown a high prevalence in adult strength and endurance athletes, particularly those who train 20 hr per week or more (284). In the 1984 Los Angeles Olympic Games, 67 of 597 (11%) U.S. athletes reported having EIA (312). Forty-one of them won medals (15 gold) in 14 different sports.

Aerobic Fitness

In nonathletic pediatric patients, maximal aerobic power ranges from normal (39; 48; 50; 157; 394; 416) to moderately low (144; 300; 354; 394; 415).

The mechanism for low aerobic fitness is not clear. Attempts to identify medical and psychological factors that contribute to poor fitness have yielded very little insight. In one study (394), three medical variables explained most of the variance in aerobic fitness: recent exacerbation of the disease, a low FEV$_1$, and a low specific air conductance. However, the three combined accounted for only 8.1% of the variance. Psychological characteristics such as relationship with family and peers, adaptation to school, and the presence of psychiatric symptoms added only 1% to the prediction. The ability of children with asthma to achieve normal maximal aerobic power through physical conditioning strongly suggests that low fitness is not inherent to their disease (243; 277).

It therefore seems that a sedentary lifestyle may be the main cause of low fitness, irrespective of the child's pulmonary function (123).

Anaerobic Fitness

Although asthma is a pulmonary disease, some studies have focused on anaerobic muscle performance in people with asthma. The rationale for such investigation is that the disease is sometimes accompanied by undernutrition and that patients, with or without chronic use of steroids, may manifest growth retardation and delayed puberty (333). Moreover, because children with asthma are often insufficiently active, their anaerobic performance may be deficient. Findings in these studies are inconsistent. While some show normal anaerobic fitness, as measured by the Wingate anaerobic test (48; 50) or the accumulated O_2 deficit method (69), others show low performance among patients with asthma using the force-velocity test (91) or the Wingate anaerobic test (92). Such inconsistencies are hard to explain, but may reflect interstudy differences in the severity of asthma, nutritional and pubertal status, and the level of physical activity.

Exercise As a Diagnostic Tool in Asthma

Exercise-induced asthma should be suspected whenever a child with asthma or a parent reports exercise-related cough (often the only complaint), shortness of breath, or wheezing. Sometimes, the only complaint is exertional chest pain. Although most cases of EIA can be diagnosed by history, an exercise challenge test is sometimes indicated.

Rationale for Exercise Testing

Exercise testing of children with asthma, or when asthma is suspected, should be considered for the following reasons.

Documentation of Exercise-Induced Asthma   Documenting EIA is especially important for the child who presents with less specific exercise-related complaints such as chest pain or cough. One should bear in mind that some patients develop EIA during spontaneous activities but not in the laboratory and that others show the reverse pattern. Exercise testing is also useful for patients who are thought to have EIA but have not responded to a therapeutic trial of inhaled bronchodilators.

Evaluation of Athletes   Athletes with no prior diagnosis of asthma, but with marked exertion-related dyspnea, may benefit from an exercise provocation test.

Evaluation of Medication   A standardized challenge test, repeated using different drug regimens, helps to determine the efficacy of a specific drug or adequacy of a certain dose (12; 150; 396). One must realize, though, that such information is specific to pharmacologic effects on EIA and not on asthma in general.

Diagnosis of Hyperreactive Airways   Exercise can be used as a nonspecific challenge test to determine asthma in patients with atypical respiratory symptoms, or in those who have extended symptomless periods. This is analogous to challenge by cold or to inhalation of histamine or methacholine. The sensitivity of an exercise challenge for this purpose is comparable to or somewhat lower than that of histamine (264) or metacholine (242). In a school-based study (168), the frequencies of EIA and of bronchial hyperreactivity to metacholine were compared. While some of the children responded positively to both challenges, others responded to only one of them. The authors concluded that the two challenges identify different abnormalities of the airways. They further suggested that exercise challenge proved to be a practical epidemiologic tool for objective measurements of bronchial responsiveness in children.

Screening for Asthma   Exercise-induced bronchoconstriction may occur in a child before asthma has been diagnosed. Epidemiologic studies have been conducted to determine whether exercise provocation can be used as a screening test for asthma (67; 168; 203; 421). For example, primary schoolchildren in Wales, with and without diagnosed asthma, performed a 6-min free-running provocation test. Among 864 pupils not known to have asthma, 60 had a positive EIA test. Of 92 children known to have asthma, 33 responded with a positive test, and 7 could not complete the run because of marked shortness of breath. This denotes a sensitivity of 43% and a specificity of 93% (203). In a 6-year follow-up, 32 of 55 (58%) children who did not have asthma in the original survey, but had a positive EIA test, were later diagnosed with asthma. The low sensitivity found in this and other studies (421), however, reduces the usefulness of an exercise test as a screening tool for asthma.

Determination of Exercise Tolerance   As in other diseases, exercise tolerance is an important aspect to consider before one constructs an exercise program. A deconditioned child may have high ventilation at a relatively low work rate, with a resulting high respiratory heat loss and EIA.

Instilling Confidence in Child and Parent   A benefit of testing that is often underestimated is its ability to instill confidence. Both parent and child can gain confidence from the realization that the child can sustain high-level activity.

The Optimal Exercise Challenge

There are two conflicting considerations in administering an exercise challenge test: (1) optimizing the provocation so that maximal bronchial response is obtained and at the same time (2) ensuring that the test is not detrimental to the child's health. The following guidelines take these considerations into account.

Pretest Preparation   The following steps should be taken:

1. Obtain medical history with emphasis on the cardiopulmonary system, habitual activity, and exercise-related symptoms.
2. Perform physical examination with emphasis on the cardiopulmonary system.
3. Formulate the rationale for exercise testing in each patient.

4. To alleviate their apprehensions, explain to the child and the parent the purpose and protocol of the test. Let the child become familiar with the procedures and equipment.

5. Obtain baseline pulmonary functions. For clinical purposes, sufficient information can be obtained with peak flowmeter or, preferably, a positive displacement one-breath spirometer.

6. *Do not proceed* with an exercise test if resting FEV_1 or peak expiratory flow is less than 60% of height-predicted level, or if the child is dyspneic at rest (115). Pre-exercise wheezing *per se is* not a contraindication.

Withdrawal of Medication Before an Exercise Test   Pretest withdrawal of anti-asthma drugs depends on the purpose of the test. If one wishes to maximize the response (e.g., to document EIA or airway hyperreactivity), all drugs, with the exception of corticosteroids, should be withdrawn. β_2-agonists, methylxanthines, and anticholinergics should be discontinued at least 8 hr prior to testing (12 hr for long-acting β_2-agonists and methylxanthines). Disodium cromoglycate should, ideally, be withdrawn 24 hr before exercise testing (115). If, on the other hand, the exercise test is conducted to assess the efficacy of medication, withdrawal of drugs will be selective.

The Exercise Protocol   Unlike tests of aerobic fitness that are of the multistage "progressive" type (see appendix II, "Prototypes of Exercise Tests"), an exercise challenge test for EIA should comprise a single 6- to 8-min stage. The intensity should be such that heart rate reaches 85% to 90% of maximum (170-180 beat · min^{-1} for children and adolescents).

When the child is tested for the first time, the exact power (or speed and slope on the treadmill) that raises heart rate to this level is still unknown. The investigator must therefore select an arbitrary initial level and, based on interim heart rate, modify it within 2 to 3 min until the target heart rate is reached (see description of the single stage with adjustments protocol in appendix II). A test is technically successful if the required load is kept for *at least* 5 min. Heart rate should be monitored periodically by a heart rate monitor. The former is preferable because it can disclose coincidental dysrhythmia or other electrocardiographic abnormalities.

The ergometer of choice is a treadmill, but a test with a cycle ergometer will yield similar results. If these are unavailable, a step with a standardized height would be a good alternative. The main advantage of treadmill running is for young children, who can master the technique and sustain the effort better than with other ergometers. Furthermore, test-retest reliability seems to be higher with the treadmill than with the cycle ergometer (11). Some clinicians use free running in or around the hospital, or up and down a flight of stairs. Although such tasks are asthmogenic, they cannot be sufficiently standardized and are not recommended.

To safeguard standardization of an exercise challenge test, one should adhere to the following guidelines:

• Have the child avoid exertion for at least 2 1/2 to 3 hr prior to testing. Failure to do so may lead to less than full bronchial response, as the child will be in a refractory period following prior activity (111). If at all possible, have the child avoid intense exercise for at least 10 hours prior to testing. This will reduce the likelihood of bronchconstriction that results from a late response to prior exercise rather than from the test itself.

• Climatic conditions, especially humidity, should be standardized. The use of commercially available air conditioners, humidifiers, and dehumidifiers is sufficient for clinical purposes. In select cases we use a climatic chamber to induce a cold environment (e.g., 5° C) for a combined exercise-plus-cold provocation. Another approach is to use a heat exchanger that cools the inspired air to around 0° C (278).

• When a test must be repeated, do not allow an interval shorter than a day. An interval of at least a day will reduce the likelihood that a late asthmatic response to the previous exercise test will affect the subsequent response. Likewise, do not wait more than a week for the second test, because test-retest variability at longer intervals is high.

• To minimize diurnal variations, attempt to retest any given child at the same time of day.

Pulmonary Function Tests   Baseline testing is done just prior to exercise. If a particular drug is being evaluated, a preliminary pulmonary function test should precede its administration. The timing of postexercise testing varies among laboratories. Some perform the test at 2, 5, and 10 min postexercise, while others repeat it also at 15 and 20 min. In the case of a positive response, testing continues every 5 min thereafter until EIA subsides (with or without administration of a bronchodilator). Note that 10% to 15% of responders may have their greatest FEV_1 drop only at 30 min postexercise (62), as summarized in figure 6.11.

Interpretation of Post-Exertional Bronchoconstriction

Assuming A = pre-exercise level of, for example, FEV_1 and B = the lowest postexercise level, the degree of EIA can be expressed in a number of ways. The most common one is

$$(1)\ EIA = (A - B)\ /\ \times 100$$

or the maximal drop in function as a percentage of pre-exercise function. Another expression of EIA is the degree of drop as a percentage of the height-predicted (P) function:

$$(2)\ EIA = (A - B)\ P\ 100$$

This approach is more relevant to the *clinical* status of the patient prior to the test. The child may have, for example, a small drop according to equation (1); but due to low A, the drop in

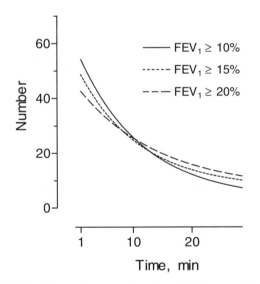

FIGURE 6.11 Time after exercise when the increase in airway resistance is maximal. A total of 397 school athletes, 12 to 18 years old, performed a 6-min treadmill test that raised their heart rate to ≥ 170 beat $\cdot$ min^{-1}. Three criteria for a positive drop in forced expiratory volume in the first second (FEV_1) were used: 10%, 15%, and 20% of baseline.

Adapted from Brudno et al. 1994 (62).

function may be sufficient to induce a major obstruction and discomfort. A third approach is to express EIA in absolute terms and not as a percentage of resting value:

$$(3)\ EIA = A - B$$

In assessing the effect of a drug, one should separately evaluate the pre-exercise bronchodilatory effect (see "Pharmacotherapy") and the amelioration of EIA. One assesses the bronchodilatory effect by comparing the pre- and postdrug functions, both measured before exercise. To assess the affect on EIA, some investigators use the predrug and others the postdrug resting value as a baseline.

Another consideration in evaluating anti-EIA drugs is their placebo effect, which is often very marked (7; 154; 360; 376). While relevant to research, this aspect also has clinical implications. To assess a placebo effect one must test the child twice, using drug and placebo in a blindfold fashion. Various mathematical approaches are available for such a comparison (for details see Johnson 1979 [202]).

The criterion for a "positive exercise test" (i.e., EIA of a clinically significant degree) is not uniform among laboratories. Using equation (1) it ranges between 10% and 25%. Most workers in the field, however, advocate a 10% or 15% maximal drop as their criterion. Our own criterion is 15%.

Management of the Child With Exercise-Induced Asthma

In spite of the risk that exercise will trigger asthmatic response, patients with asthma should be encouraged to take part in physical activities, with little or no limitation. The American Academy of Pediatrics, in a Position Statement (3), declared, *"as a general rule, every effort should be made to minimize restrictions and to invoke them only when the condition of the child makes it necessary."* Indeed, there are currently several ways to help the child with asthma lead an active lifestyle (page 155).

The main goal in managing the exercise responses of the child with asthma is the *prevention of EIA*. With today's therapeutic means, EIA is preventable in the great majority of patients. Indeed, most patients can lead an active lifestyle and successfully participate in sport at all levels. Management of EIA can be divided into pharmacologic and nonpharmacologic therapies.

Pharmacotherapy

Treatment can be initiated as prophylaxis or given once EIA has begun. As a general rule, the prophylactic approach is much preferable. Children often do not plan their activities in advance, however, and they may resort to medication only after symptoms have begun.

Figure 6.12 is an algorithm for the drug therapy strategy. The first line of drugs in current use comprises the inhaled β_2-adrenergic agonists. Another potent and widely used drug is inhaled disodium cromoglycate (cromolyn). A similar derivative is nedocromil sodium. While corticosteroids are the medication of choice for ongoing prevention of airway inflammation and edema in asthma, a single dose prior to exercise does not seem to ameliorate EIA. Also in use are methylxanthines, anticholinergics, leukotriene antagonists, and calcium channel blockers.

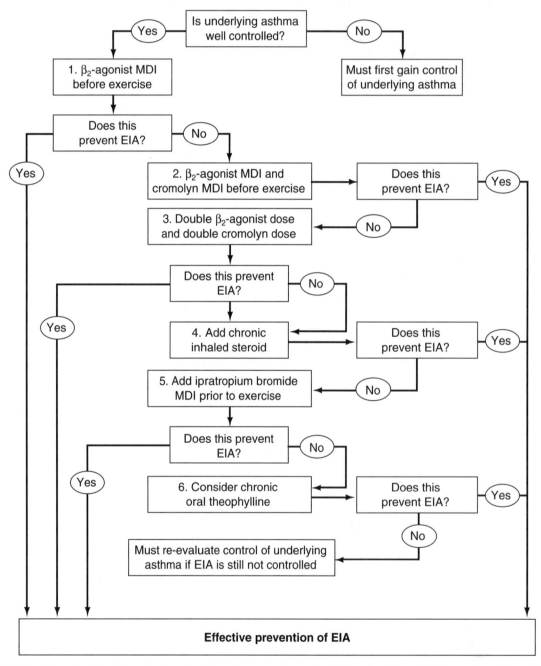

FIGURE 6.12   Flowchart for the use of medications to prevent exercise-induced asthma. MDI = metered-dose inhaler.

Adapted, with permission, from Morton and Fitch 1992 (269).

Short-Acting β₂-Agonists Clinicians worldwide have been prescribing short-acting β₂-agonists as the first line of prevention. The inhaled form of such medication is preferable to oral preparations because of its fast action and focused effect on the airways, which reduces the likelihood of systemic side effects. When used 10 to 15 min before exercise, a β₂-agonist (one to two puffs) will prevent or ameliorate EIA in almost 100% of patients (figure 6.13) (13; 152; 183). Its protective effect lasts 1 to 6 hr, depending on the specific preparation (14; 37; 344; 368; 429). The duration of the effect can be extended when a β₂-agonist is taken together with cromolyn sodium (429). Thanks to their bronchodilatory effect and direct action on the airways, the β₂-agonists can also relieve asthmatic symptoms *before* the start of exercise and potentiate bronchodilatory changes during the activity (368).

The β₂-agonists are the drug of choice for stopping EIA once it has started. While some children will respond to one puff, others may need two or even three puffs, 3 to 4 min apart. Some patients respond better to nebulized salbutamol. The basic problem in reversing EIA, particularly if

severe, is the inability of the distressed child to inhale properly while dyspneic. In addition, the narrowed airways may impede penetration of the drug even when inhalation is mechanically adequate. An alternative approach to inhaling β₂-agonists is the use of a transdermal application. While this approach makes sense, there is very little information about its efficacy in preventing EIA (327).

A word of caution: Evidence generated in the last decade (346; 375) suggests that the increase in asthma morbidity and mortality may be related to

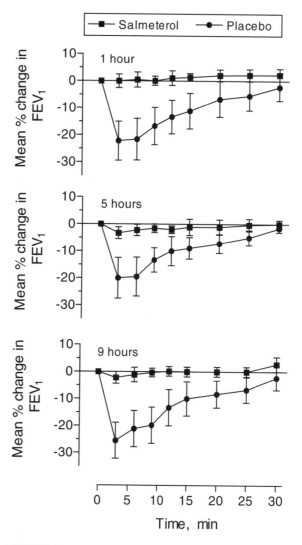

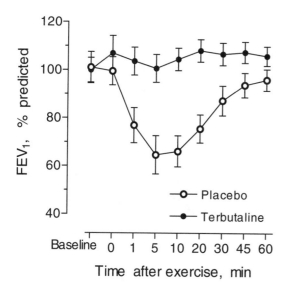

FIGURE 6.13 The prophylactic effect of a short-acting β₂-agonist on exercise-induced asthma. Ten 10- to 14-year-old children with asthma performed a 5-min cycle ergometer test at 90% of their predicted maximal heart rate. They inhaled dried air (0% relative humidity) at 20° C. Fifteen minutes prior to the test they inhaled two puffs of either terbutaline (0.5 mg) or a saline buffer. Vertical bars denote SEM.

Reproduced, with permission, from Dinh Xuan et al. 1989 (108).

FIGURE 6.14 The effect of long-acting β₂-agonists lasts at least 9 hr. Thirteen children with asthma and documented exercise-induced asthma took part. They exercised 1, 5, and 9 hr following the administration of either a single dose of 50 μg salmeterol or a placebo.

Reproduced, with permission, from C.P. Green and J.F. Price, 1992, "Prevention of exercise induced asthma by inhaled salmeterol xinafoate," *Archives of Diseases in Childhood* 67: 1014-1017.

the pervasive use of β_2-agonists as maintenance therapy. Patients and parents should be advised strongly against this practice. β_2-agonists should be taken only when asthma symptoms occur, or in prevention of EIA.

Reports exist of athletes without asthma who use inhaled β_2-agonists for their assumed cardiovascular doping effect (128; 269). Evidence suggests that while some β_2-agonists (e.g., clenbuterol) have a performance-enhancing effect (371), salbutamol (albuterol in the United States) and terbutaline do not have doping properties (263; 271; 358). A study with healthy, nonathletic prepubertal boys showed no beneficial effect of an inhaled β_2-agonist on submaximal running economy (409).

Long-Acting β_2-Agonists   Long-acting β_2-agonists such as salmeterol (160; 220; 276) and formeterol (52; 253) maintain their prophylactic effect up to 9 hr or more (figure 6.14). The benefit of such medication is that the child does not need to carry it during the day or to administer the drug just prior to exercise. The latter is an important advantage for children, whose activity often occurs spontaneously and in spurts. Long-term

β_2-agonists are suitable primarily for individuals who are physically active repeatedly throughout the day.

A concern has been raised that the efficacy of long-acting β_2-agonists diminishes with time (109; 276; 362). A study with adults (276) has shown that the protective effect of salmeterol against EIA was sustained over 30 days of daily administration. However, the duration of the effect on a given day became shorter with time. A similar pattern was shown for 12- to 16-year-old patients who were monitored during 28 days of therapy with salmeterol (362) (figure 6.15). One possible solution is to administer the medication twice a day, but more research is needed to determine the long-term effect of this approach.

Cromolyn Sodium   Cromolyn sodium is another excellent prophylactic agent, but is somewhat less effective than inhaled β_2-agonists. Approximately 70% to 80% of patients will benefit from it (152). It is particularly effective for patients with atopic asthma. Unlike β_2-agonists or methylxanthines, cromolyn does not cause bronchodilation at rest (152; 272; 367). Based on in vitro data, it apparently stabilizes the mast cell membrane and prevents

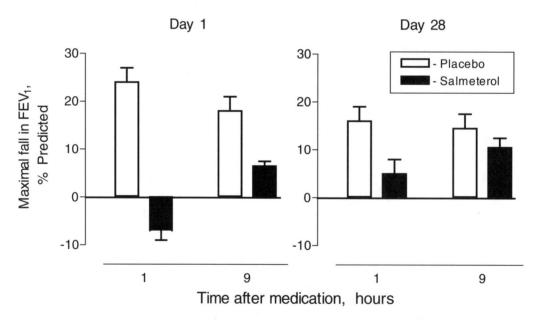

FIGURE 6.15   Changes in the protective effect of a long-acting β_2-agonist over time. Fourteen subjects (12-16 years old) took part in a randomized, placebo-controlled, double-blind crossover study in which they were treated for 28 days with two puffs of either 25 µg salmeterol or placebo. The medication was taken each morning at 8 a.m. An 8-min treadmill test was conducted 1 and 9 hr later. Intensity was set at the lower of 90% maximal heart rate or 180 beats · min^{-1}. Data are maximal postexercise fall in forced expiratory volume in the first second (FEV_1) as a percentage of predicted. Vertical bars are 1 SEM.

Modified, with permission, from Simons et al. 1997 (362).

the release of mediators from these cells. There is, however, indirect in vivo evidence that it may act through other mechanisms (119). Cromolyn seems to act mostly on the large airways and less on the small ones (397). It is especially indicated for patients who are asymptomatic prior to exercise and who do not respond well to β₂-agonists.

A major advantage of cromolyn and nedocromil (see next section) is that they have no side effects and are therefore the safest among EIA medications. These drugs can be taken orally or can be inhaled. The oral route is used for maintenance, but the drugs are ineffective for prevention of EIA when taken by mouth before exercise (97). The optimal timing for inhalation is 15 to 30 min before exercise, but there is some effect when the medication is taken just prior to the activity (360).

Cromolyn is highly effective for up to 2 hr (228), with some effect still detected after 4 hr. Its effectiveness can be extended when it is taken together with a β₂-agonist (429). It is ineffective if taken once exercise has begun (360). To maximize its effect, a dose of cromolyn should be taken before exercise, *in addition* to the regular daily dose (unless exercise happens to occur soon after the inhalation of a regular dose). Children whose EIA cannot be prevented by either β₂-agonists or cromolyn alone can benefit from a combination thereof.

Nedocromil Sodium   *Nedocromil sodium* is chemically related to cromolyn and, when taken by metered-dose inhaler (2 mg per puff), has a similar prophylactic effect on the exercising child (99). An even greater effect is elicited when the medication is taken together with the diuretic *furosemide* (282). The rationale for the use of furosemide is that it interferes with ion and water movement through the airway epithelium, thus negating exercise-induced osmotic changes. Another possible mechanism is the release of bronchodilator prostaglandins (304).

Methylxanthines   *Methylxanthines* (especially theophylline), while sometimes prescribed for the general control of asthma, are less often used for prevention of EIA. The primary reasons are the greater efficacy of the β₂-agonists and cromolyn and the greater toxicity of the methylxanthines. Like β₂-agonists they are potent bronchodilators at rest. They can reduce EIA in 65% to 80% of

patients (7; 24; 42; 152; 207). This effect is dose dependent (41), minimal therapeutic plasma concentration for theophylline being 10 µg/ml (314). There is a low correlation between the efficacy of methylxanthines in the general control of asthma and their prophylactic effect in EIA (152).

Anticholinergics   *Anticholinergics*, such as ipratropium bromide aerosol, have been efficacious in some patients and not in others (7; 55; 428; 430). They counteract the bronchoconstrictive effect of the parasympathetic system. One use of ipratropium bromide has been with patients who do not benefit from the combined regimen of β₂-agonist and cromolyn (330). Why some individuals benefit from anticholinergic drugs and others do not is unclear. One possibility that needs further study is that this reflects variability in parasympathetic tone among people. It may be that those individuals with a high tone would respond favorably while others would not (226).

Leukotriene Antagonists   The leukotriene antagonists were introduced as a result of studies that demonstrated the role of leukotrienes in the development of EIA (247) (see earlier section "Mechanisms Underlying Exercise-Induced Asthma"). The most fully studied compound is montelukast (Singulair in the United States and Canada), which is a leukotriene-receptor antagonist. It is administered orally as a chewable tablet. This gives it an advantage over inhaled medications. Its protective effect lasts as long as 20 to 24 hr in adults (240) and in children (221), particularly when the severity of asthma is mild to moderate. Data so far suggest that the efficacy of montelukast does not wane over time (figure 6.16). While montelukast is not effective in all patients, it seems to be safe and not to have interaction with other medications (175).

Corticosteroids   *Corticosteroids* are the mainstay of treating airway inflammation and edema in asthma. However, when administered in a single dose, they have no effect on EIA. In contrast, when given daily on an ongoing basis, they reduce the likelihood of EIA (418). Another suggested role for corticosteroids is that they potentiate the effect of other anti-EIA agents (9; 182; 184; 414; 418).

Nasal Breathing As Protection

The nose has long been recognized as a highly efficient conditioner of inspired air (316). Nasal

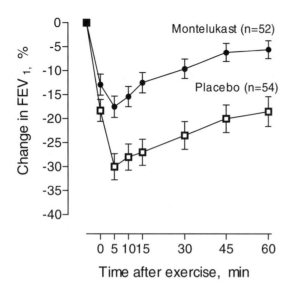

FIGURE 6.16 Leukotriene-receptor antagonist protects against exercise-induced asthma. A total of 106 adolescent and adult asthma patients were tested after 12 weeks of treatment with oral montelukast or with a placebo. Vertical lines denote 1 SEM.

Reproduced, with permission, from Leff et al. 1998 (240).

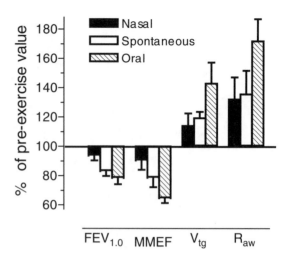

FIGURE 6.17 Post-exertional changes in pulmonary functions as related to the mode of breathing. Twelve 7- to 14-year-old asthma patients walked on three different days on a treadmill at 75% to 85% of maximal heart rate, using different inspiratory modes. Ambient conditions were 20° to 22° C and 25% to 30% relative humidity. Measurements were taken 7 to 9 min postexercise. FEV_1 = forced expiratory volume in the first second; MMEF = maximal midexpiratory flow; V_{tg} = thoracic gas volume; R_{aw} = airway resistance. Vertical lines denote 1 SEM.

Adapted from Shturman-Ellstein et al. 1978 (357).

breathing can very effectively change the characteristics of inspired air so that even in extremely low ambient humidity and temperature, the air reaching the nasopharynx is humidified to near 100% relative humidity and its temperature reaches 32° to 37° C (316; 317). As seen in figure 6.17, whereas oral breathing during a treadmill walk (even when ambient air is not cold) induces marked bronchoconstriction, an increase in airway resistance, and air trapping in the lungs, nasal breathing attenuates these responses and, in practical terms, protects the child against EIA (357). Similar effects have also been shown during free running (246) and during isocapnic hyperventilation at rest (435). In addition to its air-conditioning effect, nasal breathing seems to partially protect against the bronchoconstriction that is induced by sulfur dioxide when asthma patients exercise in a polluted environment (223).

The data in figure 6.17 refer to mild or moderate exercise intensities. There are two potential problems with nasal breathing during more intense activities, particularly when performed in the cold. One is the small size of the nasal vestibule, which can accommodate less airflow compared with oral breathing. The other is the increase in nasal airway resistance in the cold (88). On the plus side, nasal resistance to flow *decreases* with exercise (134; 209; 323; 349); the higher the exercise intensity,

the greater the nasal patency (134). The exercise-induced decrease in nasal resistance is, at least in part, due to a rise in sympathetic activity during exercise (209; 289). This spontaneous decrease in nasal resistance does not seem, in itself, to prevent EIA in patients with asthma (392).

One suggested means of artificially increasing the patency of the nasal vestibule is through *external nasal dilator strips*. These strips, which are available commercially, effect a reduction in nasal resistance during exercise and, to a lesser extent, delay the spontaneous switch from nasal to oral breathing (350).

Artificial Means for Warming and Humidifying the Inspired Air

An alternative to nasal breathing is the use of a face mask to warm and humidify the inspired air on cold days. The mask, which is available commercially, causes mixing of the inspired air with air from the preceding expiration, thus increasing its temperature and water vapor content. A slight disadvantage of this procedure is some increase in dead space. The efficacy of a mask as protection against EIA has been shown experimentally (59; 338). Air temperature inside the mask can exceed

30° C during exercise, even when the outside temperature is only 4° C (338). In the absence of a mask, one can cover the mouth and nose with a scarf and obtain similar results.

Choosing Less Asthmogenic Activities

As described earlier, some activities are more asthmogenic than others. For most patients, swimming and other water-based sports or games are the activities of choice and should be strongly recommended. Among land activities, preference should be given to the intermittent ones (1- to 3-min exercise bouts, interspersed with rest or, preferably, with mild exertion) (301). These activities are less asthmogenic than prolonged, continuous activities. Examples are circuit training, gymnastics, board diving, and downhill skiing. Among team games, water polo, American football, and baseball are suitable.

Warming Up

Using the principle of the refractory period, a warm-up should be beneficial in preventing or ameliorating EIA (see the earlier section "Refractory Period"). In one study, although a 3-min walk or jog was not found to be beneficial (270), repeated 30-sec runs 10 min prior to an exercise task did reduce the fall in peak expiratory flow by 36% (342). Other studies (259; 321) showed that the warm-up routine does not need to be intense or to induce EIA. Nor does it have to use the same muscle group that will be used subsequently for the sport routine itself (425). We and others (301) recommend a warm-up at the beginning of each exercise session. It should last 6 to 10 min, raising heart rate to not more than 150 beat · min⁻¹. However, an optimal warm-up routine may differ among individuals and among sports.

When EIA starts, moderate-intensity exercise can be used to reverse or even stop the attack. This is based on the *bronchodilation* that takes place during exercise. A short bout (up to 1 min) of activity has been found to be effective in reversing EIA (206) and it is especially effective in increasing the bioavailability of inhaled medication (9).

Physical Training and the Child With Asthma

Two aspects of the effects of training on children with asthma are clinically relevant: improvement in exercise performance and reduction in asthmatic complaints, especially EIA.

Improvement of Fitness

Numerous studies have shown that most children with asthma are trainable and, through proper programs, can increase their maximal aerobic power, muscle strength, and other fitness components (65; 126; 131; 144; 159; 241; 277; 298; 299; 308; 411; 412). In point of fact, patients with asthma can reach athletic glory at the highest elite levels (312). As in healthy children, the degree of improvement in work performance is related to the intensity and volume of training (159). Patients with severe asthma may not improve their fitness through a training program because they are often not able to withstand intense activity without triggering EIA. However, proper medication will usually help them to increase their activity level and train (326).

Psychosocial Benefits

The ability of asthma patients to participate in play and sport yields important psychosocial dividends. These include improved school attendance, increased acceptance by peers, improved self-confidence and sense of accomplishment, decrease in frequency and intensity of emotional upsets, and above all, the recognition by the patient that sickness need not be a way of life (4; 22; 116; 192; 193; 301; 339; 393). A case of a young patient is instructive. This Australian girl decided to start swimming in order "to lick asthma." Some years later she became an Olympic champion and one of the most notable figures in swimming history (129).

Effect on Exercise-Induced Asthma

To the clinician, the key question is whether physical training can improve asthmatic control. More specifically, can it prevent or reduce EIA? Some reports suggest that, in spite of improving fitness, training does not reduce the occurrence of EIA (64; 131). Others show that, once trained, the child with asthma responds to a given work rate with less bronchoconstriction (15; 187; 300; 399). Such an improvement is demonstrated in figure 6.18. The reduced EIA is most probably the result of a reduction in submaximal ventilation (326; 411) and in respiratory heat loss, rather than direct effect on EIA. Whatever the mechanism, aerobic training seems to

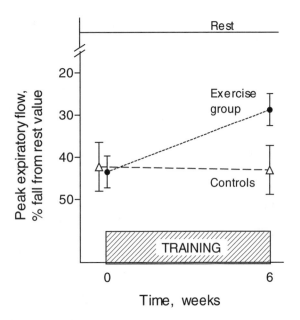

FIGURE 6.18 Training and exercise-induced asthma. Twenty children with asthma (11 ± 2 years) underwent a 6-week program of calisthenics, team games, and circuit training. Fourteen patients served as controls. Values are the lowest peak expiratory flow after a treadmill test. Vertical lines denote 1 SEM.

Reproduced, with permission, from Bar-Or 1985 (26), based on data by Henrikson and Nielsen 1983 (186).

raise the threshold for the triggering of EIA in some patients. There is, however, no proof of a relationship between the degree of reduction in EIA and the increase in fitness. A person with asthma may reach top world standards in some sports without any appreciable alleviation of EIA (126). Nor do we know whether training at a young age has any bearing on the long-range prognosis in asthma.

Children with asthma should be encouraged to take part in regular physical activities at school. There are, however, those who may first need a short (4-6 weeks) specialized initiation program during which they can improve their fitness and regain self-confidence. As part of their own education, teachers should be instructed to allow children with asthma to bring their medication to school and self-administer it whenever needed. The American Academy of Pediatrics (4), in a Position Statement, has recommended that "School children and adolescents with EIA whose physicians have prescribed metered-dose inhaled β_2-agonists, cromolyn, or both should be allowed (and encouraged) to use them before school-sponsored sport activities, including team

practices and competition, and physical education classes." (page 130)

Swimming As a Therapeutic Tool

As discussed earlier in the chapter, swimming and other water-based activities are the least asthmogenic among all physical activities. This is the main reason why swimming has become such a popular sport among asthma patients and why it is mostly swimmers who have reached elite world levels (126). This section examines the evidence that aquatic programs are beneficial to the fitness and health of the young asthma patient and also looks at possibly detrimental effects of such programs.

Beneficial Effects of Swimming

As discussed previously for land-based programs, swimming can improve aerobic fitness in children with asthma (18; 30; 131; 340; 398; 400; 402).

Swimming programs have been shown to decrease indices of asthma morbidity (e.g., frequency of attacks, rate of hospitalization, school absence days) (131; 192; 400; 402). A case in point is a study by Huang et al. (192) that included randomly assigned swimming and control groups of 6- to 12-year-old pupils with asthma. The intervention comprised in-school 2-month, three-per-week swimming classes. Indices of morbidity of both groups were then followed up for a further 12-month period. As seen in table 6.1, there was a major decrease in several morbidity indices in the swimming group, and considerably less of a decrease in the controls. It is particularly impressive to note that the improvement in morbidity remained long after the cessation of the formal intervention.

Another important question is whether swimming programs can reduce the severity and frequency of EIA. Most intervention studies that have addressed this question could not show a beneficial effect. Two studies did indicate a reduction in EIA severity, but one was noncontrolled (267) and the other did not use random allocation to intervention and control groups (252; 398).

Detrimental Effects of Swimming

In general, all aquatic activities and sports are safe for children with asthma. One exception is deep-water diving (e.g., scuba diving) because of the possibility that a mucous plug may cause air trapping. Expansion of the air pocket when the

TABLE 6.1 | **Effects of Swim Training on Asthma Morbidity**

Variable	Swimming group	Controls
Frequency of attacks	↓ 78	↓ 11
Change in PEFR at 12 months	↑ 63	↑ 25
Wheezing days	↓ 66	↓ 20
Days requiring medication	↓ 61	↓ 17
Emergency room visits	↓ 89	↓ 13
Rate of hospitalization	↓ 84	↓ 19
School absent days	↓ 82	↓ 17

PEFR = peak expiratory flow rate. The direction of the arrow reflects the change. Subjects were 90 schoolchildren with asthma, 6 to 12 years old, who were randomly assigned to swimming and control groups. Morbidity indices were monitored for 12 months after cessation of the 2-month intervention. Values are percentage change from the pretraining period. All between-group differences were significant (p < 0.01).

Reproduced, with permission, from Huang et al. 1989 (192).

nitrogen trichloride. Some people, asthma patients in particular, are sensitive to this compound, the main reaction being a coughing spell. This response is more pronounced in indoor pools where the nitrous trichloride gas may accumulate over time (306). There are no data on the degree of discomfort and harm, if any, that this response entails.

In conclusion, the benefits of a water-based program far outweigh the minor detrimental effects. Swimming and other water-based sports and games are definitely the activities of choice for the child with asthma (18; 30; 72).

diver returns to the surface may cause rupture of lung parenchyma. The likelihood that such an event will occur is very low.

There are two potentially detrimental effects of aquatic activities: enhanced parasympathetic drive and sensitivity to chlorine and its derivatives.

Enhanced Parasympathetic Drive Immersing the face in water, particularly cool water, induces a parasympathetic response. This "diving reflex" is manifested mostly by bradycardia. This response is more pronounced in people with airway hyperreactivity, asthma patients included, and is proportional to the degree of hyperreactivity (395). A related response to facial immersion in cool water is mild bronchoconstriction, mostly of the small airways, among healthy people (274). Putting these two studies together suggests that asthma patients may respond to facial immersion in cool water with even greater bronchoconstriction. More research is needed to tell whether a child with asthma is at risk for clinically meaningful bronchoconstriction while being immersed in cool water.

Sensitivity to Chlorine and Its Derivatives Chlorine is used in pool water as a disinfectant. In the water, it may combine with ammonia and urea (which are present in sweat and urine) to form

Bronchopulmonary Dysplasia

Prematurity and the development of respiratory distress syndrome in the newborn often require high positive-pressure mechanical ventilation and supplemental oxygen therapy. These may lead to bronchopulmonary dysplasia (BPD), also called *chronic lung disease of prematurity*. The diagnosis of BPD is based on a specific X-ray pattern in a baby who required oxygen supplementation beyond the 36th week of gestation and mechanical ventilation for 1 week or more. The likelihood of having BPD is inversely related to the duration of pregnancy (169). Therefore, children who were premature and of a very low or extremely low birth weight are at a special risk of developing BPD. With the increase in survival of such babies into adolescence and young adulthood (see chapter 10, section "Extremely Low Birth Weight"), the long-term impairments that accompany BPD have received increasing attention in recent years. Such impairments are manifested at rest as well as during or following exercise. The intent of this section is to outline the exercise-related outcomes of BPD. Table 6.2 lists the main physiologic functions that are impaired in BPD.

TABLE 6.2	Impaired Physiologic Functions in Children and Adolescents With a History of Bronchopulmonary Dysplasia	
At rest	**During or after exercise**	
• Low or normal forced vital capacity	• Exercise-induced bronchoconstriction	
• Low FEV_1 and other spirometric indices of airway patency	• Normal or somewhat low maximal O_2 uptake or maximal work rate	
• Airway hyperreactivity (as discerned by methacholine or histamine provocation)	• High ventilatory cost (ventilation/O_2 uptake ratio)	
• Hyperinflation (high ratio of residual volume to total lung capacity)	• Increased use of ventilatory reserve (maximal V_E/MVV)	
• High transcutaneous CO_2 tension	• High O_2 uptake at a given work rate	
• High 24-hr energy expenditure in infants, but normal resting metabolic rate in children	• Arterial O_2 desaturation	
	• High transcutaneous CO_2 tension	

Exercise-induced bronchoconstriction occurs in some children with a history of BPD (23; 336). The reported rate of EIA in such patients ranges from 17% to 50%. The etiology of this response is not clear, but it may reflect hyperreactive airways in these children. As in patients with asthma, bronchoconstriction disappears spontaneously.

While most studies (23; 179; 200) show normal aerobic performance in BPD patients, others (309; 336) indicate a somewhat low maximal O_2 uptake in comparison with term controls. Authors usually agree, however, that children with a history of BPD have high ventilation for a given O_2 uptake, or use a high percentage of their ventilatory reserve (maximal ventilation/maximal breathing capacity at rest) during exercise. This excessive ventilation is accompanied by a high respiratory rate (309).

Unpublished data from the Children's Exercise & Nutrition Centre documented a high O_2 uptake for a given mechanical work rate during cycle ergometry. This may reflect excessive ventilatory cost, but could also result from a lack of coordination among skeletal muscles as found in children with extremely low birth weight (118; 219). For details, see chapter 10, section "Extremely Low Birth Weight." Babies with BPD have a high 24-hr energy expenditure, which is associated with a high respiratory frequency (102). However, resting metabolic rate of children (age 11.8 ± 1.7) with BPD seems to be normal (145). It is unlikely, therefore, that the high metabolic cost of exercise can be explained by excessive resting metabolism.

Irrespective of the cause of a high metabolic cost of exercise and of high submaximal ventilation, the functional implication is that children and adolescents with a history of BPD may fatigue early during exercise. This is notwithstanding their normal or near-normal maximal O_2 uptake.

Arterial O_2 desaturation occurs in some patients during exercise, particularly at high intensities (17; 23; 200; 336). This, combined with a high transdermal CO_2 tension (23), may reflect an impairment in gas exchange as evidenced also by a reduced diffusion capacity (200). In spite of the arterial O_2 desaturation, maximal exercise seems to be safe for children with a history of BPD (211), particularly if they have no accompanying pulmonary hypertension. Likewise, there seems to be no contraindication for the participation of such patients in sport.

In conclusion, BPD is sometimes accompanied by pulmonary hyperinflation, exercise-induced bronchoconstriction, high ventilation, and arterial O_2 desaturation. Even though maximal aerobic power and maximal O_2 uptake are usually normal, patients with BPD may fatigue early due to a high metabolic cost of exercise. More research is needed to determine the habitual activity patterns of such patients, as well as the long-range tracking of these impairments into adult years.

Cystic Fibrosis

Cystic fibrosis (CF) is an autosomal recessive incurable disease that affects primarily the lungs, pancreas, intestinal mucosa, and sweat glands. It occurs in 1 of 1,700 to 7,500 live births among Caucasians, which makes it the most prevalent hereditary disorder in this population. The genetic defect in CF, discovered in 1989, is expressed as a disruption in the cystic fibrosis transmembrane regulator proteins (CFTR). With advances in therapy, median survival has increased dramatically, from 10.6 years in 1966 to 29.4 years in 1992 (291) and 33 years in 2001 (334). Males with CF have a longer life expectancy than females (328), possibly as a result of the females' faster decline in pulmonary function (90) and a lower fat and energy intake (89). Because of the increase in survival, clinicians now face the challenge of fostering a normal quality of life for their young patients, and not only of preventing and treating medical emergencies. The relevance of physical activity and exercise to CF is multifaceted. The disease lowers pulmonary function, which may reduce aerobic fitness; it is often accompanied by a state of undernutrition, which may affect muscle function; and it is characterized by poor heat tolerance due to a depressed thirst mechanism and a resulting *involuntary dehydration* during exercise in the heat. In addition, CF is accompanied by increased energy expenditure, which may contribute to a negative energy balance. For more details about involuntary dehydration and heat intolerance, see section "Health Hazards in Hot Climates" in chapter 3.

Habitual Physical Activity and Energy Expenditure

No studies have directly compared physical activity behavior of CF patients and healthy controls. Using an activity log (54), 7- to 15-year-old patients were rated as having a "moderate" to "high" activity level (174). None were rated as "sedentary" or "athletic." It should be emphasized that these patients had a low severity of CF as judged from their pulmonary function and nutritional status. Based on 24-hr heart rate monitoring, the ratio of total energy expenditure to resting energy expenditure was not different between 8- to 24-year-old patients and healthy controls (374). Assuming that this ratio reflects the amount of physical activity (see chapter 2), the habitual activity levels of these children were not deficient. As in healthy children, girls with CF seem to be less active than boys with CF (381). The data just cited are based on small sample sizes of Norwegian and Swiss patients. It has yet to be determined whether the same pattern of activity behavior exists for other populations.

Our experience and that of other clinicians has been that patients with CF and their parents often have a positive attitude toward physical activity. When asked to rate their attitude toward

various recreational activities, 9- to 21-year-old Norwegian patients of both genders gave scores similar to those given by healthy controls. The only difference was for endurance-type activities, which were rated lower by the patients. Attitudes improved further following a 2-week sport camp (385).

Resting (5; 63; 164; 374; 406; 410) and total 24-hr energy expenditure (60; 355; 406; 410) are 5% to 25% higher among infants and children with CF than in the general population. Such a high metabolic rate, together with limitations in intestinal absorption, may contribute to the chronic undernutrition often observed in CF. The high energy expenditure, with resulting high metabolic heat production, may also contribute to heat intolerance among some patients (for details, see chapter 3). During exercise, the metabolic cost was high in one study (100), but not in others (164; 374).

The mechanism for an elevated metabolic rate in CF is not clear. One suggested cause is a high energy cost of breathing (5; 136). However, excessive energy expenditure exists also in children who are at a preclinical stage of the disease, when lung function is not yet affected. Another possible cause is a primary energy-consuming defect at the mitochondrial level. While some authors (63; 355) subscribe to this explanation, others suggest that the high energy expenditure reflects a relatively high fat-free mass in patients with CF (60; 378). A related controversy is whether patients who are homozygous for the delta F508 mutation (see later section "Heredity and Exercise Performance") do (406) or do not (60; 136) have an especially high resting and total energy expenditure.

Treatment by aerosolized recombinant human deoxyribonucleic acid (DNA) is accompanied by an increase in forced vital capacity and a decrease in resting metabolic rate. Furthermore, as seen in figure 6.19, there is a high correlation between the two responses (5). This pattern is in line with the theory that the high energy expenditure in CF is due to a high cost of breathing.

Physiologic Causes of Deficient Exercise Performance

Exertional or post-exertional dyspnea and cough are common among CF children with advanced lung damage. Physical ability is limited in proportion to the clinical severity of the lung disease and the child's nutritional status. Table 6.3

summarizes the pathophysiologic changes that are relevant to exercise performance in the child with CF.

Factors Affecting Maximal Aerobic Power

Maximal aerobic power in CF patients is deficient, whether calculated in absolute terms or per kilogram body mass or fat-free mass (73; 76; 94; 100; 166; 189; 195; 234; 260; 280; 293; 351).

The degree of deficiency in aerobic performance is related to the severity of the pulmonary disease (75; 77; 94; 142; 153; 189; 210; 279; 297; 307), as seen in figure 6.20. It is affected by airflow limitation, parenchymal damage, hemodynamic deficiency, undernutrition, or a combination of these. Reduced aerobic performance, however, may also be explained by a reduced muscle mass and muscle force (100).

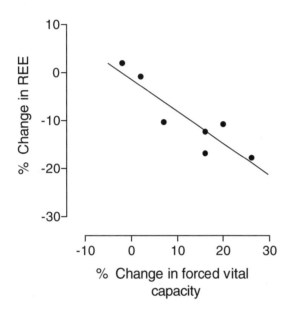

FIGURE 6.19 Relationship between the reduction in resting energy expenditure (REE) and the increase in percentage predicted forced vital capacity after 14-day treatment by aerosolized recombinant human DNA. Data are based on nine 15- to 40-year-old patients with cystic fibrosis.

Reproduced, with permission, from Amin and Dozor 1994 (5).

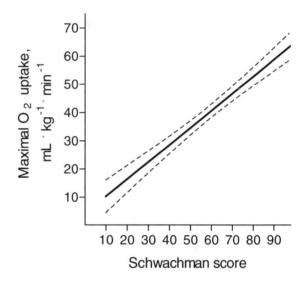

FIGURE 6.20 Aerobic performance is dependent on the severity of lung disease in cystic fibrosis. Relationship is displayed between $\dot{V}O_2$max and the Schwachman score (an index of severity of the lung disease, based on clinical and X-ray criteria) in eight 15- to 17-year-old patients. Broken lines denote 95% confidence interval. Modified from Hjeltnes et al., 1984 (189).

TABLE 6.3 Pathophysiologic Changes That Affect Physical Fitness in Children With Moderate and Advanced Cystic Fibrosis

Pathology	Physiologic outcome	Effect on physical fitness
Airflow limitation	High airflow resistance, high O_2 cost of breathing, high ventilatory equivalent, reduced alveolar ventilation and maximal ventilation	Deficient maximal O_2 uptake, exertional and post-exertional bronchoconstriction, preserved or elevated ventilatory muscle endurance
Destruction of lung parenchyma and pulmonary vascular damage	Low diffusing capacity, O_2 desaturation, CO_2 retention, low lung recoil	Deficient maximal O_2 uptake, cyanosis
Hemodynamic dysfunction	Reduced stroke volume and cardiac output, right-sided heart failure, pulmonary hypertension	Deficient maximal O_2 uptake
Protein-energy undernutrition	Reduced muscle mass and, possibly, contractile properties	Deficient anaerobic and aerobic fitness, reduced respiratory and skeletal muscle strength
Low oxidative efficiency	High O_2 cost at a given mechanical power	Early fatigability

Airflow Limitation   *Airflow limitation* results from bronchial narrowing due to the accumulation of dry mucus, swelling of the bronchial mucosa, and bronchospasm. It causes a reduction in maximal breathing capacity at rest (153) and maximal ventilation during exercise in the child with advanced lung disease (77). During maximal exercise, healthy children use only 60% to 70% of their resting maximal breathing capacity, while patients with advanced CF may reach, and even exceed, 100% of their maximal breathing capacity (153). This suggests a deficient ventilatory reserve (293). In contrast, whereas healthy children have a maximal heart rate around 200 beat · min^{-1}, patients with CF often terminate their effort when their heart rate is only 160 to 170 beat · min^{-1}. Thus, while in healthy people aerobic performance is limited mostly by maximal cardiac output and by the energy turnover capability of the exercising muscles, the patient with advanced CF is often limited by his or her ventilatory capacity (153; 195).

Airflow limitation induces an increase in dead space and a reduction in alveolar ventilation as manifested by CO_2 retention (74; 77; 86; 153; 195; 251; 260). Such CO_2 retention is particularly evident in patients with advanced disease and is associated with O_2 desaturation, low tidal volume, and low ventilation (86). During submaximal exercise, ventilation is often excessive, probably to compensate for the increased dead space (189; 195). One potential disadvantage of excessive ventilation is that a high proportion of O_2 uptake is diverted to the respiratory muscles (213) rather than to those skeletal muscles that perform the exercise. This may lead to early fatigability.

Air trapping with hyperinflation is another important pathophysiologic outcome of airflow limitation. This is reflected by a high ratio of residual volume to total lung capacity (75; 189; 232; 233; 236; 251; 297). Hyperinflation may be the cause of low inspiratory muscle strength (233; 337), because these muscles (diaphragm mostly) start contracting at a reduced length, which puts them at a mechanical disadvantage (58; 233). A high residual volume-to-total lung capacity ratio may be one factor that affects maximal aerobic power, as shown in figure 6.21 (296).

The high ventilatory demand in the child with CF may act as a training stimulus to the respiratory muscles, such that their endurance (i.e.,

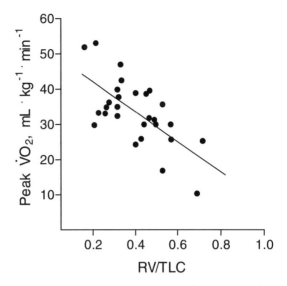

FIGURE 6.21   Peak O_2 uptake ($\dot{V}O_2$) as a function of the degree of lung hyperinflation, which is represented by the ratio residual volume/total lung capacity (RV/TLC). Individual data for 28 patients with a large range of pulmonary dysfunction.

Reproduced, with permission, from Orenstein and Nixon 1989 (296).

ability to sustain a high level of ventilation for prolonged periods) is increased. In fact, ventilatory endurance in patients with CF is sometimes *higher* than in healthy controls (218).

Destruction of Lung Parenchyma and of the Pulmonary Capillary Network   Destruction of lung parenchyma and the pulmonary capillary network is reflected by a reduced lung diffusion capacity at rest and, especially, by an inability to increase diffusion during exercise (153; 436). This pattern is shown in figure 6.22. The compromised lung performance results also in "wasted" ventilation and increased physiologic dead space (153).

A related characteristic of the moderately or severely affected patient is arterial O_2 desaturation during intense exercise (75; 77; 86; 94; 142; 181; 237; 251; 279; 307). It is sometimes accompanied by cyanosis and may reduce maximal aerobic power. Figure 6.23 shows the relationship between peak aerobic power and the degree of O_2 desaturation. The hyperbolic relationship suggests that a mild deficiency in saturation may be associated with a marked deficiency in exercise performance. Desaturation seldom occurs in patients with a mild-to-moderate severity of lung damage (e.g., FEV_1 greater than 50% of forced vital capacity), but it can become extreme

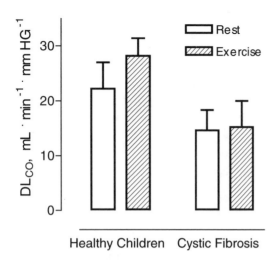

FIGURE 6.22   Lung diffusion capacity for carbon monoxide (DLco) in children with cystic fibrosis (n = 18) and in healthy controls (n = 8), 5 to 15 years old. A comparison between resting values in the sitting position and values obtained immediately after maximal cycle ergometer exercise. Vertical lines denote 1 SD.

Adapted from Zelkowitz and Giamonna 1969 (436).

when the resting FEV_1 drops to below 40% to 45% of forced vital capacity (figure 6.24) (181). Increase in O_2 saturation, as a result of intense hospital therapy, is accompanied by an increase in maximal aerobic power. Such improvement is particularly evident in patients with severe lung disease (75).

Hemodynamic Dysfunction

During mild-to-moderate stages of CF, hemodynamic responses to exercise seem normal (234; 250; 307); but advanced CF may be accompanied by hemodynamic aberrations. For example, stroke volume and cardiac output during submaximal exercise were considerably lower in children and adolescents with resting FEV_1 of less than 50% predicted, compared with patients with a less advanced disease and with healthy controls (307). Severe lung parenchymal changes may be accompanied by *cor pulmonale* and a resulting congestive right heart failure (387). Thus, exercise-induced dyspnea in advanced CF may be due not only to the lung disease but also to ventricular dysfunction.

Protein-Energy Undernutrition

Protein-energy undernutrition often occurs in advanced stages of the disease (319). It reflects a reduced intestinal absorption of nutrients due to

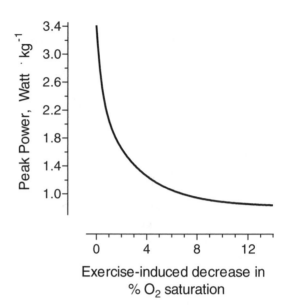

FIGURE 6.23 Arterial O_2 saturation as a possible limiting factor in peak aerobic power of children with cystic fibrosis. Twenty patients underwent a progressive continuous maximal cycling task. O_2 saturation was measured at rest and continuously during exercise.

Adapted from Cropp et al. 1982 (94).

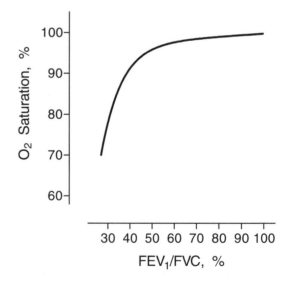

FIGURE 6.24   Relationship between O_2 desaturation during peak aerobic exercise and the severity of lung damage, expressed as the forced expiratory volume in the first second/forced vital capacity (FEV_1/FVC) relationship at rest. Ninety-one 7- to 35-year-old patients performed maximal cycling effort.

From Henke and Orenstein 1984 (181).

a deficient exocrine pancreatic function. In addition, high energy expenditure may lead to a negative energy balance. While such undernutrition affects mostly muscle performance, as discussed further on, it may also affect aerobic performance (100; 249; 273). Figure 6.25 demonstrates the relationship between maximal aerobic power and fat-free mass in children with mild or moderate CF and in healthy controls (100). While aerobic power is strongly related to fat-free mass in all three groups, it is lower in the CF patients for any given fat-free mass. The authors concluded that CF may be accompanied by abnormalities in the skeletal muscles (e.g., low efficiency of oxidative mitochondrial adenosine triphosphate synthesis, or in the recruitment of motor units). A low oxidative efficiency theory is in line with data derived from magnetic resonance spectroscopy (101). It is noteworthy in this regard that the O_2 cost for pedaling at a given mechanical power is excessive in patients with CF (100).

To summarize, as is evident from table 6.3, a reduction in maximal aerobic power may occur due to airflow limitation, decreased diffusing

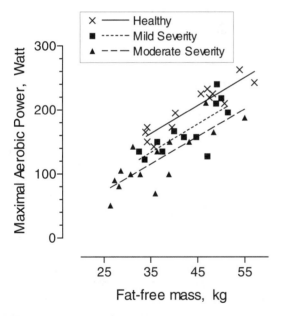

FIGURE 6.25 Relationship between maximal aerobic power, fat-free mass (estimated from skinfold thickness), and disease severity in 10- to 18-year-old cystic fibrosis patients and healthy controls. Subjects performed a progressive cycling test to volitional maximum. Triangles denote patients with moderately severe cystic fibrosis (n = 15, r = 0.82). Squares denote mild severity (n = 13, r = 0.75), and Xs denote controls (n = 13, r = 0.93).

Reproduced, with permission, from De Meer et al. 1999 (100).

capacity, reduced cardiac function, undernutrition, reduced oxidative efficiency, or a combination thereof. One should emphasize, though, that the extent of reduction in maximal O_2 uptake is related to the severity of the lung disease. Children with mild disease often have normal maximal O_2 uptake and can do very well in aerobic tasks. A case in point is a group of three male patients whose maximal O_2 uptake ranged from 42.2 to 63.7 ml · kg^{-1} · min^{-1} (382) and who completed the 1984 New York marathon (384).

Anaerobic Performance and Muscle Strength

One outcome of protein-energy undernutrition is a reduced lean body mass and muscle mass (100; 180; 199; 273; 319) and, possibly, dysfunctional contractile characteristics. Such a deficiency may then affect muscle performance, as manifested by low anaerobic peak power, local muscle endurance, and strength. It may also be the cause for reduction in respiratory muscle strength (199).

Traditionally, research on exercise and CF has focused on aerobic performance. It was only at the start of the 1990s that scientists and clinicians became interested also in anaerobic performance and, to a lesser extent, the muscle strength of patients with CF (27; 48; 51; 70; 71; 174; 233-236). Some studies have shown that anaerobic performance, as measured by the Wingate test (51; 71), a 30-sec isokinetic cycling sprint (234; 236), or exercise to exhaustion at approximately 130% of maximal O_2 uptake (351), is lower in patients with CF than in healthy controls. This deficiency is apparent when power is calculated in absolute units (51; 71; 234; 236) and sometimes also per kilogram fat-free mass (51) and per kilogram body mass (71). In contrast, other groups (48; 174; 235; 345) did not find such a deficiency. A possible reason for this discrepancy is that disease severity in the latter studies was only mild to moderate; also the subjects were not undernourished. Indeed, as seen in figures 6.26 and 6.27, anaerobic performance is related to the patient's nutritional status (51; 71) and to the severity of lung disease (71). Because undernutrition and advanced lung disease often occur simultaneously, it is hard to tell which of the two is the primary cause of deficient anaerobic performance. It is conceivable, though, that anaerobic skeletal muscle function, which is little dependent on O_2 transport, will be more directly affected by the muscles' nutritional status than by respiratory function. In point of

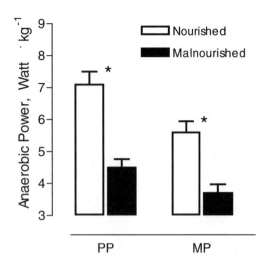

FIGURE 6.26   Peak (PP) and mean (MP) anaerobic power and nutritional status. Forty 10- to 39-year-old females and males with cystic fibrosis performed the Wingate anaerobic test. Values are presented per kilogram ideal body weight. "Nourished" reflects ≥90% or more ideal body weight. "Malnourished" reflects <90% ideal body weight. Vertical lines denote 1 SEM. * = p < 0.001.

Reproduced, with permission, from Cabrera et al. 1993 (71).

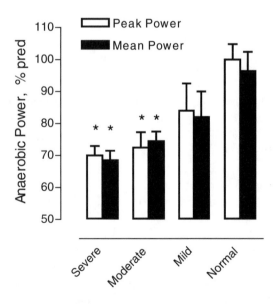

FIGURE 6.27   Peak and mean anaerobic power as a function of the severity of lung disease. Subjects and exercise protocol were as in figure 6.26. Degree of lung dysfunction is based on a pulmonary function score (94). Mechanical power values are percentage predicted from healthy populations. Vertical lines denote 1 SEM. * = p < 0.01 when compared with the normal lung function group.

Adapted, with permission, from Cabrera et al. 1993 (71).

fact, based on a study with adult CF patients (249), even aerobic performance seems to be related to the nutritional status, independently of pulmonary function.

Heredity and Exercise Performance

In the late 1980s, a genetic mutation was described that is associated with CF. This *delta F508* mutation comprised a deletion of phenylalanine at position 508 on the CFTR protein. Various studies have looked at the correlation between clinical manifestation and prognosis of CF on the one hand and the presence of delta F508 mutation on the other. As stated in the earlier section "Habitual Physical Activity and Energy Expenditure," some patients who are homozygous for delta F508 have a high resting metabolic rate (405) and total 24-hr energy expenditure as determined by doubly labeled water (406). A question remains as to whether physical performance is also affected in such patients. Evidence so far (210), based on cross-sectional analysis, suggests that the delta F508 mutation does not affect pulmonary function at rest or maximal O_2 uptake. A longitudinal follow-up at the Children's Exercise & Nutrition Centre suggests that patients who are homozygous for delta F508 do have a somewhat inferior anaerobic performance compared to those who are heterozygotes for this mutation (345).

In a recent study (348), the relationship between physical fitness and genotype was assessed in 8- to 17-year-old patients who were heterozygotes or homozygotes for delta F508 and also had a second mutation. Patients with second mutations that cause a defective CFTR production or processing had a lower maximal O_2 uptake than those with mutations that cause a defective CFTR regulation. Likewise, reduction in anaerobic performance (Wingate test) occurred with some second mutations and not others.

In conclusion, exercise performance of the child with CF, although normal in the mildly affected patient, is limited when the disease is severe. Primarily affected is maximal aerobic power, which depends on the O_2 transport system. Even though cardiac function may be deficient, the physiologic limiting factor in aerobic exercise is not circulatory, but ventilatory—both airway and parenchymal functions being affected. Muscle strength and the response to short-term, high-intensity anaerobic exercise may also be affected, particularly among patients who are undernourished. More research is needed to determine the possible role of physical hypoactivity on exercise performance. Likewise, more information is needed about gender differences in physiologic responses to exercise in CF.

Exercise As a Clinical Assessment Tool

Exercise is not relevant for the actual diagnosis of CF. It is useful, though, for the following:

- Evaluation of O_2 desaturation and alveolar hypoventilation during various intensities of exercise
- Determination of maximal heart rate and ventilatory threshold, for prescription of exercise intensity
- Prognosis regarding survival and responsiveness to therapy
- Identification of patients who respond to exercise with bronchoconstriction

The main clinical use of an exercise test is the identification of a child who responds with arterial O_2 desaturation. It is important to determine the exercise intensity (e.g., by measuring the respective heart rate) at which desaturation occurs. The patient can then be advised to limit the intensity of activities so as not to exceed that level. Centers where exercise prescription is given as therapy advocate an exercise provocation test prior to the commencement of the program (47; 73; 291). Children who respond with arterial O_2 desaturation or alveolar hypoventilation are identified and given special attention during the ensuing program. Such children should, at least initially, be prescribed only mild activities.

Another use of exercise testing is to determine each child's maximal heart rate, particularly if the child engages in competitive sport (291). The rationale is that a certain percentage of maximal heart rate is often used by coaches and athletes to determine exercise intensity during training. For most healthy children, maximal heart rate is approximately 200 beat · min^{-1}, but it is often lower in CF patients, particularly at advanced stages of the disease (see earlier section "Factors Affecting Maximal Aerobic Power"). As a result, a certain submaximal heart rate (e.g., 150 beat · min^{-1}) would denote a greater percentage of maximum than in a healthy child. It therefore is

important to determine the maximal heart rate of each individual athlete. According to the same rationale, a laboratory test can help to determine a child's ventilatory threshold, which can be used to determine training intensity. Ideally, this decision should be made based also on changes in O_2 saturation.

An inverse relationship has been found between mortality over a 26-month period and maximal aerobic power among patients with very severe CF (437). In a landmark study, Nixon and colleagues (280) determined factors that are associated with an 8-year survival rate of children, adolescents, and young adults with CF. They found that peak O_2 uptake was a powerful predictor of survival, even after adjustment to other prognostic variables such as age, gender, lung function, bacterial colonization, and nutritional status. Eighty-three percent of patients who had a "high" fitness level (≥82% of predicted peak O_2 uptake) were still alive 8 years later. This compares favorably with 51% survival of those who had a "medium" aerobic fitness level (59-81% predicted) and 28% of those with a "low" level (≤58% predicted) (figure 6.28). Obviously, these data do not prove a cause-and-effect relationship, because aerobic fitness may be a marker for the patient's overall health. Notwithstanding, the study shows that aerobic fitness

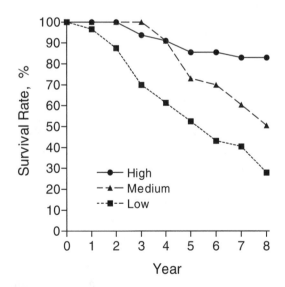

FIGURE 6.28 Eight-year survival as a function of aerobic fitness. Subjects were 109 males and females with cystic fibrosis, 7 to 35 years old. "High" = ≥82% predicted peak O_2 uptake, "medium" = 59% to 81% predicted; "low" = ≤58% predicted.

Reproduced, with permission, from Nixon et al. 1992 (280).

can be used as a strong predictor of survival. A subsequent study (268) showed an association between peak O_2 uptake and 5-year survival rate in adult CF patients.

Even though exercise performance is strongly related to the severity of the disease, there are only few data on the utility of exercise testing prior to therapy as predictor of its success. In one study (70), anaerobic performance on admission did not predict the outcome of a 2-week in-hospital regimen that included physiotherapy, intravenous antibiotics, aerosols, pancreatic enzymes, and vitamin supplementation. The program did not include exercise training or nutritional supplementation. In another study, the effect of 6-month nutritional supplementation on the increase in percentage ideal body weight was correlated to an increase in elbow flexor strength (174). The authors concluded that "changes in the strength of skeletal muscle groups, such as elbow flexors, can serve as a marker of early changes in nutritional status" (p. 584). This conclusion, however, needs further confirmation, because neither changes in the strength of other muscle groups, nor changes in anaerobic endurance and power, correlated with improvement in nutritional status.

Patients with CF often respond to a bout of exercise with transitory bronchodilation (315; 365); but some, as in asthma, respond by bronchoconstriction. The reported prevalence of bronchoconstriction has ranged from 0% to 65% of tests (98; 189; 365; 385; 433). Such a response cannot be predicted from resting pulmonary functions. Therefore, an exercise provocation test is warranted for patients who complain of post-exertional shortness of breath and persistent cough. It should be remembered, though, that unlike the situation in patients with asthma, exertional or post-exertional cough may simply reflect enhanced expulsion of respiratory mucus and is a *beneficial* effect of exercise.

Attempts have been made to use exercise provocation to identify CF patients who also have an atopic disease. Response to a treadmill test, however, was noncontributory (433); and it seems that an exercise provocation test alone is of no differential diagnostic value in separating atopic from nonatopic patients with CF.

Right ventricular function (e.g., ejection fraction) may be deficient at rest and during exercise in CF patients. Left ventricular function seems to be preserved at rest, but patients with moderate

to advanced CF may have an impaired ejection fraction with exercise (36; 83). Thus, an exercise test might be used to diagnose early stages of left cardiac involvement in CF. The predictive value of such a finding, however, has yet to be determined.

Means of Improving Response to Exercise

For patients who show O_2 desaturation during exercise, response can be improved through the use of O_2 supplementation (85; 95; 251; 279) and continuous positive air pressure (142). The improvement with O_2 supplementation (e.g., 30% instead of 21% O_2 in inspired air) includes an increase in endurance time, reduction in arterial O_2 desaturation, and a decrease in the maximal ventilatory equivalent (251). Some studies have shown an increase in maximal O_2 uptake while others have not, which poses a question as to whether maximal O_2 uptake is limited by O_2 saturation (73; 291). As seen in figure 6.29, the

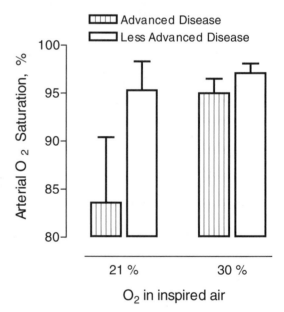

FIGURE 6.29 Beneficial effect of O_2 supplementation on arterial O_2 saturation occurs particularly in patients with an advanced lung disease. Twenty-three 13.9 ± 4.1-year-old patients with forced expiratory volume in the first second (FEV_1) <50% of FVC ("advanced disease") and thirteen 20.8 ± 4.5-year-old patients with FEV_1 >50% FVC ("less advanced disease") performed a maximal aerobic test. Values for O_2 saturation were measured at rest and at peak exercise while patients breathed room air (21% O_2) or 30% O_2. Vertical lines denote 1 SD. Data from Nixon et al. 1990 (279).

beneficial effect of O_2 supplementation on O_2 saturation occurs mostly in patients with advanced lung disease, and less so when disease severity is mild to moderate (279). O_2 supplementation is also useful in conjunction with physical conditioning among patients with advanced lung disease who otherwise would not be able to exert themselves (177; 178). The disadvantage of this approach is that patients are limited in their training repertoire to a stationary cycle or other indoor-based apparatus.

A beneficial effect for patients with advanced lung disease was achieved also by the use of continuous positive airway pressure (5 cm H_2O) during exercise. The benefits included a decrease in submaximal O_2 uptake, which, as interpreted by the authors (142), reflected reduced work of breathing; increased exercise tolerance; a reduction in a dyspnea index; an increase in arterial O_2 saturation; and a decrease in transdiaphragmatic pressure for a given tidal volume. These findings need to be confirmed by other studies.

The use of digoxin has been suggested, but it does not seem to improve aerobic exercise performance or hemodynamic response to exercise in patients who do not have congestive heart failure (87).

Beneficial Effects of Training

There is increasing evidence that enhanced physical activity and sport participation can be of benefit to the child and adolescent with CF. The following beneficial effects have been suggested:

- Improved clearance of respiratory mucus
- Increased endurance and strength of respiratory muscles
- Reduced airway resistance
- Improved exercise tolerance
- Improved sense of well-being

It is important to realize that in most intervention studies, the design was inadequate to test the efficacy of training. Most studies did not have controls. Some had controls, but without random or matching allocation. In only a few studies (20; 73; 337; 343) were subjects assigned randomly to a training or control group. The following information, therefore, should be read in this context and not be considered definitive.

Clearance of Respiratory Mucus

During a swimming program, sputum volume was 15% higher on days when the patients exercised (432). A similar benefit was reported during a walk/jog program (295; 423) and a home-based cycling program (335). In fact, during a 17-day intense activity program, sputum clearance was adequate even though aerosol inhalation and chest physiotherapy had been discontinued (431). Although these studies suggest that an acute bout of exercise, a training program, or both may enhance the clearance of respiratory mucus, none was designed specifically to address this important phenomenon. Sputum collection is a crude method that does not take into account lost or swallowed sputum. Another experimental approach was taken by Kruhlak et al. (232), who examined air trapping before and after a bout of submaximal cycling. In patients with moderate to severe airway obstruction, air trapping, mostly in the apical fields, diminished with exercise. The authors attributed this to enhanced clearance of mucus. It is interesting that airway trapping was aggravated in the two patients whose pre-exercise obstruction was only mild.

The mechanism for enhanced mucus clearance, as suggested in the studies just described, is unknown. More data are needed on the specific physical activities that may help clear the bronchi before exercise can be considered an adjunct to, or substitute for, chest physiotherapy. A more sophisticated approach (e.g., tagging the mucus with a marker and measuring the clearance of the marker) is indicated for future research. Such an approach was taken in adults with chronic bronchitis, whose mucus clearance, using aerosolized Technitium 99, was enhanced during exercise (288). A similar effect was found in healthy adults (427). This isotope, however, is radioactive, and its use with children would be unethical.

Exciting new insights regarding the mechanism for exercise-induced improvement in mucus clearance may be derived from in vivo research on the nasal epithelial mucosa (2; 176). Alsuwaidan and colleagues (2) found that although at rest the transepithelial electric potential gradient in patients with CF was more negative than in healthy controls (ages 9-24 years), the differences between the groups disappeared during exercise. In a subsequent study, A. Hebestreit and colleagues (176) confirmed this finding (figure 6.30) and then used selective ion channel blockers to

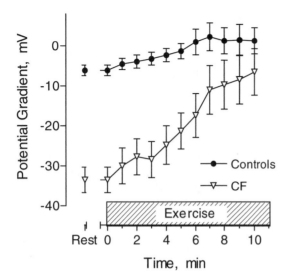

FIGURE 6.30 Exercise normalizes the electric potential gradient across the respiratory mucosa of patients with cystic fibrosis. Eleven patients (age 10-33 years) and nine controls cycled semi-supine at a moderate intensity (85% ventilatory threshold) for 10 min. The transepithelial potential gradient across their nasal mucosa was monitored while the exploring electrode was perfused with saline. Vertical lines denote 1 SEM.

Reproduced, with permission, from Hebestreit et al. 2001 (176).

determine whether exercise affects the sodium or the chloride channels. Their subjects were 10- to 33-year-old CF patients and healthy controls, who cycled for 10 min at moderate intensity. Exercise partially blocked the amiloride-sensitive sodium channel in the nasal epithelium of the patients. The authors concluded that the inhibition of luminal sodium conductance during exercise may have increased the water content of the mucus in the CF lung during exercise and reduced its viscosity. This may explain, in part, the enhancing effect of exercise on mucus clearance.

Endurance and Strength of Respiratory Muscles

Programs of walking and jogging (293), swimming and canoeing (218), and specific training of the respiratory muscles (20; 218) resulted in increased endurance of the respiratory muscles in CF patients. Likewise, inspiratory muscle strength can be increased by specific training (inspiration against resistance) (20; 337). In one study (337), allocation to high-resistance and control groups was done randomly. The high-resistance group increased its peak inspiratory pressure by 13%,

with no change in the controls. Even though the training resulted also in an increase in aerobic performance, there was no relationship between the increases in inspiratory muscle strength and performance. In another study (20) a crossover random design was used, in which some of the patients started with an inspiratory muscle training period while the others were first observed during a 4-week nontraining period. Both inspiratory muscle endurance and strength were improved by training, but there was no increase in aerobic performance.

Even though an improvement in respiratory muscle function may benefit physical performance, it is not clear whether it can also help the child fare better during exacerbation of the disease.

Decrease in Airway Resistance

With few exceptions, related to swimming (431; 432), training does not induce a decrease in airway resistance (16; 46; 110; 217; 218; 293; 335). There is little (388) or no (293; 431) change in lung volumes or in the residual volume-to-total lung capacity ratio. This suggests that hyperinflation is not reduced by training. The improved airway flow is short-lived once the program is over. As seen in figure 6.31, increase in FEV_1 and maximal midexpiratory flow following a 7-week swimming program disappeared within 10 weeks after swimming was discontinued.

Improved Exercise Tolerance

Children and adolescents with CF are trainable. Peak aerobic power, maximal O_2 uptake, and exercise tolerance time may all increase with various regimens of training (16; 73; 177; 178; 293; 335; 337; 380; 383), especially in those patients with very low fitness at the start of the program (16; 293). Some can improve markedly in their athletic performance and participate in events as demanding as marathon running (379) or triathlon (291).

Most studies with CF patients have assessed the effects of aerobic training. In a recent training study, Selvadurai et al. allocated inpatients randomly into aerobic and strength training groups, as well as controls. The 8- to 16-year-old patients, who were admitted because of intercurrent infection, exercised five times a week for approximately 3 weeks. The strength training group responded with a greater weight gain and increased fat-free mass, muscle strength, and FEV_1 (347). It thus

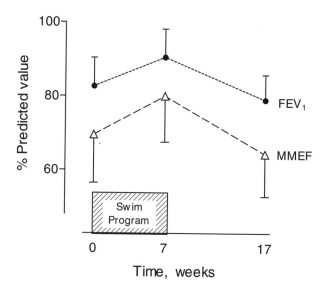

FIGURE 6.31 Durability of training-induced changes in pulmonary functions of children with cystic fibrosis. Ten 6- to 18-year-old patients participated in a 7-week swimming program, two to three sessions per week. Forced expiratory volume in the first second (FEV$_1$) and maximal midexpiratory flow (MMEF) were compared before the program, at its conclusion, and again 10 weeks later. Vertical lines denote 1 SEM. Data by Zach et al. 1981 (432).

seems that the addition of strength training to an activity program is beneficial to children and youth with CF. Because its benefits are observed within a short time, strength training can be an important motivator.

Improved Sense of Well-Being

Because of the progressive nature of CF, its inevitable outcome, the need to undergo daily therapy for life, and the sense of "being different" from other children, patients and their family members are subjected to a great emotional burden. It has been our experience and that of other clinicians (291) that the potential psychological benefits of enhanced physical activity are as important as the physiologic ones. Yet there has been only little research on the effects of an active lifestyle on the attitudes, behaviors, and emotional well-being of children with CF (110; 163; 347). In a randomized controlled study, the quality of well-being of inpatients who took part in an aerobic program improved in proportion to the increase in their aerobic fitness. Another group of patients who underwent strength training did not show any improvement in their quality of well-being (347).

A major, as yet unanswered, question is whether improvement in exercise tolerance, in endurance of the respiratory muscles, and in general well-being can slow down deterioration in clinical status. It is important, though, to reemphasize that the survival rate is higher among patients with a high aerobic fitness (see figure 6.28 and the section "Exercise As a Clinical Assessment Tool") (280; 437). In addition, enhanced physical activity is accompanied by a slowing down in the deterioration of lung functions, as shown in a randomized controlled 3-year home-based program (343).

Physical Training Versus Chest Physical Therapy

Chest physical therapy, which includes percussion and postural drainage, has been the mainstay approach to mobilizing and expelling the respiratory mucus in CF (322). With the mounting evidence on the efficacy of physical conditioning in improving lung function and possibly mucus clearance, a question arises as to whether this mode of therapy is as effective as physical therapy. Several authors have reported that during periods when their patients were physically active, there was little or no need for physical therapy (16; 46; 230; 295; 431). Others (335) reported greater sputum production with physical therapy (forced expiration combined with postural drainage) than with exercise (cycling at 50% maximal aerobic power). Based on such scant evidence it is premature to draw conclusions about whether, indeed, exercise is as effective as physical therapy in clearing the airway mucus. Until further insights are obtained, a prudent approach would be to maintain a physical therapy regimen even when children are physically active. One can consider, however, individual cases in which a consistent physical activity regimen may suffice. The individualized approach will also depend on patient compliance with the respective regimens.

Training Programs Suitable for the Child With Cystic Fibrosis

In general, children with CF can participate in all sports and physical activities. Programs ranging from swimming, cycling, and walking and jogging, to ball games, skating, and trampoline jumping have all been found to be efficacious. However, activities that include an aerobic component may yield the best physiologic dividend. Strength training of the skeletal muscle may increase fat-free mass. In recommending a program one must consider the child's own preference. We have noticed over the years that once young patients pursue their favorite activities, they often adhere to them for a long time.

An effective training program does not have to push the child to exhaustion: Exercise intensities that raise heart rate to 70% to 80% of maximum are sufficient to increase aerobic exercise tolerance (293; 347). Emphasis should be given to an individualized approach, keeping in mind the wide variability in pulmonary function, exercise capacity, nutritional status, and level of apprehension in these patients.

The Value of Home-Based Programs

A question often asked is whether an activity program should be supervised (e.g., within a clinic, camp, or fitness center) or whether it can be pursued unsupervised at home. This issue has important psychosocial and economic implications. Experience so far with young patients in the home setting has been encouraging (28). Although most home-based programs were carried out successfully with little or no supervision (16; 46; 165; 343; 380), a small number have not proven useful (190), mostly because of poor compliance. The contents of home-based programs may determine compliance: When the prescription is regimented (e.g., skipping rope for prolonged periods), compliance, with some exceptions (343), is usually poor. In contrast, patients participate more consistently when the activities vary and include recreational elements (46). It is not clear whether supervision by parents—who often are not conversant with the notions of fitness and activity—is effective. As recently suggested by Prasad and Cerny (313) factors likely to increase compliance in this population include explicit and continued encouragement and support from the family and healthcare team, and the introduction of behavior-changing strategies."

With good organization and compliance, the dividends derived from a home-based program can be impressive. An example is a study by Schneiderman-Walker et al. (343) in which 7- to 19-year-old patients were randomly assigned into an exercise group (three-a-week 20-min sessions at a heart rate of approximately 150 beat · min⁻¹) and a control group (maintaining their regular physical activity) over a 3-year period. Pulmonary functions declined more slowly in the exercise group than in the controls (figure 6.32). In addition, the exercise group showed an increase in well-being.

There is another important advantage to home-based programs. For children, especially when young, enhanced physical activity should be fun and informal rather than bear the stigma of therapy. This may be better achieved in the home environment.

Further research is needed to answer several questions. At what age can a child be given a home-based program? How can parents be taught to supervise their child and ascertain safety and compliance? How can one tailor a program to the child's physical ability, temperament, and health status? How should home-based physiotherapy and activity programs be combined? How can one ascertain adherence once the formal program is concluded?

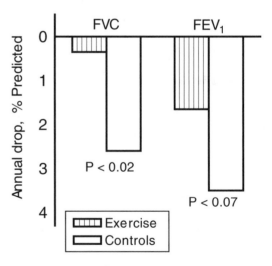

FIGURE 6.32 Effect of home-based exercise program on the decline in pulmonary functions of patients with cystic fibrosis. Data are for 65 patients, 7 to 19 years old, who were randomized into exercise and control groups. The exercise group performed three 20-min or more sessions of aerobic exercise a week for 3 years. FVC = forced vital capacity; FEV_1 = forced expiratory volume in the first second. Based on Schneiderman-Walker et al. 2000 (343).

Safety Considerations

Should any activities be prohibited for the child with CF? While there are no absolute contraindications for any activity or sport, a safe approach is to advise against scuba diving and similar underwater activities that may induce major pressure changes in the lungs (291). The rationale is that such changes, when they occur in a diseased lung, may aggravate existing bronchiectasis and bullae and induce pneumothorax. Although contact and collision sports are not contraindicated, children who have an enlarged spleen or a diseased liver should avoid such sports (291).

As discussed earlier (in the section "Factors Affecting Maximal Aerobic Power"), some patients with advanced disease (e.g., FEV_1 less than 50% predicted) respond to intense exercise with arterial O_2 desaturation. The concern expressed traditionally has been that in a child with an already compromised pulmonary function, such a repeated response might be deleterious, inducing or aggravating pulmonary hypertension and right ventricular hypertrophy. With the concomitant hypoxemia, exercise might induce cardiac dysrhythmia (94). However, there has been no evidence that brief exercise-induced O_2 desaturation causes such short- or long-term damage to the child with CF (84; 291). We propose that O_2 desaturation during exercise is a self-limiting process, because it curtails the child's ability to sustain high-intensity effort and forces termination of the activity. Soon after cessation of the activity, O_2 saturation returns to its pre-exercise levels. However, clinicians often prefer not to subject their patients to hypoxic conditions, even if for a brief period. A prudent approach therefore is to adopt an individualized policy as to whether a given patient should or should not perform certain exercise intensities. Periodic monitoring by an exercise test is a recommended precaution while the child participates in a sport program.

Pulmonary infection is an ongoing concern for all patients with CF. It therefore is important to know whether exertion suppresses their immune response. In one study (49), the cellular immune response (changes in the blood count of leukocytes, total and T subsets of lymphocytes, granulocytes, and natural killer cells, as well as in the activity of natural killer cells) was determined during and for 60 min after a maximal aerobic task. Subjects were 8- to 21-year-old patients with mild to moderate disease severity, and they had no acute exacerbation during the study. Their response to exercise was similar to that of healthy controls, and the authors concluded that there is no evidence for suppression of the immune response in these patients. Because patients with CF are prone to lung infections, more research is needed on this important issue, particularly regarding the possible effect of exercise intensity and frequency on the immune response.

The increased salt loss through sweat in patients with CF may impair their well-being and exercise tolerance, particularly in hot climates. As discussed in chapter 3 (in the section "Health Hazards in Hot Climates"), these patients are prone to heat-related illness. Even though children with CF thermoregulate effectively during short bouts of exercise in the heat (294), those exposed to the heat for several hours may dehydrate excessively (29; 231) unless given flavored NaCl-containing beverages (231). Patients who train regularly may incur salt loss unless salt is replenished in the diet. Salt tablets, however, are contraindicated.

We recommend that children with CF who exercise in a warm or humid climate be encouraged to drink repeatedly (i.e., every 15-20 min) above and beyond their thirst and include approximately 50 mmol $\cdot$ L^{-1} NaCl in their drink. As a rule of thumb when the only available drink is tap water, children below age 10 should drink approximately 75 to 100 ml beyond thirst. Older children should drink 150 to 175 ml beyond thirst. For additional recommendations regarding behavior in hot climates, see the section "Guidelines for the Conduct of Athletic Events in the Heat" in chapter 3.

In CF patients who also develop diabetes mellitus, prolonged exercise (e.g., 30 min or more) may induce hypoglycemia. In addition, high urinary output may induce dehydration. These complications should not preclude the child from pursuing an active lifestyle. The most practical precaution during prolonged activity is to include carbohydrates such as glucose and sucrose in the child's drink and make sure that he or she is given the drink every 15 to 20 min. For more details, see the section "Exercise-Induced Hypoglycemia" in chapter 8.

Cardiovascular Diseases

*T*he circulatory system plays an integral role in the physiological responses to exercise. Augmented blood flow not only satisfies the muscle's needs for oxygen supply and energy substrate but also serves to remove metabolic wastes and buffer accumulated lactic acid. Blood flow is diminished to nonexercising tissues, with redistribution of flow not only to contracting muscle but also to the skin for thermoregulation.

Given the critical nature of these responses to exercise, it is not unexpected that a close and complex relationship exists between cardiovascular health and exercise performance. In responses to repeated bouts of exercise (i.e., physical training), cardiovascular fitness improves. This enhancement of cardiac capacity is linked to salutary long-term outcomes, including decreased risk for adult coronary artery disease (see chapter 5). Depressed cardiovascular function resulting from heart disease, on the other hand, impairs exercise performance. In addition, high exercise intensity places stress on the cardiovascular system and can create risks of complications—including sudden death—for individuals with certain forms of heart disease.

Recognizing these risk-benefit relationships of exercise in patients with heart disease is important for clinicians. Exercise by adults with coronary artery disease can either increase or decrease risk of myocardial infarction, depending on factors such as extent of atherosclerotic disease and nature of the physical activity. Heart disease in children and adolescents is different. Most children and adolescents have congenital cardiac anomalies that create a different and more diversified set of issues regarding risks and benefits of physical activity. Health care providers need

to counsel their young patients into appropriate levels of exercise, promoting physical activity for its health benefits within the safe limits created by the heart disease.

Regular exercise and participation in sport activities are considered important contributors to normal growth and development of children from physical, emotional, and social standpoints. Fortunately, most children with heart abnormalities can be encouraged to enjoy the benefits of normal physical activities and do not need restriction from athletic participation. Limited exercise tolerance in some of these young patients will preclude their involvement in high-level exercise, while a very small number carry a risk for sudden cardiac death from sport play. This chapter reviews these issues in patients with the forms of heart disease most commonly encountered by clinicians caring for children and adolescents.

Congenital Heart Disease

The heart is formed early in gestational life. Beginning as looping of a primitive vascular tube, the process of demarcation of atria and ventricles, creation of atrioventricular and semilunar valves, and septation of chambers is complete by the end of the eighth week. In about 1 newborn of 100, this process has gone awry, resulting in a structural abnormality defined as congenital heart disease. As with most forms of congenital malformations, the etiology of these errors in the normal process of cardiac embryogenesis is largely unknown. Many forms of congenital heart disease exist, but such "errors" are not random. The relative uniformity of the different kinds of anomalies, besides aiding in diagnosis and treatment, suggests that critical periods of normal heart formation are particularly sensitive to teratogenic influences.

The varying forms of congenital heart disease alter normal cardiovascular function and physiology in different ways. Consequently, the clinical expression of cardiac disease in young patients can range from totally inconsequential to life-threatening hypoxemia or impairment of circulation requiring immediate surgical intervention shortly after birth. As might be expected, then, the impact of congenital heart disease on exercise capacity varies greatly, depending on the type and severity of the anomaly.

Most forms of congenital heart disease fall into one of three pathophysiologic groups: left-to-right shunts, right-to-left shunts, and obstructions. When sufficiently severe, each of these categories of anomalies can interfere with normal exercise tolerance, but by different mechanisms. Evaluating exercise tolerance, either by history-taking or by laboratory stress testing, can therefore serve as a useful means of assessing disease severity.

In patients with a *left-to-right shunt*, systemic venous blood returns normally to the right side of the heart, is pumped to the lungs and oxygenated, and returns to the pulmonary circulation again. This left-to-right shunt occurs whenever a congenital defect exists that allows communication between the two sides of the heart. This happens when the atrial septum is incorrectly formed (atrial septal defect), or a hole is present in the ventricular septum (ventricular septal defect), or a communication exists at the great vessel level (as when the ductus arteriosus does not close normally after birth). If such a defect is small, the amount of shunt causes no cardiac embarrassment. If large, the heart is faced with an increased volume load; excessive pulmonary blood flow creates pulmonary artery hypertension and ventilatory overwork; and signs of congestive heart failure (tachypnea, hepatomegaly, failure to thrive) ensue.

In a child with a *right-to-left shunt*, deoxygenated systemic venous blood finds a way to reach the aorta and systemic circulation without passing through the lungs. The resulting arterial hypoxemia is expressed as cyanosis and often by progressive acidosis, necessitating early surgical intervention. Right-to-left shunts can occur when a communication exists between the two sides of the heart in conjunction with pulmonary outflow obstruction (tetralogy of Fallot), or when an anomaly causes mixing of systemic and pulmonary venous blood (single ventricle). Severe hypoxemia also occurs when the aorta arises off the right ventricle and the pulmonary artery from the left ventricle (transposition of the great arteries), creating two separate systemic and pulmonary circulations.

Obstructions typically present as stenosis, or narrowing, of a congenitally abnormal aortic or pulmonary valve. The consequent obstruction to flow necessitates increased work by the corresponding ventricle to generate increased pressure. The same effect occurs with a congenital narrowing of the aorta (coarctation), usually just distal to the takeoff of the left subclavian artery. When severe, such obstruction can limit distal flow, particularly during exercise.

Children and adolescents with congenital heart anomalies can therefore experience limitations of exercise tolerance from different mechanisms—myocardial and ventilatory fatigue, hypoxemia, and restricted systemic blood flow—depending on the type of defect. As noted previously, in the majority of young patients the magnitude of these physiologic insults is not sufficient to preclude normal physical activities, including sport participation. In a minority, guidance is important in directing these children into appropriate exercise activities.

Ease of fatigue with exercise in patients with congenital heart disease may reflect not simply the severity of their underlying abnormality. From fear or ignorance, these children are often channeled by themselves or their families into a sedentary lifestyle, further compounding low levels of fitness. In a classic study of children in the Seattle junior high schools, Bergman and Stamm demonstrated that restriction of physical activity and parental overprotection were as likely to occur in children with minor defects, or even those with no heart disease at all (that is, children erroneously labeled with heart disease), as in children with significant abnormalities (figure 4.4) (16).

It is important that children with congenital heart defects be encouraged to participate in physical activities to the safe limits of their capabilities. The need is not simply to optimize physical and psychosocial development but also to provide the long-term benefits of regular activity in preventing adult cardiovascular disease. That may be particularly pertinent for those young patients who could face long-term cardiac complications in the adult years from their congenital heart anomalies.

Two recent trends in the surgical management of children with congenital heart disease have influenced issues regarding exercise. First, children with serious abnormalities now undergo surgical repair at an early age, usually during early infancy. No longer does the child with a large ventricular septal defect or tetralogy of Fallot wait several years before definitive repair. Instead, surgical intervention is typically achieved by several months of age. As a result, operative repair for many children with significant heart defects occurs before the age in which issues surrounding exercise become pertinent. Concerns regarding physical activities and sport participation in these patients now typically surround their postoperative status.

Second, dramatic successes have been achieved in the surgical approaches to children with complex heart disease. As a result, infants with heart anomalies that once would have precluded long-term survival (such as hypoplastic left heart syndrome and asplenia syndrome) are now living and functioning well through the pediatric years and into young adulthood. The long-term complications facing these patients (myocardial dysfunction, dysrhythmias) have created challenging new problems in patient management, including the risks and benefits of physical activities.

Left-to-Right Shunts: Atrial Septal Defect

The physiologic embarrassment created by congenital heart defects that cause left-to-right shunting includes ventricular volume overload (myocardial stress), excessive pulmonary blood flow (ventilatory work), and pulmonary hypertension. It would be expected, then, that large defects would impair ability to perform sustained exercise, which places similar demands on the cardiovascular system. Most large left-to-right shunts (ventricular septal defect, patent ductus arteriosus) become clinically apparent in infancy with signs of congestive heart failure, and these children undergo early surgical or catheter-device closure of these defects with low risk and

no residual disease. Older children who have undergone repair of these defects, as well as those with small shunts for whom surgery is not indicated, typically have normal exercise tolerance and deserve no special concerns regarding physical activities or sport play (126).

Patients born with an abnormal communication between the atria—an atrial septal defect— behave differently than those with other left-to-right shunts. Although those with large defects have excessive pulmonary blood flow similar to that occurring with other left-to-right shunts, pulmonary hypertension does not occur (due to pulmonary vasodilation). Consequently, these children do not typically demonstrate findings of congestive heart failure, and they remain asymptomatic. In addition, because the small interatrial pressure gradient causes little turbulence to flow, findings on physical examination may be minimal and the diagnosis missed.

Congestive heart failure from sustained right ventricular volume overload as well as serious pulmonary hypertension will eventually develop in many patients with a large atrial septal defect in the adult years, and, if left untreated, will significantly shorten life span. Therefore, early closure of large atrial septal defects during childhood—either surgically or by a device placed at cardiac catheterization—is warranted to prevent these long-term complications.

Since most children and adolescents with even large atrial septal defects are asymptomatic and diagnosis is sometimes difficult, it is not rare for older children and adolescents to participate in sports and physical activities with undetected atrial shunts. It is pertinent, then, to consider the effects of unoperated large atrial septal defects on exercise tolerance. Assessment of cardiac responses to exercise testing after repair of these lesions has also provided insights into the effect of relief of right ventricular volume overload.

Exercise and Atrial Shunting

Adverse effects of exercise in patients with an atrial septal defect would be expected if physical activity caused an increase in the left-to-right shunt, an abnormal rise in pulmonary artery pressure, or both. The extent and direction of shunting of blood through a large atrial septal defect depend on the relative diastolic compliances of the right and left ventricles, which in turn are dictated by the balance of pulmonary and systemic vascular resistances, respectively. An exaggerated rise in pulmonary artery pressure would occur if this balance favored an increase in left-to-right shunting.

Systemic and pulmonary vascular resistance both normally decrease from arteriolar vasodilation during exercise in healthy subjects. Thus, the effect of exercise on the balance of these factors—and the consequent degree of shunting and pulmonary blood pressure—in patients with atrial septal defects is not readily predictable.

Studies on this issue have been performed with supine cycle exercise during cardiac catheterization, with pulmonary and systemic flows estimated by the Fick technique. Reports of changes in blood flow and pulmonary-to-systemic flow ratios during exercise in patients with atrial septal defects have been conflicting. These reports have described increases (85), decreases (138; 155), or stability (13) in left-to-right shunting with exercise. It can be reasonably concluded, however, that substantial increases in such shunting or exaggerated rises in pulmonary artery pressure are not expected in most children with large atrial shunts who have no pulmonary hypertension at rest. In adults with even mild pulmonary hypertension, however, reduced exercise tolerance can be associated with abnormal increases in pulmonary artery pressure (143).

Effects on Exercise Tolerance

Although most children with atrial septal defects will claim to have no exercise-related symptoms, limitations of exercise tolerance have been documented in some with laboratory testing (123). The degree of exercise incapacity in these studies has been related to the presence and severity of resting pulmonary hypertension, not the degree of pulmonary blood flow (50; 60; 81). Children with atrial septal defects who have no elevation of pulmonary artery pressure usually demonstrate normal exercise fitness (37; 50). Data reported by Frick et al. (60) and Cumming (37) indicate that performance can be expected to be limited in patients who have a resting pulmonary artery systolic pressure over 50 mmHg. Thus, symptoms of fatigue with exercise with a large atrial septal defect are more likely in older adolescents and young adults. It has been suggested that exercise intolerance in a patient with normal pulmonary artery pressure may be due to a sedentary lifestyle (60; 159).

The adverse effects of a large atrial septal defect on exercise tolerance, then, relate principally to the presence and degree of associated pulmonary hypertension. As elevation in pulmonary artery pressure with atrial shunting is not typical during the childhood and early adolescent years, most young patients with an atrial septal defect experience no limitations in physical fitness. Consideration of this diagnosis is important, however, in any patient with otherwise unexplained exercise intolerance.

Postoperative Findings

When an atrial septal defect is eliminated by surgical or device closure, right ventricular work is normalized, pulmonary blood flow decreases, and pulmonary hypertension—if present—resolves. Consequently, any preoperative exercise intolerance would be expected to improve. While the clinical outcome of these patients following surgery is excellent, some postoperative exercise studies have indicated evidence of residual cardiovascular impairment. These reports, which include findings of reduced cardiac output and exercise-induced dysrhythmias, generally reflect surgical outcomes 30 to 40 years ago and typically involve adult patients (125; 155).

More recent studies, particularly in children undergoing early repair, are reassuring. Rosenthal et al. used cycle ergometry to test 22 children who were more than 6 months postsurgical repair (162). Maximal exercise performance was no different from that of healthy controls, but peak heart rate was lower in the atrial defect patients than the normal children. Reybrouck et al. found that in 11 patients operated on before age 5 years, the ventilatory threshold during a progressive treadmill test was equivalent to that of healthy children (159). However, in 13 children who underwent repair after age 5 years, values were below normal. In this study a reduced heart rate response to exercise was also observed.

This relative chronotropic incompetence, or failure of the sinus node to respond normally to exercise, has been noted following surgical repair of several different forms of congenital heart disease. While intrinsic sinus node dysfunction and autonomic impairment have been suggested as responsible, the findings of Perrault et al. indicated that placement of venous cannulae at the time of cardiopulmonary bypass may cause sinus node injury (153).

In summary, children who have undergone surgery or device closure of an atrial septal defect can be expected to exhibit normal exercise tolerance. This is particularly true if repair occurred at an early age and in the absence of pulmonary hypertension. Demonstration of reduced exercise tolerance in these patients may reflect a low level of habitual physical activity, or, unusually, the effects of significant chronotropic incompetence.

Obstructive Abnormalities

Congenital defects of the heart that result in obstruction to blood flow create a "pressure head" that must be overcome by increased ventricular work. The increase in blood flow with exercise magnifies the effect of these fixed obstructions, exaggerating ventricular stress. With mild-to-moderate degrees of obstruction, the increase in heart work imposed by physical activity is insufficient to limit exercise capacity or create health risk. With more significant stenosis, however, exercise performance is often restricted; and in some cases, patients may be at risk for syncope and sudden death.

Aortic Valve Stenosis

In patients with congenital aortic valve stenosis, fusion of the aortic valve commissures limits leaflet mobility and reduces the size of the valve orifice. To overcome the resulting blockage of blood flow, the left ventricle must generate increased pressure, which is accomplished by hypertrophy of the ventricular myocardium. During the childhood years, the ventricle is capable of responding to this augmented pressure work without impairment of function. However, sustained high pressure overload for many years may eventually result in myocardial decompensation and congestive heart failure in adulthood.

The pressure difference, or gradient, between the left ventricle and ascending aorta serves as a numerical indicator of the severity of outflow obstruction. Normally less than 5 mmHg, a resting gradient over 70 mmHg is indicative of severe obstruction and signals a need for surgical or balloon-catheter valvuloplasty. Such intervention is also often performed in patients who have at least a 50-mmHg gradient and who demonstrate symptoms of angina, syncope, or exercise fatigue, with or without electrocardiographic ischemic changes. Those with a gradient of less than 50 mmHg are usually asymptomatic, have a normal electrocardiogram (ECG), and need no treatment.

Exercise and Aortic Stenosis   During dynamic exercise, the increased circulatory demands of contracting skeletal muscle are met through an augmentation of cardiac output. To achieve this, frequency, velocity, and strength of myocardial contraction must be increased. The consequent rise in myocardial oxygen uptake is accomplished through dilation and increased perfusion of the coronary arteries. Because of the physical compression of coronary arteries created by myocardial contraction, volume of coronary flow is about 2.5 times greater in diastole than in systole. This becomes a critical issue during high-intensity exercise, when, because of the rising heart rate, the duration of diastole is shortened to about one-third that at rest. The task of the heart during exercise is therefore a challenging one: providing increased systemic circulation (to exercising muscle) and coronary flow (to the myocardium) as the time constraint of a decreasing diastolic time becomes progressively limiting to both ventricular filling and coronary perfusion.

The fixed left ventricular outflow obstruction created by aortic valve stenosis stresses the capabilities of the circulatory system to respond to these supply/demand challenges of exercise. Exercise in patients with aortic stenosis increases the left ventricular outflow gradient in direct relationship to the rise in cardiac output (36). Consequently, patients with significant stenosis demonstrate a blunted stroke volume and lower cardiac output at high exercise intensities (46). At the same time, myocardial and peripheral muscle metabolic demands rise.

These two issues seriously strain the relationship between the demand for myocardial (as well as systemic) blood perfusion and the supply of coronary blood flow. As children with aortic stenosis exercise, both the exaggerated rise in left ventricular pressure and the hypertrophied ventricular muscle mass escalate myocardial oxygen uptake above that observed in healthy children. This occurs concomitantly with an increasingly limited diastolic coronary flow (from the diminishing diastolic time), as well as the restriction of both systemic and coronary flow created by the valve obstruction (108).

These issues are generally not significant in patients who have aortic outflow gradients less than 40 to 50 mmHg at rest. In those with greater levels of stenosis, the resulting perfusion/demand mismatch with exercise may create symptoms of angina, syncope, and exercise fatigue and may predispose to sudden death from ischemia and ventricular dysrhythmias (36; 40).

Resistance exercise such as weightlifting triggers a different set of cardiac responses than dynamic exercise. During resistance exercise, significant increases in both systolic and diastolic pressure are observed with limited changes in heart rate and cardiac output. Intra-arterial pressures as high as 480/350 mmHg have been reported with maximal lifts in adults, but levels of 180/140 are more common in children (137). These cardiovascular responses impose a significant afterload stress on myocardial function, exaggerating that already created by aortic valve stenosis with outflow obstruction.

Risk of Sudden Death Both children and adults with significant degrees of aortic valve stenosis are at increased risk for sudden death. In untreated children the risk was initially felt to be reasonably high (about 7%), but more recent information puts the chance of such a tragedy at a lower figure of 1% (179). The mechanism for sudden death in these patients is not certain but has been presumed to mimic the same supply/demand imbalance that causes ischemic death in adults with coronary artery disease. The clinical and physiologic pictures in the two cases are similar, with angina and electrocardiographic ischemic ST-segment changes reflecting insufficiency of coronary flow in response to metabolic demands. Myocardial ischemia with terminal ventricular dysrhythmia has therefore been assumed to be the most probable explanation for sudden death in patients with severe aortic valve stenosis.

An alternative explanation for sudden demise of these children involves the triggering of left ventricular mechanoreceptors by the high cavity pressures generated by the hypertrophied ventricle, particularly with exercise. A reflex outpouring of vagal impulses in response can result in hypotension and syncope and may serve as a mechanism for sudden death.

Most cases of sudden unexpected death in patients with aortic stenosis do not occur during sport participation, but a disproportionate number of cases do happen with vigorous activities. For instance, Doyle et al. reported that among 19 patients dying suddenly, 8 were involved in vigorous physical activity (football, swimming), 6 were exerting themselves mildly (walking home from school), and 5 were at rest (sleeping, watching television) (46).

Considerable attention has been focused on clinical markers that would identify the child with aortic stenosis who is at risk for sudden death. Unfortunately, many of the details of reported cases are incomplete, but certain patterns have been identified. In the causes of sudden death reviewed by Doyle et al., the resting ECG revealed ischemic changes in 70%; 21% had left ventricular hypertrophy, and only 9% had normal tracings (46). Twenty-eight of 31 patients had described symptoms of angina, syncope, or dyspnea with exercise. Few of these patients had resting gradients measured by cardiac catheterization, but in all known cases the value was over 60 mmHg.

Other reports describe similar findings. Wagner et al. characterized the clinical course of 294 children and adolescents with aortic stenosis over an 8-year period (202). Within this time there were three sudden deaths—of a 9-year-old with ischemic ECG changes and a 55-mm gradient, a 15-year-old with a 46-mm gradient and ischemic changes on ECG, and a 16-year-old with a 68-mmHg gradient and normal ECG. In the study by Reid et al., the two boys who died suddenly with aortic stenosis both had left ventricular hypertrophy on the ECG but had never complained of symptoms (158).

Based on this information, it is reasonable to conclude that sudden unexpected death in a child or adolescent with aortic valve stenosis is highly unlikely if all three of the following conditions are met: (1) left ventricular outflow gradient less than 40 to 50 mmHg; (2) normal resting and exercise ECG; and (3) absence of symptoms of angina, dizziness, syncope, and fatigue with exercise.

Recommendations for Sport Participation Children who have aortic valve stenosis do not typically demonstrate evidence of myocardial dysfunction, and there is no evidence to suggest that intense exercise or sport training hastens the development of congestive heart failure. The potential restriction of these patients from participation in competitive athletics therefore concerns the risk of sudden death. Because such events are rare and their predictability by clinical markers is uncertain, a good deal of subjectivity has influenced the creation of recommendations for limiting exercise in these patients.

Any restriction of exercise in children with aortic stenosis needs to be weighed against the advantages that can be gained from participation in both recreational and organized sport. Inappropriate limitations of physical activities may deprive these young patients of important physical and psychosocial benefits.

The decision whether a child or adolescent with aortic valve stenosis should play competitive sport—and which type of sport—needs to be individualized for each patient in consultation with a pediatric cardiologist. Consideration must be given to factors such as the competitive level and training regimen of the sport involved, the position being played in the sport, the patient's degree and progression of stenosis, presence or absence of symptoms, and findings on laboratory testing (ECG, echocardiogram, stress testing).

A consensus conference sponsored by the American College of Cardiology and the American College of Sports Medicine in 1994 outlined recommendations for sport participation by patients with unoperated aortic stenosis (70). Under these guidelines, all forms of competitive athletics are allowed if the peak instantaneous outflow gradient is less than 20 mmHg, the patient has no exercise intolerance or symptoms (chest pain, syncope), and the ECG is normal. If a patient has a gradient between 20 and 50 mmHg, mild or no left ventricular hypertrophy on echocardiogram, normal left ventricular voltages on ECG, absence of symptoms, and a normal exercise stress test (absence of symptoms, dysrhythmias, or ischemic ECG changes), competition is allowed in certain low-intensity sports (including baseball, doubles tennis, and volleyball) but not others (football, basketball, swimming, gymnastics, running, and hockey). According to these guidelines, no patient with a resting gradient over 50 mmHg should be allowed to participate in any competitive sport.

These guidelines are highly conservative, and strict adherence to the recommendations should not put any patient with aortic stenosis at undue risk for sport participation. On the other hand, they do disallow participation in popular sports in patients with relatively mild disease, in whom the risk for sudden death should be expected to be very low.

The consensus guidelines were formulated with respect to participation of patients with aortic valve stenosis in organized competitive sport teams, a setting in which high myocardial demands can be expected. The extent to which such recommendations can be extrapolated to less intense recreational sport activities (sandlot games, intramural sport, physical education class) is not clear. It might be assumed that exercise guidelines in these conditions can be liberalized, since such activities rarely demand the high level of physical intensity of interscholastic team or individual competition.

The risks of sport play following successful surgical or balloon relief of left ventricular outflow obstruction from aortic stenosis have also not been clearly delineated. According to the consensus guidelines, the patient's post-intervention cardiac status with regard to gradient, symptoms, and laboratory findings can dictate level of sport involvement in accordance with the same recommendations as for untreated patients.

Exercise As a Diagnostic Tool   The diagnosis of aortic valve stenosis usually arises during physical examination, when typical findings of a loud harsh systolic ejection murmur with accompanying thrill are detected at the right upper sternal border with radiation into the neck. Direct pressure measurements during cardiac catheterization serve as the "gold standard" for assessing severity of obstruction, but Doppler echocardiography provides an accurate noninvasive estimate of the outflow gradient.

Assessment of blood pressure and electrocardiographic responses to treadmill or cycle exercise continues to serve as a useful adjunct in evaluating patients with congenital aortic valve stenosis. Information from the exercise testing laboratory not only adds to the estimation of valve obstruction and indications for intervention but also, as reviewed in the previous section, aids in creating recommendations for participation in sport activities.

Systolic blood pressure is expected to rise by 40 to 60 mmHg during a progressive maximal exercise test. In patients with severe aortic stenosis, blood pressure response to exercise is blunted, and the magnitude of blood pressure response is inversely related to the degree of obstruction.

Alpert et al. found that a systolic blood pressure rise of over 35 mmHg indicated there was only a 10% chance of a patient's having a left ventricular outflow gradient more than 50 mmHg at heart catheterization (6). Following successful surgical intervention, blood pressure responses to exercise in patients with aortic stenosis increase significantly.

The generally accepted electrocardiographic sign of myocardial ischemia has been horizontal or downward ST-segment depression of 0.1 mV or more lasting greater than 80 msec. Although this criterion during exercise testing of adults has been adopted for use in children, its applicability to young patients with non-coronary disease is not certain. Patients with significant aortic stenosis may demonstrate ischemic ECG changes as well as ventricular dysrhythmias with exercise testing, and the presence and extent of ST depression may be related to degree of outflow obstruction. Patients whose resting gradient does not exceed 50 mmHg can be expected to have a normal ECG without ST changes during exercise (27; 77; 89) (figure 7.1). Following surgical or balloon intervention, however, there is little predictive value of ST changes with exercise testing as an indicator of residual gradient.

Coarctation of the Aorta

Congenital narrowing or coarctation of the aorta typically involves a constriction just beyond the aortic arch, immediately distal to the takeoff of the left subclavian artery. The point of obstruction is usually discrete but can involve more diffuse narrowing of extended segments of the aortic arch and isthmus. Blood pressure is consequently elevated in the upper extremities, and a significant coarctation is indicated by a brachial-femoral artery pressure gradient at rest exceeding 20 to 30 mmHg. Patients with severe coarctation can present with congestive heart failure during infancy from the marked left ventricular afterload. Often, however, the diagnosis is made in an asymptomatic older child who is found to have systemic hypertension, left ventricular hypertrophy, and decreased femoral pulses on a routine examination.

Operation for coarctation of the aorta usually involves resection of the narrowed segment with end-to-end anastomosis. More extensive hypoplasia may necessitate interposition of prosthetic graft material to enlarge the aortic diameter. Surgery in early infancy has been associated with an increased chance of recurrent obstruction, which has been successfully treated with balloon dilation. Techniques to safely allow balloon dilation of unoperated, or "native," coarctations have been developed in some centers as well.

Given these straightforward interventions, patients with coarctation of the aorta might be expected to demonstrate normalization of blood pressure and resolution of left ventricular hyper-

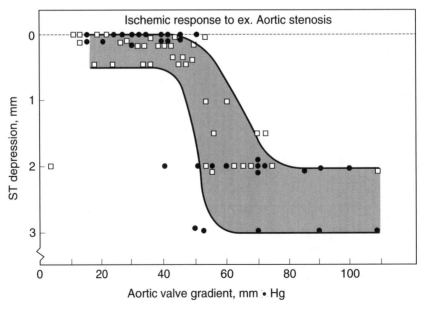

FIGURE 7.1 Electrocardiographic ST depression during exercise testing relative to degree of outflow obstruction in patients with aortic stenosis. Data from Chandramouli et al. 1975 (□) (27), Halloran 1971 (●) (77), and James 1978 (●) (89).

trophy after successful surgical or balloon relief of aortic obstruction. However, coarctation of the aorta is a more complex pathophysiologic abnormality than simply a mechanical obstruction. Findings of persistent hypertension, abnormalities in vascular reactivity, and hyperkinetic left ventricular function are not unusual following surgery, even when the aortic gradient has been successfully eliminated. The persistence of these abnormalities may contribute to the increase in long-term morbidity and mortality observed in patients following coarctectomy in adult life (114). The explanation for these findings is uncertain, but some evidence suggests that residual abnormalities may be less likely with early intervention (172).

Exercise in a healthy individual is a hypertensive event, as systolic blood pressure generally rises by approximately 40 to 60 mmHg (with little change in diastolic pressure) during a maximal treadmill or cycle test. Upper and lower extremity blood pressure responses to exercise stress have therefore proven useful in evaluating patients with coarctation of the aorta, both pre- and postoperatively.

Residual Coarctation A residual or recurrent coarctation of the aorta occurs in 10% to 20% of patients following surgery, primarily in those with intervention in the first year of life. Exercise testing can aid in the diagnosis of repeat obstruction, which is usually managed with balloon dilation.

During dynamic exercise, cardiac output and systolic blood pressure normally rise, while peripheral vascular resistance falls. Since a coarctation of the aorta represents a fixed obstruction, the upper-lower extremity pressure gradient increases with exercise. As blood pressure in the leg cannot be measured during exercise, upper and lower extremity blood pressures are determined in the immediate postexercise period. A gradient "unmasked" by exercise of over 35 mmHg has been suggested as an indicator of a significant recoarctation, especially if combined with findings of exercise-induced upper extremity hypertension (>200 mmHg), a resting arm-leg gradient over 15 mmHg, or both (33; 113; 211). Data reported by Guenthard et al. suggested that the degree of aortic obstruction is more related to the rise in exercise-induced gradient than the increase in upper extremity blood pressure (73).

Persistence of Hypertension After "Successful" Coarctation Repair Surprisingly, resting blood pressure remains elevated in 10% to 20% of young patients after an apparently successful coarctation resection. And among those with postoperative normalization of blood pressure at rest, 20% to 40% demonstrate an abnormal rise in blood pressure with exercise (9; 39; 151; 185). Pelech et al. considered this finding clinically significant, noting that "exercise blood pressure reflects the dynamic nature of the cardiovascular response to activity, and, particularly in the young patient, may be more representative of the average daily blood pressure" (p. 167)(151).

A clear explanation for this postoperative hypertension after coarctation repair, with or without exercise, remains elusive. Residual elevations in blood pressure appear to be more common with older age of repair. Sigurdardottir and Helgason compared blood pressure responses to exercise in children who had undergone coarctation resection in the first year of life to values in those with later operation (184). Those with early repair demonstrated lower resting and exercise values; hypertension with exercise was observed in 7% of the early-repair group and in 23% of those with later operation.

Decreases in compliance of precoarctation arterial vessels following surgery have been correlated with peak systolic blood pressure during exercise (63; 72). Diminished baroreceptor sensitivity, reset to a high arterial pressure level, has also been reported in patients following coarctectomy (14). Pelech et al. suggested, however, that this observation might represent the consequence rather than the cause of persistent hypertension (151).

Gradients created by aortic arch hypoplasia have been demonstrated to cause exercise-induced hypertension in some post-coarctectomy patients (100; 107; 204). This obstruction reflects a discrepancy in the growth of the aorta upstream from the surgical site rather than an actual recoarctation. Cyran et al. emphasized that these "aortic arch abnormalities thought to be minor at rest may be mistakenly underestimated as important hemodynamic determinants of hypertension during periods of increased cardiac output, e.g., exercise" (p. 986) (39).

Some data have suggested a hyperresponsiveness of the renin-angiotensin system after coarctation repair (3; 188). However, Simsolo et al. could find no differences in resting or exercise-induced changes in plasma renin activity among children who were normotensive, normotensive after coarctation repair, and hypertensive following coarctectomy (185).

Cardiac Hypercontractility   Increased cardiac work and myocardial hypertrophy are expected findings in patients with coarctation of the aorta, but these responses should resolve after surgical or balloon elimination of the coarctation gradient. Instead, postoperative patients may continue to demonstrate increases in heart mass and augmented myocardial contractility in the absence of resting hypertension or arm-leg pressure gradients. The postsurgical patients with hyperkinetic cardiac findings described by both Kimball et al. and Leandro et al. had normal blood pressure responses to exercise but gradient increases (mean values of 22 and 37 mmHg, respectively) with exercise (100; 107). This left ventricular kinesia may persist. Carpenter et al. described findings of hyperdynamic ventricular function in a group of patients an average of 15 years following surgery (26). Not all studies have demonstrated this finding, however. Murphy et al. found no differences in resting and exercise heart rate, cardiac output, or stroke volume in children with coarctation (before or after operation) compared to normal controls (134).

Cardiac hyperkinesia following coarctation repair could reflect increased sympathetic tone or failure of the heart to "remodel" after relief of pressure overload. Whether these findings are indicative of long-term cardiac risk from sustained increases in myocardial oxygen uptake or augmented arrhythmogenicity is unknown. Equally unclear is whether the appearance of these findings might be obviated by early surgical repair.

Impaired Lower Limb Flow   Using Doppler ultrasound, Johnson et al. measured flow velocities in lower extremity vessels at rest and immediately following exercise in 10 children following successful coarctation resection and compared them to values in a group of normal controls (92). The findings indicated that despite a greater dilation of the femoral artery with exercise, the post-coarctectomy patients had impaired lower limb blood flow. That these children may be characterized by limited peripheral circulation was also suggested

by the finding of Rhodes et al. that adolescent patients after surgery had a lower ventilatory anaerobic threshold with exercise than healthy subjects (160). The authors suggested that this indicates a greater reliance on anaerobic metabolism, reflecting "persistent blood flow abnormalities across the aortic arch during exercise which may be present even after apparently successful surgery" (page 213).

Right-to-Left Shunts: Postoperative Status

The often life-threatening hypoxemia from cyanotic congenital heart disease usually necessitates early surgical intervention. The astounding progress in operative techniques to deal with complex cardiac abnormalities in these small babies has permitted long-term survival with many defects that were once frequently fatal at an early age. Consequently, as these patients have grown into childhood, issues of exercise tolerance have assumed clinical importance for health care providers. From a diagnostic standpoint, exercise testing has been helpful in evaluating issues surrounding postoperative complications. Many questions regarding the natural course and prognosis for these children currently remain unanswered. Clearly, though, the increasing number of survivors with complex cardiac disease in the early 21st century will create a new set of future management problems for health care providers.

Transposition of the Great Arteries

In babies born with transposition of the great arteries, the aorta arises from the right ventricle and the pulmonary artery from the left. Since the great arteries are reversed, or transposed, with respect to the ventricles, two separate circulations are created. Desaturated blood is pumped from the right ventricle out the aorta into the systemic circulation, with venous return to the right atrium. Meanwhile, oxygenated blood from the lungs arrives in the left atrium and left ventricle, only to be pumped out to the lungs again.

During fetal life, with patency of the ductus arteriosus and oxygenation provided by maternal placental flow, this anomalous takeoff of the great arteries has no physiologic impact. However, at birth the newborn with transposition is on its own, and with closure of the ductus arteriosus and initiation of pulmonary blood flow, two independent circulations are formed. As a result, the systemic circulation becomes starved for oxygenated blood, a condition incompatible with life. Initially at birth, limited patency of the foramen ovale and ductus arteriosus permits some mixing of oxygenated blood into the systemic circulation. However, in the absence of some intervention to create adequate mixing between the two circulations, progressive hypoxemia will ensue, with increasing acidosis and death within the first days of life.

Before 1950, virtually no survival was expected in newborns with transposition of the great arteries. The subsequent development of surgical and balloon-catheter techniques to create an atrial septal defect and provide systemic-pulmonary mixing lowered the neonatal mortality rate. The introduction of the Mustard procedure in the 1960s then recreated the original plan of pulmonary and systemic circulations in series with each other by means of an "atrial switch." In this operation, the atrial septum is excised, and a pericardial baffle is placed such that systemic venous return from the superior and inferior venae cavae is directed through the mitral valve and into the left ventricle. Pulmonary venous blood then passes around the baffle and crosses the tricuspid valve into the right ventricle. This achieves a "normal" pattern of a right-sided pulmonary pump (right atrium to ventricle to pulmonary artery) supplying blood flow to the lungs and a left pump (left atrium to ventricle to aorta) delivering oxygenated blood to the systemic circulation. The difference from normal, however, is that the right ventricle (with its tricuspid valve) continues to serve as the systemic pump.

The advent of the Mustard operation drastically altered the outlook for children with transposition of the great arteries, turning a condition with nearly 100% mortality into one in which survival of over 95% was expected. Following surgery, children with transposition generally grew up without significant medical problems, eventually raising families and leading productive lives. With time, however, clinical problems began to surface that have dampened optimism for the long-term outlook for these patients. Right ventricular dysfunction, "sick sinus syndrome" requiring pacemaker placement, and difficult-to-manage supraventricular dysrhythmias have created problems for many of these patients in their adult years.

These difficulties with the Mustard procedure stimulated development of a new arterial switch operation in which the aorta and pulmonary artery are exchanged back to their appropriate ventricle by an anastomosis above their respective semilunar valves. The coronary artery ostia are then transplanted into the base of the "neo-aorta." This procedure, which is currently the operation of choice for transposition, returns the heart to a more anatomically appropriate configuration, and patients should be expected to avoid the long-term complications observed after the Mustard procedure. Early follow-up studies have been encouraging, but insufficient information has been obtained to demonstrate whether these patients will remain free of coronary artery insufficiency, valvular problems, or other complications in the future.

Exercise testing has proven useful in evaluating ventricular function, dysrhythmias, sinus node function, and coronary perfusion in patients with both atrial and arterial switch procedures. Insights from stressing the cardiovascular system in these patients can guide both medical and surgical management decisions as well as assess the natural course following these operations.

Abnormalities Following the Mustard Operation

Patients who have undergone the Mustard operation for transposition of the great arteries generally report no exercise intolerance in their normal daily activities. However, exercise stress testing has identified complications following this procedure that may have important long-term health implications.

Decreased Endurance Capacity

Endurance time on cycle or treadmill tests is typically lower in patients who have undergone the Mustard operation, but the degree of exercise intolerance is rarely dramatic. In a review of five published reports of work capacity in these patients, Paul and Wessel found that performance averaged 72% of normal (range 42-84%) (150). The cause of limited exercise capacity following the Mustard operation could be surmised to be related to any or all of the abnormal cardiac responses discussed in the following paragraphs. However, Gilljam et al. pointed out that these factors have not been demonstrated to serve as strong predictors of exercise capacity (68).

Low Aerobic Fitness ($\dot{V}O_2$max)   Patients who have undergone the Mustard procedure show a depression in maximal aerobic power similar in degree to that in exercise performance. In nine studies, Paul and Wessel found an average $\dot{V}O_2$max of 31.2 ml · kg^{-1} · min^{-1}, which represented a mean of 76% of values in healthy control subjects (150). Low aerobic fitness in the post-Mustard patients reflects a depressed cardiac output from a combination of limited heart rate response to exercise and inadequate rise (or actual decline) of stroke volume.

Inotropic Incompetence   Low resting and peak heart rates are often observed in patients after atrial switch surgery. As noted previously, this chronotropic incompetence is evident following operative repair of other forms of congenital heart disease and may reflect damage to the sinus node during cardiac bypass. Such insults to conduction tissue might be expected to be more pronounced with the atrial incisions and restructuring involved during the Mustard procedure. Average maximal heart rate during exercise testing is about 86% predicted (150). The extent to which limitations in heart rate are responsible for decreases in aerobic fitness in these patients is unclear. Failure to demonstrate a relationship between $\dot{V}O_2$max or rate of change in $\dot{V}O_2$ with heart rate in some studies suggests that ventricular dysfunction with impaired stroke volume responses to exercise may play a more key role in limiting fitness than chronotropic incompetence (68; 148).

Depressed Stroke Volume With Exercise   Stroke volume responses to progressive exercise are limited following the Mustard operation (54; 68; 145). The 17 patients (mean age 14.5 years) studied by Gilljam et al. demonstrated a mean decline of stroke index from 50 ml · m^{-2} at rest to 39 ml · m^{-2} at peak exercise (68). This might reflect inability of the right ventricular myocardium to generate the increased force of contraction necessary to sustain stroke volume as systolic ejection time shortens. However, in the report by Gilljam et al., none of the subjects with ventricular dysfunction (at rest) showed the poorest stroke volume response to exercise. The authors suggested, instead, that inadequate filling of the ventricles by small, poorly functioning atria contributed importantly to the abnormal stroke volume response.

Right Ventricular Dysfunction   Abnormal function of the systemic right ventricle has been a common finding both at rest and during exercise after the Mustard operation. Approximately three-quarters of patients exhibit an abnormally slow rise in right ventricular ejection fraction with exercise (150). Interestingly, half of patients demonstrate evidence of depressed left ventricular function as well. This suggests that factors other than sustained systemic work, such as hypoxia and surgery, might be involved.

In examining the studies reporting these data, Paul and Wessel noted that "clearly the single, almost universal observation, was the noncorrelation of clinical status and symptoms with exercise capacity or ventricular function since almost all patients were declared asymptomatic" (p. 53) (150). The high frequency of abnormal findings does, however, raise the specter that progressive ventricular dysfunction with time may eventually contribute to long-term morbidity and mortality in these patients.

Oxygen Desaturation With Exercise   Oxygen desaturation observed in some patients during exercise testing has been attributed to shunting through leaks in the atrial baffle, but decreases in oxygen saturation have been observed in patients without baffle defects (68; 122). All 16 patients exercised by Gilljam et al. had an arterial oxygen saturation over 90% at rest, and each demonstrated a fall with exercise (68). At maximal exercise, 5 had values below 90%. The authors felt that ventilation-perfusion inequality, intrapulmonary shunts, or oxygen diffusion limitation might contribute to this finding.

Outcome From the Arterial Switch Procedure
The arterial switch operation avoids atrial baffle construction and leaves the left ventricle to its usually assigned task as the systemic pump. Thus, patients who have undergone this aorta-pulmonary artery exchange should be free of complications of sinus node dysfunction, supraventricular rhythm disturbances, and depressed right ventricular function observed following the Mustard operation. Potential derangements following the arterial switch, on the other hand, include (1) kinking or torsion of the coronary arteries, which have been transplanted from the base of the original aorta (which arose off the right ventricle) to the neo-aorta arising from the left

ventricle; (2) obstructions at the supraventricular surgical anastomotic sites; and (3) dilation of the root of the neo-aorta as well as regurgitation of the systemic semilunar valve (which anatomically is the pulmonary valve).

The clinical outcome of the arterial switch operation, developed in the 1980s, has been excellent, but whether long-term issues will arise with respect to the potential complications just noted remains to be seen. These patients are now reaching the age when exercise testing can begin to provide information regarding these possibilities.

Using treadmill testing, Weindling et al. studied 23 children ages 4 to 8 years who had undergone an arterial switch operation in order to assess the prevalence of myocardial perfusion abnormalities (by radionuclide scan) and exercise capacity (206). Left ventricular contractility and regional wall motion were normal in all subjects. All had normal exercise tolerance without ECG ischemic changes with exercise. At rest, some abnormality on perfusion scan was detected in 22 (96%). When the left ventricle was divided into segments, resting scans indicated that 75% were normal and that 4%, 15%, and 6% had mild, moderate, and severe defects, respectively. With maximal exercise, 79% were normal and 8%, 11%, and 2% demonstrated mild, moderate, and severe abnormalities.

This study may be interpreted as indicating that abnormalities of left ventricular perfusion are common following the arterial switch operation, presumably from stenosis or kinking of the translocated coronary arteries. However, the authors note that these findings did not worsen (and actually improved) with exercise, and there were no other indicators of diminished coronary perfusion or myocardial ischemia. Moreover, it was acknowledged that no data exist regarding perfusion scan findings in normal healthy children. The significance of the abnormalities in this study therefore remains unclear.

Massin et al. compared treadmill endurance time, heart rate, and blood pressure responses, as well as ECGs, of 50 asymptomatic children following the arterial switch procedure with values in normal children (120). In 47 (94%), of the postoperative patients, all findings were normal. Two children with known coronary artery obstruction and another with a single right coronary ostium demonstrated myocardial ischemic changes or ventricular ectopy on the exercise ECG.

These findings suggest that endurance fitness, ventricular function, and coronary perfusion during exercise are not expected to be impaired following switch operation for transposition of the great arteries. The question of possible long-term coronary perfusion impairment needs to be addressed by future studies as these patients reach older ages.

Tetralogy of Fallot

As its name implies, tetralogy of Fallot is a complex of four cardiac anomalies (described in 1888 by the French physician) that includes a large ventricular septal defect, pulmonary stenosis, an aorta overriding the ventricular septum, and right ventricular hypertrophy. From a physiologic (and surgical) standpoint, however, this is more specifically a "duology," a right ventricular outflow obstruction creating right-to-left shunting of desaturated blood through a ventricular communication and out the aorta to produce systemic hypoxemia. The level of systemic desaturation depends on the degree of pulmonary stenosis, which can range from mild to complete obstruction (pulmonary atresia).

Surgery for tetralogy involves patch closure of the ventricular septal defect and relief of pulmonary outflow obstruction and is generally indicated for all patients, because of either hypoxemia or elevated right ventricular pressure. Before total intracardiac repair became possible in the 1960s, operative interventions were palliative, with creation of a systemic-to-pulmonary artery shunt (Blalock-Taussig, Potts, Waterston) that increased pulmonary blood flow and improved—but did not normalize—systemic oxygen saturation.

With the introduction of the heart-lung pump, total surgical repair became feasible but needed to be postponed until patients reached an adequate body size at approximately 5 years of age. Babies born with severe tetralogy in the 1960s and early 1970s, then, initially underwent a palliative shunt in infancy and then intracardiac repair of the ventricular septal defect and pulmonary stenosis several years later. More recently, improvements in surgical and anesthesia techniques permit complete repair even in early infancy with excellent outcomes.

Long-term concerns in patients following tetralogy of Fallot repair include ventricular dysrhythmias and progressive pulmonary valve insufficiency. These issues have generally arisen in the few patients who have inadequate surgical relief of pulmonary outflow obstruction or immediate postoperative excessive pulmonary valve leakage.

Understanding this evolution of surgical techniques and postoperative outcomes in patients with tetralogy of Fallot is important in interpreting research data describing their cardiac responses to exercise. Wessel and Paul reviewed 87 studies published between 1965 and 1999 that described exercise findings in patients with tetralogy (209). The great majority of these reports involved patients operated on at older ages (typically 8-10 years), often following earlier shunt surgery. While these patients generally claim to be asymptomatic, exercise testing has indicated a number of abnormalities, including diminished work performance and $\dot{V}O_2$max as well as lower heart rate and cardiac output responses. It would seem likely that the current surgical trend of earlier repair, which avoids extended periods of hypoxemia and systemic right ventricular pressure overwork, is likely to ameliorate these postoperative issues. However, few long-term data are yet available in patients who have undergone total repair of tetralogy of Fallot in the first 1 to 2 years of life.

There is more research evidence to indicate that a second factor, the success of the surgery itself, has an important bearing on clinical outcomes, including aerobic fitness and cardiac responses to exercise. The two most critical issues are the extent of relief of pulmonary outflow obstruction and the degree of postoperative pulmonary valvular insufficiency. That is, impaired exercise performance is more likely in the postoperative tetralogy patient who has persistent right ventricular pressure or volume overwork.

Low level of habitual physical activity can also serve to negatively influence physical fitness in these patients. At the time when total repair was postponed for several years, hypoactivity was not uncommon because of true limitations in exercise capacity as well as parental overprotectiveness. Contemporary surgical repair in the first few months of life may help prevent the adoption of a sedentary lifestyle in these children.

Aerobic Fitness

As would be expected from their hypoxemia, early exercise studies of children who had undergone systemic-to-pulmonary artery shunt palliation for tetralogy of Fallot showed rather profound impairment of aerobic capacity. Average $\dot{V}O_2$max was $18\ ml \cdot kg^{-1} \cdot min^{-1}$, and the level of aerobic fitness was directly related to degree of hypoxemia (21; 35). After complete repair, average $\dot{V}O_2$max in the studies reviewed by Wessel and Paul was $36\ ml \cdot kg^{-1} \cdot min^{-1}$, which represented 81% of predicted (209). These authors noted that a good surgical outcome appeared to be the most important factor responsible for improvements in fitness. With adequate relief of pulmonary obstruction, some studies revealed near normalization of exercise capacity (38; 90). Some reports did suggest, in addition, that earlier age of surgery corresponded to a better fitness outcome (90; 104).

Chronotropic Incompetence

In common with other patients who have undergone open heart surgery, children following tetralogy of Fallot repair demonstrate diminished heart rate responses to exercise compared to normal children. Typically this chronotropic incompetence is not profound, with a maximal heart rate about $175\ beat \cdot min^{-1}$ (90% predicted). Thus, although this limited heart rate response may contribute somewhat to a lower $\dot{V}O_2$max, other factors must be playing a role.

As discussed earlier, the consistent finding of chronotropic incompetence after open heart repair of various defects suggests the impact of intraoperative damage to the sinus node (153). However, Bricker et al. could find no relationship between heart rate response to exercise and sinus node function (recovery time, sinoatrial conduction time) measured during cardiac catheterization in cardiac patients (24). Driscoll et al. suggested that chronic arterial hypoxemia might be responsible for chronotropic incompetence (48). Lack of fitness itself could play a role, since heart rate response to exercise has been demonstrated to improve with physical training in these patients (22). Wessel et al. showed normal heart rate kinetics at the onset of exercise in postoperative tetralogy patients, suggesting that their vagal tone is not impaired (208).

Wessel and Paul pointed out that because of chronotropic incompetence, the use of submaximal fitness indices that rely on heart rate-work rate relationships (such as PWC_{170}) is not appropriate in children who have undergone surgical correction of tetralogy of Fallot (as well as other defects requiring open heart surgery [209]). Use of such indices will yield an overestimate of their real aerobic fitness.

Cardiac Output

Depressed stroke volume responses to exercise have accounted for low maximal cardiac output in children after tetralogy repair (21). These findings have been attributed to decreased right ventricular compliance and contractility combined with the effects of cardiopulmonary bypass, right ventriculotomy, abnormal ventricular diastolic filling, and sedentary lifestyle. Extra-cardiac factors such as skeletal muscle function may also play a role (174).

The long-term course of these findings is unknown. As noted previously, despite the abnormalities that have been reviewed, most patients following tetralogy of Fallot repair are asymptomatic and lead normal active lives. A consideration of the potential factors that influence cardiac function with exercise would suggest that early surgery, with a good anatomic result, should improve functional outlook. Some evidence exists to support this idea (90; 104), while other research does not (174).

Sarubbi et al. examined exercise performance in 41 postoperative tetralogy patients to assess the influence of age of operation, previous systemic-pulmonary shunt, and surgical approach (right ventriculotomy vs. transatrial/transpulmonary) (174). In this analysis, the patients, who were studied approximately 9 years following total correction, were divided into early-repair (age 1.4 ± 0.5 years) and late-repair (2.4 ± 1.3 years old) groups. Compared to an age-matched control group, the tetralogy patients demonstrated a lower peak work rate (68% predicted) and maximal heart rate (168 vs. 176 beat $\cdot$ min^{-1}). However, no differences were observed between the subgroups of tetralogy patients. That is, previous palliative shunt, right ventriculotomy, and a 1-year mean difference in age of surgery did not affect work performance or heart rate response.

Functional Single Ventricle: The Fontan Procedure

Congenital heart disease can be marked by severe underdevelopment of the right ventricle (tricuspid or pulmonary valve atresia) or left ventricle (hypoplastic left heart syndrome). The former typically causes diminished pulmonary blood flow and hypoxemia, while the latter is usually manifest through impairment of systemic output, or cardiogenic shock. Other children are born with both great arteries arising from a single ventricle, which results in venous-arterial blood admixture and congestive heart failure from excessive pulmonary blood flow.

All these forms of cardiac defects share the common problem of a single functional ventricle, with no surgical possibility for creating a normal two-pump heart. Babies with these anomalies are candidates for the Fontan procedure, a two-staged operation that directs systemic venous return from the superior and inferior venae cavae directly to the pulmonary arteries. Hypoxemia is thereby resolved, and the single ventricle serves as the systemic pump.

From its inception in the 1970s, methods for performing this procedure have evolved, beginning with a direct right atrial-pulmonary artery anastomosis. The most common current technique involves first separating the superior vena cava from the right atrium and then connecting it end-to-side to the right pulmonary artery (a bidirectional Glenn shunt) in the first year of life. This is followed later by a second operation in which the inferior vena cava is baffled through the right atrium and anastomosed to the underside of the right pulmonary artery. The coronary sinus is typically left to drain into the systemic circulation, resulting in mild depression of oxygen saturation (90-95%).

Given adequate function of the systemic ventricle and low pulmonary artery pressures, this redirection of plumbing is physiologically effective, and the Fontan procedure has been successful in extending survival in these patients with complex cardiac anomalies. However, eventual complications such as serious dysrhythmias and protein-losing enteropathy (from chronic mesenteric venous hypertension) have dampened optimism regarding long-term prognosis.

Patients who have undergone the Fontan procedure typically have a low tolerance for sustained exercise (49; 80; 212). For example, the total work during maximal cycle testing of 29 postoperative Fontan patients described by Driscoll et al. was 37% that of healthy control subjects (48). This is reflected in low values for maximal aerobic power as well. In the four studies reviewed by Driscoll and Durongpisitkul, $\dot{V}O_2$max averaged

57% of normal (49). The mechanisms principally responsible for this poor aerobic fitness have not been well deciphered. In this case, besides the usual suspects—myocardial dysfunction, chronotropic incompetence, and hypoxemia—a new factor, absence of a pulmonary pump, serves as a possible explanation.

Insights into the factors that limit exercise capacity in Fontan patients might prove particularly illuminating. Such information could determine the utility of exercise rehabilitation programs, improve surgical timing and techniques, and provide a better understanding of the basic physiologic mechanisms governing blood circulation during exercise.

At the same time, interpreting the existing research literature must be undertaken with some caution. Most Fontan procedures are now performed in very early childhood. The research literature, however, generally describes outcomes of patients who underwent operation at an older age, typically 8 to 20 years old. This has permitted exercise testing studies to be performed pre- and postsurgery, providing information on physiologic alterations resulting from the Fontan procedure. It is important to recognize, though, that the outcomes from present surgical intervention performed at a much earlier age may be very different.

In the years before their Fontan operation, patients had very low endurance fitness as a consequence of systemic hypoxemia, myocardial dysfunction, or both (since the single ventricle must confront increased volume work pumping to both systemic and pulmonary circulations). Because the Fontan procedure eliminates significant hypoxemia and reduces ventricular work, endurance fitness and $\dot{V}O_2$max improve postoperatively but still remain far below normal values. In 20 patients studied before and after surgery, Zellers et al. found that average total work capacity rose from 28% to 47% predicted, $\dot{V}O_2$ improved from 54% to 59%, and maximal heart rate increased from 74% to 81% (212).

The mildly depressed oxygen saturations (90-95%) in patients following the Fontan procedure typically fall by approximately 5% with exercise and may contribute to exercise intolerance (48; 87; 139; 212). This persistent hypoxemia may be surgically induced (a fenestration placed in the atrial vena caval baffle to allow a systemic venous pressure blow-off), or could be caused by pulmonary arteriovenous fistulae, atrial baffle leaks, or abnormal ventilation/perfusion relationships.

Both resting and maximal cardiac output with exercise testing are depressed in Fontan patients (48; 52; 139; 212). Chronotropic incompetence plays some role, as maximal heart rate is usually about 70% to 80% predicted (48; 139). Abnormal stroke volume responses are probably more responsible, as values change little or may even decline with increasing exercise intensity (34; 48; 139; 212).

Whether diminished stroke volume responses to exercise reflect poor myocardial contractility or diminished ventricular filling is unclear. Most patients demonstrate impaired ventricular systolic function at rest and with exercise after Fontan surgery (42; 80). Del Torso found that two-thirds of subjects had <5% increase in ejection fraction with exercise (42). However, Gewillig found that no correlation existed between resting ventricular shortening fraction and cardiac output responses to exercise (66).

Absence of a pump driving pulmonary blood flow might be responsible for decreased left ventricular filling. Although the pulmonary artery circulation is normally a low-resistance circuit (and must be so in Fontan patients), the evolutionary development of the right ventricle implies its importance in maintaining pulmonary blood flow and left atrial filling. In the absence of a pulmonary pump, patients with Fontan surgery maintain adequate pulmonary blood flow, at least at rest, although central venous pressure is elevated. Whether the absence of the pulmonary pump during exercise impairs increases in systemic venous return and pulmonary blood flow is currently an unanswered question.

Recommendations for Sport Participation As reviewed thus far in this chapter, patients who have undergone surgery for cyanotic congenital heart disease often have diminished work capacity with at least mild chronotropic incompetence and depressed cardiac function. In addition, some experience mild hypoxemia with exercise, and others may be prone to develop dysrhythmias during intense physical activities. For many of these patients, then, competitive sports are not advisable, while others may participate in certain forms of athletics after proper evaluation. The reader is referred to specific recommendations published from the 1994 Bethesda Consensus

Conference for sport participation in these postoperative children (70).

Noncongenital Heart Disease

If this chapter were being written in 1940, more than half its pages would be devoted to children with heart problems after having had rheumatic fever. Fortunately, with the virtual disappearance of this disease (for reasons not altogether clear) in Westernized societies, acquired heart disease is now unusual in the pediatric age group. There are, however, issues involving exercise surrounding some forms of noncongenital heart disease in children. This section deals with three of these: questions regarding appropriate sport play by adolescents with mitral valve prolapse, the use of exercise testing to assess coronary artery involvement in Kawasaki disease, and the physiologic impact of cardiomyopathies and heart transplantation on cardiac responses to exercise.

Mitral Valve Prolapse

Mitral valve prolapse is the bowing of one or both mitral valve leaflets back toward the left atrium during ventricular systole. Given sufficient displacement, the tips of the leaflets can become unopposed, resulting in valvular regurgitation. Prolapse is identified clinically by findings on cardiac auscultation of a midsystolic click at the apex, followed by a late systolic murmur of mitral insufficiency if regurgitation is present.

Mitral valve prolapse typically does not become manifest until the teen years but then is not an uncommon finding in young adult women, who have a frequency as high as 6%. Prolapse is a common feature of patients with Marfan syndrome (see discussion later in the chapter), and this diagnosis needs to be considered, particularly when the findings of mitral valve prolapse are discovered in males. Prolapse has also been described more commonly in patients with isolated scoliosis.

Most children and adolescents who have mitral valve prolapse have no symptoms and remain free of complications. In large groups of adults with prolapse, recurrent chest pain, endocarditis, arrhythmias, progressive mitral regurgitation, and sudden death have all been reported, but the incidence of these complications is very low in younger patients (20).

The most common consideration in patients with mitral valve prolapse is an increased risk of ventricular and atrial arrhythmias. Kavey et al. described premature ventricular contractions in 43 of 103 children and adolescents with prolapse after evaluation by exercise testing and 24-hr electrocardiographic recordings (98). In none of these patients, however, did these findings appear to have clinical significance.

Rare reports of sudden death in patients with mitral valve prolapse are presumably related to ventricular tachyarrhythmias. These tragedies have largely been limited to adults who have significant mitral valve regurgitation with previously identified heart rhythm disturbances. A review of 60 such cases of sudden death by Jeresaty revealed only 4 who were under the age of 20 years (91). Of the total group, only 3 died during vigorous physical activity (lawn mowing, tennis, football). It would appear, then, that the very small chance of sudden death in individuals with mitral valve prolapse is not increased by sport participation.

The discovery of mitral valve prolapse generally does not require any restriction from sports or vigorous physical activities. An ECG is appropriate in the initial assessment, but exercise testing and other diagnostic tests are unnecessary in the asymptomatic patient. If the patient with mitral valve prolapse gives a history of unexplained tachycardia, palpitations, dizzy spells, or syncope, however, further evaluation is indicated. In some cases an echocardiogram is important in ruling out associated aortic root dilation when Marfan syndrome is suspected.

There are certain situations in which young athletes might be at risk. Jeresaty suggested the following criteria for restricting patients with mitral valve prolapse from competitive sport play (91):

1. A history of syncope
2. A family history of sudden death with mitral valve prolaspe
3. Chest pain exacerbated by exercise
4. Repetitive forms of ventricular ectopic activity or sustained supraventricular tachycardia, particularly if worsened by exercise
5. Significant mitral valve insufficiency
6. Associated Marfan syndrome

In summary, in children and adolescents, mitral valve prolapse is generally a benign condition that should rarely affect participation in athletics. A careful history, physical examination, and resting ECG are sufficient in most cases to clear the asymptomatic patient for sport play. However, because of an increased chance of cardiac arrhythmias, the unusual symptomatic patient (chest pain, palpitations, unexplained tachycardia, syncope) or those with the risk factors just outlined require further investigation and possible restriction from athletics.

Kawasaki Disease

Kawasaki disease is an acute, generalized vasculitis that principally affects young children. The cause of this inflammatory illness is unknown, and the diagnosis is made through recognition of its constellation of clinical findings that include fever, conjunctivitis, rash, lymphadenopathy, oral erythema, and edema of the hands and feet with desquamation of fingers and toes. These cardinal features are often associated with evidence of multi-organ inflammation, such as meningitis, arthritis, perimyocarditis, and hydrops of the gallbladder.

These inflammatory changes are generally all self-limiting and reversible, and the course of Kawasaki disease was initially felt to be benign. However, it became apparent that a small percentage of children with this disease died suddenly in the convalescent phase of the illness, and autopsies indicated myocardial infarction caused by thrombosis in coronary artery aneurysms. Subsequent studies indicated that such aneurysms develop in about one in five cases, typically in the subacute stage, 2 to 3 weeks after clinical presentation. Fortunately, minor aneurysms tend to regress or resolve, but large aneurysms (>8-mm diameter) create risk for coronary artery stenosis, myocardial infarction, and sudden death. The overall chance of such a catastrophe in a child with Kawasaki disease was initially felt to be about 1%; but with the advent of gamma globulin treatment during the acute phase, this risk has substantially diminished (as low as 0.08%).

The natural course of coronary artery health at older ages in patients who had Kawasaki disease as young children, both with and without the development of coronary artery aneurysms,

is still being evaluated. Since myocardial infarction has been described many years after initial disease, it is important to identify those at risk. Echocardiography, and in some cases angiography, is the key methods for recognizing coronary artery complications in these patients. Exercise stress testing can provide a valuable adjunct in assessing coronary artery perfusion by examining evidence of myocardial ischemia under conditions of increased metabolic demand.

Several key questions surround the follow-up of patients after Kawasaki disease: What is the late outcome of those patients who appear to have normal coronary arteries or those who have small aneurysms that resolve? Specifically, are there subclinical alterations in the coronary vessels that place these patients at risk for late ischemic disease? How can the magnitude of risk be identified in those who have residual aneurysms? What are the indications for surgical intervention? For patients who have had Kawasaki disease, what guidance is appropriate as they reach the age of competitive sport participation?

Exercise Testing for Myocardial Ischemia

Exercise testing studies in patients who have had Kawasaki disease have focused on the identification of coronary artery stenosis and perfusion abnormalities in those with and without aneurysm formation. Besides endurance fitness and $\dot{V}O_2$max, techniques that have been utilized include stress echocardiography, exercise electrocardiography, and stress nuclear imaging.

Endurance Fitness   Studies assessing the relationship of coronary artery abnormalities and endurance capacity during exercise testing have produced conflicting results. Paridon et al. reported decreased cycling endurance in four of five patients with dilated or stenotic coronary arteries at cardiac catheterization (149). In another study, Paridon et al. investigated 46 Kawasaki patients, 27 with no aneurysms at any time, 11 with resolved aneurysms, and 8 with persistent lesions (147). Endurance fitness was normal regardless of the coronary artery status. Each of the 28 children tested by Pahl et al. had coronary artery abnormalities (4 with giant aneurysms), but all except 2 had normal treadmill exercise tolerance (146). Average exercise time was shorter compared to that of healthy controls in the 17 patients of Henein et al., 5 of whom had aneurysms and 12 of whom had coronary artery ectasia (82).

Ischemic Electrocardiographic Changes Surprisingly, the traditional electrocardiographic marker of myocardial ischemia during exercise testing, ST-segment depression, appears to be of limited value in identifying patients with coronary artery abnormalities following Kawasaki disease. In most exercise studies, no such changes have been reported despite significant coronary artery involvement (82; 83; 135; 146; 147; 149). Allen et al. did, however, describe ischemic ST changes with exercise in their two patients with large aneurysms (2), and Kato et al. reported similar ECG findings in 5 of 13 patients with aneurysms or stenosis by angiography (97).

Stress Nuclear Imaging Patients with coronary artery involvement after Kawasaki disease often demonstrate myocardial perfusion defects on nuclear scans during exercise (83; 147; 149). For example, all eight patients with persistent aneurysms described by Paridon et al. had perfusion abnormalities using a radionuclide scan (149). In that study, 10 of 27 post-Kawasaki patients who had no recognized coronary artery abnormalities at any time in their clinical course also had perfusion defects with exercise. These findings may represent "false positives" (no comparison information in healthy children is available for ethical reasons). Alternatively, they could indicate that potentially significant perfusion abnormalities can exist in children who have appeared to escape coronary involvement with their disease.

Stress Echocardiography Assessing abnormalities in ventricular segmental wall motion by two-dimensional echocardiography immediately after exercise is a useful means of identifying ischemic coronary artery disease in adults. This technique appears to have value in evaluating patients with coronary abnormalities following Kawasaki disease as well. Two of four patients with giant aneurysms studied by Pahl et al. showed segmental akinesis (146). Normal stress echocardiograms were seen in the 15 patients with minimal coronary abnormalities. None of the 10 children without aneurysms tested by Tomassoni et al. had postexercise echocardiographic changes (195). In the study by Henein et

al., 5 patients had aneurysms and 12 had ectasia of the coronary arteries (82). Postexercise, mean values for left ventricular long axis excursion, peak lengthening velocity in diastole, and peak shortening velocity were all decreased compared to values in healthy controls.

Hijazi et al. compared these diagnostic techniques in 11 patients with aneurysms (3.5- to 10-mm diameter) (83). With exercise, none of the patients had ischemic ECG changes, while 5 showed perfusion defects on nuclear scan and 4 had abnormal stress echocardiograms. Two patients had significant stenosis associated with their aneurysms (established by cardiac catheterization), and both of these had abnormal nuclear scans and stress echocardiograms. The findings in this study, then, appear to be representative of the available research literature; that is, nuclear scan and postexercise echocardiography appear to have value in identifying children with Kawasaki disease who have significant coronary artery disease, whereas exercise electrocardiography is of limited value.

Exercise Recommendations

Clinical experience has failed to identify risk of myocardial infarction during participation in sport by children and adolescents with earlier Kawasaki disease. Still, it is intuitively prudent, based on the recognized risks of adults with coronary artery disease, to consider restriction from high-intensity exercise in Kawasaki patients who have coronary aneurysm formation or stenosis. At the same time, no increased risk should be expected in those who have not demonstrated coronary artery involvement.

In 1994 the Committee on Rheumatic Fever, Endocarditis, and Kawasaki Disease of the American Heart Association published exercise recommendations by stratifying patients who had experienced Kawasaki disease into certain risk groups (32). As indicated in table 7.1, no restriction of sport play was felt to be necessary in patients who had no coronary artery involvement or who had transient ectasia of the coronary vessels in the acute stage. For those with coronary artery aneurysms without obstruction, decisions regarding sport participation are to be based on the results of exercise stress testing. If coronary artery obstruction is present, intensive sport play is to be avoided.

TABLE 7.1   Exercise Guidelines for Patients With Previous Kawasaki Disease

Risk level	Physical activity
No coronary artery changes at any stage of illness	No restrictions 6-8 weeks after illness
Transient coronary artery ectasia that disappears during acute illness	No restrictions 6-8 weeks after illness
Small to medium solitary coronary artery aneurysm	For patients under age 10 years, no restriction 6-8 weeks after illness. Over age 10 years, recommendations guided by annual stress testing. Strenuous athletics are strongly discouraged.
One or more giant coronary artery aneurysms, or multiple small to medium aneurysms without obstruction testing	For patients under 10 years, no restriction 6-8 weeks after illness. Over age 10 years, recommendations guided by annual stress testing. Strenuous athletics are strongly discouraged.
Coronary artery obstruction	Avoidance of contact sports, isometrics, and weight training. Other recommendations guided by stress testing or myocardial perfusion scan.

From the Committee on Rheumatic Fever, Endocarditis, and Kawasaki Disease, American Heart Association 1994 (32).

Dilated Cardiomyopathy

Primary heart muscle disease manifest as a dilated cardiomyopathy in children has many potential causes but is usually idiopathic. Acquired cardiomyopathy can follow viral myocarditis and is an unfortunate complication of anthracycline chemotherapy for malignancies. The outcome for children and adolescents with cardiomyopathies is variable, ranging from complete recovery to rapid clinical deterioration with progressive heart failure and death (121).

The physiologic hallmark of dilated cardiomyopathy is depression of myocardial function, and measures of severity of left ventricular dysfunction at rest in children are predictive of clinical outcome (121). As exercise stresses the inotropic capacity of the heart, treadmill or cycle testing should be helpful in assessing disease severity and prognosis in these patients. In fact, in adults with ischemic and primary cardiomyopathies, low aerobic fitness and diminished rise in left ventricular ejection fraction (<5%) during exercise testing are predictive of survival rate (136; 180). Interestingly, however, these features do not consistently relate to indicators of myocardial function (such as ventricular ejection fraction or shortening fraction) measured at rest (59).

Responses to Exercise

The limited information available regarding physiologic responses of children with cardiomyopathies to exercise testing suggests findings similar to those in adult patients. Rowland et al. used Doppler echocardiography to measure cardiac responses to maximal semi-supine exercise in 11 patients ages 7 to 17 years with myocardial dysfunction (shortening fraction <28% at rest), most with doxorubicin toxicity (168). Compared to values in healthy subjects, the mean values for maximal cardiac index, stroke index, heart rate, peak aortic velocity, and left ventricular shortening fraction were all significantly lower. At high work rates, stroke volume plateaued in the normal subjects but declined in the patients with cardiomyopathy (figure 7.2). The authors interpreted this finding as indicating the inability of ventricular myocardium to increase force and velocity of contraction to sustain stroke volume as the systolic ejection period shortened. As in the adult studies, neither resting ventricular shortening fraction nor endurance fitness (cycling time to exhaustion) was related to any of the exercise variables (peak or change in shortening fraction, maximal stroke index, or maximal peak aortic velocity).

Similar findings have been observed in patients who have normal indices of myocardial function at rest. Johnson et al. evaluated the cardiac fitness of 13 young patients 5 years after low-dose anthracycline therapy for cancer (93). Resting echocardiograms revealed no evidence of any abnormality in left ventricular contractility or dimensions. Compared to control subjects, however, the patients demonstrated a lower $\dot{V}O_2max$ (32 ± 6 vs. 41 ± 8 ml $\cdot$ kg^{-1} $\cdot$ min^{-1}). At similar submaximal exercise loads, the patients had a lower cardiac index relative to oxygen uptake, and this was due to a smaller rise in stroke volume. These findings suggest that exercise can unmask myocardial dysfunction in patients who have no obvious myocardial anthracycline toxicity at rest.

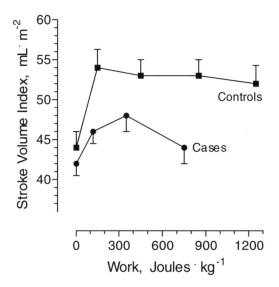

FIGURE 7.2 Changes in stroke index in patients with cardiomyopathy and healthy controls. Based on Rowland et al. 1999 (168).

Similar findings were described by Weesner et al. in eight patients aged 10 to 20 years who had received anthracycline treatment for malignancies 4 to 13 years earlier (205). Cardiac responses to cycle exercise were compared to those of 10 cancer survivors who had received no anthracyclines. Both groups had normal resting cardiac output, heart rate, blood pressure, and echocardiographic indicators of myocardial function (shortening fraction, velocity of circumferential fiber shortening, and peak aortic velocity). No significant differences were observed in maximal work performed, but immediate postexercise echocardiograms showed that the untreated subjects had a greater increase in all three measures of myocardial function compared to the treated group. The authors concluded that subtle abnormalities of myocardial function can exist in survivors of childhood cancer who have received anthracycline treatment and who demonstrate normal cardiac function at rest.

These findings suggest that variables assessed during exercise testing may serve as more sensitive indicators of myocardial health than those at rest. Additional research is needed to identify those physiological markers that might prove most valuable as predictors of clinical outcome.

An unexpected finding in adults with chronic congestive heart failure is that level of exercise intolerance does not necessarily correlate with cardiac output or muscle blood flow during exercise (8; 30). There is, in fact, evidence that

symptoms of dyspnea and fatigue that limit performance on a progressive exercise test are not related to the patient's hemodynamic status. Other factors that have been suggested to account for exercise fatigue in these patients are abnormalities in peripheral blood flow, endothelial dysfunction, abnormal skeletal muscle structure and metabolic function, and muscle wasting. This issue has not yet been evaluated in children with cardiomyopathies.

Heart Transplantation

Heart transplantation has become a standard form of treatment for children with end-stage myocardial dysfunction. Dilated cardiomyopathy, both idiopathic and anthracycline induced, is the most common indication for transplant, while about one-quarter of procedures are performed in children with severe ventricular failure after surgery for congenital heart disease (Fontan, Mustard operations) (17). The risks of complications following heart transplantation are high, including rejection, infection, hypertension, malignancies, and coronary artery disease. Still, advances in immunosuppressive therapy have resulted in a 1-year survival rate of 75% and a survival rate of 65% at 5 years (203).

Following transplantation, patients who previously showed depressed cardiac function and marked exercise fatigue experience considerable improvement in their tolerance to physical activity. However, the transplanted heart, being devoid of autonomic innervation, does not respond normally to exercise.

Studies describing cardiac responses to exercise following heart transplantation in adults have been summarized by Shephard (181). Resting heart rate is increased (90-110 beat · min⁻¹), presumably related to loss of vagal tone. With exercise, heart rate is driven by circulating catecholamines rather than by sympathetic nervous stimulation. As a result, the expected rise in heart rate is delayed and peak values are lower. At rest the stroke volume is low (due to the high heart rate); but throughout exercise, values for stroke volume tend to be greater than normal. At peak exercise, cardiac output is approximately 25% below predicted, as are $\dot{V}O_2$max and work capacity. Myocardial inotropic response to exercise appears to be normal. As Shephard noted, in adults the transplanted heart is structurally and functionally normal, but cardiac capacity during exercise is limited by lack of sympathetic drive to increase heart rate.

A limited number of studies performed in children following heart transplantation demonstrate similar findings. As outlined in table 7.2, these include diminished heart rate response to cycle exercise, with low work capacity and $\dot{V}O_2$max. In the study by Christos et al., maximal stroke volume was greater than in control subjects (29). Hsu et al. repeated exercise testing in 16 of their original 31 patients 2 years later and obtained the same results (84). These authors ascribed findings of decreased fitness in their patients to blunted heart rate responses as well as a sedentary lifestyle.

Cardiac Exercise Rehabilitation Programs

In the broad picture of children with heart disease, most have normal exercise tolerance. However, for some, particularly with the defects reviewed in this chapter, inability to participate in normal physical activities is a significant obstacle to enjoying life. The concept that a supervised, structured program of regular exercise might prove beneficial to these patients in improving their fitness is based on the success of cardiac rehabilitation programs for adults with coronary artery disease.

An examination of the various factors that limit exercise capacity in children with serious heart disease indicates that some can be potentially modified by exercise while others cannot (figure 7.3). Hypoxemia, sinus node dysfunction, and residual anatomic disease, for example, are unlikely to be altered by repetitive exercise. Based on findings in healthy subjects and athletes, myocardial contractile function is probably not altered

in training programs. On the other hand, depressed cardiac output from low stroke volume and skeletal muscle aerobic capacity can respond positively to exercise training (152).

Despite the conceptual appeal of exercise rehabilitation for children with heart disease, few long-standing programs have been developed. As noted by Wessel and Paul, these activities are difficult to conduct, hampered by the heterogeneity and the lack of geographical proximity of patients, limited funding, and necessary availability of facilities and trained personnel (209). Nonetheless, the limited data published regarding these programs highlight their potential benefits.

Insights From Exercise Rehabilitation in Adults With Coronary Artery Disease

Exercise rehabilitation programs have become standard for adult patients with coronary artery disease. These programs have focused on the goals of improving exercise capacity, increasing work rate at the threshold of angina symptoms, reducing coronary risk factors, and decreasing risk of myocardial infarction and death. Studies indicate that a period of exercise training in adults with coronary disease leads to a substantial rise in $\dot{V}O_2$max (ranging from 11% to 56%) and a better ability to tolerate physical activity (193). Moreover, most studies suggest favorable effects of these programs on decreasing long-term morbidity and mortality.

These investigations have generally been performed in patients with preserved myocardial function, and the salutary effects of exercise training have not included evidence of improvements in heart muscle performance. Increased maximal stroke volume is the principal physiologic outcome, with improvements in ventricular preload most likely due to increases in plasma volume and a more efficient skeletal muscle pump (152).

Similarly, it is doubtful that exercise training can improve myocardial contractility in patients who have significant left ventricular dysfunction. Exercise intolerance in these

TABLE 7.2	Exercise Studies in Children and Adolescents After Cardiac Transplantation		
	Maximal heart rate	Predicted work capacity	$\dot{V}O_2$max (ml · kg⁻¹ · min⁻¹)
Hsu et al. (84)	136 ± 22	61%	20 ± 6
Christos et al. (29)	154 ± 8	61%	22 ± 3
Nixon et al. (140)	134	64%	22
Schulz-Neick et al. (176)		62%	30

Untrainable	Trainable
Hypoxemia	Low physical activity
Sinus node dysfunction	Depressed skeletal muscle aerobic capacity
Residual anatomic defects	Decreased muscle capillarization
Myocardial dysfunction (?)	Low stroke volume (preload)

FIGURE 7.3 Factors influencing exercise tolerance in patients after surgery for cyanotic congenital heart disease.

individuals is caused primarily by early onset of anaerobic metabolism, which in turn results from abnormalities in skeletal muscle metabolic response to exercise (decreased aerobic enzyme content, low number of type I fibers, reduced mitochondrial volume density, and decreased capillarization) (8; 30; 189). Since these factors may be improved with exercise training, increased fitness can be achieved without primary changes in myocardial function per se. In reviewing these data, Sullivan concluded that "the concept that the skeletal muscle response to exercise plays a primary role in determining exercise tolerance provides a physiological rationale for using exercise training as a potential means to improve exertional symptoms in patients with chronic heart failure" (p. 365) (189).

Exercise Rehabilitation in Pediatric Patients

In some ways, cardiac rehabilitation programs for children are based on different goals than those for adults. The patient population and disease processes are different, and the effects of exercise on atherosclerotic vascular disease should be expected to differ from those on congenital heart abnormalities (47). Rehabilitation programs in the two groups do, however, share the common objective of improving the quality of life by allowing patients with previous low level levels of fitness to participate in the normal activities of daily living (62). Also, in addition to providing psychosocial benefits (improved social skills and enhanced self-confidence), these programs can serve as a stimulus for persistence of regular exercise habits.

Improvements seen in work capacity and $\dot{V}O_2$max following 6- to 12-week rehabilitation exercise programs in children simulate those observed in adults (table 7.3). Barber reviewed the findings in 12 published reports of such programs in patients who had undergone surgery primarily for defects such as transposition of the great arteries, tetralogy of Fallot, and single ventricle (11). The average change in work capacity was 21%, compared to a mean increase of 30% in training studies of healthy children. $\dot{V}O_2$max rose by an average of 11% in the cardiac rehabilitation studies, while an increase of 4% was found after training of normal children.

The physiologic mechanisms that account for these improvements after exercise programs in children with heart disease have not been addressed. If the physiologic basis for cardiac rehabilitation is the same in children as in adults, increases in maximal stroke volume are most likely the result of enhanced plasma volume, improved skeletal muscle pump function, and increased aerobic metabolic capacity in skeletal muscle cells.

While most authors have been impressed with positive psychosocial outcomes from exercise training programs in children with heart disease, the evidence for these effects remains largely anecdotal. Calzolari et al. noted that "at the end of the programme, the children all showed increased independence and initiative and were more self-confident in social relations" (p. 156) (25). In a questionnaire study of four patients, Donovan et al. found that an exercise program relieved mothers' fears surrounding the child's activity and reported that the program caused the child to have a better self-image (45).

Review of the research data on pediatric cardiac rehabilitation programs indicates that weaknesses in study design are common (high dropout rates, short duration, lack of nontrained controls). Driscoll (47) concluded that several important questions remain to be answered: (1) Do the improvements in physical fitness induced by an organized program persist beyond the duration of the program? (2) Are these programs effective in altering the child's lifestyle and approach to exercise and fitness? (3) Are the reported changes in exercise tolerance and $\dot{V}O_2$max an effect of the exercise rehabilitation program, or do they simply reflect participant selection and dropout bias?

Risks of Exercise

As much as regular physical activity and sport participation are healthy and beneficial pursuits,

TABLE 7.3 Responses to Cardiac Rehabilitation Programs in Children With Congenital Heart Disease

Study	Number of patients	Duration (weeks)	Increase in work capacity (%)	Increase in $\dot{V}O_2$max (%)
Vaccaro et al. (198)	21	12	19%	21%
Balfour et al. (10)	16	12	21%	20%
Sabath et al. (173)	14	12	22%	29%
Bradley et al. (22)	11	12	17%	21%
Goldberg et al. (69)	26	6	25%	
Calzolari et al. (25)	9	12	8%	

vigorous exercise does, rarely, pose a risk of sudden death in susceptible children and adolescents. Protecting such youngsters from undue jeopardy often necessitates restriction from certain forms of physical activities. The potential effectiveness of this intervention is contingent, however, on identifying those few individuals who are at risk.

Sudden Unexpected Death in Athletes With Structural Cardiovascular Disease

The sudden death of a young athlete with previously unrecognized heart disease is a rare event. Although probably underreported, the incidence of such tragedies in the United States is approximately 10 to 13 per year, creating a risk in the individual athlete of about 1 in 250,000 participants (199). Nonetheless, the sudden collapse and death of an adolescent involved in what is considered a healthy, wholesome activity is particularly poignant. It is important during the preparticipation medical assessment, then, to make efforts to recognize athletes with heart disease that poses a risk for sudden death during sport play.

Since sudden cardiac death probably occurs only in athletes with underlying heart disease, it might be anticipated that cardiac diagnostic tests should be capable of identifying those abnormalities that pose risk. To this end, recommendations have been published for screening methods to detect and prohibit athletes at risk from sport participation (118; 119). Unfortunately, not all forms of such cardiac defects can be readily identified by routine diagnostic measures, and in some cases even restriction from sport play may not substantially reduce the risk of sudden death. It is

important, though, that clinicians recognize those heart defects that can be identified and restrict activities in would-be athletes at risk.

This section reviews the structural conditions that create a risk for sudden cardiac death during sport play. Particular attention focuses on appropriate means of screening for these abnormalities in the preparticipation evaluation.

Hypertrophic Cardiomyopathy

Hypertrophic cardiomyopathy is a genetic condition characterized by dramatic thickening of the ventricular walls of the heart. The hypertrophy is usually asymmetric, involving mainly the ventricular septum (thus the alternative designation "asymmetric septal hypertrophy," or ASH), and often includes some degree of subaortic outflow obstruction (accounting for another name, "idiopathic hypertrophic subaortic stenosis" or IHSS). In some series of reported cases the risk of sudden death in individuals with this condition has been high, as much as 2% to 4% per year, but recent information suggests that the mortality rate may be substantially lower.

The mechanism for sudden death with hypertrophic cardiomyopathy is uncertain but is most likely ischemic, with relative inadequacy of coronary perfusion of the thickened, poorly compliant myocardium, particularly at times of increased metabolic demand. Most individuals who die suddenly with this condition have not been engaged in vigorous physical activity at the time, but the physiologic demands of exercise appear to increase the chance of sudden death. In the 78 adult patients described by Maron et al. (117), 37% died at rest or during sleep, 24% with mild exertion (e.g., shopping), and 29% with vigorous physical activities such as running and hiking. For this reason, patients with hypertrophic cardiomyopathy have been restricted from competitive sport play as a means of reducing risk.

Unfortunately, detection of this disease during the preparticipation examination is often difficult. As a result, unrecognized hypertrophic cardiomyopathy is the most common cause of sudden unexpected death in young athletes during sport play (118) (figure 7.4). This history may provide a clue, since a report of family members having this condition can be obtained in about 20% of cases, and at least half of individuals with hypertrophic cardiomyopathy will describe previous chest pain, syncope, or ease of fatigue with exercise.

The physical examination is often deceptively benign. A heart murmur will be audible only if subaortic obstruction or associated mitral valve insufficiency is present. The murmur may increase in intensity with standing up or during a Valsalva maneuver, but these findings are not always evident.

The ECG in those with hypertrophic cardiomyopathy is almost always abnormal, demonstrating left ventricular hypertrophy, deep Q waves, or ischemic ST changes. The echocardiogram provides a definitive diagnosis, revealing a ventricular septal thickness often exceeding 18 mm with abnormalities of mitral valve motion in systole and often some degree of subaortic obstruction.

Because the medical history and physical examination often fail to identify athletes with this disease, the routine use of echocardiography has been advocated to screen for those at risk. However, echocardiography is a prohibitively expensive test, and its routine use is confounded by the problematic identification of many false positives (that is, adolescents who have mildly thickened but normal left ventricles). As Maron et al. concluded, such findings "would generate heavy emotional, financial, and medical burdens for the athlete, family, team, and institution by virtue of the uncertainty created and the requirement for additional testing" (p. 854) (119). The same argument holds true when one considers the ECG for routine screening.

Interpreting borderline high values for ventricular septal thickness on an echocardiogram (12-15 mm) in an otherwise healthy, asymptomatic adolescent with a negative family history is difficult. Does the athlete have mild hypertrophic cardiomyopathy with risk of sudden death, signaling a need for restriction from sport play? Or is this finding a normal anatomic variant, perhaps reflecting cardiac responses to sport training (the "athlete's heart"), in which case no restriction is indicated?

Maron et al. have suggested several criteria which may be useful in making this differentiation (115). The diagnosis of hypertrophic cardiomyopathy can be assumed if (1) the ventricular septal thickness exceeds 18 mm, (2) the left ventricular diastolic dimension is small, (3) septal hypertrophy fails to regress after a period of detraining, (4) echocardiograms of family members indicate left ventricular hypertrophy, and (5) an abnormal pattern of ventricular diastolic filling (tall A waves indicative of decreased compliance) is observed by Doppler echocardiography.

Coronary Artery Anomalies

Young athletes are not at risk for atherosclerotic coronary artery disease. However, inadequate myocardial perfusion from rare but often occult congenital anomalies of these vessels can create a risk for sudden death during sport play. In the most common form, the left main coronary artery arises from the right sinus of Valsalva. Sudden death from coronary insufficiency might then occur as this vessel is compressed in its course between the base of the aorta and pulmonary artery, or, alternatively, from an obstructed os that results from its acute angle of takeoff. Other anomalies include coronary hypoplasia or stenosis, vessels buried within the heart muscle, and anomalous origin of the left coronary artery from the pulmonary artery.

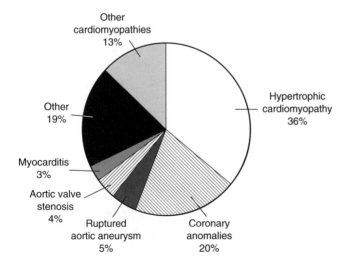

FIGURE 7.4 Causes of sudden unexpected cardiac death in young athletes.

These abnormalities are exceedingly rare, which is fortunate considering that they are often impossible to diagnose clinically. Chest pain, dizziness, and syncope with exercise have been reported as premonitory signs in adolescents with coronary anomalies, but usually the first indication of their presence is sudden collapse and death. Basso et al. described 27 cases of sudden death from coronary artery anomalies, all of which occurred during sport play (12). Mean age at death was 16 ± 5 years. Previous chest pain or syncope had occurred in 10 cases. It is troublesome to note that 9 cases had previous ECGs and 6 had undergone exercise stress tests (as part of routine preparticipation screening) and all these were normal.

Marfan Syndrome

Individuals with Marfan syndrome, an autosomal dominant inherited condition, have exaggerated laxity of connective tissue. They are tall with long extremities and fingers (arachnodactyly), and the full-blown syndrome is characterized by musculoskeletal anomalies (scoliosis, pectus excavatum), joint hyperflexibility, and dislocated ocular lenses. The most serious aspect, though, is weakening of the aortic wall at its root, with progressive dilation and aneurysm formation. This can lead to increasing aortic valve insufficiency and eventual aneurysm rupture. Preventive interventions include beta-blocker therapy (to diminish aortic pulsativity), restriction from certain forms of physical activity, and surgical replacement of the aortic valve and root when the aortic diameter exceeds 55 to 60 mm.

The risk of sport play by individuals with Marfan syndrome was highlighted by the sudden collapse and demise of the volleyball player Flo Hyman during a game in 1985. This tall star of the 1984 United States Olympic team was found at autopsy to have died of a ruptured aortic aneurysm with typical Marfan features. This tragic event heightened awareness of possible Marfan syndrome in tall athletes. Restrictions of individuals with Marfan syndrome from certain forms of sport play considered to pose a risk for acceleration of aortic dilation or aneurysm rupture were then published (23; 156). These include resistance exercise (weightlifting, wrestling), sports that might cause a blow to the chest, and endurance training.

There is currently no laboratory test that establishes the diagnosis of Marfan syndrome.

These individuals are recognized by their typical physical features, often in conjunction with a positive family history. Suspicion of Marfan syndrome on preparticipation sport screening should prompt referral to a cardiologist. Most patients with Marfan syndrome have laxity of the mitral valve (as indicated by echocardiographic findings of mitral valve prolapse) and widening of the aortic root.

Aortic Valve Stenosis

The risk of sudden death in children and adolescents with significant obstruction from aortic valve stenosis was discussed earlier in this chapter. Of the various forms of heart disease that create risk for sudden death in young athletes, this condition is by far the most common. Still, it does not rank prominently as a cause of these events, since, unlike hypertrophic cardiomyopathy and coronary anomalies, the diagnosis is easily made on physical examination. The finding on the preparticipation assessment of a prominent systolic murmur at the upper right chest, particularly if accompanied by a palpable thrill, should prompt referral to a cardiologist before clearance for sport play.

Long QT Syndrome

Long QT syndrome is an inherited electrophysiologic abnormality of heart muscle cells that delays ventricular repolarization and prolongs the QT interval on the ECG. This rare condition has attracted considerable medical and public attention because individuals with long QT syndrome are at substantial risk for ventricular tachycardia and sudden death. It has been suggested that this syndrome may account for as many as 3,000 unexpected deaths per year (200). The mortality rate in untreated cases is high, with a 10-year incidence of death reported to be as much as 50%. Approximately 85% of such events, which typically present before the end of the teen years, are triggered by exercise and emotional stimulation, creating risk for sport play. Identification and restriction from participation in athletics and vigorous physical activities are therefore critical in the management of these patients (178).

Understanding of the mechanisms, diagnosis, and treatment of individuals with long QT syndrome has undergone substantial evolution in the past decade. The prolonged QT interval and occurrence of ventricular tachyarrhythmias were initially considered to be due to "autonomic imbalance," and the condition was felt to be inherited in either an autosomal recessive (Jervell-Lange-Nielsen syndrome, with congenital deafness) or autosomal dominant (Romano-Ward syndrome) pattern.

It is now recognized that long QT syndrome is an autosomal dominant genetic disorder with variable penetrance. A considerable degree of genetic heterogeneity has been observed, with over 180 mutations on several specific gene loci; yet in half of clinically diagnosed cases, no abnormalities of these genes are evident. Nonetheless, there is early evidence that defects in certain genes are associated with characteristic risk factors, natural course, and treatment options, offering hope for gene-specific management of this disorder (200). Thirty percent of patients have no positive family history and presumably represent spontaneous mutations (207).

A prolonged QT interval with risk of sudden death can also be acquired, most particularly through heart disease (especially cardiomyopathies), electrolyte disturbances, and certain drugs. In his review of this condition, Vincent provided a list of 88 medications that have been known to prolong the QT interval (200). Since this happens in only a very few patients taking these drugs, certain individuals may be genetically predisposed to this response to medication.

The presenting symptom of individuals with prolonged QT syndrome is usually recurrent syncope and seizure-like episodes, which represent spells of self-limited polymorphic ventricular tachycardia. However, as many as 30% to 40% of sudden deaths with this condition occur as the initial event. Approximately 60% of people with the syndrome, on the other hand, remain asymptomatic.

The hallmark of this condition is prolongation of the QT interval on the ECG. The QT interval, which defines the total duration of ventricular depolarization and repolarization, normally varies inversely with heart rate. For analysis in clinical practice, values of the QT interval are corrected to a rate of 60 beat $\cdot$ min^{-1} by the Bazett formula, in which the absolute QT interval is divided by the square root of the R-R interval to obtain the QTc.

The average rate-corrected QT interval (QTc) in normal healthy individuals is 0.40 sec, with an upper limit of 0.44 sec. In the prolonged QT syndrome, the average QTc is 0.49 sec, but in 5% of cases the interval is less than 0.44 sec (28; 200). This small overlap of values for normal and affected individuals, plus the observation that the QT interval can vary with time, has created some diagnostic difficulties. While the great majority of persons with a borderline long QT interval are healthy, it is also true that the diagnosis of long QT syndrome cannot be excluded by a QTc in the upper limits of normal.

Genetic testing offers hope for patient identification and risk stratification in the future but is not yet available as a routine diagnostic tool. The diagnosis of long QT syndrome currently rests on consideration of several factors, including length of the QTc and its response to exercise (see next section); family members with long QT syndrome or unexplained sudden death; and a history of syncope, particularly with stressful events (28).

Several treatment options are available for patients with prolonged QT syndrome. Most respond to beta-blocker medication, with an 80% to 90% decrease in rate of sudden death. In other patients, pacemakers, cardiac sympathetic denervation, or an implantable defibrillator may be indicated.

QT Interval Response to Exercise

As already noted, patients with long QT syndrome typically experience events (syncope, sudden death) in response to emotional stress (loud noise, anger) or physical exercise. This observation, along with the benefits seen with beta-blocker therapy, indicates that the most likely mechanism behind these events is stimulation of the sympathetic nervous system. As exercise is accompanied by increases in both circulating catecholamines and sympathetic activity, alterations in the ECG in these patients might be expected. Reports indicate, in fact, that failure of the QT interval to shorten normally as heart rate increases with exercise is characteristic of patients with long QT syndrome; and this finding may prove useful as a diagnostic marker.

It should be recognized that measurement of the QT interval during exercise testing is not usually easy. The T wave approaches the baseline asymptotically, and identification of its termination may be difficult. This problem is confounded by the tachycardia of exercise, when the P wave of the next beat becomes superimposed on the downslope of the T wave and the baseline may be unsteady. Researchers describing QT changes with exercise and recovery have attempted to solve these difficulties by using a "tangent approach," in which the end of the QT interval is defined by the intercept of the tangent drawn from the steepest downslope of the T wave to the PQ line.

Studies in healthy subjects indicate that values for QTc are not necessarily stable with the tachycardia accompanying exercise. Bhatia et al. reported in a study of children that the average QTc rises to 0.41 to 0.42 sec in the heart rate range of 100 to 170 beat · min^{-1} but returns to 0.40 sec at higher rates (18).

This finding notwithstanding, studies that have directly compared responses in patients with long QT syndrome and normals have indicated significant differences in changes in QTc during exercise and recovery. In addition, these reports have indicated that submaximal and maximal heart rate values are lower in the patients, suggesting a chronotropic incompetence related to cardiac sympathetic deficiency.

Vincent et al. performed exercise testing in 27 patients with long QT syndrome and compared maximal and recovery QTc values to those of healthy subjects (201). At rest, mean values were 0.47 and 0.42 sec for patients and controls, respectively. Average QTc did not change significantly at maximal exercise or at 3 min into recovery in the normal subjects (0.40 and 0.43 sec); but at maximal exercise, the average QTc in the patients rose to 0.54 sec and in recovery to 0.55 sec. Maximal heart rate was 156 ± 10 versus 188 ± 9 beat · min^{-1} in patients and controls, respectively.

The 11 patients with long QT syndrome reported by Shimizu et al. demonstrated a rise in QTc from 0.48 ± 0.06 sec at rest to 0.55 ± 0.03 sec in the first 2 min of recovery from maximal treadmill exercise (182). The QTc rose in 10 of the 11 subjects. Respective values for 12 healthy controls were 0.40 ± 0.02 and 0.41 ± 0.02 sec. Small increases of 0.01 and 0.02 sec were seen in 6 of the normal subjects (figure 7.5).

Katagiri-Kawade et al. reported an average increase in QTc from 0.47 sec at rest to 0.52 sec at 1 min postexercise in six children with long QT syndrome (96). In six controls the mean value at rest was 0.40 sec and during recovery was 0.39 sec. Five of the six patients showed an increase, while QTc in all but one of the controls decreased.

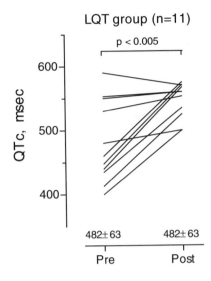

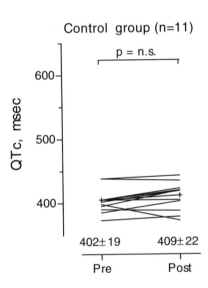

FIGURE 7.5 Changes in the corrected QT interval during exercise in patients with long QT syndrome and healthy controls (182).

Treadmill testing by Weintraub et al. in 16 young persons with long QT syndrome revealed a mean QTc of 0.47 sec at rest, 0.51 at peak exercise, and 0.52 at 2 min into recovery (207). Schwartz et al. described a 9.5% ± 0.3% shortening of the QT interval for each 100 msec of R-R interval shortening in normal subjects (177). Values in 15 patients with long QT syndrome with two different genotypes were 3.6% ± 0.3% and 2.8% ± 1.3%. On the basis of the relationship between absolute QT interval and heart rate with exercise, Swan et al. suggested that abnormal lengthening of the QTc in patients with long QT syndrome should be most obvious during recovery (190).

These findings indicate that changes in QTc with exercise testing may have diagnostic value, particularly in differentiating abnormal findings in borderline cases. The data suggest that such comparisons are best performed in early recovery. An increase in QTc above the resting value, particularly if the change is greater than 0.03 to 0.04 sec, supports the diagnosis of long QT syndrome.

At present, no predictive value with respect to clinical course has been established for exercise-induced QT interval changes in these patients. As noted by Chiang et al., there are no normal standards, and prospective accuracy of changes in QT interval with exercise in stratifying risk has not been studied (28).

Exercise Recommendations

Restriction from competitive sport and high-intensity physical activities is necessary for patients with long QT syndrome, since many experience events during exercise. Interestingly, swimming seems to create a particularly high risk (133; 177). In one series of cases, 107 of 320 patients (33%) with a known trigger experienced syncope while swimming, and this sport should be avoided (177). Patients who experience events with swimming, however, seem to possess one specific genotype, and genetic testing may be able to discriminate specific risks in the future.

Schwartz et al. suggested that it is not physical activity itself but rather the associated emotional excitement that serves as the primary stimulant of adverse events with long QT syndrome (178). They proposed that swimming stands out as a particularly risky form of exercise since this is "a physical exercise that may be accompanied by some degree of emotion, if not actual fear, in children" (p. 78).

Commotio Cordis

Commotio cordis (literally, concussion of the heart) refers to the sudden collapse and frequent death of a child or adolescent after a traumatic blow to the anterior chest. These tragedies occur in young persons with a structurally normal heart who are struck over the left side of the chest usually by projectiles such as a baseball, hockey puck, or lacrosse ball. Commotio cordis has also been described from more diffuse blows to the chest (knee in the chest during football, a karate kick). The object creating the trauma is always hard,

and no deaths from commotio cordis have been recorded after chest trauma from a pneumatic ball (football, basketball) (109).

These tragic events fortunately are rare, but their true frequency is not clear. Approximately two cases per year in the United States have been reported due to baseball blows to the chest (1), which is the most common scenario; but this number must represent an underestimate of total events. Moreover, cases in which recovery is rapid might go unrecorded.

Typically the speed of the projectile has not been considered by witnesses to be excessive, and almost always the object thrown or the blow created has been by a peer rather than an adult. Most episodes occur during competitive sport play, but many cases have happened during informal or recreational play on the playground or at home. In 25 cases of commotio cordis reviewed by Maron et al., nine of the individuals were not considered to be competitive athletes (116).

The risk of commotio cordis is largely restricted to children and young adolescents. The average age of reported cases is 11 to 12 years old, with a range of approximately 3 to 19 years (109; 116). This increased risk in younger persons has been explained by the greater compliance of the growing chest wall, which facilitates transmission of energy to the underlying heart muscle.

For reasons that are not clear, almost all cases have involved boys. In the review by Maron et al., only one event occurred in a female (116). As Link et al. have discussed, this gender difference is not explained by the male:female ratio of sport participation but may reflect biological differences in chest wall anatomy or susceptibility to heart rhythm disturbances (109).

In 72% of reported cases the anterior chest was exposed to the inciting trauma. However, in over a quarter of episodes of commotio cordis, the subject was wearing protective padding that appeared to cover the area of projectile impact. This includes cases of death in a football player, a lacrosse goalie, and a baseball catcher (116).

In the typical case, a young athlete is suddenly struck in the left anterior chest over the heart, collapses to the ground, and cannot be resuscitated. In some cases a brief recovery period is seen, with the youngster conversant or even walking, only to collapse again with apparent cardiac arrest. In about 1 in 10 cases the person survives, but even in these situations the individual may experience severe neurological sequelae. The suggested mortality rate of 90% does not consider, however, the possibility that there may be cases with rapid recovery that go unreported.

The resistance of these children to resuscitative efforts has been puzzling (1). Among 25 cases, 19 received cardiopulmonary resuscitation immediately at the scene, in seven of the situations by trained medical personnel (116). In only two subjects was resuscitation successful in restoring cardiac rhythm (and both of these youngsters eventually died within 9 days).

Findings at autopsy examination of the heart following commotio cordis have been conspicuously unrevealing. There has been no evidence of any underlying predisposing pathological conditions. No signs of myocardial injury or damage to coronary arteries or heart valves have been seen in these cases (116).

In the few cases in which an ECG has been obtained at the time of the event, ventricular fibrillation has almost always been documented. Significant elevations in ST segments consistent with myocardial injury have been observed in those who have been resuscitated. This led to early theories that commotio cordis represented a fatal ventricular dysrhythmia triggered by myocardial damage from the chest impact.

Experiments in swine by Link et al. have provided valuable insights into the mechanism of commotio cordis (110). In this model, a low-energy impact from a wooden projectile the size and weight of a baseball mimicked the scenario of commotio cordis in humans. The projectile was propelled to the chest of the animals at a speed of 30 mph (48 km · hr^{-1}), and the impact was timed to various portions of the cardiac cycle.

Ventricular fibrillation was produced in 90% of the occurrences when the impact was timed to occur from 30 to 15 msec before the peak of the T wave on the ECG. If the impact happened during the QRS complex, complete heart block occurred transiently in 40% of the tests. No dysrhythmias occurred when impact coincided with other portions of the cardiac cycle. These findings support the concept that commotio cordis results when a blow to the chest occurs during narrow vulnerable periods in the cardiac cycle, triggering ventricular fibrillation or heart block.

The study by Link et al. also provided information regarding potential means of preventing commotio cordis. The experimental protocol was repeated using a regulation baseball and three softer balls of varying degrees of hardness. The risk of provoking ventricular fibrillation was directly proportional to the ball hardness. This suggests that use of a softer ball in baseball play by children might diminish risk. Whether commercially available chest protectors can decrease risk for commotio cordis is unknown. No experimental model has been developed to test the usefulness of such equipment in humans. As noted earlier, deaths have occurred while athletes were wearing protective pads.

Cardiac Non-Disease in Children

From the preceding discussion, it is clear that irregularities in heart rhythm, cardiomegaly, and symptoms of chest pain and syncope, particularly during exercise, deserve careful diagnostic attention, especially in young athletes. Indeed, such findings may serve as the only clue to serious underlying cardiovascular disease that could pose a risk to vigorous activities and sport play.

At the same time, it is important to recognize that most irregular heart rhythms in children and adolescents are benign, that chest pain is only very rarely caused by heart disease in this age group, that syncope is usually due to other causes, and that bradycardia and mild heart enlargement can represent normal physiologic responses to endurance training. Almost always, then, these findings reflect "cardiac non-disease," raising concern over possible heart abnormalities when, in fact, none exist. Since there is a significant morbidity to this situation, with anxiety and parental overprotection, it is important that these youngsters not be erroneously labeled as carrying a risk from heart disease (16).

The highly visible occurrences of collapse in young athletes have generated a great deal of anxiety over the risk of sudden death during sport play. As necessary as it is to be alert to findings of occult cardiac disease, it is equally important to assure families when their children are normal and to keep the extremely low risk of such tragedies in perspective.

Exercise-Related Syncope

Syncopal events are not uncommon in children and adolescents and are generally benign. Sudden unconsciousness that occurs surrounding physical exercise and sport play has a higher probability of an organic cause and signals the need for diagnostic attention (142). Still, in most cases, exercise-related syncope is not an indicator of serious underlying cardiopulmonary or metabolic disease (167).

The evaluation and management of the child or adolescent who experiences syncope during sport require an understanding of the broad differential diagnosis involved (table 7.4). The cardiac-related causes have already been reviewed. To rule out these possibilities, referral to a cardiologist is warranted, and further investigations such as echocardiography and exercise stress testing are often appropriate. The central clues to explaining these episodes, however, are most often obtained from a careful history of the details surrounding the event.

While no epidemiologic analysis of the causes of exertional syncope has been performed, alterations in peripheral vascular tone, hydration status, ambient temperature, and ventilatory patterns with exercise are far more often responsible for syncope during sport play than is cardiac disease. Coaches have long emphasized, for example, that athletes should keep moving after running competition to prevent fainting from venous pooling in vasodilated leg muscles. Despite educational efforts, inadequate fluid intake is still common in young athletes, and dehydration coupled with hyperthermia is a common cause of syncope with sport play.

Hypoglycemia is not a frequent cause of exercise-related syncope in an otherwise healthy youngster but needs to be considered in the differential diagnosis; and rarely, cases of convulsive disorders triggered by sport play have been described. Syncope can also result from hypocapnia and alkalosis in athletes who hyperventilate with the excitement of sport play. These events are often preceded by a sensation of air hunger, anxiety, and numbness or paresthesias of the fingers.

In individuals with neurocardiogenic syncope, reflex vagal activity with vasodilation and bradycardia is triggered by ventricular mechanoreceptors that are stimulated by high intracavity pressures and low volumes. Athletes, who often have greater vagal tone to start with, may be particularly susceptible during sport play, when high cardiac output increases ventricular pressures in conjunction with low plasma volumes from dehydration.

An upright tilt test has commonly been utilized to identify patients who are prone to neurocardiogenic syncope. In this test, subjects are typically tilted on a table to a 60° incline to provoke a diagnostic reaction of bradycardia, hypotension, or both (192). It has been suggested that tilt testing should not be performed in the assessment of trained athletes with syncope, since their tendency for orthostatic intolerance may cause a high frequency of false positive tests (142). However, Grubb et al. demonstrated that 17 of 24 trained young athletes with recurrent unexplained syncope had positive tilt tests compared to none among 10 control subjects (71). These authors concluded that tilt testing was useful in the evaluation and management of syncope in athletes.

The management of noncardiac exertional syncope depends on the specific mechanism, but adequate hydration is a theme common to most of the causes just outlined. Neurocardiogenic syncope has been treated with supplementary fluids and salt intake, beta-blocker medications, and 9-alpha fluorocortisol, a salt-retaining steroid.

TABLE 7.4   Causes of Exertional Syncope

Cardiac	Noncardiac
Hypertrophic cardiomyopathy	Hyperthermia
Aortic stenosis	
Coronary artery anomalies	Dehydration
Dilated cardiomyopathy	
Pulmonary hypertension	Dependent venous pooling
Long QT syndrome	
Right ventricular dysplasia	Hyperventilation
Supraventricular tachyarrhythmias	
Ventricular tachycardia	Neurocardiogenic
	Hypoglycemia
	Seizure disorder

Recurrent Chest Pain

Recurrent chest pain is not an uncommon complaint heard in the pediatrician's office. It has been estimated that as many as 650,000 patients between the ages of 10 and 21 years are seen each year in the United States by physicians for this concern (56). Parents associate chest discomfort with coronary artery disease and myocardial infarction and are understandably fearful that their child may experience the same fate. Perhaps because of this, chest pain is the second most common reason a child is referred to a pediatric cardiologist (103).

In fact, symptoms of chronic recurrent chest pain in young people are rarely caused by heart disease. In the otherwise healthy child who experiences episodes of chest pain at rest and who has no associated symptoms and a normal examination, seldom can any organic cause be found at all. In a follow-up study of 31 young patients with an initial diagnosis of idiopathic chest pain, symptoms had disappeared in 81% of those who were followed for more than 3 years (169). Equally reassuring was the observation that in no case did occult disease subsequently appear to account for the initial symptoms.

When episodes of chest pain occur with exercise, an explanation is more often forthcoming. And since this complaint is consistent with the picture of angina pectoris, a careful consideration of possible cardiac disease is in order. The usual suspects are the same abnormalities as those posing a risk for syncope and sudden death outlined previously, and referral to a cardiologist is appropriate. Exercise stress testing is almost always indicated for patients who have exercise-induced chest pain, looking for ischemic ST changes, dysrhythmias, and abnormal blood pressure responses that might signal the presence of underlying heart disease. Even when this complaint is associated with exercise, however, heart disease is rarely the explanation. Exercise-induced chest pain in children and adolescents is almost always related to extra-cardiac causes.

Exercise-induced asthma is probably the most common cause of chest pain with exercise in the pediatric age group. Typically the discomfort is described as a "tightness" of the chest and is accompanied by dyspnea, cough, wheezing, or a combination of these (see chapter 6). Symptoms characteristically occur during recovery from extended periods of activity (at least 6-12 min). Most patients who have asthma will develop bronchoconstriction with exercise, but so will a high percentage of atopic patients without asthma (allergic rhinitis, atopic dermatitis), and, in a smaller number of cases, even those with no allergic history. A decline in peak expiratory flow rate or forced expiratory volume in the first second (FEV_1) of >15% with exercise is diagnostic, but a clinical therapeutic trial of pre-exercise bronchodilators in a child with typical symptoms is an appropriate first step.

Gastroesophageal reflux is commonly induced by vigorous physical activities, particularly running. Symptoms of "heartburn," chest discomfort, and belching have been described with exercise in as many as 8% to 36% of adult subjects (132). Reflux seems to be equally common in athletes and nonathletes (187). The diagnosis should be suspected in patients who complain of midline anterior pain (not necessarily burning) induced by particularly vigorous activities. In some cases, nausea, cough, vomiting, or an acid taste in the back of the throat will be reported. The patient can initially be provided a therapeutic trial of anti-reflux medication. In some cases, exercise testing with an esophageal pH probe in place can provide a more definitive diagnosis.

Musculoskeletal pain, sometimes related to previous chest trauma, is often an obvious explanation for chest discomfort with exercise. Sometimes, however, less well-defined chest wall pain is triggered by the hyperpnea of exercise. Often the label of costochondritis, or inflammation at the chondrosternal junctions, has been applied to these cases. The patient complains of pain that is exacerbated by breathing and worsened during exercise, and the diagnosis is established by eliciting point tenderness over the chest. Treatment consists of rest and analgesic/anti-inflammatory agents.

A *"stitch"* in the side is a cramping pain with exercise that has been experienced by most individuals. Typically the pain is located laterally beneath the left rib cage but can occur on the right as well. The pain worsens with breathing, and exercise can become intolerable. Originally thought to be caused by trapped gas in the gastrointestinal tract or splenic engorgement, the pain is now considered by most to be due to spasm or strain of musculoskeletal structures supporting the diaphragm. This condition is often seen in association with low levels of endurance fitness, and improving exercise capacity may alleviate symptoms. Other suggested remedies have included forceful breathing through pursed lips (i.e., "stretching out" the diaphragm), squeezing the site of the pain with one's hands, and training the abdominal muscles with sit-ups (129).

Irregular Heart Rhythm

An ECG performed on a pediatric patient with an irregular heart rhythm usually shows an exaggerated sinus arrhythmia, premature atrial beats, or premature ventricular beats. Sinus arrhythmia is the physiologic phasic change in heart rate that normally occurs with respirations. Premature atrial contractions from an ectopic supraventricular focus are benign, identified on the ECG by an early but unremarkable QRS complex with a missing or abnormal P wave. Premature ventricular contractions (PVCs) are characterized by an early bizarre QRS complex and represent cardiac depolarizations initiated by an ectopic focus in the ventricles. Premature ventricular contractions in young individuals usually bear no clinical importance but can reflect underlying heart disease or predispose to more significant ventricular tachyarrhythmias.

Premature ventricular contractions can be detected on 0.3% to 2.2% of ECGs performed on resting children with apparently normal hearts (64). They can be considered benign if (1) there is no evidence of underlying cardiac or metabolic disease, (2) no symptoms can be elicited that might indicate previous episodes of ventricular tachycardia or fibrillation (syncope, seizures), and (3) the PVCs disappear during an exercise test. Typically benign PVCs are no longer apparent after even light exercise and sometimes are abolished by the anticipatory tachycardia before testing begins. Patients in which ectopy increases with exercise or occurs as runs of ventricular tachycardia require further evaluation, usually with electrophysiologic study at cardiac catheterization.

Frequency of PVCs does not affect this analysis, as long as each PVC is isolated (i.e., there are no couplets or triplets of PVCs occurring together). Premature ventricular contractions appearing as bigeminy (every other beat) or trigeminy (every third beat) are still considered innocuous if the criteria just outlined are met. Premature ventricu-

lar contractions have been considered to be more likely benign if they are unifocal (as indicated by uniformity of their configuration) rather than multifocal (having different configurations, implying multiple foci). The latter are very uncommon in children with a normal heart, and patients with this finding require careful assessment to rule out underlying cardiac disease (64).

Athletes with PVCs that fit the cited criteria for benign ectopy can participate in all competitive sports (213). In many of these cases, PVCs will disappear over time. In the 17 patients described by Jacobsen et al., only 8 still had ectopy after an average 7-year follow-up (88). Tsuji et al. reported that PVCs disappeared in 28% of 78 children after a period of 6 ± 3 years (196).

The "Athlete's Heart"

Highly trained adult endurance athletes typically demonstrate a combination of clinical findings that has been termed the "athlete's heart" (65; 164). This consists of cardiac enlargement on X-ray and echocardiogram, sinus bradycardia, and a number of electrocardiographic changes, most commonly left ventricular hypertrophy. While these features are felt to represent physiologic adaptations to training, they also mimic findings in individuals who have heart disease. It is important, then, that clinicians be able to distinguish between normal cardiac characteristics resulting from endurance training and those that indicate cardiac abnormalities.

Findings in Adult Athletes

Enlargement of left ventricular diastolic dimension by echocardiography in adult male distance runners, swimmers, and cyclists is usually not dramatic, averaging about 54 to 56 mm (compared to the normal range in nonathletes of 40-52 mm) (55). The ventricles of adults trained in resistance sports (powerlifting, wrestling), on the other hand, are often characterized by a disproportionate wall hypertrophy rather than chamber dilation (131). The resting heart rate of the adult endurance athlete averages approximately 11 beat · min^{-1} slower than that of the nonathlete (154). Values of 35 to 50 beat · min^{-1} are not uncommon, presumably reflecting enhanced vagal tone.

The mechanisms responsible for the "athlete's heart" have been presumed to involve a combination of genetic endowment and physiologic adaptations to long-term endurance training. Studies assessing the hereditability of heart size have generally not indicated a substantial genetic influence (19; 43; 102). There is some evidence, however, that genetic factors may play a role in determining the extent of cardiac responses to exercise (19; 105).

Insights into this question have been provided by studies examining changes in left ventricular size during periods of training and detraining. In general, these indicate that with detraining, the ventricular diastolic size decreases (by an average of about 1 mm) but is still significantly greater than that of nonathletes (53; 55; 61). These reports appear to indicate, then, that both genetic and training factors influence findings of the "athlete's heart."

The mechanism by which endurance training might serve to increase left ventricular size is not certain. The response could be primarily cardiac, an adaptation to chronic, repetitive volume overload (31). Perrault and Turcotte argued, however, that the determinants were more likely extra-cardiac, related to a combination of training-induced increases in plasma volume and resting sinus bradycardia from augmented vagal tone (154).

The "Athlete's Heart" in Children

Clinicians caring for children need to know if findings such as bradycardia, heart enlargement, and left ventricular hypertrophy on the ECG are expected findings in trained endurance athletes before the age of puberty. If the answer is no, these features are more likely to indicate heart disease in the child athlete rather than physiologic adaptations. There are, in fact, reasons to suspect that the child athlete might be less likely to manifest findings of the "athlete's heart." Young athletes have not yet experienced the years of intensive training that appear to be necessary to stimulate these features. Increases in maximal stroke volume in response to aerobic training are less in prepubertal compared to mature subjects (171). Endocrine factors might also be critical. Levels of testosterone and growth hormone, two hormones that may influence myocardial responses to training, are lower in prepubertal athletes.

Cardiac features have been described in child swimmers, distance runners, and cyclists, mainly in males. These studies indicate that some but not all of the characteristics of the "athlete's heart" are observed in children as well as adults (165).

Sinus Bradycardia　Five of six studies examining resting heart rates in child endurance athletes have demonstrated lower values than in nonathletic controls (165). The average difference between the athletes and nonathletes was 15 beat · min⁻¹, similar to that observed between adult athletes and controls. In three of the pediatric studies, the resting bradycardia was observed in children who also had a larger left ventricular diastolic size.

Electrocardiographic Changes　No differences on the resting ECG (except sinus bradycardia) have been noted between prepubertal child athletes and nonathletes (141; 166; 170). Young competitors have not shown findings of left ventricular hypertrophy, conduction block, ST changes, and dysrhythmias reported in adult endurance athletes.

Cardiac Enlargement　Mild increases in left ventricular diastolic size have been seen in highly trained child athletes, comparable relative to body size to those observed in adult athletes. Interestingly, however, the reports to date seem to indicate a sport-specific response. Five studies in swimmers, for instance, all demonstrated a larger ventricular size compared to that in nonathletes (165). The mean difference in these studies was 10%, which is not different from the magnitude of increase observed in adult endurance athletes. In two studies of prepubertal male cyclists, adjusted left ventricular diastolic size averaged 8.7% greater than in nonathletic controls. However, five reports in child distance runners failed to demonstrate any enlargement of left ventricular diastolic size (165). The explanation for these differences is not readily apparent.

Complete Heart Block and Pacemakers

Complete heart block occurs when impulses initiated at the sinus node and traveling through the atria fail to be conducted down the atrioventricular node to the ventricles. A "backup" pacing focus high in the ventricles takes over the task of innervating the ventricular Purkinje fibers and myocardium, although at a rate considerably slower than that of the sinus node. In this condition, then, the heart has two pacemakers, sinus node and ventricular, firing independently of each other at different rates. As a result, the atria and ventricles do not contract in sequence. The diagnosis is easily made by an ECG in an individual with a slow pulse, as no relationship is observed between the P waves and QRS complexes.

Complete heart block in children is usually congenital, present at birth. Acquired heart block in the pediatric age group is almost always a consequence of cardiac surgery, though rarely, infectious and inflammatory illnesses such as Lyme disease, diphtheria, and rheumatic fever can be responsible. Most children with congenital complete heart block do not have underlying structural congenital heart anomalies. In many of these cases, heart block is associated with clinical or laboratory evidence of maternal connective tissue disease. Approximately 25% to 33% of cases do occur with structural cardiac abnormalities, most commonly ventricular inversion with L-transposition of the great arteries ("corrected" transposition) (163).

The principal issue for patients with complete heart block is failure to generate a normal heart rate. The firing rate of the ventricular pacemaker is about two-thirds that of the sinus node and decreases with age. The rate in older children is generally about 45 to 50 beat · min^{-1}. More importantly, this focus is poorly responsive to autonomic and humoral stimuli and does not increase appropriately with exercise. The clinical complications of the condition vary widely, depending on the adequacy of this response. Most children with congenital complete block without underlying cardiac disease lead normal lives but may experience some degree of exercise intolerance. On the other hand, infants born with heart block and a rate <55 beat · min^{-1} often present with findings of congestive heart failure (127). Children with compete heart block are at risk for syncope and sudden death, although the mechanisms and predictive signs for these events are not fully understood. Syncope in these children (Stokes-Adams attacks) may be related to ventricular tachycardia or bradycardia (130).

The specific treatment for patients with complete heart block who have congestive heart failure, significant exercise intolerance, or syncope is pacemaker implantation. Prophylactic pacemaker placement is indicated for asymptomatic young patients who demonstrate certain findings that may indicate a risk for sudden death. These include low ventricular rate (<40 beat · min^{-1} awake), frequent or multifocal ventricular ectopy, prolonged QT interval, and cardiomegaly (163; 183).

Responses to Exercise

When a normal healthy child performs a maximal exercise test, cardiac output increases three to four times over resting values, due almost entirely to increases in heart rate (typically 80 beat · min^{-1} at rest to about 200 beat · min^{-1} at peak exercise). Changes in stroke volume contribute little, since values during exhaustive exercise in the upright position are usually only about 30% greater than those at rest.

The patient with complete heart block has a dampened heart rate rise with exercise, which in some cases may be very minimal. In the average patient, atrial rate responds normally, while the ventricular rate doubles to approximately 100 beat · min^{-1}. Considerable variability has been observed, however, with peak rates reported from 50 to 145 beat · min^{-1} (194) (figure 7.6). These differences presumably reflect varying sensitivities of the ventricular pacemaker to autonomic stimulation. There appears to be no correlation between resting heart rate in these patients and extent of their heart rate response to exercise.

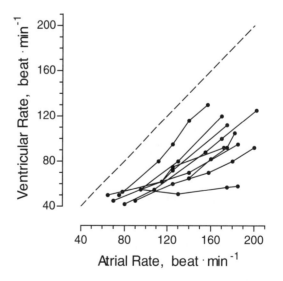

FIGURE 7.6 Comparison of atrial and ventricular rates during exercise in patients with complete heart block (194).

Considering these limitations in heart rate response, it is not difficult to appreciate why subjects with complete heart block may demonstrate limited cardiac reserve, low $\dot{V}O_2$max, and depressed endurance fitness. Intriguingly, however, many patients with congenital heart block and limited heart rate response display normal exercise tolerance (86; 191). In fact, some have even performed well in competitive athletics (86). Among the 14 patients with isolated congenital heart block exercised by Winkler et al., only 1 had an exercise intolerance (treadmill time more than 2 SD below predicted) (210). These reports illustrate that it is possible for individuals with complete block to utilize compensatory mechanisms that permit normal endurance fitness.

Patients with complete heart block often demonstrate greater stroke volumes with exercise; but this does not usually compensate for low heart rates, and cardiac output remains depressed (191). Beat-to-beat stroke volume variability with complete heart block must be considerable, since left ventricular filling volume is constantly changing with atrial-ventricular asynchrony. How this might affect stroke volume responses to exercise is unknown.

Taylor and Godfrey concluded that "the difference between the patients reaching a normal maximal work load and those failing to do so is presumably due to the response of the working muscle . . . to extract oxygen from the circulation" (p. 933). They suggested that an individual's ability to tolerate low venous oxygen content may permit greater exercise tolerance with complete heart block. An increase in maximal arterial-venous oxygen difference has been observed in some patients undergoing exercise testing, but this has not been a consistent finding (86; 191).

Ventricular ectopy is common in patients with complete heart block during exercise testing, occurring in as many as 40% to 70% of cases (86; 191; 210). The clinical significance of this finding, however, remains uncertain. Winkler et al. described complex ectopy (multiple PVCs, couplets, or ventricular tachycardia) in half of their patients (210). In that study the frequency of ectopy was age related; no ectopy was observed in patients under the age of 10 years.

Why patients with complete heart block demonstrate ventricular ectopy with exercise is unknown. Low heart rate per se is an unlikely explanation, since PVCs are also recognized in those at rest, and rate of ectopy during exercise does not correlate with rise in heart rate (210). Winkler et al. suggested that the frequent appearance of ectopy with heart block indicates that "complete heart block is a complex disease, rather than a simple problem of low heart rate" (p. 91).

The prognostic importance of ventricular ectopy in these patients is also uncertain. As noted previously, syncope with complete heart block has been associated with ventricular tachyarrhythmias (130). No clear-cut relationship between ventricular ectopy and clinical outcome has been observed (183). In formulating indications for pacing in these patients, Sholler and Walsh did include the presence of "frequent, multiform, or repetitive ventricular ectopy" (p. 197) (183).

Exercise Testing and Recommendations for Sport Play

Cycle or treadmill exercise testing in patients with complete heart block can provide information regarding level of physical fitness, heart rate response to exercise, and appearance of ventricular ectopy. As already noted, such findings are useful in making decisions regarding pacemaker placement as well as providing patients with guidelines for appropriate physical activities and sport participation.

Submaximal tests to estimate maximal aerobic power that assume a normal heart rate response to exercise are clearly inappropriate for children with complete heart block. A drastic overestimation of maximal aerobic power, for instance, will occur in a child whose maximal heart rate is 100 rather than 200 beat · min^{-1}. Use of the atrial rate, on the other hand, should be more accurate. However, assessments of aerobic fitness by measurement of $\dot{V}O_2$max, endurance treadmill time with a Bruce protocol, or peak mechanical power achieved on the cycle ergometer are preferable methods.

Recommendations from the 1994 Bethesda Consensus Conference stated that full sport participation could be approved for patients with complete heart block if they had (1) normal ventricular function and no underlying structural disease, (2) no symptoms of syncope or near-syncope, (3) a narrow QRS complex, (4) a ventricular rate at rest over 40 to 50 beat · min^{-1}

that increases with exercise, and (5) an exercise test demonstrating no more than occasional PVCs and no ventricular tachycardia (213).

According to these guidelines, athletes who have syncope, significant ventricular arrhythmias, and fatigue with exercise should have a pacemaker implanted before sport participation. Those with a pacemaker should not play sports in which the chance of body collision might endanger the pacemaker system.

Effects of Pacemaker Placement

Remarkable technological advances have made pacemaker implantation a routine intervention for children with serious heart rhythm problems (186). In older children the pacing generator is placed in a subpectoral pocket, and leads are inserted transvenously to an endocardial site near the apex of the right ventricle (67). The advent of lithium iodide batteries has extended the expected life span of pacing systems to 3 to 15 years, depending on the mode of pacing.

Complete heart block, either congenital or acquired, is one of the most common indications for pacemaker placement in children and adolescents. Patients with postsurgical block and those with underlying structural disease often require immediate pacing. In the past, pacing was delayed in patients with congenital complete block but no heart defects until the advent of symptoms (syncope, marked exercise limitations). With the recognition that the first event for these patients may be sudden death, prophylactic placement is now performed if certain premonitory findings outlined previously in this section are present. Because of the rarity of complete heart block and the limited follow-up time, however, the predictive validity of these markers has not yet been established (186).

The goal of pacing is to improve hemodynamic function by attempting to simulate the normal cardiac electrical system. Early fixed-rate ventricular pacemakers for complete heart block that increased resting rate to a specific normal level were inhibited if and when this rate was exceeded by that of the patient's own ventricular focus. These systems fail to increase heart rate with exercise for most patients and do not restore atrial-ventricular synchrony.

Newer dual-chamber pacemakers provide a more physiologic response to exercise and are currently the mode of choice for children with complete heart block (186). Since sinus node function is usually normal, these pacemaker systems, which sense atrial depolarization and trigger ventricular contractions, restore atrial-ventricular synchrony and increase ventricular rate with exercise. As might be expected, children with complete heart block show greater cardiac capacity after implantation of a dual-chamber pacemaker compared to a fixed-rate system. Karpawich found that dual-chamber pacing improved maximal cardiac output by 23% over values with fixed-rate pacing (95). Mean peak heart rates were 175 and 94 beat · min^{-1}, respectively. Interestingly, however, the mean exercise duration was similar in the two groups.

Other ventricular pacemakers increase firing rate by sensing body movement rather than sinus activity. These rate-responsive pacemakers also cause a rise in heart rate with exercise; but, as with dual-chamber pacemakers, their influence on endurance performance has not been impressive. Hanisch et al. compared maximal cycle exercise performance in seven children with heart block with their pacemakers in fixed-rate and rate-responsive modes (78). Exercise duration was 16% longer during rate-responsive pacing, with a peak heart rate of 124 beat · min^{-1} compared to 82 beat · min^{-1} with fixed-rate pacing. $\dot{V}O_2$max values were 29.7 ± 2.9 ml · kg^{-1} · min^{-1} during rate-responsive exercise and 25.7 ± 2.8 ml · kg^{-1} · min^{-1} with fixed-rate pacing.

In a similar comparison by Ragonese et al., peak heart rates in 10 young patients were 77 ± 13 and 130 ± 15 beat · min^{-1} for fixed-rate and rate-responsive pacing, respectively (157). Endurance time was longer during rate-responsive pacing (10.4 ± 2.3 vs. 9.0 ± 1.8 min), but the difference was not statistically significant. Miller et al. could find no differences in maximal treadmill times in five patients exercising in fixed-rate and rate-responsive modes (128).

Systemic Hypertension

Systemic hypertension is a major public health issue. As many as 15% to 25% of adults in developed countries have elevated blood pressure, making them three to four times more likely to develop coronary artery disease, congestive heart failure, stroke, and peripheral vascular disease (74). In most cases the specific etiology is unknown, but this *essential hypertension* is related to elevated cardiac output and peripheral vascular resistance in response to hormonal effects on blood volume and electrolyte concentrations. Essential hypertension has a strong genetic basis and hence often demonstrates familial aggregation.

Essential hypertension does not usually surface clinically until the adolescent years. Before this age, elevated blood pressure in children often has an underlying organic basis, most frequently chronic renal or endocrine disease. In general, the younger the child and the higher the blood pressure, the more likely a primary disease process can be identified. It is important to recognize, too, that body fat is a major contributor to elevations in blood pressure. In the study by Londe et al., 53% of children 4 to 15 years old with hypertension were obese (111). Lauer et al. found that among overweight children, one-quarter had a blood pressure exceeding the 90th percentile for age (106).

Both systolic and diastolic pressure increase with age, and norms are also influenced by gender and height. Accurate measurement techniques are particularly critical in avoiding false values. Hypertension is defined as pressures exceeding the 90th to 95th percentile on repeated measurements. The reader is referred to the update of the 1987 Task Force Report on High Blood Pressure in Children and Adolescents for further details regarding diagnostic criteria and methodology (197).

Several questions surround exercise in hypertensive children and adolescents:

- What is the role of physical activity and fitness in the pediatric years in preventing the development of lifelong hypertension?

- Do young subjects with hypertension respond differently to exercise than normotensive individuals?

- Is exercise helpful as a therapeutic intervention in children and adolescents with hypertension?

- Are there risks from intensive exercise or sport play in patients with elevated blood pressure levels?

The first question is addressed in chapter 5. This section deals with the remaining issues, which bear importance in the clinical management of these patients.

Response to Exercise

An acute bout of exercise in normal individuals is a hypertensive event. The mechanism and extent of blood pressure elevation depend on the type of activity being performed. In dynamic exercise (running, cycling, swimming), elevation in cardiac output drives up systolic pressure, while increases in peripheral resistance with static exercise (weightlifting) cause a rise in diastolic pressure as well. It is pertinent, then, to determine if these responses are exaggerated in patients with hypertension.

Dynamic Exercise

In the course of a maximal exercise test, systolic blood pressure in the normal older child rises from about 120 mmHg to 150 to 170 mmHg, while diastolic pressure changes little (4; 161). The same pattern is seen in hypertensive subjects. Exercise studies in children and adolescents with hypertension have consistently indicated that, although resting and peak blood pressures are increased, the magnitude of rise is similar to that of normotensive subjects (see [5] for review).

Peak systolic blood pressure in hypertensive subjects frequently exceeds 200 mmHg (30% in the report of Fixler et al. [57]), but values are seldom over 230 mmHg. These findings suggest that the hemodynamic and vascular responses to exercise that are responsible for changes in blood pressure with dynamic exercise (increased cardiac output and diminished peripheral vascular resistance) are similar in normotensive and hypertensive subjects. However, Klein et al. found that hypertensive subjects had a greater level of exercise-induced epinephrine levels and suggested that this represented an altered hemodynamic response to exercise (101).

Resistance Exercise

In contrast to dynamic activities, static exercise is characterized by sharp increases in peripheral vascular resistance, with limited rise in oxygen uptake and cardiac output. Both systolic and diastolic pressures rise to levels determined by the extent, duration, and number of repetitions of the load. Pressures as high as 480/350 mmHg have been described in a trained adult bodybuilder performing a double-leg press (112). Peak values in young subjects are lower. Nau et al. reported an average blood pressure (by intra-arterial measurement) of 162/130 mmHg in eight children ages 8 to 16 years during a single maximal arm bench press (137).

Subjects with hypertension have higher systolic and diastolic pressures in response to resistance exercise than normal individuals. In contrast, no differences in the degree of change in blood pressure with static exercise have been observed in direct comparisons of hypertensive and normotensive youth (57; 101; 175). However, these studies have all been limited to low-resistance handgrip exercise (25-30% maximal). This level of work triggered a rise in diastolic pressure to over 100 mmHg in one-quarter of the hypertensive patients described by Fixler et al. (57).

Predictive Value of Exercise Testing

Reports in adults suggest that exaggerated blood pressure changes with exercise can predict those at risk for future hypertension. In a review of 11 such studies, Benbassat found that eventual hypertension in normotensive subjects who had a higher blood pressure response to exercise was 2.1 to 3.4 times more common than in subjects with normal pressure increases (15). This predictive value of blood pressure responses to exercise has been described in subjects as young as 25 years old (44).

However, limited information suggests that blood pressure response to exercise does not have similar prognostic implications in children. Fixler et al. could not find any predictive value of blood pressure responses to either dynamic or static exercise in teenagers who had been identified 1 year before as being hypertensive (58).

Beneficial Effects of Enhanced Physical Activity

Information from studies in adults indicates that a prescription of regular aerobic exercise can be a valuable nonpharmacologic treatment for systemic hypertension. In a review of 25 studies, Hagberg found that a period of endurance training in adult hypertensive patients produced significant decreases in both systolic and diastolic pressures in a majority, amounting to an average reduction of about 10 mmHg (74). Exercise training can also reduce blood pressure in normotensive adults. Reports indicate an average

reduction of systolic blood pressure by 4 mmHg and diastolic pressure by 3 mmHg (99). Regular aerobic exercise has therefore been advocated in the management of adults with essential hypertension (144).

The number of investigations in older children and adolescents is small and limited to those with mild essential hypertension. Still, the limited available data suggest that young persons with hypertension can also benefit from regular aerobic exercise.

Hagberg et al. reported favorable effects of a 6-month aerobic training program (three sessions per week, 30-40 min per day) on 25 adolescents with mild hypertension (76). Mean systolic pressure fell from 137 to 129 mmHg with training, and average diastolic pressure decreased from 80 to 75 mmHg. Blood pressures returned to pretraining values when measured 6 months after the training stopped.

In a second study, the same investigators showed similar results in blood pressure decrements after 5 months of endurance training in six hypertensive adolescents (75). When the program was then switched to weight training, continued reductions in blood pressures were seen. The average final systolic pressure was 17 mmHg lower than at the start of the program (figure 7.7). Alpert and Fox have pointed out that these subjects underwent weight training (repetitions of submaximal lifts), which is primarily a

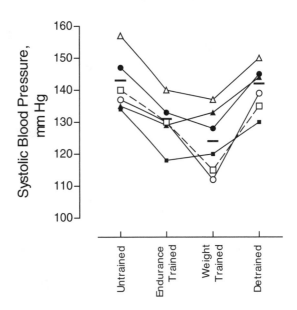

FIGURE 7.7 Changes in systolic blood pressure with endurance and weight training in adolescents (75).

dynamic exercise, rather than weightlifting (resistance training with maximal lifts) (5).

Hanson et al. added two extra sessions of physical education per week for 8 months in 69 children with hypertension and 68 who were normotensive (79). Significant reductions in systolic (6.5 and 4.5 mmHg) and diastolic (3.8 and 4.1 mmHg) blood pressures were seen in both the hypertensive and normotensive subjects, respectively. Danforth et al. enrolled 12 hypertensive black children ages 8 to 12 years in a 12-week program of walking, jogging, and cycling (3 days per week, 30 min per session) (41). Significant reductions in systolic and diastolic pressures as well as body fat were observed.

While this information is encouraging, the specific role of exercise in managing hypertensive youth needs to be clarified. Type, frequency, duration, and intensity of exercise that will prove beneficial are still unclear, and whether home-based or structured institutional programs are more likely to be effective is unknown. The role of exercise in the treatment of children with secondary hypertension from underlying chronic renal or endocrine disease has not yet been assessed.

The mechanisms by which exercise might lower blood pressure are also uncertain. Hemodynamic considerations dictate that to decrease blood pressure, repetitive exercise must reduce either cardiac output or peripheral vascular resistance (or both). Some evidence suggests that early essential hypertension in young subjects is related to a hyperkinetic cardiovascular system, which is associated with increased levels of circulating epinephrine (124). Regular exercise may decrease blood pressure by reducing epinephrine levels at rest and submaximal exercise. Supporting this idea, Duncan et al. showed that the decrease in blood pressure with exercise training in young adults was greater in those with higher catecholamine levels, and the fall in blood pressure was directly related to the decrease in catecholamine levels (51). In older individuals, exercise may serve to diminish vascular resistance by promoting peripheral vasodilation.

The reduction in blood pressure with exercise observed in obese subjects is not due to decreases in body fat alone. As indicated by Hagberg, several studies have shown that the antihypertensive effect of regular exercise is not related to the degree of weight loss (74).

Risks of Exercise

The evidence for salutary effects of exercise programs in patients with systemic hypertension is compelling. Nonetheless, concern surrounds the possible risk of repeated bouts of acute blood pressure elevations in those who are already hypertensive. The sharp increase in both systolic and diastolic pressures with resistance exercise has been considered particularly worrisome. Concern that the intensive nature of sport play might hasten complications of hypertension (heart stress, stroke, nephropathy) has led to recommendations that youth with elevated blood pressures be restricted from certain forms of athletics.

This intuitive concern has not been supported by exercise studies in hypertensive youth. Both dynamic and static exercise testing have failed to produce symptoms, dysrhythmias, or significant ischemic ST changes in these subjects. Moreover, there is a conspicuous absence of documented medical complications in hypertensive children and adolescents during sport play.

After reviewing the data on exercise in hypertensive youth, Alpert and Fox (5) concluded that "we have been unable to find data from either dynamic or isometric exercise studies which document a measurable morbidity or risk of exercise of any specific level of resting or exercise systolic or diastolic blood pressure" (p. 202). These authors and others have suggested that the principal focus should be on identifying the cause of elevations in pressure and initiating proper medical or lifestyle management.

Given this information, most recommendations for sport participation by hypertensive children and adolescents have been permissive. The 1994 Bethesda Consensus Conference recommended that the child or adolescent with mild-to-moderate hypertension who has no evidence of end-organ damage (retinal changes, left ventricular hypertrophy, renal dysfunction) should not be restricted from any athletic activity (94). The group did counsel that those with severe hypertension (systolic pressure >135-144 mmHg or diastolic pressure >90-94 mmHg at ages 10-12 years; systolic pressure >150-159 mmHg or diastolic pressure >95-99 mmHg at 13-15 years) should be restricted, particularly from resistance sports, until their blood pressure is controlled.

Some have considered an exaggerated blood pressure response to a standard maximal exercise test (systolic pressure rising above 230-250 mmHg) as an indication to limit sport participation. As Alpert and Fox pointed out, however, there are no data to specifically support the validity of this guideline for exercise restriction (5). This type of response may, however, signal a need for pharmacologic intervention (7).

CHAPTER 8

Endocrine Diseases

A strong link exists between the endocrine system and physical activity. Both a single bout of exercise and training are accompanied by hormonal changes. In addition, a child's physical performance and physiological response to exercise are often dependent on her or his hormonal status. This chapter reviews diseases of the endocrine system to which exercise is relevant.

Type 1 Diabetes Mellitus

Type 1 diabetes mellitus (DM), previously known as "insulin-dependent DM" (IDDM) or "juvenile DM," is an autoimmune disease characterized by a lack of production of insulin by the beta cells of the pancreas. As a result, patients depend on exogenous insulin. Typically, but not always, type 1 DM starts in childhood. Its prevalence varies among geographic regions and is higher in adolescents than in children. In North America, the average prevalence among the school-age population is approximately 2 per 1,000, with incidence of about 15 per 100,000 per year in Caucasians and 11 per 100,000 per year among African Americans. In Europe, prevalence and incidence are higher in northern countries than in the south.

Perhaps more so than in any other disease, the level of physical activity is of utmost importance to the daily management of patients with type 1 DM. Activity, nutrition, and insulin therapy are the pillars of this management. Because of the strong interaction among the effects of this *triad*, any change in one of its elements often requires adjustments in the other two. This section focuses on the detrimental effects of acute exercise on the metabolic control of such patients, as well as on

the benefits that the child with type 1 DM may derive from a physically active lifestyle. Further information about these issues can be found in a comprehensive 1996 review by Dorchy and Poortmans (40) and in a 2001 book published by the American Diabetes Association (132).

Habitual Activity

Because any change in physical activity may affect the nutritional and insulin requirements of a child with type 1 DM, physicians should be thoroughly familiar with the physical activity pattern of their patients. It is therefore surprising that little information is available in the medical literature regarding activity behaviors and attitudes toward physical activity among patients with type 1 DM. In general, activity patterns of the patients during childhood are similar to those of the healthy population. In contrast, pubertal and postpubertal patients are often less active than their healthy peers. Figure 8.1 summarizes the frequency of participation in physical education classes of girls and boys with type 1 DM and of healthy controls (146). While the 7- to 14-year-old patients showed good attendance, the older ones had an extremely high level of absenteeism. As for leisure time sport, the 15- to 20-year-old boys with type 1 DM were considerably hypoactive. Among the girls, both postpubertal groups, healthy and those with DM, reported a rather low level of voluntary activity.

Why would adolescent patients be reluctant to exert themselves? One possibility is their fear of hypoglycemic crisis (66). They have come to realize that by keeping their activity level consistently low, they remove one uncertainty

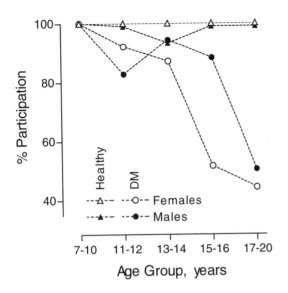

FIGURE 8.1 Physical activity in type 1 diabetes mellitus. Frequency of participation in physical education classes of 136 schoolchildren with diabetes and of 123 age-matched controls. Data from Sterky 1963 (146).

regarding metabolic control. Unfortunately, some health professionals, out of ignorance regarding the importance of an active lifestyle, concur with this and encourage a policy of physical passivity for the young person with diabetes.

In a study of attitudes toward activity among Swedish patients (89), fewer than half stated that they exercised daily. The same percentage found physical education and sports "very interesting." While less than 10% of the respondents were negative toward exertion, most felt that they just did not have enough time to get around to being active. Most patients downplayed the fear that too much exercise might disrupt their metabolic control. The authors felt that there was a definite gap between what the patients knew about the benefits of physical activity and their actual commitment to an active lifestyle. Although this is a common pattern in most segments of the child and adult population, the authors did not provide a comparison between their patient sample and healthy controls.

These findings are culture specific and cannot be deduced for all populations with DM. Even so, this survey points out the gap between *knowing* about the importance of exercise on the one hand and *being involved* in regular activities on the other.

A low level of habitual activity may also be the result of a low fitness level (see following section).

Adolescent patients, especially girls, often gain weight excessively (42; 146). If obesity develops in such individuals, it could be another factor in the vicious cycle of low fitness and hypoactivity.

Physical Fitness

Various studies (11; 12; 64; 67; 82; 83; 116; 117; 146), but not all (6; 42; 59; 61; 130; 133), indicate a low fitness level among patients with type 1 DM. This is particularly true for maximal aerobic power, although deficient performance has also been shown for motor fitness components such as strength, speed, and muscle endurance (45).

There are several possible reasons for the inconsistent data on the fitness of patients with type 1 DM. While there is little or no relationship between maximal aerobic power and the age of onset or the duration of diabetes (42; 133), the *combined* effects of early onset and long duration of the disease do contribute to the low maximal aerobic power (146).

Aerobic fitness is low in patients who have a cardiovascular autonomic dysfunction compared with those who have a normal autonomic function (14). Furthermore, maximal aerobic power is inversely related to the degree of glycemic control as assessed by glycosylated hemoglobin (6; 64; 116). Such a relationship is displayed in figure 8.2. Adolescent patients with relatively poor metabolic control (hemoglobin$_{A1}$ higher than 8.5%) had a lower maximal O_2 uptake than did patients with better metabolic control (hemoglobin$_{A1}$ lower than 8.5%). Both groups had a lower maximal O_2 uptake than did the healthy controls.

The deficient fitness and motor performance of some children with type 1 DM have been explained by their lower body stature as compared with age-matched children or adolescents (41; 146). Indeed, when corrected for body size, differences in maximal aerobic power diminish (146) or disappear altogether (30; 59; 133).

Another indication that low fitness is not due to the disease per se is that only some age groups are unfit, notably the pubertal and postpubertal patients. In contrast, the younger adolescents and children are often as fit as their healthy peers (42; 83; 130; 146). The lower exercise performance of pubertal or postpubertal patients could, in part, be due to the lower level of their habitual activity (146). Indeed, when such patients are exposed to a conditioning program, their maximal aerobic

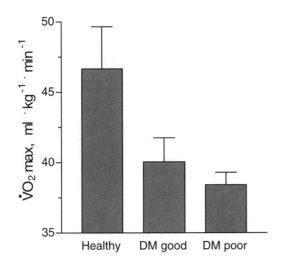

FIGURE 8.2 Relationship between maximal O₂ uptake and metabolic control of patients with type 1 diabetes mellitus. Data are presented for 17 healthy adolescents, 9 patients with relatively good metabolic control ("DM good"), and 8 patients with relatively poor metabolic control ("DM poor"). Vertical lines denote 1 SD. For details, see text.

Based on data from Poortmans et al. 1986 (116).

power and other cardiorespiratory functions improve and can equal those of healthy youths (26; 27; 29; 65; 83). Considerable numbers of elite athletes in various sports have type 1 DM (40), which further indicates that low physical fitness is not inherent to the disease.

There are several deficiencies in exercise-related cardiorespiratory functions of patients with DM, but these do not induce appreciable functional impairment. During exercise, oxygen pulse, which is proportional to the product of stroke volume and arterial-venous oxygen difference, is lower in patients with type 1 DM. This suggests reduced peripheral oxygen utilization, possibly due to subclinical microvascular involvement (11). In young adults with type 1 DM, muscle tissue capillarization is lower than normal, as is the activity of their muscle oxidative enzymes. Muscle glycogen concentration and glycogen utilization during exercise are normal (135), as is the ability to resynthesize muscle glycogen following depletion by prolonged exercise (92). The implication of such normal resynthesis is that patients with good control can train hard and recover well after prolonged intense bouts of exercise.

Although there is no direct association between type 1 DM and obesity, some patients may become obese, possibly as a result of their

diabetic management. This may occur when a patient who receives excessive dosages of insulin is advised to consume high amounts of food energy. Physical fitness may then be impaired as a result of the excess weight per se.

Perceived Exertion in Type 1 Diabetes Mellitus

As discussed in chapter 1 (section "Rating of Perceived Exertion As a Means for Production and Self-Regulation of Effort"), rating of perceived exertion (RPE) is used to assess the physical effort that a person perceives while performing a given exercise task. Compared with healthy children and youths, patients with chronic diseases and disabilities often have a high RPE (9; 10) as determined by the Borg 6 to 20 category scale (20). Such a high rating may be one reason for the sedentary lifestyle adopted by some of these patients. Little information is available regarding the RPE in children with type 1 DM. Highly trained Swedish adolescent patients rated their exercise intensity 1 to 1.5 points higher than did healthy controls (110). Nonathletic Canadian 13- to 19-year-old patients exercised for 60 min at 60% maximal O₂ uptake. At each stage of the exercise task, their RPE was 1.5 to 2 points higher than in healthy controls (figure 8.3) (122). This high rating could not be explained by differences in fitness, because the two groups did not differ in their maximal O₂ uptake. Nor could it be explained by hypoglycemia, because even when the subjects drank carbohydrate during the 60 min (thereby preventing a hypoglycemic response), the higher rating by the patients prevailed. Pediatric type 1 DM is associated with peripheral sensory nerve dysfunction (13). It is possible, therefore, that the exaggerated RPE reflects impairment in the sensory feedback mechanism in these patients. This issue needs more research.

Exercise-Induced Hypoglycemia

The metabolic responses to acute exercise in the patient with type 1 DM depend on his or her glycemic control prior to exercise; food intake; type, time, and site of insulin injection; and the intensity and duration of the activity. In a reasonably controlled patient with type 1 DM, prolonged (30 min or more) exercise induces a gradual decline in blood glucose level (17; 94; 118; 123;

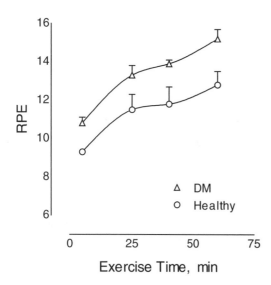

FIGURE 8.3 Rating of perceived exertion (RPE) in type 1 diabetes mellitus. Eight male patients and eight healthy controls (13-19 years of age) performed two 30-min exercise sessions (cycling at 60% maximal O₂ uptake), with a 5-min rest in between. Rating of perceived exertion was determined periodically using the Borg 6 to 20 category scale (20). Vertical lines denote SEM.

Adapted, with permission, from Riddell et al. 2000 (122).

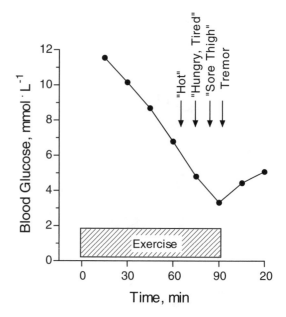

FIGURE 8.4 Hypoglycemia during prolonged moderate exercise in a boy with type 1 diabetes mellitus. Blood samples were taken through an indwelling antecubital venous catheter. For details see text.

157; 167). Such a decline, if not controlled, may reach hypoglycemic levels (3.3 mmol · L⁻¹, or 60 mg · dL⁻¹). A case in point is that of a 13-year-old boy with a 3-year history of type 1 DM, whose blood glucose response to prolonged exercise is shown in figure 8.4. This boy presented with a vague story of episodes of weakness and hunger during distance cycling—his favorite sport. He rejected the possibility that these might reflect hypoglycemia. He performed a simulation of a 90-min road ride on a cycle ergometer, choosing the resistance and pedaling rate. Exercise intensity was moderate, raising his heart rate to 140 to 150 beat · min⁻¹. Ninety minutes prior to the ride he injected into the thigh his dose of medium-acting insulin and had breakfast. His blood glucose markedly declined, reaching 3.2 mmol · L⁻¹ (58 mg · dL⁻¹) by the end of the ride. Seventy minutes after the start, he began to complain of a sensation of heat, followed by hunger, fatigue, and sore thighs. Five minutes before the conclusion of the ride, he developed a fine tremor in various muscle groups. These complaints disappeared within 10 to 15 min postexercise and the consumption of a carbohydrate drink. Other symptoms and signs of hypoglycemia include headache, cognitive dysfunction, dizziness,

blurred vision, and aggressive behavior (these probably reflect reduced glucose levels in the central nervous system). Very severe hypoglycemia (≤1 mmol · L⁻¹) may lead to convulsions and coma, but these manifestations are extremely rare in the exercising person.

There is a large variability in glycemic responses among patients (94; 121). However, as seen in figure 8.5, the intra-individual glycemic responses to exercise are quite consistent from one day to the next as long as insulin therapy and energy intake are consistent. This day-to-day consistency is encouraging for clinicians who wish to evaluate the efficacy of an intervention. If their intervention is accompanied by changes in the rate of blood glucose decline, it is likely that these changes reflect the effects of the intervention, rather than a haphazard response.

Late-Onset Hypoglycemia

Although most hypoglycemic responses occur during or immediately after a prolonged exercise bout, hypoglycemia may appear *several hours after exercise* (90), or even on the following day (4). In some patients, hypoglycemia occurs as late as 36 hr after exertion, but the most common time interval is 6 to 10 hr. Patients and parents

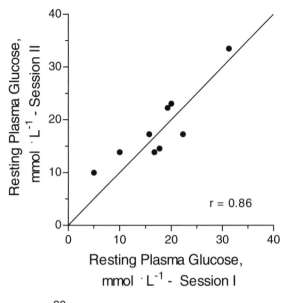

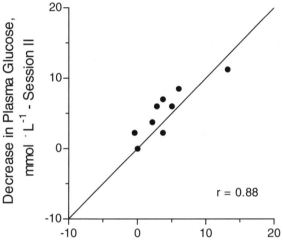

FIGURE 8.5 Consistency in the exercise-induced reduction of blood glucose. Nine adolescent boys (age 16 ± 1.8 years) with type 1 diabetes mellitus were tested twice, 7 to 15 days apart. In each session they performed six 10-min bouts of cycling at a moderate intensity (heart rate 145-150 beats · min[-1]). The top graph depicts individual values of blood glucose at rest, plotting session 2 versus session 1. The bottom graph depicts the degree of decline in blood glucose in the two visits.

Reproduced, with permission, from McNiven-Temple et al. 1995 (94).

should realize that for a child who exercises in the evening, there is a risk of hypoglycemia occurring during the sleeping hours. While the exact mechanism for late-onset hypoglycemia is not clear, it reflects insufficient replenishment of muscle and liver glycogen stores during recovery from exercise.

Ability of the Patient to Perceive Hypoglycemia

Can patients with type 1 DM perceive impending hypoglycemia during exercise? A study in which adolescent and adult patients were asked to guess their blood glucose level at rest (at least 40 times each) revealed low perceptual ability (102). Some 25% of patients (mostly adults) with type 1 DM have *hypoglycemia unawareness*. This condition is manifested by the lack of warning symptoms prior to the appearance of hypoglycemia, even if severe (98). This apparently can occur also during exercise. Figure 8.6 summarizes an experiment in which adolescent patients exercised at a moderate intensity (heart rate 145-150 beat · min[-1]) for 60 min. The patients markedly underestimated their blood glucose levels when the levels exceeded 15 mmol · L[-1] and still underestimated them when they ranged between 10 and 15 mmol · L[-1]. However, when blood glucose levels were below 10 mmol · L[-1], the patients often overestimated them (94; 120). It thus seems that young patients with type 1 DM are incapable of correctly perceiving their blood glucose level, whether high, normal, or low.

Mechanisms for Exercise-Induced Reduction in Blood Glucose

A decline in the blood glucose level reflects an imbalance between increased glucose utilization by the exercising muscle and an insufficient increase in its production by the liver (143; 167). The degree of such an imbalance in the type 1 DM patient depends on the availability of exogenous insulin and on the sensitivity of insulin receptors in the exercising muscle. When circulating insulin level is high, there is a greater suppression of hepatic glucose production, a greater uptake by the muscle, and a faster decline in blood glucose level. In addition, in patients with type 1 DM, exercise is accompanied by a blunted response of counterregulatory hormones, mostly glucagon (77) and catecholamines (138). This is particularly so in patients with diabetic neuropathy. The blunted increase in catecholamines may be a reason for hypoglycemia unawareness as described in the previous section.

Factors that can modify the level of circulating insulin during exercise include the elapsed time between insulin administration and exercise, the

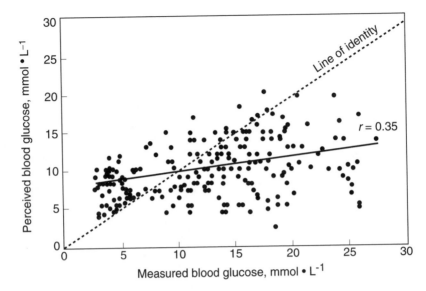

FIGURE 8.6 Ability of patients with type 1 diabetes mellitus to perceive their blood glucose levels during exercise. Individual data are shown for nine 13.8- to 19.5-year-old males with type 1 diabetes mellitus who were asked to guess their blood glucose levels during each of six 10-min cycling bouts at approximately 50% maximal aerobic power.

Reproduced, with permission, from Riddell and Bar-Or 2001 (120), based on data by McNiven-Temple et al. 1995 (94).

type of insulin, the presence of insulin antibodies, the site of injection, and route of administration.

Route and Site of Insulin Injection   Table 8.1 summarizes the effects of route and site on blood glucose level and on related variables. When insulin is injected subcutaneously (SC) to the exercising limb, there is a faster release of insulin from the depot during exercise than at rest (18). Likewise, release is faster than from a nonexercising site (78; 167). This results in increased plasma immunoreactive insulin levels, reduced glucose production by the liver, and a drop in blood glucose level. In contrast, SC injection into a nonexercising site, or a continuous intravenous (IV) infusion, does not impede glucose production by the liver and does not result in a reduction of the blood glucose level (some reduction may take place, but to a lesser extent than following SC injection to the exercising limb). Injection directly to the to-be-active muscle enhances the disappearance of insulin from the injection site, and induces higher levels of plasma insulin and a faster decline in blood glucose than with the SC route (50). To avoid accidental injection into the muscle, one can use shorter needles for SC injection or inject into a skinfold.

The faster release of insulin from a depot that overlies an exercising muscle may be due to a local increase in temperature, to hyperemia, or to enhanced lymph flow. Release is also enhanced in higher versus lower ambient temperatures (124), most probably because of higher skin blood flow during warm weather conditions. While there are no dose-response data regarding the effect of ambient temperature on insulin release, marked changes in climatic conditions (e.g., visiting a warm country during winter) may require adjustment of the insulin dosage.

Insulin Binding to Receptors   A bout of exercise causes an increase in insulin binding to receptors, located on monocytes or erythrocytes, in sedentary healthy individuals and in patients with type 1 DM (109). Under some conditions a correlation exists between in vivo insulin sensitivity and insulin binding to monocytes. One may therefore speculate that an exercise-induced rise in insulin binding to monocytes reflects a similar increased binding to receptors in exercising muscle cells. This could be another mechanism for the glucose-lowering effect of exercise and for improved glucose tolerance after exercise.

Intravenous glucose tolerance during (36) and after (37) exercise is increased in insulin-treated type 1 DM adolescents and adults but not in insulin-deprived adolescents (38). Figure 8.7 demonstrates the exercise-induced increase in glucose clearance from the blood in adolescents with type 1 DM.

TABLE 8.1	Effects of Route and Site of Insulin Injection on Blood Glucose Level and Related Variables of an Exercising Child With Insulin-Dependent Diabetes Mellitus				
Route of injection	Site of injection	Plasma insulin level	Glucose production by liver	Glucose utilization by muscle	Blood glucose level
Intravenous	Any	No change	≠≠	≠≠	No change
Subcutaneous	Nonexercising limb	No change	≠≠	≠≠	No change or mild reduction
Subcutaneous	Exercising limb	≠≠	≠	≠	▯

≠ or = some change; ≠≠ = major increase.

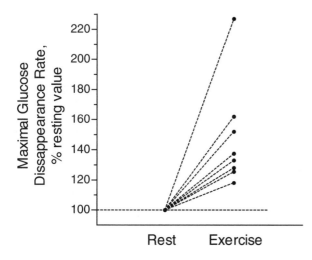

FIGURE 8.7 Intravenous glucose tolerance and exercise. Maximal disappearance rate of glucose from the blood at rest and during 35 min of cycling at 50% of maximal O_2 uptake. Subjects were eight 12- to 18-year-old non-obese males with type 1 diabetes mellitus who were given intravenous infusion of insulin and 50% glucose. Individual values are expressed as a percentage of maximal glucose disappearance at rest. Data from Dorchy et al. 1976 (36).

Means for the Prevention of Exercise-Induced Hypoglycemia

To prevent hypoglycemia in the exercising patient, one must reduce the insulin dose, increase the carbohydrate intake, or do both. These measures may not be necessary if the activity lasts less than 30 min, because relatively short-term exercise usually does not induce hypoglycemia (155). However, they should be considered whenever the patient is about to perform a more prolonged activity, such as a hike, a "field day," or a soccer tournament.

Determining optimal modifications in the insulin dose and carbohydrate intake prior to or during exercise is of particular challenge to young patients, whose activity pattern is often irregular and unpredictable (8).

Reducing the Insulin Dose Because of the synergistic action of insulin and physical activity, reducing the insulin dose in anticipation of prolonged exercise has been the most common approach to the prevention of hypoglycemia.

No studies have systematically addressed the optimal extent of dosage reduction prior to each activity. A trial and error approach may be needed for patients who take part in sport or in other prolonged activities. Some adults and well-experienced adolescents who exercise habitually can tailor their insulin dosage to the anticipated activity. An example is the postman with diabetes, cited by Joslin et al. (71), who customarily cycled 25 km daily on his rounds. Before starting in the morning he stuck his head out of the window and if the wind was blowing against him he lowered the insulin by 20 units. This approach requires much experience and assumes that the amount and intensity of the anticipated activity are predictable. However, it is only a small minority of pediatric patients whose activity is as regular and predictable as that of a mail carrier.

The extent of insulin reduction depends on the duration and intensity of exercise, the time after the last meal, and the insulin regimen of the particular patient. For example, when a prolonged, moderate-intensity activity is anticipated 1 to 3 hr after the administration of short-acting insulin and a meal, the usual bolus insulin may need to be reduced by 50% or more (137). However, an optimal reduction of the insulin dose is highly individualized and should be determined with the aid of frequent blood glucose monitoring. In our experience (123), exercise performed approximately 90 min after a meal may pose a particular danger of hypoglycemia, because circulating insulin level at that time is particularly high.

Reduction of the insulin dose as a sole strategy has a major flaw: A decision regarding the extent of such a reduction must be made *prior* to the activity and cannot be reversed even if the subsequent exercise deviates from that anticipated. This is particularly a problem with young children, whose activity pattern is often unpredictable. Another potential drawback of insulin reduction is the risk of exercise-induced *hyperglycemia* (see later section, "Exercise-Induced Hyperglycemia and Ketoacidosis").

Increasing the Carbohydrate Consumption The strategy of increasing carbohydrate consumption is more flexible than the pre-exercise manipulation of insulin dosage because the additional carbohydrate can be consumed not only before exercise, but also while the child performs the activity. This method allows the amount of carbohydrate to be modified while the activity is still in progress.

How much carbohydrate should be added? An authoritative 2000 textbook of pediatrics recommends: "In anticipation of vigorous exercise, one additional carbohydrate exchange may be taken prior to exercise, and glucose in the form of orange juice, carbonated beverage, or candy should be available during and after exercise" (p. 1783). This recommendation may be too general and does not accommodate individual differences (i.e., size of the child or the type, intensity, and amount of the activity). For further elaboration of this issue, see the next section, "The Exercise Exchange Menu."

How well can the consumption of carbohydrate before and during exercise prevent hypoglycemia? As shown in figure 8.8, when the amount of periodically ingested carbohydrate equals the amount of carbohydrate utilized by the body during prolonged exercise, the drop in blood glucose is all but eliminated (121). In that study, when the patients drank water, nearly half of them reached or approached hypoglycemia during a 60-min cycling task. When they drank glucose, only 15% approached hypoglycemia. It is important to note that when the ingested glucose was only half the amount utilized by the body, a drop in blood glucose level could not be prevented.

The efficacy of carbohydrate drinking is not limited to patients whose pre-exercise blood glucose is within normal. Even when the initial level is 20 mmol · L^{-1} or more, a precipitous reduction in blood glucose can be prevented (figure 8.9).

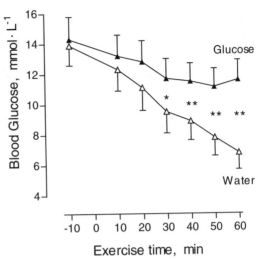

FIGURE 8.8 Ingested carbohydrate before and during exercise can prevent exercise-induced hypoglycemia in adolescents with type 1 diabetes mellitus. Seventeen male and three female patients (14 ± 1.0 years) cycled twice for 60 min (60-65% maximal O$_2$ uptake), 100 min after they had injected their regular dose of insulin and eaten their usual breakfast. In one visit, they intermittently drank water to prevent dehydration. In a subsequent visit, they drank a 6% to 8% glucose solution. The total amount of glucose for each child matched the amount of glucose utilized in the first visit, as calculated from respiratory exchange rate and O$_2$ uptake. The vertical lines denote SEM. * = P < 0.05; ** = P < 0.01.

Reproduced, with permission, from Riddell et al. 1999 (121).

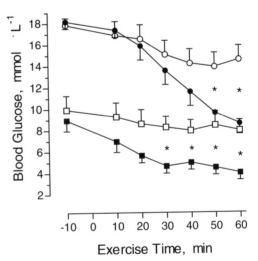

FIGURE 8.9 Ingestion of carbohydrate can prevent a precipitous decline in blood glucose level, even when the pre-exercise level is very high. Subjects, exercise, and drinking protocols were the same as in figure 8.8. The subjects were divided into two groups based on their pre-exercise blood glucose level: >15 mmol · L^{-1} (two upper lines) and <15 mmol · L^{-1} (two lower lines). The open circles and squares denote glucose ingestion; the black circles and squares, water ingestion.

Reproduced, with permission, from Riddell et al. 1999 (121).

The Exercise Exchange Menu

As most clinicians do not have access to an exercise laboratory that measures a patient's carbohydrate utilization rate, a more practical approach is to *estimate* the required carbohydrate. In the first edition of this book, we introduced the concept of the *exercise exchange menu*, which can be used in analogy to the *food exchange menu*. Each exercise exchange is equivalent to energy expenditure of 100 kcal (420 kJ). Taking the carbohydrate contribution as 60% of total energy fuel, one exercise exchange is equivalent to 60 kcal or 15 g of carbohydrate. By accepted convention, 15 g of carbohydrate is also one food exchange. Thus, in terms of carbohydrate balance, one exercise exchange equals one food exchange.

A list of exercise exchanges is given in table 8.2 for children of different body weights. Such a list can be used for educational purposes with parents and patients. Patients can create their own exercise exchange menu, combining games and other activities of their choice. Within the setting of a diabetic clinic, it is the nurse or the dietitian who should be thoroughly familiar with the exercise exchange concept. However, the physician must also understand this concept and be able to prescribe activities accordingly.

The approach just described may be the optimal one, but it may not be effective for all children, especially younger ones. A young child may not abide by overly regimented instructions. Pushing such a child to adhere to the exercise exchange concept may be counterproductive. It is for the clinician and the parents to sense how far, and how fast, they should introduce this approach with any given child. Success will be achieved once the child exercises daily within the framework of exercise exchange and still enjoys the fun of sport.

Exercise-Induced Hyperglycemia and Ketoacidosis

Exercise does not always cause hypoglycemia. Patients who exercise while in a state of hyperglycemia or ketoacidosis, or who are deprived of insulin, may respond to exercise with an elevation of their glucose level (3; 17; 46; 93; 119; 136; 157). Figure 8.10 is an example of this phenomenon in adolescent patients who skipped their morning insulin dose and then exercised. Rarely, patients may respond with hyperglycemia during or

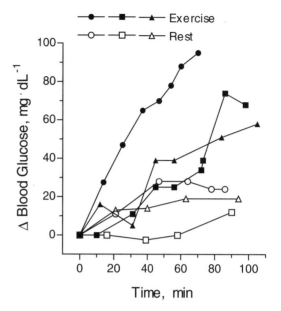

FIGURE 8.10 Insulin deprivation and exercise-induced hyperglycemia. Three 13.1- to 13.8-year-old boys were each tested on two mornings. They had not taken any insulin since the preceding evening. On one morning, blood glucose was monitored at rest, and on the other it was monitored while the patient performed two or three 5-min bouts of stair climbing. Individual changes (Δ) in blood glucose level during the experiment.

Adapted from Marble and Smith 1936 (93).

immediately following moderate-intensity exercise, even after having taken an adequate amount of insulin (39). In others, hyperglycemia may occur during brief, very intense exertion, possibly due to a marked surge in their sympathetic-adrenergic tone (164).

The cause of exercise-induced hyperglycemia in the insulin-deprived patient with type 1 DM is the increased hepatic production of glucose in excess of its uptake by the muscle (136). Such production is normally inhibited by insulin. The patient who is ketotic at rest has a greater than normal production of ketone bodies during exercise (46). Even though uptake of ketone bodies in the muscle is high in such a patient (157), there is a net increase in the blood level of these substances, especially during strenuous exercise (17; 157). Another possible cause for exercise-induced ketosis in the insulin-deprived patient with type 1 DM is a rise in plasma glucagon and cortisol (17).

To prevent exercise-induced hyperglycemia, patients with type 1 DM should not exercise if their blood glucose level is above 15 mmol · L⁻¹ (270 mg/dl), combined with ketonuria.

TABLE 8.2	"Exercise Exchanges" of 100 Kilocalories (420 KiloJoules) in Children of Various Body Weights*		
		Body weight	
Activity	**20 kg**	**40 kg**	**60 kg**
Basketball (game)	30	15	10
Calisthenics	75	40	25
Cross-country skiing (leisure)	40	20	15
Cycling			
10 km · hr⁻¹	65	40	25
15 km · hr⁻¹	45	25	15
Field hockey	35	20	15
Figure skating	25	15	10
Horseback riding			
Canter	110	60	40
Trot	45	25	15
Gallop	35	20	10
Ice hockey (ice time)	20	10	5
Judo	25	15	10
Running			
8 km · hr⁻¹	25	15	10
10 km · hr⁻¹	20	15	10
12 km · hr⁻¹		10	10
14 km · hr⁻¹			5
Sitting			
Complete rest	125	100	85
Quiet play	90	65	55
Snowshoeing	30	15	10
Soccer	30	15	10
Swimming—30 m · min⁻¹			
Breaststroke	55	25	15
Front crawl	40	20	15
Backstroke	60	30	20
Table tennis	70	35	25
Tennis	45	25	15
Volleyball (game)	50	25	15
Walking			
4 km · hr⁻¹	60	40	30
6 km · hr⁻¹	40	30	25

*Values are the number of minutes that a certain activity should be sustained.

Effect of Exercise on Renal Function

Some studies (62; 96; 106), but not all (113), suggest that children and youths with type 1 DM respond to exercise with a greater degree of albuminuria than do healthy controls. Likewise, some adolescents with type 1 DM respond to exercise with an exaggerated elevation in arterial blood pressure (28; 101). Greater albuminuria occurs mostly during mild exertion (97), but less so during moderate or maximal exercise intensities (113; 114). It is tempting to assume that the excessive exercise-induced albuminuria in DM may reflect early, subclinical stages of nephropathy. However, experts in the field reject the use of an exercise provocation as a tool for the diagnosis of future nephropathy (40; 117). Another intriguing idea, that an excessive blood pressure response to exercise in adolescents with type 1 DM signifies risk for diabetic nephropathy 5 years later (101), needs further confirmation.

Responses to Physical Training

Enhanced physical activity yields several benefits to young patients with type 1 DM. These include an increase in aerobic fitness and other fitness components, an increase in insulin sensitivity, improved lipoprotein profile, reduction of % body fat in patients who are overweight, and, most importantly, enhancement of sociability and quality of life.

Effects on Physical Fitness

As in healthy children and youths, patients with type 1 DM can improve their fitness through training. In a randomly controlled study, 5- to 11-year-old patients underwent a 12-week, three times per week aerobic program (26). The 30-min sessions were quite intense, raising the children's heart rate to ≥160 beat · min⁻¹. By the end of the

12 weeks, maximal O_2 uptake (ml $\cdot$ kg^{-1} $\cdot$ min^{-1}) increased by 7% and maximal minute ventilation by 12%. Similar responses have been reported by others (65; 82; 99). Such improvement is similar to that found in healthy children. Other physiologic changes that result from training include an increase in muscle strength, reduction of % body fat, and improved lipoprotein profile (99). A 2-week, 6 hr per day aerobic program was accompanied by a 50% reduction in exercise-induced urinary excretion of albumin and β_2 microglobulin (115).

Effects on Insulin Requirements

In the pre-insulin era, exertion was considered a major mode of treatment in diabetes. The following statements, for example, appeared in a medical journal in 1915 (3): "[Diabetic] dogs which for months had regularly shown glycosuria whenever they were given 100 grams of bread, on exercise became able to take 200 grams of bread as a regular daily ration without glycosuria." And, as to humans: "In a patient free from glycosuria with persistent hyperglycemia, one fast day with exercise may reduce the blood sugar as much as several fast days without exercise."

The synergistic effects on the blood glucose level of insulin and exercise were first reported in 1926: Exercise practiced for some weeks helped patients with diabetes double their daily carbohydrate consumption without any concomitant increase in their insulin dose. Some patients could even reduce their insulin dosage despite a greater carbohydrate intake (84). Such observations have since been reproduced in many studies (1; 43; 69; 71; 82; 83).

The reduction in insulin requirement can be quite dramatic: We have been seeing young patients who require 40 to 60 units daily during fall and winter but only 5 to 10 units during spring and summer. A similar trend is seen in figure 8.11 for a child who almost halved his insulin intake during three consecutive summer seasons, when his activity level was higher than during the school year (66).

The mechanism that enables the active child with type 1 DM to use less insulin is not entirely clear. There is strong evidence that in vivo sensitivity to insulin and in vitro binding to monocytes are higher among trained individuals and can increase in sedentary people who undergo a conditioning program (79; 87; 142; 165). Studies on rats have suggested that training increases the activity of GLUT 4 glucose transporter (55).

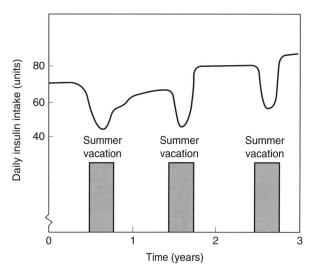

FIGURE 8.11 Insulin requirements are lower during periods when the child is active. Record of a boy with insulin-dependent diabetes mellitus.

Adapted from Jackson and Kelly 1948 (66).

The validity of these data for the child with type 1 DM has yet to be shown. Moreover, the greater sensitivity to insulin in the trained individual is short-lived and may disappear within 1 to 2 days following the last training session (5; 25). Thus, it may reflect increased glucose tolerance following a single exercise bout rather than a characteristic of the trained person. If this assumption is correct, a patient with diabetes can expect to enjoy the insulin-sparing effect of exercise only as long as he or she is regularly active.

Effects on Glycemic Control

A relationship has been suggested between glycemic control at a young age and the future risk of diabetic complications, such as neuropathy, nephropathy, and retinopathy. The assumption is that strict daily control is important for the reduction of such a risk. It is therefore pertinent to ask whether physical activity can affect the control of diabetes.

Taking the frequency of glycosuria as a rough indicator of diabetic control, regular physical activity seems to increase control (43; 71; 83; 84). An example often quoted is that of special diabetes camps, in which children who become more active retain or even improve their metabolic control in spite of a lower insulin dosage and higher calorie intake (1; 83). Figure 8.12 is a summary of observations from such a training camp: Daily calorie intake increased by 50% to 100% and yet the rate of glycosuria decreased from 65% to less than 30% of

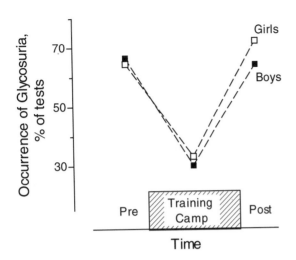

FIGURE 8.12 Physical activity and daily metabolic control of type 1 diabetes mellitus. Occurrence of glycosuria in diabetic girls and boys before, during, and following a training camp. Data from Sterky as quoted by Larsson et al. 1962 (83).

the tests. A statistical association has been reported between the *extent* of activity and an "index of control" based on two to four daily urinalyses (88). The common variance, however, between the two variables was low, suggesting that even if the role of regular activity is important, other factors strongly affect the control of diabetes.

Another index of control is the percentage of glycosylated hemoglobin (HbA$_1$, or its fraction HbA$_{1c}$) in the blood. Some studies documented a drop of HbA$_1$ in children who took part in exercise intervention programs (1; 26; 30; 65). For example, in patients who underwent a 5-month training program, HbA$_1$ dropped from 15.1% to 13.8% (30). Although statistically significant, such a reduction is small biologically and its actual benefit hard to assess. Furthermore, more recent studies could not confirm a training-induced reduction in glycosylated hemoglobin (7; 81; 131; 147). It is hard to reconcile this inconsistency. One possible factor is the length and intensity of the intervention. Another is the age or maturational stage of the subjects. In that regard, it is noteworthy that in two randomly controlled studies performed by the same group, HbA$_1$ decreased in prepubescent (26), but not in adolescent, patients (81).

In conclusion, regular physical activity is probably of benefit in the daily control of type 1 DM. Such benefit might, however, be indirect because children who become physically active may be changing other daily habits that themselves affect

metabolic control. When an increased activity is irregular, control may, in fact, be a greater challenge due to variations in day-to-day metabolism.

Exercise and Dietary Considerations in the Daily Management of Patients With Type 1 Diabetes Mellitus

In no other disease are food, activity, and medication so closely linked as in type 1 DM. A change in any one automatically calls for adjustment of the other two. Changes in insulin dosage are discussed in the earlier section "Reducing the Insulin Dose." This section focuses on recommended physical activities and on nutritional guidelines for the physically active patient.

Choice of Physical Activities

Young patients with type 1 DM can participate with their healthy peers in any physical activity or sport. For optimal metabolic control, these ideally should be physical pursuits in which the energy expenditure is predictable, such as walking, running, swimming, cycling, or cross-country skiing (40; 63). This does not preclude participation in sports that require short bursts of intense activity and thus wide variation in energy expenditure (e.g., team games), but the balancing of glycemic control will then require more frequent monitoring of blood glucose.

Special precautions should be taken with activities that may induce risk to the patient (mostly in case of hypoglycemia) or to others. Examples include skydiving, mountaineering, scuba diving, and hang gliding. As a general precaution, a patient with type 1 DM should be accompanied by another person while participating in sport.

Nutritional Considerations

Planning of nutritional modification should be done before an anticipated practice session, competition, or any other prolonged activity (31; 57; 66; 83; 121; 123; 129; 145). Food items to be added prior to and during the activity can be calculated *as fruit exchanges* and *starch exchanges*, as listed in table 8.3.

On a day of competition or major practice session, meals should be eaten not less than 3 hr before warm-up time (31) and should include a higher than normal carbohydrate content (129).

TABLE 8.3 — Fruit Starch "Exchanges" to Be Taken by a Child With Insulin-Dependent Diabetes Mellitus Before an Extra Physical Activity*

Activity	Duration (min)	20 kg Energy expended (kcal)	20 kg Carbohydrate (g)	20 kg Fruit	20 kg Starch	30 kg Energy expended (kcal)	30 kg Carbohydrate (g)	30 kg Fruit	30 kg Starch	40 kg Energy expended (kcal)	40 kg Carbohydrate (g)	40 kg Fruit	40 kg Starch
Basketball (game)	20	100	15		1	200	30		2	300	45		3
	40	200	30	1	1.5	400	60	1	3	600	90	2	4
Cross-country skiing (leisure pace)	30	70	10	.5	.66	140	20	.5	1.5	210	30	1	2
	60	140	20		1	280	40		2	420	60		3
Cycling: 10 km · hr⁻¹	30	45	7		.5	75	10		.66	120	20		1.5
15 km · hr⁻¹	30	65	10		.66	120	18		1	180	30		2
Figure skating (practice)	20	80	12	.5	.75	160	25	1	1.6	240	35	.5	2
	40	160	25			320	50		2	480	70		4
Ice hockey (on ice)	10	50	10		.66	100	20		1.5	150	30		2
	20	100	20		1.3	200	40		2.5	300	60		4
Running: 8 km · hr⁻¹	30	110	17		1	200	30		2	270	40		2.5
12 km · hr⁻¹	30	—	—	—	—	270	50		3	370	70		4.5
Snowshoeing	30	100	15	.5	1	200	30	1	2	300	45	1.5	3
	60	200	30		1.5	400	60		3	600	90		4.5
Soccer (game)	30	110	17	.5	1	215	32	1	2	320	50	2	3.5
	60	220	35		1.5	430	65		3	640	95		4.5
Swimming (breaststroke— 30 m · min⁻¹)	20	60	10	.5	.66	120	18	.5	1	230	35	1	2
	40	120	20		1	240	36		2	460	70		3.5
Tennis	30	75	12	.5	.75	130	20	1	1.5	190	30	1	2
	60	150	24		1.5	260	40		2	380	55		3
Walking: 4 km · hr⁻¹	40	75	12		.75	105	16	.5	1	135	20	.33	1.5
6 km · hr⁻¹	40	105	16	1	1	135	20		1	170	26		1.5

Purpose: To instruct children with diabetes mellitus—and their parents—about the compensatory increase in calorie intake for physical activities that are *above and beyond* those practiced daily. Examples are given for children of 20-, 40-, and 60- kg body weight.

Assumptions:

1. These added activities have not been taken into consideration in the *regular* daily food intake.
2. Carbohydrate equivalent of activities in 60% of energy expenditure, apart from ice hockey and running at 12 km · hr⁻¹, where the carbohydrate equivalent is 75% of total energy expenditure.
3. The additional food is taken *before* the activity (for activities that last more than 1 hr, periodic food intake should be practiced *during* the activity).

*Table constructed with the help of Mrs. Karen Chelswick.

For a regular practice session, a 60- to 80-min interval is sufficient. During prolonged activities, sugar-containing fluids (glucose, sucrose, or fructose) must be taken every 15 to 20 min (121; 123). Special attention should be given to proper hydration, as discussed in the section "Fluid and Electrolyte Balance" in chapter 3. When exercise is performed in late afternoon or in the evening, the addition of complex carbohydrates should be considered in order to prevent late-onset hypoglycemia.

The Insulin Pump and Exercise

An alternative approach for better control of the insulin dosage is the use of continuous subcutaneous infusion by an *insulin pump* (73; 166). The use of a pump prevents uncontrolled release of insulin from the site of injection that often occurs during exercise (see earlier section, "Mechanisms for Exercise-Induced Reduction in Blood Glucose"). It also allows a flexible pre-exercise modification of the insulin dose. Insulin pumps are suitable for children (15; 73), although some patients use them only during sleeping hours (73).

Users of an insulin pump during sport activities should take several precautions, as summarized by Zinman (166):

- **Water sports.** If the pump is waterproof, it can be worn throughout water-based activities. If not, it should be removed.

- **Contact and collision sports.** In sports such as basketball, American football, soccer, or ice hockey, the pump can be protected by the use of a sport guard case or a protective pad. Another precaution is to wear the pump on a site that is less likely to be hit in a given sport (e.g., small of the back in soccer).

- **Winter sports.** Because insulin may freeze in cold weather, the pump and tubing must be thermally protected by, for example, sufficient layers of clothing.

- **Removing the pump.** If the pump is removed for less than 1 hr, no extra insulin is required. For longer periods, one should administer a bolus of insulin (through the pump) prior to removal.

Type 2 Diabetes Mellitus

In contrast to the situation with type 1 DM, insulin production by the pancreas is normal in type 2 DM (also known as "adult-onset" DM, or non-insulin-dependent DM). In point of fact, circulating insulin levels are excessive, reflecting *insulin resistance* in body tissues, including the muscles. Insulin resistance clusters with obesity, dyslipidemia, hyperglycemia, and high blood pressure (2; 19; 58; 107). The term *syndrome X* has been coined for this cluster. As in adults, insulin resistance in children as young as 7 years is positively associated with the amount of visceral fat (154).

The first edition of this book, published in 1983, did not include an entry for type 2 DM because the condition was not considered "pediatric." This, however, has changed dramatically in the last 15 to 20 years. Type 2 DM is now seen in growing numbers of children and, in particular, adolescents (16; 75; 111; 126). For example, a study from Cincinnati showed that the incidence of type 2 DM increased 10-fold among children and adolescents between 1982 and 1994 (111). An even greater increase (30-fold) was reported from Japan, in a comparison between 1976 and 1995 (76). In North America, Native Americans are the most affected group, followed by African Americans, Hispanics, and Caucasians. A recent study, however, shows that the prevalence of type 2 DM in moderately-to-markedly obese children and adolescents is similar among Hispanics, African Americans, and Caucasians (141).

This surge in the prevalence and incidence of type 2 DM accompanies the juvenile obesity epidemic (141) (see chapter 9). Indeed, type 2 DM among young people appears mostly in obese individuals (75; 111).

Together with adequate nutrition, enhanced physical activity is a major element in the prevention and treatment of type 2 DM. Research in children and youth has concentrated on the efficacy of enhanced physical activity in prevention of type 2 DM, rather than on treatment of individuals who already have the disease. Physical training induces benefits similar to those described for obesity: decrease in body weight, % body fat, and resting blood pressure, as well as an increase in insulin sensitivity (47; 58). While there are no dose-response data on the relationship between these functions and the increase in children's activity, adult data suggest that it is the volume of added activity rather than its intensity that really matters (33).

The beneficial effect on insulin sensitivity is apparently short-lived. It disappears when the patient resumes a less active lifestyle. For example, plasma insulin levels decreased by 10% in 7- to 11-year-old obese children who took part a 4-month, five times per week aerobic program. But 4 months after the cessation of the program, their plasma insulin rebounded, increasing by 19% (47).

Unlike what occurs in type 1 DM, exercise does not induce hypoglycemia in patients with type 2 DM, nor is there a risk for hyperglycemia and ketoacidosis. Caution should be taken when a patient has retinopathy or neuropathy, but this is not likely to occur in children or adolescents.

In conclusion, an active lifestyle is highly recommended for patients with type 2 DM or insulin resistance. Specific recommendations for activities are similar to those suggested for children and youths with obesity (chapter 9).

Growth Hormone Deficiency

Growth hormone (GH) deficiency occurs either as an isolated condition or as part of general hypopituitarism. It can be congenital or can start at a later stage in life (e.g., following trauma or brain surgery). Growth hormone deficiency results in short stature, growth retardation, or both. Its diagnosis is most important, because this is the main cause of short stature for which there is an effective therapy. One of the means for reaching such a diagnosis is a provocation test by exercise, which is the focus of this section.

The Need for a Provocation Test

Growth hormone is secreted from the pituitary gland in a pulsatile pattern, particularly during the waking hours (95); and its serum levels between pulses are extremely low, barely detected by current analytical techniques. Therefore, a random blood test is unsuitable as a diagnostic tool. There is a need to "provoke" the pituitary gland to produce and release GH prior to blood testing (72; 125; 127). A child whose hypothalamic-pituitary function is adequate usually responds with an elevation of serum GH, while a child with hypopituitarism has little or no response. Pharmacologic stimuli include insulin-induced hypoglycemia; intravenous administration of arginine or GH-releasing factor; intramuscular glucagon; and oral L-dopa, clonidine, or propranolol. The two physiologic stimuli are exercise and sleep, but fasting also stimulates a rise in GH. While there are no universally accepted criteria for a normal response to a provocation test (127), it is commonly accepted that a rise in serum GH level to 7 to 8 $\mu g \cdot L^{-1}$ indicates that the child is unlikely to have GH deficiency. The term "positive test" is used to denote no rise or an insufficient rise in serum GH.

Provocation Tests at Rest

The most potent pharmacologic provocation of GH release is through insulin-induced hypoglycemia (48; 72; 86; 150). Its main drawback is discomfort to the child. Furthermore, because of the hazards of hypoglycemia, this procedure is not without risk. Especially at risk are children with pituitary insufficiency, who may respond with extreme hypoglycemia (e.g., blood glucose levels of 1 mmol $\cdot$ dL^{-1} or less). Other pharmacologic stimuli such as arginine administration, while better tolerated by the child, are less potent than insulin and as a result may have high false positive responses (48; 53; 127). Sleep serves as an effective provocation for GH production, but it requires an overnight stay in hospital and simultaneous electroencephalogram monitoring to synchronize blood sampling with specific stages of sleep. This approach is therefore financially and logistically demanding and is not widely used for clinical purposes.

Exercise As a Provocation Test

It was first shown by Roth et al. (128) that exercise induces a rise in serum GH. Since the early 1970s, numerous pediatric services have adopted exercise as a screening or diagnostic test of GH deficiency (22; 32; 68; 91; 100; 103; 108; 112; 134; 139; 156; 160; 161; 163). Figure 8.13 depicts the range of GH rise as a function of the time elapsed since the start of exercise. In some protocols, exercise is used in combination with a pharmacologic stimulus (85; 91; 140).

Attempts have been made to determine whether combining a pharmacologic provocation with exercise would increase the usefulness of the exercise test. As shown by Sutton and Lazarus (148), GH response to exercise in young healthy adults was potentiated following β-adrenergic blockade (intravenous propanolol).

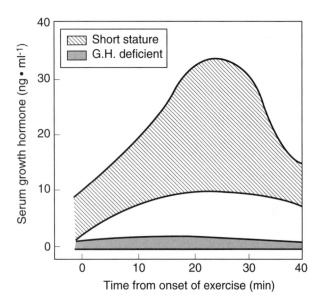

FIGURE 8.13 Changes in serum growth hormone following exercise provocation in growth-hormone-deficient (dot pattern) and non-growth-hormone-deficient (diagonal line pattern) children. A composite graph of the range of response. Based on mean data from various studies (74; 80; 100; 105; 108; 112; 139; 140; 161).

When children were given oral propranolol 2 hr prior to an exercise test, the rise in GH level was such that there were no false positive responses among the healthy children and no false negative responses among the five children with GH deficiency (figure 8.14) (140). The authors did not report the effect of an exercise test alone on the GH response of their patients. More research is needed to determine the diagnostic value of combining an exercise test with another stimulus.

The objective of any screening test is to filter out those children who do *not have the disease*, thus obtaining a smaller number of children who need a more definitive evaluation. An ideal screening test should be safe, simple, inexpensive, and highly sensitive (few false negative results). Although it is advantageous if a screening test is also specific (few false positives), this is not a prerequisite. Combining 11 studies in which the exercise test was performed on either a cycle ergometer or a treadmill (23; 32; 51; 56; 80; 100; 108; 112; 139; 161; 163), none of the patients with documented GH deficiency had a false negative result. Among 502 patients without GH deficiency, 380 had a normal rise in GH, which denotes 20% false positives. Such specificity is comparable to that of pharmacologic or other physiologic tests.

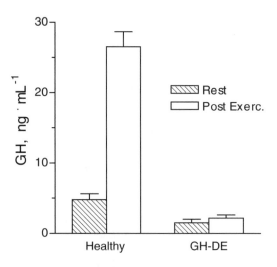

FIGURE 8.14 Combining exercise provocation with propanolol in the diagnosis of growth hormone deficiency. Thirty-two healthy, short-statured 3- to 15-year-old children and five children with growth hormone deficiency (GH-DE) performed moderate-to-intense 20-min exercise 2 hr after receiving orally 24 to 40 mg of propranolol. Blood was sampled immediately after exercise. Vertical lines denote 1 SEM. Data from Shanis and Moshang 1976 (140).

The major reasons for false positive results are as follows:

- The exercise stress is too short.
- Its intensity is too low.
- The GH peak is missed due to too few blood samples.
- The test is done at a time when the pituitary is refractory to further stimuli.

In addition, the pathways by which exercise triggers the release of GH are different from those following other stimuli (54; 134; 150; 153). As a result, some types of GH deficiency may be better diagnosed by one test, and others by another. Such a discrepancy may occur particularly in the case of partial GH deficiency (103; 127; 148; 162).

Factors That Modify the Growth Hormone Response to Exercise

A major reason for variation in the response to testing is the lack of standardization of *exercise intensity*. Intensities of 70% to 100% of maximal aerobic power yield a greater GH rise than do milder intensities (22; 35; 60; 100; 139; 149). At the Children's Exercise & Nutrition Centre we use a progressive maximal test (approximately 10 min), followed by a 10-min rest and then 10-min exercise at 75% to 80% maximal aerobic power. A dual exercise provocation has been used also by others (108). Some laboratories have been using stairway running, skipping, or walking with a parent (24; 54; 85; 105; 112). Such protocols are not recommended because exercise intensity is hard to standardize and is often too mild.

Another factor that may affect the outcome of an exercise test is the *timing of blood sampling.* Ideally, one should obtain a sample when serum GH concentration is at its peak, about 20 to 35 min after the start of the exercise (22; 34; 100; 108). Thus, when the task is very brief, a peak may be reached long after its cessation. On the other hand, when the exercise lasts more than 40 min, the peak may be reached while the individual is still exercising (22; 161). It has been suggested that one postexercise blood sample is sufficient for screening, with (160) or without (70) a pre-exercise sample. Such a practice is not advisable because the time to peak response varies among individuals. In addition, some children have a high pre-exercise GH level and a paradoxical lowering of the serum level following exercise. If only the latter is determined, a false positive conclusion may be reached. Therefore, in order to reduce the likelihood of a false positive result, one should take two to three post-exertional blood samples.

Growth hormone response depends on the *fitness level* of the patient. When an identical task is given to individuals who differ in fitness, the less fit will show a greater GH rise than the fit (22; 151; 158). The *relative* exercise intensity should therefore be equated among individuals. This can be achieved by exercising all children at a heart rate of 170 to 185 beat · min $^{-1}$.

Obese children respond to exercise and to pharmacologic stimuli with less elevation of GH than do non-obese children (52; 159). *Children with type 1 DM,* on the other hand, have a normal or even enhanced GH response to exercise (52; 152). Neither the mechanisms nor the implications of such responses are clear.

The concept of a *refractory period* has been introduced to denote the time during which a normal pituitary gland will not respond sufficiently to a stimulus for GH release. Such refractoriness can occur following a previous exercise stimulus (44), as well as following stimulus by a pharmacologic agent (34) such as growth hormone-releasing factor. The duration of the post-exertional refractory period has not been determined. It would be prudent, however, to ascertain that the child has performed no strenuous activity 4 to 6 hr prior to testing and for the child to be rested physically and emotionally for 45 to 60 min before such testing starts.

High ambient temperature is in itself a stimulus for a rise in serum GH (49; 104). A summative effect on GH of high ambient temperature and of exercise has been observed experimentally (49), and lowering of environmental temperature caused attenuation of GH release (21). One should therefore attempt to control climatic conditions to standardize the test.

Optimizing the Exercise Protocol

Based on the preceding discussion, the following protocol is recommended:

- The child should avoid intense exertion at least 4 to 6 hr before the test.
- The child should have fasted for at least 8 hr.
- Once in the laboratory, the child should rest and avoid emotional stress for at least 45 to 60 min before exercise.
- Draw venous blood for a baseline value 5 to 10 min prior to the exercise test. Use of an indwelling catheter is ideal.
- Use an ergometer (cycle, treadmill) for which the intensity can be standardized. If an ergometer is not available, one can use a step test, in which the height and cadence are standardized (see appendix II, "Step Tests" section).
- Exercise should last 10 to 15 min.
- Aim at an exercise intensity of at least 75% to 85% of maximal aerobic power (heart rate 170-185 beat · min^{-1}).
- Keep room temperature at 23° to 25° C.
- Have at least two to three postexercise blood samples, starting at 10 min postexercise.

CHAPTER 9

Nutritional Diseases

The links among nutrition, physical activity, and physical fitness are of utmost importance. Adequate nutrition is essential for normal growth and development, as well as for health, well-being, and physical performance of children and adolescents. Specifically, the balance between energy intake and energy expenditure may have major effects on growth, pubertal changes, and body composition. An excessively positive balance may lead to overweight and obesity. A negative balance during childhood and adolescence, particularly if accompanied by insufficient protein intake, may impede linear growth, induce the loss of fat and fat-free mass, and adversely affect the child's physical performance and well-being.

Several nutritional aberrations are relevant to pediatric exercise medicine. This chapter focuses on those three that are most prevalent among children and youth: anorexia nervosa, obesity, and chronic undernutrition.

Anorexia Nervosa

Anorexia nervosa (AN) is one of three main eating disorders found among adolescents. The other two are bulimia nervosa and binge-eating disorder. Issues related to physical activity and exercise performance are particularly relevant to AN. This illness is manifested by four main diagnostic elements (3):

- Body weight <85% of expected (or in young adults, body mass index <17.5 kg · m⁻²)
- Intense fear of gaining weight, or of becoming fat, even if very lean
- Inaccurate perception of own body size, weight, or shape

- Amenorrhea (absence of at least three consecutive menstrual cycles in a postmenarcheal girl)

Anorexia nervosa occurs typically in females, but approximately 5% of patients are males. The estimated prevalence of AN in the general population ranges between 0.2% and 1%; but these numbers are most probably an underestimate, because many of the patients are reluctant to seek medical advice and their condition therefore is undiagnosed. Prevalence is greater among dancers and athletes, particularly in endurance and "appearance" events such as gymnastics, rhythmic gymnastics, and figure skating (273). The incidence of AN has increased in recent decades, in conjunction with the notion that for the female body, "thin is beautiful."

Habitual Activity and Energy Expenditure

Unlike the involuntarily undernourished adolescent, the patient with AN is often more active than the healthy adolescent. Only in advanced inanition will the patient be debilitated and rendered inactive. Excessive engagement in exercise and sport is part and parcel of an overall energy-expending "strategy" of many patients with AN. Concomitantly with their food-rejecting obsession they may develop subtle, and often clandestine, techniques for enhancing their energy expenditure (54). While some concentrate on distance running or swimming, others may perform several hundred push-ups or sit-ups per day; run up and down numerous flights of stairs; or constantly move on their chair and fidget while they do homework, watch TV, or use a computer. Patients typically prefer to be active away from their peers, siblings, or parents so that they will not be accountable for such "forbidden" behavior: Some exercise in their bedroom, whereas others prefer to work out away from home. Indeed, exercise energy expenditure per day is high in patients with AN (53), while their resting metabolic rate is low (252). According to one study, the end result is that total daily energy expenditure, as measured by doubly labeled water, is similar in patients with AN and in healthy controls (53). It should be noted that the patients in that study were in good energy balance (maintaining a steady body weight). Such findings, however, may not apply to patients who are in negative balance (losing weight) or positive balance (gaining weight).

In some patients, enhanced habitual activity is apparent well before the onset of dieting or weight loss (73; 74). For example, retrospective analysis of adult female inpatients suggested that they had been more active than healthy controls from early adolescence. Sixty percent of them had been competitive athletes, and their high-level physical activity predated dieting (Figure 9.1) (74). In another study, premorbid activity levels predicted the extent of excessive activity that subsequently accompanied the disease (73). It has therefore been suggested, but not proven, that excessive physical activity may be an etiological factor in the development of AN (73).

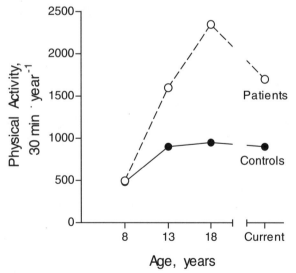

FIGURE 9.1 Excessive physical activity predates the diagnosis of anorexia nervosa. Retrospective information was obtained from young adult women in whom diagnosis of anorexia nervosa had been made around age 16 to 20 years. Units for activity are based on the amount, frequency, and intensity of the reported activities.

Reproduced, with permission, from Davis et al. 1994 (74).

At advanced stages of undernutrition and muscle wasting, patients often report a "low energy level." As a result, they moderate their activity behavior and may become even less active than their healthy peers (40). This is a major source of frustration to those who still strive to reduce their body weight.

Such dedication to, and dependence on, exercise and sport can be used by the therapist as the basis for behavior modification therapy, in which exercise is allowed as a reward whenever weight gains have been made.

Physical Fitness

The self-inflicted undernutrition of AN is accompanied by physiological and performance characteristics similar to those found during imposed undernutrition, even though patients with AN are often highly active. They may become hypoactive only at advanced stages.

A summary of physiologic characteristics at rest and during exercise is given in table 9.1. In addition to a reduction in fat mass, other dimensions such as lean body mass, mineral bone content and mass, total body potassium, intracellular water, heart volume, and blood volume are also reduced (69; 141; 208). Resting physiologic functions that are particularly low include metabolic rate, respiratory rate, heart rate, blood pressure, and cardiac output (95). The following sections address exercise-related functions that are particularly affected.

Aerobic Performance

Although a reduction in body dimensions has a direct bearing on performance, it cannot fully explain the low physical working capacity. *Peak power output and maximal O_2 uptake* are low, even when expressed per body mass or height (67; 97; 98; 149). For example, a maximal O_2 uptake of 35.0 and 34.3 ml · kg^{-1} · min^{-1}, respectively, was found in boys and girls with AN. These values are about 2 SD below the expected mean for the boys and 1 SD below mean for the girls. The deficit of maximal O_2 uptake in AN is more extreme than expected from the small heart volume (96; 97) or the low fat-free mass (69). On average, a healthy adolescent with a heart volume of 500 ml will have a maximal O_2 uptake of nearly 2 L · min^{-1}, compared with only 1.3 L · min^{-1} in a patient with the same heart volume (97). Deficiency in aerobic performance seems to be related to the severity of undernutrition (149), as shown in figure 9.2.

TABLE 9.1	Some Physiologic Characteristics of the Adolescent With Anorexia Nervosa	
At rest		
	Percent fatness	Low
	Lean body mass	Low
	Heart volume	Low
	Blood volume	Low
	Core temperature	Low
	O_2 uptake	Low
	Respiratory rate	Low
	Heart rate	Low
	Blood pressure	Low
	Cardiac output	Low
	ECG voltage	Low
	Blood lactate	High
During exercise		
	Peak aerobic power	Low
	O_2 uptake—submax and max	Low
	Heart rate—submax and max	Low
	Blood pressure—submax and max	Low
	Cardiac output—max	Low
	Cardiac output per O_2 uptake	Normal
	Blood (muscle?) lactate—max	Normal

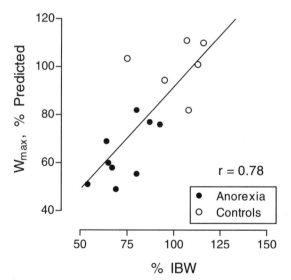

FIGURE 9.2 Maximal aerobic power (Wmax), expressed as percent predicted power, in relation to nutritional status, expressed as percentage of ideal body weight (% IBW). Nine adolescent girls with AN and nine healthy controls were tested on a cycle ergometer. All patients had reported a weight loss of at least 25% at the time of the study.

Adapted, with permission, from Lands et al. 1992 (149).

Another typical response of patients with AN is a low *maximal heart rate*. It seldom reaches 180 beat · min^{-1} (175; 245) and in some patients is as low as 150 to 160 beat · min^{-1}. This may be due to reduced muscle mass, or to high vagal activity and a low catecholamine response to exercise (175; 193). A low maximal heart rate and the small cardiac volume both result in a low *maximal cardiac output*. This, with or without iron deficiency anemia, results in a subnormal maximal O_2 uptake. It should be noted, though, that *submaximal cardiac output* and *submaximal minute ventilation* at a given O_2 output are within normal (149).

Anorexia nervosa patients also have a low resting and *submaximal heart rate*. We and others (183) have encountered adolescent patients whose resting heart rate was less than 30 beat · min^{-1}. This most probably reflects high vagal tone. One practical implication is that submaximal heart rate should not be used as a predictor of maximal O_2 uptake in individuals with AN. It will yield an overestimation of a person's real aerobic fitness.

Skeletal Muscle Function

As in other conditions of advanced malnutrition, skeletal muscle function is often deficient in patients with AN, particularly when muscle mass is depleted. In one study, the *maximal force* exerted by voluntary contraction, as well as by electrical stimulation, was low, and *muscle fatigability* was high in young adult females whose body weight ranged between 48% and 80% of ideal body weight (208). Girls with advanced AN who are tested in our clinic often show a deficient *anaerobic performance* as assessed by the Wingate anaerobic test.

Other Physiologic Functions

Patients, particularly when amenorrhea is prolonged, may have a reduction in bone mass and bone mineral content. A question often asked is whether the enhanced physical activity practiced by these patients can ameliorate the osteopenia. There is some evidence (134) that while moderate activity may protect the bone, intense activity is accompanied by bone loss. Other studies did not show a relationship between bone metabolism and the degree of physical activity in AN. The thermoregulatory capability of the exercising patient with AN is deficient, both in hot and in cold climates. For details see chapter 3.

Effects of Nutritional Repletion on Physical Fitness

Girls and boys with AN were followed up during a period of nutritional repletion (96) and markedly improved their peak aerobic power and maximal O_2 uptake (figure 9.3). These changes accompanied the return to normal of total body mass, fat-free mass, and other body dimensions. In another study (257), improvement in aerobic performance lagged behind an increase in body mass and muscle mass. One can therefore assume that the low aerobic performance of the patient with AN is, at least in part, a direct result of undernutrition. Nutritional repletion is also accompanied by improved muscle function. In one study (208), muscle fatigability reached normal levels within 4 weeks of oral refeeding, while muscle strength became normal within 8 weeks of refeeding. It is of note that these improvements preceded the return to normal of total body nitrogen, body fat mass, and the volume of intracellular water.

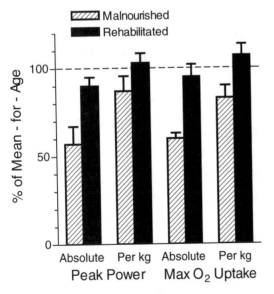

FIGURE 9.3 Effect of nutritional rehabilitation after anorexia nervosa on performance. Eight adolescent patients (five boys, three girls) were tested on a cycle ergometer before and after nutritional repletion. Values are presented as percent of expected mean. Based on data from Fohlin 1980 (99).

Exercise Perception

Patients with AN often engage in strenuous and prolonged physical activities which, to a bystander, may seem incompatible with their malnutrition and low working capacity. It has been suggested (47) that, in analogy to their distorted body image and denial of feeling tired, these patients also deny being overly active. Do these patients lose their perceptual "acuity" regarding exercise intensity? This idea has been examined experimentally (67) and found to be unjustified. The adolescent with AN does have the ability to discriminate well among various exertional intensities.

Obesity

There are several definitions for obesity and overweight among infants, children, and adolescents. Most are based on *weight for age* or *weight for height*. For infants, a weight for length of >2 SD of reference values is considered overweight. Reference values are based on the 1977 charts of the U.S. National Center for Health Statistics, and have also been adopted by the World Health Organization. These charts were updated in 2000 (146), but the following definitions and prevalence data are still based on the 1977 charts. The commonly used threshold for overweight in preschoolers is weight for height at the 95th percentile. In older children and in adolescents, the criteria are based on *body mass index (BMI)*, which is the ratio of body weight in kilograms to body height squared (m^2). The BMI threshold for overweight is the 85th percentile, while that for obesity is the 95th percentile. In adults, 25 and 30 $kg \cdot m^{-2}$ are the absolute thresholds for overweight and obesity, respectively. These values cannot be used for younger people, because the distribution of BMI in adolescents, children, and infants is different from that of adults (63). Figure 9.4 depicts the cutoff points for overweight and obesity, which are based on over 190,000 data points for infants, children, and adolescents from six countries.

The obvious drawback of using height and weight as the sole criteria for nutritional status is that these measures ignore the contribution of fat mass to total body mass. An attempt has been made to correct this gap by adding the triceps skinfold to the measurement of BMI and then using the 85th and 95th percentile of both to

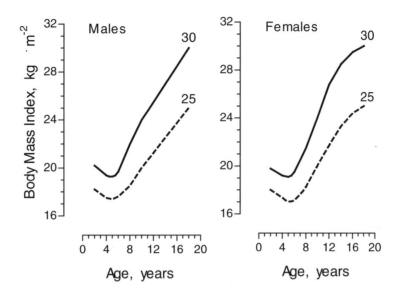

FIGURE 9.4 International cutoff points in body mass index for overweight (broken line) and obesity (full line) in males (n = 97,876) and females (n = 94,851) ages 2 to 18 years. The lines pass through body mass index of 25 and 30 kg · m^{-2}, respectively, at age 18.

Reprinted, with permission, from Cole et al. 2000 (63).

determine overweight and obesity, respectively (173). In the Children's Exercise & Nutrition Centre we have been using the following criteria for the severity of obesity, based on underwater weighing or bioimpedance analysis:

- 30% to 35% body fat = mild obesity
- >35% to 40% body fat = moderate obesity
- >40% body fat = marked obesity

Prevalence and Public Health Implications

The prevalence of childhood and adolescent obesity (juvenile obesity) has been on the rise in recent years (50; 107; 179; 248; 270) in both technologically developed and developing countries (275). In the United States, nearly one in four of all children and adolescents is considered overweight or obese. A nationwide survey conducted in 1988 to 1991 revealed an increase of some 20% in prevalence, compared with that in a similar survey from 1976 to 1980 (248). "Obesity" was defined as a BMI above the 85th percentile as determined in two nationwide surveys during the 1960s. This increase was particularly evident among adolescents and among Hispanics and African Americans. It was dramatically greater for "superobesity" (BMI above the 95th percentile). In Canada, there was more than a twofold increase in the prevalence of obesity of children and adolescents between 1981 and 1996. This was particularly evident in the 7 to 9 age group (247). In England, a 70% increase in prevalence was observed from 1989 to 1998 among 2.9- to 4.0-year-old children (50). Such a fast increase is likely due to changes in lifestyle of these young people and their families, rather than to a genetic phenomenon (14).

Although obesity is associated with risk factors for coronary artery disease (22; 52; 116) and non-insulin-dependent diabetes mellitus (116; 192), a major comorbidity of juvenile obesity, particularly among adolescents, is psychosocial (46; 101). From a public health point of view, the main concern with juvenile obesity is the likelihood that it will track into adult life. While the overall tracking of body fatness (250) and of fat distribution (254) from childhood and adolescence to adulthood is only fair, the relative risk that an obese child will become an obese adult is about 2.0. The respective relative risk for an obese adolescent may be as high as 5.0, particularly if the parents are obese (268).

Compared with those who were not overweight during adolescence, women who were overweight have an enhanced risk for atherosclerosis, and men who were overweight during adolescence face more than twice the risk of dying of coronary artery disease (174). This difference holds true even if these adults are not obese at present. Other long-term outcomes of juvenile obesity are low employability and fewer economic opportunities in later years (108).

In summary, juvenile obesity has reached *epidemic proportions* in various countries as its prevalence has risen at a fast pace. The implications to public health are immense. The highest priority should be given to prevention and treatment of juvenile obesity.

Habitual Activity and Energy Expenditure

The relationship between childhood obesity and physical hypoactivity is of unique importance: In contrast to the case with other diseases, hypoactivity may play a role in the *etiology*, as well as the *outcome*, of childhood obesity. The increase in prevalence in recent decades has occurred despite an overall reduction in fat consumption and little or no change in overall energy intake. This strongly suggests that energy expenditure has been decreasing (14). A similar pattern seems to have occurred among adults (234). In this section we first discuss the evidence that obese children and youth are hypoactive. This is followed by an analysis of the possible *causal relationships* between obesity and hypoactivity.

Adiposity and Habitual Activity

The idea that physical hypoactivity is typical of obese people and may play a role in the etiology of obesity was first suggested in 1907 (256). Various authors have subsequently discussed this relationship, but the first to systematically evaluate the activity patterns of overweight children was Bruch in 1940 (46). Among 140 overweight 2- to 14-year-old children, 68% of the girls and 76% of the boys were classified as inactive. These children were subjected to greater parental protectiveness and had considerably fewer social contacts than the more active children. Many studies since 1940 have confirmed the low activity level among obese children and adolescents, compared with their leaner peers, be it at school (13; 61; 132), after school (61; 161; 238), in summer camp (49), or at home (100; 148; 263). In a nationwide study in the United States, adiposity was inversely related to physical activity (189). Even when obese youths do participate in sport, the *intensity* of their activity is often low

(49). This is especially so in nonregimented play (65; 161), when the individual child, rather than the teacher or instructor, determines his or her level of involvement.

Other studies have indicated little or no relationship between activity level and degree of adiposity (27; 162; 210; 212; 232; 237; 261; 271). On balance, however, there is enough evidence to allow the conclusion that obese children and youth are often less active than the general population (13). The discrepancy among studies may reflect the variety of methods used to assess activity and the criteria used to define obesity, as well as the ethnic and social background of the samples. In addition, activity patterns differ when a child is exposed to different environmental stimuli. For example, obese boys, when compared with their non-obese siblings, were much less active at home, slightly less active outside home, and similarly active at school (263).

Among infants, an inverse relationship has been found between adiposity and activity. When 4- to 6-month-old babies were subdivided according to adiposity, the most obese exhibited the fewest limb motions as assessed by actometers (figure 9.5). Total energy intake and "extra" intake (i.e., energy consumed above maintenance requirements) were also inversely related to the level of adiposity (206). Likewise, direct observations, repeated at ages 6, 9, and 12 months, revealed an inverse relationship between adiposity and activity (151). In contrast, a larger-scale study (129) showed no difference in activity (according to a 1-day log kept by the mother) between obese and lean 6-month-old infants. Nor was there a relationship between 24-hr energy expenditure (doubly labeled water) and adiposity in 12-week-old healthy infants (264).

On balance, while there is evidence to suggest a reduced level of activity among obese infants, this is not reflected by a low total energy expenditure.

Another factor to consider is *parental obesity*. In one study (112), 4- to 5-year-old subjects were selected according to their parents' adiposity rather than their own. The offspring of obese parents were less active (based on heart rate monitoring) and ate less than those of leaner parents. These findings are important because, at the time of testing, the children of the two paren-

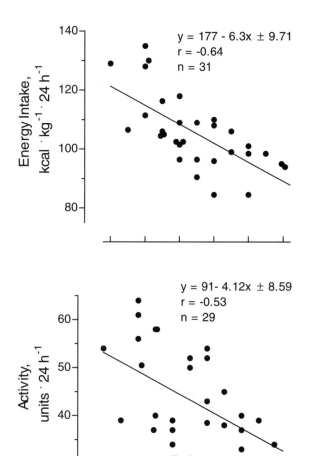

$y = 177 - 6.3x \pm 9.71$
$r = -0.64$
$n = 31$

Energy Intake, kcal · kg^{-1} · 24 h^{-1}

$y = 91 - 4.12x \pm 8.59$
$r = -0.53$
$n = 29$

Activity, units · 24 h^{-1}

Triceps Skinfold, mm

FIGURE 9.5 Physical activity and energy intake in relationship to adiposity (triceps skinfold) in 6-month-old infants. Activity is represented by actometer units. Actometers were attached to wrist and ankle. Data from Rose and Mayer 1968 (206).

tal groups had similar anthropometric and body composition characteristics. This parent-child relationship was confirmed, using direct observation of activity, for 3- to 6-year-old children (139). In that study, the strength of parental influence was particularly evident when both parents were obese. In contrast, there is no relationship between parental adiposity and the *total daily energy expenditure* of 12-week-old infants (72), or of 4- to 7-year-old children (104), as assessed by doubly labeled water. On the other hand, the resting metabolic rate of obese 8- to 12-year-old girls with two obese parents was lower than that of

obese daughters of non-obese parents (276). The reason for such a discrepancy is not clear.

Television Viewing and Juvenile Obesity

Television viewing, with or without computer games, is a dominant, and sometimes the most prevalent, sedentary pursuit of children and adolescents in many countries. In large-scale surveys, a strong relationship has been found between the likelihood of being obese and the extent of TV viewing (81; 109; 189). Based on 1990 U.S. data, the odds of being overweight ($\geqslant$85th percentile for BMI) at ages 10 to 15 were 5.3 times greater among those watching TV more than 5 hr per day compared with those watching 0 to 2 hr daily (109). The likelihood of being overweight increased as a function of hours of TV viewing, reaching 32.8% in those who watched more than 5 hr per day (figure 9.6). A 4-year follow-up of adolescents showed an increase of 1.3% in the incidence (new cases) of obesity for each additional hour of watching TV, as well as a strong negative relationship between the rate of remission from obesity and the extent of TV viewing (106). Such longitudinal observations are highly suggestive, but not evidence, of a causal relationship between juvenile obesity and the amount of TV viewing.

In contrast, some studies have shown little or no relationship between the prevalence or incidence of juvenile obesity and the extent of TV viewing (84; 204; 249). The reason for such a discrepancy is not clear, but it may reflect the smaller sample size in the latter studies.

The link between TV viewing and obesity may be related both to low energy expenditure during this sedentary pursuit, particularly in children who are obese (140), and to enhanced food intake. While there may be little (204) or no (241) relationship between the extent of TV viewing and the child's overall activity behavior, energy expenditure drops during TV watching. This is depicted in figure 9.7, which demonstrates a reduction in the sitting metabolic rate when the TV was turned on (140). Snacking while watching TV is enhanced (58), and it rises in proportion to the number of food-related commercials (241). This may reflect the large proportion of high-energy foods advertised on TV (236). The effect of TV on eating patterns is immense when one bears in mind that U.S. children see, on the average, some 10,000 food-related commercials per year, 95%

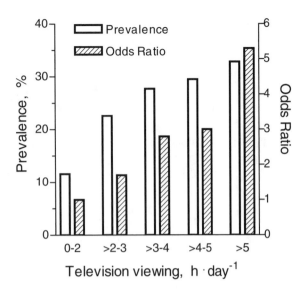

FIGURE 9.6 Prevalence and odds ratio of overweight among U.S. adolescents in relation to the number of hours per week that they watch TV. Data are based on a nationwide sample of 746 boys and girls, 10 to 15 years old, who were surveyed in 1986. "Obesity" is defined as triceps skinfold ≥85th percentile of the 1971 to 1973 NHANES. Based on data from Gortmaker et al. 1996 (109).

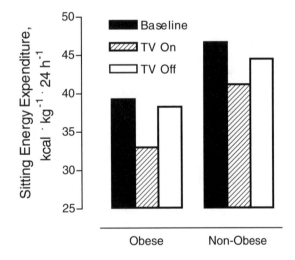

FIGURE 9.7 Effect of TV viewing on sitting metabolic rate. Fifteen obese 8- to 12-year-old children and 16 non-obese controls were monitored at the sitting position while the TV was off or on for 25 min. Based on data by Klesges et al. 1993 (140).

of which promote fast food, candy, sweetened cereals, and sport drinks (20).

Adiposity and Energy Expenditure

While most workers in the field agree that obese children are hypoactive, there is less evidence

that their daily energy expenditure is lower than in their non-obese peers. Waxman and Stunkard (263) were the first to suggest such a dichotomy, based on their study of obese and non-obese siblings. Even though the obese boys were less active than the non-obese siblings, there was no difference in energy expenditure as assessed by oxygen cost of activities. These authors did not correct for the larger body mass of the boys who were obese, which would increase the metabolic demands during activities such as walking, running, jumping, and climbing. This concept is discussed further in the section on the physical working capacity of the obese child.

Another issue to consider is that total energy expenditure (TEE) reflects two components other than activity, that is, the basal (or resting) metabolic rate (BMR) and the thermic effect of food. It is particularly important to correct for BMR because for most individuals, this constitutes the bulk of TEE. Being related to fat-free mass, BMR is higher in obese persons when calculated in absolute terms, but lower when divided by total body mass (10; 45; 240). Such a division, however, violates scaling principles (see section "Size-Dependent and Size-Independent Differences" in chapter 1 and appendix V) and would give large individuals fictitiously low values.

To more accurately estimate the energy expenditure of activities, one can correct for BMR by subtracting it from TEE. The difference reflects the energy expenditure of activity plus the thermic effect of food. Dividing TEE by BMR is another approach.

The way in which energy expenditure is calculated may determine our conclusion on the effect of adiposity. A study by Bandini and colleagues (10) is a case in point. With use of the doubly labeled water technique, TEE in absolute units was higher in obese adolescents than in lean controls. However, when calculated per body mass, TEE was lower in the adolescents who were obese (figure 9.8). A similar pattern was obtained for prepubescents (76). In 1.5- to 4.5-year-old children with a wide range of adiposity, both TEE/BMR (figure 9.9) and TEE – BMR were inversely related to % body fat (71). Based on a small sample size, obese adolescents with Prader-Willi syndrome have a lower activity-related energy expenditure than do those with exogenous obesity (216).

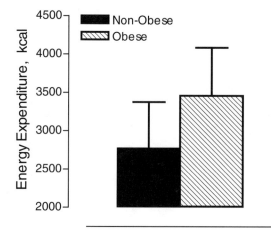

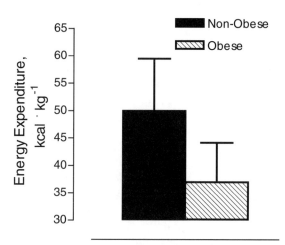

FIGURE 9.8 Absolute versus relative total daily energy expenditure and obesity. Subjects were 35 obese and 28 non-obese 12- to 18-year-old girls and boys. Total energy expenditure was assessed by doubly labeled water; % body fat by ^{18}O dilution. Based on data by Bandini et al. 1990 (10).

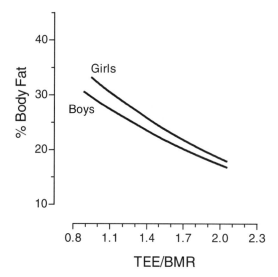

FIGURE 9.9 Activity-related energy expenditure and body fatness in preschoolers. Percent body fat (deuterium dilution technique), total 24-hr energy expenditure (TEE, doubly labeled water), and basal metabolic rate (BMR, calculated from body mass) were measured in 77 girls and boys, 1.5 to 4.5 years old. The lines denote regression.

Adapted from Davies et al. 1995 (71).

We are still left with the question of why juvenile obesity is linked to a low activity level but less so to low energy expenditure. This paradox may be due to methodological limitations in quantifying physical activity, or it may simply reflect the fact that activity behavior and energy expenditure are two different phenomena. Indeed, a dissociation between hypoactivity and low energy expenditure has been shown among preschoolers and schoolchildren (105; 274).

According to an interesting hypothesis that has been raised (80), hypoactivity covaries with other behaviors that may induce obesity (e.g., eating) rather than having a direct effect on the child's energy expenditure. Such a relationship between low activity and high food consumption has been found, for example, among 18-month-old children (255).

Energy Intake and Juvenile Obesity

Excessive eating may lead to obesity. However, contrary to common belief, once children become obese, they do not necessarily eat excessively. They often *eat less* (11; 41; 65; 132; 213; 232) and have a lower daily energy turnover than do lean children (41; 65; 132; 213; 232). An example of such low turnover is presented in figure 9.10 regarding adolescent girls. Sixty-eight percent of the overweight group expended and ate less than 8.4 mJ per day (2,000 kcal per day), compared with 11% among the lean subjects. In contrast, 53% of the lean and only 11% of the overweight subjects ate and expended more than 10.5 mJ per day (2,500 kcal per day). A recent large-scale survey in the United Kingdom showed no relationship between body mass index and food composition of children ages 1.5 to 4.5 years (70).

A word of caution regarding food intake data: Obese patients and their parents often underreport the child's food intake. In our clinic it is not uncommon to hear a parent stating with full conviction that his or her child "eats absolutely nothing." Indeed, studies have documented that

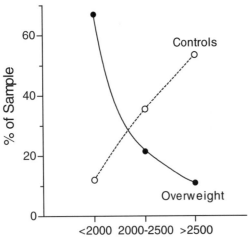

FIGURE 9.10 Frequency distribution of 29 overweight and 28 control adolescent girls according to their daily energy turnover. Based on interviewer-administered recall questionnaires. Schematic adaptation from Johnson et al. 1956 (132).

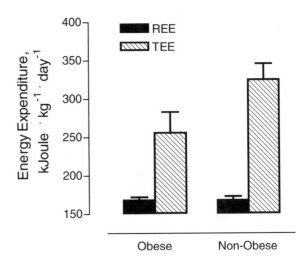

FIGURE 9.11 Low energy expenditure at age 3 months as a possible cause of overweight at age 12 months. Total energy expenditure (TEE) of 12 infants of obese mothers was determined (doubly labeled water) at age 3 months, and the infants' body adiposity (body mass index and skinfolds) was assessed at age 12 months. Data from Roberts et al. 1988 (202). REE = resting energy expenditure.

obese adults (152), adolescents (11), and children (157) markedly underreport their food intake (and often overreport their activity level!). Intake data based on reporting should therefore be regarded with skepticism.

Hypoactivity As a Possible Cause of Childhood Obesity

To assess a possibly *causal* relationship between juvenile obesity and hypoactivity, one should gather data from the early stages of life, preferably before an individual has become obese. In an early study (155), five newborns of obese mothers were observed during the first 8 weeks of their life. Weight gain in these formula-fed babies was inversely related to their activity as assessed by actometers. Gain in length, on the other hand, was positively correlated with activity. In contrast, when eating patterns of 288 babies were observed during their first year of life, neither energy intake nor the type of feeding, frequency of meals, or the age at which solids were introduced was a predictor of skinfold thickness at age 1 year (102).

Recent studies, using doubly labeled water, have revealed an interesting pattern. In one study (202), TEE was determined at age 3 months in normal-weight infants of obese mothers. By age 12 months, some of the infants had become overweight and the others had not. As seen in figure 9.11, TEE at age 3 months was 21% lower

in those who later became overweight. There was no intergroup difference in resting energy expenditure or energy intake at 3 months. The authors concluded that low activity at age 3 months may be an important factor in the rapid weight gain during the first year of life. However, in subsequent studies, using similar methodology, there was no relationship between TEE at age 12 weeks and body adiposity at age 2 years (131) and at 2.5 and 3.5 years (265). The discrepancy between these findings may be reconciled by the fact that mothers in the latter two studies were not obese. Our tentative conclusion is that, for infants of obese mothers, energy expenditure during the early months may be a cause of subsequent infant overweight. We have no data on the possible effect of energy expenditure beyond age 3 months on the subsequent development of juvenile obesity. It seems, though, that a low activity level in preschoolers is related to the development of overweight at school age (145; 172).

Causes of Hypoactivity

Why are obese youngsters insufficiently active? Although there may be still unexplained hereditary or other physiologic mechanisms that might determine the level of activity early in life, there are also well-documented psychosocial and fitness-related factors that may influence the willingness of obese children to pursue an active life. We often see patients with obesity, mostly adolescents, who are reluctant to participate in sport and games because of their "ugly body." Boys are often ashamed of their large breasts. Our explanation that this simply denotes a fat pad may be accepted on a cognitive, but not emotional, level.

Poor self-esteem, distorted body image, and depression occur in obese youth (101). Obese girls are often preoccupied with the thought that they are overweight. They show passivity and withdrawal and are found to react in a way typical of minority ethnic groups, perceiving themselves as "victims of prejudice" (169). Such girls are aware of their low levels of activity but often have no perception of the *degree* of their hypoactivity (48).

Activity behaviors seem to aggregate within families (211). Parental obesity and activity behavior, separately and in combination, affect a child's activity (139). According to one study, parents of obese children discouraged social contacts and any physical activities that connoted "risk" (46). Indeed, a relationship was found between such parental overprotection and the lack of activity among the children. Families of obese adolescent girls pursued fewer common recreational activities than lean girls. The latter seemed also to have more intersibling ties and friendships and to report a more unified family (46).

Compared to children, obese adolescents seem more influenced by their peers (176). Direct observations and questionnaires reveal that obese youths and adolescents perceive that their non-obese peers consider them "too awkward" and "physically limited" (1). And, indeed, obese children and adolescents are isolated by their peers (25), who do not invite them to take part in games and often ridicule them.

As discussed in the next section, the physical fitness of obese children and adolescents is often low, which makes participation in sport harder and less attractive. Obese schoolchildren perceive a given exercise level as harder than do their lean counterparts (260). The greater perceived strain of exercising may be another reason they are reluctant to pursue strenuous or prolonged activities.

In conclusion, hypoactivity is prevalent among obese infants, children, and adolescents of both genders, as it is also for obese adults. More research is needed regarding the association between juvenile obesity and other "screen"-based activities such as computer games and Internet surfing. Of particular concern is the association between juvenile obesity and the extent of TV viewing. Hypoactivity is not always reflected by a lower energy expenditure, which suggests that a sedentary lifestyle may covary with other behaviors (e.g., eating) that are associated with juvenile obesity. Animal studies, as well as information on infants and on preobese children, suggest that inactivity may be an *etiologic* factor in the genesis of childhood obesity. This relationship, however, may be limited to those infants who have obese parents. We still lack definitive confirmation of such a causal relationship, and no clear-cut explanation is available as to *why* babies, or fetuses, vary in their activity. Is it maternal behavior, the availability of food, a subtle interaction between baby and mother, or perhaps genetic disposition? In later childhood and in adolescence, hypoactivity could be a cause—but also a result—of obesity. Whatever the initial stage, hypoactivity and obesity create an unhappy symbiosis through which the affected child may enter a vicious cycle of inactivity-positive energy balance-obesity-reduced fitness-further inactivity.

Physical Fitness

There are many physical tasks that obese children and youth cannot perform at par with their leaner peers. This reflects a lower level of physiologic fitness. Indeed, whether measured in the laboratory or in the field, the fitness level of obese people is often lower than among the non-obese. Such a deficiency is related to the degree of obesity.

Maximal Aerobic Power

Whether expressed as maximal O_2 uptake per kilogram, anaerobic threshold, or the mechanical power at a heart rate of 170 beat $\cdot$ min^{-1} (W_{170}), the aerobic performance of obese children and youth is usually lower than in the non-obese. This has been shown in numerous (35; 59; 68; 77; 167; 170; 187; 198; 200; 201; 243; 280), but not all (156), studies. Likewise, performance in distance running is low in obese children and youth (59; 111; 262). Aerobic performance is inversely related to the degree of overweight or to % body fat (168; 200; 207; 262). This is true even when data are normalized for lean body mass (198), muscle mass (68), or heart volume (35).

One should exercise caution in interpreting the results of maximal O_2 uptake among people who vary in body mass and composition. Some authors (156; 207) have reported that compared with their non-obese peers, obese children had a *higher* maximal O_2 uptake when expressed in liters per minute. This does not denote a superior maximal aerobic power, but merely the fact that obese people have a large fat-free mass. To better reflect the ability of a child to perform a high-intensity aerobic task, one should express maximal O_2 uptake per kilogram body mass or fat-free mass or should use allometry (see appendix V). Indeed, when correction is made for body size, obese children almost invariably have low scores in aerobic performance. This is also manifested in their ability to sustain high-intensity effort over time (figure 9.12).

Low maximal aerobic power, in itself, does not denote an abnormality in the O_2 transport system (64; 207). When corrections were made for body size (by subtracting O_2 uptake during cycling at "zero load" from maximal O_2 uptake, and likewise for anaerobic threshold), cardiorespiratory responses to exercise were normal in two-thirds of obese boys and girls. The other subjects, however, did show abnormalities (64). One common pattern for the obese subjects was that, even though the on-transient time for O_2 uptake was normal, the transient times for CO_2 output and for minute ventilation were abnormally prolonged. A markedly long transient time for minute ventilation may result in low arterial pressure of oxygen during the initial stages of exercise.

Physiologic and Mechanical Cost of Locomotion

One possible reason for a deficient performance during aerobic tasks is the greater metabolic cost of locomotion (figure 9.13). In physiologic terms, obese children require a higher O_2 uptake to perform submaximal tasks, such as walking or running at a given speed, as reported by most (7; 68; 158; 201; 280), but not all (207), authors. The high metabolic cost is apparent particularly at high walking speeds (135). However, excessive energy cost is not limited to weight-bearing tasks such as walking, running, or climbing. It also occurs, albeit to a lesser extent, during cycling, in which the body is in large part supported by the seat (4; 6; 64).

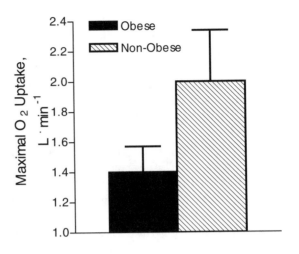

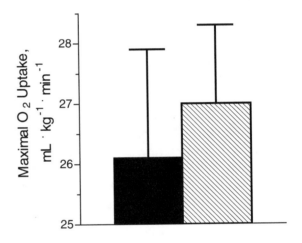

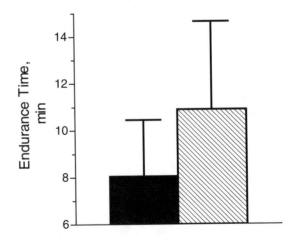

FIGURE 9.12 Aerobic performance and obesity. Fourteen 15- to 18-year-old obese girls and 13 non-obese controls performed a maximal walking test on a treadmill at 5.25 km · hr^{-1}. The slope was increased by 2% every 3 min. Data from Rowland 1991 (207).

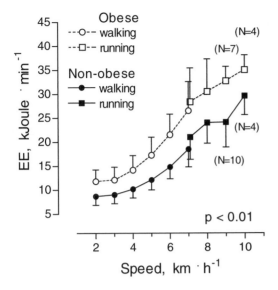

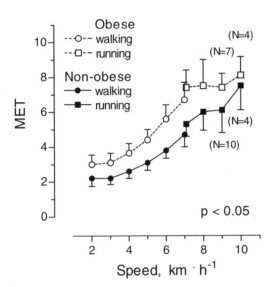

FIGURE 9.13 Energy cost of walking in obese and non-obese children. Twenty-three obese girls and boys (9.3 ± 1.1 years, 31.3 ± 6.9% body fat) and 17 controls (9.2 ± 0.6 years, 16.5 ± 5.4% fat) walked on a treadmill at several speeds. Top graph = absolute energy cost; bottom graph = energy cost expressed as METs (energy cost during exercise divided by the resting metabolic rate). The intergroup differences in the top and bottom graphs are significant.

Reproduced, with permission, from Maffeis et al. 1993 (158).

Does the high energy cost of locomotion result from a large total body mass or a large fat mass? In a study in which obese adolescent boys were compared with lean controls matched for total body mass, there were no intergroup differences in the O_2 cost of treadmill walking at slow and moderate speeds. Only at a fast walking speed (6 km · hr^{-1}) did the obese respond with a higher

metabolic cost. Fat distribution between trunk and limbs did not affect the O_2 cost or other cardiorespiratory variables (7).

The high cost of locomotion may reflect a "wasteful" walking or running style (126; 127; 166). In a study that compared mechanical gait characteristics in prepubertal obese and non-obese children, the obese children had a longer gait cycle duration, lower relative velocity (stature per second), and a longer stance period (126). In addition, they had greater asymmetry in temporal gait components (125), but there were no differences in electromyographic patterns (127). More research is needed to determine a possible relationship between biomechanical gait aberrations and the *degree* of obesity in children. Likewise, it has yet to be shown that such kinematic characteristics explain the difference in metabolic cost among people who vary in body adiposity.

Cardiorespiratory Responses to Submaximal Exercise

In line with the high metabolic cost of activity, obese children also exercise at a higher percentage of their maximal heart rate (25; 168; 197). This would suggest a low cardiac reserve. It has been estimated that among adults, each 10% increase in body fat is accompanied by a 10 beat · min^{-1} rise in heart rate (19). Whether the same is true for children has not yet been established. Nor are there data on the effects of childhood obesity on exercise cardiac output or stroke volume; but the O_2 pulse, which reflects stroke volume, is lower in children who are obese (207). Systolic arterial blood pressure is excessively elevated during exercise in the child who is obese, more so than at rest (8; 18). Some of the difference may result from the larger arm circumference in persons who are obese.

Obese children have a high minute ventilation during walking and running at a given speed and slope (158; 207). Furthermore, the O_2 cost of breathing is excessive, as found among markedly obese adults (266). Obese adults also have a somewhat reduced tidal volume, excessive ventilation and respiratory rate, and increased alveolar-arterial O_2 differences during submaximal exercise (78). There are no similar data for children.

At rest, obese children have a low ventilatory response to CO_2 (55), reduced ventilatory muscle endurance, airway narrowing, and a low diffusing capacity (130). A question remains as to whether this respiratory pattern reflects a constrictive effect of the heavy chest wall or also bronchial hyperreactivity. As we have noticed in our clinic, and as reported by others (178), the occurrence of asthma in obese children and adolescents is quite common. It is not clear, though, whether obesity is secondary to the hypoactivity that often accompanies asthma or whether there is a more subtle pathophysiologic link between the two. In an uncontrolled intervention study, weight loss was accompanied by an increase in vital capacity and forced midexpiratory flow, but there was no change in bronchial reactivity as determined by a methacholine provocation test (178).

The combined stresses of exercise and ambient heat induce a high physiologic strain on the child with obesity. For a detailed discussion see the section "Obesity" in chapter 3.

Muscle Strength

Even though they often are taller and appear to be stronger than their peers, obese children (33; 34) and young adults (138) have a lower muscle strength per kilogram body mass. As a result, when performing a task that requires carrying or lifting their own body, they are at a disadvantage (17; 26; 123; 221; 222). An example is given in figure 9.14, which summarizes performance of three groups of 50 adolescents each who were tested in the Wingate Institute laboratory. One group comprised very lean individuals; another, those of average adiposity; and the third, mildly obese subjects. Whereas the lean and the average group had similar scores in strength-related events, the performance of the obese group was dramatically inferior. Some 70% of the latter could not pull themselves up on the horizontal bar even once and could not complete a single "dip" on the parallel bars (one "dip" comprises a change of position from straight-arm support to bent-elbow support and back).

Because of the relationship between muscle mass and lean body mass, the finding just cited might suggest that a muscle of a given size generates less force in those who are obese. However, based on computed axial tomography to calculate the cross-sectional area of the knee extensors and elbow flexors (33; 34), there was no difference in isokinetic muscle strength of obese and non-obese prepubescent boys. Nor were there differences in contractile characteristics (peak twitch torque, time to peak torque, half-relaxation time). The

obese boys, however, had a lower percentage of activated motor units during maximal voluntary contraction (81.5% vs. 95.3%) (34). Whether this lower activation rate is relevant to strength and motor performance in those who are obese is unclear.

It thus seems that the lower performance of obese boys in tasks that require lifting or carrying their own body does not result from an intrinsically weak muscle or from deficient contractile characteristics, but rather from the excessive body mass that resists these actions. More research is needed to determine whether the same holds true for girls.

Motor Performance

Apart from their lower performance in activities that require carrying or lifting the whole body, as discussed in the preceding section, the motor skills of boys (26) and girls who are obese (160) are often deficient. A case in point is a study of 7- to 16-year-old Belgian obese and non-obese girls. The obese girls scored significantly lower in items that measure speed, agility, peak muscle power, abdominal muscle endurance, the ability to lift their body by the arms, and balance. A similar pattern is shown for adolescent boys in figure 9.14. The improvement in skills following a training program seems to be lower in obese children than in non-obese controls (25).

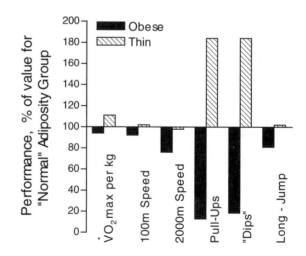

FIGURE 9.14 Maximal aerobic power (V̇O₂max), speed, muscle strength, and power of overweight and very thin male adolescents. Values expressed as percentage performance of adolescents with average adiposity. Based on unpublished data from the Wingate Institute, Israel.

Is there a linear relationship between the *level* of adiposity and impaired performance? As seen from figure 9.15, maximal O_2 uptake declines linearly with the increase in percentage of fat. Other fitness components, however, do not have a linear relationship with percent fatness. The lower half of figure 9.15 is a case in point—the "Fitness Index," as calculated from a step test, was a little lower in mildly obese adolescents than in lean ones. For those boys

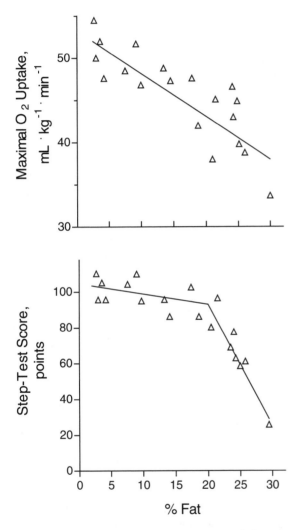

FIGURE 9.15 Maximal O_2 uptake measured during a treadmill test (top graph), and the Fitness Index in a modified Harvard Step Test (bottom graph), in relationship to body fatness. Individual performance of 19 nonathletic adolescent boys who ranged in adiposity from very thin to moderately obese. The Fitness Index is positively related to the number of steps that the individual manages to complete and inversely related to the postexercise heart rate.

Data from Bar-Or and Zwiren, unpublished report, 1972.

who had more than 23% fat, however, the score became markedly low. Thus the more obese adolescents were "penalized" twice: Their maximal aerobic power was low and they also had to lift a heavier body, which caused a clumsy up-and-down motion and early fatigue in their legs. In other fitness components in which body weight has to be supported, we also find that the decline in performance is more dramatic from mild to moderate obesity than it is from normal adiposity to mild obesity.

It has been suggested (246) that mild obesity is not necessarily accompanied by decreased performance because the extra body mass serves as a training stimulus. Such a rationale may hold true only if the child is active enough. If, on the other hand, the obese child is hypoactive, then the extra weight is only detrimental to performance.

Schoolchildren who are obese tend to achieve lower grades in physical education classes (161; 197). Among the leanest of 51 schoolchildren, more than 80% were high achievers in physical education class. The pattern for the most overweight group was practically reversed: The low achievers outnumbered the high achievers by more than three to one (figure 9.16).

Rating of Perceived Exertion

Scant data (260) indicate that children and adolescents who are obese rate the intensity of exercise higher than do their non-obese peers. This difference is equivalent to 1.5 to 2 points on the Borg 6 to 20 category scale (figure 9.17) (38). While rating of perceived exertion is not commonly considered a fitness indicator, a high rating may explain reluctance to pursue intense or prolonged activities.

Beneficial Effects of Enhanced Physical Activity

Enhanced activity is usually considered a means of reducing body fatness, but it has several other proven or presumed benefits. In this section, we address several questions. How effective is enhanced physical activity as a reducing regimen? How does it affect fat distribution? How do the effects of enhanced physical activity compare with the restriction of energy intake? Does an exercise program affect the spontaneous physical activity of the child who is obese? Does

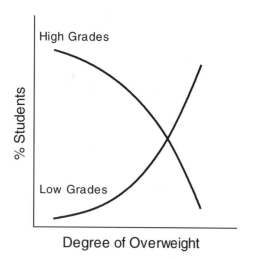

FIGURE 9.16 Relationship between grades obtained in physical education classes and the level of overweight among 518 girls and boys, 6 to 18 years old. Schematic adaptation from Rehs et al. 1973 (197).

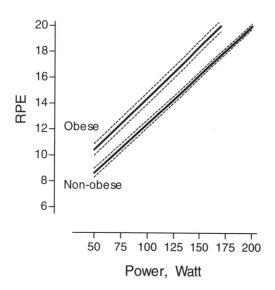

FIGURE 9.17 Rating of perceived exertion in obese and lean adolescents. Subjects were 13.0- to 17.9-year-old girls and boys. Thirty-four were obese (mean = 35.6% fat), and 50 were non-obese. All performed a multistage cycling test. Based on Ward et al. 1986 (260).

it affect appetite? What is the role of the school in prevention and treatment of juvenile obesity? Studies that address these issues have compared either an exercise group to a nonexercise group, or an exercise-plus-diet group to a diet group without added exercise. In general, the latter design yielded greater changes in body composition, physiologic variables, and fitness (13; 86).

Enhanced Physical Activity As a Reducing Regimen

Physical activity is by far the single most flexible component of TEE. A mild activity such as walking or jogging for 45 to 60 min can induce an energy expenditure of as much as 10% to 15% of daily energy expenditure (see also appendix IV for the energy equivalents of various activities). More intense or prolonged exertion, as practiced by athletes, can raise daily expenditure by 100% and more.

Are such increases sufficient to affect body weight and fatness? One often hears parents and health practitioners rationalize that a reducing regimen by exercise "isn't worth it" because of the low energy equivalence of various activities. Indeed, to lose 1 kg of adipose tissue (approximately 35 mJ, or 8,300 kcal), a 40-kg child would have to run 165 km or play tennis for 30 hr. If these tasks were to be completed in 1 to 2 weeks, such an objection would indeed be valid. However, energy expenditure has a cumulative effect. Assuming no change in energy intake, walking or running 4 km per day (200 kcal in a 40-kg child), for example, will be equivalent within 30 days to 0.7 kg of fat, or 4.2 kg in half a year.

Studies with non-obese children and adolescents have shown no effect of physical training on % body fat (13; 272). In contrast, similar regimens applied to obese persons are efficacious (87; 116). However, changes in body composition are not sustained beyond the duration of the intervention program (93) unless accompanied by behavior modification (44; 85; 88). Several studies have been conducted in which obese children and adolescents participated in conditioning programs. The rate of fat reduction markedly varies among studies, ranging between 1% and 5% of the initial fat mass per week. This rate depends neither on the intensity nor on the type of exercise. Apparently it is the overall energy expenditure that really matters (215). However, there is no evidence for a dose-response relationship between reduction in adiposity and the amount of fat loss (13).

Following an exercise bout, basal and resting metabolic rates are enhanced (194). The extent of this rise depends on the duration and intensity of the activity. A small rise may still be detected up to 15 hr following exercise. Therefore, the energy equivalent of an exercise bout will be an *underestimation* of the actual additional energy expenditure.

For reasons unknown, boys seem to benefit more than girls from weight-reducing training programs (186; 218; 224). A similar gender-related difference has been shown for rats (128). Still, weight and fat reduction are feasible also for girls who undergo short-term conditioning regimens (114; 115; 171; 186; 187; 218; 269).

The effectiveness of a reducing regimen may be enhanced by exposure to a cold climate (219). This has been suggested for young adults who were exposed to a 10-day mission in the Arctic (177) and was confirmed by a study in a climatic chamber cooled to –40° C (220). The long-term effects of this phenomenon are not known, nor has its potential as a reducing regimen been studied.

Effects of Enhanced Physical Activity on Fat Distribution

In adults, the accumulation of abdominal, and specifically visceral, fat is considered a risk factor for coronary heart disease and for type 2 diabetes mellitus (79). The same has been shown for children and adolescents (52; 110). Enhanced physical activity was efficacious in reducing the abdominal and visceral fat of 7- to 11-year-old obese children (117). Specifically, 4 months of five-per-week 45-min sessions of moderate to intense exercise (mean heart rate 157 beats · min)$^{-1}$ induced a decrease in the volume of subcutaneous abdominal adipose tissue and an attenuation of the natural increase in visceral adipose tissues (figure 9.18). The program also induced a decrease in other risk factors, such as total body fatness and insulin resistance.

Dietary Restriction Versus Exercise Therapy

Low energy intake and increased activity can, separately or in tandem, induce loss of weight and fat. The use of diet alone as a reducing regimen is not discussed here. This topic has been amply covered elsewhere. Our purpose is to compare the merits and disadvantages of diet and exercise as treatment modes, as summarized in table 9.2.

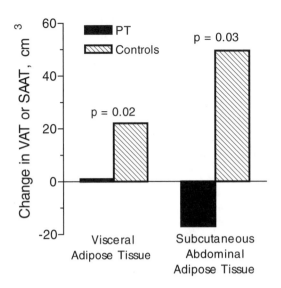

FIGURE 9.18 Effects of an aerobic exercise program on visceral and subcutaneous abdominal fat in 87 obese children, 7 to 11 years old. Fat volume was determined by magnetic resonance imaging. The training program lasted 4 months. Each of five sessions per week lasted 45 min.

Reproduced, with permission, from Gutin and Owens 1999 (117).

TABLE 9.2 A Comparison Between Dietary Restriction and Physical Exercise As Reducing Regimens in Childhood Obesity

Effect	Diet	Exercise
Weight loss	Yes	Yes
Fat loss	Yes	Yes
Fat-free mass	Loss	Gain
Growth retardation	Possibly (extreme diets)	No
Increased fitness	None or some	Yes
Reduction of adipocyte size	Yes	More than in diet (?)
Feeling of hunger	Yes	No
Rate of weight loss	Fast or slow	Slow
Carryover effect	No	No

It is easier to induce a negative energy balance through dieting than through enhanced physical activity. Therefore, short-term low-energy diets are more efficacious than short-term exercise programs in reducing body weight and adiposity. Another advantage of dietary changes is that their effect can be noticed within a few days. The change is considerably slower with use of enhanced physical activity. To avoid disappointment, one must explain the slow effect of exercise to the parents and child prior to the start of a program.

Enhanced physical activity has several advantages over dieting. A major concern in any reducing regimen is that it might interfere with normal growth. Ideally, one would like to achieve a reduction in body fat without a concomitant loss of protein, water, minerals, or vitamins. A very low energy regimen (e.g., less than 1,000 kcal per day), although effective for short-term weight reduction, may impede growth velocity (2; 42; 82; 186). A very low energy diet may induce catabolic effects. These are manifested by a negative nitrogen balance and by the loss of fat-free mass (31). As shown for adults, these effects can be reduced, or even reversed, when exercise is incorporated (142; 217; 224; 281). Such an anabolic effect is especially relevant during growth. However, there is only scant information, based on uncontrolled observations (239), to suggest that fat-free mass is preserved in obese children and adolescents undergoing a combined very low energy diet-plus-exercise regimen. Future research should determine whether resistance training would be more efficacious than aerobic training in preventing fat-free mass loss in the growing person.

As discussed in the later section "Effects Other Than on Body Composition," weight loss through exercise increases fitness level. Weight loss by diet may also improve fitness; but extremely low energy intake, such as in prolonged semistarvation (136), may induce weakness and a reduction in exercise performance. For more details, see the sections on anorexia nervosa and undernutrition earlier in this chapter.

In animals, either a low-energy diet or training may cause a decrease in adipocyte size (181), but training may induce a greater reduction in adipocyte size than diet. The practical implications of this difference are not clear.

Neither exercise nor a dietary regimen is effective if the child does not pursue both on a long-range basis. To some children who are obese, the hunger that accompanies dietary restriction is unbearable, so that even a well-motivated patient may not comply with the treatment. In this respect exercise has the advantage of not being accompanied by hunger.

The moral of all this information is that a low-energy diet and exercise *should be combined* in a reducing regimen. Both should be integrated with behavior modification. Such a combination will maximize the dividends and counteract the deficiencies of either treatment alone.

Effects of Structured Exercise on Spontaneous Physical Activity

Changes in body mass and adiposity often do not tally with the additional energy expenditure provided by an exercise program (171; 253). Such a program may be accompanied by changes in energy intake or other components of energy expenditure. The latter can occur, for example, if the child changes his or her *spontaneous* activity as a result of a regimented intervention program.

A school-based study has shown that, indeed, third to fifth graders decreased their after-school activity by 16% during a 2-year intervention by exercise and nutrition modification (83). In contrast, total daily energy expenditure of 10- to 11-year-old mildly obese boys increased considerably more than the increase expected from a laboratory-based, 4-week cycling program (32) (figure 9.19). Total energy expenditure was measured by doubly labeled water, which gives credence to these findings. Spontaneous activity, as assessed by heart rate monitoring, increased somewhat; and there may have been increases in the thermogenic effect of food and in the postexercise resting metabolic rate.

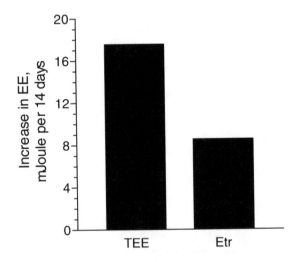

FIGURE 9.19 Total energy expenditure (TEE) increases much more than expected from the energy expenditure of a regimented training program (E_{tr}). Ten 10- to 11-year-old mildly obese boys took part in a 4-week laboratory-based cycling program. Total energy expenditure was assessed by doubly labeled water.

Reproduced, with permission, from Blaak et al. 1992 (32).

One possible reason for the discrepancy in the two studies just discussed may be the intensity of prescribed exercise. As found among moderately obese boys (143), energy expenditure and spontaneous activity on the day of a laboratory-based exercise bout, as well as on the following day, decreased when the exercise was intense (40 min at a heart rate of 150-160 beat · min^{-1}, followed by progressively increasing effort to maximum). In contrast, activity and energy expenditure increased when the laboratory-based exercise was moderate (30 min at a heart rate of 130-140 beat · min^{-1}). Until more research is done in this important area, we are left with the hypothesis that a prescribed intervention of mild-to-moderate intensity may enhance spontaneous activity and increase overall energy expenditure beyond that yielded by the program itself. In contrast, the fatigue following intense exercise may induce a reduction in spontaneous activity.

Enhanced Physical Activity and Changes in Appetite

A question commonly asked by parents is whether the effects of increased activity will not be nullified by a concomitant increase in appetite. It is true that among active people, appetite often increases with a rise in energy expenditure. Children who train extensively in sport consume more calories than nonathletic children (1; 15; 24). However, the positive relationship between the level of activity and appetite does not hold throughout the activity spectrum. Studies with rats (164; 235), a dog (188), and monkeys (9) have indicated that when a highly sedentary animal increases its level of physical activity, appetite may *decrease* rather than increase. In adult humans, cross-sectional (165) and longitudinal (150) studies confirm this pattern.

The same phenomenon has also been described for children. Overweight 8- to 10-year-old boys attended a 4-month program of one or two extra physical education periods weekly. By the end of the program, their daily energy intake had decreased by 12% (from 2,129 to 1,874 kcal per day). Those who had two extra periods weekly had a greater decrease in energy intake than those having only one extra period (35). In some training studies obese children were allowed to eat ad libitum, without any instructions for nutritional modification (114; 182; 215), and their body weight decreased. This does not prove that the subjects' appetite decreased, but it does provide indirect evidence for the lack of a sufficient compensatory increase of eating with training.

The mechanism for the apparent paradox is not clear. There may be a threshold activity level (see figure 9.20) below which appetite is not well regulated and a person eats more than is justified by his or her energy expenditure. Only when this threshold is surpassed does appetite compensate for changes in activity. It has been speculated (164) that from the perspective of evolution, a sedentary lifestyle in humans and in some animals is below the physiologic range for adequate appetite regulation.

Whatever the mechanism, it is evident that changes in appetite do not compensate for increased activity of people who are sedentary and obese. Even for individuals whose food intake does increase with exercise therapy, the end result is a negative energy balance (28; 51).

Effects Other Than on Body Composition

In addition to its effects on adiposity, a conditioning regimen induces other changes, as listed in table 9.3. While some are specific to the individual who is obese, others apply to anyone who undergoes conditioning.

Metabolic changes are evident in carbohydrate, lipid, and protein metabolism alike. Plasma insulin concentration is reduced with conditioning (93): It can be as low as half of that found

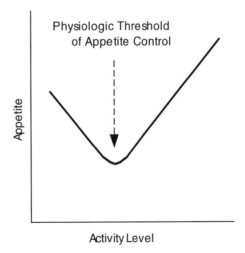

FIGURE 9.20 Appetite and physical activity. A schematic presentation of a concept. See text for details.

TABLE 9.3	Conditioning Effects in the Obese Child, Other Than on Body Composition

Function	Effect*
Metabolic	
Plasma insulin	−
Cell sensitivity to insulin	+
Glucose tolerance	+
Free fatty acid mobilization	+
Lipolysis in tissues	+
Serum triglyceride	−
Total cholesterol	−/0
Low-density lipoproteins	−/0
High-density lipoproteins	0/+
Low-density lipoproteins/Apoprotein B	+
Plasma leptin	−
Nitrogen sparing	+
24-hr energy expenditure	+
Cardiorespiratory	
Heart rate—rest and submaximal	−
Systolic blood pressure—rest	−
Ventilatory equivalent—submaximal	−
O_2 uptake—submaximal	−
O_2 uptake—maximal	+
Psychological and behavioral	
Self-image; body image	+
Rating of perceived exertion	−
Self-confidence	+
Sociability	+
Spontaneous physical activity	+

*− = decrease; 0 = no change; + = increase.

in the inactive individual. This drop, coupled with better glucose tolerance, reflects a greater cell sensitivity to insulin in various tissues (28; 29; 93; 223; 233). Improved sensitivity to insulin can result from weight loss per se. But it is also a result of an actual increase in fitness, irrespective of body weight changes (30), and is highly correlated with improvement in maximal aerobic power (223).

Changes in lipid metabolism include enhanced lipolysis, free fatty acid mobilization from adipose tissue, and a decrease in plasma low-density lipoprotein cholesterol. An increase in high-density lipoprotein cholesterol and a decrease in total cholesterol and triglycerides have been found in

some studies, but not in others (103; 269; 277). No information is available on the long-range carryover effects of conditioning on the lipid profile of children who are obese. In light of the relationship between lipid profile and the risk for coronary heart disease, this question is of paramount importance. Recent data suggest a decrease in plasma leptin levels following a 4-month training program in 7- to 10-year-old obese children (118). This effect occurred irrespective of changes in body fat, which suggests a direct response to the exercise program and the negative energy balance that it induced (113).

The nitrogen-sparing effect of exercise in juvenile obesity has already been discussed. From biochemical analysis and assessment of lean body mass, it appears that conditioning is accompanied by an anabolic process—specifically, protein synthesis (37; 133; 144; 217; 281).

Several *cardiorespiratory changes* occur with training and weight loss. These are compatible with a decreased physiologic strain during exercise. Heart rate at rest and at submaximal exercise drops (187; 233; 267; 277), as does resting arterial blood pressure (205; 267). The decrease in blood pressure is greater in obese adolescents whose blood pressure prior to the treatment is the highest (44). No data are available on changes in stroke volume or cardiac output. Ventilation at submaximal exercise drops (267), especially at work rates above the anaerobic threshold (267). The work of breathing decreases with a reduction in thoracic and abdominal wall mass, which further contributes to the lower O_2 cost of exercise.

Psychosocial changes that occur with weight reduction have been studied mostly in conjunction with dietary restriction. Training-related changes include improvement of self-image and body image (231), self-confidence, and ability to adjust to peer society (191). Hitherto-withdrawn children develop social awareness and an ability to integrate into, and enjoy, group activities. We have seen obese adolescent females who, prior to weight loss, limited their social activities to playing with children much younger than themselves—the adolescent assuming a motherly role. After successfully losing weight, these individuals actively sought company of their own age. Such individuals often take up sport and recreational activities in which they were previously reluctant to participate. They also seem to take better care of their appearance, motivated by their new (smaller size!) clothing and perception of their improved physical appearance.

In one experiment, 10- to 12-year-old obese girls and boys underwent a 6-week program in which two physical education periods were added to the regular two physical education periods at school. Rating of perceived exertion using the Borg scale (38) decreased by some two scale units, at work rates of 50 and 100 W. The inference was that a given task seemed easier to the now-conditioned subjects.

A rewarding turn of events occurs when a child's success in losing weight draws other family members, notably overweight parents, to similar activity programs. Our clinical experience suggests that such a development is good prognosis for further, more prolonged weight loss by the child.

Functional changes that result from enhanced activity in the child who is obese are similar to those found in any child who improves his or her fitness level. Observations made specifically on obese youngsters show an improvement in their maximal aerobic power (35; 185; 187; 225; 244; 277), and O_2 uptake for a given task drops in proportion to weight loss (187; 267). With such an increase in the economy of work, the child's running performance improves (62). The ability of the child to lift his or her own weight, as in pull-ups (57), improves due to the reduction of body weight and increase in muscle strength.

Adipose tissue cellularity may be affected by physical activity. The total mass of adipose tissue is a product of the number of adipocytes (fat cells) and their average mass. Understanding the growth and replication of adipocytes may shed light on the genesis of childhood obesity and aid in its prevention.

There are no data on the effect of physical training on the adipocytes of obese children. There is, however, an elegant study on the long-term effect of training on the adipocytes of rats: Starting on their fifth day of life, groups of rats were exposed to daily swimming or a diet regimen until week 28 of their life. All animals then remained sedentary until week 62. Measurements taken at that stage conclusively show (figure 9.21) that those who had swum during their growth retained a lower body

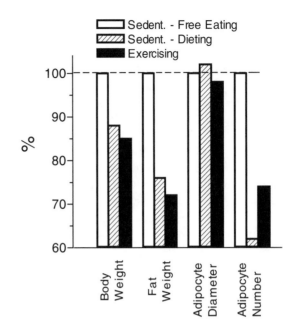

FIGURE 9.21   Long-term effect of exercise or low-energy diet during growth on body mass, body fatness, adipocyte size, and adipocyte number in the mature rat. Values were obtained at week 62 of life and are expressed as a percentage of values obtained in rats that neither exercised nor dieted during growth. Data from Oscai et al. 1974 (181).

mass, fat mass, and number of adipocytes than did the control animals. The adipocyte count was also lower than that of the rats who had dieted (180; 181). This study demonstrates that exercise during growth is effective in reducing the rate of replication of adipocytes in the mature rat and that the effect is long lasting. While such results cannot be extrapolated to humans, they pose a definite challenge for a similar intervention study in children.

Lasting Effects of Enhanced Physical Activity

A question of immense practical importance is whether activity programs have a carryover effect beyond their duration. Early work on this topic was not encouraging. Investigators from Czechoslovakia, for example, reported fat loss and increased work performance in obese children during a 7-week diet and exercise summer camp, but these benefits virtually disappeared when the children returned to their home routine (186). A school-based program was found to be effective for weight control of obese boys. However, during a 3-month vacation, the experimental group gained more weight than obese controls who had not taken part in the program (57). Another successful reducing regimen was carried out in a public school for 4 years (219). In a follow-up survey, conducted 3 years later, the beneficial effects of the program had all but disappeared (163).

These earlier studies show that, with few exceptions (186; 233), the effects of exercise and diet did not carry over beyond the duration of the programs (60). Studies with adults suggest that behavior modification techniques could prolong exercise therapy effects for up to a year beyond the duration of the treatment (122; 230), but it was not until the 1980s that similar research was done on children. The next section provides details.

Motivational and Behavioral Aspects

The key to success in treating people of any age who are obese is to maintain compliance during the intervention program and adherence to changes in lifestyle once the program is completed. Bearing in mind that, to start with, children and youth with obesity are unfit and reluctant to be active, *motivation* becomes the cardinal element in any program. Whereas adults may appreciate the health dividends of a "healthier" lifestyle, children need more concrete and immediate gratification. It is unrealistic to assume, for example, that a child (or even an adolescent) will join an activity program because it may help prevent diabetes mellitus 30 years hence. A more realistic strategy is to use a motivational approach that is extrinsic to the actual health goals. The parent can give a token prize once the child has reached a goal or a milestone. Such goals and milestones should not be based on changes in body weight but rather on activity behaviors. For example, the child can accumulate a certain walking mileage per week, stay outdoors for so many hours per week, or have one "TV-free" day each week. It is very important that milestones be realistic and cover a short time span (not more than 1 month).

In cases in which the services of a qualified behavior therapist are not available, other therapists—such as a nutritionist, kinesiologist, physician—should incorporate a behavioral approach as well. Elements within this approach might include the following:

• **Self-monitoring** of activities. The child keeps an activity log or attaches stickers to a calendar.

• **Positive reinforcement.** The child collects points for minutes of daily activities that, as in a "frequent flyer" program, will be redeemed for an award. A formal "contract" can be signed between child and parent to determine the rules and awards.

• **Modeling.** The child and the parent are taught how to set an example for each other.

• **Contingency contracting.** The parent deposits money (e.g., at the clinic or the school) that is returned to the parent once the child has reached a certain milestone or has attended a certain number of activity sessions. This approach is geared toward parental compliance, but it can also be used to motivate the patients. It is more suitable for adolescents than for young children.

One can take similar approaches regarding nutritional changes. The role of parents and other family members in the child's program cannot be overemphasized (44; 89; 90). In a classical experiment, Brownell and colleagues (44) showed that behavior modification in an activity-plus-diet program was most efficacious when both mother and child participated, but in separate groups (figure 9.22). Moreover, the effect

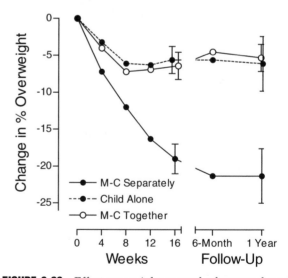

FIGURE 9.22 Effect on weight control of parental participation in a behavior modification program. Subjects were 42 obese girls and boys, 12 to 16 years old, divided into three treatment conditions: child alone, child and mother together, child and mother in separate sessions. Treatment lasted 16 weeks. M = mother; C = child. Vertical lines denote SEM.

Reproduced, with permission, from Brownell et al. 1983 (44).

of this approach was still maintained 1 year after the end of the intervention.

L. Epstein and his colleagues have added particular insights to our understanding of behavioral principles for a successful activity program (86). One important message is that "lifestyle" activities (e.g., walking to and from school, taking the dog out for a walk), as well as activities that include recreational elements, are more efficacious than regimented, supervised programs. Moreover, people adhere to these less structured interventions long after the actual program has been completed (92). More recent observations have shown that reduction in the time spent on sedentary activities such as watching TV, *without any exercise prescription*, is more efficacious than a prescribed activity (regimented or lifestyle) (91). To facilitate this, the parent together with the child should set up a reward system.

Epstein's observations were made in a research setting, with select groups of motivated subjects and cooperative families. More experience is needed to tell whether this approach is also applicable to a real-life situation in a clinic, club, or school. The initial impression in our clinic is that a reduction of TV viewing time is indeed effective for most patients, and is quite easy to adhere to.

In conclusion, one must remember that any intervention program that does not include changes in attitudes and behavior, of the child and family alike, is not likely to be effective. Nor will children adhere to changes in activity and nutrition without behavior modification.

Criteria for Success of Weight Control

How does one determine whether a weight control program has been successful? Changes in body weight may themselves serve as a useful gauge in adults, but not in a person who is still growing. A child, for example, can gain weight yet still slim down. One obviously needs to consider also changes in body height and, ideally, body composition. Figure 9.23 summarizes three

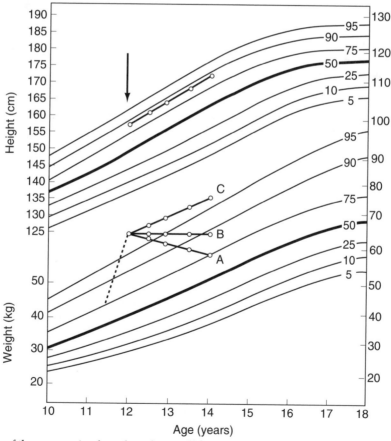

FIGURE 9.23 Illustration of three scenarios, based on changes in body weight and height, that denote success in a weight control program. The arrow indicates the start of the program. The charts are based on those developed by the U.S. National Center for Health Statistics. For details, see text.

scenarios based on changes in body weight and height. Line A reflects an ideal scenario in which height increases and weight decreases. However, line B shows a good response as well, because the child maintains a constant body weight while growing. Does line C reflect a good response? Based on values obtained during the intervention program, it does not seem to. However, if one bears in mind the growth data obtained prior to the start of the program, the rate of increase in weight has slowed down during the program, which indicates a positive response.

While changes in weight versus height are useful gauges of success, one should pay special attention to changes in lifestyle behavior, attitude, and mood. For long-term success these may be more important than changes in weight or adiposity.

Recommended Activities Within a Weight Control Program

Research so far has not provided us with sufficient know-how for optimizing physical activity, prescribed separately or together with dietary changes. This section is therefore based mostly on experience accumulated in the author's (O. B.-O.) Children's Exercise & Nutrition Centre.

Table 9.4 lists the most important elements to be considered in an activity program. Three principles should guide the selection of activities for the obese child. A program should be *effective* for fat reduction, *feasible* for the child and family, and *fun*.

Effectiveness

Activities for fat reduction should emphasize energy expenditure, which is a function of *work* rather than power. As large a body mass as possible, preferably the whole body, should be transported from one place to another—as in walking, running, cycling, dancing, skating, cross-country skiing, or swimming. Activities that are less suitable for weight control are, for example, strength training, gymnastics, or downhill skiing. These involve a high power output, but the overall energy expenditure is not high because the actual time of sustaining the effort is brief.

In walking or running, it is distance rather than speed that determines overall energy expenditure. Thus, one should emphasize to the child and parents that walking 1 km is almost as effective as running 1 km, the only difference being that walking takes more time. As with diet, activities can be exchanged based on their energy equivalent. Table 9.5 shows examples of activities that require 200 kcal (840 kJ) when performed by a 40-kg child. The heavier the child, the greater is the energy expenditure in a given task. This is especially true for tasks that involve transporting the whole body. Appendix IV presents a more comprehensive list of activities and their energy equivalents for children whose body weight ranges between 20 and 60 kg. Such a list can be used by the practitioner for exercise prescription.

To maximize weight reduction by exercise, one should strive to yield high-energy expenditure for any given task. An interesting question is whether a particular exercise yields the same energy expenditure at different times of the day. Circadian rhythms of energy expenditure have been reported for adult obese women, who expended some 20% more energy in afternoons and evenings than in the mornings while performing an identical cycling task.

TABLE 9.4	Components of an Optimal Enhanced Activity Program for Juvenile Obesity
Use large muscle groups (to achieve high-energy expenditure).
Move the whole body over distance (e.g., walking, skating, dancing, swimming).
Emphasize duration; deemphasize intensity.
Aim at energy equivalent of 10% to 15% total daily expenditure (e.g., 200-300 kcal) per session.
Include resistance training (particularly if program includes low-energy dieting).
Gradually increase frequency and volume (strive for daily activity, 30-45 min per day).
Consider the child's preference for activities.
Emphasize water-based sports and games.
Consider lifestyle changes, not merely regimented activities.
Let parents contract with the child to reduce sedentary activities (e.g., TV).
Build in token remuneration.
Include other obese children in group activities (this reduces inhibition).
Remember, key to compliance is *fun, fun, fun!*

TABLE 9.5	Time Required by a 40-Kilogram Child to Expend 200 Kilocalories (840 Kilojoules) During Various Activities	
Activity		**Duration (min)**
Cross-country skiing (leisure)		50
Cycling—10 km · hr^{-1}		80
Ice hockey (on ice)		20
Running—8 km · hr^{-1}		30
Running—10 km · hr^{-1}		25
Soccer (game)		30
Swimming (breaststroke, 30 m · min^{-1})		50
Tennis		45
Walking—4 km · hr^{-1}		80
Walking—6 km · hr^{-1}		60

These differences were above and beyond those found in their resting metabolic rate (278). It is not clear whether similar circadian rhythms exist in children and adolescents.

Feasibility

Not all sports are equally feasible for the obese child. Some "penalize" obese persons because of their physique and a priori put them at a disadvantage. For example, in jumping, rope climbing, or maneuvers on the parallel bars, excess body weight is a distinct handicap (see also figures 9.14 and 9.15). To prevent embarrassment and disappointment, such activities should not be prescribed.

Some of the physical characteristics of children and adolescents with obesity are actually advantageous for sport. These individuals are often taller and have a large lean body mass, which is reflected by high absolute strength. Their high body fat increases their buoyancy in water and provides subcutaneous thermal insulation. These characteristics may enhance their performance in such activities as shot put, discus, some positions in football and basketball, and especially in swimming and water polo.

Many adolescents who are obese choose to swim not only because of their high buoyancy and thermal comfort, but also because their body is submerged and not exposed to the watchful eyes of their peers. We have seen, for example, an obese 13-year-old boy who refused to take part in physical education classes and other sports because his "large breasts" could be seen when he wore a T-shirt. This boy happily joined a swimming program. A year later, following fat reduction, he joined the football team at school.

Ideally, one would like to integrate children who are obese into the physical activities of their non-obese peers. This approach unfortunately has not been found to be successful, because obese children are often rejected by other children. More success has been shown in programs in which all the participants were obese. It seems that, as with adults, obese children or adolescents accept each other quite readily and are less inhibited than when in the company of the non-obese (191; 218). Programs exclusively for obese children have been tried, with some success, in summer camp (187; 191; 225; 267; 269) and in hospitals (230) and schools (35; 39; 57; 171).

Fun

To increase motivation and compliance, activities must be enjoyable. Those preferred by obese children and adolescents include sailing, dancing, horseback riding, archery, ice-skating, cycling, and, particularly, water-based games. Such sports share a number of common features: (1) They usually involve the individual rather than a team, thus avoiding peer disapproval; (2) they are recreational rather than competitive, which is particularly important to adolescent females (191); (3) in each of these sports, the intensity and overall amount can be individually tailored according to inclination and ability. This is especially critical at the start of a new program, when the emphasis should be put on gradual progression. Dance-oriented exercise has been found to be highly successful with people of all ages who are obese. If performed correctly it can induce high levels of energy expenditure combined with pleasant recreation. It can be done within a group, but also in the privacy of the home (153).

Use of Rating of Perceived Exertion to Prescribe Exercise Intensity

Even though activities need not be intense in weight control programs, one sometimes needs to explain to the child how to select a certain intensity. One approach, practiced by athletes, is to monitor heart rate. However, palpation of pulse (e.g., carotid) during exercise requires skill and, in our experience, is unreliable and impractical when used by children. With the popularization of electronic heart rate monitoring, one can now easily obtain reliable and valid information on heart rate, but the costs preclude the use of these devices by most families.

An alternative approach is to use rating of perceived exertion (RPE). For details, see the section "Rating of Perceived Exertion" in appendix II. Briefly, one should first teach the child the concept of RPE and let the child experience what it feels like to exercise at several intensities on an RPE scale. Our own experience has been with the 6 to 20 Borg category scale (38). This can be administered in the laboratory during a multistage exercise test, or outdoors where the therapist performs with the child activities that require a wide range of efforts (e.g., walking and running at various speeds). Once the child has learned and experienced the notion of RPE, he or she is ready for the actual prescription stage, that is, to receive instruction to perform activities at given intensities based on RPE. One might say, for example, "Walk for 5 minutes at number 8 on the scale and then run for 5 minutes at number 12, and then increase your running speed until it

feels like number 15 and keep it up for 2 minutes." Once they go through this procedure, children and adolescents can discriminate among various cycling or running tasks as prescribed by RPE (figure 9.24).

Role of the School in Prevention and Treatment of Juvenile Obesity

Within what framework should an enhanced activity program for obese children be conducted? Much of the practical experience about managing juvenile obesity has been gleaned from clinics and summer camps. However, while medical services provide expertise and sophisticated equipment, their use is expensive and they can reach only a small fraction of the obese population, or of those at risk of becoming obese. The disadvantage of summer camps is that they present an environment that cannot be sustained year-round. As a result, dividends gained in the camp are short-lived.

With the current epidemic of juvenile obesity, one should seek alternatives that provide large-scale, yet efficacious, preventive and management programs. A major player for such a role is the school. Schools are particularly suited to accommodate large-scale juvenile obesity programs (table 9.6).

Because 95% of North American children ages 5 to 17 attend school, school-based programs can potentially reach most juveniles with obesity and the at-risk population. Contact with students occurs almost daily for 8 to 9 months of the year, and there are proven, streamlined means of communication with parents. While expertise varies among schools, many have specialists in physical education and health education, as well as counselors. Some schools, especially in European countries, have a resident nurse. With their ability to conduct activity programs and to monitor anthropometric changes, physical educators can play a unique role. They can identify children who are obese, as well as those who gain weight excessively and are at risk of becoming obese.

Because schools often have food services that provide lunch, they are in an excellent position to control the composition of this important meal. Furthermore, the availability of a kitchen can provide a valuable hands-on educational environment regarding food selection and cooking.

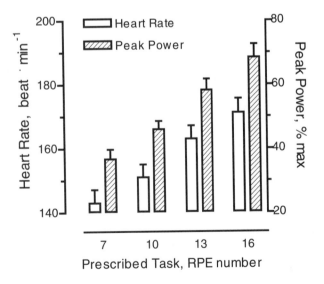

FIGURE 9.24 Ability of children and adolescents with obesity to discriminate among exercise intensities prescribed to them as numbers on the Borg 6 to 20 category scale. Data are for 20 girls and boys, 7 to 15 years old, with mild to moderate obesity. Based on Ward and Bar-Or 1990 (259).

TABLE 9.6	Characteristics of Schools That Are Conducive to Effective Prevention and Management of Juvenile Obesity
Pooling of obese children and adolescents	
Pooling of at-risk children and adolescents	
Pooling of experts (nurse, physical and health educator, counselor, nutritional services)	
Available volunteer role models among the students	
Available indoor and outdoor sport facilities	
Daily contact with students most months of the year	
Effective lines of communication with parents	
No "medical" stigma	

Adapted from Bar-Or 1995 (12).

Most schools have indoor and outdoor sport facilities and equipment, geared for use by children. These are often underutilized after school hours and could provide excellent venues for obesity-related programs.

A streamlined means of communication with the parents is important, both in order to inform parents about their child's progress and to involve them in the program. Parental involvement is often (43; 137), though not always (154), instrumental to success.

How efficacious are school-based programs? Studies addressing this question have been summarized in several reviews (184; 199; 209; 258). In general, programs that combined physical activity and a nutrition education program had a greater impact on body weight and adiposity than did those using a single intervention modality. This effect was sustained as long as the intervention was in effect, but very seldom (279) did it last beyond the duration of the program. The addition of behavior modification (ideally, for both children and parents) seems efficacious (43), but its feasibility for large-scale programs is yet to be tested.

While most programs attempted to induce a reduction in adiposity, others focused on changes in attitudes and health behaviors. Noteworthy among these is the CATCH (Child and Adolescent Trial for Cardiovascular Health) project. This largest-ever randomized, controlled trial comprised a 3-year intervention that started in grade three at 56 schools in California, Louisiana, Minnesota, and Texas (154). The program included enhanced physical education, school food service modifications, and classroom health curricula. Some schools also conducted family education. Major outcomes included an increase in the intensity of activities during physical education classes, more time spent on daily vigorous activities, and a reduced fat intake during school lunch and other meals. There was no change, however, in adiposity or blood lipids. The addition of family education did not modify the outcomes.

In a smaller-scale 2-year project (83), physical activity in the classroom and high-density lipoprotein cholesterol increased, and there was a reduction in energy and fat content of lunches. However, reported activity outside of school decreased, and energy intake outside of school increased. This compensatory behavior may explain why the intervention did not affect body adiposity.

A recent randomized controlled study demonstrated a unique potential for school-based programs: Ninety-two third and fourth graders in one school received a classroom-based 7-month behavior modification program aimed at reducing the time spent at home on TV, video games, and videotapes. One hundred students in another school served as controls (203). The intervention group indeed decreased the time spent watching TV and the number of meals eaten while watching. Compared with the controls, they responded with significantly lower increases in BMI, triceps skinfold thickness, waist circumference, and waist-to-hip ratio.

Undernutrition

In spite of a global increase in the prevalence of juvenile obesity, many children worldwide are still exposed to *insufficient food intake*. Undernutrition, especially the protein-energy type, is a childhood disorder highly prevalent in various geographic areas and is by far the most common cause of growth failure during infancy and childhood. Undernutrition can occur also when a specific element (e.g., a vitamin or mineral) is deficient. Discussion of such specific deficiencies is beyond the scope of this book. We focus here on the protein-energy type.

There are several categories of protein-energy undernutrition, based on measurement of weight and height (or length). The three categories suggested by the World Health Organization are as follows:

- **Underweight:** weight for age >2 SD below the 50th percentile of the U.S. National Center of Health Statistics/WHO reference
- **Stunting:** height (length) for age >2 SD below the same 50th percentile
- **Wasting:** weight for height <2 SD below the same 50th percentile

The exact prevalence of undernutrition is not known, but in developing countries it seems to exceed 40% of children under 5 years of age (251). The greatest percentage of undernourished children in this age group has been documented in Africa and Southeast Asia. Based on World Health Organization data, prevalence in the *underweight* category was 27.8% in 1995. The respective prevalence was 34.9% for the *stunted* group and 8.4% for the *wasted* group (75).

While undernutrition refers to dietary deficiencies, one should bear in mind that it is often combined with parasitic and other infectious diseases, as well as with poverty and specific societal/cultural customs. These may affect a child's health, lifestyle, and behavior. It is therefore quite impossible to tease out the effects of dietary deficiency per se. Much exercise-related research has focused on *marginal undernutrition*, which denotes weight for age and for height of less than 95% of expected (228).

Habitual Activity and Energy Expenditure

In a classic study by Keys and colleagues (136), adults who were put on a starvation regimen reduced their spontaneous activity. For obvious ethical reasons, such an intervention protocol cannot be conducted with children or infants. Several observational studies, however, have shown that infants and children with protein-energy undernutrition are less active than their well-nourished peers. Based on direct time-and-motion observations, the activity level of stunted (height <2SD below expected) 12- to 24-month-old Jamaican infants was lower than that of their nonstunted controls (165a). The dif-

ference was particularly apparent in moderate-to-vigorous activities. Using heart rate monitoring, the calculated daily energy expenditure of 6- to 16-year-old Colombian children with marginal undernutrition was lower than in well-nourished controls (figure 9.25). A similar difference was observed for activity-related energy expenditure (228). The difference in energy expenditure was further magnified when the children attended a summer day camp, in which all participants were encouraged to take part in sport and active games (227).

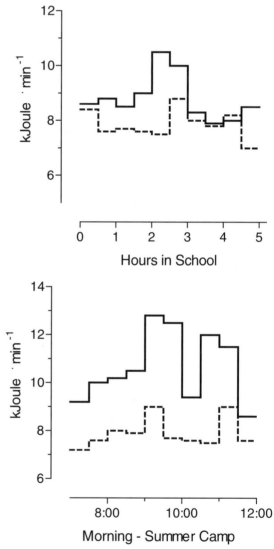

FIGURE 9.25 Average energy expenditure over 30-min periods, assessed through minute-by-minute heart rate monitoring, during a morning in school (top) and in a day camp. Subjects were 19 marginally undernourished (broken line) and 14 nutritionally normal (solid line) 10- to 12-year-old boys.

Adapted, with permission, from Spurr and Reina 1988 (227).

One of the most important sources on activity and undernutrition is a longitudinal study of two groups of Mexican infants who were followed from birth until 2 years of age. In one group, both mother and child received a supplement to their food, while the other group followed the regular diet consumed by low-socioeconomic families in a poor rural area. Physical activity was monitored by direct observation and by interviewing the mothers. The supplemented babies increased their activity level with age, while the less nourished ones had a consistently lower activity level. The difference between the groups grew with time so that, by the age of 24 months, the well-nourished children were six times as active as the undernourished ones (56). In the same study, the undernourished infants spent less time outdoors and were carried more frequently on their mothers' backs. For details, see figure 9.26.

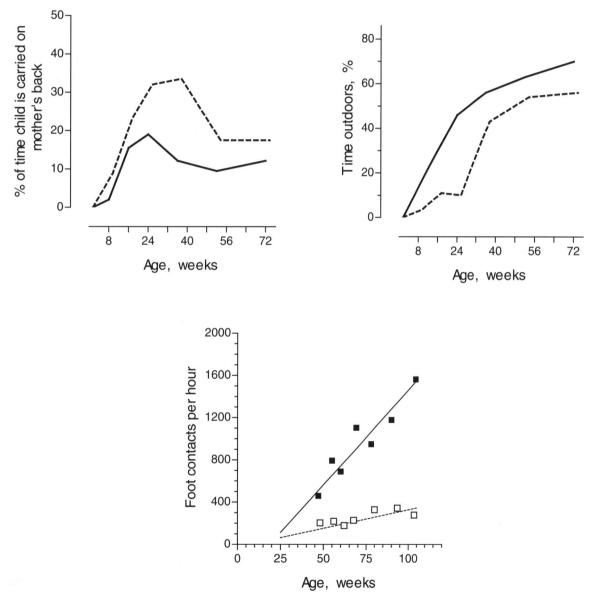

FIGURE 9.26   Longitudinal follow-up of activity patterns in two groups of infants from rural Mexico. In one group (broken line and ■), the pregnant mothers received 350 kcal per day supplements from day 45 of pregnancy. Their infants started being supplemented from the third or fourth month of life in amounts that would make their diet comparable to that of an urban family. The other group (solid line and □) received no supplements. The bottom graph displays the number of foot contacts with a surface of support (e.g., crib, bed, ground) as directly observed every 10 min.

Adapted, with permission, from Chavez and Martinez 1984 (56).

Chavez and Martinez (56) concluded that study with the statement that "malnutrition depresses activity which, in turn, isolates the individual from contact with the environment, from necessary interaction with the mother and the family, and from all sources of stimuli that are of a vital importance to the functional development of the brain" (p. 319). A similar hypothesis, put forth in 1971 by Heywood et al. (124), suggests that the subpar intellectual performance of undernourished, or previously undernourished, children is induced by restricted activity and the resulting limited stimuli and learning opportunities. Because of the immense numbers of undernourished children worldwide, such a hypothesis deserves a thorough test and should be a challenge to anthropologists, nutritionists, and exercise physiologists.

In conclusion, aberrations in nutritional status seem to affect the activity behavior of infants, children, and adolescents. Obese as well as undernourished children are less active than their well-nourished but lean counterparts (figure 9.27).

Exercise Performance and Growth

Chronic undernutrition during infancy and early childhood often results in growth failure. This is particularly evident after age 6 to 9 months, when the mother's milk becomes insufficient. In the absence of nutritional supplementation, growth velocity remains slow throughout childhood and adolescence. In addition to stunted growth, muscle mass and lean body mass develop

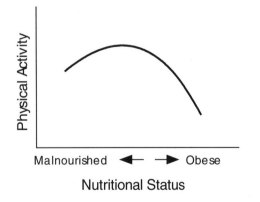

FIGURE 9.27 Habitual activity in relation to the level of nutrition in children and infants. A schematic representation.

slowly. The severely malnourished child may display edema, flabby muscles, and postural abnormalities.

Will subpar growth and muscle development affect the physical working capacity of an undernourished child? Based on simple physical principles, a small body size should be a handicap for fitness components that depend on height or absolute muscle mass. For example, when a small and a large child are to push an equal resistance (e.g., a wheelbarrow), throw an object (put a shot or pitch a baseball), or elevate the body (e.g., high jump, volleyball), the small child will be at a disadvantage. The same goes for more prolonged activities, in which a given task (e.g., rowing) will be sustained longer by a larger individual. Indeed, various studies have shown that the small body size of undernourished children and adolescents is the best predictor of their low exercise capacity and deficient motor performance. As might be expected, most data on this important topic are derived from populations in underdeveloped countries.

Aerobic Performance

Whether measured as maximal mechanical power, maximal O_2 uptake, W_{170}, or performance in distance running, aerobic performance is often low in children and youth who are undernourished or whose growth has been stunted due to earlier undernutrition (5; 36; 66; 226; 229). A study of 6- to 16-year-old Colombian girls and boys (229) showed that absolute (liters per minute) maximal O_2 uptake was smaller in those who were marginally undernourished than in their well-nourished counterparts. However, when the calculation was per kilogram body mass, the difference between the groups disappeared. A similar pattern was shown for Ethiopian (5) and other East African (66) groups. In some studies (36; 226), maximal O_2 uptake, relative to body mass, remained lower in undernourished children (6-year-olds) than in well-nourished ones. In severely undernourished adults (serum albumin <2.5 g · dL^{-1}, creatinine excretion <450 mg · L^{-1} · hr^{-1} per square meter skin area), the deficiency in maximal aerobic power was greater than could be explained merely by body mass or muscle cell mass (16).

One may conclude that, in most cases, the low aerobic performance of an undernourished child is directly related to reduced body mass and, specifically, the lean mass. In especially severe levels of undernutrition, low maximal aerobic power may be related also to a deficient O_2 transport system, which, for example, may accompany iron deficiency anemia. This aspect is discussed in chapter 11.

Which form of undernutrition is most detrimental to physical working capacity? Due to their retrospective nature, studies on humans cannot easily be designed to compare specific types of undernutrition. It seems, though, that the major type of undernutrition that induces stunted growth and reduced working capacity is a combined protein-energy undernutrition. In a study on mature rats (121), endurance time of swimming was found to be *higher* in those given a 10-week low-protein diet than in those given a normal diet. This seeming paradox may be explained by greater starch consumption in the protein-deficient animals, which may have increased the glycogen content of their muscle, thus improving endurance time. A similar phenomenon has been suggested for humans (16).

Muscle Strength and Anaerobic Performance

While most research on the effect of undernutrition on physiologic fitness has focused on aerobic performance, some studies suggest that muscle strength, peak anaerobic power, and muscle endurance are also affected. For example, prepubescent girls of a low socioeconomic class in Bolivia, classified as having marginal undernutrition, had a lower peak power (force-velocity test) and mean power (Wingate test) than did their higher-socioeconomic counterparts. This deficiency prevailed when power was expressed per body mass (36), which suggests that there are *qualitative differences* in the muscles or in neuromotor control. For example, in the same Bolivian girls (21) there was a relationship between peak power and the level of insulin-like growth factor, which is known to stimulate glycolysis.

As stated in the section on anorexia nervosa, muscle fatigability increases with undernutrition and decreases with nutritional supplementation. In a study of children and adolescents with cystic fibrosis, we found that nutritional status was one of the predictors of force, contractility, and fatigability of larger muscle groups (120).

Motor Performance

Studies by anthropologists and exercise scientists have documented deficiency in motor development (159) and in motor performance (23) of children with current or previous malnutrition.

Motor tasks that have been found to be deficient include sprinting, agility, throwing of an object for distance, and long jump. Deficiency has also been documented in neuromotor coordination components such as catching a ball, hopping on one foot, standing on tiptoes, and performing an accuracy throw. Once the data are corrected for body size and age, the motor tasks do not differ among various nutritional groups, but coordination components remain deficient in the undernourished children (23). It is not clear how undernutrition affects coordination, but there are indications that children with severe undernutrition have a slower nerve conduction velocity and that this deficit remains at least during the preschool years (147). Rat studies have shown a delay in nerve myelinization as a result of undernutrition.

There are limited data from developed countries on the effects of undernutrition on physical working capacity in young people. A study performed at the Wingate Institute in Israel compared fitness of male high school graduates who differed in adiposity. When the 50 leanest boys in a population of 2,000 graduates were compared to 50 boys with average adiposity, the former had as good a performance in 100- and 2,000-m runs and an even better performance in pull-ups and "dips." As expected, the absolute maximal O_2 uptake of the thin boys was lower, but not when expressed per kilogram of body mass. A similar trend was observed in a comparison of the physical performance of Italian preschool children from the south with better-nourished preschool children from the center of the country (94). These data suggest that low body weight and marked leanness among children and adolescents who are generally well nourished are not per se detrimental to working capacity. Furthermore, adolescents with average adiposity within a well-nourished population are probably *overnourished*, when physical working capacity is the criterion.

Effects of Nutritional Supplementation on Fitness

The question whether nutritional rehabilitation in later years can improve the physical working capacity of the previously undernourished child is of clinical and public health importance. Figure 9.28 summarizes data on 6-year-old Colombian children (226). Group A had been undernourished at age 3 and beyond. Group B had been undernourished at age 3 and since that time had been nutritionally rehabilitated. Group C had been well nourished throughout their first 6 years of life. In spite of nutritional rehabilitation, Group B still had stunted growth and reduced W_{170}. Other observations (214) also indicate that the eventual maximal aerobic power of the previously undernourished child depends on whether his or her body size can reach normal standards. A similar improvement following nutrition supplementation has been observed for patients with AN (see earlier section).

It should be realized, though, that undernutrition-induced low performance in early age is not fully reversible with subsequent nutritional supplementation. Studies on rats (195; 196) show that in spite of supplementation, animals that had been undernourished in the first weeks of life still had deficient muscle endurance and glycolytic activity (figure 9.29), as well as reduced endurance time for swimming.

Effects of Training on Fitness and on Growth

To what extent can the physical working capacity of nutritionally deficient children be improved by physical training? There are no available intervention studies in which undernourished children underwent a training regimen. There is, however, evidence that undernourished boys who are physically active have better exercise performance than their less active undernourished peers (5; 214). Undernourished rats, when given a swimming training program, showed a distinct improvement in muscle endurance and in swimming endurance time (196). No concomitant increase was observed in their body weight.

From the practical point of view, these data suggest that the exercise performance of previously undernourished children can be improved by training even if their body dimensions are still diminished.

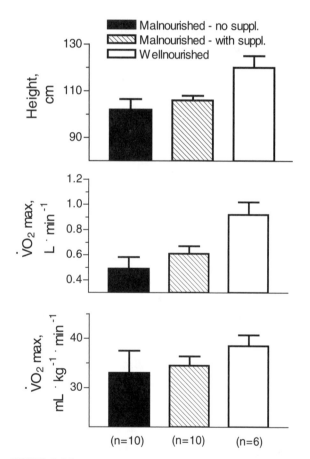

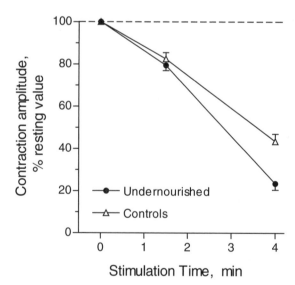

FIGURE 9.28 Nutrition and working capacity. Body height and maximal aerobic power ($\dot{V}O_2$max) of 6-year-old Colombian children. Group A had been undernourished at age 3 and not been given nutritional supplement since then. Group B had been undernourished at age 3 and then been given a nutritional supplement, health care, and psycho-educational stimulation in a special school. Group C had been well nourished throughout their childhood.

Data from Spurr et al. 1978 (226).

FIGURE 9.29 Muscle fatigability and nutritional history. The graph displays the reduction in amplitude of repeated contractions of the *in vitro* gastrocnemius of mature rats. One group ("undernourished") had been given a reduced diet for the first 13 weeks of life and then was nutritionally rehabilitated for 15 to 20 weeks. The other group ("controls") had been given a normal diet. Values are expressed as percent of the initial contraction amplitude. Mean ± 1 SEM.

Data from Raju 1974 (195).

Neuromuscular and Musculoskeletal Diseases

By definition, physical activity involves skeletal muscle action, which is initiated and controlled by the nervous system. Muscle action, in turn, applies mechanical forces to a bone through its tendon or ligament. The integrity of the neuromuscular and musculoskeletal systems is paramount, therefore, for successful and effective physical performance.

Several pediatric diseases affect muscles, nerves, bones, joints, or a combination of these. Children with these diseases often are physiologically impaired and have some physical disability. In some diseases, Duchenne muscular dystrophy for example, the disability progresses as the child grows. This may induce extreme limitation of physical activity and further deconditioning. In other diseases such as epilepsy, the disability usually does not progress and the child may sustain a reasonable activity and fitness level over the years. In yet other conditions, such as juvenile idiopathic arthritis, the disease may enter full remission and the child may regain a normal activity pattern and fitness level. Understanding the interactions among the disease, the child's activity behavior, and his or her physical performance is therefore most important.

The intent of this chapter is to highlight the more prevalent pediatric neuromuscular and musculoskeletal diseases and analyze such interactions. The chapter also addresses the physiologically and clinically relevant benefits that the child may reap from enhanced physical activity or from training of a specific fitness component. Whenever relevant, we discuss potentially detrimental effects of exercise and how to prevent them.

Pulmonary and cardiovascular diseases, discussed in chapters 6 and 7, affect primarily the child's O_2 transport capacity, which in turn impedes aerobic performance. In contrast, neuromuscular and musculoskeletal diseases affect mostly the performance of the muscles, as manifested by reduced strength, peak anaerobic power, local muscle endurance, and neuromuscular coordination. In the following discussion we therefore give special attention to these fitness components.

Cerebral Palsy

Cerebral palsy (CP) is a nonprogressive encephalopathy that affects posture and movement, mostly locomotion. It occurs in 4 out of 1,000 live births and is associated with gestational and perinatal insults to the central nervous system (e.g., maternal infection and perinatal asphyxia, as well as cerebral hemorrhage and periventricular leukomalacia in those born at less than 1,000 g weight). The disease is classified according to the limbs affected (e.g., diplegia, hemiplegia, quadriplegia), as well as the type of motor abnormalities (e.g., spasticity, athetosis, ataxia).

Habitual Physical Activity and Energy Expenditure

One of the criteria for success in the rehabilitation of the physically handicapped child is the achievement of verticalization and walking (231). In the child with CP, a great deal of surgical, physiotherapeutic, pharmacological, and educational effort is invested in the early years to achieve these goals. The end result is that, with or without the aid of crutches, orthoses, or braces, the child can often be helped to walk and perform other useful motor activities. It has therefore been a frustrating clinical experience that such success is often short-lived. When these children reach adolescence they often become progressively reluctant to walk, and eventually regress to wheelchair status. A case in point is a group of adolescents who were followed up for 1 to 2 years. Their heart rate at a given power became higher with age, at a rate of 10 beat · min⁻¹ per year (149). This suggests progressive deconditioning and deterioration of cardiovascular adaptation to exercise. Superimposed on the high metabolic cost of ambulation (159; 171; 245), these are soon followed by deterioration in mobility. While assistive devices such as electric wheelchairs, ramps, and hydraulic beds are a convenience, they reduce the patient's reliance on muscle work and induce further deconditioning. A similar regression in ambulation occurs in adolescents with other motor disabilities, such as myelomeningocele, muscular atrophy, and muscular dystrophy.

A quantitative assessment of activity among patients with CP was first made by heart rate telemetry (27). Energy expenditure ranged from 950 to 3,500 kcal · 24 hr⁻¹, obese and spastic patients being considerably less active (and eating less) than the leaner patients and those with dyskinesia. Clinical experience shows that, indeed, severely affected children with CP are considerably less active than their healthy peers. It is also clear that habitual activities (including regular physical education classes) that are not administered within special supervised programs are seldom intense enough to induce conditioning changes (26).

Using the doubly labeled water technique (249), total 24-hr energy expenditure of five girls and five boys (8.0 ± 1.4 years) with spastic CP was compared to that of matched controls. Sleeping metabolic rate was monitored in a respiration chamber. Even though only one of the CP subjects used a wheelchair, their mean 24-hr energy expenditure was 21% lower than in the controls (7.0 vs. 8.5 mJoule per day). Likewise, the ratio of total to sleeping energy expenditure, taken as an index of physical activity, was significantly lower in the patients (1.56 vs. 1.95) (figure 10.1). In point of fact, for the patients this ratio was similar during free-living conditions and during a full-day stay in the respiration chamber, which further suggests that they used few of the opportunities for physical activities that were available to them. A low total-to-basal energy expenditure ratio was also found among nonambulatory adolescents with CP, but there was no difference between the ambulatory patients and their healthy controls (14).

More recent observations have confirmed and extended the evidence that children with CP are hypoactive compared with their able-bodied peers, and that their activity level deteriorates during adolescence. In a province-wide survey in Ontario, 6- to 20-year-old patients and their families responded to a questionnaire regarding activity behavior and attitudes (146; 147). Based on a point score, the great majority of them were classified as "sedentary" or "moderately active." Eighty percent of the patients felt that they were

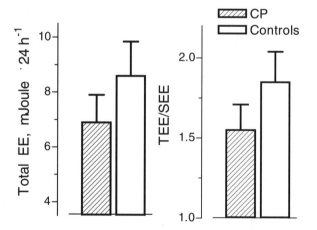

FIGURE 10.1 Daily energy expenditure is low in children with cerebral palsy. Total energy expenditure (TEE) in 24 hr and sleeping energy expenditure (SEE) were measured in 10 children with CP (ages 8.0 ± 1.4 years) and 9 healthy controls (8.4 ± 1.0 years), using the doubly labeled water technique and respiratory chamber, respectively. The ratio TEE/SEE reflects the energy expenditure of activity. Vertical lines denote 1 SD. Based on data by Van Den Berg-Emons et al. 1995 (249).

limited in the amount of physical activity that they could do. Half of the respondents perceived their fitness level to be lower than in their peers, while only 10% thought they were fitter than their peers.

In the past it was commonly believed that because skeletal muscles remain spastic even during sleep and rest, the metabolic rate of a child with CP would be high even when he or she is not active. This, however, has been proven wrong. While some authors (249) found no difference in sleeping metabolic rate between patients with CP and able-bodied controls, others (14) have actually reported lower than normal resting or basal metabolic rates among the patients.

Physiologic Fitness

The child with CP, whether spastic, athetotic, or a combination thereof, is limited in his or her physical abilities. Muscle spasticity is often accompanied by low muscle strength, muscle endurance, peak power, and stamina; but patients with spasticity can perform fine movements better than children with athetosis. The latter can better perform activities that require strength, and they have a greater walking ability (108). Children with mixed spastic-athetotic manifestations are the most limited. This section focuses on specific components of physiologic fitness and on the relationship between fitness deficiencies and the limitation in motor ability of young patients with CP.

Aerobic Fitness

Directly measured maximal aerobic power of children and adolescents with CP is 10% to 30% lower than in controls (20; 29; 105; 152; 153). Figure 10.2 displays considerably lower individual maximal O_2 uptake of patients compared with able-bodied controls.

When maximal aerobic power is indirectly assessed from submaximal heart rate, children with CP score only some 50% of normal (61; 62; 149). Submaximal heart rate is disproportionately high in children with CP because of their low economy of movement, apparently because of spasticity. Peak heart rate, on the other hand, is low (20; 105; 248). The use of heart rate to predict maximal aerobic power, therefore, is not recommended.

An alternative index of maximal aerobic power that does not require expired gas collection is *the highest mechanical power* achieved through a progres-

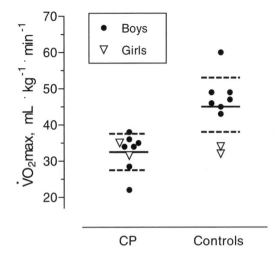

FIGURE 10.2   Individual maximal O_2 uptake of 10- to 16-year-old adolescents with cerebral palsy and able-bodied controls who walked on a treadmill. The solid lines denote mean and the broken lines denote 1 SD.

Reproduced, with permission, from Hoofwijk et al. 1995 (105).

sive "all-out" protocol (e.g., McMaster Progressive Continuous Cycling Test, appendix II). Used with CP patients in one study, this test has been found to be feasible and reliable (250). The patients in that study, who had spastic diplegia or quadriplegia, scored approximately 50% of expected when tested with the leg version of the protocol (250).

It is unlikely that the low aerobic fitness of patients with CP is due to a deficiency in their oxygen transport system. A more plausible explanation is that the skeletal muscles cannot generate enough metabolic drive to fully tax the cardiorespiratory system; combined with physical hypoactivity, this results in low aerobic fitness.

Anaerobic Fitness

Anaerobic performance, manifested by short-term muscle endurance and peak muscle power, is markedly deficient in most children and adolescents with CP (184; 185; 242). Figure 10.3 displays the mean power during the 30-sec arm cranking Wingate anaerobic test, which is an index of muscle endurance, in patients with spastic CP (183). All 27 patients scored 2 to 6 SD below the mean for a comparison group of able-bodied children. Performance was particularly low among the patients with hemiplegia and quadriplegia compared to those with diplegia. With such marked differences between the anaerobic power of children with CP and their healthy peers, it is not useful for exercise laboratories and clinicians to relate the performance of a CP patient to normative data.

The mechanism for low anaerobic performance in children with CP is unclear. One possibility is that because of spasticity and low coordination between antagonist muscles (24; 244), such patients exert extra energy that is not registered as mechanical work during pedaling or arm cranking. For more details, see the later section on economy of movement. The opposing forces by antagonist muscles seem to further increase with the rise in movement velocity (70). Moreover, patients with CP are particularly limited in their ability to produce muscle contractions at high frequencies (176), which is the required task in the Wingate test. Finally, muscle spasticity may be accompanied by a selective reduction in fast-twitch fibers (42). This in itself may impair performance in the Wingate test, which is dependent on the preponderance of fast-twitch fibers (18).

Muscle Strength

Children and adolescents with CP have low muscle strength, particularly in the affected limbs. Strength impairment may be mild or extreme. In one study (54), isometric strength of the knee extensors in 6- to 14-year-old girls and boys with spastic diplegia was only 37% that of able-bodied controls at 30° flexion. Respective values for 60° and 90° were 53% and 69% (figure 10.4). Likewise,

hamstring strength at 90° flexion was 52% that of the controls (54). Isokinetic strength of patients with CP is also quite low (figure 10.5) (248).

Metabolic Cost and Economy of Movement

The metabolic cost of movement is high in

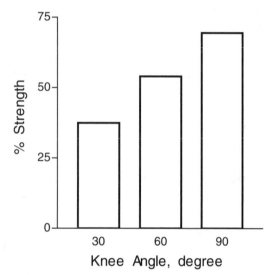

FIGURE 10.4 Isometric quadriceps extension strength in children with cerebral palsy as a percentage of the value in able-bodied controls. Fourteen 6- to 14-year-old patients with spastic diplegia were compared with 25 healthy controls. Measurement of strength was performed at 30°, 60°, and 90° knee flexion. Data from Damiano et al. 1995 (54).

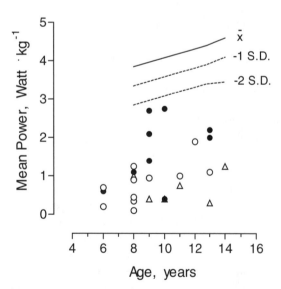

FIGURE 10.3 Mean anaerobic power in the arm cranking Wingate test, as a function of age, in 27 boys with spastic cerebral palsy. The lines represent data for able-bodied boys, tested in the Children's Exercise & Nutrition Centre.

Adapted, with permission, from Parker et al. 1992 (183).

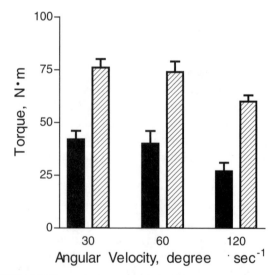

FIGURE 10.5 Isokinetic strength of the knee extensors of children with spastic cerebral palsy (black bars) compared with able-bodied controls (shaded bars). Eight subjects with spastic cerebral palsy and 38 controls were tested, at three angular velocities, using a Cybex II dynamometer. Vertical lines denote 1 SEM.

Data from Van Den Berg-Emons 1996 (248).

patients with CP, particularly those with muscle spasticity. This is true whether measurements are taken during pedaling (26; 150; 151), arm cranking (20; 158), or walking (159; 245).

As shown in figure 10.6, the average mechanical efficiency during pedaling in children with spastic CP is only some 50% to 60% that of able-bodied children. Efficiency in patients with athetosis is approximately halfway between values for those with spastic CP and those without disease (150). We have observed patients who exert force on the pedals in two opposite directions, causing a disruption in the continuous flow of pedaling or arm cranking. Such effort does not contribute to overcoming the pedal resistance and is not included in the calculation of mechanical power output. O_2 cost during cycling may vary considerably from one test to another (26). Such variability suggests that a higher metabolic cost among patients with CP is not due to an intrinsic biochemical aberration but rather to less economical movement.

The net O_2 cost (exercise minus resting VO_2) at various walking speeds is higher in most children with spastic CP than in able-bodied controls (200; 245) (figure 10.7). It is particularly high in those who have quadriplegia and, to a lesser extent, hemiplegia. In some patients, the metabolic cost of walking may be two to three times that observed in able-bodied controls (200; 245).

It is unlikely that the adenosine triphosphate yield per unit metabolic fuel is different between patients and healthy children. Unnithan and colleagues have shown that the high metabolic cost in CP patients is due to a "wasteful" activity pattern of their spastic or dyskinetic muscles (245). This is manifested by a high total body mechanical power (247) and by incoordinated action of the antagonist muscles in the calf and the thigh (244; 246). As seen in figure 10.8, the metabolic cost of walking is high in patients who have a high total mechanical power. This is in contrast to what occurs in able-bodied children, whose metabolic cost is less dependent on mechanical power (247). A high mechanical power results from aberrations in posture and in walking style (e.g., crouching, feet in equinovarus position, tiptoe walking).

A normal pattern of muscle activation during walking and running comprises alternating relaxation and contraction in the antagonist muscles of the limbs. During most of the normal gait cycle, the antagonist muscle relaxes while the agonist contracts, as displayed by electromyography. The time period during which both an agonist and its antagonist are active has been termed "co-activation," also referred to as "co-contraction" when related to the actual contraction of the muscles. As seen in figure 10.9, the *co-activation index* (a composite expression of the length of the co-activation period and the intensity of electrical activity; see chapter 1, section on locomotion) in the thigh and

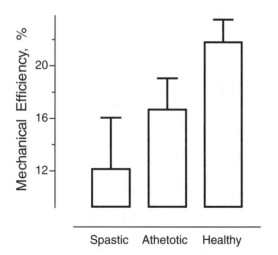

FIGURE 10.6 Mechanical efficiency in cerebral palsy. Comparison among patients with spastic cerebral palsy, athetotic cerebral palsy, and able-bodied controls. Data from Lundberg 1975 (150).

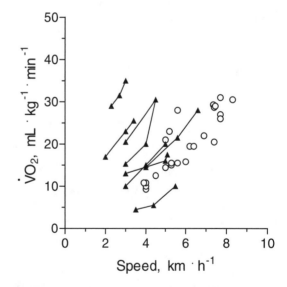

FIGURE 10.7 Net O_2 cost of locomotion of boys with spastic cerebral palsy compared with able-bodied controls. Subjects walked on the treadmill without holding on to the side bar. The lines connect data points for each patient.

Reproduced, with permission, from Unnithan et al. 1998 (245).

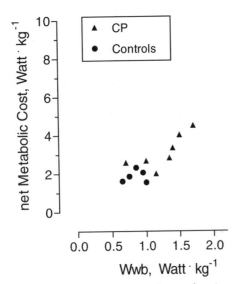

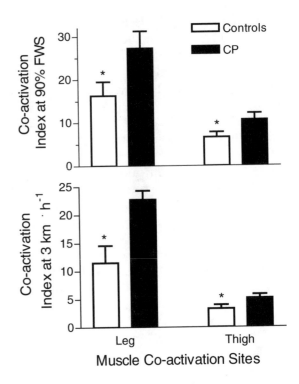

FIGURE 10.8 Net metabolic cost of walking of patients with spastic cerebral palsy increases with the increase in total body mechanical cost. Eight children with spastic cerebral palsy (ages 12.7 ± 2.8 years) and able-bodied controls (13.6 ± 2.1 years) walked on a treadmill at 3 km · hr^{-1}. Wwb denotes total body mechanical power, including power transfer between segments of the limbs.

Reproduced, with permission, from Unnithan et al. 1999 (247).

FIGURE 10.9 Co-activation index of muscle antagonists during treadmill walking in nine patients with spastic cerebral palsy (CP) (ages 12.7 ± 2.8 years) and able-bodied controls (13.6 ± 2.1 years). Measurements were taken at 90% of fastest walking speed (FWS) and at 3 km · hr^{-1}.

Reproduced, with permission, from Unnithan et al. 1996 (244).

calf muscles is considerably greater in children with spastic CP than in able-bodied controls during a treadmill walk (244). Furthermore, this excessive co-activation is an important cause of the patients' high metabolic cost of locomotion (figure 10.10), as those who have a high O$_2$ uptake also have a high co-activation index (246).

In conclusion, it seems that both the wasteful mechanical walking style and the co-activation of antagonists play a role in the low economy of locomotion in CP patients. A considerably high co-activation occurs also during cycle ergometry in children with CP (116).

Relationship Between Functional Ability and Physiologic Fitness

Can the low physiologic fitness of children with CP explain their poor functional motor ability? If so, which fitness components are the limiting factors in motor ability? There are no universally accepted indices for "functional ability" of children with CP. One index that has been gaining international recognition is the Gross Motor Function Measure (GMFM), which was developed specifically for children with CP or head injuries (202; 206) and subsequently refined

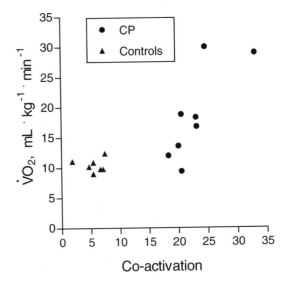

FIGURE 10.10 Relationship between metabolic cost of locomotion and the co-activation index of the leg antagonists during treadmill walking (3 km · hr^{-1}) in eight patients with spastic cerebral palsy (CP) (ages 12.7 ± 2.8 years) and seven able-bodied controls (13.6 ± 2.1 years).

Reproduced, with permission, from Unnithan et al. 1996 (246).

(205). The test components were selected to represent normal developmental milestones. Ability of the child to perform the following functions is assessed sequentially: lying prone and supine, rolling, sitting, kneeling, crawling, standing, walking, running, and jumping. In one study (185) the authors determined the extent to which performance in the GMFM could be explained by aerobic or anaerobic fitness of children with spastic CP. Maximal aerobic power was determined by arm ergometry. Anaerobic performance was determined for the upper and the lower limbs using the Wingate test. Neither the aerobic nor the anaerobic scores were strongly associated with functions such as lying, rolling, sitting, crawling, or kneeling (r = 0.65 or less, n = 23). However, when related to the standing, walking, running, and jumping components, as well as the total GMFM score, anaerobic fitness of the lower limb had considerably higher correlation coefficients than did the aerobic fitness or the anaerobic upper limb fitness (table 10.1 and figure 10.11).

A similarly low association (r = –0.47 or lower) has been reported between aerobic fitness and walking time at maximal speed, or up and down a ramp, in adolescents with severe CP or paresis due to poliomyelitis (17). Even lower correlations (r = –0.30 or lower) were found between maximal O_2 uptake of young adults with CP and their ability to wheel their chair to volitional fatigue (29). It thus seems that anaerobic fitness limits the motor ability of children with CP to a greater extent than does their aerobic fitness.

Among fitness components, low muscle strength is a more important limiting factor of motor function in CP. Deterioration in gait as the child grows (39), and the assumption of "crouch gait," seem to be related to muscle weakness (52) and to strength imbalance between agonists and

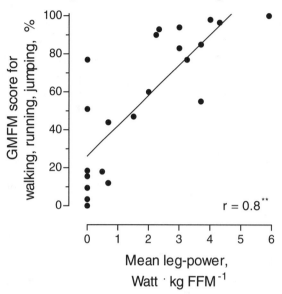

FIGURE 10.11 Relationship between a composite score for walking, running, and jumping in the Gross Motor Function Measure (GMFM) and the muscular endurance of the lower limb. The latter is expressed as mean power per kilogram fat-free body mass in the Wingate anaerobic test (lower limb protocol). Subjects were 7- to 14-year-old ambulatory girls and boys with spastic cerebral palsy. FFM = fat-free mass. ** = p<0.01

Adapted, with permission, from Parker et al. 1993 (185).

TABLE 10.1 Correlation Coefficients Between Scores in the Gross Motor Function Measure (GMFM) and the Aerobic and Anaerobic Fitness of Children With Cerebral Palsy

	GMFM COMPONENTS					
Fitness component	LR	Sit	CK	Stand	WRJ	Total
Peak power, arms	0.41	0.36	0.36	0.30	0.28	0.34
Mean power, arms	0.50*	0.44*	0.38	0.36	0.32	0.40
Peak power, legs	0.65*	0.42	0.59*	0.75**	0.83**	0.80**
Mean power, legs	0.62*	0.44	0.55*	0.72**	0.80**	0.73**
Aerobic power	0.43	0.48˙	0.40	0.38	0.31	0.55*

Twenty-three 7- to 14-year-old girls and boys with spastic cerebral palsy performed a maximal aerobic power test (arm cranking) and a Wingate anaerobic test (arm cranking and leg pedaling). Fitness was calculated in Watts per kilogram fat-free mass. LR = lying and rolling; Sit = sitting; CK = crawling and kneeling; Stand = standing; WRJ = walking-running-jumping.

* = p<0.05

** = p<0.01

Adapted, with permission, from Parker et al. 1993 (185).

antagonists (53; 54). Weakness of the knee extensors is an important limiting factor in a patient's spontaneous walking velocity (54). Finally, muscle strength is positively associated with scoring in the GMFM (130).

In conclusion, several physiologic fitness components, mostly strength and anaerobic power, have an effect on the motor ability of the child with CP. It is important to emphasize, though, that motor ability depends also on range of motion, degree of spasticity, and neuromotor control. More research is needed, using multiple regression techniques, to determine the relative importance of these components.

Exercise Testing

Testing of children with CP often requires special protocols and equipment, as well as highly skilled and devoted personnel. Testing in a laboratory setting may be necessary for some patients.

Laboratory-Based Testing

Many patients cannot use their legs and must be tested by *arm cranking ergometry* (1; 21; 68; 158; 243). The shoes of those who can cycle may need to be strapped to the pedals (184; 185; 243; 248). This is especially indicated for the child with athetosis. Because of their spasticity and incoordination, many patients find it hard to pedal at a constant pace (116); and, when a mechanically braked ergometer is used, one should continuously count the actual number of pedal revolutions (248). Whereas able-bodied children are usually tested at cycling or arm cranking speeds of 50 or 60 rpm, this speed is too fast for patients with CP. In one study (158), arm cranking velocity of 30 rpm was found most comfortable and metabolically more economical than 20 or 40 rpm. Some young patients have an extremely short concentration span. Others may find it impossible to tolerate a mouthpiece and a nose clip while exercising (164). As a result, it is sometimes impossible to obtain direct measurements of O_2 uptake, especially at high exercise intensities. A maximal aerobic test without gas collection (e.g., the McMaster protocol, appendix II) can be used as an alternative. This is preferable to a submaximal test, which is based on heart rate.

There have been few attempts to measure $\dot{V}O_2max$ *during treadmill walking* in children with CP (105; 152; 195). No study has involved attempts to determine it during treadmill running. It is a

challenge for a child with any neuromuscular disease to master the technique of treadmill walking. However, with proper learning and practicing, even CP patients with marked limitation in their walking ability (e.g., those who must resort to the use of crutches or a walker) can learn how to walk on the treadmill without holding on to the hand railing (105). In point of fact, such patients sometimes have a higher maximal walking speed on the treadmill than on a carpeted floor. While able-bodied children typically need only 2 to 4 min to master the walking or running technique, patients with CP may require as much as 15 to 20 min (105; 159; 245). No studies have systematically determined how much practice is required for full habituation.

Physiologic responses to *submaximal treadmill walking* or *ground walking* can be used as an indicator of a child's exercise ability. As discussed in the earlier section on physiologic fitness (see also figure 10.7), the metabolic cost of walking differs markedly among patients with CP. A higher metabolic cost usually denotes a lower fitness level. Therefore, measurement of submaximal O_2 uptake during walking at standardized speeds can be used as a fitness test. It can also be used to determine the effect of treatments. In one study (159), the net O_2 uptake of 10 children with spastic CP (age 9.0 ± 2.1 years) who walked at 3 km · hr^{-1} dropped by 8.9% when they wore ankle-foot orthoses compared with no orthoses. The drop at 90% fastest walking speed was 5.9%. A similar pattern was observed for minute ventilation. In another study (161), O_2 uptake of 15 patients, age 4 to 13 years, dropped by 2 ml · kg^{-1} · min^{-1} 2 months after intramuscular injection of botulinum toxin. In most of the patients, O_2 uptake returned to preinjection levels four months later.

When expired gas collection is not feasible, one can *monitor heart rate*, instead, as a surrogate measure of the physiologic cost of walking (38; 113; 155; 199; 200). The equation used to calculate the Physiologic Cost Index (PCI), also called Energy Expenditure Index, is as follows:

PCI (beats per meter) = (walking HR – resting HR) / walking speed in meters per minute

The rationale for using this index is that the submaximal walking heart rate is well correlated with the respective O_2 uptake (ml · kg^{-1} · min^{-1}) in children with CP (r = 0.84, n = 13) (200). The lower the index, the higher the walking performance. For example, while the average PCI in healthy subjects, ages 7 to 17 years, was 0.41 beat · m⁻¹ (range 0.25-0.64), it was 1.38 (range 0.24-2.58) in those with CP (199). Based on anecdotal data, the PCI seems useful in assessing the functional outcome of orthotic treatments. For example, an 11-year-old girl with spastic diplegia had a PCI of 1.53 beat · m⁻¹. When she used an ankle-foot orthosis, her index dropped to 0.98 beat · m⁻¹ (38).

An alternative ergometer, particularly for patients with diplegia or quadriplegia, is the *wheelchair ergometer.* Various methods have been proposed, but the general principle is to position the subject's wheelchair on rollers that move with or without friction. Variables measured include maximal distance covered over a time period, the highest friction that can be tolerated at a predetermined wheeling speed, or an actual measurement of maximal O_2 uptake. In one study (29), a reliability of r = 0.89 (n = 6) was obtained when young adult athletes with CP performed a wheeling maximal O_2 uptake test twice over a 48-hr period. Comparison with arm cranking ergometry yielded a validity coefficient of only r = 0.31. The maximal O_2 uptake during wheelchair ergometry was 97% of that obtained with arm cranking. More research is needed to show whether this approach is suitable for younger subjects who are not athletes.

Testing of anaerobic performance was ignored until the last decade, when the suitability and utility of the Wingate anaerobic test were documented for children with CP and other neuromuscular disabilities (16; 69; 184; 185; 243; 250; 253). The Wingate test is feasible for the great majority of CP patients. For example, 100% of 27 girls and boys with spastic CP performed the upper limb protocol successfully. The lower limb protocol was performed successfully by 89% of the patients. The youngest child who performed the test was 5 years old. Feasibility for patients with athetotic CP was lower (91% and 46% for the upper and lower limbs, respectively) (243).

As shown for able-bodied populations, the Wingate test is highly reliable also when used with children who have CP (69; 243; 250). Figure 10.12 displays test-retest scores for patients with CP and other neuromuscular diseases. Testing was performed 1 to 2 weeks apart. Reliability coefficients for arm cranking and pedaling were 0.98 and 0.96, respectively, for all groups com-

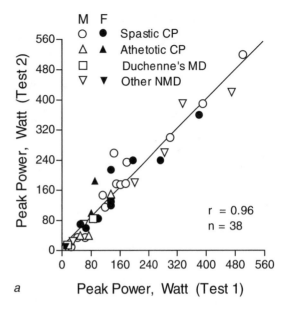

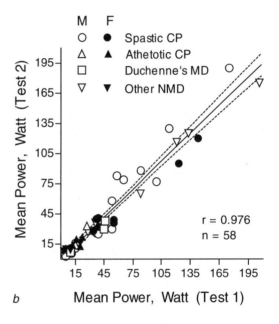

FIGURE 10.12   Test-retest reliability in the Wingate test, performed by 5- to 18-year-old patients with various neuromuscular disabilities (NMD) using the lower (*a*) and upper (*b*) limbs. MD = muscular dystrophy; CP = cerebral palsy.

Reproduced, with permission, from Tirosh et al. 1990 (243).

bined. For the spastic group alone, the respective r values were 0.96 and 0.95. They were 0.92 and 0.82 for the patients with athetosis. Similarly high test-retest correlations were shown for 6- to 12-year-old Dutch patients (69).

Mechanical power is the product of velocity and force. Theoretically, one can obtain a given power by any combination of force and velocity. However, based on force-velocity contractility considerations, there is an optimal force during pedaling or arm cranking in anaerobic exercise that will elicit the highest mechanical power. As discussed in appendix II, such a force is usually selected according to the subject's body weight, assuming a normal relationship between muscle mass and total body mass (15). Because muscle mass is often abnormally low for patients with a neuromuscular disability, including CP, the challenge is to select a braking force based on characteristics other than total body weight. Van Mil and colleagues (253) recently validated two alternative approaches for children and adolescents with CP and other neuromuscular diseases who performed the arm cranking test. One is based on the lean arm volume and the other on each subject's optimal force during the force-velocity anaerobic test (212; 254).

Testing in the Field

The relative complexity and cost of laboratory exercise testing of children with CP warrant the use of fitness tests that can be administered in a clinic or a sport club. The use of simple and familiar motor tasks will also have a strong motivational effect. Various field tests have been used to assess fitness of patients with CP. Very few, however, have been validated against laboratory tests. For example, in a study of adolescents with severe CP and post-poliomyelitis (231), several walking tasks were performed periodically during a 2-year conditioning program, and the results were related to those obtained in the laboratory. The tasks included a 25-m walk on the level at a comfortable speed, a 25-m walk on the level at maximal speed, walking up a 5-m-long ramp with a 15° incline, and walking down the same ramp. The subjects used their regular assistive devices (crutches or braces), and there was no attempt to make them modify their habitual gait. Neither the walk at normal speed nor climbing up or down the ramp had any relationship to maximal O_2 uptake. Walking at maximal speed did show significant correlation with maximal O_2 uptake, but only 30% of the walking speed variance could be explained by maximal aerobic power. Thus success in such walking tasks apparently is not limited by the maximal aerobic power of the severely affected patient. It may be related more to walking skill and perhaps to local muscle strength and endurance.

Field-based evaluation can include motor ability tasks, such as catching a ball, throwing a ball to a target, or static and dynamic balance. Muscle endurance can be assessed, for example, by the number of sit-ups performed per time unit (197).

For patients who use a wheelchair, or those with severe functional limitation of the lower limbs, tests based on wheeling ability are of practical relevance. One approach is to document the time that is needed to cover a predetermined distance, or the distance covered during a given time period. Such protocols can include short-term "anaerobic" tasks, or longer-term "aerobic" tasks (180). They can also include more complex tasks such as circling an object, wheeling backward, or wheeling in a zigzag pattern (19).

In conclusion, even though the physiologic correlates of field tests are not yet fully established, such tests are recommended for their simplicity and motivational impact.

Response to Physical Training

Children and, in particular, adolescents with CP often become progressively immobile and deconditioned. Rehabilitation by training can slow down and, it is hoped, reverse such deterioration. An increase in physiologic fitness may also improve well-being, social integration, and occupational potential.

Effect on Maximal Aerobic Power

The hemodynamic, respiratory, and metabolic responses to training of the child with CP are similar to those of healthy youngsters (20; 25; 68; 76; 153; 154; 194; 231; 251).

Increases in maximal O_2 uptake or in maximal mechanical power following a training program have ranged from 8% to 35%. There is only one study, by Van Den Berg-Emons and colleagues, that included a randomly selected control group (251). In that project, patients were matched by severity of disability and then randomly assigned into either a training or a control group. Training consisted of four-per-week 45-min aerobic exercise sessions over 9 months. These sessions were in addition to school physical education and regular physical therapy sessions. Activities included cycling, wheeling, running, swimming, mat exercises, and training on a "flying saucer." The program elicited a 35% increase in maximal aerobic power (Watts per fat-free mass), compared with no change in the controls (figure 10.13). In a subsequent school year, a similar program, at two-per-week frequency, also induced an increase in aerobic fitness, but only by 21%. The 35% increase that occurred in the first year is much higher than the improvement commonly observed among nondisabled children (see chapter 1). This may reflect the very low activity level of such patients prior to the program and the resulting deconditioning effect. Less intense programs did not induce an improvement in directly measured maximal O_2 uptake, although there was a decrease in heart rate at a given work rate (e.g., [20]).

One training response described for patients with CP is an increase in blood flow to the exercising muscles during a submaximal task (154). The mechanism for such an increase is unclear. Spastic muscles have subnormal blood flow during exercise (135), and the posttraining increase in blood flow could reflect a decrease in spasticity. However, such a decrease has yet to be clinically and objectively confirmed.

Effect on Mechanical Efficiency

Training effects on mechanical efficiency (O_2 cost of cycling) have been assessed in four studies (20; 62; 153; 248). While three of these showed no change in mechanical efficiency, one study (62) reported a significant 10% decrease in the O_2 cost of cycling at 30 Watts following a 10-week program of enhanced physical education. That study included a group of matched, randomly assigned controls, but so did one of the studies that did not show a change in mechanical efficiency (248). It is hard to reconcile these differences in the outcomes.

Effect on Muscle Endurance and Strength

The only attempt to document the effect of physical training on anaerobic performance is the study by Van Den Berg-Emons and colleagues (251), who found no change in anaerobic indices. This may reflect the aerobic nature of the intervention, but it has yet to be shown whether anaerobic characteristics are trainable in children or adolescents with CP.

Effects of resistance training on muscle strength have been documented in various studies. Most of these, however—a frequent occurrence in training studies with patients—did not include a randomly assigned control group (55). Figure 10.14 summarizes the response of 6- to

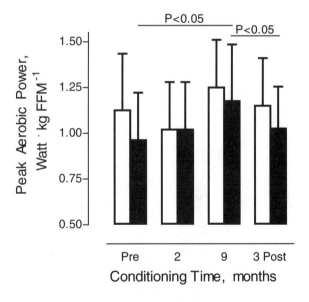

FIGURE 10.13 Effect of aerobic training on maximal aerobic power of 7- to 13-year-old children with spastic cerebral palsy. Twenty patients were randomly assigned to a four-per-week aerobic training program (in addition to physical education and physical therapy, black bars) and controls (physical education and therapy only, white bars). Pair matching was done for physical ability and mental function. Training lasted 9 months, and the subjects were retested at 12 months.

Based on Van Den Berg-Emons et al. 1998 (252).

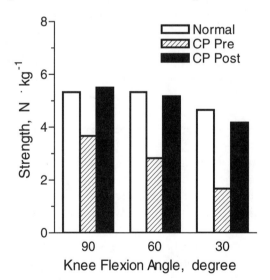

FIGURE 10.14 Effect of resistance training on knee extensor strength in children with spastic cerebral palsy (CP). Fourteen 6- to 14-year-old patients underwent a three-per-week, 12-week quadriceps-strengthening program, using ankle weights. Values are isometric strength as determined by handheld dynamometer at three knee angles. Vertical lines denote 1 SEM.

Adapted, with permission, from Damiano et al. 1995 (54).

14-year-old girls and boys with spastic diplegia to a 6-week, three-per-week program (54). The program focused on the knee extensors of both legs and yielded a considerable increase in the isometric strength of the quadriceps muscle. In point of fact, most subjects reached strength values comparable to those of able-bodied controls. There was very little increase in the strength of the hamstrings, which points at the specificity of the intervention.

Even though the four-per-week, 9-month intervention by Van Den Berg-Emons et al. (252) consisted of aerobic activities, it resulted in an increase in peak isometric knee flexion torque of 32% within 2 months and 39% within 9 months. These values were achieved by the less affected limb. In the more severely affected limb, the increase after 2 months was 28%, with no further increase at 9 months.

Effect on Muscle Spasticity

The possible effect of physical training on muscle spasticity is of both practical and theoretical value. Assessed semiquantitatively by the Ashworth clinical scoring protocol (10), muscle spasticity did not change following a 4-week aerobic program among adults (269) and an 8-week strength program among adolescents with CP (156). Likewise, using integrated electromyography as a more objective criterion for spasticity, a 1-year program of vertical jumps, trampoline jumping, and neuromuscular activation by the Bobath method did not reduce muscle spasticity (203). Nor did a 12-month mild enrichment program of games and calisthenics, in which the H:M reflex ratio (8) was taken as an index of spasticity (231).

There is one study (248) in which the effects of aerobic training on spasticity were assessed during dynamic exercise (cycling and one-arm cranking). Nine months of two-per-week, 45-min sessions resulted in a trend (nonsignificant) for a reduction in spasticity, as assessed by integrated electromyography, and in co-activation of antagonist muscles during cycling, but not during one-arm cranking.

In conclusion, the possible beneficial effect of exercise on muscle spasticity in CP has been suggested by clinicians but not proven by objective criteria. More refined testing methods are needed for further study of this phenomenon.

Effects on Motor Function

Physical educators or physiotherapists who use sport as a means of rehabilitation do not need scientific tools to discern improvement in the mobility of children with CP. The ability of these youths to train and compete in running, swimming, volleyball, hockey, and many other sports is testimony to their potential for improvement in mobility and other motor tasks.

Functional improvement along with physiologic adaptation to training has been widely documented (25; 53; 54; 203; 229; 231). Even the very severely disabled individual may enjoy functional improvement. Figure 10.15 summarizes the walking performance of 19 severely disabled adolescents with spastic and mixed spastic-athetotic CP who took part in a 2-year sport rehabilitation program (231). The improvement in walking speed at comfortable and at maximal pace was evident. Even more impressive was the 45% to 55% improvement in the speed of walking up or down a 5-m ramp (15° slope). The patients were subdivided into those who at the start of the program could perform short distance walks for daily needs ("functional walkers") and those who could not take more than a few steps, which were insufficient to serve their functional needs ("physiologic walkers"). As seen in figure 10.16,

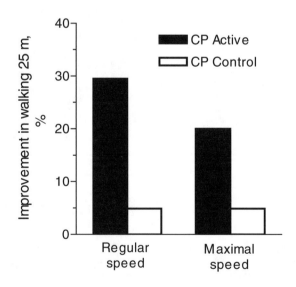

FIGURE 10.15 Effect of a 2-year training program on the walking speed of adolescents with cerebral palsy (CP). A comparison between active participants and sedentary controls. Values are percentage improvement.

Adapted from Spira and Bar-Or 1975 (231).

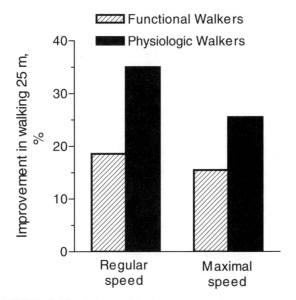

FIGURE 10.16 Relationship between the improvement in walking and the severity of disability in adolescents with cerebral palsy. Comparison between "functional" and "physiologic" walkers (see text for definitions) who took part in a 2-year training program. Values are percentage improvement.

Adapted from Spira and Bar-Or 1975 (231).

the latter group derived the greater functional benefit from the program, with a dramatic improvement in walking ability. When given a questionnaire to evaluate the program, 70% of the participants reported some "relief of their spasm" and of muscle clonus while walking.

An increase in quadriceps muscle strength following a 6-week resistance training program was accompanied by a decrease in the *crouch gait* pattern of 6- to 14-year-old patients with spastic diplegia (53; 54)—specifically, an increase in stride length and a decrease in the knee flexion angle at heel strike. The latter is an index of crouch. Interestingly, the improvement in crouch gait is similar to that achieved through surgical lengthening of the hamstring muscle. This is an important example of the potential of exercise treatment to obviate the need for more aggressive interventions.

Recommended Activities for the Child With Cerebral Palsy

Neither activities administered at regular or special schools nor habitual activities at home seem sufficient to induce optimal energy expen-

diture or fitness level in patients with CP (248). Reinforcement of these activities by special training programs is needed to induce an increase in physiologic fitness. Activities that have been found to be effective in eliciting higher maximal aerobic power among mildly or moderately affected patients include exercising with medicine balls and pulleys, pedaling or arm cranking a cycle ergometer, riding a tricycle, wheelchair sprinting, wheelchair slalom, and swimming. More severely affected patients can be given mat exercises (individualized according to ability); floating, ducking, and moving in the water; crawling; and pedaling a cycle ergometer in the supine position—all tailored to their own ability. Improvement in muscle strength can be achieved through the use of free weights, as well as exercise machines that provide isometric and isokinetic routines (55). Modest improvement in motor skills can be achieved through ball throwing and catching and, to a lesser extent, dynamic balance exercises (197).

Unless highly motivated, children with CP will not adhere to prolonged, intense activities. It is important to intersperse activities like those just mentioned with games that emphasize recreation and fun. Because of their relatively short concentration span, such patients should frequently change from one game to another. Intense activities need not last more than 10 to 15 min. Activities that are less intense and include a recreational element can last up to 30 to 45 min. Two sessions per week are sufficient to induce a noticeable training effect within a few weeks. Because of the variety in anatomic distribution of the disease and in the severity of the impairment, training interventions in CP must be individualized.

Enhanced physical activity must be sustained and become part of the child's lifestyle. Cessation of the activity will be followed by a rapid deterioration in performance. Figure 10.17 demonstrates changes in maximal aerobic power among schoolchildren with CP who had stopped their intervention program for the summer vacation (25). Following the vacation, peak O_2 uptake deteriorated in five of the eight subjects, almost nullifying the effect of the preceding activities. This further emphasizes the inability of children and youth with CP to retain their fitness unless they engage in special programs throughout the year.

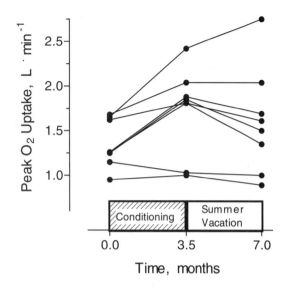

FIGURE 10.17 Individual changes in aerobic fitness (peak O₂ uptake) following a 3.5-month activity program at school and after a summer vacation. Subjects were 10- to 18-year-old patients with cerebral palsy.

Adapted from Berg 1970 (25).

Chronic Fatigue Syndrome

Chronic fatigue syndrome (CFS) is manifested by marked physical and mental fatigue following minimal physical exertion, as well as by muscle and joint pain at rest and following exercise. While most often encountered in adults, CFS occurs also in children and adolescents. Some patients complain of chronic headache and exhibit symptoms and signs of depression. In some, the fatigue and pain first appear following viral infection. There are three main diagnostic criteria for CFS: (1) The symptoms have lasted at least 6 months; (2) fatigue and lassitude are present for at least 50% of the day; and (3) there are no clear characteristics of another muscular, metabolic, or inflammatory disease.

Response to Acute Exercise

Patients with CFS complain of exercise-induced fatigue that forces them to reduce their habitual physical activity. Objective measurement (accelerometry over 2 weeks) of daily activity patterns revealed that the actual decline in activity following an exhausting treadmill task occurred only 12 to 14 days later (225). There is no clear pathophysiologic pattern that underlies CFS (65). Magnetic resonance spectroscopy suggests that repeated muscle contractions may lead to excessive muscle cell acidity in some patients (23). Muscles display normal contractile properties, as well as a normal recovery of maximal voluntary contraction within 24 hr (91). Aerobic fitness is normal or somewhat low, but peak heart rate is below normal. Some patients respond to exercise with high blood lactate (136) or low submaximal and maximal heart rate (109). An important feature, as shown for adult patients, is a high rating of perceived exertion at a given submaximal heart rate (87; 91). There are no similar data for children with CFS.

Beneficial Effects of Enhanced Physical Activity

Clinical experience has shown that even though exercise is the main trigger of fatigue and pain in CFS, patients often benefit from an activity program that targets those muscle groups in which the pain occurs and that is based on a gradual increase in intensity and volume. Such programs have been found to be efficacious in randomly controlled trials, and they do not induce any adverse effects (87; 189; 219; 263; 268). A recent comprehensive evidence-based review of available therapies shows that graded exercise interventions yield the most consistent beneficial effect (reduction of fatigue and pain and improved quality of life) for patients with CFS (263). However, there are no data on the long-term effects of such interventions once the program is over.

Epilepsy

Epilepsy is a common neurologic pediatric disorder, occurring in 0.5% to 1% of children. It is manifested by seizures of various kinds, which are not induced by fever or acute insult to the central nervous system. The classification of epileptic seizures has evolved over many years. The current international classification includes *partial seizures, generalized seizures,* and *unclassified seizures.* Some types are accompanied by loss of consciousness and others are not. This section addresses epilepsy in general, without attempting to cover its various subgroups. We give major emphasis to safety issues as related to exercise.

Habitual Physical Activity

There are very few documented surveys of the habitual physical activity of children and adolescents with epilepsy. It seems, though, that the activity level of such patients is often curtailed. Limitations are often imposed by parents, teachers, and other adults, who are concerned about the child's safety. For example, in a survey among Japanese school nurses, 24.5% reported that they had imposed limitations on participation in swimming for pupils with epilepsy even if their attending physician had permitted the child to participate in the lessons (178). In a survey of Hong Kong parents, 20% thought swimming should be prohibited even if seizures could be controlled (134). Adolescents and adults with epilepsy seem to be less active than their healthy peers. In one study (30), less than 25% of 16- to 60-year-old Norwegian patients practiced sports regularly, compared with 50% in the general population. The main self-reported reason for the sedentary lifestyle was a fear of seizures. The authors of that study stated that patients who had been overprotected as children subsequently became overcautious as adults.

Physiologic Fitness

In a study of adolescents and adults with various kinds of epilepsy, maximal aerobic power was only some 75% to 80% of that reported for the general population (30). Likewise, young adults with epilepsy had impaired aerobic fitness, muscle endurance, flexibility (232), and strength (112). It is unlikely that the disease, in itself, causes a deficiency in the oxygen transport system or muscle performance. The explanation most probably is the sedentary lifestyle practiced by some patients.

Does Exercise Induce Seizures?

Very few epidemiologic data address the question whether exercise brings on seizures. For example, in a survey among 200 Norwegian adults with epilepsy, approximately 10% reported that exertion triggered a seizure. This was particularly apparent among those with symptomatic partial epilepsy (173). The following comments are based on the cumulative experience of clinicians and on very few laboratory observations.

Although patients and their parents often fear that physical exertion may induce a seizure, there are conflicting data as to whether this indeed is the case. One reason for the discrepancy in reports is that epilepsy encompasses a variety of clinical manifestations that may be induced by more than 40 factors, some of which interact with each other (e.g., state of hydration, sleep deprivation, and acid-base balance) (4). Several authors have reported cases in which sports, or other strenuous activities, induced a clinical seizure, usually accompanied by electroencephalographic (EEG) abnormalities (115; 129; 182; 214). The paucity of such reported cases limits our ability to suggest a clinical profile of the child who would respond to exertion with a seizure. However, Schmitt et al. (214) proposed the following common characteristics of such children:

- Epilepsy is diagnosed in the first 1 to 3 years of life.
- Exercise-induced seizures start 1 or more years later.
- EEG is normal during clinical remissions.
- Hyperventilation does not trigger a seizure.
- Brain magnetic resonance imaging and computed tomography are normal.
- Epilepsy is of an unknown origin.

Nakken et al., who studied responses of children to maximal exertion (175), suggest that children who respond during exercise with a diminution of the rate of EEG epileptiform pattern are unlikely to experience seizures during or after exercise. In contrast, those who do not have this EEG response during exercise report exercise-related seizures. These authors further suggest that a laboratory-based exercise test with monitoring of EEG can serve as a diagnostic test to identify children who are at risk for exercise-induced seizures.

No available studies have documented the effect of *exercise intensity* or *duration* on the induction of a seizure. It seems, though, that in the majority of cases, a seizure occurs when the activity is prolonged (e.g., more than 10 min). We have tested in our clinic two children with a history of exercise-induced seizures. In the laboratory, while cycling continuously, they both developed a clinical seizure some 10 min into a moderate-intensity exercise (heart rate of 140-150 beat · min^{-1}) session. In a subsequent visit, when the child cycled intermittently (2 min on and 2 min off) at the same intensity, there were no seizures even after 25 to 30 min. We repeated these two routines in subsequent visits, and the phenomenon recurred.

The pathophysiology of exercise-induced seizure is unknown. Schmitt et al. (214) suggest that exercise-induced seizures are most likely a form of reflex epilepsy. Much research is needed to elucidate possible mechanisms.

Does Exertion Protect Against Seizures?

Several authors have reported a reduction in seizure frequency during periods of enhanced physical activity (57; 72; 173-175). For example, 20 of 26 children with intractable partial and generalized epilepsy responded to maximal exercise with a reduction in the rate of epileptiform EEG discharge. However, this rate showed a rebound above baseline during a 10-min recovery (175). In a survey among Norwegian adults with epilepsy, 36% reported that regular exercise contributed to better seizure control (173). Rats with induced epilepsy responded to 45 days of aerobic training with fewer seizures than did nontraining control rats (9).

Would prior exertion negate the detrimental effect of hyperventilation? The first study that addressed this question (93) included 30 adolescents with epilepsy (primary generalized and other forms) whose EEG was monitored while they first hyperventilated at rest, then exercised (20-50 deep knee bends), and again hyperventilated. The initial hyperventilation induced an increase of EEG abnormalities in all patients. In contrast, exercise was accompanied by diminution of slow-frequency waves and of seizure discharge and by a decrease in wave amplitude. Such normalization of the EEG pattern lasted 1 to 2 min after exercise. Furthermore, deliberate hyperventilation 10 to 15 sec following exercise did not induce as much wave abnormality as did the initial hyperventilation. This study suggests that exercise may suppress the hyperventilation-induced wave abnormality and seizures in some, but not in all, patients. A similar beneficial effect of exertion has since been confirmed (72; 106; 132; 133). Of special importance is the study by Esquivel et al. (72), who monitored the plasma pH and the EEG at rest and during maximal exer-

cise, recovery, and voluntary hyperventilation in children and adolescents with *absence* epilepsy. The occurrence of seizures was lowest during exercise and highest during hyperventilation. As seen in figure 10.18, there was a tight positive association between the frequency of seizures and plasma pH. It is thus reasonable to assume that the mechanism for raising the seizure threshold by exercise is through metabolic acidosis, which counteracts the respiratory alkalosis of hyperventilation. This postulated mechanism, however, does not explain why the reduction in seizure frequency starts immediately after the beginning of exercise, before plasma pH has declined.

Does Fatigue Induce Seizures?

Fatigue and sleep deprivation have both been implicated as factors that increase the likelihood of a seizure (4). There is no documented evidence, however, that fatigue per se induces a seizure. On the contrary, most seizures occur during sleep or rest, as shown clinically and by continuous EEG monitoring (93; 129; 142). Nor is there any need for a child with epilepsy to sleep or rest longer hours than his or her healthy peers. As aptly summarized by Lennox in 1941 (cited by Livingston in

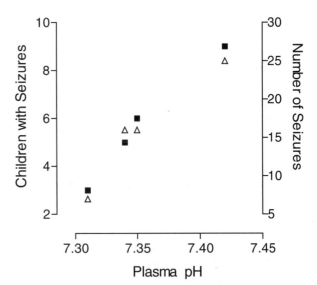

FIGURE 10.18   Relationship between plasma pH and the occurrence of seizures. Twelve 6- to 17-year-old patients with *absence epilepsy* rested for 15 min, exercised to volitional maximum, recovered for 30 min, and then hyperventilated for 3 min. An *absence* seizure was counted whenever the electroencephalogram showed a burst of spikes and waves at a frequency of 3 Hz, combined with clinical *absence*.

Reproduced, with permission, from Esquivel et al. 1991 (72).

1971 [142]), "Physical and mental activity seems to be an antagonist of seizures. Enemy epilepsy prefers to attack when the patient is off-guard, sleeping, resting or idling."

Does Head Trauma Induce Seizures?

An objection often raised against the participation of children with epilepsy in collision or contact sports is that repeated mechanical blows to the head may aggravate the state of the disease. While head trauma, especially penetrating injury, may exacerbate epilepsy, there are no epidemiologic data that evaluate the risk of collision or contact sports. Livingston (142) reported in 1971 that out of 15,000 young patients under his supervision during 34 years, there was not a single case of seizures due to athletics-related head trauma.

Accidents to the Child and to Others

It is commonly accepted that a potential hazard exists when a child with epilepsy has a seizure while performing such activities as cycling, horseback riding, rope or tree climbing, skin diving, and swimming (6; 207). Similarly, if a seizure occurs while the patient is engaged in a sport that involves throwing objects or shooting (e.g., javelin, discus, archery), there is a definite danger to people nearby (145). Precautions should be taken to reduce the likelihood of such accidents. Patients who are not well controlled by drugs should be barred temporarily from engaging in such sports.

While these precautions make sense, epidemiologic data on the accidental injury rate among young patients with epilepsy are scarce. In a recent study, injury rates were compared between 5- to 16-year-old patients of a Canadian hospital and their healthy controls (124). Subjects were matched for gender and age. Interviews indicated that there was no difference in the injury rate or severity between the two groups. Children in both groups who had attention deficit hyperactivity disorder were more prone to accidents than those without this disorder.

A Physician's Dilemma

As discussed in the preceding sections, intense exertion per se should not pose a danger to the child with epilepsy. In point of fact, it may reduce the risk of seizures. Yet there are young patients who do have a seizure during exercise and not while at rest. Should the physician then "play it safe" by instructing all children with epilepsy to refrain from intense activities, or from collision or contact sports? Before such a simplistic and "safe" conclusion is reached, physicians and parents must realize that, to any child, activity can be a primary avenue for self-expression and social acceptability. Children with epilepsy are not inherently inferior in their physical potential and should therefore be allowed to cultivate this means of physical and psychosocial expression. A striking case is that of a 13-year-old girl with a history of psychomotor seizures who took up distance running as a hobby. This girl, to show her ability to overcome her handicap, successfully completed approximately 2,000 km during a 40-day run and took part in other ultramarathon races (50). As pointed out repeatedly (5; 34; 35; 43; 49; 95; 129; 142-144; 201), the damage caused by inactivity due to overprotection often outweighs the remote risk of a sport injury or a sport-induced seizure.

In a 1983 position statement, the American Academy of Pediatrics (47) recommended that participation in contact and collision sports be determined individually for each child with epilepsy and that patients and parents should be taught that, indeed, sport activities may entail risk. A trial period should be initiated during which the response of the child to specific sports can be evaluated. In a 1994 position statement (6), the academy recommended that patients who are well controlled can participate in all sports. Those with poorly controlled epilepsy should be assessed individually for participation. As a general guideline, they should avoid archery, riflery, swimming, weight- or powerlifting, strength training, and sports involving heights.

Final decisions should, therefore, be made for each individual case, taking into consideration the level of seizure control, the ambitions of the child, the availability of other (nonexertional) means of expression, and the likelihood of compliance with an imposed regimen.

Recommendations for Physical Activity

Given the considerations just outlined, it is impossible to set forth rules that will be relevant for all children with epilepsy. The following recommendations should be treated as general guidelines, based on clinical experience and physiologic reasoning.

- A child with medically controlled epilepsy should be encouraged to do as much physical activity as he or she desires.

- Strenuous activity (distance running, a prolonged tennis match) is not contraindicated, even if it causes marked fatigue.

- Activities such as horseback riding, rock climbing, swimming, or diving should not be practiced without supervision.

- Bicycle riding should be limited only if the state of epilepsy is not well controlled by medication.

- Collision (football, ice hockey, lacrosse, rugby) and contact (basketball, soccer, wrestling) sports can be practiced in the medically controlled patient.

- Boxing, involving repeated mechanical impact on the head, should not be allowed.

- For unexplained reasons, certain activities may be more epileptogenic than others to a given child. If a certain activity repeatedly triggers a seizure of any kind, it should be avoided.

- Activities that may cause damage to spectators should not be allowed in a child with uncontrolled epilepsy.

- Physicians should give individualized consideration to each patient. Recommended activities should be determined with the cooperation and consent of parents and, preferably, the child.

Extremely Low Birth Weight

With the improvement in perinatal care, survival of infants who were born prematurely, particularly those at extremely low birth weight (ELBW = less than 1,000 g) and short gestational age (less than 26 weeks), has increased dramatically. For those born at 23 weeks gestation, survival ranges from 2% to 35%. At 24 weeks, survival is 17% to 58%, and at 25 weeks gestation it rises to 35% to 85% (96).

Such survival is not without cost: While many survivors of ELBW have reached the third decade of life, a high percentage of them are affected by neurodevelopmental disabilities such as deficient cognitive function, CP, blindness, and deafness (267). Another complication is bronchopulmonary dysplasia (BPD). This section focuses on the exercise performance of children and adolescents who were ELBW. Discussion of BPD is included in chapter 6.

Habitual Physical Activity

While physical activity and daily energy expenditure are limited in children with CP (see preceding section) or mental retardation (see chapter 12), there is little published information on the habitual activity patterns of those who were ELBW and do not show overt neuromotor or mental limitation. Unpublished observations from the Children's Exercise & Nutrition Centre show that parents of 5- to 7-year-old ELBW survivors reported no differences in activity patterns compared with controls who had normal birth weight (NBW = more than 2,500 g). There are no data on the 24-hr energy expenditure of ELBW children.

Physiologic Fitness

Children and adolescents with ELBW often have a low physiologic fitness (101). Specifically, they often have a low maximal aerobic power (188; 208; 211), anaerobic muscle power, and muscle endurance (73; 121; 226).

Aerobic Performance

Maximal O_2 uptake in children born prematurely ranges from normal (13; 22) to very low (208; 211). It is reasonable to assume that performance may be related to birth weight. In one study, the children with the lowest reported birth weight (800 g or less) had a low maximal O_2 uptake: 30 ml $\cdot$ kg^{-1} $\cdot$ min^{-1} in the girls and 34 ml $\cdot$ kg^{-1} $\cdot$ min^{-1} in the boys (208). However, the lowest maximal O_2 uptake (25 ml $\cdot$ kg^{-1} $\cdot$ min^{-1}) was reported for 6- to 12-year-old children whose mean birth weight was 1,400 g. It is possible that the low aerobic performance in ELBW is related to pulmonary functional impairment (see section on BPD in chapter 6), but it may also result from a low activity level.

It is not known whether the low aerobic performance in children with low birth weight tracks to adult years. In a study of 19-year-old Swedish army recruits, aerobic performance (peak mechanical power in cycle ergometry) was determined in 218 individuals whose birth weight was less than 1,500 g (i.e., very low birth weight, VLBW) (71). The odds ratio for this group not to reach a normal maximal power was 3.3, which suggests that, as a group, these low birth weight adults had somewhat impaired aerobic performance.

Anaerobic Performance and Muscle Strength

Even when they show no overt manifestations of a neuromuscular disability, 5- to 8-year-old girls and boys with ELBW have low peak muscle power and muscle endurance, as measured by the Wingate test, compared to those with VLBW and controls with NBW (>2,500 g) (73; 121). This pattern still exists when performance is corrected for total body mass (figure 10.19*a*) or fat-free mass. In one study, the power generated during a vertical jump and the height of the jump were impaired in children with ELBW compared with NBW controls (73). To determine whether there is a "catch-up" of anaerobic performance in subsequent years, 11- to 17-year-old ELBW adolescents were tested in another study (226). Again, the low birth weight subjects had no overt neurological or mental deficiency, yet their anaerobic performance was low (figure 10.19*b*).

It is less clear whether *muscle strength* is affected by ELBW. Handgrip strength of 5-year-old children who were VLBW was lower than in term controls (83). While peak torque in isokinetic knee extension and flexion was not different between 11- to 17-year-old girls and boys with ELBW and their term controls (226), the composite strength (handgrip, knee extension, and elbow flexion) of 19-year-old men who were VLBW was reduced (71). It is not clear whether the inconsistent reports on strength impairment reflect differences in muscle quality or in size.

Metabolic Cost of Exercise

Three studies showed a high O_2 uptake for a given mechanical task in children born prematurely. In one, 7- to 12-year-old children with VLBW who had been small for gestational age had a higher O_2 uptake during treadmill running than did VLBW children whose weight was appropriate for gestational age and a control group of term children (22). Two other studies have indicated a

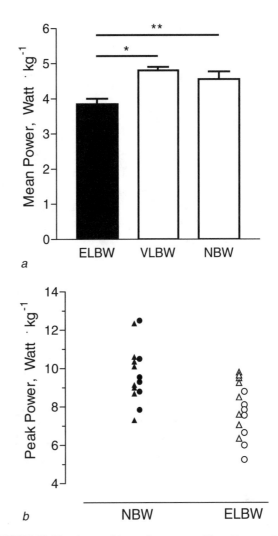

FIGURE 10.19 Anaerobic performance of 5- to 7-year-old children (*a*) and of 11- to 17-year-old adolescents (*b*) who were extremely low birth weight (ELBW). Subjects were tested by the Wingate anaerobic test. VLBW = very low birth weight; NBW = normal birth weight. The triangles represent individual data for boys, and the circles represent girls. Vertical lines denote 1 SD.

Graph in figure 10.19*a* is reproduced, with permission, from Keller et al. 2000 (121). Graph in figure 10.19*b* is reproduced, with permission, from Small et al. 1998 (226).
Graph in 10.19*b* is reprinted, by permission, from E. Small et al., 1998, "Muscle function of 11- to 17-year-old children of extremely low birthweight," *Pediatric Exercise Science* 10: 327-336.

high O_2 cost during cycling among children born prematurely (102; 131). The cause of low economy (treadmill walks) and low efficiency (cycling tasks) is not clear, but it presumably reflects impaired development of the central nervous system. Such impairment may interfere with normal neuromotor control, muscle coordination (see following section), and metabolic efficiency. The end result of a high metabolic cost is that such children work at a relatively high percentage of their maximal aerobic power and thus may fatigue early (see the section "Mechanical Efficiency and Economy of Movement" in chapter 1).

Neuromotor Function

As stated previously, it is quite clear that anaerobic performance is distinctly lower in children and adolescents with ELBW, while isometric or isokinetic strength is inconsistently affected. One possible reason for this discrepancy is that "sprint" pedaling on a cycle ergometer—as in the Wingate test—or a vertical jump on a force platform requires much more neuromotor coordination than does a static test of isometric or isokinetic strength. Indeed, various motor functions have been found to be impaired in children born prematurely, particularly those who were ELBW. Functions that are impaired include total body coordination (120), the sequence of force and velocity generation during a vertical jump (figure 10.20a) (73), reaction time (120), maximal pedaling speed against "zero" resistance (figure 10.20b) (120), balance (160), and gross motor functions and visual-motor integration (209; 210). The high energy cost of exercise (see preceding section) may be another indication of incoordinated motor function.

It is important to note that the impairments outlined in this section are found not only among survivors with obvious disabilities such as CP, but also among those who have no overt manifestation of any neurological or cognitive deficiency. While some improvement in motor function may occur between ages 6 and 8 years (160), there seems to be no further "catching up" from age 8 to 12 years (190). It is likely that perinatal insults to the central nervous system, such as periventricular hemorrhage, prolonged hypoxemia, and leucomalacia, contribute to these impairments. However, there is no evidence for such cause-and-effect relationship.

Juvenile Idiopathic Arthritis

With an incidence of 4 to 20 new cases every year per 100,000 children and a prevalence of over 100 cases per 100,000 children in various countries, juvenile idiopathic arthritis (JIA) (previously

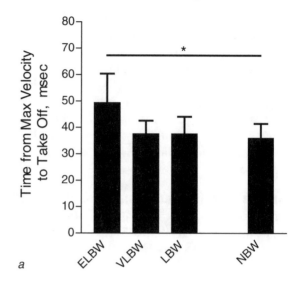

a

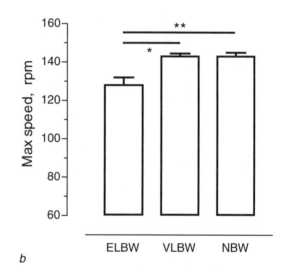

b

FIGURE 10.20 Motor function is impaired in children born at extremely low birth weight (ELBW). (*a*) The time elapsed from maximal velocity to takeoff in the vertical jump as performed by 5- to 8-year-old children. (*b*) Maximal pedaling speed against "zero" resistance in 5- to 7-year-old children. VLBW = very low birth weight; LBW = low birth weight; NBW = normal birth weight.

(*a*) Reproduced, with permission, from Falk et al. 1997 (73). (*b*) Reproduced, with permission, from Keller et al. 1998 (119).

called *juvenile rheumatoid arthritis, JRA*) is the most common pediatric arthritic disease. It is a major disabling condition in childhood and adolescence, and the disability tracks to adulthood in as many as 35% to 55% of the patients. While there is still some controversy regarding the classification of the various forms of childhood arthritis (230), the clinical entities usually agreed upon are pauciarticular JIA (one to four affected joints), polyarticular JIA (five or more affected joints), psoriatic JIA, and systemic JIA (also known as Still's disease).

Habitual Physical Activity

Symptoms such as pain, stiffness, and fatigue, often associated with their disease, are likely to limit the spontaneous physical activity of patients with JIA. However, there are only scant data about the activity patterns of such children.

In a Canadian study of 8- to 17-year-old patients with a mild disability, 53% participated in all physical education classes, 29% participated "some of the time," and 18% did not participate at all (157). It thus seems that nearly half of all patients in this study were limited in their physical activity at school. Within a province-wide survey of activity behaviors in children and adolescents with a disability or a chronic disease in Ontario, Canada, 25 individuals had arthritis. Based on responses to a questionnaire, their overall physical activity was calculated to be "moderate." While 63% perceived their fitness to be similar to that of their peers, 37% stated that they were less fit. The main reason for a low activity level was the patients' perceived physical limitation (147).

In a study that combined a questionnaire and objective monitoring of body movements by motion sensors, mean physical activity and participation in organized sport were lower in 5- to 11-year-old patients with mild and moderate JIA than in controls. However, the number of movements per day was similar in the two groups (104). Among 6- to 17-year-old Austrian patients, 74% cycled and 63% swam regularly. It is likely that these two sports were preferred because of their non-weight-bearing nature. Third in popularity came soccer (13%) (123).

In conclusion, while some patients with JIA have a limited activity repertoire, most of them conduct an active lifestyle. The factors that limit physical activity in JIA are not clear.

Physiologic Fitness

A small number of studies have documented exercise-related physiologic functions of patients with JIA. A reduction in *muscle strength* has been documented in several studies (90; 126; 181). Muscle weakness is secondary to disuse atrophy, which lasts long after the child has entered

remission, particularly when the disease started before age 3 years (258).

Anaerobic fitness, as determined by the Wingate test, did not differ between patients 8 to 17 years old and healthy controls, nor was there any relationship between anaerobic fitness and disease severity (157). In contrast, a study of Dutch patients revealed an impairment in mean power and peak power of those with poliarticular JIA and systemic JIA, and less so in those with pauciarticular JIA (figure 10.21) (240). One possible mechanism for the low anaerobic performance, in addition to general disuse atrophy of skeletal muscles, is a selective reduction in type II muscle fibers. This has been documented in adults with rheumatoid arthritis, but not in children with JIA.

Results regarding *aerobic fitness* are inconsistent. For example, in one study (89), maximal O_2 uptake was lower in JIA patients compared with controls, but there was no relationship between aerobic fitness and the severity of the disease. In contrast, compared with healthy controls, most 8- to 17-year-old patients with JIA had a normal maximal aerobic power, but there was an inverse relationship between aerobic fitness and the severity of the disease (157) (figure 10.22). Neither the disease duration nor the active joint count seemed to affect the fitness of

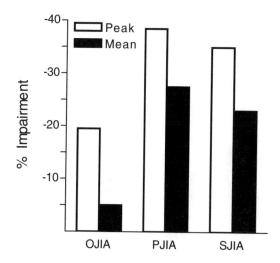

FIGURE 10.21 Impairment in the anaerobic performance (Wingate test) of 8- to 13-year-old patients with juvenile idiopathic arthritis. Subjects were divided into three diagnostic groups: pauciarticular (OJIA), polyarticular, (PJIA) and systemic arthritis (SJIA). Their mean values are presented as percentage of impairment compared with values for healthy controls.

Adapted from Takken et al. 2001 (240).

this sample. A recent meta-analysis of five studies (a total of 144 patients) suggests that their peak O$_2$ uptake, on the average, is 21.8% lower than in healthy controls (239). It is not clear why JIA may be accompanied by impairment in aerobic fitness. One possible cause is the anemia that often accompanies the disease, especially the systemic type, which would result in a reduction in the O$_2$-carrying capacity of arterial blood. Disuse atrophy of skeletal muscles, along with a resulting decrease in O$_2$ extraction from arterial blood, may be another cause. The most likely cause, however, is an overall deconditioning that accompanies reduced physical activity.

Does aerobic fitness remain low during remission? In a study of HLA-B27-positive patients with spondyloarthropathy, those under 18 years of age had a maximal O$_2$ uptake and maximal peak aerobic power similar to those of matched controls. However, in subjects older than 18 years, the aerobic fitness was deficient (103). It is not clear whether the impairment of aerobic fitness among the older patients reflects deconditioning due to a sedentary lifestyle or a deficiency in muscle performance related to the disease process itself.

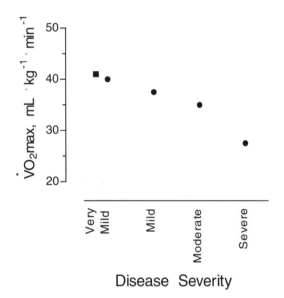

FIGURE 10.22 Maximal O$_2$ uptake ($\dot{V}O_2$max) is inversely related to the severity of juvenile idiopathic arthritis. Thirty-one 8- to 17-year-old patients (circles), who varied in disease severity (global assessment by a physician), and 16 healthy controls (square) performed cycle ergometry. Data are mean values.

Adapted from Malleson et al. 1996 (157).

Relationship Between Physical Function and Fitness

While muscle-strengthening and range of motion programs are included in the therapy of the child with JIA, there is little information regarding the extent to which muscle strength, or other physiologic functions, affects daily functional ability. In one study, scores in the Childhood Health Assessment Questionnaire were moderately correlated with quadriceps torque of 6- to 16-year-old girls with JIA, mostly of the polyarticular and systemic types (75). However, in another study, there was no relationship between several fitness items and indices of disease activity of patients with polyarticular JIA (126).

The implication is that the somewhat low fitness of many patients with JIA, particularly those with a mild-to-moderate severity, does not pose much limitation to their daily functioning. For example, as reasoned by Hebestreit and colleagues (103), even a maximal O$_2$ uptake as low as 27 ml · kg^{-1} · min^{-1} is sufficient to allow adolescent or young adult patients normal performance in daily, nonathletic physical activities.

Beneficial Effects of Enhanced Physical Activity

Until recent years, rest has been recommended as an important component of therapy for patients with JIA. The rationale for limiting patients' activity was that inflammatory conditions are best treated by rest. It was further believed that intense exertion might induce pain, swelling, extreme fatigue, and flare-ups. In the last decade, however, several studies have suggested that activity programs can increase a patient's physical performance and well-being (125; 170; 241), with minimal, if any, deleterious affects.

A well-designed 8-week, three-per-weeek conditioning program of low-impact aerobics, dance, stepping up and down a platform, resistance training, and range of motion exercise resulted in improvement in aerobic performance (9-min run, figure 10.23) and a reduction in affected joint count and in disease severity scores (figure 10.24) (125). The importance of this study was that most of the activities were weight bearing, hitherto looked upon as unsafe for patients with JIA. Improvement in aerobic fitness was achieved also through laboratory-based cycling in a 12-week

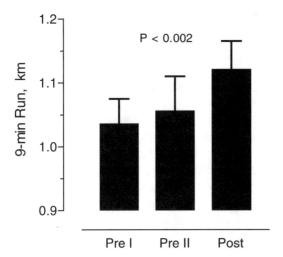

FIGURE 10.23 Effect of an 8-week conditioning program on the aerobic performance of patients with polyarticular juvenile idiopathic arthritis. Twenty-five 8- to 17-year-old patients performed a 9-min running test three times: at baseline, 8 weeks later just before the start of the intervention, and 8 weeks after the intervention. Data are mean and SEM.

Based on Klepper 1999 (125).

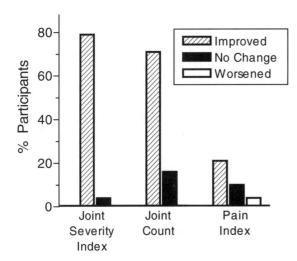

FIGURE 10.24 Effect of an 8-week conditioning program on disease severity and joint pain. Subjects and study design are the same as in figure 10.23.

Based on Klepper 1999 (125).

intervention. The patients had an 11% increase in maximal O_2 uptake and a 41% increase in exercise time to exhaustion, as well as reduction in affected joint count (172).

Safety Considerations

Traditionally, patients with JIA were instructed to avoid intense weight-bearing exercise for fear of exacerbation of inflammation and of symptoms (127). However, studies over the last decade suggest that exertion, even when quite intense, does not trigger deleterious effects in these patients. For example, children with polyarticular disease did not report increased pain, excessive fatigue, or any other problems during and after intense tests of aerobic and anaerobic endurance, muscle strength, or repetitive muscle contraction such as in sit-ups (126; 157). Furthermore, an 8-week *weight-bearing* intervention program (low-impact aerobics, dance steps, and stepping up and down a 10-cm platform) was found to be safe when conducted by professionals (125). While these findings are encouraging, it would be prudent to limit the activity of patients *during periods of acute inflammation* of their weight-bearing joints until more safety data are available.

There is limited information about the *long-term* safety of enhanced physical activity. An 8-year prospective follow-up of 62 patients, ages 6 to 17 years, revealed no differences in the total joint scores (an index of severity of the disease) of those who took part regularly in sport and those who did not. The two most popular sports were cycling and swimming, both of which are non-weight bearing (123). More research is needed regarding the long-term safety of weight-bearing exercise programs.

McArdle's Disease

In the rare autosomal recessive myopathy known as McArdle's disease (also called glycogenosis type V), skeletal muscle phosphorylase is absent or deficient. As a result, the muscle is limited in its ability to generate energy anaerobically through glycolytic pathways, and must resort to blood supply of free fatty acids and glucose. As described originally in 1951 by McArdle (162), patients fatigue early during exercise and often experience muscle cramps, tenderness, and swelling. Their work performance can be extremely low. This disability occurs mostly upon transition from rest to exercise and during anaerobic exercise. If mild exercise is sustained for a few minutes, the patient can usually increase the intensity without undue fatigue, probably due to an enhanced supply of free fatty acids. This phenomenon has been referred to as "second wind" (186).

McArdle's syndrome is often diagnosed during adult years, but careful history-taking reveals the beginning of symptoms in childhood. In some patients the clinical picture becomes apparent in infancy or early childhood. Initial manifestations include suckling and swallowing difficulties, hypotonia, clumsiness, exercise intolerance, severe post-exertional muscle pain, and myoglobinuria. Affected children are often described as "lazy" (204).

Typical metabolic and cardiopulmonary responses to exercise include little or no increase in blood lactate or pyruvate levels, excessive increase in plasma free fatty acid levels (204), high minute ventilation, a high ratio of ventilation to O_2 uptake (97), a depressed ventilatory threshold (196), and enhanced cardiac output increase for a given increase in O_2 uptake (139).

There are several ways to increase the exercise tolerance of patients with McArdle's disease. Infusion of glucose before and during exercise provides an alternative energy source, which alleviates the symptoms and helps to normalize the response to exercise (140). Glucose taken orally, particularly if combined with branch-chained amino acids (46), can also increase exercise performance. The same can be achieved, albeit to a lesser extent, by fasting prior to exercise, which increases the availability of free fatty acids (97). Local cooling has also been reported to delay fatigue and muscle cramps (192).

A recent crossover controlled study has shown that treatment by oral creatine over 5 weeks improves skeletal muscle performance. Specifically, there was an increase in maximal voluntary contraction during an ischemic condition, as well as an increase in endurance time for submaximal muscle contractions. Magnetic resonance spectroscopy revealed an increased use of intramuscular phosphocreatine (257).

Myelomeningocele

Myelomeningocele (MMC, also called *spina bifida*) is found in 1 of 1,000 live births. It is the most severe form of neural tube closure defect, often resulting in neurological deficits that include bladder and bowel incontinence, hydrocephalus, and flaccid paralysis of the lower limbs. The paralysis occurs especially when the lesion is at a midlumbar level or higher. This section focuses on the physiologic fitness of patients with MMC and the role played by physical conditioning.

Habitual Physical Activity and Energy Expenditure

There is a wide range of activity levels among patients with MMC, much of which depends on their walking ability. In one study, the reported daily walking distance ranged from 100 to 5,000 m (79). A person's ambulatory status clearly depends on the level of the spinal lesion. It is extremely rare that a child with a thoracic or high lumbar lesion will be a community walker (i.e., be able to use walking, with or without aids, as the sole means of ambulation in and out of home). In contrast, when the lesion is sacral or low lumbar, the majority of patients are community walkers (e.g., [3]).

Total energy expenditure among adolescents with MMC, based on doubly labeled water, is considerably lower than in healthy controls (14). To a great extent, this reflects a low mobility due to dependence on a wheelchair, walker, braces, or crutches.

In a survey of Ontario children and adolescents with MMC, 85% reported that they were limited in the type or amount of activity that they could do, even though 52% perceived their fitness level to be similar to that of their peers. Overall, the calculated activity score of those with MMC was lower than that in the general Canadian population, but considerably higher than in adolescents with CP or muscular dystrophy.

Physiologic Fitness

Physiologic fitness of patients with MMC depends largely on the level of the spinal lesion. The higher the lesion, the lower the fitness. Information is available mostly regarding muscle strength, aerobic performance, and the energy cost of locomotion.

Muscle Strength

Although the strength of the upper limb muscles is usually normal in patients with MMC, their lower limb strength is deficient (3). Such impairment is a major cause of ambulation disability (166). Figure 10.25 demonstrates the strong relationship between maximal ambulation velocity and the strength of the knee extensors in 10- to 15-year-old patients. Seventy-two percent of the variance in ambulation velocity was explained by the maximal force of the hip extensors relative to body weight (3).

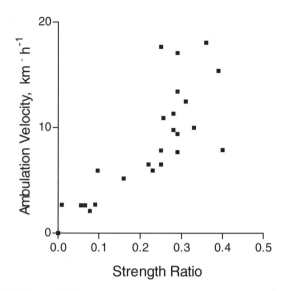

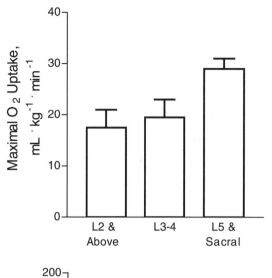

FIGURE 10.25 Relationship between maximal ambulation velocity and knee extensor strength in patients with myelomeningocele. Thirty-three patients, ages 10 to 15 years, participated. Walking speed was measured over 30 m. Strength ratio is the ratio of maximal isometric force (in kilograms) to body mass (in kilograms).

Modified from Agre et al. 1987 (3).

Aerobic Fitness

Maximal O_2 uptake varies markedly in patients with MMC. It can range from normal levels for patients with no neural deficit to as low as 10 to 15 ml $\cdot$ kg^{-1} $\cdot$ min^{-1} in those with a high lumbar or a thoracic lesion (3). As seen in figure 10.26, maximal O_2 uptake is strongly related to the level of the spinal lesion, as are maximal heart rate and maximal minute ventilation. The low maximal heart rate and ventilation in patients with high spinal lesions suggest that maximal O_2 uptake is limited not by the cardiopulmonary system, but by the mass and function of the lower limb muscles. Lower aerobic fitness has been described in patients with associated hydrocephalus, but this difference disappears when the level of the lesion and the level of ambulation are considered (3).

Energy Cost of Ambulation

O_2 uptake at a given walking or wheeling speed is excessive in those MMC patients who have a neural deficit (3; 63; 78), particularly in those with a high spinal lesion. In one study, the O_2 cost per meter walking at a comfortable speed was 0.55 ml $\cdot$ kg^{-1} $\cdot$ m^{-1} in 5- to 12-year-old patients with a lesion at L3/L4, compared with 0.39 ml $\cdot$ kg^{-1} $\cdot$ m^{-1} when the lesion was at L5/S1. The respective cost for healthy controls was 0.24 ml $\cdot$ kg^{-1} $\cdot$ m^{-1}

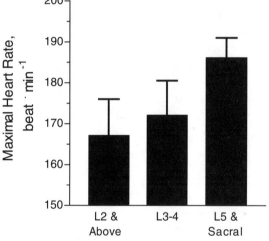

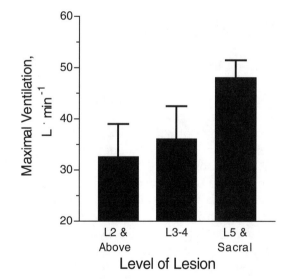

FIGURE 10.26 Relationship between maximal physiologic functions and the level of the spinal lesion in myelomeningocele. Patients, 10 to 15 years old, walked on a treadmill, using a progressive-continuous protocol to exhaustion. Vertical lines denote 1 SD. Based on Agre et al. 1987 (3).

(63). The energy cost in patients who use a walker because of a thoracic lesion can be as high as 1.5 to 2.0 ml · kg⁻¹ · m⁻¹ (148).

These data indicate, then, that excessive energy cost of ambulation, combined with weak lower limb muscles, contributes significantly to walking disability in patients with MMC.

Beneficial Effects of Training

Like healthy children, patients with MMC are physically trainable (7; 179). For example, in a randomized controlled study, eight patients underwent a 10-week, 1 hr per week program of aerobic and strengthening elements. Their performance in a 9-min walk/run test improved by 46%, compared with a decrease of 11% in the controls. Likewise, the strength of the shoulder flexors increased by 19% in the training group and decreased by 2% in the controls (7).

Strengthening of upper body muscles is an important means of improving wheelchair propulsion, which for many patients is the major or exclusive means of ambulation. For example, an 8-week, three-per-week program of circuit resistance training (shoulder girdle, elbow extensors and flexors, and bench pressing) was accompanied by a 29% increase in the distance covered during a 12-min test (179). A strong upper body is also important for patients who use crutches for walking.

Muscular Dystrophy

Several skeletal myopathies that appear during childhood have a progressively debilitating course. This section focuses primarily on the most prevalent in the pediatric age group—Duchenne muscular dystrophy (MD). Duchenne MD is an X-linked genetic disease that occurs in 1 of 3,000 to 4,000 live-born males. A very rare variation of Duchenne MD can occur in females. The disease is characterized by the absence of *dystrophin*, a protein essential to the structural integrity of the muscle cell membrane (187). The pathologic changes and functional deterioration occur first in the proximal muscle groups (e.g., at the shoulder and pelvis) and gradually affect the smaller, distal ones.

Duchenne MD is invariably progressive, and it causes premature death. Muscle weakness is usually not identifiable before the second year of life, and the diagnosis is made by age 4 to 5 years. Most patients can still walk at age 10 to 12 years, but some must use a wheelchair as early as age 6. At a later stage, the patients become bedridden. The great majority of them die by age 18. Death occurs mostly due to cardiomyopathy and, less frequently, respiratory failure.

Becker MD occurs less frequently than Duchenne MD and progresses more slowly. The boys are still ambulatory during adolescence, and some can still walk at age 20 years. While most patients die during the third decade of life, some survive until their early 40s.

Habitual Physical Activity

There is little epidemiologic information on the habitual physical activity and attitudes toward activity of children with muscular dystrophy. In a survey of 342 Ontario children and adolescents with a physical disability, 6 to 20 years old, those with Duchenne MD reported the lowest level of activity compared with responders who had CP, head injury, or myelomeningocele. Only 18% of those with Duchenne MD were physically active. Seventy-nine percent of them perceived their fitness level to be lower than in their peers, and all reported that they had a "physical limitation" (147).

The physical activity of patients with MD parallels the natural history of their disease. The functional level of an MD patient who can still walk (often with the aid of braces or a walker) can be divided into four stages (256):

1. Normal ambulation pattern, with only a slight functional deficit

2. Definite decrease in the ability to perform strenuous tasks and a concomitant reduction in habitual activity

3. Decrease in all types of physical activity; daily duration of walking and standing less than 2 hr

4. Daily duration of walking and standing less than 30 min

The decline to wheelchair status results from the following factors:

- Diminished residual muscle strength, especially of the knee and hip extensors
- Contractures in the lower limb joints, especially the knee and hip

- Bed rest that may follow illness, surgery, or injury
- Development of overweight or obesity, which causes an excessive load on the weak muscles
- Psychological factors, such as anxiety due to fear of falling, or withdrawal from the outside world
- Decision by well-meaning relatives, teachers, and health care providers to "make it easy" for the child by providing him with a wheelchair

Additional factors that enhance the loss of walking ability include prolonged sitting with bent knees; plantar flexion contracture, which forces the child to walk on the toes; and asymmetry of lower limb deficit, which induces "favoring" of one side and unilateral contractures. Tests of timed motor functions, such as the time needed to walk a certain distance, can be used to predict how near a child is to resorting to wheelchair status. This was shown by prospective observations of 51 patients. Of those who could walk 30 ft in less than 6 sec, 89% were able to avoid wheelchair status for over 2 years. In contrast, all those who needed 9 to 12 sec to walk that distance were using a wheelchair within 2 years (165). Whatever the reason for nonambulation, a child with MD who is put to bed for a few weeks may never be able to walk again. Contractures and atrophy of residual muscle develop so fast during bed rest that even surgery or braces may not reambulate such a child (36).

In conclusion, although transition from an ambulatory to a more sedentary life is inevitable, a premature decline in activity status may well reflect some failure of management.

Physical Fitness

Four fitness components are deficient in the child with MD: muscle strength, peak mechanical power, muscle endurance, and maximal aerobic power.

Muscle Strength

The fitness component muscle strength is crucial to the child's ability to rise, walk, and perform other daily functions. Muscle strength has been assessed mostly by clinical, ordinal rating (81; 165; 256), but also by more quantifiable and objective methods (37; 56; 85; 107; 165; 167; 215; 228; 237).

Either approach has shown a continuous decline in muscle strength. This is exemplified in figure 10.27. Figure 10.27a depicts the deterioration in manually determined strength in children and adolescents with Duchenne MD (165); figure 10.27b relates isometric strength to body height of patients and controls. While the latter show a continuous increase in strength, the patients have a negligible change in absolute strength, so that their scores at age 16 are similar to or lower than those at 5 years (86). Such nonprogression in absolute muscle strength during growth is equivalent to a marked drop in functions such as walking speed (215) or the ability to negotiate stairs. This deficiency is further aggravated if the child becomes overweight.

The extent of such regression may be appreciated if we compare the strength of children with Duchenne MD to that of the 5th percentile of size-matched controls, as shown in figure 10.28 for six muscle groups. Almost invariably, patients taller than 120 cm score below the 5th percentile. Thus, strength norms established for healthy children are of little use for the evaluation of a child with MD.

Anaerobic Performance

Clinical observations show that children with MD have a low muscle endurance. This can be judged from their easy fatigability when walking or climbing stairs. An attempt was made to assess muscle endurance objectively by measuring the time for which a supine child could hold his neck or leg at 45° from the ground (107). Ninety-two percent of the patients scored below the 5th percentile of healthy children. The drawback of such a measurement is its low reproducibility and questionable standardization (82).

In the last decade, the 30-sec Wingate anaerobic test has been used to assess peak muscle power and muscle endurance of children with MD. This test has been found to be highly reliable for such patients (test-retest $r = 0.99$ for mean power and peak power, $n = 13$ with DMD). (See also figure 10.12 [243].) Their ability to complete the 30-sec pedaling or arm cranking test depends on the severity of the disease. In general, the test is much more feasible for use in arm cranking, while the child sits in his own wheelchair, than in pedaling. Patients with Duchenne MD, and some with Becker MD, often find it easier to perform the 30-sec anaerobic test than a longer cycling or arm cranking aerobic protocol.

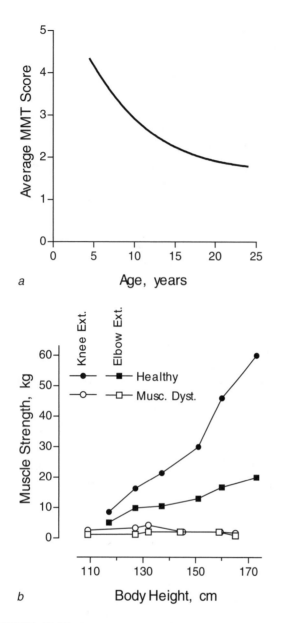

FIGURE 10.27 Deterioration of muscle strength in Duchenne muscular dystrophy. (*a*) Summary of data for 70 patients whose strength was assessed by a manual muscle test (MMT). The MMT score is a composite value for 34 muscle groups. Reproduced, with permission, from McDonald et al. 1995 (165). (*b*) Summary of data for 43 patients and 45 controls, ages 5 to 16 years. Their isometric strength was measured by cable tensiometry.

Based on data from Fowler and Gardner 1967 (85).

Invariably, children with Duchenne MD score extremely low (as low as 5 to 7 SD below the mean for healthy children) in both muscle endurance and peak anaerobic power. Figure 10.29 displays longitudinal changes in peak power of three patients, compared with normative data. The figure depicts the futility of trying to compare the performance of such patients to normal values. It is impossible to provide "typical" values for anaerobic performance in children with MD because of the progressive nature of the disease. However, as seen in figures 10.12 and 10.29, scores for a child with Duchenne MD can be as low as 15 W for peak power and 10 Watts for mean power (242). These values are close to the lowest levels detected by most cycle ergometers.

The low anaerobic performance of children with MD is linked to their deteriorating strength. It may result also from a selective loss of fast-twitch muscle fibers (187), particularly of the IIb type (260).

Aerobic Performance

Peak aerobic power output and maximal oxygen uptake are markedly low in patients with MD (228) and other myopathies (40). This is mostly due to the small number of functional motor units in the muscles of these patients and, to a lesser extent, their compromised respiratory (12; 94; 110; 193; 198; 233) or cardiac (88) function. One indicator that the circulatory system is not taxed to its potential during "aerobic" activities is the very low peak heart rate reached by MD patients during ergometry. They often are forced to terminate a test while their heart rate is only 120 to 130 beat · min^{-1}. It is thus safe to assume that the main limiting factor in these children's physical ability is not their oxygen transport system, but rather their muscle endurance, anaerobic power, and strength. Based on limited information, mechanical efficiency and anaerobic threshold are not different between patients with myopathies and healthy controls (40; 61). Hamsters with MD, however, have a low mechanical efficiency and possibly reduced oxidative capacity (137).

Response to Training

Residual, unaffected muscle tissue is trainable. If the strength and endurance of the residual motor units can be improved, one can also expect some arrest of functional deterioration and even a temporary functional improvement (213). The end result may be a prolongation of the period in which the child can still maintain an upright position and be ambulatory. The effects of training on limb muscles have been evaluated in various studies in humans (1; 56; 60; 86; 117; 168; 169; 177; 216; 233; 255) and in animals (41; 64; 84; 100; 265). While most human studies focused on the

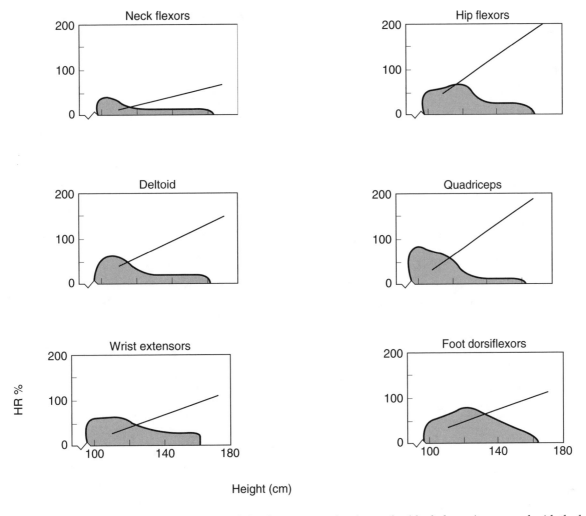

FIGURE 10.28 Muscle strength of children with Duchenne muscular dystrophy (shaded areas) compared with the lower limit of normal in healthy children. Measurements were made with the Hammersmith myometer. Regression lines are "near fit" 5th percentile of 215 healthy children.

Adapted from Hosking et al. 1976 (107).

effect of resistance training on upper or lower limb muscles, the aim in some was to train the respiratory (128; 198; 259) or masticatory (117) muscles. Other studies used electrical stimulation of the muscles (217; 218; 271).

An example of the possible benefits of strength training is shown in figure 10.30. Ambulatory boys with Duchenne MD, 6 to 10 years old, underwent a 1-year program that included resistive and assistive exercise of the hip abductors, hip extensors, knee extensors, arm flexors, and abdominal muscles (255). As expected, there was a decrease in their composite strength index of about 15% during the year preceding the program. But there was no further decline in strength during the year of conditioning. This was in contrast to findings for age-matched patients who did not train and

whose strength kept declining. One should note, however, that the decline in strength during the year prior to training was considerably slower in the controls, which suggests that the two groups were not matched for disease severity. This detracts from the validity of this study.

Indeed, obtaining a proper control group is a difficult challenge for investigators. The major reason is a paucity of available subjects. Another constraint is the inconsistent rate of functional deterioration, which makes matching difficult. Most patients are given medications and other therapies, which makes it hard to tease out the specific effect of physical training. The main constraint, however, is ethical: How can one justify a program in which the controls are denied physical rehabilitation?

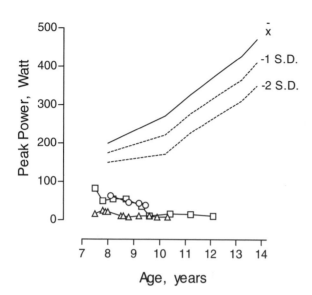

FIGURE 10.29 Anaerobic performance of children with Duchenne muscular dystrophy. Longitudinal data are displayed for three male patients who were tested periodically during clinic visits. The lines represent cross-sectional norms established at the Children's Exercise & Nutrition Centre, using identical equipment and protocol for the Wingate anaerobic test.

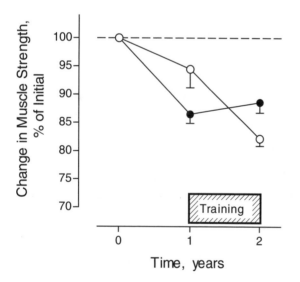

FIGURE 10.30 Changes in isometric muscle strength of boys with Duchenne muscular dystrophy (n = 14) before and during a 1-year strength training program (black circles). Comparison with a sedentary group (n = 14, open circles). "Muscle strength" is a composite index, based on the strength of several muscle groups. Weighted mean was obtained according to the assumed mass of each muscle group.

Adapted from Vignos and Watkins 1966 (255).

One solution is to compare two modes of training, without the use of nonexercising controls. Scott et al. (216) allocated boys with Duchenne MD into strength versus muscle endurance training groups. There was no intergroup difference in the changes in muscle strength and locomotive function by the end of the home-based 15 min per day, 6-month program. In the absence of nonexercising controls it was impossible to determine whether either training regimen was helpful.

An alternative solution is a design in which one limb is trained while the contralateral limb serves as control. De Lateur and Giaconi (56) trained the knee extensors of one limb for 6 months in four 4.8- to 11.1-year-old boys with Duchenne MD. Three of the subjects improved their peak isokinetic torque. This effect was still noticeable 18 months after the intervention. A similar design yielded an increase in strength of the training arm, but not of the contralateral, nontraining arm, of adults with MD who underwent a strength training program (163). Trainability has been documented also for masticatory muscles of patients with Duchenne MD (117), as well as for limb muscles of *mdx* mice that lack dystrophin (41; 64; 84; 100; 265).

Strengthening of skeletal muscles can be achieved also by electrical stimulation, particularly of the low-frequency type. For example, a 3-month stimulation of the tibialis anterior muscle resulted in an increase in muscle torque compared with that in the nonstimulated leg (272). This intervention did not improve muscle fatigability. In a subsequent study, electrical stimulation over 9 months was more efficacious than after 3 months (271).

In conclusion, based on human and animal studies, it seems that the residual functional muscle is indeed trainable. The dividends of muscle training are particularly evident in patients who are at early stages of the disease and still have abundant functional muscle fibers (213). Moreover, the progressive loss of strength can be slowed down and, for a while, reversed. An interesting, as yet unexplained, finding is the improvement in short-term associative learning of boys with MD following a bout of exercise (60; 67).

Training of the Respiratory Muscles

Deterioration of muscle function in MD includes also a progressive reduction in respiratory functions (12; 110; 193; 198; 233). The changes are of the restrictive type, reflecting a weakness of the respiratory muscles and a decrease in chest wall compliance. Some patients develop scoliosis, which further impedes their respiratory function. Slowing down this deterioration is of importance, since respiratory insufficiency is a major cause of death in these young patients (193). The question is whether specific training of the respiratory muscles is beneficial. Some studies that have suggested this were not designed to exclude a possible placebo effect (99). In a randomized, placebo-controlled, double-blind crossover design, 9- to 14-year-old boys with Duchenne MD underwent inspiratory training (20 inspirations against resistance daily for 18 days), but there was no improvement in lung volumes and flows or in maximal inspiratory and expiratory pressures (198). In another randomized, crossover design (233), training of 10- to 23-year-old patients with Duchenne MD lasted 6 months. Compliance was encouraged by computer games, which operated only when the patient inhaled with sufficient force. Even though there were no changes in forced vital capacity or respiratory muscle endurance, there were near-significant increases in maximal inspiratory and expiratory pressures. Beneficial effects of inspiratory muscle training were still evident several months after the cessation of training, as shown in a controlled study of 15 boys with Duchenne MD (259). The authors concluded that such a benefit is particularly apparent at early stages of the disease. More research is needed in this important field to identify optimal training programs, compliance, and the age at which intervention might be most efficacious.

Prevention and Management of Obesity

Overweight and obesity often occur in children with MD (264), particularly in early stages of the disease. The exact mechanism is unclear, but overweight is most likely an outcome of a positive energy balance due to low resting (98) and total energy expenditure. Typically, fat accumulates in the dystrophic muscles (138). Because of the weak skeletal muscles, excessive body mass further impedes the child's ability to ambulate. At advanced stages of the disease, overweight also interferes with respiratory function, as the heavy chest increases the mechanical and metabolic cost

of breathing (262). Prevention and management of obesity in these patients are therefore of major importance. Ideally, body adiposity should be monitored periodically, and nutritional counseling should be given to the child and parents even before body fat mass starts to increase. Helping the child is more difficult once he is already obese. One potentially detrimental effect of a low-energy diet is the loss of fat-free mass, muscle in particular. Edwards et al. (66) treated two markedly obese boys, one with Duchenne MD and the other with Becker MD, with a very low-energy diet for approximately 1 year. Both boys lost considerable amounts of body mass, but there was only a transitory mild negative nitrogen balance. The authors concluded that the low-energy treatment was safe. More research is needed to determine the efficacy, effectiveness, and safety of various dietary changes, but prevention of obesity is still an appropriate approach.

Deleterious Effects of Immobilization

To prevent a precipitous decline in fitness and motor function, children with MD should avoid immobilization. It has been a common clinical experience that once a patient is immobilized for several weeks, for whatever reason, his ability to ambulate may be jeopardized. A case in point is a boy with Duchenne MD who was first seen in our clinic at age 7.5 years. At that point he was still able to walk unassisted. The patient complied fairly well with a combined activity-nutrition program during the following 6 months and, as seen in figure 10.31, his anaerobic muscle performance remained stable during that period. He then underwent a bilateral release of his heel cord, which was accompanied by bed rest for several weeks. When he was subsequently seen at the laboratory, the peak and mean anaerobic power of his legs had declined by more than 50%, and it did not rise in subsequent tests. A similar pattern occurred 2 months later in the performance of his arms. In point of fact, after the surgery this boy never walked again. Obviously, this example cannot be generalized to all instances, but it does demonstrate the potentially detrimental effects of immobilization.

More information is needed on the optimal training methods for children with MD. Special attention must be given to improvement

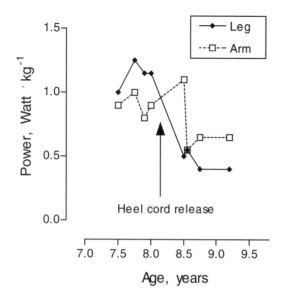

FIGURE 10.31 Changes over 20 months in peak mechanical power (Wingate anaerobic test) of the arms and legs in a boy with Duchenne muscular dystrophy. The arrow denotes time of surgery for a bilateral release of the Achilles tendon. Data from the Children's Exercise & Nutrition Centre.

Reproduced, with permission, from Bar-Or 1993 (16).

of muscle strength and endurance. Based on current knowledge, one can draw the following conclusions:

1. The rate of decrease in muscle strength, muscle endurance, and anaerobic power of children with Duchenne dystrophy, and especially in more slowly progressing muscular dystrophies, can be reduced by strength training of a few months' duration. In some cases, these functions can *improve* transiently.

2. The degree of improvement is a function of residual muscle mass. Therefore, the more advanced the disease, the less apparent the improvement in muscle function.

3. Strength training per se is ineffective in slowing down the loss of ambulation. There is a need for additional activities that specifically make the patient rise, stand, and walk, even if helped by assistive devices or by others.

4. Specific attention should be given to the prevention and management of contractures. A muscle that is strong enough to move a limb may prove too weak when a contracture is superimposed.

5. One should guard against obesity. The weak muscles may not be able to carry an overweight individual.

Even though wheelchair status is inevitable for all patients with Duchenne MD and for many patients with other MDs, it always represents a major setback to child and parents alike. In giving himself up to physical passivity, the patient is inviting obesity, cardiorespiratory dysfunction, scoliosis, lower extremity edema, accelerated muscular atrophy, weakness, contractures, and emotional crises (224). It is therefore the goal of the therapist to turn such a change of status into an opportunity for enhanced activities rather than to let the patient become resigned to a passive lifestyle. Examples of specific wheelchair-based exercises are available in textbooks of physiotherapy or adapted physical education (51; 266).

Safety Considerations

Concern has been raised that enhanced physical activity may be detrimental to the child with MD. This has been based on two arguments: (1) Clinical and histopathologic data suggest that deterioration is faster in the more active muscle groups, and (2) exertion is followed by an excessive rise in creatine kinase (CK). How valid are these claims?

The argument regarding activity-induced deterioration is based mostly on anecdotal, uncontrolled observations of "overwork weakness" of active muscle groups (e.g., [114]). However, reports of controlled interventions (56; 216) with follow-up periods as long as 24 months (56) have not confirmed excessive weakness in the training muscles. A similar lack of overwork weakness has been reported for adults with slowly progressing neuromuscular diseases who underwent a low-intensity 12-week aerobic program (270).

Another basis for the concern that activity accelerates deterioration of active muscles in these patients emanated from a postmortem study of one child with Duchenne MD (32). His proximal limb muscles and some postural muscles were more affected histologically than were other muscle groups. The author concluded that the affected muscles had been more active during the boy's life and therefore that physical training might accelerate the pathologic process. It is not clear, though, whether this histologic difference truly reflected differences in habitual activity. If it did, other muscle groups that are even more active (e.g., intercostal respiratory muscles) would be expected to show deterioration. In this context, muscle groups of mice with MD that had been trained daily for 3 weeks (low-intensity, high-repetitive exercise) showed less degeneration than the respective muscle groups in untrained dystrophic mice (84). This suggests a protective, rather than detrimental, effect of exertion. Children with MD often respond to exertion with excessive levels of serum CK. However, there is no proof that such a response reflects damage to their muscles (80; 111).

In conclusion, enhanced physical activity, particularly at mild-to-moderate intensities, appears to be safe for children with MD. However, one should address each case individually. In general, patients should be encouraged to increase their level of activity; but if a certain task induces pain or excessive fatigue, its intensity, or the frequency of exercise sessions, or both should be decreased. Additional research is needed to further refine safety-related recommendations.

Scoliosis

Scoliosis indicates abnormal lateral curvature of the spine. It may be secondary to other conditions, such as muscular dystrophy or myelomeningocele, but the most common type is *idiopathic scoliosis* of the thoracic spine, which is the topic of this section. This condition may develop in otherwise healthy children and adolescents—mostly the latter—and is more common among females than among males. The degree of malalignment often progresses with growth, particularly among female patients. This serves as the main rationale for early treatment. The severity of scoliosis is determined by the angle (using the *Cobb* method) of the lateral curvature as measured by radiographic anteroposterior imaging of the chest.

Habitual Physical Activity

There are few quantitative data on the physical activity or daily energy expenditure of patients with scoliosis. It seems, though, that patients with scoliosis are often reluctant to take part in games and other physical activities because of their perceived nonaesthetic appearance and poor self-image (92). Some limit their activity because of the need to wear a spinal brace (2). In advanced scoliosis, exercise-induced dyspnea is another barrier to physical activity. A vicious cycle of further deconditioning and functional deterioration may ensue.

Functional and Physiologic Impairments

When scoliosis is advanced, exertional dyspnea is common and the patients have a low maximal aerobic power and impaired cardiorespiratory function (31; 222; 223). Maximal O_2 uptake as low as 11 ml $\cdot$ kg^{-1} $\cdot$ min^{-1}, and usually not exceeding 25 ml $\cdot$ kg^{-1} $\cdot$ min^{-1}, has been found among severely affected patients (compared with 40-55 ml $\cdot$ kg^{-1} $\cdot$ min^{-1} in healthy children). Subsequent studies (44; 58; 59; 77; 118; 122) have shown that even mild-to-moderate scoliosis (e.g., less than 60-70°) is accompanied by reduced aerobic fitness and cardiopulmonary function. For example, when 38 adolescents (mean age 15.1 years) with scoliosis performed a maximal treadmill test, even mild curvatures of 20° or 30° were associated with a decline in aerobic performance (44).

A low maximal aerobic power is related to the chest deformity, undersized lung, small peripheral musculature, hypoactivity, and generalized deconditioning. The main pulmonary functional abnormalities at rest and exercise are summarized in table 10.2. Chest deformity of the child with scoliosis may result in reduced total lung and vital capacities at rest (even when allowance is made for reduced body stature); increased work of the respiratory muscles, which contract at a mechanical disadvantage due to distorted position of the ribs (28; 31; 48); ventilation-perfusion imbalance in some lung regions, which is manifested by a widened alveolar-arterial PO_2 gradient, particularly during exercise (gradients as high as 35-40 mmHg have been reported); low tidal volume; high physiologic dead space during exercise; and a high pulmonary artery blood pressure (but not pulmonary wedge pressure) during rest and exercise (221; 227). The O_2 cost of walking is excessive among patients with advanced scoliosis (141), probably due to mechanically distorted gait but also because of the high O_2 cost of breathing.

In mild-to-moderate scoliosis, a sedentary lifestyle may be more of a limiting factor than are the direct respiratory deficiencies, which become limiting mostly when the deformity has reached an advanced stage (31; 118). According to another opinion (220), ventilation is the limiting factor for the majority of exercising children and adolescents with scoliosis.

TABLE 10.2	Pulmonary and Metabolic Function During Rest and Exercise in Adolescents With Moderate or Severe Scoliosis, Compared With Healthy Individuals

Function	Compared to healthy individuals
Rest	
Total lung capacity	Lower
Vital capacity	Lower
Work of breathing	Higher
Alveolar ventilation	Lower
Pulmonary artery pressure	Higher
Submaximal exercise	
Minute ventilation	Higher
Breathing rate	Higher
Minute ventilation/CO_2 output	Normal
O_2 cost of locomotion	Normal to higher
Tidal volume/vital capacity	Normal to lower
Arterial O_2 saturation	Normal
(Aveolar-arterial) PO_2 gradient	Higher
Pulmonary artery pressure	Higher
Maximal exercise	
O_2 uptake	Lower
Minute ventilation	Lower
Minute ventilation/CO_2 output	Normal to lower
Breathing rate	Normal to higher
Tidal volume	Normal to lower
Tidal volume/vital capacity	Normal
Arterial O_2 saturation	Normal to lower
Pulmonary artery pressure	Higher

Effects of Treatments on Physiologic Fitness

The three most common treatments of scoliosis are surgery (usually spinal fusion), thoracic bracing, and strengthening exercise for specific muscle groups. The *benefit of surgery* to exercise performance is marginal. A slight increase in maximal ventilation (235) and a small decrease in submaximal ventilation (222) have been noted. There has also been some decrease in O_2 cost during a treadmill walk at slow speeds (141). Such changes (or their absence) are hard to interpret because of differences in body dimensions, level of maturity, activity patterns, and fitness status between pre- and postsurgical patients. Differences in response may also result from the variability in surgical corrections.

Bracing of the chest is used in order to reduce the angle of scoliosis or prevent its further progression. A side effect of this therapy is a restriction in the mobility of the chest and the abdomen. This may lead to respiratory impairment during rest and exercise and a mild reduction in exercise performance when the patient wears the brace (77; 191). For example, 10- to 16-year-old girls with mild-to-moderate scoliosis (Cobb angle 19-60°) showed a decrease in resting lung functions such as vital capacity (reduction of 20%), forced expiratory volume in the first second (20%), and maximal voluntary ventilation (16%). They had an increase of 32% in the ventilatory cost and of 20% in the metabolic cost during submaximal cycling exercise (191). The exercise-related impairment may reflect an excessive anatomical dead space and metabolic cost of breathing. Interestingly, the high ventilatory and metabolic cost diminished gradually over 6 months of bracing, which suggests a gradual habituation to the brace and, possibly, a greater reliance on diaphragmatic breathing (191). In another study (141), the wearing of a brace was not accompanied by an increase in the metabolic cost of walking. It is possible that the restrictive effect of the brace was compensated for by a more favorable posture during the walks.

An additional effect of bracing may be an increase in dyspnea during exercise because of the extra inspiratory effort. Using the Borg 0 to 10 scale (see appendix II), six 12- to 15-year-old girls with mild scoliosis perceived increased dyspnea when they were wearing a Boston or a Milwaukee brace at peak and at 50% peak cycling exercise intensity (figure 10.32) (77).

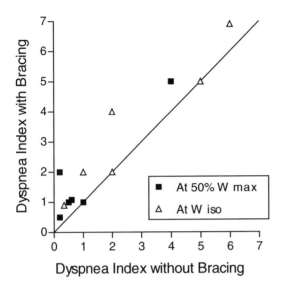

FIGURE 10.32 Effect of chest bracing on the dyspnea index of girls (12-15 years) with idiopathic thoracic scoliosis. Six girls performed a maximal test on a cycle ergometer. Their dyspnea index was determined by the Borg 0 to 10 scale (33) and is presented at two exercise intensities for each girl.

Based on Ferrari et al. 1997 (77).

Effects of Training

Exercise therapy is recommended in scoliosis for two reasons. The first is the mobilization and strengthening of the trunk and leg muscles that are related to posture. Training is expected, in conjunction with other modes of therapy, to slow down and possibly reverse the progress of the spinal curvature. Second, through training of specific muscle groups in the trunk, pulmonary functions such as vital capacity can be improved (261). It is not within the scope of this book to discuss in detail those "remedial exercises" that have been suggested to arrest the postural deterioration of the spine. This topic is covered in textbooks of physiotherapy and adapted physical education. The results of such exercise programs have been equivocal, especially when a control group is also observed (236).

An increase in maximal aerobic power and in ventilatory efficiency (pulmonary ventilation/O_2 uptake) can be obtained through training. This has been shown for patients with various degrees of scoliosis (11; 31; 74; 234; 238). For example, in a randomly controlled intervention, twenty 12- to 14.5-year-old girls with mild scoliosis (27.4° ± 1.9°) took part in a 2-month four-per-week cycling program. Each session lasted 30 min. Their W_{170} (mechanical power at a heart rate of 170 beat · min^{-1}) improved by 48%, compared with a decline of 9% in the 20 controls (11). As is true for other training programs, compliance is better when the sessions are supervised. When the program is home based, there may be no appreciable benefit (223).

Does the effectiveness of training programs vary with the severity of the curvature? There is no definitive study comparing large enough groups of patients, selected according to the angle of the curvature. It does seem, though, that the least trainable are those with very advanced scoliosis (e.g., 130-150° curvature). Some of these patients also develop high pulmonary arterial blood pressure during exercise, and for these persons intense exertion is not advisable (221). Among the less severely affected patients there has been no correlation between the improvement in maximal aerobic power and the initial spinal curvature (31; 234).

Hematologic, Oncologic, and Renal Diseases

*T*he production of motor activity is an extraordinarily complex event that brings into play the actions or reactions of virtually every body system. It follows that disturbed function or disease in any component of the exercising "machine" can be expected to interfere with normal muscle function, provision of energy substrate, or maintenance of body homeostasis. In addition, the chronically ill individual is likely to adopt a sedentary lifestyle, further compromising physical fitness and the health benefits of regular exercise. These patients also experience secondary psychosocial effects that can negatively affect quality of life. Understanding the relationship between specific chronic diseases and exercise may therefore bear importance from physiologic, preventive health, and therapeutic perspectives.

This chapter explores these aspects of exercise in children and adolescents with hematologic diseases, cancer, and chronic renal disorders.

Anemia

The energy demands of contracting muscle during sustained exercise—even that lasting as briefly as 30 sec—require an ongoing source of oxygen. The body meets this need through a transport of oxygen from the ambient air to its interior milieu through the integrated function of lungs, heart, and circulatory system. Within this chain of oxygen delivery, the transport of O_2 bound to hemoglobin in circulating erythrocytes serves as a critical link in providing the cellular needs for aerobic metabolism. Any decrement in the number or function of red cells in the blood, or anemia, can therefore be expected to impair oxygen delivery, limit aerobic metabolism, and consequently reduce level of physical fitness.

According to the Fick equation (see chapter 1), oxygen uptake is the product of cardiac output and the arterial-venous oxygen difference, or the difference in O_2 content (defined by the hemoglobin concentration and oxygen-binding capacity) between the arterial and venous blood. In the resting adult male, the oxygen contents of arterial and venous blood are approximately 19 and 14 mL $O_2 \cdot$ 100 mL^{-1} blood, respectively, creating an arterial-venous oxygen difference of 5 mL $O_2 \cdot$ 100 mL^{-1}. With increasing exercise intensity, little change occurs in the arterial oxygen content, but that of the venous effluent progressively falls with greater cellular extraction of oxygen to levels as low as 2 to 3 mL $O_2 \cdot$ 100 mL^{-1}. That is, the arterial-venous oxygen difference in the adult male at peak exercise is typically about 16 to 17 mL $O_2 \cdot$ 100 mL^{-1} (5).

In the individual with anemia, the arterial oxygen content—and consequently oxygen delivery—is diminished. Since venous oxygen content cannot be further reduced by a substantial degree, the individual with decreased red cell mass experiences a reduction in maximal arterial-venous oxygen difference and limitation of $\dot{V}O_2$max. This critical dependence of oxygen transport on a sufficient number of red blood cells is demonstrated by the close relationship

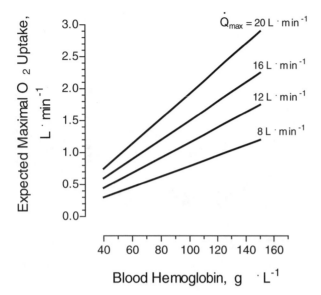

FIGURE 11.1 Maximal oxygen uptake as a function of hemoglobin concentration and maximal cardiac output ($\dot{Q}$max).

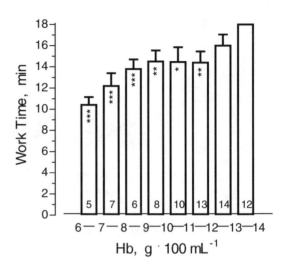

FIGURE 11.2 Maximal treadmill endurance time for individuals with different blood hemoglobin (Hb) concentrations ($P < 0.05*$, $P < 0.01**$, $P < 0.001***$).

Reprinted, with permission, from Gardner et al. 1977 (29).

observed between absolute VO_2max and total body hemoglobin (figure 11.1) (4).

It takes little decrease in the normal amount of circulating erythrocytes (expressed as the hematocrit, or percent of red blood cells) to negatively affect exercise performance (24; 42). For instance, Gardner et al. reported a 20% decrease in treadmill exercise time in adult women with a hemoglobin concentration between 11.0 and 11.9 g · dL^{-1} compared to those with a value over 13.0 g · dL^{-1} (figure 11.2) (29). A decrease in hemoglobin concentration of 0.3 g · dL^{-1} is associated with a decline in maximal aerobic power of approximately 1% (32).

The principles just outlined hold for children as well as adults, but quantitative changes in hematocrit and hemoglobin concentrations occur during the growing years. Both values rise slowly during childhood, with little gender difference (figure 11.3) (18). For instance, the typical 2-year-old boy or girl has a hemoglobin concentration of 12.6 g · dL^{-1}, but this level rises to 13.7 g · dL^{-1} by age 12 years. At puberty the actions of testosterone cause a rise in hematocrit and hemoglobin concentrations in males (34). At age 16 years the average male has an 11% greater hemoglobin concentration than a female (15.2 vs. 13.7 g · dL^{-1}).

Because of these normal alterations in hemoglobin concentration with biological maturation, the definition of anemia changes depending on age and gender. Table 11.1 outlines the average as well as the lower limits of normal hemoglobin

concentration and hematocrit during the prepubertal, adolescent, and adult years (66). Values indicative of anemia rise during childhood but are not different in boys and girls until puberty.

Consistent with this information, no differences are seen in maximal arterial-venous oxygen difference in prepubertal boys and girls matched for chronological age. On the other hand, males after puberty demonstrate values for maximal arterial-venous oxygen difference that are about 4 mL · 100 mL^{-1} greater than for prepubertal subjects and postpubertal females (figure 11.4) (79; 80).

Anemia can reflect a decrease in red cell production by the bone marrow, premature or increased rate of destruction of red cells (hemolysis), or blood loss from acute or chronic hemorrhage. A decline in hemoglobin concentration, hematocrit, or both can occur acutely or over time on a chronic basis, which may affect the degree of hemodynamic compromise. This chapter addresses issues surrounding exercise and chronic anemia created by sickle cell disease, thalassemia, and iron deficiency. Exercise considerations in persons with sickle cell trait are also discussed, as well as the "pseudoanemia" of highly trained athletes.

Sickle Cell Anemia

Patients with sickle cell disease possess a homozygous gene that causes formation of an

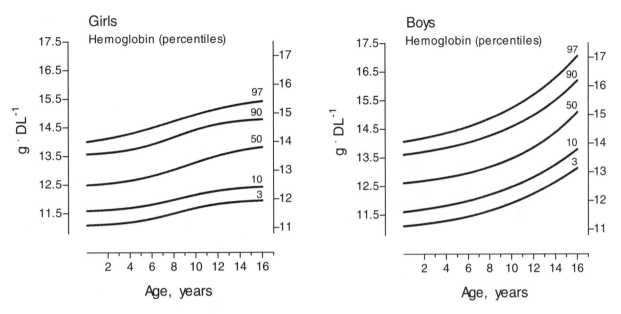

FIGURE 11.3　Changes in hemoglobin concentration with growth during childhood.
Reprinted, with permission, from Dallman and Siimes 1979 (18).

TABLE 11.1	Normal Values for Hemoglobin and Hematocrit by Gender and Age			
	Hemoglobin (gm · dl)		**Hematocrit (%)**	
Age (years	**Mean**	**Lower limit**	**Mean**	**Lower limit**
2-4	12.5	11.0	38	34
5-7	13.0	11.5	39	35
8-11	13.5	12.0	40	36
12-14 Females	13.5	12.0	41	36
Males	14.0	12.5	43	37
15-17 Females	14.0	12.0	41	36
Males	15.0	13.0	46	38

Lower limits are defined by two standard deviations below the mean value. From Oski 1993 (66).

abnormal form of hemoglobin (hemoglobin S) in their circulating erythrocytes. The presence of hemoglobin S causes the red cells to take on a sickled shape, particularly in an environment of low oxygen tension and reduced pH. Sickling of red cells increases their mechanical fragility, leading to a chronic hemolytic anemia. The increased number of sickled erythrocytes also produces greater blood viscosity with clinical manifestation of microvascular obstruction.

Sickle cell disease occurs almost exclusively in the black population, in which the incidence is 0.3% to 1.3%. The hemoglobin concentration in these patients usually ranges between 6 and 9 g · dL^{-1}. The course of the disease is often marked by painful and aplastic crises, skeletal deformities, growth retardation, and increased risk of infection, all complications of chronic anemia and vascular obstruction and stasis.

Signs of sickle cell disease usually become apparent in early childhood. While risk of death is increased, particularly between the ages of 1 and 3 years, the probability of patients with sickle cell disease surviving to age 20 years is approximately 85% (50).

Besides the detrimental limitation of oxygen delivery from chronic anemia, exercise performance in children and adolescents with sickle cell disease may be impaired by the direct effects of this condition on cardiovascular and pulmonary function. These patients are characterized by a high cardiac output state in response to their low hemoglobin concentration, often with associated left ventricular enlargement and hypertrophy (51). It is possible, too, that heart muscle function may be impaired by microvascular occlusion in the coronary circulation from sickled red blood cells (a so-called sickle cardiomyopathy) (26).

Considerable controversy and conflicting data have surrounded the existence of intrinsic myo-

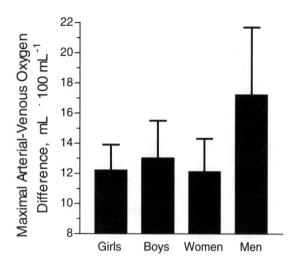

FIGURE 11.4 Maximal arterial-venous oxygen difference by gender and maturity. Data from Rowland et al. 2000 and 1997 (79; 80). Vertical lines denote ISD.

cardial disease in sickle cell patients beyond the effects of their high-output state. In reviewing this information, Willens et al. commented that such a distinction may be difficult to make, since signs and symptoms of congestive heart failure from myocardial dysfunction may mimic those of a long-standing chronic high-output state (100). Noninvasive studies of these patients at rest have indicated both normal myocardial function (30; 100) and abnormalities in contractility (20; 76), while others have concluded that left ventricular dysfunction develops with increasing age (6; 96).

Cardiac function may be further compromised by myocardial deposition of iron, or hemosiderosis, resulting from chronic hemolysis as well as recurrent blood transfusions. Also, in adults with long-standing disease, pulmonary hypertension may develop, causing cor pulmonale and right-sided heart failure. These patients may also demonstrate pulmonary dysfunction, with restrictive lung disease and increased alveolar dead space (61). Sproule et al. reported that the oxygen saturation of arterial blood in subjects with sickle cell disease was lower than in healthy individuals, possibly from intrapulmonary shunting (91).

Considering the characteristic occlusion of small blood vessels by sickled cells in this condition, it is surprising that sickle cell disease has not been associated with an increased risk of myocardial infarction. In addition, autopsy specimens in young adults generally indicate an obvious but unexplained lack of atherosclerotic lesions in the coronary arteries (44).

Effects on Exercise Performance

Multiple factors, then, may be responsible for abnormalities observed in patients with sickle cell disease during exercise testing. While decreased oxygen delivery from low arterial oxygen content is likely to be the most important, the negative effects of a long-term high-output cardiac state as well as myocardial and pulmonary dysfunction from microvascular occlusion may also contribute. As might be expected from these influences, endurance fitness is typically depressed in sickle cell patients (2; 17), and fitness improves following blood transfusion (60).

Electrocardographic ST-segment depression indicative of myocardial ischemia with exercise testing has been described in 15% to 30% of patients with sickle cell disease (2; 11; 57). The clinical significance of this finding is unclear, however, since symptoms of angina with exercise are rare (36). McConnell et al. reported that patients who developed ST changes with exercise were more likely to have lower hemoglobin levels and higher peak blood pressure and heart rate (indicative of higher myocardial oxygen consumption) than those without (57).

Covitz et al. studied cardiac performance at rest and during exhaustive upright cycle exercise using radionuclide angiography in 22 adolescents with sickle cell disease (17). Heart rate and left ventricular ejection fraction at rest were not significantly different when compared to values in healthy control subjects. Work capacity was lower in the sickle cell disease patients and was directly related to hematocrit (r = 0.50). Similarly, maximal cardiac output, heart rate, and ejection fraction were less in the patients.

Alpert et al. performed longitudinal exercise testing in 74 patients with sickle cell anemia 1 to 7 years following initial study (2). Depressed levels of maximal heart rate, maximal blood pressure, and aerobic work capacity were evident on both tests, but there was no evidence of deterioration of these variables over time. Correlational analysis revealed that hemoglobin concentration was the most critical determinant of exercise responses.

Pianosi et al. measured cardiac responses in 30 children with sickle cell anemia during exercise to 50% of maximum (69). Compared to healthy controls, patients demonstrated a greater cardiac output and stroke volume, a finding interpreted as indicating a compensatory mechanism for lower blood oxygen-carrying capacity. This increase in cardiac output was greatest in the older adolescents who had lower hemoglobin concentrations. Similar findings were described in adults with sickle cell disease by Lonsdorfer et al. during supine exercise (53). However, Sproule et al. found no differences in cardiac output and stroke volume between adult sickle cell disease patients and controls at maximal exercise (91).

In a spirometric study of 28 youth, ages 6 to 19 years, who had sickle cell disease, Pianosi et al. reported low VO_2max (30.1 ± 6.6 mL · kg^{-1} · min^{-1}) but no difference in %VO_2max at ventilatory threshold compared to values in healthy controls (70). The patients with sickle cell disease had an increased ventilatory response to exercise caused partially by augmented physiologic dead space, possibly as a result of impaired alveolar capillary perfusion.

Although myocardial ischemia is not an expected complication of sickle cell disease, Acar et al. found ventricular perfusion defects by thallium scanning at maximal exercise in a small group of eight asymptomatic pediatric patients (1) . Subsequent coronary angiography in those with perfusion defects disclosed no abnormalities. The authors suggested that myocardial ischemia may be an underestimated complication in patients with sickle cell anemia because of the low sensitivity of other diagnostic techniques.

Exercise Recommendations

The ability of patients with sickle cell disease to tolerate exercise and participate in sport varies according to the severity of their anemia and frequency of complications. Some are highly capable athletes, while others have little tolerance for strenuous physical activities (89). Recommendations for sport play, then, need to be individualized.

No formal guidelines for participation in specific forms of physical activity and sport have been formulated for patients with sickle cell disease. However, a number of considerations have led to recommendations that certain limitations be placed on these individuals. Given the nature of this condition, with anemia, microvascular obstruction, increased blood viscosity, pulmonary disease, and possible myocardial dysfunction, it is not difficult to recognize limitations and potential risks that might be incurred from vigorous sport play. Given this observation, it is interesting that, in fact, very few complications of this illness related to exercise have been reported. As previously noted, there appears to be no increased risk for sudden death from myocardial infarction. There have been no clear-cut reports of life-threatening rhabdomyolysis from exertion in patients with sickle cell disease (see next section on sickle cell trait). Kark and Ward have suggested that "it is possible that people with sickle cell disease have a risk of unexpected death [and other complications] . . . but their restriction from military duty, competitive athletics, and heavy labor has been sufficiently protective to make such complications extremely rare" (p. 221) (44).

The American Academy of Pediatrics has given a "qualified yes" for sport participation in those with sickle cell disease, based on assessment of the individual athlete (3). The academy concluded that "in general, if status of the illness permits, all but high exertion, collision/contact sports may be played. Overheating, dehydration, and chilling must be avoided" (p. 760). This caution reflects the tendency for some patients to develop increased sickling in conditions of hypoxia, dehydration, and acidosis that may be created by vigorous physical activities. This may occur particularly in cold air or at high altitudes.

Sigler and Zinkham noted that selecting sports for patients with sickle cell anemia is hampered by lack of scientific knowledge regarding the effects of different activities on this disease (89). They suggested that those who have known cardiorespiratory abnormalities should be advised against aerobic sports such as basketball and swimming. Patients with splenomegaly, frequently observed in this condition, should not participate in contact sports. These authors also considered contact or collision sports potentially hazardous because of the possible risk—although unproven—that trauma might trigger painful crises.

Sickle Cell Trait

Persons with sickle cell trait are heterozygous for the gene coding for hemoglobin S. Consequently, the amount of sickle cell hemoglobin is only 25%

to 50% of the total hemoglobin concentration. These individuals are healthy, and their life expectancy is normal (83). They are not anemic, have no symptoms, and demonstrate no abnormalities on physical examination or in hematologic laboratory studies. Sickle cell trait is present in approximately 7% to 9% of the American black population, amounting to approximately 3 million individuals.

Despite the generally benign course of sickle cell trait, evidence of a tendency for microvascular obstruction can be detected. Many with the trait demonstrate microscopic infarction of the renal medulla, possibly because of the local conditions of hypoxemia and acidosis that exist deep in the kidney, and this may explain the hyposthenuria (inability to fully concentrate the urine) and episodes of hematuria occasionally observed in persons with sickle cell trait. Of greater concern, a report of increased risk of "sudden death" in black military recruits with sickle cell trait during vigorous physical training has raised concern regarding the risks for these individuals during sport play (43). The validity of the association between sudden death with exercise and sickle cell trait, as well as the extent of any risk to persons with this condition, has since generated considerable debate.

In 1987, Kark et al. found that among 2 million American military recruits, the risk of unexplained sudden death between the years 1977 and 1981 was 40 times higher in blacks with sickle cell trait than in the general recruit population (43). The overall risk of sudden death in blacks with sickle cell trait was 1 in 3,200. In a subsequent review article, Kark and Ward examined the features not only of these deaths but also of those described by others (44). All deaths occurred in persons between the ages of 16 and 32 years, and most of these events happened surrounding highly strenuous physical activity (most frequently distance running) in the heat. In some cases, high altitude may have been contributory.

The majority of the deaths were, in fact, not "sudden" but rather involved survival for at least 6 to 24 hr following the provoking exercise. The clinical picture was often that of severe rhabdomyolysis (skeletal muscle breakdown) with accompanying renal failure, acidosis, and hyperkalemia. In general, there was no evidence of increased sickling or microvascular complications that are seen in patients with sickle cell

disease (see earlier section). Kark and Ward concluded that most of the deaths in blacks with sickle cell trait were related to exertional heat illness in undertrained, nonacclimated persons performing highly strenuous exercise (44).

The relevance of these data to risk for young athletes with sickle cell trait has come under close scrutiny. Most have agreed that "people with sickle cell trait are not at a clinically significant level of risk from sports" (p. 221) (44). No excessive mortality has been observed in black military personnel after graduation from basic training (44). Murphy reported no morbidity or mortality related to sickle cell trait in National Football League players, many of whom competed frequently at an altitude of 1,600 m (63).

Pearson concluded that ". . . past observations and common sense suggest that if there is an increased risk of sudden death associated with student athletic competition due to sickle cell trait, it is of a very small order of magnitude. Millions of young people with sickle cell trait have participated in sport. One would have anticipated that if there were a significant risk, a disproportionate number of sudden deaths of young black athletes should have been recognized" (p. 614) (68). Eichner agreed, stating that "there is still no cogent evidence that sickle cell trait per se increases the risk or consequences of exertional rhabdomyolysis or plays any role in exercise-related deaths" (p. 436) (21).

Effects on Exercise Fitness

Exercise testing studies of subjects with sickle cell trait have consistently revealed no ill effects of this condition on measures of physical fitness (see [44] for review). Variables including maximal oxygen uptake, cardiac function, peak work rates, and metabolic recovery have all been normal in this population. Electrocardiograms during cycle or treadmill exercise have revealed no ischemic changes or dysrhythmias. The presence of sickled red cells in venous blood has been reported to increase at high-intensity exercise but to levels no more than clinically insignificant 1%.

Exercise testing of recruits with sickle cell trait has been conducted at simulated altitudes as high as 13,000 ft (3,962 m) via breathing of a hypoxic gas (55). Again, no abnormalities in aerobic fitness or cardiopulmonary function were observed. However, the number of sickled cells rose to a mean of 8.5%.

Exercise Recommendations

Despite some concerns, it is generally considered appropriate to permit persons with sickle cell trait unrestricted sport play. Certain considerations, however, may be important. The excessive mortality in military recruits reported by Kark et al. appeared to be related to several factors, including heat illness, dehydration, poor fitness, inadequate heat acclimatization, associated viral illness, and extremes of physical exertion (43). Therefore, athletes with sickle cell trait should be counseled to avoid these conditions through adequate training, hydration, and attention to signs of heat illness. Extremes of exercise should be avoided, particularly during viral infections. Symptoms of collapse or chest pain during training or competition should be fully evaluated by a physician. As noted by several authors, these recommendations are no different from those that should be emphasized to all athletes (22; 44).

Thalassemia

Thalassemia is an inherited hemoglobinopathy occurring often in individuals of Mediterranean descent in which an abnormal hemoglobin causes hemolysis and disturbed erythropoiesis. The disease population has been broadly divided into persons with thalassemia major, the homozygous or more severe form, and those with heterozygous, less severe thalassemia minor. However, within each category exists a wide spectrum of clinical manifestations.

Patients with thalassemia major typically demonstrate a hypochromic, microcytic anemia with a hemoglobin concentration from 5 to 9 g · dL⁻¹. Repeated blood transfusions may be necessary to maintain adequate hemoglobin concentrations, resulting in liver and cardiac damage from iron deposition (hemosiderosis). Severe cases are marked by growth retardation, splenomegaly, and cardiac enlargement.

Cardiac dysfunction from iron overload is the major cause of death in these patients. The extent of cardiac disease can be directly related to number of transfusions, ranging from mild left ventricular thickening to decreased ejection fraction, life-threatening ventricular ectopy, and overt congestive heart failure (58). At the other end of the clinical spectrum, patients with thalassemia minor may have normal hemoglobin concentrations and lead healthy lives.

Exercise tolerance in patients with thalassemia therefore depends on the severity of the clinical expression of this disease. Those with significant anemia and cardiac complications of iron overload tolerate physical activities poorly and may be at risk from high-intensity exercise. Exercise studies of these patients predictably indicate low maximal oxygen uptake, pulmonary function abnormalities, hypoxemia, and elevated cardiac output and heart rate related to the amount of oxygen consumed (35; 91; 97). Maximal arterial-venous oxygen difference is reduced from the anemia (35). Hypoventilation is observed during exercise, with an increased alveolar PCO_2.

Transfusion therapy in these patients may improve maximal oxygen uptake. Villa et al. reported a rise in $\dot{V}O_2$max from an average of 38.5 to 45.7 mL · kg⁻¹ · min⁻¹ in 13-year-old patients with thalassemia major whose mean hemoglobin was increased by blood transfusion from 10.9 to 13.7 g · dL⁻¹ (97). However, those with past pulmonary symptoms demonstrated an increase in their dyspnea index and end-tidal PCO_2 after transfusion.

Exercise recommendations for patients with thalassemia need to be tailored to disease severity. People with minor forms and little or no anemia may be able to tolerate all sports effectively and safely. On the other hand, the low hemoglobin concentration, cardiac involvement, and splenomegaly often observed in patients with thalassemia major require curtailment of many forms of physical activity. Sigler and Zinkam recommended that youngsters with thalassemia be limited to low-intensity aerobic forms of exercise (89).

Iron Deficiency Anemia

Approximately two-thirds of body iron content is incorporated in the form of hemoglobin. Iron deficiency resulting from inadequate iron intake or loss decreases hemoglobin concentration and causes a microcytic hypochromic anemia. In the pediatric years, iron deficiency anemia most often becomes a clinical issue during infancy (typically between the ages of 9 and 24 months) and is usually due to inadequate dietary iron consumption. Iron deficiency may also appear during adolescence, when the iron requirements of accelerated growth may not be satisfied by marginal eating habits. This may become an issue particularly in females whose iron losses become exaggerated with onset of menses. Given the potential impact of iron deficiency on sport performance, adequate iron stores are a particular concern for teenage athletes.

There is no question that iron deficiency anemia, defined in teenagers and adults as a hemoglobin concentration less than $12.0 \text{ g} \cdot \text{dL}^{-1}$ in females and $13.0 \text{ g} \cdot \text{dL}^{-1}$ in males, impairs exercise performance. When adult volunteers have undergone repeated phlebotomies, the decline in hemoglobin concentration was closely correlated with a decrease in treadmill endurance time and maximal aerobic power (24; 101). The percentage fall in $\dot{V}O_2$max in these studies approximates that of the decline in hemoglobin concentration (see also figure 11.2).

The mechanism for the decline in aerobic fitness with iron deficiency anemia may not be simply that of a decline in oxygen transport. The depression in hemoglobin concentration obviously impairs oxygen delivery to exercising muscles, but iron has other roles in the body that can contribute to defining exercise performance. For example, iron within the muscle cells plays a key role in several aerobic processes. Iron is a critical component of myoglobin, which acts as a reservoir for oxygen; and iron is essential in the action of several key enzymes important in energy metabolism (aconitase, cytochrome oxidase, succinate dehydrogenase, tyrosine hydroxylase). That is, iron deficiency may inhibit biochemical functions within the cell that are important in the utilization of oxygen for aerobic metabolism. There is also evidence, less well defined, that iron is important in cognitive function and might play a critical role in defining motivation and thresholds of central nervous system-mediated fatigue (12; 65).

In the early stages of iron deficiency, iron stores (as indicated by blood levels of ferritin, the storage form of iron) can be low ($<12 \text{ ng} \cdot \text{mL}^{-1}$) without a decrease in hemoglobin concentration. Studies in animals clearly indicate that iron deficiency without anemia significantly impairs endurance exercise performance (27). This does not appear to be the case in humans. Most studies of individuals with non-anemic iron deficiency have failed to indicate a decrement in performance or $\dot{V}O_2$max (78). However, non-anemic iron deficiency is common in adolescent athletes, particularly in the female population, in which the frequency

may be as high as 25% to 50%. The frequency reported in males is typically much lower, about 0% to 17%. It would appear, then, that although overt iron deficiency anemia is unusual in these athletes (0-3% in most studies), the number who are at risk for depression in hemoglobin concentration from iron deficiency is substantial.

It is possible, too, that some persons with low ferritin levels and borderline hemoglobin concentrations (12-$12.5 \text{ g} \cdot \text{dL}^{-1}$ in females) may actually be impaired in their sport performance by true mild iron deficiency anemia. Lamanca and Haymes studied the effect of iron supplementation in women 19 to 35 years old who had an initial ferritin concentration of less than $20 \text{ ng} \cdot \text{mL}^{-1}$ and mean hemoglobin of $12.7 \text{ g} \cdot \text{dL}^{-1}$ (48). With iron treatment, the average hemoglobin concentration rose to $13.5 \text{ g} \cdot \text{dL}^{-1}$ while submaximal cycle endurance time increased by 38%.

The high frequency of hypoferritinemia in adolescents reflects to some extent low dietary iron intake, since foods rich in iron such as liver, lima beans, and green vegetables are not typically favorites, and red meats may be avoided because of concern over fat content. As noted previously, menstrual losses place females at particular risk. There is also some evidence that sport training and competition itself may augment iron losses through sweat, hemolysis, and gastrointestinal bleeding (78).

These data indicate that maintenance of adequate body iron stores is important for young athletes, many of whom—particularly females—may be at significant risk for iron deficiency. These athletes should be encouraged to consume iron-rich foods such as lean red meat, poultry, iron-enriched breakfast cereals, and green vegetables. Determination of serum ferritin and hemoglobin concentration is indicated in any athlete whose performance is declining. Overt or borderline low hemoglobin concentration can be treated with oral iron, and some have suggested that those with a ferritin level less than $12 \text{ ng} \cdot \text{mL}^{-1}$, even with a normal hemoglobin concentration, should also receive iron supplementation.

"Pseudoanemia" in Athletes

Highly trained adult endurance athletes often demonstrate mildly depressed hemoglobin and hematocrit values compared to nonathletes. This usually does not represent a true anemia but rather a dilutional phenomenon from increases in plasma volume. Athletic training is associated with a 5% to 10% increase in plasma volume, while cross-sectional studies have demonstrated differences as great as 20% to 25% between elite athletes and nonathletes (15). The expanded plasma volume offers the advantages of increased cardiac stroke volume, improved thermoregulation, and decreased blood viscosity. The total red cell mass in highly trained athletes is also often increased. The hematocrit may be reduced, however, because the rise in plasma volume often exceeds the increase in red cell mass (21).

Limited information suggests that this "pseudoanemia" may not be expected in child athletes. Eriksson described an average 12.3% rise in blood volume after 16 weeks of endurance training in 12 boys 11 to 13 years old (25). While total body hemoglobin mass rose from 389 to 428 g (10% increase), blood hemoglobin concentration did not change.

No significant differences in hemoglobin concentrations were observed by Sundberg and Elovainio between 12- to 16-year-old endurance runners and untrained controls (93). Rowland et al. found similar average hemoglobin concentrations in female high school cross country runners (13.3 g · dL^{-1}), swimmers (13.3 g · dL^{-1}), and nonathletes (13.4 g · dL^{-1}) (82). Whether this failure to observe "pseudoanemia" in young athletes compared to adults reflects a difference in red cell mass or plasma volume responses to training is unknown.

Hemophilia

Patients with hemophilia lack adequate amounts of circulating factor VIII or IX, important proteins in the normal mechanism triggering the clotting of blood. Inherited as a sex-linked recessive, hemophilia affects only males, while females who carry the gene are asymptomatic. Because of their clotting deficiency, persons with hemophilia experience recurrent episodes of bleeding, particularly in response to trauma. Consequently, what does or does not constitute safe participation in physical activities and sport for these individuals needs to be carefully considered.

The extent of the clinical risk of hemorrhage with hemophilia depends on the quantitative level of factor VIII or IX, which varies widely from patient to patient. A normal individual has 50 to 200 units per milliliter plasma. Patients with mild hemophilia demonstrate 5% to 40% of normal levels; those with moderate disease have about 5%, and severely affected patients have 1% to 2%. Recognizing the risk of hemorrhage from clotting factor levels, as well as a past history of bleeding episodes, is therefore useful in counseling the individual patient regarding sport participation.

The characteristic hemorrhage in patients with hemophilia is within the joints—hemarthrosis—and recurrent joint bleeds can lead to a crippling chronic hemophilic arthropathy. In those with severe disease, such bleeding can occur with even little obvious trauma. Bleeding elsewhere, particularly intracranial, is less common but still a risk, with potential tragic outcomes. With onset of signs of tissue bleeding, patients with hemophilia receive immediate intravenous infusion of deficient clotting factors to restore normal clotting function. In some cases, prophylactic treatment, or regular infusion of clotting factors every 2 to 3 days, has been used to prevent hemorrhage.

Hemophilia does not, by itself, negatively influence physical fitness or ability to perform in sport. However, those who have developed an arthropathy from recurrent joint hemorrhage may have limitations in neuromuscular function of the involved joints, with resulting diminished muscle strength and endurance (71). The more specific issue surrounding patients with hemophilia with regard to sport play is risk of hemorrhage, particularly intramuscular, articular, and intracranial. One must individualize selection of appropriate physical activities for patients with hemophilia, taking into consideration hemorrhagic tendency, nature of the sport, and past history of bleeding episodes.

Sports particularly appropriate for patients with hemophilia include swimming, table tennis, walking, golf, dancing, and archery. Those forms of athletics that create highest risk and are generally banned are boxing, martial arts, hockey, football, and wrestling (64). Some patients are prone to hemorrhage in specific joint sites, and this may dictate choice of sport. For instance, a patient with recurrent wrist hemorrhages is unlikely to tolerate tennis or bowling. As Beardsley has noted, "the best sport for one child with hemophilia may not be wise for another with the same diagnosis . . . [and] in many cases the only way to determine whether a particular sport will be tolerated is by trial participation" (p. 307) (8). In some cases protective equipment such as extra joint padding, or modification of the sport itself, may be appropriate (i.e., no jump dismounts in gymnastics). A plan for immediate management of hemorrhage needs to be devised with parents and coaches before any child with hemophilia participates in sport activities.

Beardsley has emphasized the importance of encouraging patients with hemophilia to participate as much as is appropriate in regular physical education as well as in recreational and team sports (8). Besides the psychosocial benefits of such participation, improving joint strength may help protect against the occurrence and severity of hemarthroses. To this end, swimming and controlled-resistance weight training have been considered particularly useful for these patients.

It may also be important that acute bouts of exercise cause significant increases in blood levels of factor VIII (92). This is particularly true with high-intensity running, during which elevations ranging from 20% up to as high as 336% have been described that persist for up to 10 hr afterward. This response is considered secondary to β-adrenergic receptor stimulation. Other factors are not affected by exercise.

Bone Marrow Transplantation

Bone marrow transplantation is a therapeutic procedure in which bone marrow from a healthy donor is transfused intravenously into a patient with one of a variety of malignant and nonmalignant hematologic diseases. The concept behind this technique is that stem cells in healthy marrow will grow within the patient's marrow and replace diseased or malignant cells. Success of this procedure is contingent upon a number of variables, including donor-recipient gene matching and adequate recipient immunosuppression by drugs or radiation to prevent graft rejection.

Bone marrow transplantation has been of particular benefit in selected patients with aplastic anemia, severe combined immunodeficiency syndrome, thalassemia, acute and chronic leukemias, and malignant lymphomas (99). More limited use has also involved those with congenital hypo-

plastic anemia (Diamond-Blackfan syndrome), osteopetrosis, Gaucher's disease, and other storage abnormalities. The major problem limiting success in these patients is the occurrence of graft-versus-host disease; finding a histocompatible donor can reduce the risk of this. However, such successful matches are found in only 20% to 40% of cases.

Children and adolescents have demonstrated low levels of endurance fitness following bone marrow transplantation (38; 49). While the mechanisms responsible remain unclear, this finding has led to concern regarding the possible cardiovascular sequelae of high-dose chemotherapy and radiation treatments. Specifically, evidence exists that cardiac function may be suppressed in these patients. Resting echocardiograms and radionuclide angiography have demonstrated depressed myocardial contractility (72; 77); and overt congestive heart failure, pericardial effusions, and dysrhythmias have been reported (14; 37).

Evidence for depression of cardiac function has been borne out with exercise testing as well. Larsen et al. performed maximal cycle testing on 20 children and young adults, most with acute leukemia, before transplant and a second group of 31 patients after transplant (49). Compared to healthy controls, both groups exhibited reduced exercise times, $\dot{V}O_2$max, cardiac output, and ventilatory anaerobic threshold. While this study showed a low level of cardiorespiratory fitness in these patients, the effect of transplantation itself could not be ascertained, since the same patients were not studied before and after the procedure.

Hogarty et al. performed a longitudinal assessment of the natural course of aerobic fitness and cardiac function in bone marrow transplant recipients (38). Exercise testing with a cycle ramp protocol was performed in 33 children at 1, 2, and 5 years after transplant. Mean age at transplantation was 11.3 years. At the initial posttransplant testing, maximal cardiac index, $\dot{V}O_2$max, $\dot{V}O_2$ at the ventilatory threshold, and maximal work rate were all low, with 62%, 62%, 75%, and 63% of predicted normal values, respectively. On serial testing, maximal cardiac index did not change, but $\dot{V}O_2$max rose by 4% per year to 69% predicted, and maximal work had increased to 77% predicted by 6 years following transplantation. These data were interpreted as indicating that significant cardiac dysfunction does not improve in short-term follow-up after bone marrow transplantation. However, aerobic fitness may increase over time, perhaps because of greater efficiency of oxygen extraction at the musculoskeletal level.

Exercise and Cancer

Compared to the situation in adults, cancer is not a common affliction in the pediatric age group. Approximately 6,500 children under the age of 15 years in the United States are diagnosed with a malignancy each year. Most of these cases are hematologic or lymphoproliferative in nature (leukemia, lymphoma), while solid tumors such as brain tumors, neuroblastomas, Wilms' tumor, and osteogenic sarcomas make up a minority of cases. Recent dramatic improvements in chemotherapy, surgery, and radiation have drastically improved the outlook for many of these types of malignancies; and overall, more than 50% of children with cancer are cured (84).

Studies addressing the association of physical activity and fitness with cancer have generally involved adult populations. These reports have suggested a role of physical activity in the prevention of certain forms of malignancies as well as the efficacy of exercise programs in rehabilitation following treatment. Little information is available regarding similar issues in the pediatric age group. Nonetheless, there is evidence to suggest that there are a number of ways in which exercise might play similar roles in children and adolescents with cancer. This section addresses three such areas: physical activity in the prevention of cancer, the contributions of exercise to the physical and psychosocial well-being of patients during cancer treatment, and the role of exercise testing in assessing the impact of therapeutic interventions.

Exercise and Cancer Prevention

The 1996 U.S. Surgeon General's report examined the scientific literature to that date regarding the role of physical inactivity as an etiologic factor for certain forms of cancer (95). As concluded in the report, good evidence existed to indicate that physical activity is associated with a reduced risk of cancer of the colon (a neoplasm of adults), possibly from its effect in decreasing gastrointestinal transit time. Data regarding the association of exercise with other forms of cancer were considered too inadequate to allow any conclusions.

In one area, however, some interesting information suggested the possibility that physical activity by females during adolescence might serve to diminish the risk of future breast cancer. Of the five studies identified that examined this question, two reported a significant reduction in risk, one found a lower risk that was statistically insignificant, and two described no association. Bernstein et al. examined the relationship of leisure activities after menarche with breast cancer in 545 women over 40 years old (9). The relative risk of those who participated in over 3.8 hr per week in the first 10 years after menarche, compared to that of women describing no regular activity, was 0.41. Similarly, Mittendorf et al. reported a relative risk of 0.50 in women with breast cancer who were involved in daily strenuous activities at ages 14 to 22 years compared to those with none (62).

Studies published subsequently to that review have continued to support the possibility that regular involvement in physical activity during the teenage years might offer some protection against future breast cancer. Marcus et al. questioned 864 women ages 20 to 74 years who had breast cancer and 790 healthy controls regarding participation in four activities at age 12 (walking or cycling to school, competitive athletic training, and regular household chores) (54). Those who reported any of the four had a modest reduction in risk for breast cancer (odds ratio 0.8).

Carpenter et al. found that regular exercise habits needed to be continued into the adult years if risk of breast cancer was to be reduced (13). No effect of early activity was observed on risk unless the postmenopausal women in their study (ages 55-64 years) continued to exercise at least 4 hr per week. Moreover, the risk reduction effect of exercise disappeared in women who gained considerable weight during adulthood.

Shoff et al. reported similar findings (86). Lowering risk of future breast cancer through exercise habits at ages 14 to 22 years was directly linked to frequency of activity and inversely to weight gain in the adult years. Most of this information suggests, then, that teenagers who are involved in regular vigorous physical activity and avoid weight gain as adults may have a reduced chance of developing breast cancer.

Such a protective effect of physical activity may be related to a reduction in the frequency of ovulatory cycles in adolescents who exercise regularly (9). Risk of breast cancer appears to reflect cumulative exposure of breast tissue to ovarian hormones (early menarche and later menopause are both associated with greater breast cancer risk). Carpenter et al. hypothesized that early and persistent exercise, by reducing exposure to estrogen, should be expected to lower this risk (13). Maintaining normal body weight is also important, since adipose tissue is a primary source of estrogen after menopause (86). Weight gain in adults, by itself, has been related to an elevated breast cancer risk (13).

Exercise in Rehabilitation: Psychosocial and Physical Effects

While surgical, radiation, and drug treatments for cancer have enhanced survival, patients often demonstrate a lack of both physical and pyschosocial well-being following these interventions. Courneya has identified depression, anxiety, concern over body image, lowered self-esteem, and social isolation as particularly anticipated outcomes in adults following cancer treatment (16). These emotional issues are often intertwined with physical complaints of weakness, weight loss, insomnia, anorexia, and gastrointestinal disturbances. As part of these complications, participation in regular physical activity often diminishes.

It is the goal of cancer rehabilitation programs to counter these issues that significantly affect quality of life. Given the potential for physical activity to ameliorate emotional disorders (see chapter 12), regular exercise has been proposed as a useful component of rehabilitative efforts. Courneya reviewed 11 studies in the adult literature that examined the role of exercise as a rehabilitative modality in patients undergoing cancer treatment (16). All 11 reports described positive effects. With exercise interventions, improvements were observed in physical issues such as sleeping, eating, strength, and pain, as well as positive emotional responses of improved functioning, mood, and overall quality of life.

While very limited research has addressed this issue in children and adolescents, there is no reason to expect that exercise interventions would not have similar benefits in the pediatric age group. Shore and Shephard examined the physical and psychologic effects of a 12-week exercise program in three children ages 13 to 14 years who had been successfully treated for lymphoblastic leukemia (87). Previously low $\dot{V}O_2max$, excessive body fat, and high anxiety scores on psychologic testing all improved with training. However, the exercise program reduced immune function, though not to levels that were considered a concern for health. These authors suggested that exercise programs have benefit in the posttreatment management of children with cancer but that immune responses need to be monitored.

Sharkey et al. assessed the effects of a 12-week, twice-weekly, hospital-based aerobic conditioning program on 10 postpubertal patients who were free of cancer and had been off chemotherapy (>100 mg $\cdot$ m^{-2} of anthracyclines) and radiation treatments for at least 1 year (85). With this program, treadmill endurance time improved 13% and $\dot{V}O_2max$ improved 8%, but values remained well below predicted for normals. It was concluded that deconditioning from a sedentary lifestyle may contribute to some, but not all, of the decrease in aerobic fitness observed in childhood cancer survivors.

Smith et al. found that a summer camp experience for 18 children with cancer and their families produced effects of improved social, physical, and self-initiated activities (90). By maternal report, many of these positive outcomes were still evident 1 month after the camp.

Physical Fitness Posttreatment

Given the negative physiologic and behavioral impacts of malignancy, it is not unexpected that children who are cancer survivors often demonstrate a diminished level of physical fitness. For example, Jenney et al. studied cardiorespiratory function at rest and during exercise in 70 survivors (mean age 14.6 years) of lymphoblastic leukemia (40). Mean age at diagnosis was 5.8 years, and average time since completion of chemotherapy was 4.2 years. Treatment had consisted of several combinations of chemotherapeutic agents, radiation, or both. Echocardiography at rest indicated a lower mean left ventricular ejection fraction (71.6%) compared to that of healthy controls (76.6%). No significant differences in the maximal heart rate were seen between the two groups with exercise testing; but $\dot{V}O_2$max, maximal work rate, and maximal minute ventilation were lower in the patients (89%, 91%, and 91% of predicted, respectively).

Warner et al. reported similar findings in a group of 35 survivors of acute lymphoblastic leukemia ages 7 to 19 years (98). Mean $\dot{V}O_2$max was lower in these patients compared to their healthy siblings (30.5 vs. 41.3 mL · kg^{-1} · min^{-1} for the girls and 39.9 vs. 47.6 mL · kg^{-1} · min^{-1} for the boys). Energy expenditure during low-intensity exercise was negatively related to body fat (after adjustment for body weight), causing the authors to speculate that reduced exercise capacity might contribute to the excessive adiposity in leukemia survivors.

The issue of drug toxicity in these survivors has drawn particular clinical and research attention. Anthracycline antitumor drugs, a main component in the treatment of many forms of childhood cancer, are toxic to the myocardium. Late-onset cardiomyopathy, typically nonreversible, can therefore be an unfortunate complication of these medications, the risk increasing with cumulative dosage. The ill effect is the reduction of myocardial mass, with progressive decrement in both systolic and diastolic function (39). These abnormalities can be demonstrated by measures obtained in the resting state both by echocardiography and by nuclear myocardial scans.

Information obtained from exercise testing in these patients may serve as a sensitive indication of anthracycline cardiac toxicity. Depressed endurance fitness has been described in up to one-half of patients who received high-dose anthracycline treatment 1 to 15 years previously (52). Fukazawa et al. studied echocardiographic changes during both handgrip and supine cycle exercise in 13 children, ages 9 to 18 years, who were in complete remission after treatment of acute leukemia that had included anthracycline chemotherapy (28). These subjects were divided into three risk groups based on cumulative anthracycline dose. Exercise echocardiographic findings of left ventricular fractional shortening, end-systolic stress-volume index (ESS/ESVI), left ventricular diastolic filling velocity ratio, and normalized peak rate of diastolic increase in left ventricular internal dimensions (dLVDt · dt) were compared to findings in 10 healthy con-

trol children. During cycle exercise the ESS/ESVI was decreased in the patients in a dose-dependent fashion. However, dLVDt/dt was decreased at rest only in the high-risk group compared to controls. These findings suggest that the end-systolic stress-volume index might serve as a useful index of severity of anthracycline myocardial toxicity during exercise testing.

Johnson et al. performed maximal and submaximal cycling tests on 13 patients (mean age 13 ± 4 years) in remission who had received 292 ± 119 mg · m^{-2} of adriamycin at least 2 years earlier (41). Resting echocardiograms were all normal. $\dot{V}O_2$max was 32.0 ± 6.3 and 41.3 ± 8.4 mL · kg^{-1} · min^{-1} for patients and healthy controls, respectively (p < .05). During submaximal exercise, increases in stroke index were smaller in the patients (33% vs. 54% at 33% of $\dot{V}O_2$max; 33% and 69% at 66% of $\dot{V}O_2$max). These findings suggest a limited inotropic reserve following anthracycline therapy.

Rowland et al. studied cardiac responses to semi-supine maximal cycle testing with Doppler echocardiography in 11 patients with myocardial dysfunction (left ventricular shortening fraction <28% at rest), 8 of whom had anthracycline toxicity (figure 7.3) (81). Maximal cardiac index, stroke index, heart rate, peak aortic velocity, and left ventricular shortening fraction were all significantly lower in the patients compared to healthy controls. However, the resting shortening fraction in the patients bore no relationship to any of these exercise variables. Stroke volume declined at high exercise intensities in the patients but remained stable in the controls.

These data suggest that exercise testing might serve as a valuable tool in detecting and assessing severity of anthracycline cardiomyopathy. However, additional research will be needed to determine which techniques and variables are most sensitive to changes in myocardial function.

Chronic Renal Disease

Normal kidney function is critical in the maintenance of fluid, electrolyte, and acid-base homeostasis. The kidneys excrete nitrogenous waste products (urea, creatinine), the by-products of cellular metabolism, and maintain normal blood concentrations of sodium, potassium, and other electrolyte constituents. The kidney tubules are the sites of action of hormones such as aldosterone and vasopressin, which control body water content and osmolality. Blood pH is regulated in the kidneys through selective reabsorption or buffering of hydrogen ion. The kidneys produce renin, which leads to the formation of angiotensin II, a potent vasoconstrictor and stimulant of aldosterone formation. Erythropoietin from kidneys stimulates red blood cell production in response to hypoxic conditions (33).

Given this wide array of functions, it is not surprising that disturbed renal function resulting from disease of the kidneys can be manifest through a multitude of significant complications (45). Patients with chronic renal disease often demonstrate fluid accumulation, hypertension, acidosis, anemia, electrolyte disturbances, bone disease, growth failure, and malnutrition. With full-blown renal failure, these aberrations are reflected in laboratory findings of elevated blood urea nitrogen (BUN) and creatinine, reduced urinary creatinine clearance, anemia, hyperkalemia, hypocalcemia, hypoalbuminemia, and hyperuricemia.

Renal failure can occur in end-stage chronic renal disease from a variety of causes (59). In children, most cases of chronic renal disease result from inflammatory diseases of the kidney (acquired glomerulonephritis) or congenital renal malformations. Other causes include systemic inflammatory disease (such as systemic lupus erythematosus) and renal parenchymal damage from obstruction to urine flow.

Fortunately, chronic renal disease is not common in children. The incidence is approximately 1 case per 100,000 each year in the pediatric population of the United States (about 830 patients) (45). Among those affected, 85% are over the age of 5 years.

With worsening of renal function, three interventions can be considered in these patients: hemodialysis, peritoneal dialysis, and kidney transplantation. Hemodialysis, which involves regular filtration of toxic solutes from the blood by means of an arterial-venous fistula in the arm, is the most common form of therapy for end-stage renal disease in adults (67). With peritoneal dialysis, the membranes within the abdominal cavity serve to filter out abnormal substances from the blood after instillation and subsequent removal of dialysis fluid by a catheter. In children, kidney transplantation is the preferred treatment, since

it allows a more active lifestyle. Sixty percent of adolescents in the United States who are 3 years of age after onset of end-stage renal disease have received a transplant, while a third are on hemodialysis and 10% are receiving peritoneal dialysis treatment (45).

Cadaver kidneys can be used for transplantation; but an organ from a family member, particularly when immunologically compatible with the recipient, improves graft survival. Data from the North American Pediatric Renal Transplant Cooperative Study indicated that graft survival was 74% at 1 year and 62% at 3 years among recipients of cadaver kidneys (59). Patients receive intensive immunosuppressive therapy following transplant to prevent rejection. These medications, which include cyclosporin, corticosteroids, and antilymphocytic globulin, improve graft survival but unfortunately are also responsible for most of the complications observed following transplantation (infection, neoplasia, hypertension, atherosclerotic vascular disease) (67).

Effects on Physical Fitness

Considering the multiple organ system complications of chronic renal disease, it is not surprising that these patients demonstrate a poor capacity for exercise (figure 11.5). Anemia, muscle wasting (from protein malnutrition), hypertension, fluid and electrolyte abnormalities, emotional disturbances, and complications of therapy combine to significantly impair skeletal muscle performance, cardiac and pulmonary function, and oxygen transport.

Painter has reviewed results of the greater experience of fitness testing in adults with end-stage renal disease, most of whom were being managed with hemodialysis at the time (67). These reports indicate that endurance fitness in older patients is approximately 50% that of healthy persons, with typical values of $\dot{V}O_2$max of 15 to 25 mL $\cdot$ kg^{-1} $\cdot$ min^{-1}.

Exercise training programs in adults with chronic renal disease, ranging in duration from 10 weeks to 12 months, have demonstrated improvements in aerobic fitness, with typical increases in $\dot{V}O_2$max of about 25%.

Kennedy and Siegel have noted that exercise test findings in adults may not necessarily be applicable to children, since (1) most pediatric

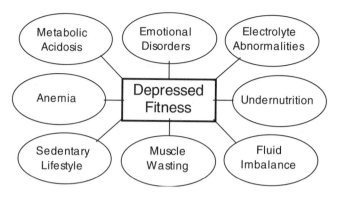

FIGURE 11.5 Complications of chronic renal disease that influence physical fitness.

patients are not treated with hemodialysis, (2) children and adolescents have different forms of kidney disease than adults, and (3) older subjects often have comorbid conditions such as atherosclerotic vascular disease (45). Nonetheless, the five studies that have examined physical fitness in children and adolescents with chronic renal disease demonstrate a similar picture of impaired performance.

Krull et al. performed maximal cycle exercise tests on 33 females and 37 males an average of 3.6 years after they had received a transplanted kidney (47). Average age of the patients was 17.6 years, with a range of 8.2 to 23.8. Mean values of $\dot{V}O_2$max for the males and females were 23.2 ± 5.9 and 28.3 ± 5.9 mL $\cdot$ kg^{-1} $\cdot$ min^{-1}, respectively. Peak heart rate was blunted in those on beta-blocker treatment, with an overall average of 153 ± 21 beat min^{-1} for the males and 166 ± 23 for the females. Resting and peak blood pressures were normal in this study.

The 10 posttransplant patients, ages 6 to 22 years, tested by Matteucci et al., however, demonstrated a systolic hypertensive response to treadmill exercise (56). Peak systolic pressure averaged 175 mmHg in the renal disease subjects compared to 130 mmHg in healthy controls matched for body surface area. No differences in diastolic pressures were seen between the two groups. Endurance time of the patients and controls averaged 10.8 and 12.8 min, respectively.

In a similar study, Giordano et al. assessed exercise tolerance and blood pressure responses to treadmill exercise in 45 children (mean age 14.3 $\pm$ 4.2 years), all of whom were at least 6 months posttransplant and taking multiple immunosuppressive drugs (31). All had a hemoglobin con-

centration over 10 mg $\cdot$ dL^{-1}. Treadmill endurance time was reduced in the patients—10.1 ± 2.1 min compared to 15.1 ± 1.7 min in healthy controls. The posttransplant children demonstrated a greater systolic blood pressure response to exercise (150 ± 26 vs. 134 ± 13 mmHg).

Bonzel et al. compared physiologic responses to maximal cycle exercise in young patients on conservative treatment, hemodialysis, or posttransplant with those of healthy controls (10). $\dot{V}O_2$max averaged 16.5, 25.1, and 35.9 mL $\cdot$ kg^{-1} $\cdot$ min^{-1} for the three renal disease groups, respectively, while the value in the controls was 44.1 mL $\cdot$ kg^{-1} $\cdot$ min^{-1}. The greater levels of aerobic fitness in the posttransplant patients compared to the others with chronic renal disease may have reflected their higher mean hemoglobin concentration (9.7, 7.2, and 14.6 g $\cdot$ dL^{-1} for the conservative treatment, dialysis, and posttransplant patients, respectively).

The findings of Ulmer and colleagues highlight the importance of anemia in reducing physical fitness in patients with chronic renal disease (94). Forty children with chronic renal failure underwent serial submaximal cycle testing for determination of aerobic fitness, defined as the work rate at a heart rate of 170 beat $\cdot$ min^{-1} (PWC$_{170}$). Fitness was inversely related to the degree of renal functional impairment, and a positive correlation was observed between PWC$_{170}$ and hemoglobin concentration (figure 11.6).

Evidence indicates, as well, that correction of anemia in children with chronic renal failure can improve their exercise tolerance. The seven patients reported by Baraldi et al. (mean age 13.9 years) performed maximal treadmill exercise before and after erythropoietin treatment for anemia (7). Pre- and posttreatment hemoglobin concentrations were 6.3 ± 0.9 and 11.2 ± 1.2 g $\cdot$ dL^{-1}. With correction of anemia, mean $\dot{V}O_2$max rose from 24.1 ± 7.1 to 32.6 ± 12.7 mL $\cdot$ kg^{-1} $\cdot$ min^{-1}, compared to 44.7 ± 7.1 mL $\cdot$ kg^{-1} $\cdot$ min^{-1} in healthy control subjects.

Exercise Recommendations

Neither acute bouts of exercise nor physical conditioning appears to influence renal functional capacity in healthy individuals. Consequently, exercise programs in persons with kidney disease are not expected to improve or cause deterioration in renal function itself (33). Nonetheless,

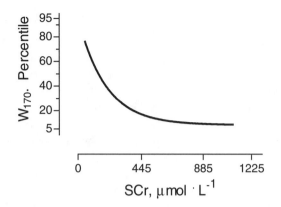

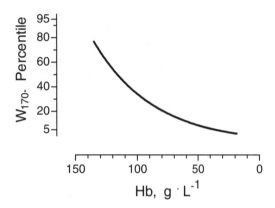

FIGURE 11.6 Relationship of serum creatinine concentration (SCR) and blood hemoglobin (Hb) to percentile of physical fitness (as defined by W_{170}) in 20 children with chronic renal failure and 22 control children.

From Ulmer et al. 1978 (94).

there are a number of potential salutary effects of regular physical activity on the complications of chronic renal disease that might prove beneficial for these patients. Aerobic exercise programs can lower blood pressure in those with hypertension, as well as improve serum lipid profiles. Weight-bearing activities can improve bone density. Importantly, too, exercise can provide a means of enhancing self-esteem and social confidence while combating the depression often observed in patients with chronic renal disease (46).

At the same time, it needs to be recognized that participation in sport and vigorous physical activities can pose risks to these patients. Most problematic is the risk of trauma to a subcutaneous peritoneal dialysis catheter, the arterial-venous fistula for hemodialysis, or the transplanted kidney in the lower abdomen. Patients with advanced renal disease, particularly before transplant, are often in a delicate

fluid-electrolyte balance, with risks of dehydration, hyperkalemia, and hyponatremia that can be exaggerated by exercise. Those with bone weakness (osteodystrophy) may incur fractures in certain activities. Exercise can trigger cardiac dysrhythmias in patients with renal disease or produce unacceptably high levels of systolic blood pressure (45).

With these considerations in mind, specific forms and intensities of exercise can be recommended for young patients with chronic renal disease. Factors that need to be weighed in making decisions regarding sport participation by individual patients include severity of disease and nature of complications, presence of a transplanted kidney or dialysis catheter, and basic level of fitness. Some children and adolescents are severely handicapped by their disease and can be expected to engage in little activity without exhaustion, while others can perform certain sports at a high level of proficiency.

Kennedy and Siegel reported findings in a questionnaire, returned by 54 pediatric nephrologists in the United Sates, that sought attitudes regarding exercise in patients with chronic renal disease (45). Most respondents strongly encouraged exercise within the limitations of safety identified for individual patients. Highly acceptable sports included cycling, swimming, aerobics, dance, golf, running, tennis, and baseball. Most recommended that children with a transplanted kidney avoid contact sports such as football and karate as well as events that involve isometric exercise (such as weightlifting).

Findings from laboratory cycle or treadmill exercise testing may be useful in making these recommendations for physical activity. Level of fitness and complications such as dysrhythmias, blood pressure responses to exercise, and exercise-induced bronchospasm can be evaluated.

School administrators and physical education teachers may be reluctant to have their students with chronic renal disease participate in physical activities. Close communication between physicians caring for these patients and families and school personnel is important in assuring that optimal exercise and safe activities can be performed. Kennedy and Siegel emphasized that while promotion of some form of exercise is generally appropriate, these children should not be placed in situations in which success in performing certain physical activities is unlikely (45).

Exercise-Induced Proteinuria and Hematuria

The appearance of protein or red blood cells in the urine often signifies the presence of renal disease, but each can occur as a benign finding following vigorous physical activity. Proteinuria and hematuria triggered by exercise have generally been considered to carry no long-term health implications. It is important to recognize that these findings can occur in a urine sample taken as a medical test in an individual who has participated in athletics in the preceding 24 to 48 hr (33). On the other hand, persistence of either of these findings beyond 2 days of vigorous exercise should prompt consideration of other causes.

Increased protein excretion in the urine with exercise may occur from increased glomerular permeability, elevated hydrostatic pressure within the glomerulus, or inhibition of tubular reabsorption of protein (33; 73). Studies in adults indicate that the extent of protein content in the urine depends on the intensity of exercise (as indicated by serum lactate level) rather than its duration (figure 11.7) (75). Postexercise proteinuria has been reported to be more pronounced following maximal short-burst activities than in endurance sports such as cycling and swimming (74). Effects of biological maturation on this phenomenon have not been described.

Benign hematuria following exercise is most likely caused by mechanical trauma to the posterior wall of the bladder. Microscopic (>1,000 red cells per milliliter of urine) or overt gross hematuria is not uncommon in highly competitive athletes, having been described in 20% of marathon runners and 50% to 70% of those engaged in ultramarathon running (19; 88). Prevention

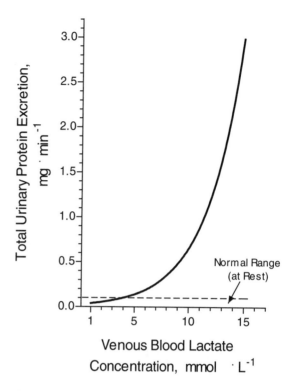

FIGURE 11.7 Relationship between venous blood lactate concentration and total urinary protein excretion in subjects following running exercise.

From Poortmans 1985 (73).

through the maintenance of adequate hydration and avoidance of urinating just prior to exercise has been suggested (23).

A red urine from hemoglobinuria has been considered a consequence of destruction of red cells caused by repeated pounding of the feet ("march hemoglobinuria," "footstrike hemolysis"). Microscopic examination of the urine (indicating no red cells) will distinguish this entity from exercise-induced hematuria.

CHAPTER

12

Emotional and Mental Disorders

Undoubtedly, mental function is closely linked to physical exercise. Most people who play tennis, go for walks, or take a swim do so simply because it's enjoyable—there is mental pleasure in physical activity. To the ancients, however, this mind-exercise link took on a more holistic, spiritual meaning. *Orandum est sit mens sana in corpore sano* (we should pray for a sane mind in a sound body), declared Juvenal in the first century A.D., indicating that we should break away from the television set to gain human completeness in exercise activities. (For the skeptical, Spinoza provided an alternative perspective of the mind-body relationship in the 1600s when he noted that "everyone is fond of relating his own exploits and displaying the strength of both his body and mind" and that "men are on this account a nuisance to one the other.")

In recent times, scientific research has offered more pragmatic, mechanistic roles for the mind in influencing muscular exercise. Habits of regular exercise are important to health; and, since motivation for participation in physical activity is, of course, a central nervous system function, factors that influence the decision to exercise or not have come under close scrutiny. There is evidence to suggest, in fact, that not all these influences are external ones (such as peer support, time constraints, and accessibility to exercise facilities). A biological control center within the central nervous system may govern, to some extent, one's level of physical activity as part of homeostatic regulation of body energy balance (77).

Increasing evidence indicates that the brain may play a critical role in determining the limits of exercise performance. That is, one's finish time in a 5-mi (8-km) road race may be governed as much by mental as by physical factors. Åstrand concluded that "the feeling of discomfort and fatigue in muscular exercise is very differently tolerated by different individuals, and the same individual does not sustain the work similarly from day to day" (p. 121) (3). This idea has been more formally constructed as the *central fatigue hypothesis,* which holds that (1) mental processes rather than muscular or metabolic ones can serve as the principal determinants of exercise performance, and (2) the central limitations are biologic (such as increases in brain serotonin levels) rather than psychologic (27).

Psychogenic factors can influence metabolic rate, both at rest and during exercise. Under the hypnotic suggestion during exercise, subjects at rest have been demonstrated to significantly increase heart rate, cardiac output, ventilation, and oxygen consumption (64). Morgan et al. described physiologic changes when perceived exertion was altered by hypnotic suggestion during steady state exercise (65). When subjects thought they were pedaling up a hill, ventilation rose 11 L · min⁻¹, and when they "reached the top" and started "downhill," ventilation diminished.

This chapter deals with an even more pragmatic aspect of the mind-exercise connection, the potential role of exercise in treating young patients who have emotional or mental disorders. The observation that many individuals "feel good" after physical activity has historically stimulated efforts to create a therapeutic role for this positive mental effect of exercise on persons with emotional disease. Somewhat disappointingly, the success in scientifically documenting such a salutary action of exercise has been limited, and positive effects have been reasonably well documented for only a few disorders. Attempts to define the true therapeutic role of exercise have been thwarted by serious methodological flaws in these research studies. Most information has been acquired in the adult population, and the potential for exercise interventions in the management of mental disorders in children and adolescents has just begun to be explored.

Scope of the Problem

The "carefree days of youth" may be anything but. The reported prevalence of clinical mental disorders in U.S. youth under the age of 18 years is 12%—one in eight—and it has been estimated that only about a third of those needing treatment receive it. Depressive disorders occur in as many as 2.5% of children and 8.3% of adolescents, numbers that have risen over the past generation. Similar frequencies have been observed for thymic disorder. The morbidity of these disorders is significant, interfering with normal social and academic function (8).

In the United States, the number of suicides in adolescents has quadrupled since 1950 (12). In 1991, suicide was the second most common reason for death in the age group 15 to 19 years, with a mortality rate of 13 per 100,000. In addition, it has been estimated that there are 50 attempted suicides for each completed event. Younger children are at risk as well, with a mortality rate from suicide in the 5- to 14-year-old age group of 0.7 per 100,000 in 1991.

Hidden behind these statistics are those youngsters—presumably a much larger number—who experience significant emotional stress without the diagnosis of overt clinical mental illness. Changes in modern society have produced increasingly stressful lifestyles, and children and adolescents may need to cope with family instabilities, frequent moves, physical intimidation by peers, and pressures for early academic or athletic success (52). The frequent appearance of young patients in the pediatrician's office with psychogenic complaints of chronic fatigue, headaches, dizzy spells, and chest pain attests to the effects of these stressors on quality of life.

Equally concerning, emotional disorders during the childhood years may serve as precursors to a lifetime of mental illness. For example, the prevalence over a lifetime for major depression with onset during adolescence has been estimated to be as high as 15% to 20%, which is comparable to that observed in adults (8).

Compounding this morbidity, depression in youth is often associated with other mental disorders, such as anxiety and bipolar disorder as well as substance abuse. As many as a third of adolescents with depression also manifest signs of anxiety disorder (54).

Considering the high frequency of mental disorders, their negative impact on well-being, and the potential for lifetime impairment, the importance of recognition and management in the young is evident. While psychotherapy and pharmacologic interventions serve as the mainstays of treatment, outcomes are not always successful. Consequently, the potential role of physical activity in the management of emotional and mental disorders bears particular significance. If exercise interventions are, in fact, effective in treating these children, they would prove a welcome adjunct to conventional therapies.

The current status of exercise in the management of emotional disorders has been summarized by Brown (17):

"To a large extent the existing situation in the mental health field is analogous to what is occurring in the physical health field. There are currently millions of people who are disabled and suffering to some degree from mental health problems, and, as with cardiovascular disease, these problems are reaching pandemic proportions in many countries. . . ."

"It therefore seems critical that efforts continue to be made to identify cost-effective interventions that can be delivered to people in their natural environments and that have the potential to prevent or lessen the severity of mental health problems in the population at large. In this regard, exercise holds considerable promise."

An Overview of Exercise and Mental Health in Adults

The bulk of research addressing the impact of exercise on emotional health has been performed in adults. It is instructive, then, to begin with a brief review of this information, with no particular assurance that such influences of physical activity or fitness can necessarily be translated to younger subjects.

Investigations in adults have focused principally on the role of exercise in promoting self-esteem and well-being in the general population and its impact on patients with anxiety disorders and depression in the clinical setting. Little is known regarding response of patients with more profound mental illness (schizophrenia, bipolar disorders) to exercise programs or that of patients with other clinical disturbances such as sleep disorders or addictive behaviors.

A major proportion of studies have been cross-sectional in nature. That is, the aim in most has been to determine if within a given group of subjects an association exists between level of physical activity or fitness and some measure of mental health. As will be discussed in more detail later in the chapter, such investigations are of limited value since identifying such a relationship does nothing to clarify cause and effect. Other studies have assessed the effect of an exercise intervention in a prospective manner, and few others have addressed the predictive value of exercise on future mental health outcomes.

Aerobic activities such as running and walking have typically been utilized as the exercise intervention in these studies. Whether other types of activities (resistance exercise or short-burst activities such as table tennis) might show more or less of a salutary effect remains unknown.

Studies can be divided into those that examine responses to single acute bouts of exercise and those that measure effects of chronic exercise over a period of time. These outcomes can be defined as "state" measures ("how I feel now") or "trait" measures ("how I usually feel"), corresponding to responses to acute and chronic exercise interventions, respectively. Psychological outcomes have generally been assessed by questionnaire, although in some cases, particularly in studies of anxiety, physiological measures such as blood pressure and electromyograms have been employed.

Methodological Considerations

Unfortunately, the quality of the research design in these studies has often been wanting, and the dearth of scientifically credible data examining the link between exercise and mental health has been recognized as a particular obstacle to progress in understanding this relationship (17; 42; 80). It is important, then, in reviewing and analyzing these data—both in adults and in children—that one recognize these problems.

1. Studies often involve a small number of participants not randomly assigned to experimental conditions, include no control subjects, or fail to provide a blind or double-blind design.

2. Initial differences in psychological or physiologic measures between interventional groups may not be considered.

3. Subjects may be permitted to self-select among the conditions or activities involved in the study.

4. The "Hawthorne effect"—the idea that simply participating in a study, regardless of type of intervention, can change a subject's behavior—may not have been considered in the study design.

5. No information is provided regarding dose-response relationships.

6. Cross-sectional studies provide limited insights, since cause-and-effect relationships cannot be determined. For instance, an association identified between self-concept and amount of habitual physical activity could mean that regular exercise promotes feelings of self-worth, or, conversely, that those who feel better about themselves are more likely to exercise (or, that both variables are independently related to a third factor) (see figure 12.1).

7. In some studies, nonvalidated self-report questionnaires are different than standard tools utilized in clinical practice.

8. The distinction between clinical and nonclinical groups with emotional or mental disorders may be vague and superficial.

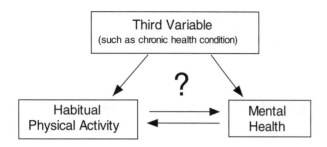

FIGURE 12.1 Cause-and-effect issues complicate understanding of the relationship between exercise and mental health.

9. Measures of physical fitness, activity, or both—before and after exercise interventions—are often lacking or insufficient.

Care is necessary, then, in interpreting the large number of published and unpublished data that have been used to assess the relationship between exercise and mental health. While a number of meta-analyses have attempted to draw conclusions from this information, such an approach does not alter the serious methodologic weakness of these reports (17).

Self-Esteem

Self-esteem—how one feels about oneself—is multidimensional (how one perceives one's effectiveness as a parent can differ from one's idea of how well one drives an automobile) and is clearly influenced by numerous variables (parental or spouse expectations, social comparisons, performance outcomes). Physical attributes have long been recognized as a key aspect of self-esteem, and this concept has served as the impetus for investigations into the role that regular exercise might play in promoting feelings of self-worth in the general population.

Sonstroem reviewed studies published before 1984 that assessed the causal effect of an exercise intervention program on self-esteem in adults (80). While these reports generally indicated that physical training is associated with improved self-esteem scores, it was concluded that serious methodologic shortcomings in these studies prevented any assumptions (1) that increases in fitness, per se, affected these scores, or (2) that change in scores was any indication of "enduring aspects of self-conception." In fact, Sonstroem considered the scientific quality of these studies to be so mediocre ("incomplete . . . vague . . . [and]

poorly presented") that "a majority of the positive results can be explained by alternative interpretations and experimental limitations" (p. 131).

Two subsequent meta-analyses of studies of exercise interventions and self-concept demonstrated effect sizes of 0.56 and 0.23 (that is, the self-concept score of a subject involved in chronic exercise was about one-half to one-fourth SD above that of nonexercising controls) (59; 82). The former study considered self-concept only in the physical realm, while the latter defined it as global self-concept. Considering these results, Gauvin et al. suggested that "the benefits for physical activity in regard to global self-concept may be overstated in the physical activity literature" (p. 962) (42).

Depression

Considerable research supports a beneficial effect of exercise for adults with depression—feelings of sadness, lethargy, and withdrawal—both in clinical and in general populations. Epidemiologic studies indicate that those who exercise regularly have reduced levels of depression compared to sedentary individuals, with an odds ratio of 1.3 to 1.8. Interventional studies support this beneficial effect of exercise, with effect sizes ranging from −0.46 to −0.97 (42).

In patients with clinical depression, exercise interventions have been demonstrated to be as effective as psychotherapy (70). Those with the greatest levels of depression appear to receive the most benefit from increased physical activity (26; 67). Getting these patients involved in a program of regular physical activity has therefore been considered a valuable adjunct to management, along with antidepressant medication.

The evidence is not as strong in general populations, but many studies have indicated that regular exercise can improve mood in people without clinical depression. However, most benefit has been observed in those who are depressed at the beginning of the exercise program. In reviewing these data, Brown concluded that "it seems that the greatest potential for chronic exercise [in the general population] lies in its use as an early intervention strategy. Those who are mild-moderately depressed and who may be at risk for experiencing greater clinical depression have benefited from involvement in aerobic types of chronic exercise" (p. 607) (17).

There is some evidence to suggest that regular exercise habits might serve to prevent the development of depression. Previously nondepressed, sedentary adults have been found to exhibit an increased chance for developing depression in the future compared to those who were initially active (31).

Anxiety

Reduction of psychic stress from physical activity is a common experience, and one that has been documented by abundant research information. Studies utilizing a questionnaire or measuring physiologic variables (blood pressure, electromyogram) demonstrate that a single bout of vigorous exercise can decrease anxiety in both normal and pathologically anxious adults. The effect is temporary but sustained, with levels of anxiety reduced for 2 to 5 hr following physical activity.

Dose-response relationships for such exercise interventions have been documented. The ability of most acute exercise to reduce anxiety levels is moderate, being equivalent to the anxiolytic effects of meditation or sedative medication.

The research literature examining the effects of chronic physical activity or exercise training on trait anxiety is less compelling. The magnitude of decline in anxiety with a program of exercise has been small, with an overall effect size of about −0.33 SD (the anxiety levels of a subject in such a program is about one-third SD below that of control subjects) (42). Given the methodologic weaknesses of these studies, it has been concluded that a definite influence of repeated bouts of exercise on chronic levels of anxiety cannot be considered established (17). It has been suggested that any such effects might represent simply the carryover responses to multiple single bouts of exercise.

Several studies have described the role of exercise in patients who suffer from panic disorder, or disabling acute episodes of severe anxiety (70). Acute bouts of activity may, on the other hand, trigger symptoms of anxiety in such individuals (who fear they may suffer a heart attack). Consequently, many refrain from regular exercise and demonstrate low levels of physical fitness (14). Broocks et al. have demonstrated, however, that a 10-week aerobic exercise program in 18-to 50-year-old patients with panic disorder decreased symptoms compared to the effect in a nontraining control group (13).

Potential Mechanisms

The means by which physical activity or training might alter emotional health are unknown. Generally, speculation has been polarized between biologic and psychologic mechanisms, and evidence exists to support the plausibility of either etiologic perspective.

Biologic

Changes in mental status following acute or chronic exercise may reflect biochemical alterations in the central nervous system that are triggered by motor activity. Several different mechanisms have been suggested, principally, as might be expected, from research involving animal models.

Alterations in Neurotransmitters   Low levels of the neurotransmitters norepinephrine and serotonin in the central nervous system have been identified in stress-induced depression in animals (63). This psychologic state has been considered similar to human depression and can be reversed by medications known to increase brain levels of these transmitters. A "therapeutic" mechanism for exercise is suggested by the observation that acute bouts of swimming as well as run training by animals can augment central nervous system neurotransmitter concentrations (5; 15). As indicated by Morgan (63), however, the notion that such data can explain mental alterations with exercise in humans can only be considered speculative.

Production of Endorphins   Based on limited animal research, the concept that exercise can stimulate the production of endorphins—narcotic-like chemicals that can create feelings of euphoria—within the brain has captured popularity. Opiate receptor occupancy has been demonstrated to occur in the brains of rats following swimming and running (71). In humans, studies of the effect of naloxone, a narcotic antagonist, on positive emotional responses to exercise have been conflicting and perhaps related to naloxone dose (58). Farrell et al. reported increased levels of plasma beta-endorphin immunoreactivity after treadmill running in humans, but changes were small and not always related to mood intensity (32). It has been suggested that "while the endorphin hypothesis has not been supported unequivocally, it does remain tenable" (p. 98) (63).

Cerebral Lateralization Hypothesis

Physical exercise may stimulate differential electrical activity in the cerebral cortex, causing a greater effect of alpha-wave activation on the right hemisphere. This theory has been developed from previous psychologic studies indicating that diminished right-sided alpha activity (produced by medication or central nervous system damage) is related to decreases in anxiety and depression. Supporting this idea, Petruzzello and Landers showed that electroencephalographic activity of the left frontal area increased with respect to the right in 19 adult males who ran on a treadmill at an intensity of 75% VO_2max for 30 min (72). While acknowledging that "in addition to its predominantly atheoretical nature, this literature has been plagued by numerous methodological shortcomings which make its interpretation difficult," the authors concluded from their study that "the cerebral lateralization hypothesis remains reasonable for explaining anxiety reductions associated with exercise" (p. 1033).

Psychologic

A number of psychologic mechanisms may contribute to the improved emotional well-being experienced with exercise. As with the proposed biologic causes, however, there exist few solid research data to specifically support any of these concepts, and most remain intuitive (70).

Distraction Hypothesis

Subjects might "feel better" after physical activity simply because the process of exercising distracts them from the serious problems surrounding their lives. The observation that exercise is equivalent in its effects on anxiety to a period of quiet meditation, for example, supports this idea.

Self-Efficacy Theory

Participation in physical activity may prove emotionally supportive for sedentary individuals who view involvement in exercise as particularly challenging. This can result in improved mood and feelings of self-confidence.

Mastery Hypothesis

Similar to the self-efficacy theory, the mastery hypothesis states that achieving success in exercise activities may provide a sense of control over one's life that improves emotional state. Such positive effects would be expected to be more evident in individuals who place importance on physical fitness and "being an athlete." This theory suggests that types of exercise in which improvement can be most easily measured and achieved might provide the most beneficial psychologic effects.

Social Interaction

Exercising with others can provide social support and improvements in self-identity. Such social interaction might be expected to underscore the importance of physical activities and potentiate feelings of mastery and self-efficacy just outlined. However, according to Ransford, experimental evidence suggested that individual rather than team sports are more likely to improve emotional well-being (76).

Studies in Children and Adolescents

There is no reason to expect a priori that the influences of exercise on mental health in children and adolescents should mimic those observed in adult subjects. Young persons differ, of course, in their immature psychosocial development. They participate in different forms of physical activity than adults, experience separate emotional disorders, and can be expected to view the importance of exercise and sport play in a different light. Even within the pediatric population, considerable differences in the exercise-mental health link can be anticipated. The impact of a program of regular aerobic exercise on the emotional well-being of a 16-year-old with depression, for instance, might be expected to differ greatly from that on a 7-year-old with attention deficit disorder.

As outlined earlier, the significance of the role of exercise in the treatment of mental illness in adults remains clouded due to the prominent weaknesses in research study design. The same problems exist in the pediatric literature, with many fewer studies. Still, there is sufficient information to suggest that properly constructed exercise interventions may prove useful in managing certain forms of emotional disorders in young patients. This is a possibility that deserves future research attention. The potential significance of *any* modality that is effective in optimizing mental health in children is very clear, since psychologic development in the growing years provides a foundation for emotional health in adulthood (21).

Several reviews have examined the link of physical activity, physical fitness, sport participation, or a combination thereof with the emotional health of children and adolescents (21; 87; 89). Studies have involved general pediatric populations as well as youngsters with clinically diagnosed mental disorders. Most have occurred in educational settings and have utilized aerobic exercise (aerobics classes, running) for older children (who have been studied most often) and motor activities and sports for younger subjects.

As with adult studies, reports of the effects of exercise on mental health outcomes in children and adolescents have been varied. In 1999, Calfas reviewed all studies in subjects under the age of 18 years during the previous 30 years that described psychologic outcomes from an exercise intervention (21). Among the total of 21 studies, approximately one-half showed a significant improvement in psychologic variables after a program of physical activity. Importantly, however, with the exception of a single study, none of these reports described the opposite—a negative influence on mental health from increased physical activity.

Self-Esteem

To a psychologist or psychiatrist, characteristics such as self-esteem, self-concept, and self-image have specific and separate meanings. In general, however, these characteristics belong to a family of psychologic constructs that have to do with feelings of self-worth and confidence to deal with one's social environment and control one's own personal destiny. The issue is a complex one in the pediatric age group, since the ideas of self-esteem and self-concept not only change with age but are invariably influenced by myriad extrinsic variables, including social interactions and comparisons, perceived competence, importance placed on a particular activity, and the influence of significant others (parents, teachers, peers).

A general perception exists that participation in physical activity benefits self-esteem in children and adolescents. If so, that's important, since the outcome should be a sense of happiness, well-being, and satisfaction with life. Optimizing a sense of self-worth through exercise in youth

might also be expected to stimulate a persistence in regular physical activity with all its positive long-term health benefits.

In examining the research literature, it is important to distinguish between those studies assessing the effects of *achievement* in sports and physical activities in children versus those evaluating the outcomes of *participation* in regular exercise on self-esteem. The former construct is a traditional tenet that has driven physical education programs and coaches of youth sport: The development of physical abilities with improvement in motor skills will heighten level of self-regard. Scientific support for this conclusion, however, is wanting. In the studies that have supported this idea, the dilemma of cause and effect is again apparent. What begets what? Are children with more self-confidence able to perform better in sport? Or does success in physical activities generate improvements in self-esteem?

In reviewing these studies, Weiss concluded that "despite these claims [that improved motor skills will improve self-esteem in children], surprisingly little empirical research exists to substantiate them. An increasingly accepted view is that success experiences are not sufficient to enhance self-esteem if the child does not perceive that he or she was responsible for that success. Actual competence is not the best predictor of future development and motivated behavior. One's perceptions of ability are more influential" (p. 105) (89).

The second issue, whether an intervention of regular exercise can enhance self-esteem, is more pertinent to the focus of this chapter. The most extensive review of this question was provided in 1986 by Gruber, who assessed the findings of 65 articles and doctoral dissertations relating to elementary school children that had appeared in the previous 20 years (44). These were divided into 10 "belief" articles, 17 small-scale studies in "disturbed" children with no data analysis, 27 controlled interventional studies (which provided the basis for a meta-analysis), and 11 correlational reports. Of the aggregate 65 studies, the authors of 53 had come to the conclusion that play and physical activity were effective in improving self-esteem in children and adolescents.

In the meta-analysis of the 27 interventional studies, Gruber found an overall average effect size of 0.41, supporting a positive effect of exercise, since self-esteem scores in subjects with exercise interventions exceeded those of 66% of subjects in control non-interventional groups. When the analysis was broken down by different categories that characterized these studies, some interesting observations emerged (figure 12.2). For example, the effect size of exercise interventions in "handicapped children" (defined as those who were emotionally disturbed, trainable or educable mentally retarded, economically disadvantaged, or perceptually handicapped) was considerably higher (0.57) than in studies involving normal children (0.34). Gruber explained this finding by the baseline low levels of self-esteem in the handicapped children. "When placed into enrichment programs conducted by trained and understanding teachers who provide individual attention, the handicapped begin to feel important and experience success in activity programs, [leading] to large improvements in their self-concept" (p. 44).

In this analysis, the type of exercise also appeared to be important. Aerobic activities were found to be clearly most useful, with an effect size of 0.89, far exceeding the outcomes from programs focusing on motor development, sport skills, or creative motions such as dance (effect sizes 0.29, 0.40, and 0.32, respectively). Length of intervention, on the other hand, did not influence

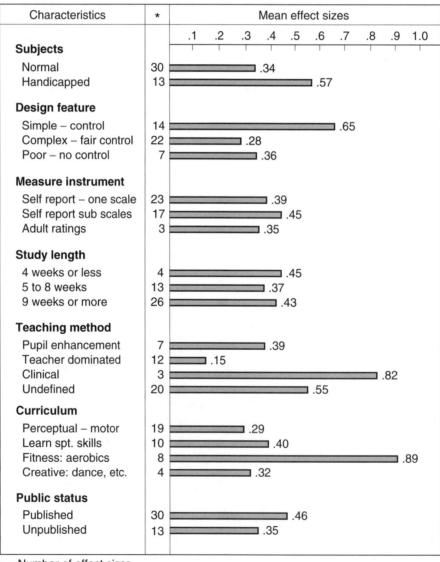

* = Number of effect sizes

FIGURE 12.2   The extent of influence of various factors regarding the effect of exercise on self-esteem in children. From Gruber 1986 (44).

effect size, nor did means of self-esteem measurement. Information from these studies was limited, but variations in subject age, gender, or race did not appear to significantly affect psychologic measurements after exercise interventions.

In considering the data in this analysis, Gruber stated that "participation in directed play and/or physical education programs contributes to the development of self-esteem in elementary school-age children." Moreover, he contended that the findings "clearly demonstrate the superiority of physical fitness activities in developing self-esteem when compared with other components of the elementary physical education curriculum." This is particularly true "for those children who feel insecure and in time could develop behavior problems" (p. 43).

Additional studies performed since that time, mostly in normal populations, have produced mixed findings (21; 87). Aine and Lester found no relationship between self-reported levels of physical activity and self-concept in their cross-sectional study of 90 subjects 15 to 24 years old (1). Pregnant adolescents improved their self-esteem after an exercise program intervention (53). Holloway et al. found that a 12-week school-based weight training program improved not only strength but also self-efficacy scores in a group of 16-year-old girls (49). Feelings of self-concept and confidence in physical appearance improved after 6 weeks of a combined exercise-education program in the adolescent girls reported by Boyd and Hrycaiko (11).

Covey and Feltz administered the Offer Self-Image Questionnaire and the Bem Sex Role Inventory to 149 high school females to assess the relationship between their psychologic health and the amount of participation in regular physical activity (25). The findings indicated that the physically active girls described a more positive self-image and more coping characteristics than those who were physically inactive.

Hay et al. examined the relationship between aerobic fitness and self-efficacy in 140 children in grades seven and eight (46). $\dot{V}O_2$max was estimated with a 1,600-m run, and self-efficacy was measured with the Children's Self-Perceptions of Adequacy and Predilection for Physical Activity. Correlational analysis indicated that aerobic fitness and self-efficacy scores were significantly and positively related (r = 0.49).

Taken collectively, these data support the concept that involvement in a program of aerobic exercise may improve self-reported levels of self-esteem in children and adolescents. Importantly, this research suggests that such interventions might be most valuable in youth with preexisting emotional disorders. The extent and duration of improvement in self-esteem that might be expected from regular exercise have not been established. Similarly, whether changes in self-report scores following exercise can be translated into meaningful improvements in clinical measures of emotional well-being has not been clarified.

Depression

The few studies that have examined either the association of exercise with, or the effect of exercise on, depressive symptoms in youth have dealt mostly with adolescents. Tortolero et al. identified 10 such reports published since 1983 (87). While the studies are plagued with methodologic flaws, these data generally support the salutary influence of physical activity, physical fitness, or both on depression, similar to that observed in adult populations.

Three of four cross-sectional studies in the general population indicate lower depressive symptoms in highly active youth (18; 60; 86). For example, using parent and teacher report of physical activity, Michaud-Tomson found that among 933 children, 8 to 12 years old, those who were sedentary were three times more likely to report depressive symptoms on questionnaires than those who were active (60). Glyshaw et al., on the other hand, could find no relationship between self-reported levels of physical exercise and depression in a group of 530 adolescents (43).

When Oler et al. administered the Children's Depression Inventory and the Suicidal Behavior Questionnaire to 823 high school students, depression and suicidal thoughts were significantly less in athletes compared to nonathletes (68). Other cross-sectional studies have failed to find a relationship between suicidal ideation or attempts and sport participation (7; 28). Thorlindssen et al. described a negative correlation between hours of sport participation and depressive symptoms in high school students (86).

Steptoe and Butler examined the relationship between self-reported participation in sports and vigorous physical activities and "emotional well-being" in a large group of high school boys and girls (83). Emotional health was estimated by questionnaire, which included issues of anxiety as well as depression. A positive relationship was observed between participation in exercise and emotional well-being that was independent of the influences of sex, social class, heath status, and the use of hospital services. While concluding that participation in vigorous activities and sport was related to mental health, the authors noted that the findings could also be interpreted as indicating that adolescents with psychologic difficulties might choose not to play sports.

Exercise intervention studies in the general pediatric population are limited and have produced mixed results. Norris et al. divided a group of 80 girls, 13 to 17 years old, into high-, moderate, and low-intensity 10-week exercise programs and compared depressive symptoms (by a multiple-adjective checklist) to those in nontraining controls (66). While exercise interventions lowered levels of anxiety, no significant differences in feelings of depression or hostility were observed between the four groups. Milligan et al. found lower levels of depression following an exercise program in boys but not girls (62). Dua and Hargreaves found that long-term exercising youth had lower (but statistically insignificant) levels of depression than sedentary subjects (29).

There is scant information on the effect of exercise interventions on young patients with clinical depression. Brown reported the effect of a 9-week, three times a week aerobic exercise program on psychologic variables in a group of institutionalized adolescents with a variety of forms of mental illness (mainly dysthymia and conduct disorders) (16). Using the Beck Depression Inventory and the Profile of Mood States, the authors demonstrated that by midway in the program the training subjects showed decreases in depression and anger that did not occur in nontraining control subjects. For most of the measures, psychologic changes occurred before improvements were noted in physical fitness. All changes with training had disappeared at the 4-week follow-up assessment.

In evaluating such studies, it is necessary to recognize the existence of evidence that the amount of physical activity itself can be a sign of depression. Aronen et al. measured 72 hr of motor activity in 27 hospitalized prepubertal children with emotional disorders using an ambulatory activity monitor (2). Reduced levels of physical activity were significantly related to severity of depressive symptoms.

Stress and Anxiety

Some information in studies of adolescents supports the role of chronic exercise in reducing levels of anxiety in the general population. As previously noted, Norris et al. found that a 10-week program of school-based aerobic training caused a decrease in anxiety and perceived stress in 13- to 17-year-old subjects (66). Thorlindssen et al. described a significant relationship between hours playing sports and lower levels of anxiety in a group of Icelandic teenagers (86). In 220 girls in a private school, ages 11 to 17 years, Brown and Lawton demonstrated that life stressors had a greater negative effect on the physical and emotional health of those who reported infrequent exercise compared to those who participated more frequently in physical activities (18).

Little research has examined the tranquilizing effect of an acute bout of exercise in healthy young people. Bahrke and Smith divided 65 children (mean age 10.6 years) into those partaking in a 15-min session of either running/walking, resting, or performing crafts (4). Using the State Trait Inventory for Children, the authors found no significant effect of the acute bouts of exercise on anxiety level compared to effects in control subjects.

Attention Deficit Hyperactivity Disorder

Children with attention deficit hyperactivity disorder (ADHD) are characterized by high levels of short-burst physical activity that is often socially disruptive and coupled with poor impulse control and learning problems. While seemingly counterintuitive, a theoretical basis exists for expecting that exercise interventions might actually improve the physical behavior of these patients and decrease their need for medication. Satterfield et al. and others (45; 78) have suggested that hyperactive children suffer from a state of depressed central nervous system arousal. These researchers have hypothesized that a defect in the reticular activating system in children with ADHD results in low brain arousal, and, consequently, that the high levels of physical activity in these patients represent a stimulus-seeking maneuver. If so, increasing structured physical activity might serve to diminish their hyperkinetic behavior.

Little experimental work has been done to test this concept. In his review of exercise and mental health in children, Brown cited as "personal communication," information regarding a study by Shipman of 30 hyperkinetic children ages 7 to 11 years whose behavior was controlled by medication (19). The experimental group participated in a running program three times a week for 5 months while control subjects simply played field games. Aggression before and after intervention was assessed while subjects played with dolls. Both groups reportedly showed a reduced amount of aggression, but those in the running program required less medication than the controls.

Medical treatment of ADHD may alter physiologic responses to exercise. Methylphenidate, which is commonly used in the management of hyperkinetic children, may increase submaximal heart rate by approximately 10 beat · min^{-1} without appreciably altering maximal rate (10).

Cognitive Function

Considering the potential influences on thought processes, it is not surprising that research attention has also addressed the ability of motor activity to optimize intellectual function. The effect of exercise on intelligence and creativity in the general pediatric population has come under scrutiny, as has the capacity of physical activity to potentiate cognitive abilities in those with learning disorders or mental retardation.

Creativity

Evidence exists that aerobic exercise can facilitate creativity in children. Herman-Tofler and Tuckman studied the responses of aerobic fitness (estimated by a timed 800-m run), perceived self-worth, and figural creativity to an 8-week aerobics training program in 26 third graders compared to controls in a conventional physical education class (47). The "creativity index score," determined by fluency, originality, and detail elaboration in figure drawing, increased significantly in those in the aerobic program. However, no significant changes in either aerobic fitness or self-perception were observed from training.

Hinkle et al. found increased figural creativity in middle school children in association with improved aerobic fitness from a running program (48). These authors hypothesized that running might serve to stimulate visual imagery by enhancing right brain activity. Tuckman and Hinkle reported that improvements in standard youth fitness tests in children in grades four to six were associated with level of creative skills (88).

Intelligence

Can increased physical activity or involvement in organized sport improve intelligence in youth? The question has long drawn research attention, particularly as it relates to concerns that inclusion of physical education in the school curriculum might detract from academic progress. Indeed, in his review of the early research literature on this subject, Kirkendall remarked that "it seemed that virtually every physical education researcher during the period from 1950 to 1970 at some time conducted a study to explore the relationship between motor performance and academic achievement or intelligence" (p. 59) (51).

In reviewing these data, Kirkendall concluded that there was a "generally positive relationship" between physical performance and intellectual abilities, and that this association was most obvious in younger children. Because of the small number of experimental intervention studies (which showed mixed results), he felt there was no conclusive evidence one way or the other on the effects of increased exercise on cognitive function. More recent studies have done little to cast more light on this question (87).

The issue of whether participation in athletics stimulates intellectual functioning is beset with a large number of variables that are difficult to control. In evaluating studies in which athletes score higher on intelligence tests or receive higher grades than nonathletes, one needs to consider several possible explanations. Athletes may enroll in easier classes than nonathletes or receive preferential treatment from teachers (the "halo effect"). Also, students with low grades may fail to qualify for participation on athletic teams. A true effect of sport play on cognitive function might reflect increased cerebral blood flow, improved nutrition and body build, hormonal changes, or alterations in self-esteem (79).

Shephard reviewed three studies that assessed the influence of daily physical education class on academic performance of elementary school students (79). Taken together, these reports indicated that when 14% to 26% of curricular time is spent in physical activity, "learning seems to proceed more rapidly per unit of classroom time, so that academic performance matches, or may even exceed, that of control students" (p. 120). Shephard considered this information as indicating that daily physical education class can be introduced into the school curriculum without fear of undermining academic achievement.

Learning Disorders

Children with learning disorders demonstrate problems in specific areas of cognitive function despite normal measured levels of intellect. Often such youngsters have difficulties in motor abilities as well, particularly muscular coordination. It has been suggested that improving motor skills in these children might improve self-esteem and social adjustment.

MacMahon and Gross assessed the effects of an aerobic exercise program on psychologic variables and physical fitness in 27 boys with learning disabilities, ages 7 to 13 years (56). The boys had average or above-average intelligence scores on standard testing but had been diagnosed as having specific learning disorders. Measures of self-concept, aerobic fitness (PWC_{170} test), academic proficiency, and anthropometric variables were obtained before and after a 20-week aerobic training program of distance running, aerobic dance, and soccer. Findings were compared to those in a control group that participated in less vigorous activities. The aerobic exercise program was effective in improving physical fitness and self-concept, but there were no significant differences between the two groups in either academic achievement or motor proficiency.

However, Bluechardt et al. concluded from their review of exercise programs designed to enhance motor skills in children with learning disorders that such interventions were "uniformly unsuccessful" (9). Some of these studies have provided evidence that an activity program with a social skills training component can enhance both motor ability and self-perception of physical and academic competence. However, such programs seem to be no more effective in this regard than "other forms of special attention."

Mental Retardation

Youngsters with mental retardation often demonstrate low levels of habitual activity, depressed physical fitness, and excessive body weight (30; 37). From both a general health and a psychosocial standpoint, therefore, these children may therefore obtain significant benefit from improving exercise habits. Such interventions can be highly effective. Programs of physical activity for mentally retarded children and adolescents have been successful in improving both physical fitness and body composition (33).

Physical Activity and Fitness

Youth with mental retardation are generally considered to be sedentary, an observation that has prompted concern regarding their risks for obesity, depressed physical fitness, and reduced potential for vocational productivity. As Fernhall and Pitetti have pointed out, however, there are few scientific data to substantiate this anecdotally based conclusion (35). Still these authors suggested that "personal observations by us and other professionals in the field are in agreement that the majority of persons with mental retardation lead a very sedentary lifestyle" (p. 176).

There exists, on the other hand, a considerable body of research information regarding levels of physical fitness in young persons who are mentally retarded. Most studies of laboratory and field-based fitness in youth with mental retardation have been performed in those with mild-to-moderate intellectual impairment. These investigations have been reviewed by Fernhall and Pitetti, who concluded that, as a group, children with mental retardation—and particularly those with Down syndrome—have depressed levels of physical work capacity (35).

Typical laboratory measurements of peak $\dot{V}O_2$ in children with mental retardation are outlined in table 12.1. Although these values are usually lower than those observed in normal children, there is a wide variability. Particularly low levels of aerobic fitness are apparent in studies of subjects with Down syndrome, who typically have a $\dot{V}O_2$max of approximately 25 mL · kg^{-1} · min^{-1} regardless of age.

Children with mental retardation also demonstrate reduced endurance capacity on field tests

TABLE 12.1 Studies of Maximal Aerobic Power ($\dot{V}O_2$max) in Youth With Mental Retardation

Study	Number of subjects (mean age)	Modality	$\dot{V}O_2$max (ml · kg^{-1} · min^{-1})	Peak heart rate (beats · min^{-1})
Bar-Or et al. (6)	89M, 32F MR 7-15 years	Tread	48-51M 42-47	195-205
Fernhall et al. (36)	9M, 8F MR, DS 14 years	Tread	39	182
Fernhall and Pitetti (34)	15M, 11F MR	Tread	35	
Fernhall et al. (38)	13M, 10F MR, DS 15 years	Tread	33M 26F	174M 180F
Fernhall et al. (39)	22M, 12F MR, DS 14 years	Tread	37	186
Maksud and Hamilton (57)	62 MR 10-13 years	Cycle	39	187
Pitetti and Fernhall (73)	17M, 12F MR, DS 14 years	Tread	37M 30F	183M 188F
Pitetti et al. (74)	12M, 11F MR	Tread	46M 32F	187M 182F
Teo-Koh and McCubbin (85)	45 MR	Tread	41	189
Yoshizawa et al. (90)	74M, 53F MR	Cycle	41-44M 33-36F	184-187M 181-186F
Fernhall and Tymeson (40)	11M, 3F DS	Tread	27	171
Millar et al. (61)	15 DS	Tread	26	166-173

MR = mental retardation; DS = Down syndrome. Modified from reference 35.

(such as walk/run) compared to normal youth (34; 36; 38), and they score lower on tests of muscular strength as well (55; 75) (figure 12.3). Fernhall and Pitetti demonstrated the interesting finding that measures of leg strength, $\dot{V}O_2$max, and endurance performance were significantly related to each other in children with mental retardation (34). They proposed that poor leg strength might be a particularly limiting factor to work performance in this group of individuals.

A sedentary lifestyle and excessive body fat may contribute to low levels of fitness in some youth with mental retardation, but biological differences in cardiovascular function may also play a role. This concept is suggested by observations of a dampened heart rate response (chronotropic incompetence) in exercise studies of children with mental retardation, especially those with Down syndrome. As indicated in table 12.1, peak heart rate in subjects with Down syndrome is usually 15 to 20 beats lower than in mentally retarded youth without Down

syndrome. As discussed by Fernhall and Pitetti, such differences are not considered to be related to lack of testing effort (35; 41). It has been suggested that chronotropic incompetence might reflect autonomic dysfunction in children with Down syndrome.

Effects of Physical Training

A period of physical training has been demonstrated to improve field endurance performance in children with Down syndrome (22; 24). While some laboratory studies have indicated similar training-induced increases in $\dot{V}O_2$max, others have failed to show changes in maximal aerobic power (see [33] for review). Fernhall noted that these training studies were typically weakened by significant methodologic shortcomings (small number subjects, lack of controls) (33).

Improvements in muscle strength and endurance have also been documented following training in children with mental retardation (33). These programs have typically been administered in the physical education class setting. Corder, for

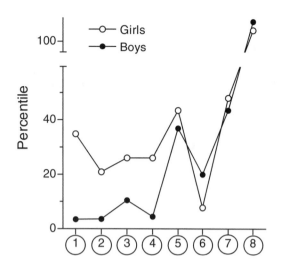

FIGURE 12.3 Mean scores (expressed in percentiles of those in normal subjects) in various motor tasks in educable mentally retarded boys (solid circles) and girls (open circles). Tests: (1) 300-yd run; (2) 50-yd run; (3) shuttle run; (4) broad jump; (5) pull-ups; (6) sit-ups; (7) handgrip; and (8) balance walk.

example, reported a 33% increase in number of sit-ups and 180% improvement in pull-ups after a 4-week training period in 24 boys with "educable" mental retardation (24).

On the other hand, there is no convincing evidence that an exercise intervention can be expected to improve intellectual functioning in mentally retarded patients. Solomon and Pangle found no increases in academic performance after an 8-week program of structured physical education in adolescent retarded boys (81). Similar results were obtained by Chasey and Wyrick in 12-year-old institutionalized mentally retarded children (23).

Others have reported small increases in intelligence scores in retarded children after physical activity interventions (20; 24; 69). It has been suggested, however, that other factors, such as improved physical fitness and social skills, may have contributed to these changes (84).

Summary

From the information presented in this chapter it is clearly not possible to draw strong conclusions

regarding the efficacy of exercise as a therapeutic modality for children and adolescents with emotional or mental disorders. Nonetheless, the data can be considered highly suggestive that improving physical activity might serve as an effective intervention in certain forms of mental illness. Given the hesitance to employ psychotropic medications in young people, exercise as an effective nonpharmacologic treatment strategy would be highly welcome.

Progress in this understanding, in both children and adults, has been seriously hampered by the lack of scientific rigor in studies assessing the link between exercise and mental health. To permit new insights, future investigations will need to conform to appropriate study designs. In this regard, the difficulty of accounting for the multiple variables that can influence behavioral studies in natural settings creates a particular challenge.

Whereas data exist that suggest the usefulness of exercise in managing mental illness, virtually nothing is known regarding the type, intensity, duration, and frequency of physical activities most likely to be effective. The importance of achieving gains in physical fitness, as opposed to volume of physical activity, is not clear. The role of social structures in developing effective exercise programs needs to be delineated. The forms of emotional illness for which exercise is most likely to be helpful have not been clarified, although persons with depressive or stress-related symptomatology appear to be most responsive to exercise interventions.

Given the rise in frequency of emotional disorders in the pediatric population, identification of useful preventive and therapeutic modalities is increasingly relevant. The negative impact of these diseases on well-being, not only during the childhood years but also as precursors of lifelong mental illness, lends a particular importance to understanding the therapeutic role of physical activity and fitness in emotionally disturbed youth.

Norms

The inclusion of norms of physical performance in this book might imply that (1) we know what is normal in children's performance; (2) the samples chosen to establish these norms are representative of their populations; (3) the norms of one population apply to others, irrespective of habitual activity, body size and composition, nutritional status, health status, climate, altitude, ethnic origin, and sociocultural conditions; and (4) the variables follow a known distribution (e.g., Gaussian) such that our choice of limits of normality is correct. None of these assumptions is entirely valid. We do not have valid criteria for normality in the various components of physical performance. Furthermore, if a range of "normal" values were to exist, it would vary among populations, depending on their level of activity (which, in itself, cannot be categorized into "normal" or "abnormal"), state of health and nutrition, prevailing climate, altitude, and ethnic origin.

Norms of one ethnic group may not apply to others. Oded Bar-Or observed Burmese children performing physical fitness tests constructed in Europe and North America. While most of the rural Burmese girls performed above the Western 95th percentile of flexed-arm hang (testing elbow flexor and shoulder girth strength and endurance), the great majority of them could not complete a single sit-up (testing abdominal muscle strength and endurance). The Burmese hosts attributed this pattern to specific traditions that encourage rural girls to perform heavy lifting tasks but forbid them from performing activities that would require dynamic abdominal muscle contraction.

Few of the samples included in the following graphs are representative of their respective populations. Most include subjects who happened to be tested in various laboratories for a specific research project or for clinical purposes. These graphs reflect the fitness of selected groups of well-nourished, nonathletic, European, North American, or Israeli children and adolescents, mostly of Caucasian origin. Subjects had no overt manifestation of a disease. As such, the graphs in this appendix should be regarded merely as guidelines for those who cannot establish norms for their own populations.

The choice of chronological age as the independent variable is not ideal. *Biological age* would have been more appropriate. However, from a practical point of view, while biological age is seldom available to a clinician, chronological age is easily ascertained. Whenever available, data are also presented per kg of body weight. While scaling considerations (see appendix V) would suggest that this might not be the ideal way to correct for differences in body size, weighing a child is easily done in a clinical environment.

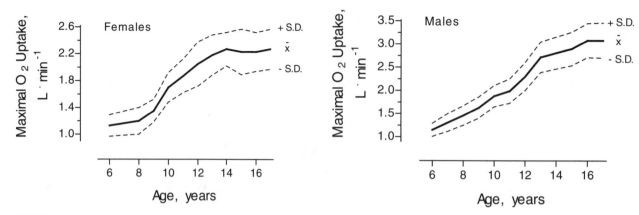

FIGURE I.1 Maximal O$_2$ uptake in children and adolescents. One hundred seventy-nine girls and 178 boys performed an all-out progressive-continuous protocol on a treadmill. Subjects were healthy nonathletes.

Data from the Wingate and the Children's Exercise & Nutrition laboratories.

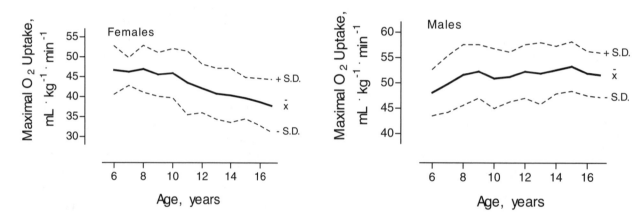

FIGURE I.2 Maximal O$_2$ uptake per kg body weight in children and adolescents. Subjects and protocols as in figure I.1.

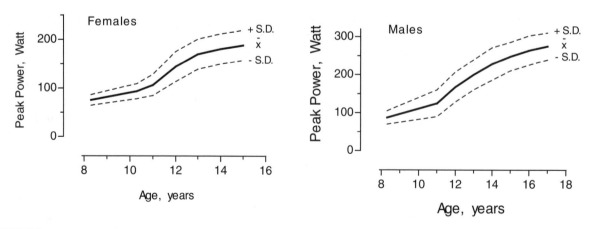

FIGURE I.3 Peak aerobic mechanical power in children and adolescents. The highest mechanical power output achieved during a progressive upright cycling test by healthy girls and boys.

Based on data from Andersen et al. 1974 (1), Seliger and Bartunek, 1976 (6), Wirth et al. 1978 (8).

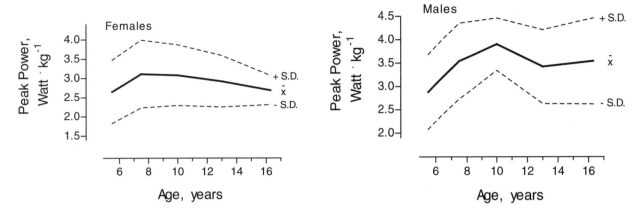

FIGURE I.4 Peak aerobic mechanical power per kg body weight in children and adolescents. The highest mechanical output achieved during a progressive upright cycle ergometer test. Subjects were girls and boys of an outpatient population who had an "innocent" murmur but no organic heart disease.

Data from Cumming 1977 (2).

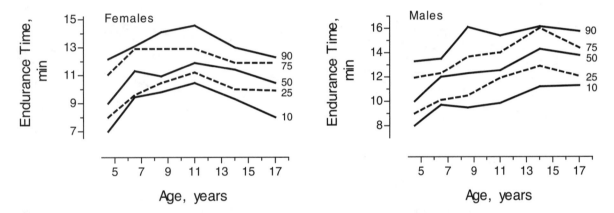

FIGURE I.5 Endurance time in the Bruce treadmill test in children and adolescents. Lines represent percentiles, which are based on performance of 160 girls and 167 boys who had an "innocent" murmur, but no organic heart disease.

Data from Cumming et al., 1978 (3).

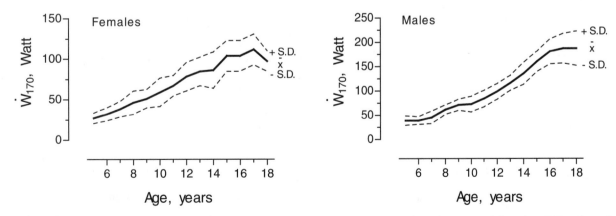

FIGURE I.6 Mechanical power during upright cycle ergometry, at heart rate of 170 beat · min⁻¹ (W_{170}) in 727 girls and boys.

Data from Rutenfranz et al. 1973 (5)

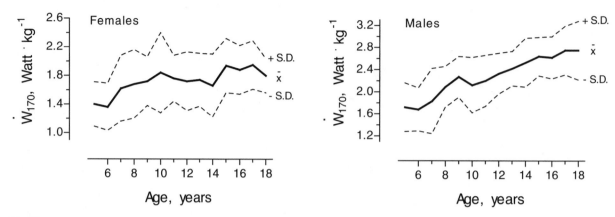

FIGURE I.7 Mechanical power per kg body weight during upright cycle ergometry, at heart rate of 170 beat · min^{-1} (W_{170}) in 727 girls and boys.

Data from Rutenfranz et al. 1973 (5).

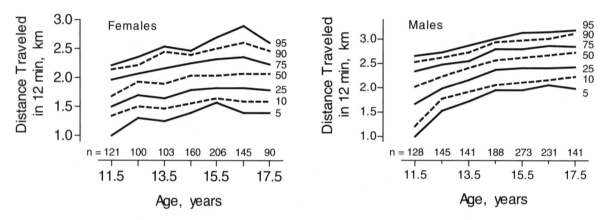

FIGURE I.8 Distance traveled by girls and boys during a 12-min run-walk test. Lines represent percentiles. Subjects were schoolchildren randomly selected from 43 schools in Halton County, Ontario, Canada.

Based on data by Roche 1980 (4).

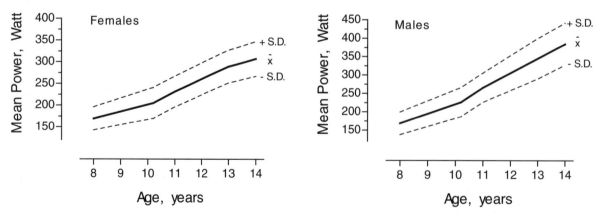

FIGURE I.9 Mean anaerobic power of the legs in children and adolescents, as determined by the Wingate anaerobic test. Subjects were 144 girls and 145 boys, all healthy nonathletes.

Based on data from the Wingate Institute and the Exercise Physiology Laboratory, McMaster University.

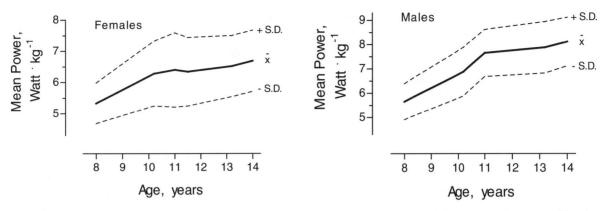

FIGURE I.10 Mean anaerobic power of the legs per kg body weight in children and adolescents, as determined by the Wingate anaerobic test. Subjects and source of data are as in figure I.9.

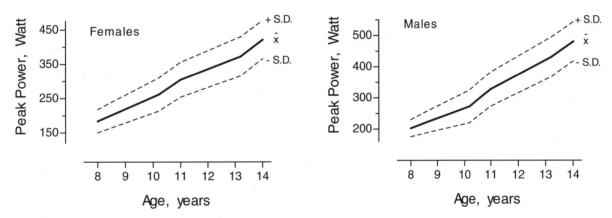

FIGURE I.11 Peak anaerobic power of the legs in children and adolescents, as determined by the Wingate anaerobic test. Subjects and source of data are as in figure I.9.

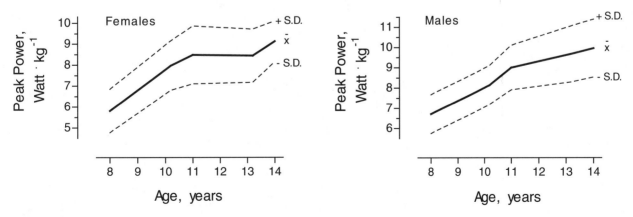

FIGURE I.12 Peak anaerobic power of the legs per kg body weight in children and adolescents, as determined by the Wingate anaerobic test. Subjects and source of data are as in figure I.9.

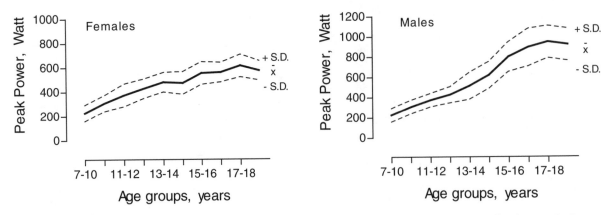

FIGURE I.13 Peak anaerobic power of the legs in children and adolescents, as determined by the force-velocity test. Subjects are 536 French girls and 510 French boys.

Based on Van Praagh 2000. (7).

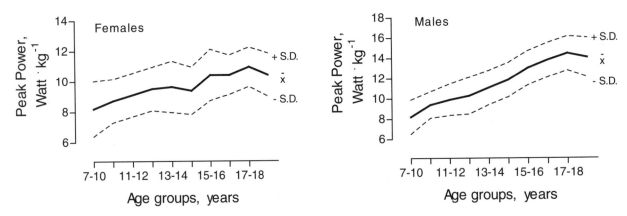

FIGURE I.14 Peak anaerobic power of the legs per kg body weight in children and adolescents, as determined by the force-velocity test. Subjects are as in figure I.13.

Based on data from Van Praagh 2000 (7).

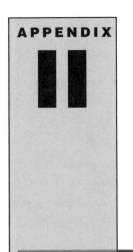

Procedures for Exercise Testing in Children

*W*hen the first edition of this text was written, there was only one available book, by S. Godfrey, on pediatric exercise testing (52). With improved technology and a growing interest in the exercising child, two more books are now available with a major focus on testing (42; 104). A reader interested in detailed discussion of exercise testing methods can consult these texts. This appendix highlights the principles and main characteristics of ergometric tests that may have a clinical relevance. Three areas to be discussed are the ergometer, the exercise protocol, and the physiologic or perceptual measurements.

Choice of Ergometer

The two most commonly used ergometers in an exercise laboratory are the cycle ergometer and the motor-driven treadmill. A step can also be used, especially in a physician's office. For some neuromuscular disabilities, one must resort to testing the arms using specially designed arm-cranking ergometers or modified cycles. There are also specialized arm ergometers to be used by wheeling a wheelchair. A comparison of the characteristics of the cycle ergometer, treadmill, step, and arm ergometer is presented in table II.1.

In spite of its greater cost and the need for safety measures and for more technicians, the treadmill is the ergometer of choice for determining maximal exercise performance of children, especially those ages 7 years and younger. Children as young as 3 years can learn to walk and run on a treadmill (89).

We often see a young child who cannot keep pedaling on a cycle ergometer even though his heart rate is only 160 to 170 beat · min^{-1}. On a treadmill the same child can reach a heart rate of 200 to 210 beat · min^{-1} and a maximal O_2 uptake that is 20% to 30% higher than the peak uptake obtained on a cycle ergometer. The apparent cause of such a discrepancy is the relatively undeveloped mass of the knee extensors (which take the brunt of the load during cycling) in the young child (83). This results in local fatigue and premature termination of the test. In contrast, during walking or running a greater muscle mass is activated, resulting in a higher cardiorespiratory and metabolic contribution. Some children, especially those who are very young and those who are mentally retarded, cannot conform to the cadence of a metronome, which is needed for mechanically braked cycle ergometers and for step testing. Furthermore, the attention span of such children is low, and they cannot maintain the required pedaling rate for the duration of the test. In contrast, a treadmill forces them to maintain a certain speed, and they can easily compensate for a momentary change of pace. Another alternative to the mechanically braked ergometer is the electronically braked cycle ergometer. It elicits a constant power, even if the child does not keep a constant pace while pedaling at 50 to 70 rev · min^{-1}.

Many laboratories still prefer the cycle ergometer to a treadmill because of its greater portability and safety. In addition, when using a treadmill one technician is fully dedicated to spotting the child, which is not needed with the cycle ergometer.

TABLE II.1 Characteristics of Ergometers in Use for Pediatric Exercise Testing

Characteristics	Cycle ergometer	Treadmill	Step	Arm ergometer
Cost	Low to medium	High	Low	Medium
Portability	High	No	High	High
Number of staff needed	1 to 2	2 to 3	1 to 2	1 to 2
Noise level	Low to medium	Medium to high	Low	Low to medium
Special safety measures	None	Harness, padding	None	None
Skill required by patient	Too hard for <5 yr or mentally retarded	Little skill	Some skill	Too hard for <5 yr or mentally retarded
Muscle mass involved	Small	Large	Large	Small
Maximal O_2 uptake	Underestimated	Achieved	Achieved	Much underestimated
Determination of mechanical power	Accurate	Estimated	Fairly accurate	Accurate
Feasibility of obtaining physiologic measurements	Easy	Less easy	Less easy	Easy
Anaerobic testing	Suitable	Not suitable	Not suitable	Suitable

In addition, certain procedures such as sphygmomanometry, oximetry, rebreathing, Doppler measurement of cardiac output, or echocardiography are easier to perform on a seated child than on a walking, running, or stepping child. For obvious reasons, a treadmill is unsuitable for the cardiac catheterization laboratory.

If a cycle ergometer is chosen, models used for adults fit most children, ages 8 years or older. For younger children, however, one needs special pediatric models or a modified existing cycle with the seat height and the length of the pedal crank reduced and the handlebar lengthened. The optimal (i.e., inducing the lowest O_2 cost of pedaling) pedal length for 6-year-old children is 13 cm, as compared with 15 cm for 8- and 10-year-old children (71) and 17.5 cm for adults. When seat height is optimal, the angle at the knee joint during extension is 15° (71). A useful indication that the seat is too high is a lateral tilt of the pelvis, as seen from behind. Most important, ergometers used for children with very weak muscles (e.g., muscular dystrophy, spinal muscular atrophy) must have the capability of eliciting very low power outputs and very small power increments. Some of these patients have *maximal* outputs as low as 10 Watts (117) (see chapter 10, figure 10.12). This is extremely low, considering that the mechanical power loss between the pedals and the flywheel in most ergometers is more than 5 Watts (some-times as much as 15 Watts). Put differently, when the dial shows "zero" resistance the child is actually cranking at more than 5 Watts.

In mechanically braked ergometers, the rate must be kept constant. Recommended rates for children and adolescents are 50 to 60 rev · min^{-1}. Among healthy 6- to 10-year-old children, 50 rev · min^{-1} yield higher mechanical efficiency than 30 or 70 rev · min^{-1} (71). Some patients, such as those with spastic cerebral palsy, need a cranking cadence slower than 50 rev · min^{-1}. Our own experience (84) has been that the optimal cranking cadence (yielding the lowest O_2 uptake at any given power) for such patients is 40 rev · min^{-1}.

On the treadmill the child should be tested only after having practiced and gained confidence in walking, running, stepping on, and stepping off. While most healthy children can master walking and running on the treadmill within 2 to 3 min of practice, children with motor disabilities may take as long as 20 min (58). In addition to learning the walking and running technique, the child should be habituated by practicing further. The amount of practice required for habituation varies markedly among children, even when they have no motor disabilities (50). We recommend that, when a child participates in a scientific experiment, a special session be devoted for habituation to the treadmill and to other instruments to be used in subsequent sessions. Habituation is less crucial for clinical testing. Most clinical laboratories perform the test without formal habituation. Economy of gait depends on stride length, the most economical being that selected by each individual. Children should therefore be allowed to use their preferred stride length. A body harness is used as a safeguard in some laboratories. In our experience, this is not needed as long as an experienced investigator stands close to the child. We do recommend, however, padding of the rear of the treadmill to prevent injury in case of a fall.

Exercise Protocol

A variety of exercise protocols is available for children. The choice of a protocol depends primarily on the specific questions to be answered but also on the abilities and limitations of the patient.

Prototypes of Exercise Tests

Protocols are schematically presented in figure II.1. These can be divided into "all-out" ones, in which the patient is expected to reach the highest possible power ("aerobic peak" and "anaerobic peak" in the diagram), and submaximal ones, in which the peak is not reached. The latter type is further subdivided into progressive and single-stage protocols. All the protocols, excluding the anaerobic ones, can be performed with a cycle ergometer, arm ergometer, treadmill, or step. The anaerobic protocols are designed for leg cycling or arm cranking only. Even though running can be done at supramaximal levels, there are still unresolved methodologic issues in this approach. As a result, very few laboratories have resorted to anaerobic testing by treadmill (48, 114).

For *direct determination of maximal aerobic power*, the most commonly used protocol is that shown in figure II.1*a*. Resistance (cycle, arm ergometer, wheelchair ergometer), slope and speed (treadmill), or height and frequency of ascents (step) are increased every 1 to 3 min, without interruption of the test, until the child can no longer maintain the activity. When the stages last 1 min or less, the term *ramp protocol* is used.

The all-out interrupted protocol (figure II.1*b)* is used whenever measurements must be taken *after* the end of each stage (e.g., blood lactate, echocardiogram) or when the investigator needs time to decide whether to continue with a higher load or to terminate the test. Sometimes, one wishes, through a single test, to collect data when the child has reached a steady state (e.g., to assess energy cost of locomotion) and to determine the child's maximal aerobic power. For such requirements, each stage should last 2 to 3 minutes. In general, children reach a steady state faster than adults do.

Submaximal progressive protocols in which maximal aerobic power is predicted rather than directly measured (figure II.1*c, d)* are resorted to when the investigator is interested only in response to submaximal exercise or is reluctant to exhaust the child by a maximal test. This is done, for example, in conjunction with heart catheterization studies or in situations where a physician is not present in the laboratory.

Single-stage protocols (figure II.1*e, f)* will be chosen when the purpose of testing is not physical working capacity, but a specific pathophysiologic response to an exercise provocation—for example, assessment of exercise-induced asthma, diagnosis of growth hormone deficiency, or provocation of exercise-induced hypoglycemia in type 1 diabetes mellitus (13). The duration of a single-stage test varies from 6 to 60 min and the intensity is submaximal. Occasionally, the exact intensity cannot be determined before testing. The investigator then starts the test at *approximately* the required intensity and makes adjustments during the initial 1 to 2 min based, for example, on heart rate response. This protocol is termed "single-stage with adjustments" (see figure II.1*f).*

The tests depicted in figure II.1*g* and *h* are designed to determine *anaerobic characteristics of a muscle group*. In the G protocol, the child performs a single supramaximal task (i.e., a task that is more intense than the aerobic peak power), which can be sustained for not more than 45 sec, as in the Wingate anaerobic test. The H protocol comprises several supramaximal tasks of 5 to 7 sec each, as in the force-velocity test.

Exercise Protocols for Aerobic Performance

There are several exercise protocols for aerobic performance. We will discuss them here.

Bruce All-Out Progressive Continuous Treadmill Test

See table II.2 for the Bruce All-Out Progressive Continuous Treadmill Test. In this test, both

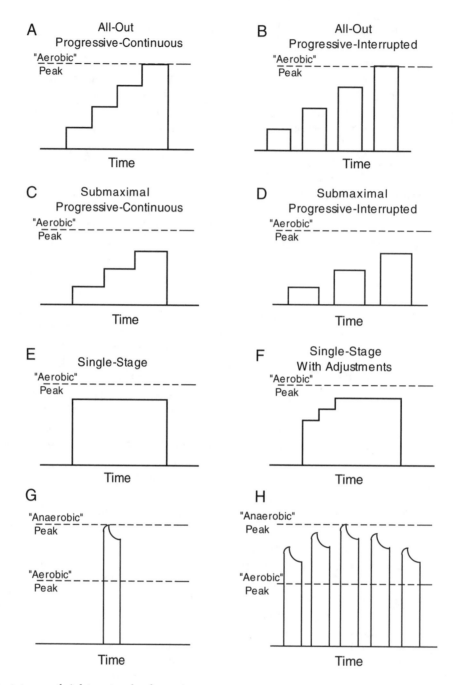

FIGURE II.1 Prototypes of eight protocols of exercise tests.

speed and slope increase from one stage to the next. The end point is the inability of the child to carry on walking or running in spite of verbal encouragement. Performance is graded by the number of minutes completed, or according to the calculated O_2 cost at the highest stage. The test was introduced for screening adult cardiac patients (27) and later adopted for use with children (38). Reproducibility among children is high and the test is feasible for 4- to 18-year-old girls

and boys. This test is not inherently better than other all-out treadmill tests (95; 111). Its basic drawback is that the initial load and the increments may be too strenuous for markedly unfit children. Another disadvantage is that highly fit children may exercise 18 min or more to determine their maximal aerobic power, by which time local fatigue and thermal load may interfere with their performance. In spite of the drawbacks, the Bruce test has been used in a clinical setup for the

TABLE II.2 The Bruce Treadmill Protocol				
Stage	**Speed**		**Grade, %**	**Duration (min)**
	Km · hr⁻¹	**Miles · hr⁻¹**		

Stage	Km · hr⁻¹	Miles · hr⁻¹	Grade, %	Duration (min)
1	2.7	1.7	10	3
2	4.0	2.5	12	3
3	5.5	3.4	14	3
4	6.8	4.2	16	3
5	8.0	5.0	18	3
6	8.8	5.5	20	3
7	9.7	6.0	22	3

assessment of maximal aerobic power. Norms are available, based on the performance of healthy children in an outpatient clinic (38) (See figure I.5 in appendix 1) and at a school (39).

Balke All-Out Progressive Continuous Treadmill Test

In the Balke protocol, progression is achieved through an increase in slope while the speed is kept constant. Various versions of this protocol have been used with children. Riopel et al. (95), for example, used 3 or 3.5 mi · hr⁻¹ and increased the slope by 2% each minute. Skinner et al. (111) increased the slope by 2.5% every 2 min. In other versions, the child is tested during running (e.g., at 5 mi · hr⁻¹) rather than walking. Peak levels are thus reached at lower slopes than with the walking versions (3; 103).

McMaster All-Out Progressive Continuous Cycling Test

See table II.3 for the McMaster All-Out Progressive Continuous Cycling Test. In this test, progression is achieved by an increase in resistance every 2 min. As in the Bruce protocol, the child is verbally encouraged to keep exercising until he or she can no longer adhere to the required pedaling rate (50 rev · min⁻¹, 60 rev · min⁻¹ in some laboratories). Performance is graded by the peak mechanical power, or the directly measured maximal O_2 uptake. When the child cannot complete 2 full min of the final load, peak power is prorated, based on the period that he did manage to complete. The following is an example:

Penultimate stage = 100 Watts
Final stage = 125 Watts
Time completed in final stage = 60 sec
Calculated peak power = 112.5 Watts

The protocol is constructed according to body height such that the total exercising time will range from 8 to 12 min for most children. The initial load and increments must sometimes be reduced to suit children with marked disability, such as muscle dystrophy or advanced cystic fibrosis.

James All-Out Progressive Continuous Cycling Test

See table II.4 for the James All-Out Progressive Cycling Test. In this test, cycling cadence is 60 to 70 rev · min⁻¹, and each stage lasts 3 minutes. The initial stage and the increments are determined according the patient's body surface area (67).

Godfrey All-Out Progressive Continuous Cycling Test

See table II.5 for the Godfrey test. Since its development in the early 1970s (52), this protocol is the first to be used systematically with pediatric patients. As in the McMaster protocol, the starting stage and the increments are based on the patient's body height. Each stage lasts 1 min, and cycling cadence is kept at 60 rev · min⁻¹.

The various cycling all-out protocols described previously are all suitable for obtaining peak power and maximal O_2 uptake. However, the Godfrey protocol is less suitable for submaximal data, because the duration of each stage is too short for obtaining a steady state.

Pulse-Activated Test

A variation of the all-out progressive continuous protocols is the *pulse-conducted exercise test*, in which the mechanical power continuously, or almost continuously, increases to induce a continuous rise in heart rate by 5 beat · min⁻¹ each minute (94; 120). The increments in power, therefore, are determined by the cardiovascular response to exercise, being slower for less fit children. Although such flexibility is an advantage, the system requires a computerized feedback system that is not readily available in clinical laboratories.

TABLE II.3 The McMaster All-Out Progressive Continuous Cycling Protocol by Body-Height Groups			
Body height (cm)	Initial stage (Watt)	Increments (Watt)	Duration of each stage (min)
≤119.9	12.5	12.5	2
120-139.9	12.5	25	2
140-159.9	25	25	2
≤160	25	♀25 ♂50	2 2

TABLE II.4 The James All-Out Progressive Continuous Cycling Protocol by Body-Surface-Area Groups			
Body surface area (m²)	Initial stage (Watt)	Increments (Watt)	Duration of each stage (min)
<1.0	33	16.5	3
1.0-1.2	33	33	3
>1.2	33	49.5	3

TABLE II.5 The Godfrey All-Out Progressive Continuous Cycling Protocol by Body-Height Groups			
Body height (cm)	Initial stage (Watt)	Increments (Watt)	Duration of each stage (min)
<120	10	10	1
120-150	15	15	1
>150	20	20	1

The McMaster All-Out Progressive Continuous Arm Test

See table II.6 for the McMaster test. This procedure is similar to that described for the McMaster cycling test, but the child uses the arms for cranking while sitting on a wheelchair or a regular chair. The axle of the pedals is kept at shoulder height. The child should be seated comfortably, supported by the back of the seat or by a pillow. At the farthest point of the pedal the arm should be fully extended without forward bending of the trunk. Strapping in required for stabilizing some patients (e.g., those with high paralysis) and to minimize trunk motion. As in the cycling test, loads must be tailored to the disability of the child. Patients with very weak muscles, as in muscular dystrophy or spinal muscular atrophy, require very low initial load and smaller incre-

ments. Not all arm ergometers are compatible with these requirements (see previous section, "Choice of Ergometer"). Likewise, while the recommended cranking cadence is 50 rev · min^{-1} for able-bodied children, it may have to be reduced for children with a neurological or muscular disability. For example, for children with spastic cerebral palsy, the optimal cadence was found to be 40 rev · min^{-1} (84).

The Cumming All-Out Progressive Intermittent Cycling Test

See table II.7 for the Cumming test. This protocol was introduced in the mid-1970s for use with cardiac catheterization; the patient cycles in the supine position (37). The wide range of power loads recommended at each stage reflects the need to individualize the test according to age, cardiac defect, and functional ability. Other protocols of interrupted exercise, to be used in conjunction with cardiac catheterization, are available (77).

Submaximal Progressive Continuous Cycling Protocols

These protocols involve three or more submaximal stages. Each stage lasts 3 to 6 min, at the end of which heart rate is determined. Performance is graded according to the mechanical power that the child produces at a heart rate of 170 beat · min^{-1} (W_{170}). This value is reached by interpolation or extrapolation of the individual regression line of heart rate over power. For more details, see the section titled "Indirect Determination: Submaximal Tests" later in this appendix. Various protocols have been used for this purpose with children (e.g., 1; 22; 88). Table II.8. is an example. Each stage in this Adams protocol lasts 6 min.

Submaximal Progressive Intermittent Step Tests

Stepping up and down on a step or a bench of a given height at a given pace is one of the oldest aerobic tests. Originally introduced for adults, step tests were modified in the 1940s for adolescents (68) and later for children (7; 53). In some tests the subjects perform a single stage, but most protocols include several stages at a progressively increasing sequence. The increase is achieved by

TABLE II.6	The McMaster All-Out Progressive Continuous Arm Cranking Protocol by Body-Height Groups			
Body height (cm)	Initial stage (Watt)	Increments (Watt)	Duration of each stage (min)	
≤119.9	8	8	2	
120-139.9	8	16.5	2	
140-159.9	16.5	16.5	2	
≤160	16.5	♀16.5 ♂33	2 2	

TABLE II.7	The Cumming All-Out Progressive Intermittent Cycling Protocol	
Stage	Mechanical power (Watt per kg body weight)	Duration
1 Rest	0.64-1.14	3 3
2 Rest	1.31-1.96	3 5-10
3	2.12-4.08	To exhaustion

TABLE II.8	The Adams Submaximal Progressive Continuous Cycling Protocol by Body-Weight Groups			
Body weight (kg)	1st-stage power (Watt)	2nd-stage power (Watt)	3rd-stage power (Watt)	Duration of each stage (min)
<30	16.5	33	50	6
30-39.9	16.5	50	83	6
40-59.9	16.5	50	100	6
≥60	16.5	83	133	6

an increase in the stepping cadence. Heart rate is measured at the end of each stage or at a precise time of recovery.

Step tests are inexpensive and require little expertise to administer. As such, they have been used in a clinical setting (53) as well as in large-scale surveys (7). The protocol used in a clinical setting, first introduced at the Wingate Institute by Hanne in 1971, is summarized in table II.9. It is based on a feasibility study with 7- to 12-year-old girls and boys (53). Stepping is performed up and down a 30-cm step to a cadence given by a metronome. Each stage lasts 5 min with a 5- to 10-min rest in between. Grading is done accord-

ing to heart rate response at the end of each stage, from which W_{170} can be calculated. Calculation of mechanical work (per one ascent) and power is done as follows:

$$\text{Work (Joule)} = \text{body mass (kg)} \times \text{step height (m)} \times 9.81 \quad (1)$$

Work *per one descent* is taken as one-third of the work per one ascent. Thus the total work of one ascent and descent in a 30-cm step equals the following:

$$\text{Work (Joule)} = \text{body mass (kg)} \times 3.92 \quad (2)$$

Assuming N ascents and descents per minute, the mechanical power output in a 30 cm step is as follows:

$$\text{Power (Watt)} = [\text{body mass (kg)} \cdot 3.92 \text{ N}]/60 \quad (3)$$

In the Canadian Home Fitness Test, stepping is done on a double 20.3-cm step, at 19 ascents per min. Scoring is done according to heart rate, measured during recovery. Special records are available, which include instructions and music at the required cadence (7).

Testing Protocols for Anaerobic Performance

There are several testing protocols for anerobic performance. They include the Margaria Step-Running test of Peak Anaerobic Power, the Wingate Anaerobic Cycling Test, the Single-Leg Wingate Anaerobic Test, the Force-Velocity Test, the McMaster Muscle Endurance Test (McMET), and the Accumulated O_2 Deficit Test (AOD.)

The Margaria Step-Running Test of Peak Anaerobic Power

This test (85) is of historical importance. With this test di Prampero and Cerretelli (41) were the first to show that children's anaerobic performance is lower than that in adults. The child is required to run upstairs at maximal speed, usually two steps at a time, after a short sprint on a flat surface. The time taken to make two full strides (i.e., from the

TABLE II.9 The Hanne Submaximal Progressive Intermittent Step Test				
Stage	Step height (cm)	Rate (ascents/ min)	Duration (min)	Approximate power (Watts/kg)
1	30	15	5	1.0
2	30	22.5	5	1.5
3	30	30	5	2.0

time that a foot takes off until it lands again on the target step) is registered. Power (P) is calculated based on body mass (m), vertical distance (h), and the time needed to cover this distance (t), as follows:

$$P \ (kgm \cdot sec^{-1}) = m \ (kg) \ 9.81 \ h(meter) \ t^{-1}(sec)$$

Young children, who cannot run two steps at a time, can use a modified protocol of one step at a time (40). The main advantage of the Margaria test is the low cost; the only equipment needed is a photoelectric timer. Although the test yields higher power values than peak power of the Wingate test (see the following paragraph), it requires some skill and cannot be applied to disabled children. Nor can arm performance be measured separately. In addition, because of the contribution of various muscle groups to the step-running task, it is difficult to tell which muscles specifically contributed to the power generation.

The Wingate Anaerobic Cycling Test

See figure II.2 and table II.10 for the Wingate Anaerobic Cycling Test. This test has been used and evaluated extensively since the late 1970s, and much has been written about its reliability, validity, and other characteristics. The interested reader can consult related books (63; 123) and review articles (12; 14; 62). We focus here on practical principles of using this test with healthy children and those with a neuromuscular disability.

The test can be performed by the legs or by the arms and with an appropriate ergometer. It lasts 30 sec, during which time the child pedals at maximal speed against a high constant resistance. The power output is therefore a function of pedaling velocity, and the mechanical work during 30 sec is a function of the total number of pedal revolutions. These can be counted by a mechanically or magnetically triggered counter.

One usually determines three indices:

- Peak power
- Total work (or mean power) over 30 sec
- Percent fatigue

Point A in figure II.2 is peak power, which is the highest power at any 3-sec period. (Depending on the system used to count pedal revolutions, one can sample at frequencies other than every 3 sec and define peak power accordingly.) The total work is the product of the total pedal revolutions, the resistance force on the flywheel, and the distance traveled by the flywheel perimeter in each pedal revolution. Mean power is the average of all power values measured at each sampling interval (10 points in the example of figure II.2). Point B in figure II.2 is the lowest power at any 3-sec period. One needs to determine this point to calculate % fatigue:

$$Percent \ fatigue = 100(A-B)/A$$

Peak power and total work (or mean power) are both considered fitness indices. Peak power represents the explosiveness of the muscle group tested. Because it is usually reached within the first 5 to 10 sec of the test, peak power is considered to reflect the rate of the ATP-CP turnover, even though there is no validation of this assumption. In fact, as found in adults, already after 10 sec after the start of the test, there is a marked accumulation of lactate in the muscle (66), suggesting that anaerobic glycolysis has taken place at that early stage. Total work and mean power are taken as indices of muscle endurance, reflecting the rate of anaerobic glycolysis (65; 66). The % fatigue index has not been linked to a specific fitness characteristic. However, it is particularly high in power athletes and sprinters (63) and it has a fairly strong association with the predominance of fast-twitch muscle fibers in the vastus lateralis muscle (18).

Ergometers suitable for the Wingate test must have a constant resistance mode. This is available in all mechanically braked ergometers and in some electrically braked ones. In the mechanically braked ergometers, one should select a model in which the resistance is applied by hanging

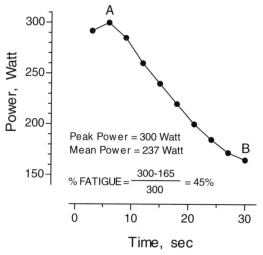

FIGURE II.2 The Wingate anaerobic test, leg performance of an 11-year-old boy. Power is measured continuously but averaged every 3 seconds.

$$\text{Work (kpm)} = \text{resistance (kp)} \times 6 \quad (1)$$

Total work in 30 sec is therefore

$$\text{Work (kpm)} = \text{resistance (kp)} \times 6 \times \text{total rev} \quad (2)$$

or

$$\text{Work (Joule)} = \text{resistance (kp)} \times 58.8 \times \text{total rev} \quad (3)$$

Mean power for the whole test is:

$$\text{Mean power (kpm} \cdot \text{min}^{-1}) = \text{resistance (kp)} \times 12 \times \text{total rev} \quad (4)$$

or

$$\text{Mean power (Watt)} = \text{resistance (kp)} \times 1.96 \times \text{total rev} \quad (5)$$

Power at each 3-sec period is calculated to determine peak power and percent fatigue:

$$\text{Power (watt)} = \text{resistance (kp)} \times 11.76 \times \text{rev in 3 s} \quad (6)$$

By increasing the number of intervals for which power is calculated (e.g., each 1 sec), one can achieve a higher calculated peak power and a higher % fatigue, but this will not affect the total work or the mean power (34).

An important refinement of the Wingate test is to include the work done in overcoming the inertia of the flywheel during acceleration at the start of the test and the frictional loss due to internal resistance of the ergometer (34; 73). When the inertia component is included, the corrected peak power increases by approximately 20% and the mean power by about 7% to 8%, as found for 9- to 10-year-old girls and boys, who were tested on a Monark model 814E ergometer (34). While such a correction is important, it should be determined separately for each ergometer and its flywheel, using expertise and equipment not readily available in most laboratories.

The resistance setting that yields the highest mean power (or total work) for various body

| Body weight (kg) | Force (kp) | | Work (J · rev⁻¹) |
	Monark	Fleisch	
20-24.9	1.75	1.05	1.02
25-29.9	2.10	1.26	1.23
30-34.9	2.45	1.47	1.44
35-39.9	2.80	1.68	1.65
40-44.9	3.15	1.89	1.86
45-49.9	3.50	2.10	2.07
50-54.9	4.00	2.40	2.40
55-59.9	4.50	2.70	2.73
60-64.9	5.00	3.00	3.06
65-69.9	5.50	3.30	3.39

TABLE II.10 Optimal Resistance for the Wingate Anaerobic Leg Test by Body-Weight Groups

Values for arm cranking are 67% of those recommended for the leg test.

weights (e.g., a Monark model 814, or a Fleisch-Metabo ergometer) rather than a pendulum. In studies at the Wingate Institute and at McMaster University, we have been using the Fleisch-Metabo ergometer (Switzerland), in which the respective distance is 10 m · rev⁻¹. The following calculation of work and power is given for the commonly available Monark ergometer, in which the distance traveled by the flywheel circumference is 6 m for each pedal revolution. Work done in one pedal revolution is as follows:

weight groups is shown in table II.10. Values are presented for the Monark and the Fleisch ergometers. Optimal resistance for other ergometers can be calculated accordingly, based on the distance traveled by the flywheel at each pedal revolution.

The Wingate anaerobic test is feasible for children as young as 5 years. It can also be applied to most children and adolescents with a neuromuscular disability (93; 117; 121). For such patients, the arm-cranking test is usually easier to perform than the leg pedaling test, which reflects the anatomic distribution of the affected muscles (e.g., in muscular dystrophy).

A 3- to 4-min warm-up is recommended before the test. It is a means of preventing injury, and it improves the child's performance (61). The warm-up should include pedaling (or arm cranking) at submaximal intensity, interpolated by several 5- to 7-sec sprints. For details see the section in chapter 1 titled "Warm-Up Effect." Once the test is completed, the child should cool down by pedaling slowly against low resistance for about 2 minutes. Some people complain of dizziness and nausea several minutes after having completed the test (this is very rare in children). They should be advised to lie down for several additional minutes.

Unlike aerobic tests, the Wingate test does not require a thermoneutral environment. That is, neither air temperature nor humidity affects children's performance in the Wingate test (44). Nor is it important to ensure that the person to be tested is not dehydrated (64). As discussed in chapter 1 (section titled "Anaerobic Performance of Children and Adolescents"), children recover very quickly after the Wingate test (55). If the test has to be repeated, the subject need not wait more than 30 minutes. Norms for peak power and mean anaerobic power of legs and arms are given in appendix I.

Single-Leg Wingate Anaerobic Test

In this variation of the Wingate test, the child pedals with one leg only, while the other leg rests. This allows one to determine the anaerobic performance of one leg versus the other, which is useful in clinical conditions that affect the legs asymmetrically (e.g., tethered spinal cord, hemiplegia, status after knee surgery). This test was studied extensively by Hebestreit et al. in healthy children and those with a disability. The author found very high test-retest reliability

and proposed a way of determining the optimal braking force that will yield the highest power (54). Another study has shown that the sum of performance (e.g., total mechanical power) of the right and left legs is greater than the performance on the Wingate test. This difference was evident in girls more so than in boys (46).

The Force-Velocity Test

This test is intended to determine the peak mechanical power during cycling or arm cranking. It involves 5 to 8 sprints (5 to 7 sec each) at maximal speed, with a 2- to 5-min rest in between. Each sprinting bout is done against a different, but constant, braking force. Based on force-velocity characteristics of muscle contraction, and as seen in figure II.3, each bout yields a different power. The highest power is taken as peak power, and the respective braking force is the "optimal force." Peak power in this test is somewhat higher than that achieved by the Wingate test. The latter is done against a single braking force, which often will not be the optimal force for that child.

The force-velocity test is easily performed by healthy children (21; 87; 122) and those with a

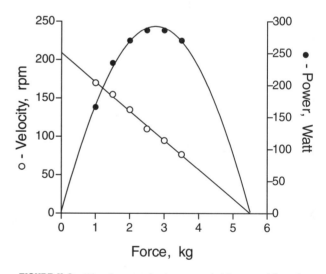

FIGURE II.3 The force-velocity test. A 10-year-old male performed six cycling sprints, each against a different braking force. The velocity-force regression line (open circles) is extrapolated to determine the force at zero velocity, which is the theoretically highest possible force. The velocity against zero force is the highest possible cycling velocity. Power values (black circles) are calculated for each sprint and form part of an ellipse. The highest power, determined by interpolation, would be elicited at an optimal force, which is approximately 50% of F_0. Reproduced, with permission, from Bar-Or 1996 (14).

neuromuscular disability (121) or pulmonary disease (36). It has been used less frequently than the Wingate test within a clinical context, mostly because it yields information on peak power only but not on muscle endurance. Like the Wingate test, the force-velocity test can be done using an ergometer in which the force is constant, but it can also be done on an isokinetic cycle ergometer, in which pedaling velocity is controlled by the investigator (108).

The McMaster Muscle Endurance Test (McMET)

This test involves 25 maximal knee extensions and flexions at an angular velocity of $180° \cdot sec^{-1}$, using an isokinetic dynamometer. Each cycle of extension and flexion lasts 1.1 to 1.2 sec so that the total duration of the test is 27 to 30 sec. The McMET yields information about peak power as well as muscle endurance (measured as total work) and fatigability (measured as the rate of power drop from its peak). These variables are calculated separately for the knee extensors and flexors. The test was found valid when compared with the Wingate test (30). A clinical use of the test is a comparison of muscle endurance of the right and left limbs. While the test was developed specifically for the knee extensors and flexors, it can also be used for other muscle groups, such as the elbow flexors and extensors.

The Accumulated O$_2$ Deficit Test (AOD)

First introduced by Hermansen and Medbo (56), the concept of accumulated oxygen deficit has been used to assess anaerobic capacity in adults (86) and, to a lesser extent, in children (31; 32). For details on the pediatric use of the AOD test, see Carlson and Naughton (32). The test is based on measuring the O_2 uptake during an all-out supramaximal exercise task (i.e., an exercise intensity that is higher than the lowest intensity that would yield maximal O_2 uptake), and subtracting it from the predicted O_2 requirements for that task. To predict such requirements, one needs to measure the child's O_2 uptake at 4 to 5 submaximal exercise intensities, as well as the maximal O_2 uptake. Furthermore, one must assume that the mechanical efficiency during supramaximal exercise is the same as that obtained during submaximal exercise.

While the notion of knowing a person's anaerobic capacity is appealing, the AOD test has been accepted with skepticism among the scientific community. The method has several shortcomings:

- The assumption about constancy of mechanical efficiency is most probably incorrect (8).
- If, indeed, O_2 deficit at supramaximal exercise reflects the *capacity* of anaerobic energy turnover, one would expect to find an identical O_2 deficit at various supramaximal intensities. This, however, seldom occurs in children, particularly during treadmill running (32).
- The test requires the child to perform several steady state submaximal stages, a direct measurement of maximal O_2 uptake, as well as an all-out supramaximal task. This is time- and effort-consuming for the subjects and for the laboratory.

Testing Protocols of Muscle Strength

A weak muscle is often a limiting factor in a child's physical performance, particularly if the child has a neuromuscular disability (11; 15; 59) or other causes of muscle atrophy. Obese individuals, while having a normal absolute strength, may be limited in strength-related fitness components that require lifting the body (see chapter 9, figure 9.14).

Strength is defined as the highest force generated by a muscle or a muscle group at a given joint angle during a single contraction. Traditionally, strength has been determined during *isometric* action (i.e., when the overall length of the muscle does not change). With currently available dynamometers, one can measure strength also during *concentric* action (i.e., force generated while the muscle length is reduced) and *eccentric* action (force generated while muscle length increases). Another term to recognize is *isokinetic* action, which denotes contraction (concentric or eccentric) at a constant angular velocity. While isokinetic action is seldom used in real-life situations, it is a useful testing mode in which the tested limb (or other body part) is allowed by the dynamometer to move only at a predetermined, near-constant speed

Clinicians often use the term *muscle power* when they mean *muscle strength*. While strength is an expression of force (or torque), power is force times velocity, or work divided by time. The distinction between force and power is important:

Though the two may be related in a given child, they do denote two distinct performance functions.

For a detailed description of measuring strength in children, see reviews by Gaul (51) and by Blimkie and Macauley (24). This section is limited to the rationale and principles of strength measurement in children and the most commonly used testing protocols.

Rationale for Strength Testing Within Clinical Context

Physiologists use strength testing as a means of understanding contractile characteristics of muscle (e.g., force-velocity relationships, rate of contraction and relaxation); the neural control of muscle action (e.g., recruitment of motor units); and the effects of growth, maturation, aging, gender, and training on muscle performance (see chapter 1, "Effects of Growth and Maturation on Muscle Strength"). To the clinician, knowing the strength of a specific muscle group can yield objective, quantifiable information on the progression of a disease process (e.g., muscular dystrophy) or on changes in muscle function through medication (e.g., deflazocort in muscular dystrophy, Botox injections in cerebral palsy), physical therapy, occupational therapy, rehabilitation, and surgery.

Principles of Strength Testing

In this section we will review the principles of strength testing, including muscle specificity, joint angle specificity, specificity of angular velocity, restriction and stabilization, and cooperation and motivation.

Muscle Specificity   When measuring muscle strength, one should realize that the effects of disease and of treatment may be specific to a given muscle group. An example is Duchenne muscular dystrophy, in which the proximal muscles of the limbs (e.g., thigh and shoulder) lose function faster than do the more distal muscles. Measuring handgrip strength, for example, at early stages of the disease may yield little clinically relevant information. Likewise, the effects of physical therapy, rehabilitation, and training are specific to the muscle group that has been treated.

Joint Angle Specificity   Strength generated through muscle action depends on the length of the muscle, relative to its maximal length. While the relative muscle length cannot be readily measured, it is represented by the joint angle. To standardize a strength test, one must therefore determine the joint angle, or angles, at which the action should be monitored. This is of particular importance when testing of a patient is repeated over time.

Specificity of Angular Velocity   Likewise, during isokinetic action, strength depends markedly on angular velocity. The forces generated during concentric action are greater at lower than at higher speeds (being greatest at zero velocity, which is the isometric status). Inversely, with eccentric action the forces get stronger with rising velocity. It is thus imperative to standardize the velocity used in a testing protocol. Many laboratories select several angular velocities for each muscle group.

Restriction and Stabilization   Some movements can be executed by more than one muscle group. For example, raising a weight by the upper limb can be done by elbow flexion but also by movement at the shoulder girdle. Therefore, to focus on a specific muscle group, one must restrict the action of other muscles by stabilizing body segments that are not being tested. This can be done, for example, by restricting the trunk movement with shoulder straps when testing elbow flexors and extensors, or by affixing straps around the waist and thigh when testing the knee extensors and flexors.

Cooperation and Motivation   Strength testing calls for maximal effort by the child. The outcome, therefore, is most dependent on the child's cooperation and motivation. In general, children like strength testing. Their cooperation depends on understanding the importance of the test and the required procedures. One should provide ample time for learning and practicing each routine. The learning phase is also used to habituate the child to the laboratory environment and to the staff administering the test.

Motivation can be further enhanced by providing feedback about the child's performance. This can be done verbally after each trial or, even more effective, by a display on a computer screen. A promising approach that needs further technological development is to construct testing routines that, to the child, will look like a computer game.

Recovery Time Most tests require several repetitions. To yield an optimal response, the child should be fully recovered from one trial to the next. There are no formal studies that have identified optimal recovery times in children. Furthermore, recovery time may differ among diseases and among muscle groups. A safe approach is to have rest periods of 90 to 120 sec between trials. However, some patients may require longer periods. This can be determined by trial and error for those children whose measured strength declines from one trial to the next.

Availability of Equipment While low-cost apparatus, such as a myometer (see the following section), can be used at the physician's office (60), more sophisticated machines, such as an isokinetic dynamometer, may be available only in exercise laboratories. Such laboratories are usually found within a physical education or a sport science department and less so in a hospital or a clinic. However, dynamometers are often available in physical therapy and rehabilitation services.

Modification of Equipment Commercially available dynamometers are made for adults. Because of children's smaller trunk and limb size, some modifications are needed for pediatric use, mostly as related to the location of restraining straps, back support, and the degree of increments of resistance forces.

Warming up As with other exercise tests, strength testing should be preceded by a warm-up. There is no single routine that has been identified as optimal.

Procedures for Strength Testing

As part of a physical examination, clinicians often assess strength by a *manual method*, in which the child is asked to push or pull against resistance provided by the clinician. This measurement yields, at best, semiquantitative information and is quite subjective. A refined approach is the use of an apparatus that senses and displays the force generated between the child's limb and the clinician's hand. The *myometer* is an example of such a device.

If all one wishes to measure is handgrip strength, one can use a *handgrip dynamometer*. Grip strength is a simple and inexpensive measurement. It has been used in numerous studies with healthy children (23) and in some studies with children who have chronic diseases such as juvenile idiopathic arthritis (132) or who had low birth weight (49). Its main disadvantage is that, based on the principle of muscle group specificity, grip strength seldom represents other muscle groups, particularly in children with a neuromuscular disability.

Cable tensiometers measure isometric strength of various muscle groups. The force sensor can be either mechanical or electronic. Much of the original information on the effects of age, maturation, and gender on isometric muscle strength (4; 33) was generated with these devices. Their main advantage is that they can be used for a variety of muscle groups, with little or no need for special attachments. In addition, they are much cheaper than the isokinetic dynamometers. The main disadvantage of cable tensiometers is that their use is limited to isometric action.

Isokinetic dynamometers are now used in many exercise laboratories and in physical therapy and rehabilitation clinics. The three most common are various models of the Cybex, Kin Com, and Biodex dynamometers. For a detailed description of the mode of action of these dynamometers, as well as the recommended testing protocols, see reviews by Gaul (51) and Sale (107). A limb segment (and sometimes the trunk) is attached by a strap to a lever arm. While a motor moves the lever arm at a predetermined speed, the child exerts force either to push the lever arm faster or to resist its movement. This force is sensed by a force transducer attached to the lever arm. The motor is strong enough to counter this force. As a result, the lever arm keeps moving at a constant (or near constant) velocity. Because the contraction velocity of the muscle itself may change somewhat as it moves through a range of motion, some investigators use the term *isodynamic* to denote the action of the muscle, as opposed to *isokinetic*, which refers to the movement of the lever arm. When the force exerted by the subject is in the same direction as the lever arm movement, the muscle action is *concentric*. If the subject exerts force to oppose the lever arm motion, then the muscle action is *eccentric*.

Because of the angular motion during iso-kinetic dynamometry, muscle strength is defined as peak *torque* rather than peak force. Torque is the product of force and the distance between the axis of rotation and the location of the strap that connects the limb to the lever arm. The unit commonly used for torque is the newton-meter.

$$1 \text{ newton-meter} = 9.81 \text{ kg meter}$$

To match the torque detected by the dynamometer with the torque generated by the muscle, the limb joint around which the motion occurs and the axis of rotation of the lever arm must be positioned at the same level. To calculate torque, one must record the location along the lever arm of the strap that connects it to the limb.

The machine records continuously the torque throughout the whole range of motion. Peak torque is defined as the highest torque during this motion. One usually records the peak torque as well as the joint angle at which it has been achieved. Other variables that can be recorded include average torque, total work, peak power, average power, and *impulse* (torque multiplied by time). Ideally, when serial measurements are made over time to assess the progression of a disease or the effect of therapy, one should try to replicate the range of motion from one test to the next. However, some patients, because of joint contractures, pain, or other reasons, cannot go through the same range of motion each time. It is likely that in such cases the angle at which peak torque is reached may be different as well. This detracts from the ability to compare tests along time, but there is no valid way of circumventing this limitation.

The force transducer on the lever arm senses the weight of the limb segment attached to it, even when the child is not exerting any force. This causes an error in the measurement, and it should be corrected. During testing of strength around the knee joint, for example, this gravitational component will decrease the torque sensed by the machine during extension and increase it during flexion. The error is greater whenever the mass of the limb is relatively high compared with the forces exerted by it. It may reach 15% in healthy adults (107) and be considerably greater in a child with advanced Duchenne muscular dystrophy (where strength is small compared with the weight of the limb) or when the leg is heavy because of obesity. Likewise, the gravity component is relatively higher, in all children, during flexion than during extension, because forces exerted during flexion are only 50% to 80% those exerted by extension (28; 43; 113). Routines for correction due to gravity are provided by the manufacturer of each dynamometer.

Measurements Taken During Aerobic Exercise Tests

Numerous variables can be measured during an exercise test. These range from the simple determination of peak power output (which does not require any instruments apart from an ergometer) to the measurement of intracardiac pressures or myocardial perfusion imaging (which require catheterization and sophisticated equipment). Methodologic details of such measurements are available in books and reviews on exercise testing (16; 24; 52; 104; 124; 130). In this section we focus on those variables that can be measured by noninvasive procedures. We also present a brief description of methods to determine cardiac output and related hemodynamic variables.

Heart Rate (HR). This variable is conveniently determined by the use of an HR monitor. This measurement has become popular and simple with the development of telemetric methods, in which a lightweight transmitter of the heartbeats is placed on the chest and a receiver is placed on the wrist. Measurements by an apparatus such as the Sport Tester are valid for children during steady state (76) and non-steady-state (19) conditions, alike. The instrument is socially acceptable by children as young as 4 years (19). Heart rate can be averaged at various intervals. This makes the method useful for testing during short bouts of exercise as well as for continuous HR monitoring over several days.

Within a clinical context, one often wishes to determine electrocardiographic (ECG) changes, which can also be used for HR determination. One can count the number of R-R intervals during a certain time period or measure the time needed for a given number of R-R intervals. Special rules are available for the latter method. Because of sinus dysrhythmia, which appears particularly at rest and during low exercise intensities, one should include at least six R-R intervals in each measurement. In some metabolic carts, the ECG is fed online to the cart, and HR is calculated automatically. Irrespective of the method of counting, it is important to periodically calibrate the apparatus or the paper speed of the electro-cardiograph.

Heart rate is sensitive to environmental changes and other factors. The utmost care must be taken to standardize room temperature and humidity, to reduce emotional stress, and to ensure that the child is well hydrated and not fatigued by prior activity. (See chapter 1, section titled "Heart Rate and Exercise," for further details.)

Minute Ventilation. Ventilatory function is often monitored to evaluate the limiting factors in performance of a dyspneic child or in a child with known pulmonary or cardiac disease (see, for example, chapter 6, section titled "Cystic Fibrosis"). Ventilation is also monitored for the measurement of O_2 uptake, as well as of the ventilatory threshold (see related section that follows). Historically, expired air used to be collected into Douglas bags, or a Tissot spirometer. Most laboratories at present use a flow meter at the inspiratory or expiratory end, or a pneu-motachograph, in conjunction with a metabolic cart. Minute ventilation is presented in volume per minute, corrected for BTPS (air at body temperature, ambient pressure, and saturated with water). However, for the calculation of O_2 uptake, ventilation is corrected for STPD (air at $0°$ C, 760 mmHg, dry). For children the dead space of the mouthpiece and one-way valve must be small (not more than 50 ml for a child under 10 years old).

Ventilatory Threshold. As discussed in chapter 1 (section titled "Ventilation and Ventilatory Threshold"), ventilatory threshold is an important variable that reflects aerobic fitness, without the need to directly measure maximal oxygen uptake. It is often used as a guide to establish intensity of exercise during conditioning programs. There are several ways of determining a person's threshold, all of which involve measurement of relative changes in different gas exchange variables as exercise intensity increases. These include the point of nonlinear increase in minute ventilation or $\dot{V}CO_2$, the upward deflection of $\dot{V}E/\dot{V}O_2$ without a concomitant increase in $\dot{V}E/\dot{V}CO_2$, and the V-slope method, in which computer-generated regression lines indicate when $\dot{V}CO_2$ begins to rise disproportionately to the linear increase in $\dot{V}O_2$.

These approaches to estimating the ventilatory threshold can be used during treadmill (walk, run) or cycling protocols. Evidence exists that measurement of ventilatory threshold in this manner provides an accurate indication of the lactate anaerobic threshold (the point in a progressive exercise test when serum lactate levels begin to rise) (92). The ventilatory threshold, then, provides a noninvasive means of estimating both anaerobiasis during exercise testing as well as aerobic fitness. Measurement and interpretation of the ventilatory threshold in children have been reviewed by Mahon and Cheatham (81) and Washington (129).

Systemic Arterial Blood Pressure. Measuring a child's blood pressure during and after exercise is commonly done within a clinical exercise testing laboratory and in physiologic research. Guidelines for blood pressure measurement in the clinical context have been reviewed elsewhere (70; 130). This section focuses on major methodologic aspects.

The use of a sphygmomanometer during exercise is feasible with children, as it is with adults. However, this measurement is prone to error resulting from lack of objectivity, insufficient skill, or nonstandardized procedures. The two difficulties in the use of the auscultation method during exercise are movement of the arm and noises other than the Korotkoff sounds. Stethoscopes are available in which the diaphragm is stabilized to the cuff. These can be helpful with a child who keeps moving the arm during exercise. Noise and vibration are transferred, during cycling, from the handlebar to the arm. It is useful therefore if the child lets go of the handlebar and keeps the arm extended during the time of measurement. Likewise, while a measurement is taken, the child should avoid performing isometric action with the arm or forearm.

The monitoring of diastolic blood pressure by auscultation is quite unreliable during exercise because one can sometimes hear the Korotkoff sounds at as low as zero pressure. The systolic pressure, on the other hand, can be determined with relative ease during moderate and intense activities because of the high intensity of the sounds.

Sphygmomanometers are available in which inflation and deflation of the cuff are done automatically and at a predetermined rate and timing. The Korotkoff sounds are detected by a microphone, with filtering of other noises. Such an apparatus increases the objectivity and standardization of the measurement and is recommended for clinics with a high patient load. We are not aware of validation studies in which blood pressure, using this apparatus, was compared with a direct intra-arterial measurement in children.

In testing children, special attention must be paid to the cuff size. The bladder should, as a rule, cover at least two-thirds of the arm length and completely encircle it (95; 115).

O_2 Uptake. A direct determination Of O_2 uptake requires the use of O_2 and CO_2 analyzers to determine the fractional concentration of these gases in the expired air. Such analyzers are usually available in exercise physiology laboratories but not in clinics or in a physician's office. With the current use of metabolic carts, one does not need to calculate the O_2 uptake, because this variable is calculated by the computer within the cart.

O_2 uptake can be assessed indirectly when the mechanical power output is known, as in pedaling. Assuming that the O_2 cost of pedaling is not markedly different among individuals (2), one can use table II.11 to assess the O_2 uptake equivalent during cycling. For many children with a neuromuscular disability (e.g., cerebral palsy) the O_2 cost of pedaling is above normal (see chapter 9, section titled "Physiologic and Mechanical Cost of Locomotion"). For such children, table II.11 should not be used.

Equations are also available for the O_2 uptake equivalents of walking, running, or stepping up and down stairs. Such equations, however, neglect to consider the marked differences in the O_2 cost of these tasks among children at different ages and even within a certain age group (see chapter 1, section titled "Mechanical Efficiency and Economy of Movement," and figure 1.8). Marked differences in the energy cost of locomotion exist among children with a neuromuscular disability (118).

Electrocardiogram (ECG)

The exercise ECG is an important tool in the assessment of children with a proven, or suspected, cardiovascular disease. The following are comments on some practical aspects of exercise ECG.

Skin Preparation. To increase the electric conductivity of the skin, hyperemia is induced through cleansing and rubbing with alcohol or acetone, and the skin is then slightly abraded.

Choice of Electrodes. A straight isoelectric baseline is essential for detection and interpretation of ST-T changes. To achieve a straight baseline during exercise, one must ascertain that the skin-to-electrode electrical impedance remains unchanged in spite of body movement. This can be achieved by the use of "floating" electrodes in which the metal is not in direct contact with the skin but is separated from it by a layer of electrolyte cream or gel. Various electrodes of this type are available commercially. Cables connecting the electrode to the ECG must be lightweight and flexible.

Choice of Leads. These can range from a single bipolar chest lead to 14 leads that combine the conventional 12-load ECG with Frank orthogonal leads. Although a single CM5 bipolar lead (reference electrode at the manubrio-sternal junction, exploratory electrode at C5, and a ground electrode on the back) can detect dysrhythmias and most ST-T changes; one should attempt to use at least three leads simultaneously (e.g., II, V2, and V5).

Cardiac Output

Estimation of cardiac output in the exercising child is feasible using invasive (direct Fick, dye-dilution) or noninvasive techniques. For ethical reasons the latter approach is preferable. Several noninvasive methods have been used, each with specific advantages and weaknesses, to estimate cardiac output during exercise in young persons. Consequently, no gold standard technique exists for this measurement. Means of determining

TABLE II.11	Oxygen Uptake Equivalents of Mechanical Power Output During Submaximal Upright Cycling	
Mechanical power (Watts)	**Oxygen uptake (L · min^{-1})**	
25	0.62	
50	0.94	
75	1.26	
100	1.58	
125	1.90	
150	1.22	

Mean values of 88 girls and 83 boys, 8 to 16 years old. There were no sex- or age-related differences in O_2 cost of pedaling.

Based on Andersen et al. 1974 (2).

cardiac output in children have been reviewed by Barber (20), Rowland (102), and Driscoll et al. (45).

The carbon dioxide rebreathing technique is an indirect means of estimating cardiac output by the Fick principle, using measures of VCO_2, venous carbon dioxide content, and arterial carbon dioxide content after inhalation of CO_2-enriched gas. Estimation of cardiac output by this technique is limited to steady state conditions, making measurement at maximal exercise problematic. In the acetylene rebreathing method, cardiac output is determined by the rate of disappearance of an inert gas (in this case, acetylene) from an inhaled gas mixture. This technique requires expensive measurement equipment (i.e., mass spectrometer) for accurate measurement of expired gas concentrations.

Doppler echocardiography (using standard echocardiographic equipment) has been used nonobtrusively to estimate cardiac output with exercise in children. This method does not require steady state conditions and can estimate cardiac output at maximal exercise. Several potential weaknesses may affect values, including variation in sampling angle, disturbance of laminar aortic flow, and change of aortic size with exercise (101). Measurement of changes in thoracic bioimpedance have been used to estimate cardiac output with exercise in youth. Motion and respiratory artifact may limit usefulness of this method at high exercise intensities.

Rating of Perceived Exertion (RPE)

The previously described measures of response to exercise are all based on the objective physiologic strain that takes place in response to a physical exercise. It is also of value, however, to assess the *subjective strain*, as perceived by the child. One way of obtaining such information is through the use of an RPE scale. The original scale, based on psychophysic principles, was devised in 1962 by G. Borg (25). As shown in table II.12, it has 15 numerical categories (6 to 20), 7 of which also have a verbal descriptor. The numerical categories and their respective descriptors were selected so that, when young adults or middle-aged people rate their effort, the numbers they choose will be approximately one-tenth of their heart rate. Hundreds of studies were done subsequently to validate this category scale, describe its uses, and analyze through it the psychophysiologic processes of effort rating. Borg and others have designed other scales, but many investigators and clinicians still use the original 6 to 20 scale. For detailed review of this topic see Noble and Robertson (90) and Robertson and Noble (100). See also chapter 1, section titled "Rating of Perceived Exertion," for a review of the characteristics of perceived exertion in children. Some of the clinical chapters include a discussion of rating characteristics in the child with a chronic disease or disability.

RPE has been used for two major purposes. One is the *estimation mode*, in which one determines the way a person perceives the exercise intensities that he is performing. The other is the *production mode*, in which RPE is used as a tool to prescribe exercise intensities.

Estimation Mode of RPE. The scale is explained to the child before the start of an exercise test. It is important that the explanation be standardized. Some investigators (80) let the child read the instructions. More commonly, one reads the instructions to the child. We use the following spoken introduction:

TABLE II.12 — **The Borg 6 to 20 Category Scale for Rating of Perceived Exertion Table II.13**

6	
7	Very, very light
8	
9	Very light
10	
11	Light
12	
13	Somewhat hard
14	
15	Hard
16	
17	Very hard
18	
19	Very, very hard
20	

"You are going to perform an exercise test in which the effort will be changed from time to time [one should not suggest to the child that the exercise test is progressive, with equal increments]. This scale describes different efforts: 6 is the lightest possible effort that you can think of, and 20 is the hardest possible effort. Therefore, all efforts that you will be making must be between 6 and 20."

"I'll show this scale to you during the test and ask you how hard you are exercising. Please answer by saying a number that best describes your feeling of the exercise at that moment. There are words opposite some of the numbers. These will help you remember what the numbers mean, but your answer must be by a number and not by a word. You can choose any number between 6 and 20 and not only those that have words opposite them. Remember, there is no correct or incorrect answer. We just want to know how you feel during the exercise."

This introduction is followed by examples and by ascertaining that the child understands the procedure. During the exercise test, one shows the scale to the child just before the end of each stage (after heart rate determination) and each time asks, "How hard are you exercising?" (To children younger than 10 years of age we say, "Guess how hard you are exercising.")

We have been using the RPE scale routinely as part of most progressive exercise tests (17). In healthy children and in many sick ones, 8 to 9 years of age or older, the subjective rating is closely correlated with physiologic indices, mostly heart rate. We have found the RPE scale of special value in children with neuromuscular disabilities, handicapping dyspnea, chronic fatigue or pain, or a high degree of apprehension; in these children there may be a discrepancy between exercise performance and physiologic strain. In such children the subjective rating may be high at a low power output, even when heart rate is low. The exercise performance of a child with, for example, muscular dystrophy is often more related to his subjective rating than to objective physiologic strain.

Another scale that has been used, mostly with adults, is a combination of a category scale and a ratio scale. In addition to its use for rating of exercise intensities,

it has been found useful for rating of specific symptoms during exercise, such as breathlessness, aches, and pains (26). This scale has yet to be evaluated for children.

The immense contribution of Gunnar Borg in establishing and developing the RPE concept was originally intended for adults. Additional methodological challenges exist when one uses this approach with children. An optimal instrument for young children should address developmental constraints such as reading proficiency, the ability to process numbers, and the ability to assign numbers to exercise-related feelings. There is also a potential semantic limitation: Words used for adults may not mean the same to a child. Furthermore, their meaning may vary according to the child's social and cultural background.

In the mid-1990s, a scale was developed specifically for children (47; 75), with an attempt to address some of these issues. This Children's Effort Rating Table (CERT), as shown in table II.13, comprises 10 numerical categories and their respective verbal descriptors. According to the developers of the CERT, it is more suitable for children than is the 6 to 20 scale, for the following reasons:

- It has only 10 numbers.
- Children are more familiar with the connotation of a 1 to 10 range.
- The wording of the verbal descriptors is more child friendly.

TABLE II.13 **The Children's Effort Rating Scale (CERT)**

1	Very, very easy
2	Very easy
3	Easy
4	Just feeling a strain
5	Starting to get hard
6	Getting quite hard
7	Hard
8	Very hard
9	Very, very hard
10	So hard I'm going to stop

More experience and research are needed to ascertain the characteristics and utility of this scale for use with children. Studies suggest good correlation between the CERT and heart rate, both in the estimation mode (47; 74; 134) and the production mode (75; 134).

Another approach for using RPE in young children is through scales based on drawings rather than numbers and words. The first of these had a group of stick diagrams of a person, which substituted for the words in Borg's 6 to 20 scale (91). Another is based on a heart-shaped person riding a bicycle (78). A more ambitious approach is the OMNI Scale of Perceived Exertion, which combines numbers, words, and pictures (figure II.4). This scale has been validated for a variety of tasks, such as cycling, treadmill walking and running, and resistance exercise (97-99; 99a; 119).

The success of the pictorial approach will be judged by its ability to expand the age range for which the more conventional scales have been found valid. Specifically, will it be useful for children younger than 8 years of age?

So far we discussed the properties of RPE as a tool for the *estimation* of subjective strain. Another use is for prescribing exercise intensities. The latter has been referred to as the *production mode* of RPE.

Production Mode of RPE. When prescribing an activity program to a child, it is relatively simple to describe the content (e.g., baseball, swimming), frequency (e.g., 3 per week) and duration (e.g.,

45 min) of each session. It is more challenging to describe exercise intensity. Several studies have shown that this can be done by using the RPE concept (75; 126-128; 133; 134). The child is first introduced to the RPE scale, usually during a progressive exercise test, using the estimation procedure as described previously. Once familiar with the RPE notion, the child can be given a prescription of exercise intensities by using numbers from the scale. For example, "I want you to run for 5 minutes at number 10 and then for another 5 minutes at number 13."

The mental processes of estimating exercise intensity and of producing an exercise task are quite different. When a person is asked to estimate exercise intensity, she must interpret physiologic and mechanical cues that come from the muscles, tendons, joints, lungs, and possibly other body organs. These stimuli are sent to the sensory cortex through afferent pathways during or immediately after exercise. In addition to this feedback process there may also be a feed-forward process, where information from the motor cortex is fed directly to the sensory cortex without going through the periphery. In contrast, in the production mode, the child has to recall, based on past experiences, how she rated such cues and then try to reproduce an exercise intensity that would create the same cues.

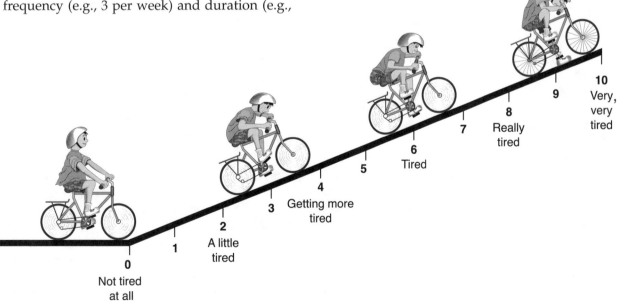

FIGURE II.4 The OMNI scale for rating of perceived exertion.

Reproduced, with permission, from Robertson et al. 2000 (99).

Differential RPE. The preceding sections discuss RPE in relation to the body as a whole, but one can use the same estimation mode to rate sensations that arise from parts of the body. For example, RPE can be partitioned into a ventilatory component and a lower limb component. This approach has been used for adults by scientists and clinicians alike (29; 69; 100). We have been using the differential RPE sporadically in our clinic. It helps shed insight into what may limit the physical activity of a patient with, for example, a neuromuscular disability or an advanced lung disease. More basic research is needed to determine the suitability of this approach for children (82).

Determination of Maximal Aerobic Power

Maximal aerobic power can be determined directly by the measurement of maximal O_2 uptake. It can also be assessed indirectly by measuring heart rate during submaximal exercise, anaerobic and ventilatory thresholds, peak power output during cycling or stepping, peak effort in a progressive treadmill test, and by field tests such as the distance covered in a 20-min shuttle run.

Direct Determination

The traditional main criterion indicating that maximal O_2 uptake has been reached during a progressive protocol is that an increase in power load is accompanied by no increase, or by an increase of less than $2 \text{ ml} \cdot \text{kg}^{-1} \cdot \text{min}^{-1}$, in O_2 uptake. Such a plateau, although common among adults, is less often reached in children, particularly during continuous-progressive protocols. According to 14 studies, as summarized by Rowland (102), only 56% of children (range 21-95%) have reached a plateau. Maximal O_2 uptake is similar among children who do reach a plateau and those who do not reach it (5; 96). This suggests that the plateau criterion is of little consequence for children.

It has become common among exercise physiologists to use the term *peak O_2 uptake,* instead of *maximal O_2 uptake,* whenever a plateau has not been reached or confirmed. While this distinction may be important in the adult literature, it has little practical value for children. Secondary

criteria have been used to decide whether a child has reached maximal O_2 uptake: heart rate of at least 195 beat min^{-1}, blood lactate concentration of 9 mmol $\cdot$ L^{-1}, or a respiratory gas exchange ratio (CO_2 output/O_2 uptake) that exceeds 1.0. These criteria have been used with healthy children, but they may not be valid in some diseases. A boy with muscular dystrophy, for example, may reach his peak effort at a low heart rate (e.g., 130-140 beat $\cdot$ min^{-1}), as will a child with complete heart block or advanced cystic fibrosis, or an adolescent with anorexia nervosa. It thus seems that a decision during the test as to whether a child has reached maximal aerobic power cannot be based on physiologic variables per se. Experienced personnel often use such *subjective* criteria as blanching of the skin (mostly around the mouth and at the neck and shoulders), widening of the pupils, or a change in gait style, which shows that the child is struggling to keep walking or running. On the cycle ergometer, a reasonably motivated child who is attempting to sustain the prescribed pedaling or cranking cadence and cannot do so, in spite of encouragement, has probably reached his maximum. As a rule, when using an "all-out" protocol, we do not terminate the test until the child can no longer continue, or there are symptoms and signs, as summarized in the section that follows, "Terminating an Exercise Test." We assign much importance to repeated verbal encouragement. Such statements as, "We know that you are terribly tired but you've almost finished" or, "We are now in the last minute" are most valuable and often elicit that all-important extra effort from the child.

Indirect Determination: Submaximal Tests

Some investigators and clinicians prefer not to bring the child to a self-imposed maximum. This approach has been taken for children who are sick or unmotivated, in conjunction with cardiac catheterization, or in large-scale surveys where a test must be short and simple. To derive information about the child's maximal aerobic power, one then has to resort to an indirect estimation based on submaximal data.

Prediction of Maximal O_2 Uptake. Various methods are available for the prediction of maximal O_2 uptake from one or more values of submaximal heart rate. The most commonly used method is based on a nomogram by Åstrand and Rhyming (6). With this nomogram, heart rate must be measured at a known mechanical power (on a cycle ergometer or by a step test) or together with a measurement of O_2 uptake (on a treadmill).

The use of the Åstrand-Rhyming nomogram is based on the following assumptions:

- Heart rate is linearly related to O_2 uptake.
- Exercise at a certain power output requires the same O_2 uptake in all individuals.
- Individuals of a certain age group have the same maximal heart rate.

Even though such assumptions have some theoretical basis, none of them holds true in practice. The result is a low reproducibility and validity when the nomogram is used with children (53; 88; 109; 112; 131). Moreover, because children's maximal heart rate is higher than in adults, for whom the nomogram was originally designed, the nomogram underestimates a child's real maximal O_2 uptake by 10% to 25% (57; 109; 110; 112; 131). To reduce this underestimation, various correction factors have been suggested. These, however, do not reduce the marked scatter in results obtained by this method of prediction. We (as do the authors of the original nomogram) therefore do not recommend the use of the Åstrand-Rhyming nomogram for the assessment of individual patients. Its use should be limited to population studies.

Physical Working Capacity. This index, commonly known as W_{170} or PWC_{170}, which is the mechanical power at which heart rate is 170 beat · min^{-1}, was introduced in 1948 by Wahlund (125) and has been in use since for children (22; 105; 106) and adults. To calculate W_{170} one does not assume a certain maximal heart rate. The only assumption is that the heart rate is still linearly related to power at about 170 beat · min^{-1}. Two or more measurements of heart rate are needed and the values are plotted against mechanical power, as shown in figure II.5. The line thus established is extrapolated (or interpolated) to heart rate = 170 beat · min^{-1}, and the corresponding power is W_{170}. The drawback of this index is that, as with other submaximal tests, it is based on heart rate, which is dependent on numerous factors in addition to aerobic fitness. Another cause for an error is that the heart rate–power relationship is not always linear.

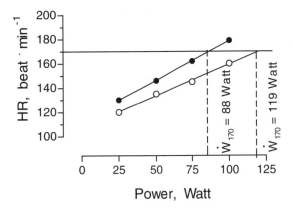

FIGURE II.5   A schematic diagram displaying the method by which $\dot{W}_{170}$ is derived from submaximal heart rate in two adolescents.

To reduce the error in this measurement, one of the heart rate values should be as close as possible to 170. Another refinement is the use of more than two values to plot the line. If one point is obviously out of line, it can be discarded. This occurs sometimes with the lowest or with the highest value. The former is often too high, because the child is emotionally excited at the start of the test. The latter may be too low if the load approaches maximum (heart rate may reach a plateau while O_2 uptake is still rising).

$\dot{W}_{17}$. An alternative index, not based on heart rate, is the mechanical power that the child produces when he *perceives* the exercise to be "hard," or 17 on the Borg category scale (table II.13). We have found this index to correlate well with other indices of maximal aerobic power and to serve as a valid fitness test in children (9; 10). As seen in figure II.6, the method of assessing $\dot{W}_{17}$ is similar to that for $\dot{W}_{170}$, with RPE being assessed instead of heart rate.

Indirect Determination of Maximal Aerobic Power: All-Out Tests

When direct measurement of O_2 uptake is not feasible, one can assess maximal aerobic power using the *performance* of the child as a criterion. One example is the highest stage reached during the Bruce treadmill test, which is highly correlated with directly measured maximal O_2 uptake. Another example is the peak power output in the McMaster cycling or arm-cranking tests.

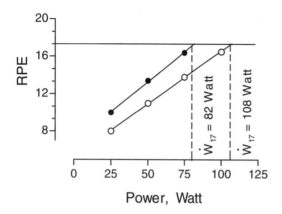

FIGURE II.6 A schematic diagram displaying the method by which $\dot{W}R_{17}$ is derived from the rating of perceived exertion in two adolescents.

Other indices are the speed of running a certain distance (72) and the distance covered during a predetermined period (79; 116). A test that can be performed indoors, even in a hospital corridor, is the 20-m multistage shuttle run (35). In this test, the child runs at a speed determined by the investigator between two markers, 20 m apart. The speed increases gradually until the child can no longer perform the run. The final speed (or the final stage) is taken as the child's aerobic performance. This has been validated against directly measured maximal O_2 uptake, and equations have been developed to calculate children's maximal O_2 uptake, based on the shuttle run.

Safety Precautions

In adults, the two main untoward effects of exercise testing are myocardial ischemia and ventricular fibrillation. Although epidemiologic data are not available on the occurrence of such complications in the pediatric age group, the risk seems to be extremely low (130). It can be kept that way if one follows certain precautions. The American Heart Association has published guidelines for exercise testing in children (130). While the guidelines address primary cardiovascular aspects, they are also relevant to other medical conditions. Their main premise is that "the need to the information to be obtained by exercise testing must be carefully weighed against its risks. However, even for children whose diagnoses place them in a high-risk group there may be compelling reasons to perform exercise testing." The guidelines further state the following: "A carefully observed laboratory exercise test

performed under controlled circumstances can be safer than the unsupervised activity of high physical intensity on the playground."

To create and maintain a safe environment for testing, one should ascertain that the laboratory staff are proficient in exercise testing techniques and in basic and advanced life support, and that they can recognize symptoms and signs that may require termination of a test. The equipment (e.g., treadmill, mouthpiece) should be structured such that it would not cause injury. A physician does not need to attend each exercise test, but it is important that a physician be available to the laboratory staff if they need further advice or supervision of a test.

Contraindications for Exercise Testing. Although the American Heart Association guidelines do not recommend absolute contraindication for testing, they list the following examples of diagnoses in which the risk of an exercise test may outweigh its benefits (130):

- Severe pulmonary vascular disease
- Poorly compensated congestive heart failure
- Recent myocardial infarction
- Active rheumatic fever with carditis
- Acute myocarditis or pericarditis
- Severe aortic or mitral stenosis
- Unstable dysrhythmia, especially when the hemodynamic function is compromised
- Marfan syndrome with suspected aortic dissection
- Uncontrolled, severe hypertension
- Evidence of hypertrophic cardiomyopathy with a history of syncope

Other, noncardiologic conditions in which the need for exercise testing should be weighed carefully against the risk include the following:

- Forced expired volume in the first second (FEV_1) lower than 60% of height-predicted in a child with asthma
- Severe arterial O_2 desaturation at rest
- Uncontrolled epilepsy with a history of exercise-induced seizures
- Elevated intracranial pressure
- Acute febrile condition
- Acute renal disease (e.g., acute glomerulonephritis)

- Acute hepatitis during 3 months since onset
- Insulin-dependent diabetes in a child who did not take his prescribed insulin or who has ketoacidosis

Terminating an Exercise Test. A patient's safety during and after an exercise test is paramount. The laboratory staff should use their discretion to decide when continuing an exercise test may present danger to the child. Staff should watch for the following symptoms and signs in particular: light-headedness, confusion, staggering gait, a severe cramp or pain (such as headache), dyspnea that seems excessive in light of the exercise intensity, and a progressive decrease in blood pressure in spite of an increase in exercise intensity. The following ECG changes merit termination of a test (130): ST segment depression or elevation greater than 3 mm, significant dysrhythmia precipitated or aggravated by the test (e.g., increasing frequency of ventricular premature beats, supraventricular or ventricular tachycardia, atrioventricular conduction block). In testing of a cardiac patient, the test should be terminated if the ECG monitoring system fails.

Methods of Determining Physical Activity and Energy Expenditure

*C*hapter 2 discusses the relevance of physical activity (PA) and energy expenditure (EE) to pediatric exercise medicine and science. This appendix reviews methodologic aspects in the determination or assessment of PA and EE.

General Principles

Several general principles should be considered before one chooses a method, or a combination of methods, to determine or assess PA and EE of children.

Objectives

Documentation of PA or EE in a patient is done for various reasons. A clinician may wish to determine whether the child is sufficiently active and what factors enhance or impede the activity behavior. Documenting the PA patterns of a patient, and the attitudes toward an active lifestyle, may help to explain exercise-induced complaints and serve as a basis for prescribing exercise. Once enhanced PA is prescribed, documentation of changes made by the patient is essential in order to determine compliance and to explain changes that may occur in the child's health. It is seldom that a clinician will need to determine the child's daily EE. In some conditions, however, this may be useful. Examples are obesity, anorexia nervosa, and type 1 diabetes mellitus.

Scientists usually have other objectives. Valid assessments of PA and EE are essential in epidemiologic surveys with the intent of understanding relationships between children's current or future health and their activity behavior. It is also important to document changes in PA and EE whenever an intervention or a disease may modify them over time. Because of the interaction between PA and fitness, the activity pattern of subjects should be documented and reported in any study where physical fitness is a dependent variable.

Validity

As with any other measurement, a method chosen for assessing PA or EE should be valid. Over the years there has been inconsistency in the literature about how to validate any given method. As a result, studies have used, for example, heart rate monitoring to validate an accelerometer, whereas others have used accelerometry to validate heart rate monitoring. To avoid such confusion, one must determine a certain hierarchy in the various methods so that a method chosen as a validation criterion will indeed be suitable. Sirard and Pate divided the various available methods into three categories using the following ascending hierarchy: (1) *subjective measures* (questionnaire, interview, diary), (2) *secondary measures* (heart rate monitoring, pedometry, acelerometry), and (3) *criterion standards* (direct observation, doubly labeled water, indirect calorimetry) (66). According to this division, a method chosen as a validation criterion must be selected from a higher hierarchical group. Furthermore, a method chosen to validate a measure of EE is unsuitable for validating a measure of PA, and vice versa. Nor will a method suitable for validation of EE (or PA) in the laboratory be suitable as a validation criterion under free-living conditions. Table III.1 summarizes the various possibilities for legitimate validation choices.

Reactivity

With some methods, the very act of documenting the activity may induce changes in the child's behavior. This phenomenon is called *reactivity*. It may occur, for example, during direct observation when the child becomes aware of the presence of an observer or when she notices the existence of cameras. Another scenario is when the child is asked to fill up a diary at short intervals. Reactivity may occur also when the child wears an instrument, such as a pedometer, accelerometer, or heart rate monitor. The extent of reactivity to a given method is not known. Our experience with preschoolers has been that reactivity to direct observations, accelerometry, and heart rate monitoring occurs only during the first few minutes. The children then resume a spontaneous activity pattern, ignoring the presence of

an observer or an instrument. Likewise, when second to sixth graders carried sealed pedometers for several days, there was little or no reactivity (74). Still, when designing a protocol to assess habitual PA or EE in children and adolescents, one should keep reactivity to a minimum. To minimize reactivity, one can repeat the measurement over several days and discard the information from the first 1 to 2 days.

Determining the Energy Cost of Activities

Energy expenditure measured during exercise (EEE) includes metabolic processes that occur at rest (resting energy expenditure—REE), which are not related to the demands of the exercise task itself. Examples are brain and bone metabolism, digestion, and hepatic and renal functions. There are several ways to quantify the energy that is actually expended by a specific physical task. The most commonly used method is the ratio EEE/REE. This ratio is called MET (short for metabolism equivalent). In the fully rested person MET = 1. During the years, tables have been constructed that assign MET values to various tasks, mostly based on measurements in adults. EEE (per kg body mass) during various activities is often considerably higher in children than in adults (see chapter 1). Likewise, the REE

TABLE III.1 Hierarchy for Validation of Methods Used to Assess Physical Activity (PA) or Energy Expenditure (EE)

| Method | Assessing | Validated by | | | | | |
		HR	Motion analyzer	VO₂ metabolic cart	Direct observation	Doubly labeled water	VO₂ respiration chamber
Questionnaire	PA, free living		X		X		
	EE, free living	X				X	
Interview	PA, free living		X		X		
	EE, free living	X				X	
Diary	PA, free living		X		X		
	EE, free living	X				X	
Heart rate monitoring	EE, laboratory			X			X
	EE, free living					X	
Motion analyzers	PA, laboratory				X		
	PA, free living				X		
	EE, laboratory			X			X
	EE, free living					X	

Free living = nonlaboratory conditions; motion analyzers = pedometers or accelerometers; HR = heart rate; VO₂ = O₂ uptake.

of children (per kg body mass or per body surface area) is usually higher than in adults. As a result, MET values of specific activities, constructed for adults, will often underestimate those for children. MET values per task also vary among children of different ages; the younger the child, the higher the MET. Few attempts have been made to determine MET specifically for children. Table III.2 gives an example.

TABLE III.2	MET Values of Selected Activities in 9- to 12-Year-Old Children	
Activity	**MET value**	
Sitting quietly	1.1	
Reading while sitting	1.2	
Watching television while sitting	1.3	
Doing puzzles while sitting	1.5	
Standing quietly	1.5	
Singing while standing	1.8	
Dressing	2.6	
Eating	1.4	
Walking at a slow pace	2.8	
Walking at a firm pace	3.5	
Cycling slowly	2.5	
Cycling at a firm pace	5.0	
Spontaneous playing outdoors	4.5	
Ballet	4.4	
Gymnastics	5.0	
Judo	6.3	
Soccer game	6.0	

Adapted from Saris 1986 (64).

Specific Methods

Over the years, several methods have been suggested for the measurement and assessment of PA and EE. They differ in sophistication, cost, and applicability (clinical or scientific) to individuals, small groups, and large populations. Table III.3 summarizes the strengths and drawbacks of these methods. The list starts with the simplest, least expensive techniques and ends with the most sophisticated, expensive ones. It is important to note that these methods can assess either PA or EE, but not both combined. However, by assigning a certain metabolic equivalent to a physical task, one can estimate the energy expended during a certain PA routine. This can be obtained, for example, from direct observation and subsequent quantification of the child's activities over a given period. This section discusses each of the methods listed in table III.3.

Recall Questionnaires

Questionnaires are used to estimate PA for research and clinical purposes, alike. Because of their relative simplicity and low cost, they are suitable for large-scale surveys. The validity of this method depends on the responder's cognitive functioning (e.g., recall ability, reading and comprehension skills), which is usually lower in children than in adults (60). As a result, one often needs to rely on a *proxy report* by a parent, teacher, or other adults who are familiar with the child's activity behavior. Unfortunately, proxy reports often have a low validity when compared with direct observation (56). This is particularly so for activities performed away from home. In another study, however, a good rank correlation (r = 0.72-0.82) was found when a proxy report by a parent was validated against 3-day heart rate monitoring (48).

Questionnaires have been validated against direct observation, heart rate monitoring, motion sensors, and doubly labeled water. In a large-scale study, responses of third and fourth graders had a high agreement (83.6%) with direct observation of 10-min bouts of moderate to vigorous activity. The reliability and validity of recall questionnaires increase with the age of the child (60).

Various questionnaires have been constructed for healthy children (30-32; 49; 63) and those with a physical disability or a chronic disease (46). They range in recall time from one day to several years (66). In general, the shorter the recall time, the higher the validity. Questionnaires vary also in the specific questions used to describe activities. Some yield merely a description of the activities, whereas others also address environmental factors that enhance or impede a child's activity behavior. Very few questionnaires address the *attitudes* of the child (31) and family members toward PA. An example of a questionnaire used in a clinical context at the Children's Exercise and Nutrition Centre is given at the end of this appendix.

TABLE III.3   Methods Used for the Assessment of Physical Activity (PA) and Energy Expenditure (EE)

Method	Function assessed	Advantages	Drawbacks	Suitable for clinical use	Comments
Questionnaire	PA	Simple, low cost; suitable for large-scale studies	Relies on memory; hard to quantify; low validity	Yes	The shorter the recall period, the higher the validity
Interview	PA	More valid than a questionnaire	Relies on memory	Yes	Interviewer can corroborate information
Diary	PA	Short recall time	Interactive	Yes	Depends on a child's interpretation
Direct observation	PA, (EE?)	No need for recall; context documented	Expensive; depends on observer's skill	No	Gold standard for behavioral aspects
Time-lapse or video photography	PA, (EE?)	Objective, hard record available	Child is limited to predetermined area	No	Less expensive than direct observation
Movement counters	PA (EE?)	Objective, little interaction; low cost	Does not detect specific movements	Yes	May be used as a motivational tool
Accelerometry	PA, EE(?)	Same as counters, plus acceleration	Does not detect specific activities	Yes, with help of an exercise lab	Some validity vs. measurements of EE
HR monitoring	EE	Little interaction; inexpensive	HR affected not only by metabolism	Yes, with help of an exercise lab	Needs individual "calibration" vs. $\dot{V}O_2$
$\dot{V}O_2$ metabolic cart	EE	Measures metabolism	Limited activities; need for mouthpiece or face mask	Rarely (e.g., to assess effect of therapy on energy cost of locomotion)	Useful for ergometry and $\dot{V}O_2$-HR calibration
$\dot{V}O_2$ portable equipment	EE	Measures metabolism away from the lab	Highly interactive; expensive	Rarely (e.g., to assess effect of therapy on energy cost of locomotion)	Limited pediatric use in free-living observations
$\dot{V}O_2$ canopy	EE	Measures metabolism	RMR only	For REE	Used in conjunction with HR monitoring
Respiration chamber	EE	Precise measurement of EE	Very limited quarters; expensive	No	Validating other tests; ideal for BMR
Doubly labeled water	EE	Best measure of EE; not interactive	Very high cost; requires at least one week	No	Gold standard for total EE, but not for the profile of EE

RMR = resting metabolic rate, BMR = basal metabolic rate. A question mark denotes uncertain validity.

Can one derive information about the child's EE from a recall questionnaire? This has been attempted with the use of tables that convert each activity into its MET value (see chapter 2) or to its equivalent energy expenditure (tables III.2 and III.4, respectively). This approach has a low to fair validity, particularly when used with children (61). One reason is that the energy cost of activities may differ between children and adults, particularly when the activity includes walking or running (see chapter 1, section titled "Mechanical Efficiency and Economy of Movement"). Moreover, such tables are based on the steady state energy cost of an activity, whereas a child's typical activity pattern of short bursts of several seconds' duration (5) does not result in a steady state. Ideally, a conversion table for use with children and adolescents should be divided into several

TABLE III.4 — Energy Expenditure of Various Activities and Sports, When Performed by Children and Adolescents of Various Body Weights

Activity	20	25	30	35	40	45	50	55	60	65
Basketball (game)	35	43	51	60	68	77	85	94	102	110
Calisthenics	13	17	20	23	26	30	33	36	40	43
Cross-country skiing (leisure)	24	30	36	42	48	54	60	66	72	78
Cycling 10 km · hr^{-1}	15	17	20	23	26	29	33	36	39	42
15 km · hr^{-1}	22	27	32	36	41	46	50	55	60	65
Field hockey	27	34	40	47	54	60	67	74	80	87
Figure skating	40	50	60	70	80	90	100	110	120	130
Horseback riding canter	8	11	13	15	17	19	21	23	25	27
trot	22	28	33	39	44	50	55	61	66	72
gallop	28	35	41	48	50	62	69	76	83	90
Ice hockey (on-ice time)	52	65	78	91	104	117	130	143	156	168
Judo	39	49	59	69	78	88	98	108	118	127
Running 8 km · hr^{-1}	37	45	52	60	66	72	78	84	90	95
10 km · hr^{-1}	48	55	64	73	79	85	92	100	107	113
12 km · hr^{-1}	--	--	76	83	91	99	107	115	123	130
14 km · hr^{-1}	--	--	--	--	--	113	121	130	140	148
Sitting										
Complete rest	8	8	9	9	10	10	11	11	12	12
Quiet play	11	12	14	15	15	16	17	18	19	20
Snowshoeing	35	42	50	58	66	74	82	90	98	107
Soccer (game)		36	45	54	63	72	81	90	99	108
Squash		--	--	64	74	85	95	106	117	127
Swimming 30 m · min^{-1} breast	19	24	29	34	38	43	48	53	58	62
freestyle		25	31	37	43	49	56	62	68	74
back		17	21	25	30	34	38	42	47	51
Table tennis		14	17	20	24	28	31	34	37	41
Volleyball (game)	20	25	30	35	40	45	50	55	60	65
Walking 4 km · hr^{-1}	17	19	21	23	26	28	30	32	34	36
6 km · hr^{-1}	24	26	28	30	32	34	37	40	43	48

body weight categories (table III.4) and, as far as possible, should use energy equivalents generated for the respective age groups. Another proof of the low validity of recall questionnaires as a tool for assessing EE is the low correlation with doubly labeled water data (14; 28). An interesting approach has been developed to display the question by use of video technology with children (mean age 7.7 years). To visually display the questions, each was accompanied by a short video display of activity intensity. Responses were correlated with accelerometry data and heart rate monitoring. The respective correlation coefficients were r = 0.40 and r = 0.50 (70).

In conclusion, despite their limitations, recall questionnaires may be the only practical tool in a clinical setting for assessing PA behaviors, especially when complemented by an interview (see interviews section that follows).

Interviews

Using an interviewer-administered questionnaire is much more expensive than a self-administered questionnaire, but it has a higher reliability and validity. Through an interview, one can elicit much more information than the child or an adult might volunteer. While children can usually report the *type* of an activity (e.g., soccer or playing tag), it is more challenging for them to accurately describe its *intensity* and *duration*. An interviewer can refresh the child's memory by first creating "anchors" for a day's events (for example, by establishing the time that the child woke up, went to school, returned from school, had dinner, and went to bed). Once established, such context-specific anchors can help to quantify the time spent on a given activity. Because of their high cost, face-to-face interviews are rarely used in large-scale surveys. The use of telephone-based interviews reduces such costs.

For clinical purposes and small-scale studies we often combine a self-administered questionnaire with a subsequent interview. Through the interview we clarify written responses and corroborate information. This approach can also be used to explore exercise-induced complaints.

Diaries

In this method, subjects record their activities as they occur or at predetermined intervals throughout the day (10; 25). Information logged in the diary can include the type of activity and its intensity and duration. The data are quantified offline to estimate the level of activity. By using conversion tables, as described in the section on questionnaires, one can also estimate the child's EE (16). Diaries have been used successfully with children as young as 10 years of age, but their validity with younger children is not clear.

The main advantage of the diary method is that it does not depend on memory. A disadvantage is its high degree of reactivity, because the need to repeatedly record activities, in itself, may induce a change in a person's spontaneous behavior. This is particularly so if the frequency of logging data is high.

Direct Observation

This method is the gold standard for documenting the behavioral aspects of PA. An investigator observes the child continuously and, using a coding system, records the activities at a high frequency (e.g., each minute). The main advantage over questionnaires, interviews, and diaries is that the data acquisition does not depend on the child's recall ability but on the skill of the observer. Several methods have been used with children:

- Behaviors of Eating and Activity for Children's Health Evaluation System (BEACHES) (51)
- Fargo Activity Time sampling Survey (FATS) (42)
- Children's Activity Rating Scale (CARS) (39)
- Activity Patterns and Energy Expenditure (19)
- Children's Physical Activity Form (CPAF)
- System for Observing Fitness Instruction Time (SOFIT) (50; 59)

The following are some of the items that are coded: the activity, who initiated it (the child, a parent, another child), where it was performed (playground, in front of the TV), whether the child interacted with toys or with others. The interobserver reliability of these methods, when performed by trained observers, is high (84-99%) (66).

The main drawbacks of direct observation are the high cost of an observer's time and the considerable effort needed to train a team of observers to become skillful. Another potential drawback is reactivity caused by the presence of the observer. Some children are likely to modify their activity pattern merely because of the observer's presence. In one study (58), only one in six children, ages 5 to 6 years, reacted to the presence of an observer. In our experience, the younger the child, the lesser the reactivity. For preschoolers it takes only a few minutes until they ignore the presence of an observer.

Time-Lapse or Video Photography

This method removes concerns about subjectivity. By installing well-positioned cameras, one can document a child's activity in a manner similar to direct observation. The main strength of photography is that a record is kept and can be analyzed offline (13; 17). Furthermore, this method induces less reactivity than direct observation does. The disadvantage of photography is that the child is confined to certain areas (e.g., schoolyard, gymnasium) within view of the camera.

Pedometers and Accelerometers

The methods described previously focus on the behavioral aspects of PA with indirect derivation of energy expenditure. In contrast, movement counters, such as pedometers and accelerometers, document the *mechanical* aspects of an activity. The device is affixed to a body part, such as the waist, wrist, or ankle. Through a mechanical or electronic mechanism that is sensitive to motion, it counts and registers the number of movements of that body part.

Pedometers are the most commonly used movement counters. They are usually attached to the waist and, by registering vertical motion, count strides. Assuming a certain stride length, the distance covered over the duration of the measurement can be calculated. The original pedometers used a mechanical system. More recent ones have an electronic mechanism, which has increased their validity. Pedometers have been proven valid when their output is correlated with O_2 uptake measured during treadmill walking or running ($r = 0.62 - 0.93$) (21; 47). They are not recommended, though, as a tool for assessing EE in a free-living situation. Correlation with direct observations of PA has ranged from 0.80 to 0.97 (40).

The main drawback of pedometers is that they ignore stride intensity or length and, therefore, cannot distinguish between slow and fast walking or between walking and running. They also cannot detect whether locomotion occurs on the level, uphill, or downhill. Another limitation is that pedometers usually do not sense movements of body parts other than the waist during, for example, cycling. As such, while useful for adults, whose main leisure activity may be walking or jogging, they are less suitable for children, who have a larger activity repertoire.

Accelerometers sense and register acceleration of each movement and not merely the number of movements. Like pedometers, they are affixed to one or more body parts. By sensing acceleration or deceleration, an accelerometer can yield important information about mechanical events that occur in that body segment.

One of the first commercially available accelerometers is the Caltrac, which senses and integrates the number of movements and their acceleration in one plane. It yields a single value over the time of observation, thus ignoring changes in the rate of counts during this period. The Caltrac has been validated against direct observations (41; 43; 54), observation by video photography (6), doubly labeled water (38), O_2 uptake during treadmill walks (11), and O_2 uptake in a respiration chamber (12). There was a wide variability in correlation coefficients, which ranged from 0.16 to 0.95. In general, validity of the Caltrac, particularly in preschoolers, was lower when the child was outdoors. This may reflect the larger variability of body movements in young children than in adults, many of which may not have been sensed by the Caltrac.

The Tritrac accelerometer senses movements in all three planes, compared with a single plane sensed by the Caltrac (20; 55). As a result, the Tritrac captures more body movements. Another important advantage is that it stores the counts sequentially, which provides a profile of the amount and intensity of body movements at various points in time. This profile can then be superimposed on the activity profile achieved by other methods, such as direct observation or heart rate monitoring. The combination of three profiles would yield information on the behavioral, metabolic, and mechanical elements of PA, combined. Attempts have been made to derive EE levels from Tritrac data (35). More research is needed, however, to validate this approach. A major disadvantage is that the Tritrac is much more expensive than the Caltrac, which makes it less suitable for large-scale studies. Figure III.1 is an example of the output obtained by the Tritrac accelerometer, comparing a sedentary child with a physically active child.

Another accelerometer that has been used with children is the CSA (Computer Science and Applications). It is lightweight (40-45 g) and small ($5 \times 4 \times 1.5$ cm) and, like the Tritrac, it can store information sequentially. Intra-instrument reproducibility is high (52). An advantage over the Tritrac is that the CSA can be programmed to turn itself on and off at predetermined intervals. However, the CSA senses movement in one plane only. Correlation coefficient of 0.87 was documented when the CSA was validated against direct observations of preschoolers (22). Correlation was somewhat lower when validation was made against O_2 uptake measured in the laboratory (21) or heart rate monitoring (36).

In spite of the theoretical advantage of three-plane over single-plane accelerometry, direct comparison between the two yields a high agreement (57; 76). Moreover, when compared with O_2 output during treadmill walking and running, the CSA and the Tritrac RD3 accelerometer yielded similar results (24). In contrast, when activities include a greater variety than treadmill exercise, the Tritrac RD3 yields a higher correlation with O_2 uptake than does the CSA (21). From a practical point of view, the light weight of the CSA compared with the Tritrac RD3 is an important advantage, particularly for preschool children.

Heart Rate (HR) Monitoring

This is the most commonly used method for estimating EE under free-living conditions. It has been used extensively in pediatric research, yielding important information about the EE of various populations (2; 23; 37; 44; 66-68; 73). The method is based on the linear relationship between $\dot{V}O_2$ and HR at a wide range of exercise intensities. eq $\dot{V}O_2$, in turn, is a surrogate indicator of the rate of the aerobic energy expenditure.

Because the $\dot{V}O_2$-HR regression varies among individuals (and in the same individual, with changes in aerobic fitness), it has to be established in each subject. Such calibration requires the simultaneous measurement of $\dot{V}O_2$ and HR. Measurements should be taken at rest (e.g., subject lying, sitting, and standing) and then at several submaximal exercise intensities (see figure III.2). Including a $\dot{V}O_2$-HR line at rest is important because children and youth spend many of the 24 hr at rest (see chapter 2), and using only the exercise regression line will not reflect their overall O_2 uptake.

Once the calibration lines have been established, heart rate is continuously monitored over 24 hr or more (ideally two or more weekdays and one

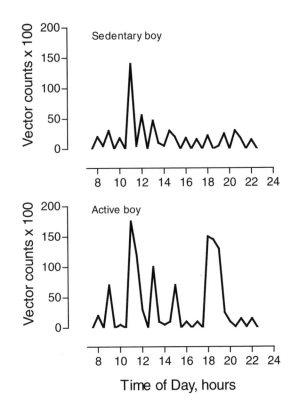

FIGURE III.1   Output (vector counts) generated by a Tritrac accelerometer. A comparison is made between the activity profiles of a sedentary boy and a physically active boy. The only exercise performed by the sedentary boy was during one physical education class. The active boy walked to and from school, was active during recess, took part in a physical education class, and had a training session in the evening.

Data from the Children's Exercise and Nutrition Centre.

weekend day). The HR profile is then transformed to $\dot{V}O_2$ and to EE profiles (figure III.3). A question remains about which of the two regression lines should be used during this transformation. One needs to establish a transition point of HR, below which the resting line is used, and above which the exercise line is used. Researchers in the field often take the standing HR as the transition point, also called the *flex point* (figure III.2) (45; 67).

One reason for the popularity of the HR-monitoring method is the availability of small, lightweight telemetric units that allow continuous monitoring in an unobtrusive way. The commonly used systems (e.g., Sport Tester, Polar, Vantage XL) are socially acceptable and highly valid for measuring HR in children as young as 4 years, compared with a direct electrocardiographic measurement (7).

As discussed in chapter 1, exercise HR is affected by factors other than metabolic rate (e.g., emotional state, climatic conditions, hypohydration). While the emotional effects are apparent mostly at rest or during low-intensity activities, the effect of warm climate occurs also at higher exercise intensities. When children exercise at 35°C, their HR can be higher by as many 25 beat · min⁻¹ than the HR at 22°C (33; 34). Because of these climatic effects, HR monitoring often overestimates the actual metabolic EE, particularly if outdoor activities are performed in a hot or cold environment. A nomogram has been constructed that corrects for the effect of climatic heat stress on HR (34).

When equipment is not available to measure $\dot{V}O_2$, one can still use HR monitoring to document a person's physical activity level. Instead of converting heart beats to metabolic rate, one can simply present raw HR values to reflect the intensity of PA (1; 3; 26; 27; 37). For example, in a 10-year-old child, a HR of 180 beat · min⁻¹ or more reflects a very intense effort, whereas 150 to 169 beat · min⁻¹ and 130 to 149 beat · min⁻¹ reflect intense and moderate efforts, respectively. Using such ranges as general guidelines, one can then determine the total time over 24 hours that the child was active at any given intensity.

A high HR may not reflect merely an intense activity but also a high basal (ideally determined in the sleeping child) HR. To correct for this, one can calculate *net HR*, which is the difference between the HR measured during an activity and that person's basal HR. When sleeping HR is not available, one can subtract the lowest values of the 24-hr HR from each of the other values (53). While a correction for resting HR has its merit, it is important to realize that there is no universally agreed-on method for the measurement of resting HR in children (71).

Measuring O₂ Uptake: Metabolic Cart

$\dot{V}O_2$ is an excellent surrogate method for determining EE. For practical purposes, one can assume that 1 L of O_2 consumed is equivalent to 21 kJ (5 kcal). To refine the conversion, one can correct for the respiratory exchange ratio (CO_2 produced/O_2 consumed). Using a metabolic cart in the laboratory limits the child's activities repertoire to, for example, cycling, step climbing, and treadmill running or walking, and it does not allow for monitoring EE under free-living conditions. Determining EE in the laboratory is useful, however, for establishing $\dot{V}O_2$-HR regression lines. It has also been used to validate heart rate monitoring and motion sensors (8; 18; 21; 62; 72).

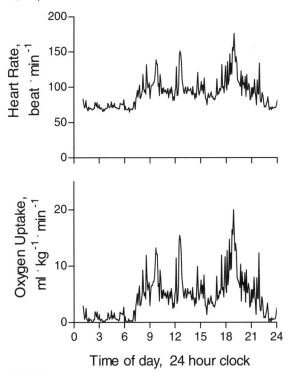

FIGURE III.3 A profile of heart rate and the corresponding O₂ uptake over 24 hr in a 10-year-old girl. Heart rate was monitored by the Sport Tester, Polar Vantage XL.

Data from the Children's Exercise and Nutrition Centre.

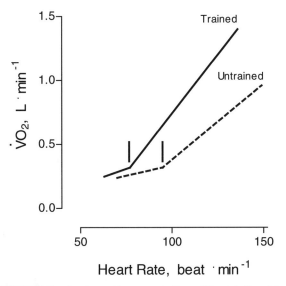

FIGURE III.2 A schematic presentation of the relationship between O₂ uptake (VO₂) and heart rate in a trained and an untrained child. The vertical lines denote a flex point (transition between resting and exercise values).

Measuring O₂ Uptake: Portable Device

Unlike a metabolic cart, portable devices can be used to measure $\dot{V}O_2$ while the child is moving freely, away from the laboratory. Currently available units are lightweight but still too heavy for many children to pursue their daily activities. Another challenge for children is that they have to wear a face mask or breathe through a mouthpiece throughout the activity.

Measuring O₂ Uptake: Canopy Method

This method, also called the ventilated hood method, is used to measure resting EE in a supine child (9). The transparent canopy is placed over the child's head. Room air is pumped through it at a measured flow rate. The outgoing air is directed to O_2 and CO_2 analyzers. By knowing the airflow and the difference between the concentration of the incoming and outgoing O_2 and CO_2, the O_2 uptake

and CO_2 output can be calculated, but not minute ventilation or other ventilatory variables.

With the canopy method there is no need for a mouthpiece, nose clip, or face mask, which makes it easier for a child who must lie still for 20 to 30 min. This is particularly useful for infants and young children.

Measuring O₂ Uptake: Respiration Chamber

These chambers are large enough to accommodate children or adults while their metabolic rate is determined continuously. They are sealed hermetically, with one exception: A pipe system provides a continuous, measurable inflow of outside air into the chamber and a similar outflow of air that is channelled into gas analyzers. By knowing the flow rate and the difference in O_2 and CO_2 concentration between the incoming and the outgoing air, one can calculate O_2 uptake and CO_2 output, but not minute ventilation or other ventilatory variables. This principle is similar to that used with the canopy described previously.

An experiment can last as long as several weeks. The chambers are furnished and equipped with amenities such as bed, toilet, telephone, television, and ergometers so that a subject can conduct normal routines or choreographed activities, albeit in very limited quarters. This method has been used successfully with children to determine sleeping EE overnight, or the metabolic rate of daily indoor-living routines (9; 18). This gold standard method has been used to validate other methods that estimate EE, such as accelerometry, HR monitoring, and doubly labeled water (18). The obvious limitation of this method is that the child is restricted to a small space and the daily routine seldom reflects a spontaneous activity behavior. Construction and operation of a respiration chamber system are rather expensive.

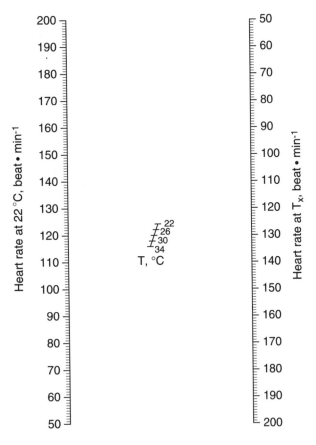

FIGURE III.4   Nomogram for the correction of heart rate as observed at ambient temperature T_x to heart rate at 22°C, assuming equal O_2 uptake. Based on data for 8- to 11-year-old girls and boys.

From Hebestreit et al. 1995 (34).

Doubly Labeled Water (DLW) Analysis

This has been the gold standard method for measuring EE under free-living conditions in infants (15; 39), children (29; 45), and adults. The subject drinks water that contains a measured amount of two stable isotopes: ^2H and ^{18}O. Within 4 to 8 hr, the isotopes become fully diluted in the total body water compartment. This is followed by periodic measurements of the ^2H and ^{18}O concentration in a body fluid (e.g., urine, saliva) for 5 to 14 days to determine their elimination rate from the body. While the ^{18}O is eliminated both through H_2O and CO_2, the ^2H is eliminated only through H_2O. Therefore, the difference in elimination rate of the two isotopes reflects the elimination rate of CO_2, which is proportional to EE.

This is the least reactive method because it does not interfere with the child's spontaneous activity behavior. Its main limitation is that the ^{18}O isotope is hard to obtain and very expensive (several hundred dollars per dose). Another constraint is the need for a high-precision isotope-ratio mass spectrometer. Indeed, there are only a few exercise laboratories worldwide that can afford to use the DLW technique routinely. Another limitation is that the measurement yields a single value of EE over the whole observation period, without any data regarding changes in EE over that period.

Advantage of Combining Methods

There are approximately 20 methods to assess the behavioral, mechanical, and physiologic aspects of PA, as well as the various components of EE. However, none of these *alone* can cover every aspect of PA and EE. There is some evidence that the combination of HR monitoring with acceler-ometry has a high validity when compared with $\dot{V}O_2$ determined in a respiration chamber (70). To obtain a comprehensive picture of a child's PA and EE within a research project, we and others (76) recommend the combined use of two or more methods (e.g., HR monitoring, accelerometry, and a recall interview by the end of the observation day). The choice of such combinations depends on the specific objectives of the study and on the availability of equipment and personnel.

Activity Questionnaire Used in a Clinical Context

The following is an example of a questionnaire to be filled out by a parent on the child's first visit to the clinic. Questions are designed to elicit information on the child's activities at home, at school, and after school; exercise-induced complaints; the activity habits of other family members; and parental attitude toward physical activity. The questionnaire is not designed to quantify the daily energy expenditure of the patient.

Dear Parent:

   The purpose of the following questionnaire is to help us evaluate the activity habits of your child. Please be as accurate as possible in your answers. Feel free to add any details that seem relevant.

1. How would you compare the physical activity of your child with that of her/his friends?
    Child is as active as her/his friends._____

    Child is more active than her/his friends._____

    Child is less active than her/his friends._____

    It is difficult to make such a comparison._____

    Details_____

2. How would you compare the activity of this child with that of your other children?
    There are no other children in the family._____

    This child is as active as my other children._____

    This child is more active than my other children._____

    This child is less active than my other children._____

    It is difficult to make such a comparison._____

    Details_____

3. Does this child take part in physical education classes at school?
    Child participates in all activities with no exception._____

    Child participates in some activities only._____

    Child does not participate in physical education classes._____

    Child does not attend school._____

    Details (especially activities the child does not take part in)_____

    _____

4. If this child is limited in activity at school, for what reason? (You may fill in more than one answer.)
    Advice of physician_____

    Advice of teacher_____

    Decision of parents_____

    Child does not want to participate _____

    Other reasons (please specify)_____

5. Is the child a member of a sports team at school or elsewhere?
    No_____

    Yes, within school (intramural)_____

    Yes, representing the school_____

    Yes, other_____

    Yes, in the past but no more_____

6. If the child is a member of a team, in which sport or sports?_____

7. If child trains regularly, what is the nature of his or her training?

	Hours per week	Time of year	Comments
1.			
2.			
3.			

8. How many hours on a typical day is the child engaged in the following:

Activity	Less than 1 hour	1–2 hours	2–3 hours	3–4 hours	4–5 hours	More than 5 hours
1. TV						
2. Video						
3. Computer						
4. Phone						

9. Are there any other members who participate in *competitive* sports?

   Yes_____

   No one_____

10. If yes, please specify.

Family member	Types of sport	Trains regularly?
1.		
2.		
3.		
4.		

11. Does the child participate in any *recreational* (noncompetitive) activity that requires physical effort (for example, skiing, cycling, dancing, swimming)? Please specify.

Type of activity	Time of year	Hours per week
1.		
2.		
3.		
4.		

12. Does any member of the family participate in *recreational* activities that require physical effort?

   Yes_____

   No one_____

13. If yes, please specify.

Family member	Type of activity	Time of year
1.		
2.		
3.		
4.		

14. Does this child complain of any difficulty *during* or *after* physical exertion?
No complaint_____

Shortness of breath_____

Coughing_____

Wheezing_____

Pains or cramps_____
Where?_____Dizziness_____

Fainting_____

Fatigue_____

Other (please specify)_____

Details_____

15. Does this child *often* sustain bruises, injuries, or other damage when physically active?
Yes_____

No_____

16. If yes, please specify. _____

17. In your opinion, is this child as active as she/he should be?
Yes_____

Child is too active_____

Not sufficiently active_____

18. If this child is not as active as she/he should be, what, in your opinion, is the reason? (You can select more than one answer.)
Lack of interest_____

Disease_____

Lack of suitable conditions_____

Other_____

I don't know_____

Details_____

19. Please check off any of the following statements that you agree with (you can check more than one statement):
Physical activity is important because it is fun._____

Physical activity is necessary for keeping fit._____

Physical activity is good for health reasons._____

Physical activity may be dangerous to one's health._____

Physical activity can prevent a person from becoming overweight._____

Physical activity is important mostly to those who wish to become professional athletes._____

Energy Expenditure During Various Activities and Sports When Performed by Children and Adolescents

TABLE

IV.1 **Caloric Equivalents of Activities by Body Weight Categories**

	Body weight in kilograms									
	20	**25**	**30**	**35**	**40**	**45**	**50**	**55**	**60**	**65**
Activity	**Kcal per 10 minutes**									
Basketball (game)	35	43	51	60	68	77	85	94	102	110
Calisthenics	13	17	20	23	26	30	33	36	40	43
Cross-country skiing (leisure)	24	30	36	42	48	54	60	66	72	78
Cycling 10 km · hr⁻¹	15	17	20	23	26	29	33	36	39	42
15 km · hr⁻¹	22	27	32	36	41	46	50	55	60	65
Field hockey	27	34	40	47	54	60	67	74	80	87
Figure skating	40	50	60	70	80	90	10	110	120	130
Horseback riding canter	8	11	13	15	17	19	21	23	25	27
trot	22	28	33	39	44	50	55	61	66	72
gallop	28	35	41	48	50	62	69	76	83	90
Ice hockey (on-ice time)	52	65	78	91	104	117	130	143	156	168
Judo	39	49	59	69	78	88	98	108	118	127

(continued)

TABLE
IV.1 *continued*

Activity	Kcal per minute									
Running 8 km · hr⁻¹	37	45	52	60	66	72	78	84	90	95
10 km · hr⁻¹	48	55	64	73	79	85	92	100	107	113
12 km · hr⁻¹	--	--	76	83	91	99	107	115	125	130
14 km · hr⁻¹	--	--	--	--	--	113	121	130	140	148
Sitting complete rest	8	8	9	9	10	10	11	11	12	12
quiet play	11	12	14	15	15	16	17	18	19	20
Snowshoeing	35	42	50	58	66	74	82	90	98	107
Soccer (game)	36	45	54	63	72	81	90	99	108	117
Squash	--	--	64	74	85	95	106	117	127	138
Swimming (30 m · min⁻¹) breaststroke	19	24	29	34	38	43	48	53	58	62
front crawl	25	31	37	43	49	56	62	68	74	80
backstroke	17	21	25	30	34	38	42	47	51	55
Table tennis	14	17	20	24	28	31	34	37	41	44
Tennis	22	28	33	39	44	50	55	61	66	72
Volleyball (game)	20	25	30	35	40	45	50	55	60	65
Walking 4 km · hr⁻¹	17	19	21	23	26	28	30	32	34	36
6 km · hr⁻¹	24	26	28	30	32	34	37	40	43	48

APPENDIX V

Scaling for Size Differences

Much has been written on the relationship between function and size within and between species (20; 26; 31). There is also a plethora of reports devoted to this question in humans, with a focus on exercise and energy expenditure (2; 5-10; 13-15; 17-19; 21-25; 27; 29; 30; 32; 33; 35; 36). The most common approach for "correcting" or "normalizing" for body size has been to divide function by a variable that reflects body size. However, the use of scaling by ratio, in spite of its popularity, has been criticized, and several alternatives have been suggested. Table V.1 summarizes the strengths and weaknesses of the various scaling approaches, which are described in the following sections. For more details on scaling strategies, see reviews by Nevill (22), Winter (37), Welsman (33), and Welsman and Armstrong (34).

Scaling by Ratio   The ratio standard has been used traditionally by exercise physiologists to "normalize" a physiologic function (Y) for body size (X). Indices of size commonly used have been body mass, fat-free mass, height, and body surface area. The correction is simply based on the ratio Y/X. For example, maximal O_2 uptake is divided by body mass (expressed as $ml \cdot kg^{-1} \cdot min^{-1}$) and resting metabolic rate by body surface area (expressed as $ml \cdot m^{-2} \cdot min^{-1}$). This approach has been found useful because of its simplicity. This is particularly so for body mass, the measurement of which is simple, highly reliable, and valid. Intuitively, division by mass makes good biological sense for activities such as walking and running, in which energy expenditure is strongly

dependent on the mass carried. It is less relevant to cycling or swimming, in which much of the body mass is supported.

Another concern with scaling by ratio is that it violates an important mathematical principle. By adopting the Y/X ratio, we are assuming that the relationship between function and size is represented by the equation Y = aX, in which *a* represents a constant. Such an equation implies that the intercept is zero. This, however, is not the case for most physiologic functions.

Linear Regression Standards   The linear regression approach, first suggested by Tanner in 1949 (30), is based upon the equation:

$$Y = a + bX \pm \epsilon$$

where *a* is the intercept, *b* the slope, and ϵ an error above and below the line. It is possible to include in the equation more than one predictor.

A major limitation of the linear regression approach is that it assumes a constant error (ϵ) at all levels of body mass, which is often not the case. The actual pattern for many physiologic functions (e.g., maximal O_2 uptake and ventilation) is the "fanning" of values as body size increases. In addition, when using linear regressions one should be cautious not to extrapolate the line beyond the actual data.

Dimensionality Principle   On theoretical and empirical grounds, the relationship between function and size is often nonlinear, but exponential. The dimensionality principle offers a

TABLE V.1 **Strengths and Weaknesses of Methods Used to Scale for Body Size**

Method	Strengths	Weaknesses
Ratio	Simple to use	Falsely assumes zero Y intercept
	Large volume of historical data	Overestimates function of small people
		Falsely assumes a linear relationship between function and size
Linear regression	Discerns size-related intergroup differences not found by ratio scaling	Falsely assumes equal spread of data along regression line
	Does not assume zero Y intercept	
Dimensionality principle	Firm theoretical basis	Assumes constant body proportions
		Discrepancy between theoretical and observed exponent
Allometric modeling	Best statistical fit for data sets	Low interstudy consistency of exponents
	Mathematically best method to discern size-related differences	Low validity for small samples
	Used successfully in interspecies comparisons	
Multilevel modeling	Mathematically best approach for longitudinal studies	Complex mathematics

Note: With few exceptions, these methods ignore variability in body composition. None of them can explain causality between size and function.

theoretical approach to predict the exponent n in the equation

$$Y = bX^n$$

It is based on the assumption that proportions among body segments remain fairly constant during late childhood, adolescence, and adulthood (5). Using this theory one can predict growth-related changes of physiologic functions by monitoring the respective changes in body length (L). For example, the length of body segments and organs will be proportional to L, their surface area or cross-sectional area proportional to L^2, and their volume or mass proportional to L^3. The same rationale applies to physiologic functions. Muscle strength, for example, should be scaled to L^2 (strength being proportional to muscle cross-sectional area), and lung volume, stroke volume, or work to L^3. Time is proportional to L because of the following reasoning. According to Newton's second law, force (F) equals the product of mass (M) and acceleration (A):

$$F = M \cdot A$$

Being proportional to surface area (e.g., force exerted by a muscle), force has the dimension L^2. Mass is proportional to L^3 and acceleration is distance over time squared (T^2). Therefore,

$$L^2 \alpha L^3 L T^{-2}$$

$$T^2 \alpha L^2, \text{ and } T \alpha L$$

Power, which is work/time, should therefore be scaled to $L^3 / L = L^2$, and so should O_2 uptake, ventilation, and cardiac output, which are volume $\cdot$ time^{-1}.

Will scaling of maximal O_2 uptake to L^2 modify our interpretation of the growth-related changes in maximal aerobic power? An answer can be illustrated by the following example. The maximal O_2 uptake of a 117-cm-tall 6-year-old boy is 1.0 L $\cdot$ min^{-1}. Scaled to L^2, his expected maximal O_2 uptake at age 18, when 176 cm tall, will be 2.3 L $\cdot$ min^{-1}. This is contrasted with a value of 3.0 L $\cdot$ min^{-1} commonly found in a healthy nonathletic young adult. Thus, if we were to infer the maximal aerobic power of children from the principles of dimensionality, it would be *lower* than that of adults.

While the dimensionality principle is appealing, reality has shown that the actual exponents often differ from the expected. For example, longitudinal studies (6; 8; 27-29) have shown that the height exponent for maximal O$_2$ uptake ranged from 1.51 to 3.21. Moreover, this approach does not account for the possible effects of age or maturation. Thus the application of dimensionality principles to the prediction of maximal aerobic power (and probably other fitness items) has not been confirmed experimentally. An alternative approach then is to *actually calculate* the exponent for each study group, rather than use a predicted value. This can be done through allometric modeling.

Allometric Modeling   Similar to the dimensionality principle, allometric modeling assumes a nonlinear relationship between function and size. The basic equation used for such models is as follows:

$$Y = a \, X^b \, \epsilon$$

in which *a* is a constant and *b* is the size exponent. Transformed to a logarithmic form, the equation becomes:

$$\ln Y = \ln a + b \ln X \pm \ln \epsilon$$

This equation, when plotted on log-log coordinates, is now a linear regression, where the slope of the line (*b*) is the size exponent. A *b* of 1.0 indicates that the function Y is related linearly to size X, as shown, for example, for mean anaerobic power (11). However, the size exponent is often different from 1.0. Large-scale cross-sectional studies on children of identical age and different sizes have shown, for example, an exponent of 0.65 to 0.66 when maximal O$_2$ uptake was related to body mass (1; 4). Note that this exponent is very close to the theoretical mass exponent of 0.67 (or body length exponent of 2.0) derived from dimensionality principles. Including more than one independent variable (e.g., height, weight, and adiposity) will result in different exponents.

An important message regarding allometry is that one can calculate an *actual* size exponent for each variable rather than predict it based on dimensionality principles. This is important because, for reasons not yet elucidated, exponents may vary among studies even for the same variables (3; 12).

Although it is the best approach to discerning the effect of body size, allometry has some inher-

ent weaknesses. It seems to yield "logical" results for large samples of subjects, say 100 or more, and less explainable results for small samples. Furthermore, it requires a sophisticated and somewhat complex statistical approach. For this reason, allometry may be less practical for clinical use than for research.

Multilevel Modeling   This mathematical method is a refinement of allometric modeling, and it seems to explain a greater portion of the variance in the relationship between function and size. It is suitable for studies in which measurements are taken repeatedly, as in longitudinal design (16), and may prove useful in future research in which independent variables other than size (e.g., % body fat, blood hemoglobin level) are to be incorporated. Because of its relative complexity, few researchers have been using this method (34).

Practical Implications   The choice of a scaling approach depends on the specific question in mind and on the required level of sophistication. Clinicians often wish to know whether the performance of a patient is within normal. To answer this question, one must resort to available norms, which at present are usually based on ratio scaling (e.g., per kilogram body mass for O$_2$ uptake, per muscle cross-sectional area for strength, or per body surface area for resting metabolic rate). If the question is about the effects of a certain therapy (e.g., muscle strengthening) over time, a linear regression approach may be taken, in which the patient's longitudinal data are plotted against size (e.g., body mass) and compared to a reference sample.

For scientists, the use of allometry may open new insights into the function-size relationships and thus a better understanding of changes that occur with growth and development. Those who wish to explore this field are strongly advised to attempt an allometric approach in addition to describing their data in the traditional ratios.

It is important to realize that none of the scaling methods explains *causality* between function and size, nor the mechanisms underlying certain relationships. Breaking with tradition in scaling may help, however, to generate hypotheses about such issues.

APPENDIX VI

Glossary of Terms

Acute Exercise. A bout of physical exercise of any intensity that may last from a few seconds to a few minutes.

Adapted (Adaptive) Physical Education. Subspecialty of physical education using movement, physical skills, and sports for the education and therapy of children and adolescents who are disabled emotionally, mentally, or physically.

All-Out Exercise. A bout of exercise that the child performs to exhaustion.

Anaerobic Exercise. A high-intensity exercise that can be sustained for not more than 30-60 sec. Such exercise depends predominantly on non-oxidative energy turnover.

Anaerobic Power. Mechanical power generated during anaerobic exercise.

Anaerobic Threshold. Exercise intensity during a progressive exercise protocol at which the rate of lactate appearance in the blood exceeds the combined rates of its metabolism and elimination from the blood (cf. Ventilatory Threshold).

Chronic Exercise. Repeated bouts of acute exercise, spreading over a period of weeks, months, or years.

Collision Sports. Sports in which hitting or ramming the opponent is integral to the sport. Examples: American football, boxing, ice hockey, rugby, karate (cf. Contact Sports).

Concentric Muscle Action. Muscle shortening during its action. This occurs when the force generated by the muscle is greater than the opposing resistance.

Conditioning (Physical). The process by which repeated bouts of exercise induce morphologic and functional changes in the body as a whole (cf. Training).

Contact Sports. Sports in which body contact between opponents is frequent but incidental to the sport. Examples: basketball, European handball, lacrosse, soccer (cf. Collision Sports).

Dehydration. The process of incurring fluid deficit, which leads to hypohydration.

Dynamic (Rhythmic) Exercise. An exercise pattern in which muscle action and relaxation alternate.

Eccentric Muscular Action. Lengthening of a muscle during its action. This occurs when the opposing resistance is greater than the force generated by the muscle.

Economy of Movement. The relationship between the degree of body movement (e.g., a given speed of running) and the metabolic energy (or power) needed to perform this movement (cf. Mechanical Efficiency.

Ergometer. An apparatus used for exercise in which the mechanical power or the walking or running conditions are quantifiable and reproducible. The most commonly used ergometers are the cycle, treadmill, and step.

Euhydration. A state of normal body hydration.

Heat Strain. Physiologic and perceptual responses to heat stress. Variables commonly included are: heart rate, rectal temperature, sweating rate, and rating of perceived exertion.

Heat Stress. A combination of environmental conditions that stresses the heat-dissipation mechanisms. The main environmental components are humidity, radiation, ambient temperature, and wind velocity. Heat produced as a result of body metabolism is another component of heat stress.

Hypoactivity. Physical activity level that is lower than in healthy individuals of similar age, gender, cultural, and socioeconomic background.

Hypohydration. A state of body hydration that is below optimal (cf. Dehydration). It is usually quantified as a percentage of body weight.

Isometric Muscular Action. Action that does not result in a change in the muscle length, nor in any skeletal movement. This occurs when the force generated by the muscle is equal to the opposing resistance.

Maximal Aerobic Power. The highest mechanical power that can be attained during an aerobic exercise test. (This term should be used instead of Aerobic Capacity.)

Maximal Exercise. The lowest exercise intensity at which O_2 uptake is maximal (cf. Supramaximal Exercise).

Maximal O_2 Uptake. The level of O_2 uptake that cannot be surpassed even when exercise intensity is increased. Serves as a measure of aerobic fitness.

Mechanical Efficiency. The ratio between external mechanical work (or power) produced by the muscle and the chemical energy (or power) utilized during the action. Usually expressed as percentage. (cf. Economy of Movement)

Muscle Endurance. The ability of a muscle, or a muscle group, to sustain prolonged (static) or repeated (dynamic) actions of high intensity.

Muscle Strength. The maximal force, torque, or moment that a muscle, or muscle group, can exert.

Oxygen Uptake. The volume of O_2 that is consumed during a time unit by a tissue, an organ, or the whole body.

Peak Aerobic Power. The highest mechanical power achieved in a progressive maximal test.

Peak Anaerobic Power. The highest mechanical power achieved over a very short time (up to 5-7 sec) during a supramaximal task.

Physical Working Capacity (PWC). A general term in common use to describe the working performance of an individual (often used as a synonym for maximal aerobic power). PWC is assessed by peak power output, total work, maximal O_2 uptake, or $\dot{W}_{170}$.

Progressive Exercise Protocol. An exercise test in which the intensity increases gradually.

Prolonged Exercise. A continuous exercise bout lasting 30 min or more.

Resistance Training. Activities designed to improve muscle strength by highly forceful actions.

Rating of Perceived Exertion (RPE). The rating of exercise intensity, as perceived by the person who performs the exercise.

Respiratory Exchange Ratio (RER). The ratio between CO_2 production and O_2 uptake. Also called Respiratory Quotient (RQ).

Static Exercise. An exercise pattern in which muscle action (usually isometric) is continuous and not alternating with relaxation.

Submaximal Exercise. An exercise level at which O_2 uptake is below maximum.

Supramaximal Exercise. An exercise level higher than maximal exercise.

Training (Physical). The process by which repeated bouts of exercise induce morphologic and functional changes in a specific tissue, organ, body system, or the body as a whole (cf. Conditioning).

Ventilatory Equivalent. The ratio between pulmonary ventilation and O_2 uptake (i.e., the number of liters of ventilated air that facilitate 1 L of O_2 uptake).

Ventilatory Threshold (also called Ventilatory Anaerobic Threshold). An exercise intensity beyond which ventilation starts increasing in an accelerating rate as related to O_2 uptake or CO_2 output. Ventilatory threshold is often used to reflect anaerobic threshold.

$\dot{W}_{170}$. The mechanical power at which heart rate would be 170 beat $\cdot$ min^{-1}. An index of maximal aerobic power.

REFERENCES

Chapter 1

1. Adams, F.H., E. Bengtsson, and H. Berven. Determination by means of a bicycle ergometer, of the physical working capacity of children. *Acta Paediatr Scand Suppl* 118: 120-122, 1959.

2. Allen, H.D., S.J. Goldberg, and D.J. Sahn. A quantitative echocardiographic study of champion childhood swimmers. *Circulation* 55: 142-145, 1977.

3. Alpert, B.S., N.L. Flood, and W.B. Strong. Responses to ergometer exercise in healthy biracial population of children. *J Pediatr* 101: 538-545, 1982.

4. American Academy of Pediatrics. Strength training, weight and power lifting, and body building by children and adolescents. A position statement [abstract]. *Pediatrics* 86: 801-803, 1990.

5. Amigo, N., J.A. Cadefau, I. Ferrer, N. Tarrados, and R. Cusso. Effect of summer intermission on skeletal muscle of adolescent soccer players. *J Sports Med Phys Fitness* 38: 298-304, 1998.

6. Andersen, B., and K. Froberg. Circulatory parameters and muscular strength in trained and normal boys during puberty [abstract]. *Acta Physiol Scand* 105: D36, 1980.

7. Andersen, B., and K. Froberg. Maximal oxygen uptake and lactate concentration in highly trained and normal boys during puberty [abstract]. *Acta Physiol Scand* 105: D37, 1980.

8. Andersen, K.L., and J.R. Magel. Physiological adaptation to a high level of habitual physical activity during adolescence. *Int Z Angew Physiol* 28: 209-227, 1970.

9. Andersen, K.L., V. Seliger, J. Rutenfranz, and I. Berndt. Physical performance capacity of children in Norway. Part II. Heart rate and oxygen pulse in submaximal and maximal exercises—population parameters in a rural community. *Eur J Appl Physiol* 33: 197-206, 1974.

10. Andersen, K.L., V. Seliger, J. Rutenfranz, and S. Messel. Physical performance capacity of children in Norway. Part III. Respiratory responses to graded exercise loadings—population parameters in a rural community. *Eur J Appl Physiol* 33: 265-274, 1974.

11. Andersen, K.L., V. Seliger, J. Rutenfranz, and R. Mocellin. Physical performance capacity of children in Norway. Part I. Population parameters in a rural inland community with regard to maximal aerobic power. *Eur J Appl Physiol* 33: 177-195, 1974.

12. Anderson, S.D., and S. Godfrey. Cardio-respiratory response to treadmill exercise in normal children. *Clin Sci* 40: 433-442, 1971.

13. Anderson, S.D., and S. Godfrey. Transfer factor for CO_2 during exercise in children. *Thorax* 26: 51-54, 1971.

14. Ariens, G.A., W. van Mechelen, H.C. Kemper, and J.W. Twisk. The longitudinal development of running economy in males and females aged between 13 and 27 years: the Amsterdam Growth and Health Study. *Eur J Appl Physiol Occup Physiol* 76: 214-220, 1997.

15. Armon, Y., D.M. Cooper, R. Flores, S. Zanconato, and T.J. Barstow. Oxygen uptake dynamics during high-intensity exercise in children and adults. *J Appl Physiol* 70: 841-848, 1991.

16. Armon, Y., D.M. Cooper, and S. Zanconato. Maturation of ventilatory responses to one-minute exercise. *Pediatr Res* 29: 362-368, 1991.

17. Armstrong, N., B.J. Kirby, A.M. McManus, and J.R. Welsman. Prepubescents' ventilatory responses to exercise with reference to sex and body size. *Chest* 112: 1554-1560, 1997.

18. Armstrong, N., and J. Welsman. *Young people and physical activity.* Oxford: Oxford University Press, 1997, 1-369.

19. Armstrong, N., J. Welsman, and B.J. Kirby. Peak oxygen uptake and maturation in 12-year-old boys. *Med Sci Sports Exerc* 30: 165-169, 1998.

20. Armstrong, N., J.R. Welsman, and B.J. Kirby. Performance on the Wingate anaerobic test and maturation. *Pediatr Exerc Sci* 9: 253-261, 1997.

21. Armstrong, N., J.R. Welsman, A.M. Nevill, and B.J. Kirby. Modeling growth and maturation changes in peak oxygen uptake in 11-13 year olds. *J Appl Physiol* 87: 2230-2236, 1999.

22. Asano, K., and K. Hirakoba. Respiratory and circulatory adaptation during prolonged exercise in 10-12 year-old children and in adults. In Ilmarinen, J. and I. Valimaki, eds., *Children and sport.* Berlin: Springer-Verlag, 1984.

23. Asmussen, E. Growth in muscular strength and power. In Rarick, L., ed., *Physical activity, human growth and development.* New York: Academic Press, 1973, 60-79.

24. Åstrand, P.O., and K. Rodahl. *Textbook of work physiology.* 3rd ed. New York: McGraw-Hill, 1986.

25. Åstrand, P.O. *Experimental studies of physical working capacity in relation to sex and age.* Copenhagen: Munksgaard, 1952.

26. Åstrand, P.O., L. Engstrom, and B.O. Eriksson. Girl swimmers. With special reference to respiratory and

circulatory adaptation and gynecological and psychiatric aspects. *Acta Paediatr Scand Suppl* 147: 1-75, 1963.

27. Åstrand, P.O., and I. Rhyming. A nomogram for calculation of aerobic capacity (physical fitness) from pulse rate during submaximal work. *J Appl Physiol* 7: 218-221, 1954.

28. Ayub, B., and O. Bar-Or. Energy cost of walking in boys who differ in adiposity but are matched for body mass. *Med Sci Sports Exerc* 35: 669-674, 2003.

29. Bailey, D.A., H.A. McKay, R.L. Mirwald, P.R. Crocker, and R.A. Faulkner. A six-year longitudinal study of the relationship of physical activity to bone mineral accrual in growing children: the University of Saskatchewan bone mineral accrual study. *J Bone Miner Res* 14: 1672-1679, 1999.

30. Bailey, D.A., W.D. Ross, R.L. Mirwald, and C. Wesse. Size dissociation of maximal aerobic power during growth in boys. In Borms, J. and M. Hebbelinck, eds., *Pediatric work physiology.* Basel: Karger, 1978, 140-151.

31. Bailey, R.C., J. Olson, S.L. Pepper, J. Porszasz, T.J. Barstow, and D.M. Cooper. The level and tempo of children's physical activities: an observational study. *Med Sci Sports Exerc* 27: 1033-1041, 1995.

32. Bal, M.E.R., E.M. Thompson, E.H. McIntosh, C.M. Taylor, and G. Macleod. Mechanical efficiency in cycling of girls six to fourteen years of age. *J Appl Physiol* 6: 185-188, 1953.

33. Banister, E.W. A comparison of fitness training methods in a school program. *Res Q Am Assoc Health Phys Educ* 36: 387-392, 1965.

34. Bar-Or, O. Arm ergometry vs. treadmill running and bicycle riding in men with different conditioning levels. In Hansen, G. and H. Mellerowicz, eds., *Internationale Seminar fur ergometrie.* Berlin: Sports Med. Inst., 1972.

35. Bar-Or, O. Age-related changes in exercise perception. In Borg, G., ed., *Physical work and effort.* Oxford and New York: Pergamon Press, 1977, 255-266.

36. Bar-Or, O. Physiologische gesetzmassigkeiten sportlicher aktivitat beim kind. In Howald, H. and E. Han, eds., *Kinder im Leistungssport.* Basel: Birkhauser, 1982, 18-30.

37. Bar-Or, O. The growth and development of children's physiological and perceptional responses to exercise. In Ilmarinen, Y. and I. Välimäki, eds., *Pediatric work physiology.* Berlin: Springer-Verlag, 1983.

38. Bar-Or, O. Temperature regulation during exercise in children and adolescents. *Perspectives in exercise science and sports medicine.* Vol. 2. *Youth, exercise, and sport.* Indianapolis: Benchmark Press, 1989, 335-367.

39. Bar-Or, O. Trainability of the pre-pubescent child. *Phys Sportsmed* 17: 65-66, 1989.

40. Bar-Or, O. Exertional perception in children and adolescents with a disease or a physical disability: assessment and interpretation. *Int J Sport Psychol* 32: 127-136, 2001.

41. Bar-Or, O., and E.R. Buskirk. The cardiovascular system and exercise. In Johnson, W.R. and E.R. Buskirk, eds., *Science and medicine of exercise and sport.* 2nd ed. New York: Harper & Row, 1974, 121-136.

42. Bar-Or, O., and O. Inbar. Relationships among anaerobic capacity, sprint and middle distance running of schoolchildren. In Shephard, R.J. and H. Lavallée, eds., *Physical fitness assessment.* Springfield, IL: Charles C Thomas, 1978, 142-147.

43. Bar-Or, O., and S.L. Reed. Rating of perceived exertion in adolescents with neuromuscular disease. In Borg, G., ed., *Perception of exertion in physical work.* Stockholm: Wenner-Gren, 1987, 137-148.

44. Bar-Or, O., R.J. Shephard, and C.L. Allen. Cardiac output of 10- to 13-year-old boys and girls during submaximal exercise. *J Appl Physiol* 30: 219-223, 1971.

45. Bar-Or, O., J.S. Skinner, V. Bergsteinová, C. Shearburn, C.W. Bell, D. Royer, and E.R. Buskirk. Maximal aerobic capacity of 6- to 15-year-old girls and boys with subnormal intelligence quotients. *Acta Paediatr Scand Suppl* 217: 108-113, 1971.

46. Bar-Or, O., and D.S. Ward. Rating of perceived exertion in children. In Bar-Or, O., ed., *Advances in pediatric sports sciences.* Champaign, IL: Human Kinetics, 1989, 151-168.

47. Bar-Or, O., and L.D. Zwiren. Physiological effects of increased frequency of physical education classes and of endurance conditioning on 9- to 10-year-old girls and boys. In Bar-Or, O., ed., *Pediatric work physiology.* Natanya: Wingate Institute, 1973, 183-198.

48. Baraldi, E., D.M. Cooper, S. Zanconato, and Y. Armon. Heart rate recovery following 1 minute exercise in children and adults. *Pediatr Res* 29: 575-579, 1991.

49. Baxter-Jones, A., and P.J. Helms. Effects of training at a young age: a review of the Training of Young Athletes (TOYA) Study. *Pediatr Exerc Sci* 8: 310-327, 1996.

50. Becker, D.M., and P. Vaccaro. Anaerobic threshold alterations caused by endurance training in young children. *J Sports Med Phys Fitness* 23: 445-449, 1983.

51. Bedu, M., N. Fellmann, H. Spielvogel, G. Falgairette, E. Van Praagh, and J. Coudert. Force-velocity and 30-s Wingate tests in boys at high and low altitude. *J Appl Physiol* 70: 1031-1037, 1991.

52. Bell, R.D., J.D. MacDougall, and R. Billeter. Muscle fiber types and morphometric analysis of skeletal muscle in six-year-old children. *Med Sci Sports Exerc* 12: 28-31, 1980.

53. Bengtsson, E. The working capacity in normal children, evaluated by submaximal exercise on the bicycle ergometer and compared with adults. *Acta Med Scand* 154: 91-109, 1956.

54. Berg, A., J. Keul, and G. Huber. Biochemische Akutveranderungen bei Ausdauerbelastungen im Kindes-und Jugendalter. *Mschr Kinderheilkd* 128: 490-495, 1980.

55. Bernhardt, D.T., J. Gomez, M.D. Johnson, T.J. Martin, T.W. Rowland, E. Small, C. Leblanc, R. Malina, C. Krein, J.C. Young, F.E. Reed, S.J. Anderson, B.A. Griesemer, and O. Bar-Or. American Academy of Pediatrics policy statement. Strength training by children and adolescents. *Pediatrics* 107: 1470-1472, 2001.

56. Beunen, G., and R.M. Malina. Growth and physical performance relative to the timing of the adolescent spurt. *Exerc Sport Sci Rev* 16: 503-540, 1988.

57. Beunen, G., R.M. Malina, M.A. Van't Hof, J. Simons, M. Ostyn, R. Renson, and D. Van Gerven. *Adolescent growth and motor performance: a longitudinal study of Belgian boys.* Champaign, IL: Human Kinetics, 1988.

58. Beunen, G., and M. Thomis. Muscular strength development in children and adolescents. *Pediatr Exerc Sci* 12: 174-197, 2000.

59. Blaak, E.E., K.R. Westerterp, O. Bar-Or, L.J. Wouters, and W.H. Saris. Total energy expenditure and spontaneous activity in relation to training in obese boys. *Am J Clin Nutr* 55: 777-782, 1992.

60. Blimkie, C.J., S. Rice, C.E. Webber, J. Martin, D. Levy, and C.L. Gordon. Effects of resistance training on bone mineral content and density in adolescent females. *Can J Physiol Pharmacol* 74: 1025-1033, 1996.

61. Blimkie, C.J.R. Age- and sex-associated variation in strength during childhood: anthropometric, morphological, neurologic, biomechanical, endocrinologic, genetic and physical activity correlates. In Gisolfi, C.V. and D.R. Lamb, eds., *Perspectives in exercise science and sports medicine.* Vol. 2. *Youth, exercise, and sport.* Indianapolis: Benchmark Press, 1989, 99-161.

62. Blimkie, C.J.R. Benefits and risks of resistance training in children. In Cahill, B.R. and A.J. Pearl, eds., *Intensive participation in children's sports.* Champaign, IL: Human Kinetics, 1993, 133-165.

63. Blimkie, C.J.R. Resistance training during preadolescence: issues and controversies. *Sports Med* 15: 389-407, 1993.

64. Blimkie, C.J.R., and O. Bar-Or. Trainability of muscle strength, power and endurance during childhood. In Bar-Or, O., ed., *The child and adolescent athlete.* Oxford: Blackwell Scientific, 1996, 113-129.

65. Blimkie, C.J.R., D.A. Cunningham, and F.Y. Leung. Urinary catecholamine excretion during competi-

tion in 11- to 23-year-old hockey players. *Med Sci Sports* 10: 188-193, 1978.

66. Blimkie, C.J.R., and S. Kriemler. Bone mineralization pattern during childhood: relationship to fracture and fracture risk. In Froberg, K., O. Lammert, H. St. Hansen, and C.J.R. Blimkie, eds., *Exercise fitness—benefits and risks.* Odense: Odense University Press, 1997, 111-137.

67. Blimkie, C.J.R., J. Martin, J. Ramsay, D. Sale, and D. MacDougall. The effects of detraining and maintenance weight training on strength development in prepubertal boys [abstract]. *Can J Sport Sci* 14: 102, 1989.

68. Blimkie, C.J.R., J. Ramsay, D. Sale, D. MacDougall, K. Smith, and S. Garner. Effects of 10 weeks of resistance training on strength development in prepubertal boys. In Oseid, S. and K.-H. Carlsen, eds., *Children and exercise XIII.* Champaign, IL: Human Kinetics, 1989, 183-197.

69. Blimkie, C.J.R., P. Roche, and O. Bar-Or. The anaerobic-to-aerobic power ratio in adolescent boys and girls. In Rutenfranz, J., R. Mocellin, and F. Klimt, eds., *Children and exercise XII.* Champaign, IL: Human Kinetics, 1986, 31-37.

70. Blimkie, C.J.R., P. Roche, J.T. Hay, and O. Bar-Or. Anaerobic power of arms in teenage boys and girls: relationship to lean tissue. *Eur J Appl Physiol* 57: 677-683, 1988.

71. Blimkie, C.J.R., and D.G. Sale. Strength development and trainability during childhood. In Van Praagh, E., ed., *Pediatric anaerobic performance.* Champaign, IL: Human Kinetics, 1998, 193-224.

71a. Boas, S.R., M.J. Danduran, A.L. McBride, S.A. McColley, and M.R.G. O'Gorman. Postexercise immune correlates in children with and without cystic fibrosis. *Med Sci Sport Exerc* 32: 1997-2000, 2000.

71b. Boas, S.R., M.J. Danduran, S.A. McColley, K. Beaman, and M.R.G. O'Gorman. Immune modulation following aerobic exercise in children with cystic fibrosis. *Int. J. Sports Med* 21: 294-301, 2000.

71c. Boas, S.R., M.L. Joswiak, P.A. Nixon, G. Kurland, M.J. O'Connor, K. Bufalino, D.M. Orenstein, and T.L. Whiteside. Effects of anaerobic exercise on the immune system in eight-to seventeen-year-old trained and untrained boys. *J Pediatr* 129: 846-855, 1996.

72. Boileau, R.A., J.E. Ballard, R.L. Sprague, E.K. Sleator, and B.H. Massey. Effect of methylphenidate on cardiorespiratory responses in hyperactive children. *Res Q Am Assoc Health Phys Educ* 47: 590-596, 1976.

73. Borg, G. *Physical performance and perceived exertion.* Lund: Gleerup, 1962.

74. Borg, G. Perceived exertion as an indicator of somatic stress. *Scand J Rehab Med* 2: 92-98, 1970.

75. Borg, G. *Borg's perceived exertion and pain scales.* Champaign, IL: Human Kinetics, 1998.

76. Borg, G. Psychophysical bases of perceived exertion. *Med Sci Sports Exerc* 14: 377-381, 1982.

77. Borjeson, O. Overweight children. *Acta Paediatr Scand Suppl* 132, 1962.

78. Bouchard, C., R.M. Malina, W. Hollmann, and C. Leblanc. Submaximal working capacity, heart size and body size in boys 8 to 18 years. *Eur J Appl Physiol* 36: 115-126, 1977.

79. Bouchard, C., L. Perusse, C. Leblanc, et al. Inheritance of the amount and distribution of human body fat. *Int J Obesity* 12: 205-215, 1998.

80. Bouchard, C., A.-L. Simoneau, G. Lortie, M.R. Boulay, M. Marcotte, and M.C. Thibault. Genetic effects in human skeletal muscle fiber type distribution and enzyme activities. *Can J Physiol Pharmacol* 64: 1245-1251, 1986.

81. Bradney, M., G. Pearce, G. Naughton, C. Sullivan, S. Bass, T. Beck, J. Carlson, and E. Seeman. Moderate exercise during growth in prepubertal boys: changes in bone mass, size, volumetric density, and bone strength: a controlled prospective study. *J Bone Miner Res* 13: 1814-1821, 1998.

82. Brown, C.H., J.R. Harrower, and M.F. Deeter. The effects of cross-country running on pre-adolescent girls. *Med Sci Sports* 4: 1-5, 1974.

83. Burkhard-Jagodzinska, K., K. Nazar, M. Ladyga, J. Starczewska-Czapowska, and L. Borkowski. Resting metabolic rate and thermogenic effect of glucose in trained and untrained girls age 11-15 years. *Int J Sports Med* 9: 378-390, 2001.

84. Burmeister, W., J. Rutenfranz, W. Stresny, and H.G. Randy. Body cell mass and physical performance capacity (W170) of school children. *Int Z Angew Physiol* 31: 61-70, 1972.

85. Cadefau, J., J. Casademont, J.M. Grau, J. Fernandez, A. Balaguer, M. Vernet, R. Cusso, and A. Urbano-Marquez. Biochemical and histochemical adaptation to sprint training in young athletes. *Acta Physiol Scand* 140: 341-351, 1990.

86. Cafarelli, E. Peripheral contributions to the perception of effort. *Med Sci Sports Exerc* 14: 382-389, 1982.

87. Cahill, B.R., ed. *Proceedings of the conference on strength training and the prepubescent.* Chicago, IL: American Orthopedic Society for Sports Medicine, 1988.

88. Caird, S.J., A.D. McKenzie, and G.G. Sleivert. Biofeedback and relaxation techniques improve running economy in sub-elite long distance runners. *Med Sci Sports Exerc* 31: 717-722, 1999.

89. Carlson, J.S., and G.A. Naughton. An examination of the anaerobic capacity of children using maximal accumulated oxygen deficit. *Pediatr Exerc Sci* 5: 60-71, 1993.

90. Carron, A.V., and D.A. Bailey. Strength development in boys from 10 through 16 years. *Monog Soc Res Child Develop* 39: 1-37, 1974.

91. Cerny, F.J., and H.W. Burton. *Exercise physiology for health care professionals.* Champaign, IL: Human Kinetics, 2001, 1-384.

92. Chatterjee, S., P.K. Banerjee, P. Chatterjee, and S.R. Maitra. Aerobic capacity of young girls. *Indian J Med Res* 69: 327-333, 1979.

93. Chausow, S.A., W.F. Riner, and R.A. Boileau. Metabolic and cardiovascular responses of children during prolonged physical activities. *Res Q Exerc Sport* 55: 1-7, 1984.

94. Cheatham, C.C., A.D. Mahon, J.D. Brown, and D.R. Bolster. Cardiovascular responses during prolonged exercise at ventilatory threshold in boys and men. *Med Sci Sports Exerc* 32: 1080-1087, 2000.

95. Clarke, D.H. Muscular strength and endurance as a function of age and activity level. *Res Q Exerc Sport* 28: 105-108, 1992.

96. Clarke, D.H., and P. Vaccaro. The effect of swimming training on muscular performance and body composition in children. *Res Q* 50: 9-17, 1979.

97. Clarke, H.H. *Physical and motor tests in the Medford boy's growth study.* Englewood Cliffs, NJ: Prentice Hall, 1971.

98. Clarke, H.H., and G.C.E. Harrison. Differences in physical and motor traits between boys of advanced, normal and retarded maturity. *Res Q* 33: 13-25, 1962.

99. Cooke, C.B., M.J. McDonagh, A.M. Nevill, and C.T. Davies. Effects of load on oxygen intake in trained boys and men during treadmill running. *J Appl Physiol* 71: 1237-1244, 1991.

100. Cooper, D.M., and T.J. Barstow. Magnetic resonance imaging and spectroscopy in studying exercise in children. *Exerc Sport Sci Rev* 24: 475-499, 1996.

101. Cooper, D.M., M.R. Kaplan, L. Baumgarten, D. Weiler-Ravell, B.J. Whipp, and K. Wasserman. Coupling of ventilation and CO_2 production during exercise in children. *Pediatr Res* 21: 568-572, 1987.

102. Cooper, D.M., D. Weiler-Ravell, B.J. Whipp, and K. Wasserman. Aerobic parameters of exercise as a function of body size during growth in children. *J Appl Physiol* 56: 628-634, 1984.

103. Courteix, D., P. Obert, A.M. Lecoq, P. Guenon, and G. Koch. Effect of intensive swimming training on lung volumes, airway resistance and on the maximal expiratory flow-volume relationship in prepubertal girls. *Eur J Appl Physiol Occup Physiol* 76: 264-269, 1997.

104. Cowden, R.D., and S.A. Plowman. The self-regulation and perception of exercise intensity in children in a field setting. *Pediatr Exerc Sci* 11: 32-43, 1999.

105. Crielaard, J.M., and F. Pirnay. Anaerobic and aerobic power of top athletes. *Eur J Appl Physiol Occup Physiol* 47: 295-300, 1981.

106. Cullen, D.M., V.T. Iwaniec, and M.J. Barger-Lux. Skeletal responses to exercise and training. In Garrett, W.E. and D.T. Kirkendall, eds., *Exercise and sport science*. Philadelphia: Williams & Wilkins, 2000, 227-237.

107. Cumming, G.R. Hemodynamics of supine bicycle exercise in 'normal' children. *Am Heart J* 93: 617-622, 1977.

108. Cumming, G.R. Supine bicycle exercise in pediatric cardiology. In Borms, J. and M. Hebbelinck, eds., *Pediatric work physiology*. Basel: Karger, 1978, 82-88.

109. Cumming, G.R. Correlation of physical performance with laboratory measures of fitness. In Shephard, R.J., ed., *Frontiers of fitness*. Springfield, IL: Charles C Thomas, 1971, 265-279.

110. Cumming, G.R. Recirculation times in exercising children. *J Appl Physiol: Respir Environ Exerc Physiol* 45: 1005-1008, 1978.

111. Cumming, G.R., A. Goodwin, G. Baggley, and J. Antel. Repeated measurements of aerobic capacity during a week of intensive training at a youths' track camp. *Can J Physiol Pharmacol* 45: 805-811, 1967.

112. Cumming, G.R., D. Goulding, and G. Baggley. Failure of school physical education to improve cardiorespiratory fitness. *Can Med Assoc J* 101: 69-73, 1969.

113. Daniels, J., and N. Oldridge. Changes in oxygen consumption of young boys during growth and running training. *Med Sci Sports* 3: 161-165, 1971.

114. Daniels, J., N. Oldridge, F. Nagle, and B. White. Differences and changes in $\dot{V}O_2$ among young runners 10 to 18 years of age. *Med Sci Sports* 10: 200-203, 1978.

115. Davies, C.T.M. Strength and mechanical properties of muscle in children and young adults. *Scand J Sports Sci* 7: 11-15, 1985.

116. Davies, C.T.M., C. Barnes, and S. Godfrey. Body composition and maximal exercise performance in children. *Hum Biol* 44: 195-214, 1972.

117. de Bisschop, C., H. Guenard, P. Desnot, and J. Vergeret. Reduction of exercise-induced asthma in children by short, repeated warm ups. *Br J Sports Med* 33: 100-104, 1999.

118. Desouza, M., M.S. Schaffer, D.L. Gilday, and V. Rose. Exercise radionuclide angiography in hyperlipidaemic children with apparently normal hearts. *Nucl Med Commun* 5: 13-17, 1984.

119. di Prampero, P.E., and P. Cerretelli. Maximal muscular power (aerobic and anaerobic) in African natives. *Ergonomics* 12: 51-59, 1969.

120. Docherty, D., and C.A. Gaul. Relationship of body size, physique, and composition to physical performance in young boys and girls. *Int J Sports Med* 12: 525-532, 1991.

121. Docherty, D., H.A. Wenger, and M.L. Collins. The effects of resistance training on aerobic and anaerobic power of young boys. *Med Sci Sports Exerc* 19: 389-392, 1987.

122. Docherty, D., H.A. Wenger, M.L. Collins, and H.A. Quinney. The effects of variable speed resistance training on strength development in prepubertal boys. *J Hum Mov Studies* 13: 377-382, 1987.

123. Dore, E., O. Diallo, N.M. Franca, M. Bedu, and E. Van Praagh. Dimensional changes cannot account for all differences in short-term cycling power during growth. *Int J Sports Med* 21: 360-365, 2000.

124. Dore, E., O. Diallo, N.M. Franca, M. Bedu, and E. Van Praagh. Dimensional changes cannot account for all differences in short-term cycling power during growth. *Int J Sports Med* 21: 360-365, 2000.

125. Dotan, R., B. Falk, and A. Raz. Intensity effect of active recovery from glycolytic exercise on decreasing blood lactate concentration in prepubertal children. *Med Sci Sports Exerc* 32: 564-570, 2000.

126. Dotan, R., A. Rotstein, G. Tenenbaum, and O. Bar-Or. Is the anaerobic threshold preferable to maximal O_2 uptake or other indicators for evaluating the effect of aerobic conditioning in prepubescent boys? Research Report [in Hebrew]. Natanya: Wingate Institute, 1982.

127. Drinkwater, B.L., I.C. Kupprat, J.E. Denton, J.L. Crist, and S.M. Horvath. Response of prepubertal girls and college women to work in the heat. *J Appl Physiol: Respir Environ Exerc Physiol* 43: 1046-1053, 1977.

128. Drinkwater, B.L., I.C. Kupprat, J.E. Denton, J.L. Crist, and S.M. Horvath. Response of prepubertal girls and college women to work in the heat. *J Appl Physiol: Respir Environ Exerc Physiol* 43: 1046-1053, 1977.

129. Duché, P., G. Falgairette, M. Bedu, N. Fellmann, G. Lac, A. Robert, and J. Coudert. Longitudinal approach of bio-energetic profile in boys before and during puberty. In Coudert, J. and E. Van Praagh, eds., *Children and exercise XVI*. Paris: Masson, 1992, 43-45.

130. Duda, M. Prepubescent strength training gains support. *Phys Sportsmed* 14: 157-161, 1986.

131. Dunaway, G.A., T.P. Kasten, G.A. Nickols, and J.A. Chesky. Regulation of skeletal muscle 6-phosphofructo-1-kinase during aging and development. *Mech Ageing Dev* 36: 13-23, 1986.

132. Ebbeling, C.J., J. Hamill, P.S. Freedson, and T.W. Rowland. An examination of efficiency during walking in children and adults. *Pediatr Exerc Sci* 4: 36-49, 1992.

133. Ekblom, B. Effect of physical training in adolescent boys. *J Appl Physiol* 27: 350-355, 1969.

133a. Eliakim, A., B. Wolach, E. Kodesh, R. Gavrieli, J. Radnay, T. Ben Tovim, Y. Yarom, and B. Falk. Cellular and humoral immune response to exercise among gymnasts and untrained girls. *Int J Sports Med* 18: 208-212, 1997.

134. Ellenbecker, T.S., G.J. Davies, and M.J. Rowinski. Concentric versus eccentric isokinetic strengthening of the rotator cuff. Objective data versus functional test. *Am J Sports Med* 16: 64-69, 1988.

135. Emons, H.J.G., and M.A. van Baak. Effect of training on aerobic and anaerobic power and mechanical efficiency in spastic cerebral palsied children [abstract]. *Pediatr Exerc Sci* 5: 412, 1993.

136. Engström, I., B.O. Eriksson, P. Karlberg, B. Saltin, and C. Thorén. Preliminary report on the development of lung volumes in young girl swimmers. *Acta Pediatr* 217: 73-76, 1971.

137. Eriksson, B.O. Cardiac output during exercise in pubertal boys. *Acta Paediatr Scand Suppl* 217: 53-55, 1971.

138. Eriksson, B.O. Physical training, oxygen supply and muscle metabolism in 11 to 15 year-old boys. *Acta Physiol Scand* 384: 1-48, 1972.

139. Eriksson, B.O. Muscle metabolism in children—a review. *Acta Paediatr Scand Suppl* 283: 20-27, 1980.

140. Eriksson, B.O., K. Berg, and J. Taranger. Physiological analysis of young boys starting intensive training in swimming. In Eriksson, B. and B. Furberg, eds., *Swimming medicine*. Baltimore: University Park Press, 1994, 147-160.

141. Eriksson, B.O., P.D. Gollnick, and B. Saltin. Muscle metabolism and enzyme activities after training in boys 11-13 years old. *Acta Physiol Scand* 87: 485-497, 1973.

142. Eriksson, B.O., P.D. Gollnick, and B. Saltin. The effect of physical training on muscle enzyme activities and fiber composition in 11 year-old boys. *Acta Paediatr Belg* (Suppl) 28: 245-252, 1974.

143. Eriksson, B.O., J. Karlsson, and B. Saltin. Muscle metabolites during exercise in pubertal boys. *Acta Paediatr Scand Suppl* 217: 154-157, 1971.

144. Eriksson, B.O., and G. Koch. Cardiac output and intra-arterial blood pressure at rest and during submaximal and maximal exercise in 11 to 13 year-old boys before and after physical training. In Bar-Or, O., ed., *Pediatric work physiology*. Natanya: Wingate Institute, 1973, 139-150.

145. Eriksson, B.O., and G. Koch. Effect of physical training on hemodynamic responses during submaximal and maximal exercise in 11 to 13 year old boys. *Acta Physiol Scand* 87: 27-39, 1973.

146. Eriksson, B.O., and B. Saltin. Muscle metabolism during exercise in boys aged 11 to 16 years compared to adults. *Acta Paediatr Belg* (Suppl) 28: 257-265, 1974.

147. Eriksson, B.O., and C. Thorén. Training girls for swimming from medical and physiological points of view, with special reference to growth. In Eriksson, B. and B. Furberg, eds., *Swimming medicine IV*. Baltimore: University Park Press, 1978, 3-15.

148. Eston, R., and K.L. Lamb. Effort perception. In Armstrong, N. and W. van Mechelen, eds., *Paediatric exercise science and medicine*. Oxford: Oxford University Press, 2000, 85-91.

149. Eston, R.G., G. Parfitt, L. Campbell, and K.L. Lamb. Reliability of effort perception for regulating exercise intensity in children using a Cart and Load Effort Rating (CALER) scale. *Pediatr Exerc Sci* 12: 388-397, 2000.

150. Eston, R.G., and M. Thompson. Use of ratings of perceived exertion for predicting maximal work rate and prescribing exercise intensity in patients taking atenolol. *Br J Sports Med* 31: 114-119, 1997.

151. Eston, R.G., and J.G. Williams. Reliability of ratings of perceived effort for regulation of exercise intensity. *Br J Sports Med* 22: 153-155, 1988.

152. Faigenbaum, A.D., L.A. Milliken, R.L. Loud, B.T. Burak, C.L. Doherty, and W.L. Westcott. Comparison of 1 and 2 days per week of strength training in children. *Res Q Exerc Sport* 73: 416-424, 2002.

153. Faigenbaum, A.D., L.D. Zaichkowsky, W.L. Westcott, L.J. Micheli, and A.F. Fehlandt. The effects of a twice-a-week strength training program on children. *Pediatr Exerc Sci* 5: 339-346, 1993.

154. Falgairette, G., M. Bedu, N. Fellmann, E. Van-Praagh, and J. Coudert. Bio-energetic profile in 144 boys aged from 6 to 15 years with special reference to sexual maturation. *Eur J Appl Physiol* 62: 151-156, 1991.

155. Falgairette, G., P. Duché, M. Bedu, N. Fellmann, and J. Coudert. Bioenergetic characteristics in prepubertal swimmers: comparison with active and non-active boys. *Int J Sports Med* 14: 444-448, 1993.

156. Falk, B., and O. Bar-Or. Longitudinal changes in peak aerobic and anaerobic mechanical power of circumpubertal boys. *Pediatr Exerc Sci* 5: 318-331, 1993.

157. Falk, B., O. Bar-Or, and J.D. MacDougall. Thermoregulatory responses of pre-, mid-, and late-pubertal boys to exercise in dry heat. *Med Sci Sports Exerc* 24: 688-694, 1992.

158. Falk, B., R. Dotan, D.G. Liebermann, R. Regev, S.T. Dolphin, and O. Bar-Or. Birth weight and physical ability in 5-8 year old healthy children born prematurely. *Med Sci Sports Exerc* 29: 1124-1130, 1997.

159. Falk, B., and G. Mor. The effects of resistance and martial arts training in 6- to 8-year-old boys. *Pediatr Exerc Sci* 8: 48-56, 1996.

160. Falk, B., E. Sadres, N. Constantini, L. Zigel, R. Lidor, and A. Eliakim. The association between adiposity and the response to resistance training among pre- and early-pubertal boys. *J Pediatr Endocrinol Metab* 15: 597-606, 2002.

161. Falk, B., and G. Tenenbaum. The effectiveness of resistance training in children. A meta-analysis. *Sports Med* 22: 176-186, 1996.

162. Ferretti, G., M.V. Narici, T. Binzoni, L. Gariod, J.F. Le Bas, H. Reutenauer, and P. Cerretelli. Determinants of peak muscle power: effects of age and physical conditioning. *Eur J Appl Physiol Occup Physiol* 68: 111-115, 1994.

163. Fixler, D.E., W.P. Laird, R. Browne, V. Fitzgerald, S. Wilson, and R. Vance. Response of hypertensive adolescents to dynamic and isometric exercise stress. *Pediatrics* 64: 579-583, 1979.

164. Fournier, M., J. Ricci, A.W. Taylor, R.J. Ferguson, R.R. Montpetit, and B.R. Chaitman. Skeletal muscle adaptation in adolescent boys: sprint and endurance training and detraining. *Med Sci Sports Exerc* 14: 453-456, 1982.

165. Freedson, P.S., T.B. Gilliam, S.P. Sady, and V.L. Katch. Transient $\dot{V}O_2$ characteristics in children at the onset of steady-rate exercise. *Res Q Exerc Sport* 52: 167-173, 1983.

166. Friman, G. Effect of clinical bed rest for seven days on physical performance. *Acta Med Scand* 205: 389-393, 1979.

167. Froberg, K., and O. Lammert. Development of muscle strength during childhood. In Bar-Or, O., ed., *The child and adolescent athlete*. London: Blackwell Scientific, 1996, 25-41.

168. Frost, G., O. Bar-Or, J. Dowling, and K. Dyson. Explaining differences in the metabolic cost and efficiency of treadmill locomotion in children. *J Sports Sci* 20: 451-461, 2002.

169. Frost, G., O. Bar-Or, J. Dowling, and C. White. Habituation of children to treadmill walking and running: metabolic and kinematic criteria. *Pediatr Exerc Sci* 7: 162-175, 1995.

170. Frost, G., J. Dowling, O. Bar-Or, and K. Dyson. Ability of mechanical power estimations to explain differences in metabolic cost of walking and running among children. *Gait and Posture* 5: 120-127, 1997.

171. Frost, G., J. Dowling, K. Dyson, and O. Bar-Or. Cocontraction in three age groups of children during treadmill locomotion. *J Electromyog Kinesiol* 7: 179-186, 1997.

172. Fuchimoto, T., and M. Kaneko. Force, velocity and power relationships in different age groups. *Res J Phys Educ* 25: 274-279, 1981.

173. Fuchs, R.K., J.J. Bauer, and C.M. Snow. Jumping improves hip and lumbar spine bone mass in prepubescent children: a randomized controlled trial. *J Bone Miner Res* 16: 148-156, 2001.

174. Fukunaga, T., K. Funato, and S. Ikegawa. The effects of resistance training on muscle area and strength in prepubescent age. *Ann Physiol Anthropol* 11: 357-364, 1992.

175. Funato, K., T. Fukunaga, T. Asami, and S. Ikeda. Strength training for prepubescent boys and girls. *Proceed Dept Sports Sci, College Arts Sci, University of Tokyo* 21: 9-19, 1987.

176. Gadhoke, S., and N.L. Jones. The responses to exercise in boys aged 9-15 years. *Clin Sci* 37: 789-801, 1969.

177. Gaisl, G., and J. Buchberger. The significance of stress acidosis in judging the physical working capacity of boys aged 11 to 15. In Lavallée, H. and R.J. Shephard, eds., *Frontiers of activity and child health*. Quebec: Pélican, 1977, 161-168.

178. Gatch, W., and R. Byrd. Endurance training and cardiovascular function in 9- and 10-year-old boys. *Arch Phys Med Rehab* 60: 574-577, 1979.

179. Geenen, D.L., T.B Gilliam, C. Steffens, D. Crowley, and A. Rosenthal. The effects of exercise on cardiac structure and function in prepubescent children [abstract]. *Med Sci Sports Exerc* 13: 93, 1981.

180. Gilliam, T.B., and P.S. Freedson. Effects of a 12-week school physical fitness program on peak $\dot{V}O_2$ body composition and blood lipids in 7 to 9 year old children. *Int J Sports Med* 1: 73-78, 1980.

181. Gilliam, T.B., P.S. Freedson, D.L. Geenen, and B. Shahraray. Physical activity patterns determined by heart rate monitoring in 6- to 7-year-old children. *Med Sci Sports Exerc* 13: 65-67, 1981.

182. Gilliam, T.B., S. Sady, W. Thorland, and A.L. Weltman. Comparison of peak performance measures in children ages 6 to 8, 9 to 10 and 11 to 13 years. *Res Q Am Assoc Health Phys Educ* 48: 695-702, 1977.

183. Girandola, R.N., R.A. Wiswell, F. Frisch, and K. Wood. Metabolic differences during exercise in pre- and post-pubescent girls [abstract]. *Med Sci Sports Exerc* 13: 110, 1981.

184. Glaser, R.M., L.L. Laubach, M.N. Sawka, and A.G. Suryaprasad. Exercise stress, fitness evaluation and training of wheelchair users. In Leon, A.S. and G.J. Amundson, eds., *Proceedings First International Conference on Lifestyle Health*. Minneapolis: University of Minnesota, 1978.

185. Godfrey, S. The growth and development of the cardio-pulmonary responses to exercise. In Davis, J.A. and J. Dobbing, eds., *Scientific foundations of paediatrics*. Philadelphia: Saunders, 1974, 271-280.

186. Godfrey, S., C.T.M. Davies, E. Wozniak, and C.A. Barnes. Cardio-respiratory response to exercise in normal children. *Clin Sci* 40: 419-431, 1971.

187. Gratas-Delamarche, A., J. Mercier, M. Ramonatxo, J. Dassonville, and C. Préfaut. Ventilatory response of prepubertal boys and adults to carbon dioxide at rest and during exercise. *Eur J Appl Physiol* 66: 25-30, 1993.

188. Gratas, A., J. Dassonville, J. Beillot, and P. Rochcongar. Ventilatory and occlusion-pressure responses to exercise in trained and untrained children. *Eur J Appl Physiol* 57: 591-596, 1988.

189. Greenleaf, J.R., and S. Kozlowski. Reduction in peak oxygen uptake after prolonged bed rest. *Med Sci Sports Exerc* 14: 477-480, 1982.

190. Grodjinovsky, A., and O. Bar-Or. Influence of added physical education hours upon anaerobic capacity, adiposity, and grip strength in 12-to 13-year-old children enrolled in a sports class. In Ilmarinen, Y., and I. Välimäki, eds., *Pediatric Work Physiology X*. Berlin: Springer-Verlag, 1983.

191. Grodjinovsky, A., O. Bar-Or, R. Dotan, and O. Inbar. Training effect on the anaerobic performance of children as measured by the Wingate anaerobic test. In Borg, K. and B.O. Eriksson, eds., *Children and exercise IX*. Baltimore: University Park Press, 1980, 139-145.

192. Gutin, B., S. Owens, F. Treiber, and G. Mensah. Exercise haemodynamics and left ventricular parameters in children. In Armstrong, N., B. Kirby, and J. Welsman, eds., *Children and exercise XIX*. London: Spon, 1997, 460-464.

193. Guy, J.A., and L.J. Micheli. Strength training for children and adolescents. *J Am Acad Orthop Surg* 9: 29-36, 2001.

194. Haapasalo, H., P. Kannus, H. Sievanen, M. Pasanen, K. Uusi-Rasi, A. Heinonen, P. Oja, and I. Vuori. Effect of long-term unilateral activity on bone mineral density of female junior tennis players. *J Bone Miner Res* 13: 310-319, 1998.

195. Hagberg, J.M., A.A. Ehsani, D. Goldring, A. Hernandez, D.R. Sinacore, and J.O. Holloszy. Effect of weight training on blood pressure and hemodynamics in hypertensive adolescents. *J Pediatr* 104: 147-151, 1984.

196. Hagberg, J.M., D. Goldring, A.A. Ehsani, G.W. Heath, A. Hernandez, K. Schechtman, and J.O. Holloszy. Effect of exercise training on the blood pressure and hemodynamic features of hypertensive adolescents. *Am J Cardiol* 52: 763-768, 1983.

197. Hakkinen, K., A. Mero, and H. Kauhanen. Specificity of endurance, sprint and strength training on physical performance capacity in young athletes. *J Sports Med Phys Fitness* 29: 27-35, 1989.

198. Hamilton, P., and G.M. Andrew. Influence of growth and athletic training on heart and lung functions. *Eur J Appl Physiol* 36: 27-38, 1976.

199. Hausdorff, J.M., P.L. Purdon, C.K. Peng, Z. Ladin, J.Y. Wei, and A.L. Goldberger. Fractal dynamics of human gait: stability of long-range correlations in stride interval fluctuations. *J Appl Physiol* 80: 1448-1457, 1996.

200. Hausdorff, J.M., L. Zemany, C.-K. Peng, and A.L. Goldberger. Maturation of gait dynamics: stride-to-stride variability and its temporal organization in children. *J App Physiol* 86: 1040-1047, 1999.

201. Hebestreit, H., and O. Bar-Or. Influence of climate on heart rate in children: comparison between intermittent and continuous exercise. *Eur J Appl Physiol* 78: 7-12, 1998.

202. Hebestreit, H., and O. Bar-Or. Exercise and the child born prematurely. *Sports Med* 31: 591-599, 2001.

203. Hebestreit, H., O. Bar-Or, C. McKinty, M. Riddell, and P. Zehr. Climate-related corrections for improved estimation of energy expenditure from heart rate in children. *J Appl Physiol* 79: 47-54, 1995.

204. Hebestreit, H., S. Kriemler, R.L. Hughson, and O. Bar-Or. Kinetics of oxygen uptake at the onset of exercise in boys and men. *J Appl Physiol* 85: 1833-1841, 1998.

205. Hebestreit, H., F. Meyer, H. Htay, G.J. Heigenhauser, and O. Bar-Or. Plasma metabolites, volume and electrolytes following 30-s high-intensity exercise in boys and men. *Eur J Appl Physiol Occup Physiol* 72: 563-569, 1996.

206. Hebestreit, H., K. Mimura, and O. Bar-Or. Recovery of muscle power after high-intensity short-term exercise: comparison between boys and men. *J Appl Physiol* 74: 2875-2880, 1993.

207. Heeboll-Nielsen, K. Muscle strength of boys and girls, 1981 compared to 1956. *Scand J Sports Sci* 4: 37-43, 1982.

208. Hermansen, L., and S. Oseid. Direct and indirect estimation of maximal oxygen uptake in prepubertal boys. *Acta Paediatr Scand Suppl* 217: 18-23, 1971.

209. Hettinger, T.L. *Physiology of strength*. Springfield, IL: Charles C Thomas, 1961.

210. Hewitt, D. Sib resemblance in bone, muscle and fat measurement of the human calf. *Ann Hum Genetics* 22: 213-221, 1957.

211. Hollmann, W., and C. Bouchard. The relationship between chronological and biological ages and spiroergometric values, heart volume, anthropometric data and muscle strength in 8- to 18-year-old youth [in German]. *Zeitschrift Fur Kreislaufforschung* 59: 160-176, 1970.

212. Hollmann, W. The preventive and rehabilitative role of sport in internal medicine. *Das Medizinische Prisma* 2: 4-28, 1978.

213. Holmes, J.R., and G.J. Alderink. Isokinetic strength characteristics of the quadriceps femoris and hamstring muscles in high school students. *Phys Ther* 64: 914-918, 1984.

214. Ienna, T.M., and D.C. McKenzie. The asthmatic athlete: metabolic and ventilatory responses to exercise with and without pre-exercise medication. *Int J Sports Med* 18: 142-148, 1997.

215. Ignico, A.A., and A.D. Mahon. The effects of a physical fitness program on low-fit children. *Res Q Exerc Sport* 66: 85-90, 1995.

216. Ikai, M. Training of muscular endurance related to age. *FIEP Bull* 3-4: 19-27, 1969.

217. Ikai, M., M. Shindo, and M. Miyamura. Aerobic work capacity of Japanese people. *Res J Phys Educ* 14: 137-142, 1970.

218. Inbar, O., and O. Bar-Or. The effects of intermittent warm-up on 7-9 year-old boys. *Eur J Appl Physiol* 34: 81-89, 1975.

219. Inbar, O., and O. Bar-Or. Relationships of anaerobic and aerobic arm and leg capacities to swimming performance of 8-12 year old children. In Shephard, R.J. and H. Lavallée, eds., *Frontiers of activity and child health*. Quebec: Pélican, 1977, 283-292.

220. Inbar, O., and O. Bar-Or. Anaerobic characteristics in male children and adolescents. *Med Sci Sports Exerc* 18: 264-269, 1986.

221. Inbar, O., and O. Bar-Or. Anaerobic characteristics in male children and adolescents. *Med Sci Sports Exerc* 18: 264-26, 1986.

222. Inbar, O., O. Bar-Or, and J.S. Skinner. *The Wingate anaerobic test: development and application*. Champaign, IL: Human Kinetics, 1996.

223. Issekutz, B. Jr., J.J. Blizzard, N.C. Birkhead, and K. Rodahl. Effect of prolonged bed rest on urinary calcium output. *J Appl Physiol* 21: 1013-1020, 1966.

224. Jacobs, I., B. Sjödin, and B. Svane. Muscle fiber type, cross-sectional area and strength in boys after 4 years' endurance training [abstract]. *Med Sci Sports Exerc* 14: 123, 1982.

225. Jacobs, I., P.A. Tesch, O. Bar-Or, J. Karlsson, and R. Dotan. Lactate in human skeletal muscle after 10 and 30 s of supramaximal exercise. *J Appl Physiol* 55: 365-367, 1983.

226. Jacobs, J.C., H.M. Dick, and J.A. Downey. Weight bearing as a treatment for damaged hips in juvenile rheumatoid arthritis. *N Eng J Med* 205: 409, 1981.

227. James, F.W., S. Kaplan, C.J. Glueck, J.Y. Tsay, M.J.S. Knight, and C.J. Sarwar. Responses of normal children and young adults to controlled bicycle exercise. *Circulation* 61: 902-912, 1980.

228. Jones, D.A., and J.M. Round. Strength and muscle growth. In Armstrong, N. and W. van Mechelen, eds., *Paediatric exercise science and medicine*. Oxford: Oxford University Press, 2000, 133-142.

229. Jones, D.A., J.M. Round, R.H. Edwards, S.R. Grindwood, and P.S. Tofts. Size and composition of the calf and quadriceps muscles in Duchenne muscular dystrophy. A tomographic and histochemical study. *J Neurol Sci* 60: 307-322, 1983.

230. Jones, H.E. *A developmental study of static dynamometric strength*. Berkeley: University of California Press, 1949, 34-52.

231. Kanaley, J.A., and R.A. Boileau. The onset of the anaerobic threshold at three stages of physical maturity. *J Sports Med Phys Fitness* 28: 367-374, 1988.

232. Kanehisa, H., S. Ikegawa, N. Tsunoda, and T. Fukunaga. Strength and cross-sectional area of knee extensor muscles in children. *Eur J Appl Physiol* 68: 402-405, 1994.

233. Kaneko, M., A. Ito, T. Fuchimoto, and J. Toyooka. Mechanical work and efficiency of young distance runners during level running. In Morecki, A., ed., *Biomechanics VII*. Baltimore: University Park Press, 1981, 234-240.

234. Karlsson, J. Muscle ATP, CP and lactate in submaximal and maximal exercise. In Pernow, B. and B. Saltin, eds., *Muscle metabolism during exercise*. New York: Plenum Press, 1971, 383-393.

235. Katzmarzyk, P.T., R.M. Malina, T.M.K. Song, and C. Bouchard. Television viewing, physical activity, and health-related fitness of youth in Quebec Family Study. *J Adolesc Health* 23: 318-325, 1998.

236. Keller, H., B.V. Ayub, S. Saigal, and O. Bar-Or. Neuromotor ability in 5- to 7-year-old children with very low or extremely low birthweight. *Dev Med Child Neurol* 40: 661-666, 1998.

237. Keller, H., O. Bar-Or, S. Kriemler, B.V. Ayub, and S. Saigal. Anaerobic performance in 5- to 7-yr-old children of low birthweight. *Med Sci Sports Exerc* 32: 278-283, 2000.

238. Kemper, H.C.G. Skeletal development in children and adolescents and the relations with physical activity. In Froberg, K., O. Lammert, H. St. Hansen, and C.J.R. Blimkie, eds., *Exercise and fitness—benefits and risks*. Odense: Odense University Press, 1997, 139-160.

239. Kemper, H.C.G., and C. Niemeyer. The importance of a physically active lifestyle during youth for peak bone mass. In Blimkie, C.J.R. and O. Bar-Or, eds., *New horizons in pediatric exercise science*. Champaign, IL: Human Kinetics, 1995, 77-95.

240. Kemper, H.C.G., R. Verschuur, and L. de Mey. Longitudinal changes of aerobic fitness in youth age 12 to 13. *Pediatr Exerc Sci* 1: 257-270, 1989.

241. Kemper, H.C.G., ed. *The Amsterdam Growth Study: a longitudinal analysis of health, fitness, and lifestyle*. Champaign, IL: Human Kinetics, 1995, 1-278.

242. Killian, K.J. Dyspnea and leg effort during incremental cycle ergometry. *Am Rev Respir Dis* 145: 1339-1345, 1992.

243. Kindermann, V.W., G. Huber, and J. Keul. Anaerobic capacity in children and adolescents in comparison with adults [in German]. *Sportarzt und Sportmedizin* 6: 112-115, 1975.

244. Kindermann, W., J. Keul, and M. Lehmann. Prolonged exercise in adolescents, metabolic and cardiovascular changes. *Fortsch Med* 97: 659-665, 1979.

244a. Klentrou, P., J. Hay, and M. Plyley. Habitual physical activity levels and health outcomes of Ontario youth. *Eur J Appl Physiol* 89: 460-465, 2003.

245. Kobayashi, K., K. Kitamura, M. Miura, H. Sodeyama, Y. Murase, M. Miyashita, and H. Matsui. Aerobic power as related to body growth and training in

Japanese boys: a longitudinal study. *J Appl Physiol: Respir Environ Exerc Physiol* 44: 666-672, 1978.

246. Koch, G. Muscle blood flow after ischemic work and during bicycle ergometer work in boys aged 12 years. *Acta Paediatr Belg* (Suppl) 28: 29-39, 1974.

247. Koch, G. Muscle blood flow in prepubertal boys—effect of growth combined with intensive physical training. In Borms, J. and M. Hebbelinck, eds., *Pediatric work physiology*. Basel: Karger, 1978, 39-46.

248. Koch, G. Aerobic power, lung dimensions, ventilatory capacity and muscle blood flow in 12- to 16-year-old boys with high physical activity. In Berg, K. and B.O. Eriksson, eds., *Children and exercise X.* Baltimore: University Park Press, 1980, 99-108.

249. Koch, G., and B.O. Eriksson. Effect of physical training on anatomical R-L shunt at rest and pulmonary diffusing capacity during near-maximal exercise in boys 11-13 years old. *Scand J Clin Lab Invest* 31: 95-105, 1973.

250. Koch, G., and L. Rocker. Plasma volume and intravascular protein masses in trained boys and fit young men. *J Appl Physiol: Respir Environ Exerc Physiol* 43: 1085-1088, 1977.

251. Komi, P.V., and S.L. Karppi. Genetic and environmental variation in perceived exertion and heart rate during bicycle ergometer work. In Borg, G., ed., *Physical work and effort*. Stockholm: Pergamon Press, 1977, 91-99.

252. Komi, P.V., V. Klissouras, and E. Karvinen. Genetic variation in neuromuscular performance. *Int Zeitschrift fur Angewandte Physiol* 31: 289-330, 1973.

253. Kondo, I., M. Fukuda, M. Souma, A. Oda, M. Iwata, S. Saito, and O. Bar-Or. A method of measuring finger force with the Kin-Com dynamometer. *J Hand Ther* 12: 263-268, 1999.

254. Krahenbuhl, G.S., R.P. Pangrazi, and E.A. Chomokos. Aerobic responses of young boys to submaximal running. *Res Q Am Assoc Health Phys Educ* 50: 413-421, 1979.

255. Kriemler, S., H. Hebestreit, and O. Bar-Or. Temperature-related overestimation of energy expenditure, based on heart-rate monitoring in obese boys. *Eur J Appl Physiol* 87: 245-250, 2002.

256. Krotkiewski, M., J.G. Kral, and J. Karlsson. Effects of castration and testosterone substitution on body composition and muscle metabolism in rats. *Acta Physiol Scand* 109: 233-237, 1980.

257. Kulangara, R.J., and W.B. Strong. Exercise stress testing in children. *Compr Ther* 5: 51-61, 1979.

258. Kuno, S., H. Takahashi, K. Fujimoto, H. Akima, M. Miyamura, I. Nemoto, Y. Itai, and S. Katsura. Muscle metabolism during exercise using phosphorus-31 nuclear magnetic resonance spectroscopy in adolescents. *Eur J Appl Physiol* 70: 301-304, 1995.

259. Kurowski, T.T. Anaerobic power of children from ages 9 through 15 years. 1977. MSc thesis. Florida State University.

260. Lakhera, S.C., T.C. Kain, and P. Bandopadhyay. Changes in lung function during adolescence in athletes and non-athletes. *J Sports Med Phys Fitness* 34: 258-262, 1994.

261. Lamb, K.L. Children's ratings of effort during cycle ergometry: an examination of the validity of two effort rating scales. *Pediatr Exerc Sci* 7: 407-421, 1995.

262. Lamb, K.L. Exercise regulation during cycle ergometry using the CERT and RPE scales. *Pediatr Exerc Sci* 8: 337-350, 1996.

263. Lamb, K.L., and R.G. Eston. Effort perception in children. *Sports Med* 23: 139-148, 1997.

264. Lefevre, J., G. Beunen, G. Steens, A. Claessens, and R. Renson. Motor performance during adolescence and age thirty as related to age at peak height velocity. *Ann Hum Biol* 17: 423-434, 1990.

265. Lengyel, M., and I. Gyarfas. The importance of echocardiography in the assessment of left ventricular hypertrophy in trained and untrained schoolchildren. *Acta Cardiol* 34: 63-69, 1979.

266. Lind, A.R. Cardiovascular responses to static exercise. *Circulation* 41: 173-176, 1970.

267. Lowry, A. The development of a pictorial scale to assess perceived exertion in school-children. 1995. University College, Chester, England. Thesis.

268. Lumley, M.A., B.G. Melamed, and L.A. Abeles. Predicting children's presurgical anxiety and subsequent behavior changes. *J Pediatr Psychol* 18: 481-497, 1993.

269. Lussier, L., and E.R. Buskirk. Effects of an endurance training regimen on assessment of work capacity in prepubertal children. *Ann N Y Acad Sci* 301: 734-747, 1977.

270. Lynn, R., and D.L. Morgan. Decline running produces more sarcomeres in rat vastus intermedius muscle fibers than does incline running. *J Appl Physiol* 77: 1439-1444, 1994.

271. MacDougall, J.D., R.S. McKelvie, D.E. Moroz, D.G. Sale, N. McCartney, and F. Buick. Factors affecting blood pressure during heavy weight lifting and static contractions. *J Appl Physiol* 73: 1590-1597, 1992.

272. MacDougall, J.D., P.D. Roche, O. Bar-Or, and J.R. Moroz. Oxygen cost of running in children of different ages: maximal aerobic power of Canadian schoolchildren [abstract]. *Can J Appl Sport Sci* 4: 237, 1979.

273. MacDougall, J.D., P.D. Roche, O. Bar-Or, and J.R. Moroz. Maximal aerobic capacity of Canadian school children: prediction based on age-related oxygen cost of running. *Int J Sports Med* 4: 194-198, 1983.

274. MacDougall, J.D., D. Tuxen, D.G. Sale, J.R. Moroz, and J.R. Sutton. Arterial blood pressure response to heavy resistance exercise. *J Appl Physiol* 58: 785-790, 1985.

275. Máček, M., V. Seliger, and J. Vávra. Physical fitness of the Czechoslovak population between the ages of 12 and 55 years. Oxygen consumption and pulse oxygen. *Physiol Bohemoslov* 228: 75-82, 1979.

276. Máček, M., and J. Vávra. Cardiopulmonary and metabolic changes during exercise in children 6-14 years old. *J Appl Physiol* 30: 202-204, 1971.

277. Máček, M., and J. Vávra. Prolonged exercise in children. *Acta Paediatr Belg* 28: 13-18, 1974.

278. Máček, M., and J. Vávra. The adjustment of oxygen uptake at the onset of exercise: a comparison between prepubertal boys and young adults. *Int J Sports Med* 1: 75-77, 1980.

279. Máček, M., J. Vávra, and J. Novosadová. Prolonged exercise in prepubertal boys. I. Cardiovascular and metabolic adjustment. *Eur J Appl Physiol* 35: 291-298, 1976.

280. Máček, M., J. Vávra, and J. Novosadová. Prolonged exercise in prepubertal boys. I. Cardiovascular and metabolic adjustment. *Eur J Appl Physiol* 35: 291-298, 1976.

281. Máček, M., J. Vávra, and J. Novosadová. Prolonged exercise in prepubertal boys. II. Changes in plasma volume and in some blood constituents. *Eur J Appl Physiol* 35: 299-303, 1976.

282. MacKelvie, K.J., K.M. Khan, and H.A. McKay. Is there a critical period for bone response to weight-bearing exercise in children and adolescents? A systematic review. *Br J Sports Med* 36: 250-257, 2002.

283. Macková, J., M. Sturmová, and M. Máček. Prolonged exercise in prepubertal boys in warm and cold environments. In Ilmarinen, Y. and I. Välimäki, eds., *Pediatric work physiology X.* Berlin: Springer-Verlag, 1983.

284. Mahon, A.D. Assessment of perceived exertion during exercise in children. In Welsman, J., N. Armstrong, and B. Kirby, eds., *Children and exercise XIX.* Vol. II. Exeter: Washington Singer Press, 1997, 25-39.

285. Mahon, A.D. Exercise training. In Armstrong, N. and W. van Mechelen, eds., *Paediatric exercise science and medicine.* Oxford: Oxford University Press, 2000, 201-222.

286. Mahon, A.D., G.E. Duncan, C.A. Howe, and P. Del Corral. Blood lactate and perceived exertion relative to ventilatory threshold: boys versus men. *Med Sci Sports Exerc* 29: 1332-1337, 1997.

287. Mahon, A.D., J.A. Gay, and K.Q. Stolen. Differentiated ratings of perceived exertion at ventilatory threshold in children and adults. *Eur J Appl Physiol Occup Physiol* 78: 115-120, 1998.

288. Mahon, A.D., K.Q. Stolen, and J.A. Gay. Differentiated perceived exertion during submaximal exercise in children and adults. *Pediatr Exerc Sci* 13: 145-153, 2001.

289. Mahon, A.D., and P. Vaccaro. Ventilatory threshold and $\dot{V}O_2$ max changes in children following endurance training. *Med Sci Sports Exerc* 21: 425-431, 1989.

290. Mahon, A.D., and P. Vaccaro. Cardiovascular adaptations in 8 to 12 year-old boys following a 14 week running program. *Can J Appl Physiol* 19(2): 139-150, 1994.

291. Malina, R.M., C. Bouchard, and O. Bar-Or. *Growth, maturation, and physical activity.* Champaign, IL: Human Kinetics, 2004.

292. Malina, R.M., and W.H. Mueller. Genetic and environmental influences on the strength and motor performance of Philadelphia school children. *Hum Biol* 53: 163-179, 1981.

293. Maliszewski, A.F., and P.S. Freedson. Is running economy different between adults and children? *Pediatr Exerc Sci* 8: 351-360, 1996.

294. Mandigout, S., A. Melin, L. Fauchier, L.D. N'Guyen, D. Courteix, and P. Obert. Physical training increases heart rate variability in healthy prepubertal children. *Eur J Clin Invest* 32: 479-487, 2002.

295. Mandigout, S., A. Melin, A.M. Lecoq, D. Courteix, and P. Obert. Effect of two aerobic training regimens on the cardiorespiratory response of prepubertal boys and girls. *Acta Paediatr* 91: 403-408, 2002.

296. Marcus, R. Exercise and the regulation of bone mass. *Arch Intern Med* 149: 2170-2171, 1989.

297. Maresh, N.M. Measurements from roentgenograms: heart size; long bone length, bone, muscle and fat widths; skeletal maturation. In McCammon, R.W., ed., *Human growth and development.* Springfield, IL: Charles C Thomas, 1970, 155-200.

298. Martin, J.C., and R.M. Malina. Developmental variations in anaerobic performance associated with age and sex. In Van Praagh, E., ed., *Pediatric anaerobic performance.* Champaign, IL: Human Kinetics, 1998, 45-64.

299. Martin, W.H. III, E.F. Coyle, S.A. Bloomfield, and A.A. Ehsani. Effects of physical deconditioning after intense endurance training on left ventricular dimensions and stroke volume. *J Am Coll Cardiol* 7: 982-989, 1986.

300. Martinez, L.R., and E.M. Haymes. Substrate utilization during treadmill running in prepubertal girls and women. *Med Sci Sports Exerc* 24: 975-983, 1992.

301. Massicotte, D.R., and R.B.J. MacNab. Cardiorespiratory adaptions to training at specified intensities in children. *Med Sci Sports* 6: 242-246, 1974.

302. Matějková, J., Z. Kopřivová, and Z. Plachetá. Changes in acid-base balance after maximal exercise.

In Placheta, Z., ed., *Youth and physical activity*. Brno: J.E. Purkyne University, 1980, 191-199.

303. Mayers, N., and B. Gutin. Physiological characteristics of elite prepubertal cross-country runners. *Med Sci Sports* 11: 172-176, 1979.

304. McManus, A.M., N. Armstrong, and C.A. Williams. Effect of training on the aerobic power and anaerobic performance of prepubertal girls. *Acta Paediatr* 86: 456-459, 1997.

305. Mercier, J., A. Varray, M. Ramonatxo, B. Mercier, and C. Préfaut. Influence of anthropometric characteristics on changes in maximal exercise ventilation and breathing pattern during growth in boys. *Eur J Appl Physiol* 63: 235-241, 1991.

306. Mero, A., H. Kauhanen, E. Peltola, and T. Vuorimaa. Changes in endurance, strength and speed capacity of different prepubescent athletic groups during one year of training. *J Hum Mov Studies* 14: 219-239, 1988.

307. Mersch, F., and H. Stoboy. Strength training and muscle hypertrophy in children. In Oseid, S. and K.-H. Carlsen, eds., *Children and exercise XIII*. Champaign, IL: Human Kinetics, 1989, 165-182.

308. Michael, E., J. Evert, and K. Jeffers. Physiological changes of teenage girls during five months of detraining. *Med Sci Sports* 4: 214-218, 1972.

309. Miyamura, M., and Y. Honda. Maximal cardiac output related to sex and age. *Jpn J Physiol* 23: 645-656, 1973.

310. Miyashita, M., K. Onodera, and I. Tabata. How Borg's RPE-scale has been applied to Japanese. In Borg, G. and D. Ottoson, eds., *The perception of exertion in physical work*. Stockholm: Wenner-Gren, 1986, 27-34.

311. Mocellin, R. Youth and sport [in German]. *Med Klin* 70: 1443-1457, 1975.

312. Mocellin, R., and J. Rutenfranz. Investigations of the physical working capacity of obese children. *Acta Paediatr Belg* (Suppl) 217: 77-79, 1971.

313. Mocellin, R., W. Sebening, and K. Buhlmeyer. Cardiac output and oxygen uptake at rest and during submaximal loads in 8 to 14 year-old boys. *Z Kinderheilkd* 114: 323-339, 1973.

314. Mocellin, R., and U. Wasmund. Investigations on the influence of a running-training programme on the cardiovascular and motor performance capacity in 53 boys and girls of a second and third primary school class. In Bar-Or, O., ed., *Pediatric work physiology*. Natanya: Wingate Institute, 1973, 279-285.

315. Morgan, D.W. Economy of locomotion. In Armstrong, N. and W. van Mechelen, eds., *Paediatric exercise science and medicine*. Oxford: Oxford University Press, 2000, 183-190.

316. Morgan, D.W., P.E. Martin, and G.S. Krahenbuhl. Factors affecting running economy. *Sports Med* 7: 310-330, 1989.

317. Morgan, D.W., W. Tseh, J.L. Caputo, I.S. Craig, D.J. Keefer, and P.E. Martin. Sex differences in running economy of young children. *Pediatr Exerc Sci* 11: 122-128, 1999.

318. Morris, F.L., G.A. Naughton, J.L. Gibbs, J.S. Carlson, and J.D. Wark. Prospective ten-month exercise intervention in premenarcheal girls: positive effects on bone and lean mass. *J Bone Miner Res* 12: 1453-1462, 1997.

319. Morse, M., F.W. Schultz, and D.E. Cassels. Relation of age to physiological responses of the older boy (10-17 years) to exercise. *J Appl Physiol* 1: 683-709, 1949.

320. Mosher, R.E., E.C. Rhodes, H.A. Wenger, and B. Filsinger. Interval training: the effects of a 12-week programme on elite, pre-pubertal male soccer players [abstract]. *J Sports Med Phys Fitness* 25: 5-9, 1985.

321. Nagle, F.J., J. Hagberg, and S. Kamei. Maximal O_2 uptake of boys and girls ages 14 to 17. *Eur J Appl Physiol* 36: 75-80, 1977.

322. Naughton, G., J. Carlson, and I. Fairweather. Determining the variability of performance on Wingate anaerobic tests in children aged 6-12 years. *Int J Sports Med* 13: 512-517, 1992.

323. Neder, J.A., and B.J. Whipp. Kinetics of pulmonary oxygen uptake. In Armstrong, N. and W. van Mechelen, eds., *Paediatric exercise science and medicine*. Oxford: Oxford University Press, 2000, 191-200.

323a. Nemet, D., Y. Oh, H.S. Kim, M. Hill, and D.M. Cooper. Effect of intense exercise on inflammatory cytokines and growth mediators in adolescent boys. *Pediatrics* 110: 681-689, 2002.

323b. Nemet, D., C.M. Rose-Gottron, P.J. Mills, and D.M. Cooper. Effect of water polo practice on cytokines, growth mediators, and leukocytes in girls. *Med Sci Sports Exerc* 35: 356-363, 2003.

324. Nielsen, B., K. Nielsen, M. Behrendt Hansen, and E. Asmussen. Training of 'functional muscular strength' in girls 7-19 years old. In Berg, K. and B. Eriksson, eds., *Paediatric work physiology IX*. Baltimore: University Park Press, 1980, 69-78.

324a. Nieman, D.C. Exercise immunology: practical applications. *Int J Sports Med* 18 (supplement 1): 91-100, 1997.

325. Noble, B., and R. Robertson. *Perceived exertion*. Champaign, IL: Human Kinetics, 1996.

326. Nordstrom, P., U. Pattersson, and R. Lorentzon. Type of physical activity, muscle strength, and pubertal stage as determinants of bone mineral density and bone area in adolescent boys. *J Bone Miner Res* 13: 1141-1148, 1998.

327. Nottin, S., A. Vinet, F. Stecken, L.D. N'Guyen, F. Ounissi, A.M. Lecoq, and P. Obert. Central and peripheral cardiovascular adaptations to exercise in endurance-trained children. *Acta Physiol Scand* 175: 85-92, 2002.

328. Nystad, W., S. Oseid, and E.B. Mellbye. Physical education for asthmatic children: the relationship between changes in heart rate, perceived exertion, and motivation for participation. In Oseid, S. and K. Carlsen, eds., *Children and exercise XIII*. Champaign, IL: Human Kinetics, 1989, 359-377.

329. Obert, P., D. Courteix, A.M. Lecoq, and P. Guenon. Effect of long-term intense swimming training on the upper body peak oxygen uptake of prepubertal girls. *Eur J Appl Physiol Occup Physiol* 73: 136-143, 1996.

330. Obert, P., M. Mandigout, A. Vinet, and D. Courteix. Effect of a 13-week aerobic training programme on the maximal power developed during a force-velocity test in prepubertal boys and girls. *Int J Sports Med* 22: 442-446, 2001.

331. Obert, P., S. Mandigout, A. Vinet, L.D. N'Guyen, F. Stecken, and D. Courteix. Effect of aerobic training and detraining on left ventricular dimensions and diastolic function in prepubertal boys and girls. *Int J Sports Med* 22: 90-96, 2001.

332. Oertel, G. Morphometric analysis of normal skeletal muscles in infancy, childhood and adolescence. An autopsy study. *J Neurol Sci* 88: 303-313, 1988.

333. Ohuchi, H., H. Suzuki, K. Yasuda, Y. Arakaki, S. Echigo, and T. Kamiya. Heart rate recovery after exercise and cardiac autonomic nervous activity in children. *Pediatr Res* 47: 329-335, 2000.

334. Oyen, E.M., S. Schuster, and P.E. Brode. Dynamic exercise echocardiography of the left ventricle in physically trained children compared to untrained healthy children. *Int J Cardiol* 29: 29-33, 1990.

334a. Osterback, L., and Y. Qvarnberg. A prospective study of respiratory infections in 12-year-old children actively engaged in sports. *Acta Physiol Scand* 76: 944-949, 1987.

335. Ozmun, J.C., A.E. Mikesky, and P.R. Surburg. Neuromuscular adaptations following prepubescent strength training. *Med Sci Sports Exerc* 26: 510-514, 1994.

336. Pařizková, J. *Body fat and physical fitness*. The Hague: Nijhoff, 1977.

337. Parker, D.F., J.M. Round, P. Sacco, and D.A. Jones. A cross-sectional survey of upper and lower limb strength in boys and girls during childhood and adolescence. *Ann Hum Biol* 17: 199-211, 1990.

338. Pate, R.R., and D.S. Ward. Endurance exercise trainability in children and youth. *Adv Sports Med Fit* 3: 37-55, 1990.

339. Pate, R.R., and D.S. Ward. Endurance trainability of children and youths. In Bar-Or, O., ed., *The child and adolescent athlete*. Oxford: Blackwell Scientific, 1996, 130-152.

340. Pauer, M., V. Sobolová, V. Zelenka, S. Bartunková, Z. Bartunek, V. Seliger, L. Havlicková, J. Heller, and J. Melichna. The effects of intensified school physical education on physical fitness [abstract]. *Physiol Bohemoslov* 29: 272, 1980.

341. Payne, V.G., and J.R. Morrow Jr. Exercise and VO$_2$ max in children: a meta-analysis. *Res Q Exerc Sport* 64: 305-313, 1993.

342. Payne, V.G., J.R. Morrow Jr., L. Johnson, and S.N. Dalton. Resistance training in children and youth: a meta-analysis. *Res Q Exerc Sport* 68: 80-88, 1997.

342a. Pedersen, B.K., and L. Hoffman-Goetz. Exercise and the immune system: regulation, integration, and adaptation. *Physiol Rev* 80: 1055-1081, 2000.

343. Pels, A.E., T.B. Gilliam, P.S. Freedson, D.L. Geenen, and S.E. MacConnie. Heart rate response to bicycle ergometer exercise in children ages 6 to 7 years. *Med Sci Sport Exerc* 13: 299-302, 1981.

343a. Perez, C.J., D. Nemet, P.J. Mills, T.P. Scheet, M.G. Ziegler, and D.M. Cooper. Effects of laboratory versus field exercise on leukocyte subsets and cell adhesion molecule expression in children. *Eur J Appl Physiol* 86: 34-39, 2001.

344. Petray, C.K., and G.S. Krahenbuhl. Running training, instruction on running technique, and running economy in 10-year old males. *Res Q Exerc Sport* 56: 251-255, 1985.

345. Pfeiffer, R.D., and R.S. Francis. Effects of strength training on muscle development in prepubescent, pubescent, and postpubescent males. *Phys Sportsmed* 14: 134-143, 1986.

346. Pikosky, M., A. Faigenbaum, W. Westcott, and N. Rodriguez. Effects of resistance training on protein utilization in healthy children. *Med Sci Sports Exerc* 34: 820-827, 2002.

347. Polgar, G., and V. Promadhat. *Pulmonary function testing in children: techniques and standards*. Philadelphia: Saunders, 1971.

348. Quanjer, P.H., G.J. Borsboom, B. Brunekreff, M. Zach, G. Forche, J.E. Cotes, J. Sanchis, and P. Paoletti. Spirometric reference values for white European children and adolescents: Polgar revisited. *Pediatr Pulmonol* 19: 135-142, 1995.

349. Ramos, E., W.R. Frontera, A. Llopart, and D. Feliciano. Muscle strength and hormonal levels in adolescents: gender related differences. *Int J Sports Med* 19: 526-531, 1998.

350. Ramsay, J.A., C.J.R. Blimkie, K. Smith, S. Garner, J.D. MacDougall, and D.G. Sale. Strength training effects in prepubescent boys. *Med Sci Sports Exerc* 22: 605-614, 1990.

351. Ratel, S., P. Duché, A. Hennegrave, E. Van Praagh, and M. Bedu. Acid-base balance during repeated cycling sprints in boys and men. *J Appl Physiol* 92: 479-485, 2002.

352. Reybrouck, T., M. Weymans, H. Stijns, J. Knops, and L. Van der Hauwaert. Ventilatory anaerobic threshold in healthy children: age and sex differences. *Eur J Appl Physiol* 54: 278-284, 1985.

353. Rians, C.B., A. Weltman, B.R. Cahill, C.A. Janney, S.R. Tippett, and F.I. Katch. Strength training for prepubescent males: is it safe? *Am J Sports Med* 15: 483-489, 1987.

354. Riddell, M.C., O. Bar-Or, H.P. Schwarcz, and G.J. Heigenhauser. Substrate utilization in boys during exercise with [^{13}C]-glucose ingestion. *Eur J Appl Physiol* 83: 441-448, 2000.

355. Riopel, D.A., A.B. Taylor, and A.R. Hohn. Blood pressure, heart rate, pressure rate product and electrocardiographic changes in healthy children during treadmill exercise. *Am J Cardiol* 44: 697-704, 1979.

356. Robertson, R.J., J.E. Falkel, A.L. Drash, A.M. Swank, K.F. Metz, S.A. Spunqun, and J.R. Le Boeuf. Effects of blood pH on peripheral and central signals of perceived exertion. *Med Sci Sports Exerc* 18: 114-122, 1986.

357. Robertson, R.J., F.L. Goss, J.A. Bell, C.B. Dixon, K.I. Gallagher, K.M. Lagally, J.M. Timmer, K.L. Abt, J.D. Gallagher, and T. Thompkins. Self-regulated cycling using the Children's OMNI Scale of Perceived Exertion. *Med Sci Sports Exerc* 34: 1168-1175, 2002.

358. Robertson, R.J., F.L. Goss, N.F. Boer, J.A. Peoples, A.J. Foreman, I.M. Dabayebeh, N.B. Millich, G. Balasekaran, S.E. Riechman, J.D. Gallagher, and T. Thompkins. Children's OMNI scale of perceived exertion: mixed gender and race validation. *Med Sci Sports Exerc* 32: 452-458, 2000.

359. Robertson, R.J., and B.J. Noble. Perception of physical exertion: methods, mediators, and applications. *Exerc Sport Sci Rev* 25: 407-453, 1997.

360. Robinson, S. Experimental studies of physical fitness in relation to age. *Z Angew Physiol Einschl Arbeitphysiol* 10: 251-323, 1938.

361. Rode, A. Some factors influencing the fitness of a small Eskimo community. 1972. University of Toronto. Dissertation.

362. Rode, A., O. Bar-Or, and R.J. Shephard. Cardiac output and oxygen conductance. A comparison of Canadian Eskimo and city dwellers. In Bar-Or, O., ed., Pediatric Work Physiology, proceedings of the IVth International Symposium. Natanya: Wingate Institute, 1973, 45-57.

363. Rohmert, W. Rechts-link Vergleich bei isometrischem armmuskeltraining mit verschiedenem Trainingsreiz bei achtjahringen Kindern. *Int Z Angew Physiol* 26: 363-393, 1968.

364. Rost, R., H. Gerhardus, and W. Hollmann. Untersuchungen zur Frage eines Trainingseffektes bei Kindern im Alter von 8-10 Jahren in kardiopulmonalen System. Unpublished. 1978.

365. Rotstein, A., R. Dotan, O. Bar-Or, and G. Tenenbaum. Effect of training on anaerobic threshold, maximal aerobic power and anaerobic performance of preadolescent boys. *Int J Sports Med* 7: 281-286, 1986.

366. Round, J.M., D.A. Jones, J.W. Honour, and A.M. Nevill. Hormonal factors in the development of differences in strength between boys and girls during adolescence: a longitudinal study. *Ann Hum Biol* 26: 49-62, 1999.

367. Round, J.M., A.M. Nevill, J. Honour, and D.A. Jones. Testosterone and the developmental difference in strength between boys and girls [abstract]. *J Physiol* 491: 79, 1996.

368. Rowland, T. On being a metabolic nonspecialist. *Pediatr Exerc Sci* 14: 315-320, 2002.

369. Rowland, T., D. Goff, L. Martel, and L. Ferrone. Influence of cardiac functional capacity on gender differences in maximal oxygen uptake in children. *Chest* 117: 629-635, 2000.

370. Rowland, T., D. Goff, B. Popowski, P. DeLuca, and L. Ferrone. Cardiac responses to exercise in child distance runners. *Int J Sports Med* 19: 385-390, 1998.

371. Rowland, T., G. Kline, D. Goff, L. Martel, and L. Ferrone. Physiological determinants of maximal aerobic power in healthy 12-year-old boys. *Pediatr Exerc Sci* 11: 317-326, 1999.

372. Rowland, T., B. Popowski, and L. Ferrone. Cardiac responses to maximal upright cycle exercise in healthy boys and men. *Med Sci Sports Exerc* 29: 1146-1151, 1997.

373. Rowland, T., J. Potts, T. Potts, J. Son-Hing, G. Harbison, and G. Sandor. Cardiovascular responses to exercise in children and adolescents with myocardial dysfunction. *Am Heart J* 137: 126-133, 1999.

374. Rowland, T.W. Aerobic response to endurance training in prepubescent children: a critical analysis. *Med Sci Sports Exerc* 17: 493-497, 1985.

375. Rowland, T.W. *Developmental exercise physiology.* Champaign, IL: Human Kinetics, 1996, 1-268.

376. Rowland, T.W. Cardiovascular function. In Armstrong, N. and W. van Mechelen, eds., *Paediatric exercise science and medicine.* Oxford: Oxford University Press, 2000, 163-171.

377. Rowland, T.W. Pulmonary function. In Armstrong, N. and W. van Mechelen, eds., *Paediatric exercise science and medicine.* Oxford: Oxford University Press, 2000, 153-161.

378. Rowland, T.W., J.A. Auchinachie, T.J. Keenan, and G.M. Green. Physiologic responses to treadmill running in adult and prepubertal males. *Int J Sports Med* 8: 292-297, 1987.

379. Rowland, T.W., J.A. Auchinachie, T.J. Keenan, and G.M. Green. Submaximal aerobic running economy and treadmill performance in prepubertal boys. *Int J Sports Med* 9: 201-204, 1988.

380. Rowland, T.W., and A. Boyajian. Aerobic response to endurance exercise training in children. *Pediatrics* 96: 654-658, 1995.

381. Rowland, T.W., and L.N. Cunningham. Oxygen uptake plateau during maximal treadmill exercise in children. *Chest* 101: 485-489, 1992.

382. Rowland, T.W., and L.N. Cunningham. Development of ventilatory responses to exercise in normal white children. A longitudinal study. *Chest* 111: 327-332, 1997.

383. Rowland, T.W., and G.M. Green. Physiological responses to treadmill exercise in females: adult-child differences. *Med Sci Sports Exerc* 20: 474-478, 1988.

384. Rowland, T.W., L. Martel, P. Vanderburgh, T. Manos, and N. Charkoudian. The influence of short-term aerobic training on blood lipids in healthy 10-12 year old children. *Int J Sports Med* 17: 487-492, 1996.

385. Rowland, T.W., and T.A. Rimany. Physiological responses to prolonged exercise in premenarcheal and adult females. *Pediatr Exerc Sci* 7: 183-191, 1995.

386. Rutenfranz, J. *Entwicklung und Beurteilung der korperlichen Leistungsfahigkeit bei Kindern und Jugendlichen*. Basel: Karger, 1964.

387. Rutenfranz, J., K.L. Andersen, V. Seliger, F. Klimmer, J. Ilmarinen, M. Ruppel, and H. Kylian. Exercise ventilation during the growth spurt period: comparison between two European countries. *Eur J Pediatr* 136: 135-142, 1981.

388. Rutenfranz, J., F. Klimt, J. Ilmarinen, and H. Kylian. Blood lactate concentration during triangular and stepwise loadings on the bicycle ergometer. In Lavallee, H. and R.J. Shephard, eds., *Frontiers of activity and child health*. Quebec: Pelican, 1977, 179-187.

389. Sady, S. Transient oxygen uptake and heart rate responses at the onset of relative endurance exercise in prepubertal boys and adult men. *Int J Sports Med* 2: 240-244, 1981.

390. Sailors, M., and K. Berg. Comparison of responses to weight training in pubescent boys and men. *J Sports Med* 27: 30-37, 1987.

391. Sale, D., and J.D. MacDougall. Specificity in strength training: a review for the coach and athlete. *Can J Appl Sport Sci* 5: 87-92, 1980.

392. Sale, D.G. Strength training in children. In Gisolfi, G.V. and D.R. Lamb, eds., *Perspectives in exercise science and sports medicine*. Vol 2: *Youth, exercise and sport*. Indianapolis: Benchmark Press, 1989, 165-222.

393. Sallis, J.F., M.J. Buono, and P.S. Freedson. Bias in estimating caloric expenditure from physical activity in children: implications for epidemiological studies. *Sports Med* 11: 203-209, 1991.

394. Sallis, J.F. ed. Physical activity guidelines for adolescents. *Pediatr Exerc Sci* 6: 299-463, 1994.

395. Saltin, B. Physiological effects of physical conditioning. *Med Sci Sports* 1: 50-56, 1969.

396. Sargeant, A.J. Effect of muscle temperature on leg extension force and short-term power output in humans. *Eur J Appl Phyiol* 56: 693-698, 1987.

397. Sargeant, A.J. Short-term muscle power in children and adolescents. In Bar-Or, O., ed., *Biological issues*. Champaign, IL: Human Kinetics, 1989, 41-63.

398. Sargeant, A.J., and P. Dolan. Effect of prior exercise on maximal short-term power output in humans. *J Appl Physiol* 63: 1475-1480, 1987.

399. Sargeant, A.J., P. Dolan, and A. Thorne. Effects of supplemental physical activity on body composition, aerobic and anaerobic power in 13-year old boys. In Binkhorst, R.A., H.C.G. Kemper, and W.H.M. Saris, eds., *Children and exercise XI*. Champaign, IL: Human Kinetics, 1985, 135-139.

399a. Scheet, T.P., P.J. Mills, M.G. Ziegler, J. Stoppani, and D.M. Cooper. Effect of exercise on cytokine and growth mediators in prepubertal boys. *Pediatr Res* 46: 429-434, 1999.

400. Schepens, B., P.A. Willems, and G.A. Cavagna. The mechanics of running in children. *J Physiol* 509: 927-940, 1999.

401. Schieken, R.M., and D.F. Geller. The cardiovascular effect of isometric exercise in children [abstract]. *Clin Res* 26: 741A, 1978.

402. Schmucker, B., J. Dordel, and W. Hollmann. The aerobic power of 7-9 year old children participating in a rehabilitative sport programme. In Lavallee, H. and R.J. Shephard, eds., *Frontiers of activity and child health*. Quebec: Pelican, 1977, 307-312.

403. Schmucker, B., and W. Hollmann. The aerobic capacity of trained athletes from 6 to 7 years of age on. *Acta Paediatr Belg* (Suppl) 28: 92-101, 1974.

404. Schneider, E.C., and C.B. Crampton. The respiratory responses of pre-adolescent boys to muscular activity. *Am J Physiol* 117: 577-586, 1936.

405. Schrader, P.C., P.H. Quanjer, G. Borsboom, and M.E. Wise. Evaluating lung function and anthropometric growth data in a longitudinal study on adolescents. *Hum Biol* 56: 365-381, 1984.

406. Seliger, V. The influence of sports training on the efficiency of juniors. *Int Z Angew Physiol Einschl Arbeitsphysiol* 26: 309-322, 1968.

407. Seliger, V. Physical fitness of Czechoslovak children at 12 and 15 years of age. IBP results of investigation 1968-1969. *Acta Univ Carol Gymnica* 5: 6-196, 1970.

408. Sewall, L., and L.J. Micheli. Strength training for children. *J Pediatr Orthop* 6: 143-146, 1986.

409. Shephard, R.J., C. Allen, O. Bar-Or, C.T. Davies, S. Degre, R. Hedman, K. Ishii, M. Kaneko, J.R. Lacour, P.E. di Prampero, and V. Seliger. The working capacity of Toronto schoolchildren. II. *Can Med Assoc J* 100: 705-714, 1969.

410. Shephard, R.J., C. Allen, O. Bar-Or, C.T.M. Davies, S. Degre, R. Hedman, K. Ishii, M. Kaneko, J.R. Lacour, P.E. di Prampero, and V. Seliger. The working capacity of Toronto schoolchildren. Part I, II. *Can Med Assoc J* 100: 560-567, 705-714, 1969.

411. Shephard, R.J., and O. Bar-Or. Alveolar ventilation in near maximum exercise. Data on pre-adolescent children and young adults. *Med Sci Sports* 2: 83-92, 1970.

412. Shephard, R.J., H. Lavallée, J.C. Jequier, C. Beaucage, and R. Labarre. Influence of added activity classes upon the working capacity of Quebec schoolchildren. In Lavallee, H. and R.J. Shephard, eds., *Frontiers of activity and child health*. Quebec: Pélican, 1977, 237-245.

412a. Shore, S., and R.J. Shephard. Immune responses to exercise and training: a comparison between children and adults. *Pediatr Exerc Sci* 10: 210-226, 1998.

412b. Shore, S., and R.J. Shephard. Immune responses to exercise in children treated for cancer. *J Sports Med Phys Fitness* 39: 240-243, 1999.

413. Siegel, J.A., D.N. Camaione, and T.G. Manfredi. The effects of upper body resistance on prepubescent boys. *Pediatr Exerc Sci* 1: 145-154, 1989.

414. Simoneau, J.-A., and C. Bouchard. Genetic determinism of fiber type proportion in human skeletal muscle. *FASEB J* 9: 1091-1095, 1995.

415. Simoneau, J.-A., and C. Bouchard. The effects of genetic variation on anaerobic performance. In Van Praagh, E., ed., *Paediatric anaerobic performance*. Champaign, IL: Human Kinetics, 1998, 5-21.

416. Simoneau, J.-A., G. Lortie, M.R. Boulay, M. Marcotte, M.-C. Thibault, and C. Bouchard. Inheritance of human skeletal muscle and anaerobic capacity adaptation to high-intensity intermittent training. *Int J Sports Med* 7: 167-171, 1986.

417. Sjödin, B., and J. Svedenhag. Oxygen uptake during running as related to body mass in circumpubertal boys: a longitudinal study. *Eur J Appl Physiol Occup Physiol* 65: 150-157, 1992.

418. Skinner, J.S., O. Bar-Or, V. Bergsteinová, C.W. Bell, D. Royer, and E.R. Buskirk. Comparison of continuous and intermittent tests for determining maximal oxygen uptake in children. *Acta Paediatr Scand Suppl* 217: 24-28, 1971.

419. Slemenda, C.W., J.Z. Miller, S.L. Hui, T.K. Reister, and C.C. Conrad Jr. Role of physical activity in the development of skeletal mass in children. *J Bone Miner Res* 6: 1227-1233, 1991.

420. Sprynarová, S. Development of the relationship between aerobic capacity and the circulatory and respiratory reaction to moderate activity in boys 11-13 years old. *Physiol Bohemoslov* 15: 253-264, 1966.

421. Sprynarová, S., J. Pářizková, and I. Irinová. Development of the functional capacity and body composition of boy and girl swimmers aged 12-15 years.

In Borms, J. and M. Hebbelinck, eds., *Pediatric work physiology*. Basel: Karger, 1978.

422. Sprynarová, S., and R. Reisenauer. Body dimensions and physiological indicators of physical fitness during adolescence. In Shephard, R.J. and H. Lavallée, eds., *Physical fitness assessment*. Springfield, IL: Charles C Thomas, 1978, 32-37.

423. Stewart, K.J., and B. Gutin. Effects of physical training on cardiorespiratory fitness in children. *Res Q Am Assoc Health Phys Educ* 47: 110-120, 1976.

424. Strong, W.B., M.D. Miller, M. Striplin, and M. Salehbhai. Blood pressure response to isometric and dynamic exercise in healthy black children. *Am J Dis Child* 132: 587-591, 1978.

425. Suei, K., L. McGillis, R. Calvert, and O. Bar-Or. Relationships among muscle endurance, explosiveness and strength in circumpubertal boys. *Pediatr Exerc Sci* 10: 48-56, 1998.

426. Sundberg, S., and R. Elovainio. Cardiorespiratory function in competitive endurance runners aged 12-16 years compared with ordinary boys. *Acta Paediatr Scand* 71: 987-992, 1982.

427. Sunnegardh, J., L.E. Bratteby, L.O. Nordesjo, and B. Nordgren. Isometric and isokinetic muscle strength, anthropometry and physical activity in 8 and 13 year old Swedish children. *Eur J Appl Physiol* 58: 291-297, 1988.

428. Tanner, J.M. *Growth at adolescence*. Oxford: Blackwell, 1962, 69-87.

429. Tanner, S.M. Weighing the risks. Strength training for children and adolescents. *Phys Sportsmed* 21: 105-116, 1993.

430. Taylor, C.M., M.E.R. Bal, M.W. Lamb, and G. McLeod. Mechanical efficiency in cycling of boys seven to fifteen years of age. *J Appl Physiol* 2: 563-570, 1950.

431. Thorén, C. Effects of beta-adrenergic blockade on heart rate and blood lactate in children during maximal and submaximal exercise. *Acta Paediatr Scand Suppl* 177: 123-125, 1967.

432. Thorén, C.A.R., and K. Asano. Functional capacity and cardiac function in 10-year-old boys and girls with high and low running performance. In Ilmarinen, J. and I. Välimäki, eds., *Children and sport*. Bedin, Italy: Springer-Verlag, 1984, 182-188.

433. Thorstensson, A. Effects of moderate external loading on the aerobic demand of submaximal running in men and 10 year-old boys. *Eur J Appl Physiol Occup Physiol* 55: 569-574, 1986.

434. Timmons, B.W., and O. Bar-Or. RPE during prolonged exercise with and without carbohydrate ingestion in boys and men. *Med Sci Sports Exerc*, submitted 2003.

434a. Timmons, B.W., M.A. Tarnopolsky, and O. Bar-Or. Immune responses to strenuous exercise and carbohydrate intake in boys and men. *Pediatr Res* In Press: 2004.

435. Timmons, B.W., O. Bar-Or, and M.C. Riddell. Oxidation rate of exogenous carbohydrate during exercise is higher in boys than in men. *J Appl Physiol* 94: 278-284, 2003.

436. Turley, K.R., D.E. Martin, E.D. Marvin, and K.S. Cowley. Heart rate and blood pressure responses to static handgrip exercise of different intensities: reliability and adult versus child differences. *Pediatr Exerc Sci* 14: 45-55, 2002.

437. Turley, K.R., and J.H. Wilmore. Cardiovascular responses to submaximal exercise. *Med Sci Sports Exerc* 29: 824-832, 1997.

438. Turley, K.R., and J.H. Wilmore. Cardiovascular responses to submaximal exercise in 7- to 9-yr-old boys and girls. *Med Sci Sports Exerc* 29: 824-832, 1997.

439. Turley, K.R., and J.H. Wilmore. Cardiovascular responses to treadmill and cycle ergometer exercise in children and adults. *J Appl Physiol* 83: 948-957, 1997.

440. Turner, M.C., E.J. Ruley, K.M. Buckley, and orthopedic immobilization. *J Pediatr* 95: 989-992, 1979.

441. Ulbrich, J. Individual variants of physical fitness in boys from the age of 11 up to maturity and their selection for sports activities. *Medicina Dello Sports* 24: 118-136, 1971.

442. Unnithan, V.B., and R.G. Eston. Stride frequency and submaximal treadmill running economy in adults and children. *Pediatr Exerc Sci* 2: 149-155, 1990.

443. Unnithan, V.B., L.A. Murray, J.A. Timmons, D. Buchanan, and J.Y. Paton. Reproducibility of cardiorespiratory measurements during submaximal and maximal running in children. *Br J Sports Med* 29: 66-71, 1995.

444. Utter, A.C., R.J. Robertson, D.C. Nieman, and J. Kang. Children's OMNI Scale of Perceived Exertion: walking/running evaluation. *Med Sci Sports Exerc* 34: 139-144, 2002.

445. Vaccaro, P., and D.H. Clarke. Cardiorespiratory alterations in 9-11 year old children following a season of competitive swimming. *Med Sci Sports* 10: 204-207, 1978.

446. Vaccaro, P., and A. Mahon. Cardiorespiratory responses to endurance training in children. *Sports Med* 4: 352-363, 1987.

447. Vaccaro, P., and A. Poffenbarger. Resting and exercise respiratory function in young female child runners. *J Sports Med Phys Fitness* 22: 102-107, 1982.

448. Van Huss, W.D., K.E. Stephens, P. Vogel, D. Anderson, T. Kurowski, J.A. Janes, and C. Fitzgerald. Physiological and perceptual responses of elite age group distance runners during progressive intermittent work to exhaustion. In Weiss, M.R. and D. Gould, eds., *Sport for children and youth.* Champaign, IL: Human Kinetics, 1986, 239-246.

449. van Mechelen, W., H. Hlobil, H.C. Kemper, W.J. Voorn, and H.R. de Jongh. Prevention of running injuries by warm-up, cool-down, and stretching exercises. *Am J Sports Med* 21: 711-719, 1993.

450. Van Praagh, E., ed. *Pediatric anaerobic performance.* Champaign, IL: Human Kinetics, 1998, 1-375.

450a. Van Praagh, E. Development of anaerobic function during childhood and adolescence. *Pediatric Exercise Science* 12:150-173, 2000.

451. Van Praagh, E., and E. Dore. Short-term muscle power during growth and maturation. *Sports Med* 32: 701-728, 2002.

452. Van Uytvanck, J and J. Vrijens. Experimentelle Untersuchungen uber Anpassungserscheinjungen von Adolezenten mit schwacher Konstitution bei Kurzfristiger, genau dosierter Arbeit. *Int Z Angew Physiol Einschl Arbeitphysiol* 25: 310-313, 1968.

453. Vanden Eynde, B., D. Van Gerven, D. Vienne, M. Vuylsteke-Wauters, and J. Ghesquière. Endurance fitness and peak height velocity in Belgian boys. In Oseid, S. and K.-H. Carlson, eds., *Children and exercise XIII.* Champaign, IL: Human Kinetics, 1989, 19-26.

454. Verhaaren, H.A., R.M. Schieken, P. Schwartz, M. Mosteller, D. Matthys, H. Maes, G. Beunen, R. Vlietinck, and R. Derom. Cardiovascular reactivity in isometric exercise and mental arithmetic in children. *J Appl Physiol* 76: 146-150, 1994.

455. Von Ditter, H., P. Nowacki, E. Simai, and U. Winkier. Das Verhalten des Saure-Basen-Haushalts nach erschopfender Belastung bei untrainierten und trainierten Jungen und Madchen im Vergleich zu Leistungssportlern. *Sportarzt und Sportmedizin* 28: 45-48, 1977.

456. Von Döbeln, W., and B.O. Eriksson. Physical training, maximal oxygen uptake and dimensions of the oxygen transporting and metabolizing organs in boys 11 to 13 years of age. *Acta Paediatr Scand* 61: 653-660, 1972.

457. Vrijens, J. Muscle strength development in the pre- and post-pubescent age. *Med Sport (Basel)* 11: 152-158, 1978.

458. Ward, D.S., and O. Bar-Or. Use of the Borg scale in exercise prescription for overweight youth. *Can J Sport Sci* 15: 120-125, 1990.

459. Ward, D.S., O. Bar-Or, P. Longmuir, and K. Smith. Use of rating of perceived exertion (RPE) to prescribe exercise intensity for wheelchair-bound children and adults. *Pediatr Exerc Sci* 7: 94-102, 1995.

460. Ward, D.S., J.D. Jackman, and F.D. Galiano. Exercise intensity reproduction: children versus adults. *Pediatr Exerc Sci* 3: 209-218, 1991.

461. Wasmund, U., P.E. Nowacki, H. Ditter, and F. Klimt. Radiotelemetric studies on heart rates of 10 year old boys and girls during a 3000m run on the sports field and on the treadmill. *Mschr Kinderheilkd* 126: 198-204, 1978.

462. Waters, R.L., H.J. Hislop, L. Thomas, and J. Campbell. Energy cost of walking in normal children and teenagers. *Dev Med Child Neurol* 25: 184-188, 1983.

463. Weber, G., W. Kartodihardjo, and V. Klissouras. Growth and physical training with reference to heredity. *J Appl Physiol* 40: 211-215, 1976.

464. Welsman, J.R., N. Armstrong, B.J. Kirby, R.J. Winsley, G. Parsons, and P. Sharpe. Exercise performance and magnetic resonance imaging-determined thigh muscle volume in children. *Eur J Appl Physiol* 76: 92-97, 1997.

465. Welsman, J.R., N. Armstrong, and S. Withers. Responses of young girls to two modes of aerobic training. *Br J Sports Med* 31: 139-142, 1997.

466. Weltman, A. Strength training in prepubertal children. In Laron, Z. and H.D. Rogol, eds., *Hormones and sport.* New York: Raven Press, 1989, 217-245.

467. Weltman, A., C. Janney, C.B. Rians, K. Strand, B. Berg, S. Tippitt, J. Wise, B.R. Cahill, and F.I. Katch. The effects of hydraulic resistance strength training in pre-pubertal males. *Med Sci Sports Exerc* 18: 629-638, 1986.

468. Weltman, A., S. Tippett, C. Janney, K. Strand, C. Rians, B.R. Cahill, and F.I. Katch. Measurement of isokinetic strength in prepubertal males. *J Orthop Sports Phys Ther* 9: 345-351, 1988.

469. Williams, J.G., R. Eston, and B. Furlong. CERT: a perceived exertion scale for young children. *Percept Mot Skills* 79: 1451-1458, 1994.

470. Williams, J.G., R.G. Eston, and C. Stretch. Use of rating of perceived exertion to control exercise intensity in children. *Pediatr Exerc Sci* 3: 21-27, 1991.

471. Williams, J.G., B. Furlong, C. MacKintosh, and T. Hockley. Rating and regulation of exercise intensity in young children [abstract]. *Med Sci Sports Exerc* 25: S8, 1993.

472. Williams, J.R., and N. Armstrong. The influence of age and sexual maturation on children's blood lactate responses to exercise. *Pediatr Exerc Sci* 3: 111-120, 1991.

473. Wilmore, J.H., and D.L. Costill. *Physiology of sport and exercise.* Champaign, IL: Human Kinetics, 1999.

474. Wilmore, J.H., and P.O. Sigerseth. Physical work capacity of young girls 7-13 years of age. *J Appl Physiol* 22: 923-928, 1967.

475. Winters, W.G., D.M. Leaman, and R.A. Anderson. The effect of exercise on intrinsic myocardial performance. *Circulation* 48: 50-55, 1973.

476. Wirth, A., E. Trager, K. Scheele, D. Meyes, K. Diehm, K. Reischle, and H. Weicker. Cardiopulmonary adjustment and metabolic response to maximal and submaximal physical exercise of boys and girls at different stages of maturity. *Eur J Appl Physiol* 29: 229-240, 1978.

477. Witzke, K.A., and C.M. Snow. Effects of plyometric jump training on bone mass in adolescent girls. *Med Sci Sports Exerc* 32: 1051-1057, 2000.

478. Yamaji, K., M. Miyashita, and R.J. Shephard. Relationship between heart rate and relative oxygen intake in male subjects aged 10 to 27 years. *J Hum Ergol* 7: 29-39, 1978.

479. Yoshida, T., I. Ishiko, and I. Muraoka. Effect of endurance training on cardiorespiratory functions of 5-year-old children. *Int J Sports Med* 1: 91-94, 1980.

480. Yoshizawa, S., H. Honda, N. Nakamura, K. Itoh, and N. Watanabe. Effect of an 18-month endurance run training program on maximal aerobic power in 4- to 6-year-old girls. *Pediatr Exerc Sci* 9: 1997.

481. Yoshizawa, S., H. Honda, M. Urushibara, and N. Nakamura. Effects of endurance run on circulorespiratory system in young children. *J Hum Ergol* 19: 41-52, 1990.

482. Yoshizawa, S., T. Ishizaki, and H. Honda. Physical fitness of children aged 5 and 6 years. *J Hum Ergol* 6: 41-51, 1977.

483. Zanconato, S., S. Buchthal, T.J. Barstow, and D.M. Cooper. 31P Magnetic resonance spectroscopy of leg muscle metabolism during exercise in children and adults. *J Appl Physiol* 74: 2214-2218, 1993.

484. Zanconato, S., D.M. Cooper, and Y. Armon. Oxygen cost and oxygen uptake dynamics and recovery with 1 min of exercise in children and adults. *J Appl Physiol* 71: 993-998, 1991.

Chapter 2

1. Andersen, L.B. Tracking of risk factors for coronary heart disease from adolescence to young adulthood with special emphasis on physical activity and fitness—a longitudinal study. *Dan Med Bull* 43: 407-418, 1996.

2. Andersen, L.B., and J. Haraldsdottir. Tracking of cardiovascular disease risk factors including maximal oxygen uptake and physical activity from late teenage to adulthood. An 8-year follow-up study. *J Int Med* 234: 309-315, 1993.

3. Atkins, S., G. Stratton, L. Dugdill, and T. Reilly. The free-living physical activity of schoolchildren: a longitudinal study. In Armstrong, N., B. Kirby, and J. Welsman, eds., *Children and exercise XIX.* London: Spon, 1997, 145-150.

4. Bailey, R.C., J. Olson, S.L. Pepper, J. Porszasz, T.J. Barstow, and D.M. Cooper. The level and tempo of children's physical activities: an observational study. *Med Sci Sports Exerc* 27: 1033-1041, 1995.

5. Bar-Or, O., J. Foreyt, C. Bouchard, K.D. Brownell, W.H. Dietz, E. Ravussin, A.D. Salbe, S. Schwenger, S. St. Jeor, and B. Torun. Physical activity, genetic, and nutritional considerations in childhood weight management. *Med Sci Sports Exerc* 30: 2-10, 1998.

6. Blanchard, S. Effects of ambient temperature and relative humidity on 8- and 10-year-old children involved in endurance activities working at 60% of maximal oxygen consumption. 1987. Dissertation. University of Maryland.

7. Bouchard, C., R.J. Shephard, T. Stephens, J.R. Sutton, and B.D. McPherson. *Exercise, fitness, and health: a consensus of current knowledge.* Champaign, IL: Human Kinetics, 1990.

8. Butcher, J. Socialization of adolescent girls into physical activity. *Adolescence* 18: 130-143, 1983.

9. Cale, L., and L. Almond. Children's activity levels: a review of studies conducted on British children. *Phys Educ Rev* 15: 111-118, 1992.

10. Centers for Disease Control and Prevention. Vigorous physical activity among high school students—United States, 1990. *MMWR* 41: 91-94, 1992.

11. Centers for Disease Control and Prevention. Participation in school physical education and selected dietary patterns among high school students—United States, 1991. *MMWR* 41: 698-703, 1999.

12. Centers for Disease Control and Prevention. Guidelines for school and community programs to promote lifelong physical activity among young people. *MMWR* 46(RR-64): 1-24, 1997.

13. Ilmarinen, J., and J. Rutenfranz. Longitudinal studies of the changes in habitual physical activity of school children and working adolescents. In Berg, K. and B.O. Eriksson, eds., *Children and exercise IX.* Baltimore: University Park Press, 1980, 149-159.

14. Jéquier, J.C., H. Lavallée, and M. Rajic. The longitudinal examination of growth and development: history and protocol of the Trois-Rivières regional study. In Lavallée, H. and R.J. Shephard, eds., *Frontiers of activity and child health.* Quebec: Pélican, 1977.

14a. Kemper, H.C.G., J.W.R. Twisk, L.L.J. Koppes, W. Van Mechelen, and G.B. Post. A 15-year physical activity pattern is positively related to aerobic fitness in young males and females. *Eur. J. Appl. Physiol.* 84: 395-402, 2001.

15. Kelder, S.H., C.L. Perry, K.I. Klepp, and L.L. Lytle. Longitudinal tracking of adolescent smoking, physical activity, and food choice behaviors. *Am J Pub Health* 84: 1121-1126, 1994.

16. Kemper, H.C.G., J. Snel, R. Verschurr, and L. Storm-van Essen. Tracking of health and risk indicators of cardiovascular disease from teenager to adult: Amsterdam Growth and Health Study. *Prev Med* 19: 642-655, 1990.

17. Klesges, R.C., L.H. Eck, C.L. Hanson, C.K. Haddock, and L.M. Klesges. Effects of obesity, social interactions, and physical environment on physical activity in preschoolers. *Health Psych* 9: 435-449, 1990.

18. Longmuir, P.E., and O. Bar-Or. Physical activity of children and adolescents with a disability: methodology and effects of age and gender. *Pediatr Exerc Sci* 6: 168-177, 1994.

19. Longmuir, P.E., and O. Bar-Or. Factors influencing the physical activity levels of youths with physical and sensory disabilities. *Adapted Phys Act Q* 17: 40-53, 2000.

20. Malina, R.M. Tracking of physical activity and physical fitness across the lifespan. *Res Q Exerc Sport* 67: 48-57, 1996.

21. Malina, R.M., and C. Bouchard. *Growth, maturation, and physical activity.* Champaign, IL: Human Kinetics, 1991, 1-501.

22. Mimura, K., H. Hebestreit, and O. Bar-Or. Activity and heart rate in preschool children of low and high motor ability: 24-hour profiles. *Med Sci Sports Exerc* 23: S12, 1991.

23. Morrow, J.R., and P.S. Freedson. Relationship between habitual physical activity and aerobic fitness in adolescents. *Pediatr Exerc Sci* 6: 315-329, 1994.

24. Pate, R.R., T. Baranowski, M. Dowda, and S.G. Trost. Tracking of physical activity in young children. *Med Sci Sports Exerc* 28: 92-96, 1996.

25. Pate, R.R., B. Long, and G. Heath. Descriptive epidemiology of physical activity in adolescents. *Pediatr Exerc Sci* 6: 434-447, 1994.

26. Perusse, L., C. Leblanc, and C. Bouchard. Familial resemblance in lifestyle components: results from the Canada Fitness Survey. *Can J Pub Health* 79: 201-205, 1988.

27. Perusse, L., A. Tremblay, C. Leblanc, and C. Bouchard. Genetic and environmental influences on habitual physical activity and exercise participation. *Am J Epidemiol* 129: 1012-1022, 1989.

28. Raitakari, O.T., K.V.K. Porkka, S. Taimela, R. Telama, L. Rasanen, and S.A. Viikari. Effects of persistent physical activity and inactivity on coronary risk factors in children and young adults. The cardiovascular risk in young Finns study. *Am J Epidemiol* 140: 195-205, 1994.

29. Ross, J.G., C.O. Dotson, and G.G. Gilbert. Are kids getting appropriate activity? The National Children and Youth Fitness Study. *JOPERD* 82: 40-43, 1985.

30. Ross, J.G., C.O. Dotson, G.G. Gilbert, and S.J. Katz. After physical education . . . physical activity outside of school physical education programs.

The National Children and Youth Fitness Study. *JOPERD* 56: 35-39, 1985.

31. Ross, J.G., C.O. Dotson, G.G. Gilbert, and S.J. Katz. What are kids doing in school physical education? The National Children and Youth Fitness Study. *JOPERD* 73: 31-34, 1985.

32. Ross, J.G., and R.R. Pate. The National Children and Youth Fitness Study II: a summary of findings. *JOPERD* 58: 51-56, 1987.

33. Ross, J.G., R.R. Pate, C.J. Casperson, C.L. Damberg, and M. Svilar. Home and community in children's exercise habits. *JOPERD* 58: 85-92, 1987.

34. Russell, S.J., C. Hyndford, and A. Beaulieu. *Active living for Canadian children and youth: a statistical profile.* Ottawa: Canadian Fitness and Lifestyle Research Institute, 1992.

35. Sallis, J.F., C.C. Berry, S.L. Broyles, T.L. McKenzie, and P.R. Nader. Variability and tracking of physical activity over 2 yr in young children. *Med Sci Sports Exerc* 27: 1042-1049, 1995.

36. Sallis, J.F., M.J. Buono, J.J. Roby, F.G. Micale, and J.A. Nelson. Seven-day recall and other physical activity self-reports in children and adolescents. *Med Sci Sports Exerc* 25: 99-108, 1993.

37. Saris, W.H. Habitual physical activity in children: methodology and findings in health and disease. *Med Sci Sports Exerc* 18: 253-263, 1986.

38. Saris, W.H.M., J.W.H. Elvers, M.A. Van't Hof, and R.A. Binkhorst. Changes in physical activity of children aged 6 to 12 years. In Rutenfranz, J., R. Mocellin, and F. Klimt, eds., *Children and exercise XII.* Champaign, IL: Human Kinetics, 1986, 121-130.

39. Scarr, S. Genetic factors in activity motivation. *Child Devel* 37: 663-671, 1966.

40. Shephard, R.J. *Fitness of a nation: lessons from the Canada Fitness Survey.* Basel: Karger, 1986.

41. Shephard, R.J., J.C. Jéquier, H. Lavallée, R. La Barre, and M. Rajic. Habitual physical activity: effects of sex, milieu, season, and required activity. *J Sports Med Phys Fitness* 20: 55-66, 1980.

42. Stephens, T., and C.L. Craig. *The well-being of Canadians: highlights of the 1988 Campbell's survey.* Ottawa: Canadian Fitness and Lifestyle Research Institute, 1990, 1-95.

43. Taylor, W.C., S.N. Blair, S.S. Cummings, C.C. Wun, and R.M. Malina. Childhood and adolescent physical activity patterns and adult physical activity. *Med Sci Sports Exerc* 31: 118-123, 1999.

44. Telama, R., J. Viikari, I. Valimaki, H. Siren-Tiusanen, H.K. Akerblom, M. Uhart, M. Dahl, E. Pesonen, P.L. Lahde, M. Pietikainen, and P. Suoninen. Atherosclerosis precursors in Finnish children and adolescents. X. Leisure-time physical activity. *Acta Paediatr Scand* 318: 169-180, 1985.

45. Torun, B., P.S.W. Davies, M.B.E. Livingstone, M. Paolisso, R. Sackett, and G.B. Spurr. Energy requirements and dietary energy recommendations for children and adolescents 1 to 18 years old. *Eur J Clin Nutr* 50: S37-S81, 1996.

46. Van Den Berg-Emons, H.J.G., W.H. Saris, D.C. de Barbanson, K.R. Westerterp, A. Huson, and M.A. van Baak. Daily physical activity of schoolchildren with spastic diplegia and of healthy control subjects. *J Pediatr* 127: 578-584, 1995.

47. Van Mechelen, W., and H.C.G. Kemper. Habitual physical activity in longitudinal perspective. In Kemper, H.C.G., ed., *The Amsterdam Growth Study: a longitudinal analysis of health, fitness, and lifestyle.* Champaign, IL: Human Kinetics, 1995, 135-158.

48. Verschuur, R., and H.C.G. Kemper. Habitual physical activity in Dutch teenagers measured by heart rate. In Binkhorst, R.A., H.C.G. Kemper, and W.H.M. Saris, eds., *Children and exercise XI.* Champaign, IL: Human Kinetics, 1985, 194-202.

49. Willerman, L. Activity level and hyperactivity in twins. *Child Devel* 44: 288-293, 1973.

Chapter 3

1. Allen, T.E., D.P. Smith, and D.K. Miller. Hemodynamic response to submaximal exercise after dehydration and rehydration in high school wrestlers. *Med Sci Sports* 9: 159-163, 1977.

2. Alter, B.P., E.E. Czapek, and R.D. Rowe. Sweating in congenital heart disease. *Pediatrics* 41: 123-129, 1968.

3. American Academy of Pediatrics Committee on Sports Medicine and Fitness. Climatic heat stress and the exercising child and adolescent. Position statement. *Pediatrics* 106: 158-159, 2000.

4. American College of Sports Medicine. Position stand on weight loss in wrestlers. *Med Sci Sports* 8: xi-xiii, 1997.

5. Araki, T., Y. Toda, K. Matsushita, and A. Tsujino. Age differences during sweating during muscular exercise. *Jpn J Fitness Sports Med* 28: 239-248, 1979.

6. Araki, T., J. Tsujita, K. Matsushita, and S. Hori. Thermoregulatory responses of prepubertal boys to heat and cold in relation to physical training. *J Hum Ergol* 9: 69-80, 1980.

7. Armstrong, L.E., W.C. Curtis, and R.W. Hubbard. Symptomatic hyponatremia during prolonged exercise in the heat. *Med Sci Sports Exerc* 25: 543-549, 1993.

8. Armstrong, L.E., J.P. De Luca, E.L. Christensen, and R.W. Hubbard. Mass-to-surface area index in a large cohort. *Am J Phys Anthropol* 83: 321-329, 1990.

9. Armstrong, L.E., and C.M. Maresh. Exercise-heat tolerance of children and adolescents. *Pediatr Exerc Sci* 7: 239-252, 1995.

10. Asmussen, E., and O. Boje. Body temperature and capacity for work. *Acta Physiol Scand* 10: 1-22, 1945.

11. Bar-Or, O. Distribution of heat activated sweat glands in 10- to 12-year-old Israeli girls and boys who differ in ethnic origin [in Hebrew]. Natanya: Wingate Institute, 1976, 1-28.

12. Bar-Or, O. Climate and the exercising child—a review. *Int J Sports Med* 1: 53-65, 1980.

13. Bar-Or, O. Physiologische Gesetzmassigkeiten sportlicher Aktivitat beim Kind. In Howald, H. and E. Han, eds., *Kinder im Leistungssport*. Basel: Birkhauser, 1982, 18-30.

14. Bar-Or, O. Thermoregulation, fluid and electrolytes in the young athlete. In Smith, N.J., ed., *Sports related health concerns in pediatrics*. Evanston, IL: American Academy of Pediatrics, 1983.

15. Bar-Or, O. Temperature regulation during exercise in children and adolescents. In Gisolfi, C.V. and D.R. Lamb, eds., *Perspectives in exercise science and sports medicine*. Vol. 2. *Youth, exercise, and sport*. Indianapolis: Benchmark Press, 1989, 335-367.

16. Bar-Or, O. Temperature regulation during exercise in children and adolescents. In Gisolfi, C.V. and D.R. Lamb, eds., *Perspectives in exercise science and sports medicine*. Vol. 2. *Youth, exercise, and sport*. Indianapolis: Benchmark Press, 1989, 335-367.

17. Bar-Or, O., C.J.R. Blimkie, J.A. Hay, J.D. MacDougall, D.S. Ward, and W.M. Wilson. Voluntary dehydration and heat intolerance in cystic fibrosis. *Lancet* 339: 696-699, 1992.

18. Bar-Or, O., R. Dotan, O. Inbar, A. Rothstein, and H. Zonder. Voluntary hypohydration in 10- to 12-year-old boys. *J Appl Physiol: Respir Environ Exerc Physiol* 48: 104-108, 1980.

19. Bar-Or, O., D. Harris, V. Bergstein, and E.R. Buskirk. Progressive hypohydration in subjects who vary in adiposity. *Isr J Med Sci* 12: 800-803, 1976.

20. Bar-Or, O., and O. Inbar. Relationship between perceptual and physiological changes during heat acclimatization in 8- to 10-year-old boys. In Lavallée, H. and R.J. Shephard, *Frontiers of activity and child health*. Quebec: Pélican, 1977, 205-214.

21. Bar-Or, O., H.M. Lundegren, and E.R. Buskirk. Heat tolerance of exercising obese and lean women. *J Appl Physiol* 26: 403-409, 1969.

22. Bar-Or, O., H.M. Lundegren, L.I. Magnusson, and E.R. Buskirk. Distribution of heat-activated sweat glands in obese and lean men and women. *Hum Biol* 40: 235-249, 1968.

23. Barcenas, C., H.P. Hoeffler, and J.T. Lie. Obesity, football, dog days and siriasis: a deadly combination. *Am Heart J* 92: 237-244, 1976.

24. Bergh, U., B. Ekblom, I. Holmer, and L. Gullstrand. Body temperature response to a long distance swimming race. In Eriksson, B. and B. Furberg, *Swimming medicine IV*. Baltimore: University Park Press, 1978, 342-344.

25. Blanchard, S. Effects of ambient temperature and relative humidity on 8- and 10-year-old children involved in endurance activities working at 60% of maximal oxygen consumption. 1987. Dissertation. University of Maryland.

26. Bosco, J.S., R.L. Terjung, and J.E. Greenleaf. Effects of progressive hypohydration on maximal isometric muscular strength. *J Sports Med Phys Fitness* 8: 81-86, 1968.

27. Boulze, D., P. Montastruc, and M. Cabanac. Water intake, pleasure and water temperature in humans. *Physiol Behav* 30: 97-102, 1983.

28. Bridger, C.A., F.P. Ellis, and H.L. Taylor. Mortality in St. Louis, Missouri during heat waves in 1936, 1953, 1954, 1955 and 1966. *Environ Res* 12: 38-48, 1976.

29. Brooke, O.G. Thermal insulation in malnourished Jamaican children. *Arch Dis Child* 48: 901-905, 1973.

30. Brooke, O.G., and C.B. Salvosa. Response of malnourished babies to heat. *Arch Dis Child* 49: 123-127, 1974.

31. Burch, G.E., and N.P. DePasquale. *Hot climates, man and his heart*. Springfield, Illinois: Charles C. Thomas, 1962.

32. Buskirk, E.R., O. Bar-Or, and J. Kollias. Physiological effects of heat and cold. In Wilson, N.L., ed., *Obesity*. Philadelphia: Davis, 1969, 119-139.

33. Buskirk, E.R., H. Lundegren, and L. Magnusson. Heat acclimatization patterns in obese and lean individuals. *Ann NY Acad Sci* 131: 637-653, 1965.

34. Cardullo, H.M. Sustained summer heat and fever in infants. *J Pediatr* 35: 24-42, 1949.

35. Choremis, K., C. Danelatou, F. Maounis, B. Basti, P. Lapatsanis, and K. Kiossoglou. Paper chromatography for amino-acid in thirst fever. *Helv Paediatr Acta* 14: 44-53, 1959.

36. Claremont, A.D., D.L. Costill, W. Fink, and P. Van Handel. Heat tolerance following diuretic induced dehydration. *Med Sci Sports* 8: 239-243, 1976.

37. Convertino, V.A., L.E. Armstrong, E.F. Coyle, G.W. Mack, M.N. Sawka, L.C. Senay Jr., and W.M. Sherman. Exercise and fluid replacement. *Med Sci Sports Exerc* 28: i-vii, 1996.

38. Convertino, V.A., L.E. Armstrong, E.F. Coyle, G.W. Mack, M.N. Sawka, L.C. Senay, and W.M. Sherman. Exercise and fluid replacement. ACSM position stand. *Med Sci Sports Exerc* 28: 1-8, 1996.

39. Costill, D.L. Sweating: its composition and effects on body fluids. In Milvy, P., ed., *The long distance runner*. New York: Urizen Books, 1978, 290-303.

40. Costill, D.L., and J.M. Miller. Nutrition for endurance sports: carbohydrate and fluid balance. *Int J Sports Med* 1: 2-14, 1980.

41. Costill, D.L., and B. Saltin. Factors limiting gastric emptying during rest and exercise. *J Appl Physiol* 37: 679-683, 1974.

42. Danks, D.M., D.W. Webb, and J. Allen. Heat illness in infants and young children: a study of 47 cases. *Br Med J* 2: 287-293, 1962.

43. Davies, C.T.M. Thermal responses to exercise in children. *Erg* 24: 55-61, 1981.

44. Davies, C.T.M., L. Fohlin, and C. Thorén. Temperature regulation in anorexia nervosa. *J Physiol (London)* 268: 8P-9P, 1977.

45. Davies, C.T.M., L. Fohlin, and C. Thorén. Thermoregulation in anorexia patients. In Borms, J. and M. Hebbelinck, *Pediatric work physiology.* Basel: Karger, 1978, 96-101.

46. Delamarche, P., J. Bittel, J.R. Lacour, and R. Flandrois. Thermoregulation at rest and during exercise in prepubertal boys. *Eur J Appl Physiol* 60: 436-440, 1990.

47. Dill, D.B., F.G. Hall, and W. Van Beaumont. Sweat chloride concentration: sweat rate, metabolic rate, skin temperature, and age. *J Appl Physiol* 21: 99-106, 1966.

48. Docherty, D., J.D. Eckerson, and J.S. Hayward. Physique and thermoregulation in prepubertal males during exercise in a warm, humid environment. *Am J Phys Anthropol* 70: 19-23, 1986.

49. Dotan, R., and O. Bar-Or. Climatic heat stress and performance in the Wingate anaerobic test. *Eur J Appl Physiol* 44: 237-243, 1980.

50. Drinkwater, B.L., and S.M. Horvath. Heat tolerance and aging. *Med Sci Sports* 11: 49-55, 1979.

51. Drinkwater, B.L., I.C. Kupprat, J.E. Denton, J.L. Crist, and S.M. Horvath. Response of prepubertal girls and college women to work in the heat. *J Appl Physiol: Respir Environ Exerc Physiol* 43: 1046-1053, 1977.

52. Editorial. Dehydration and fat babies. *Br Med J* 1: 125, 1971.

53. Ellis, F.P. Mortality from heat illness and heat-aggravated illness in the United States. *Environ Res* 5: 1-58, 1972.

54. Falk, B., M. Bar-Eli, R. Dotan, M. Yaaron, Y. Weinstein, S. Epstein, B. Blumenstein, M. Einbinder, Y. Yarom, and G. Tenenbaum. Physiological and cognitive responses to cold exposure in 11-12-year-old boys. *Am J Hum Biol* 9: 39-49, 1997.

55. Falk, B., O. Bar-Or, R. Calvert, and J.D. MacDougall. Sweat gland response to exercise in the heat among pre-, mid-, and late-pubertal boys. *Med Sci Sports Exerc* 24: 313-319, 1992.

56. Falk, B., O. Bar-Or, R. Calvert, and J.D. MacDougall. Sweat gland response to exercise in the heat among pre-, mid-, and late-pubertal boys. *Med Sci Sports Exerc* 24: 313-319, 1992.

57. Falk, B., O. Bar-Or, and J.D. MacDougall. Aldosterone and prolactin response to exercise in the heat in circumpubertal boys. *J Appl Physiol* 71: 1741-1745, 1991.

58. Falk, B., O. Bar-Or, and J.D. MacDougall. Thermoregulatory responses of pre-, mid-; and late-pubertal boys to exercise in dry heat. *Med Sci Sports Exerc* 24: 688-694, 1992.

59. Falk, B., O. Bar-Or, J.D. MacDougall, C.H. Goldsmith, and L. McGillis. Longitudinal analysis of sweating response of pre-, mid-, and late-pubertal boys during exercise in the heat. *Am J Hum Biol* 4: 527-535, 1992.

60. Falk, B., O. Bar-Or, J.D. MacDougall, L. McGillis, R. Calvert, and F. Meyer. Sweat lactate in exercise in children and adolescents of varying physical maturity. *J Appl Physiol* 71: 1735-1740, 1991.

61. Fitzsimons, J.T. The physiological basis of thirst. *Kidney Int* 10: 3-11, 1976.

62. Fortney, S.M., V.A. Koivisto, P. Felig, and E.R. Nadel. Circulatory and temperature regulatory responses to exercise in a warm environment in insulin-dependent diabetics. *Yale J Biol Med* 54: 101-109, 1981.

63. Foster, K.G., E.N. Hey, and B. O'Connell. Sweat function in babies with defects of the central nervous system. *Dev Med Child Neurol* 20: 94, 1969.

64. Fox, E.L., D.K. Mathews, W.S. Kaufman, and R.W. Bowers. Effects of football equipment on thermal balance and energy cost during exercise. *Res Q Am Assoc Health Phys Educ* 37: 332-339, 1966.

65. Geist, M., and N. Barzilai. Dilutional hyponatremia and seizures following exercise [in Hebrew]. *Harefuah* 122: 420-421, 1992.

66. Gisolfi, C.V. Work-heat tolerance derived from interval training. *J Appl Physiol* 35: 349-354, 1973.

67. Gullestad, R. Temperature regulation in children during exercise. *Acta Paediatr Scand* 64: 257-263, 1975.

68. Hadland, D.G., J.F. Stock, and M.I. Hewitt. Heat and cold tolerance: relation to body weight. *Postgrad Med* 55: 75-80, 1974.

69. Hanson, P.G., and S.W. Zimmerman. Exertional heatstroke in novice runners. *JAMA* 242: 154-157, 1979.

70. Haymes, E.M., E.R. Buskirk, J.L. Hodgson, H.M. Lundegren, and W.C. Nicholas. Heat tolerance of exercising lean and heavy prepubertal girls. *J Appl Physiol* 36: 566-571, 1974.

71. Haymes, E.M., R.J. McCormick, and E.R. Buskirk. Heat tolerance of exercising lean and obese prepubertal boys. *J Appl Physiol* 39: 457-461, 1975.

72. Horswill, C.A. Applied physiology of amateur wrestling. *Sports Med* 14: 114-143, 1992.

73. Hubbard, R.W., P.C. Szlyk, and L.E. Armstrong. Influence of thirst and fluid palatability on fluid ingestion during exercise. In Gisolfi C.V., and D.R. Lamb, eds. *Perspectives in Exercise and Sports Medicine.* Fluid hemostasis during exercise. Indianpolis: Benchmark Press, 1990, 39-95.

74. Huebner, D.E., C.C. Lobeck, and N.R. McSherry. Density and secretory activity of eccrine sweat glands in patients with cystic fibrosis and in healthy controls. *Pediatrics* 38: 613-618, 1966.

75. Inbar, O. Acclimatization to dry and hot environment in young adults and children 8-10 years old. 1978. Dissertation. Columbia University.

76. Inbar, O., O. Bar-Or, R. Dotan, and B. Gutin. Conditioning versus exercise in heat as methods for acclimatizing 8- to 10-year-old boys to dry heat. *J Appl Physiol: Respir Environ Exerc Physiol* 50: 406-411, 1981.

77. Inbar, O., R. Dotan, O. Bar-Or, and B. Gutin. Passive versus active exposures to dry heat as methods of heat acclimatization in young children. In Binkhorst, R.A., H.C.G. Kemper, and W.H.M. Saris, *Children and exercise XI*. Champaign, IL: Human Kinetics, 1985, 329-340.

78. Jacobs, I. The effects of thermal dehydration on performance of the Wingate anaerobic test. *Int J Sports Med* 1: 21-24, 1980.

79. Jokinen, E. *Children's physiological adjustment to heat stress during Finnish sauna bath*. Turku: University of Turku, 1989, 1-101.

80. Jokinen, E., I. Välimäki, K. Antila, A. Seppanen, and J.T. Philic. Children in sauna: cardiovascular adjustment. *Pediatrics* 86: 282-288, 1990.

81. Kark, J.A., and F.T. Ward. Exercise and hemoglobin S. *Semin Hematol* 31: 181-225, 1994.

82. Kawahata, A. Sex differences in sweating. In Yoshimura, H., K. Ogata, and S. Itoh, eds., *Essential problems in climatic physiology*. Nankodo: Kyoto, 1960, 169-184.

83. Keatinge, W.R. Body fat and cooling rates in relation to age. In Folinsbee, L.J., ed., *Environmental stress. Individual human adaptations*. New York: Academic Press, 1978.

84. Keatinge, W.R., and R.G.E. Sloan. Effect of swimming in cold water on body temperatures of children. *J Physiol* 226: 55P-56P, 1972.

85. Kenney, W.L. A review of comparative responses of men and women to heat stress. *Environ Res* 37: 1-11, 1985.

86. Kessler, W.R., and D.H. Andersen. Heat prostration in fibrocystic disease of the pancreas and other conditions. *Pediatrics* 8: 648-656, 1951.

86a. Knöpfli, B.H., and O. Bar-Or. Vagal activity and airway response to ipratropium bromide prior to and after exercise in ambient and cold conditions, in healthy cross-country runners. *Clin J Sport Med* 9: 170-176, 1999.

87. Kondo, N., M. Shibasaki, K. Aoki, S. Koga, Y. Inoue, and C.G. Randall. Function of human eccrine sweat glands during dynamic exercise and passive heat stress. *J Appl Physiol* 90: 1877-1881, 2001.

88. Kriemler, S., B. Wilk, W. Schurer, W.M. Wilson, and O. Bar-Or. Preventing dehydration in children with cystic fibrosis who exercise in the heat. *Med Sci Sports Exerc* 31: 774-779, 1999.

89. Kuno, Y. *Human perspiration*. Springfield, IL: Charles C Thomas, 1956.

90. Landing, B.H., T.R. Wells, and M.L. Williamson. Studies on growth of eccrine sweat glands. In Cheek, B.D., ed., *Human growth, body composition, cell growth, energy and intelligence*. Philadelphia: Lea & Febiger, 1968, 382-395.

91. Langdeau, J.-B., H. Turcotte, P. Desgagne, J. Jobin, and L.-P. Boulet. Influence of sympatho-vagal balance on airway responsiveness in athletes. *Eur J Appl Physiol* 83: 370-375, 2000.

92. Leibowitz, H.W., C.N. Abernathy, E.R. Buskirk, O. Bar-Or, and R.T. Hennessy. The effect of heat stress on reaction time to centrally and peripherally presented stimuli. *Hum Factors* 14: 155-160, 1972.

93. Leithead, C.S., and A.R. Lind. *Heat stress and heat disorders*. Philadelphia: Davis, 1964.

94. Luck, P., and A. Wakeling. Increased cutaneous vasoreactivity to cold in anorexia nervosa. *Clin Sci (Colch)* 61: 559-567, 1981.

95. Luck, P., and A. Wakeling. Set-point displacement for behavioural thermoregulation in anorexia nervosa. *Clin Sci (Colch)* 62: 677-682, 1982.

96. MacDougall, J.D. Thermoregulatory problems encountered in ice hockey. *Can J Appl Sport Sci* 4: 35-38, 1979.

97. Mackie, J.M. Physiological responses of twin children to exercise under conditions of heat stress. 1982. Dissertation. University of Waterloo.

98. Macková, J., M. Sturmová, and M. Máček. Prolonged exercise in prepubertal boys in warm and cold environments. In Ilmarinen, Y. and I. Välimäki, eds., *Pediatric work physiology X*. Berlin: Springer-Verlag, 1983.

99. Mahloudji, M., and K.E. Livingston. Familial and congenital simple anhidrosis. *Am J Dis Child* 113: 477-479, 1967.

100. Main, K., K.O. Nilsson, and N.E. Skakkebaek. Influence of sex and growth hormone deficiency on sweating. *Scand J Clin Lab Invest* 51: 475-480, 1991.

101. Malamud, N., W. Haymaker, and R.P. Custer. Heat stroke: a clinico-pathologic study of 125 fatal cases. *Milit Surg* 99: 397-449, 1946.

102. Mansell, P.I., I.W. Fellows, I.A. Macdonald, and S.P. Allison. Defect in thermoregulation in malnutrition reversed by weight gain. Physiological mechanisms and clinical importance. *Q J Med* 76: 817-829, 1990.

103. Martin, H.L., J.H. Loomis, and W.L. Kenney. Maximal skin blood vascular conductance in subjects aged 5-85 yr. *J Appl Physiol* 79: 297-301, 1995.

104. Matsushita, K., and T. Araki. The effect of physical training on thermoregulatory responses of preadolescent boys to heat and cold. *J Phys Fitness-Japan* 29: 69-74, 1980.

105. McConnell, C.S., S. Rostan, and F.A. Puyau. Heat dissipation in children with congenital heart disease. *South Med J* 63: 837-841, 1970.

106. McCormick, R.J., and E.R. Buskirk. Heat tolerance of exercising lean and obese middle-aged men. *Fed Proc* 33: 441, 1974.

107. Mecklenburg, R.S., L. Loriaux, R.H. Thompson, A.E. Andersen, and M.B. Lipsett. Hypothalamic dysfunction in patients with anorexia nervosa. *Medicine* 53: 147-159, 1974.

108. Meyer, F., O. Bar-Or, J.D. MacDougall, and G.J.F. Heigenhauser. Sweat electrolyte loss during exercise in the heat: effects of gender and maturation. *Med Sci Sports Exerc* 24: 776-781, 1992.

109. Meyer, F., O. Bar-Or, A. Salsberg, and D. Passe. Hypohydration during exercise in children: effect on thirst, drink preferences, and rehydration. *Int J Sports Nutr* 4: 22-35, 1994.

110. Meyer, F., O. Bar-Or, and B. Wilk. Children's perceptual responses to ingesting drinks of different compositions during and following exercise in the heat. *Int J Sports Nutr* 5: 13-24, 1995.

111. Morimoto, T., K. Miki, H. Nose, S. Yamada, S. Hirakawa, and C. Matsubara. Changes in body fluid volume and its composition during heavy sweating and the effect of fluid and electrolyte replacement. *Jpn J Biometeorol* 18: 31-39, 1981.

112. Nadel, E.R., K.B. Pandolf, M.F. Roberts, and J.A.J. Stolwijk. Mechanisms of thermal acclimation to exercise and heat. *J Appl Physiol* 37: 515-520, 1974.

113. Nielsen, B. Physiology of thermoregulation during swimming. In Eriksson, B. and B. Furberg, *Swimming medicine IV.* Baltimore: University Park Press, 1978, 297-303.

114. Nose, H., G.W. Mack, X. Shi, and E.R. Nadel. Role of plasma osmolality and plasma volume during rehydration in humans. *J Appl Physiol* 65: 325-331, 1988.

115. Nose, H., T. Yawata, and T. Morimoto. Osmotic factors in restitution from thermal dehydration in rats. *Am J Physiol* 249: R166-R171, 1985.

116. Oppliger, R.A., H.S. Case, C.A. Horswill, G.L. Landry, and A.C. Shelter. American College of Sports Medicine position stand. Weight loss in wrestlers. *Med Sci Sports Exerc* 28: ix-xii, 1996.

117. Oppliger, R.A., R.D. Harms, D.E. Herrmann, C.M. Streich, and R.R. Clark. The Wisconsin wrestling minimum weight project: a model for weight control among high school wrestlers. *Med Sci Sports Exerc* 27: 1220-1224, 1995.

118. Oppliger, R.A., G.L. Landry, S.W. Foster, and A.C. Lambrecht. Wisconsin minimum weight program reduces weight-cutting practices of high school wrestlers. *Clin J Sport Med* 8: 26-31, 1998.

119. Orenstein, D.M., K.G. Henke, D.L. Costill, C.F. Doershuk, P.J. Lemon, and R.C. Stern. Exercise and heat stress in cystic fibrosis patients. *Pediatr Res* 17: 267-269, 1983.

120. Orenstein, D.M., K.G. Henke, and C.G. Green. Heat acclimation in cystic fibrosis. *J Appl Physiol: Respir Environ Exerc Physiol* 57: 408-412, 1984.

121. Paterson, D.H., D.A. Cunningham, D.S. Penny, M. Lefcoe, and S. Sangal. Heart rate telemetry and estimated energy metabolism in minor league ice hockey. *Can J Appl Sport Sci* 2: 71-75, 1977.

122. Piekarski, C., P. Morfeld, B. Kampamm, R. Ilmarinen, and H.G. Wenzel. Heat-stress reactions of the growing child. In Rutenfranz, J. and F. Klimt, eds., *Children and exercise XII.* Champaign, IL: Human Kinetics, 1986, 403-412.

123. Pitts, G.C., R.E. Johnson, and F.C. Consolazio. Work in the heat as affected by intake of water, salt and glucose. *Am J Physiol* 142: 253-259, 1944.

124. Pugh, L.G.C.E., J.L. Corbett, and R.H. Johnson. Rectal temperatures, weight losses and sweat rates in marathon running. *J Appl Physiol* 23: 347-352, 1967.

125. Puyau, F.A. Evaporative heat losses of infants with congenital heart disease. *Am J Clin Nutr* 22: 1435-1443, 1969.

126. Randell, W.C. Quantitation and regional distribution of sweat glands in man. *J Clin Invest* 25: 761-767, 1946.

127. Redfearn, J.A. Jr. History of heat stroke in a football trainee. *JAMA* 208: 699, 1969.

128. Rees, J., and S. Shuster. Pubertal induction of sweat gland activity. *Clin Sci* 60: 689-692, 1981.

129. Rivera-Brown, A., R. Gutierrez, J.C. Gutierrez, W.R. Frontera, and O. Bar-Or. Drink composition, voluntary drinking, and fluid balance in exercising, trained, heat-acclimatized boys. *J Appl Physiol* 86: 78-84, 1999.

130. Robinson, S., S.L. Wiley, L.G. Bondurant, and S.J.R. Mamlin. Temperature regulation of men following heatstroke. *Isr J Med Sci* 12: 786-795, 1976.

131. Rodriguez Santana, J.R., A.M. Rivera-Brown, W.R. Frontera, M.A. Rivera, P.M. Mayol, and O. Bar-Or. Effect of drink pattern and solar radiation of thermoregulation and fluid balance during exercise in chronically heat acclimatized children. *Am J Hum Biol* 7: 643-650, 1995.

132. Rodriguez, J.R., A.M. Rivera-Brown, W.R. Frontera, M.A. Rivera, P. Mayol, and O. Bar-Or. Effect of drink pattern and solar radiation on thermoregulation and fluid balance during exercise in chronically acclimatized children. *Am J Hum Biol* 7: 643-650, 1995.

133. Rothstein, A., E.F. Adolph, and J.H. Wills. Voluntary dehydration. In Adolph, E.F., ed., *Physiology*

of man in the desert. New York: Interscience, 1947, 254-270.

134. Rohstein, A., O. Bar-Or, and R. Dlin. Hemoglobin, hematocrit, and calculated plasma volume changes induced by a short, supramaximal task. *Int J Sports Med* 3: 230-233, 1982.

135. Rowell, L.B. Human cardiovascular adjustments to exercise and thermal stress. *Physiol Rev* 54: 75-159, 1974.

136. Saltin, B. Aerobic and anaerobic work capacity after dehydration. *J Appl Physiol* 19: 1114-1118, 1964.

137. Sargeant, A.J. Effect of muscle temperature on leg extension force and short-term power output in humans. *Eur J Appl Physiol* 56: 693-698, 1987.

138. Sato, K., R. Leidal, and F. Sato. Morphology and development of an apoeccrine sweat gland in human axillae. *Am J Physiol* 252: 166-180, 1987.

139. Sawka, M.N., and K.B. Pandolf. Effect of body water loss on physiological function and exercise performance. In Gisolfi, C.V. and D.R. Lamb, eds., *Perspectives in exercise and sport medicine: fluid homeostasis during exercise.* Indianapolis: Benchmark Press, 1990, 1-30.

140. Schickele, E. Environment and fatal heat stroke. *Milit Surg* 100: 235-256, 1947.

141. Shaker, Y. Thirst fever, with a characteristic temperature pattern in infants in Kuwait. *Br Med J* 1: 586-588, 1966.

142. Shapiro, Y., A. Magazanik, R. Udassin, G. Ben-Baruch, E. Shvartz, and Y. Shoenfeld. Heat intolerance in former heatstroke patients. *Ann Int Med* 90: 913-916, 1979.

143. Shibasaki, M., Y. Inoue, and N. Kondo. Mechanisms of underdeveloped sweating responses in prepubertal boys. *Eur J Appl Physiol* 76: 340-345, 1997.

144. Shibasaki, M., Y. Inoue, and N. Kondo. Mechanisms of underdeveloped sweating responses in prepubertal boys. *Eur J Appl Physiol* 76: 340-345, 1997.

145. Shibasaki, M., Y. Inoue, N. Kondo, and A. Iwata. Thermoregulatory responses of prepubertal boys and young men during moderate exercise. *Eur J Appl Physiol* 75: 212-218, 1997.

146. Shibasaki, M., Y. Inoue, N. Kondo, and A. Iwata. Thermoregulatory responses of prepubertal boys and young men during moderate exercise. *Eur J Appl Physiol* 75: 212-218, 1997.

147. Sloan, R.E.G., and W.R. Keatinge. Cooling rates of young people swimming in cold water. *J Appl Physiol* 35: 371-375, 1973.

148. Smolander, J., O. Bar-Or, O. Korhonen, and J. Ilmarinen. Thermoregulation during rest and exercise in the cold in pre- and early pubescent boys and young men. *J Appl Physiol* 72: 1589-1594, 1992.

149. Smolander, J., J. Oksa, M. Tervo, K. Savonen, H. Pekkarinen, and O. Bar-Or. Thermal responses to cross-country skiing in prepubescent boys and in young men. In Johannsen, B.N. and R. Nielsen, eds., *Thermal Physiology.* Copenhagen: August Krogh Institute, 1997, 339-342.

150. Sohar, E., and Y. Shapiro. The physiological reactions of women and children marching during heat. *Proc Isr Physiol Pharmacol* 1: 50, 1965.

151. Spickard, A. Heat stroke in college football and suggestions for prevention. *South Med J* 61: 791-796, 1968.

152. Stiene, H.A. A comparison of weight-loss methods in high school and collegiate wrestlers. *Clin J Sport Med* 3: 95-100, 1993.

153. Sutton, J., M.J. Coleman, A.P. Millar, L. Lazarus, and P. Russo. The medical problems of mass participation in athletic competition. The "city-to-surf" race. *Med J Austr* 2: 127-133, 1972.

154. Sutton, J.R. Heat illness. In Strauss, R.H., ed., *Medicine in sports and exercise: non-traumatic aspects.* Philadelphia: Franklin Institute Press, 1983.

155. Sutton, J.R., and O. Bar-Or. Thermal illness in fun running. *Am Heart J* 100: 778-781, 1980.

156. Szlyk, P.C., I.V. Sils, R.P. Francesconi, R.W. Hubbard, and L.E. Armstrong. Effects of water temperature and flavoring on voluntary dehydration in men. *Physiol Behav* 45: 639-647, 1989.

157. Taj-Eldin, S., and N. Falaki. Heat illness in infants and small children in desert climates. *J Trop Med Hyg* 71: 100-104, 1968.

158. Tcheng, T.K., and C.M. Tipton. Iowa wrestling study: anthropometric measurements and the prediction of a "minimal" body weight for high school wrestlers. *Med Sci Sports* 5: 1-10, 1973.

159. Tipton, C.M., and T.-K. Tcheng. Iowa wrestling study. Weight loss in high school students. *JAMA* 214: 1269-1274, 1970.

160. Van Beaumont, W. Thermoregulation in desert heat with respect to age. *Physiologist* 8: 294, 1965.

161. Vroman, N.V., E.R. Buskirk, and J.L. Hodgson. Cardiac output and skin blood flow in lean and obese individuals during exercise in the heat. *J Appl Physiol* 55: 69, 1983.

162. Wagner, J.A., S. Robinson, and R.P. Marino. Age and temperature regulation of humans in neutral and cold environments. *J Appl Physiol* 37: 562-565, 1974.

163. Wagner, J.A., S. Robinson, S.P. Tzankoff, and R.P. Marino. Heat tolerance and acclimatization to work in the heat in relation to age. *J Appl Physiol* 33: 616-622, 1972.

164. Wilk, B., and O. Bar-Or. Acclimation and body fluid balance in children drinking carbohydrate-electrolyte beverage ad libitum during intermittent exercise in the heat [abstract]. *Can J Appl Physiol* 20: 58P, 1995.

165. Wilk, B., and O. Bar-Or. Effect of drink flavor and NaCl on voluntary drinking and rehydration in boys exercising in the heat. *J Appl Physiol* 80: 1112-1117, 1996.

166. Wilk, B., S. Kriemler, H. Keller, and O. Bar-Or. Consistency of preventing voluntary dehydration in boys who drink a flavored carbohydrate-NaCl beverage during exercise in the heat. *Int J Sports Nutr* 8: 1-9, 1998.

167. Williams, A.J., J. McKiernan, and F. Harris. Heat prostration in children with cystic fibrosis. *Br Med J* 2: 297, 1976.

168. Yamashita, Y., T. Ogawa, N. Ohnishi, R. Imamura, and J. Sugenoya. Local effect of vasoactive intestinal peptides on human sweat-gland function. *Jpn J Physiol* 37: 929-936, 1987.

169. Zambraski, E.J., D.T. Foster, P.M. Gross, and C.M. Tipton. Iowa wrestling study: weight loss and urinary profiles of collegiate wrestlers. *Med Sci Sports* 8: 105-108, 1976.

Chapter 4

1. Bar-Or, O. Rating of perceived exertion in children and adolescents: clinical aspects. In Ljunggren, G. and S. Dornic, eds., *Psychophysics in action.* Berlin: Springer-Verlag, 1989, 105-113.

2. Bar-Or, O., and S.L. Reed. Rating of perceived exertion in adolescents with neuromuscular disease. In Borg, G., ed., *Perception of exertion in physical work.* Stockholm: Wenner-Gren, 1987, 137-148.

3. Bergman, A.B., and S.J. Stamm. The morbidity of cardiac nondisease in schoolchildren. *N Eng J Med* 276: 1008-1013, 1967.

4. *Canada fitness survey. Canadian youth and physical activity.* Ottawa: Fitness Canada, 1983, 1-52.

5. Dominguez, R.H. Shoulder pain in age group swimmers. In Eriksson, B. and B. Furberg, eds., *Swimming medicine IV.* Baltimore: University Park Press, 1978, 105-109.

6. Grafe, M.W., G.R. Paul, and T.E. Foster. The preparticipation sport examination for high school and college athletes. *Clin Sports Med* 16: 569-591, 1997.

7. Longmuir, P.E., and O. Bar-Or. Physical activity of children and adolescents with a disability: methodology and effects of age and gender. *Pediatr Exerc Sci* 6: 168-177, 1994.

8. O'Neill, D.B., and L.J. Micheli. Overuse injuries in the young athlete. *Clin Sports Med* 7: 591-610, 1988.

9. Outerbridge, A.R., and L.J. Micheli. Overuse injuries in the young athlete. *Clin Sports Med* 14: 503-516, 1995.

10. Ranick, G.L. *Exercise and growth. Science and medicine of exercise and sport.* Harper & Row, 1974, 306-321.

11. Riddell, M.C., O. Bar-Or, H.C. Gerstein, and G.J.F. Heigenhauser. Perceived exertion with glucose ingestion in adolescent males with IDDM. *Med Sci Sports Exerc* 32: 167-173, 2000.

12. Rowland, T.W.E. *Pediatric laboratory exercise testing: clinical guidelines.* Champaign, IL: Human Kinetics, 1993.

13. Sullivan, J.A., and S.J. Anderson. *Care of the young athlete.* Chicago, IL: American Academy of Pediatrics and American Academy of Orthopedic Surgeons, 2000, 1-524.

14. Torg, J.S. The little league pitcher. *Am Fam Physician* 6: 71-76, 1972.

15. U.S. Department of Health and Human Services. *Physical activity and health. A report of the Surgeon General.* Rockville, MD: The President's Council on Physical Fitness and Sports, 1996.

16. Unnithan, V.B., C. Clifford, and O. Bar-Or. Evaluation by exercise testing of the child with cerebral palsy. *Sports Med* 26: 239-251, 1998.

17. Ward, D.S., C.J.R. Blimkie, and O. Bar-Or. Rating of perceived exertion in obese adolescents [abstract]. *Med Sci Sports Exerc* 18: S72, 1986.

Chapter 5

1. Abraham, S., and M. Nordsieck. Relationship of excess weight in children and adults. *Pub Health Reports* 75: 263-273, 1960.

2. Al-Hazzaa, H.M., M.A. Sulaiman, A.J. Al-Matar, and K.F. Al-Mobaireek. Cardiorespiratory fitness, physical activity patterns and coronary risk factors in pre-adolescent boys. *Int J Sports Med* 15: 267-272, 1994.

3. Alpert, B.S., and J.H. Wilmore. Physical activity and blood pressure in adolescents. *Pediatr Exerc Sci* 6: 361-380, 1994.

4. Armstrong, N., and B. Simons-Morton. Physical activity and blood lipids in adolescents. *Pediatr Exerc Sci* 6: 381-405, 1994.

5. Armstrong, N., and W. van Mechelen. Are young people fit and active? In Biddle, S., J. Sallis, and N. Cavill, eds., *Young and active? Young people and health-enhancing physical activity—evidence and implications.* London: Health Education Authority, 1998, 69-97.

6. Armstrong, N., and W. van Mechelen. *Paediatric exercise science and medicine.* Oxford: Oxford University Press, 2000.

7. Armstrong, N., and J. Welsman. *Young people and physical activity.* Oxford: Oxford University Press, 1997.

8. Arroll, B., and R. Beaglehole. Does physical activity lower blood pressure? A critical review of the clinical trials. *J Clin Epidemiol* 45: 439-447, 1992.

9. Baer, J.T., L.J. Taper, F.G. Gwazdauskas, J.L. Walberg, M.A. Novascone, S.J. Ritchey, and F.W. Thye.

Diet, hormonal and metabolic factors affecting bone mineral density and adolescent amenorrheic and eumenorrheic runners. *J Sports Med Phys Fitness* 32: 51-58, 1992.

10. Bailey, D.A. Prevention of osteoporosis: a pediatric concern. In Rippe, J.M., ed., *Lifestyle medicine.* Malden, MA: Blackwell Scientific, 1999, 578-584.

11. Bailey, D.A., R.A. Faulkner, K. Kimber, A. Dzus, and K. Yong-Hing. Altered loading patterns and femoral bone mineral density in children with unilateral Legg-Calve-Perthes disease. *Med Sci Sports Exerc* 29: 1395-1399, 1997.

12. Bailey, D.A., and A.D. Martin. Physical activity and skeletal health in adolescents. *Pediatr Exerc Sci* 6: 330-347, 1994.

13. Ballor, D.L., and R.E. Keesey. A meta-analysis of the factors affecting exercise-induced changes in body mass, fat mass, and fat-free mass in males and females. *Int J Obesity* 15: 717-726, 1991.

14. Bao, W., S.R. Srinivasan, and G.S. Berenson. Plasma fibrinogen and its correlates in children from a biracial community: the Bogalusa Heart Study. *Pediatr Res* 33: 323-326, 1993.

15. Barnard, R., and S.J. Wen. Exercise and diet in the prevention and control of the metabolic syndrome. *Sports Med* 18: 218-228, 1994.

16. Bar-Or, O., and T. Baranowski. Physical activity, adiposity, and obesity among adolescents. *Pediatr Exerc Sci* 6: 348-360, 1994.

17. Bar-Or, O., and R.M. Malina. Activity, fitness, and health of children and adolescents. In Cheung, L.W.Y. and J.B. Richmond, eds., *Child health, nutrition, and physical activity.* Champaign, IL: Human Kinetics, 1995, 79-123.

18. Bass, S., G. Pearce, M. Bradney, E. Hendrich, P.D. Delmas, A. Harding, and E. Seeman. Exercise before puberty may confer residual benefits in bone density in adulthood: studies in active prepubertal and retired female gymnasts. *J Bone Miner Res* 13: 500-507, 1998.

19. Becque, M.D., V.L. Katch, A.P. Rocchini, C.R. Marks, and C. Moorehead. Coronary risk incidence of obese adolescents: reduction by exercise plus diet intervention. *Pediatrics* 81: 605-612, 1988.

20. Ben-Ezra, V., and K. Gallagher. Blood profiles of 8-10 year old children: the effect of diet and exercise [abstract]. *Med Sci Sports Exerc* 21: 584, 1989.

21. Berenson, G.S., S.R. Srinivasan, W. Wattigney, L.S. Webber, W.P. Newman, and R.E. Tracy. Insight into a bad omen for white men: coronary artery disease—the Bogalusa Heart Study. *Am J Cardiol* 64: 32C-39C, 1989.

22. Bergstrom, E., O. Hernell, and L.A. Persson. Endurance running performance in relation to cardiovascular risk indicators in adolescents. *Int J Sports Med* 18: 300-307, 1997.

23. Berlin, J.A., and G.A. Colditz. A meta-analysis of physical activity in the prevention of coronary heart disease. *Am J Epidemiol* 132: 612-628, 1990.

24. Biddle, S., J. Sallis, and N. Cavill, eds. *Young and active? Young people and health-enhancing physical activity—evidence and implications.* London: Health Education Authority, 1998.

25. Blair, S.N. Physical activity, fitness, and coronary heart disease. In Bouchard, C., R.J. Shepard, and T. Stephens, eds., *Physical activity, fitness, and health: international proceedings and consensus statement.* Champaign, IL: Human Kinetics, 1994, 579-590.

26. Blair, S.N., D.G. Clark, K.J. Cureton, and K.E. Powell. Exercise and fitness in childhood: implications for a lifetime of health. In Gisolfi, C.V. and D.R. Lamb, eds., *Perspectives in exercise science and sports medicine.* Vol. 2. *Youth, exercise, and sport.* Indianapolis: Benchmark Press, 1989, 401-430.

27. Blair, S.N., N.N. Goodyear, L.W. Gibbons, and K.H. Cooper. Physical fitness and incidence of hypertension in healthy normotensive men and women. *JAMA* 252: 487-490, 1984.

28. Blair, S.N., H.W. Kohl, C.E. Barlow, R.S. Paffenbarger, Jr., L.W. Gibbons, and C.A. Macera. Changes in physical fitness and all-cause mortality: a prospective study of healthy and unhealthy men. *JAMA* 273: 1093-1098, 1995.

29. Blessing, D.L., R.E. Keith, H.N. Williford, M.E. Blessing, and J.A. Barksdale. Blood lipid and physiological responses to endurance training in adolescents. *Pediatr Exerc Sci* 7: 192-202, 1995.

30. Blimkie, C.J.R., S. Rice, C.E. Webber, J. Martin, D. Levy, and C.L. Gordon. Effects of resistance training on bone mineral content and density in adolescent females. *Can J Physiol Pharmacol* 74: 1025-1033, 1996.

31. Blumenthal, J.A., W. Jiang, and R.A. Waugh. Mental stress-induced ischemia in the laboratory and ambulatory ischemia during daily life: association and hemodynamic features. *Circulation* 92: 2102-2108, 1995.

32. Boreham, C., J. Twisk, L. Murray, M. Savage, J.J. Stran, and G. Cran. Fitness, fatness, and coronary heart disease risk in adolescents: the Northern Ireland Young Hearts Project. *Med Sci Sports Exerc* 33: 270-274, 2001.

33. Bradney, M., G. Pearce, G. Naughton, C. Sullivan, S. Bass, T. Beck, J. Carlson, and E. Seeman. Moderate exercise during growth in prepubertal boys: changes in bone mass, size, volumetric density and bone strength: a controlled prospective study. *J Bone Miner Res* 12: 1814-1821, 1998.

34. Brownell, K.D., and A.J. Stunkard. Physical activity in the development and control of obesity. In Stunkard, A.J., ed., *Obesity.* Philadelphia: Saunders, 1980, 300-324.

35. Burke, G., L. Webber, S. Srinivasan, B. Radhakrishnamurthy, D. Freedman, and G. Berenson. Fasting plasma glucose and insulin levels and their relationship to cardiovascular risk factors in children: Bogalusa Heart Study. *Metabolism* 35: 441-446, 1986.

36. Corbin, C. The 'untracking' of sedentary living: a call for action. *Pediatr Exerc Sci*, in press, 2001.

37. Corbin, C.B., and R.P. Pangrazi. Toward an understanding of appropriate physical activity levels for youth. *President's Council on Physical Fitness and Sports Activity Fitness Research Digest* 2: 1-8, 1994.

38. Corbin, C.B., and R.P. Pangrazi. *Physical activity for children. A statement of guidelines.* Reston, VA: National Association for Sport and Physical Education, 1998.

39. Courteix, D., E. Lespessailles, S.L. Peres, P. Obert, P. Germain, and C.L. Benhamou. Effect of physical training on bone mineral density in prepubertal girls: a comparative study between impact-loading and non-impact sports. *Osteopor Int* 8: 152-158, 1998.

40. Dollman, J., T. Olds, K. Norton, and D. Stuart. The evolution of fitness and fatness in 10-11 year old Australian schoolchildren: changes in distributional characteristics between 1985 and 1997. *Pediatr Exerc Sci* 11: 108-121, 1999.

41. Drinkwater, B.L. Osteoporosis and exercise. In Rippe, J.M., ed., *Lifestyle medicine.* Malden, MA: Blackwell Scientific, 1999, 237-241.

42. Epstein, L.H., R.A. Paluch, L.E. Kalakanis, G.S. Goldfield, F.J. Cerny, and J.N. Roemmich. How much activity do youth get? A quantitative review of heart-rate measured activity. *Pediatrics* 108: e344, 2001.

43. Epstein, L.H., and R.R. Wing. Aerobic exercise and weight. *Addictive Behav* 5: 371-388, 1980.

44. Ewart, C.K., D.R. Young, and J.M. Hagberg. Effects of school-based aerobic exercise on blood pressure in adolescent girls at risk for hypertension. *Am J Pub Health* 88: 949-951, 1998.

45. Fain, J.A. National trends in diabetes: an epidemiologic perspective. *Diabetes* 28: 1-7, 1993.

46. Folsom, A.R., K.K. Wu, and W.D. Rosamund. Prospective study of hemostatic factors and incidence of coronary heart disease: the Atherosclerosis Risk in Communities (ARIC) Study. *Circulation* 96: 1102-1108, 1997.

47. Forwood, M., and D. Burr. Physical activity and bone mass: exercise in futility? *Bone Miner* 21: 89-112, 1993.

48. Freedson, P.S., and T.W. Rowland. Youth activity versus youth fitness: let's redirect our efforts. *Res Q Exerc Sport* 63: 133-136, 1992.

49. Fripp, R.R., and J.L. Hodgson. Effect of resistive training on plasma lipid and lipoprotein levels in male adolescents. *J Pediatr* 11: 926-931, 1987.

50. Fulton, J.E., and H.W. Kohl. The epidemiology of obesity, physical activity, diet, and type 2 diabetes mellitus. In Rippe, J.M., ed., *Lifestyle medicine.* Malden, MA: Blackwell Scientific, 1999, 867-883.

51. Goldberg, R.J., and J.L. Yarzebski. Coronary artery disease: epidemiology, risk factors, and temporal trends in mortality rates. In Rippe, J.M., ed., *Lifestyle medicine.* Malden, MA: Blackwell Scientific, 1999, 3-24.

52. Guillaume, M., L. Lapidus, P. Björntop, and A. Lambert. Physical activity, obesity, and cardiovascular risk factors in children. The Belgian Luxembourg Child Study II. *Obesity Res* 5: 549-556, 1997.

53. Gunnes, M., and E.H. Lehmann. Physical activity and dietary constituents as predictors of forearm cortical and trabecular bone gain in healthy children and adolescents: a prospective study. *Acta Paediatr* 85: 19-25, 1996.

54. Gutin, B., C. Basch, S. Shea, I. Contento, M. DeLozier, J. Rips, M. Irigoyen, and P. Zybert. Relations among changes in blood pressure, fitness and fatness in 5-8 year old children [abstract]. *Med Sci Sports Exerc* 23: S83, 1991.

55. Gutin, B., S. Owens, G. Slavens, S. Riggs, and F. Treiber. Effect of physical training on heart-period variability in obese children. *J Pediatr* 130: 938-943, 1997.

56. Gutin, B., S. Owens, F. Treiber, S. Islam, W. Karp, and G. Slavens. Weight independent cardiovascular fitness and coronary risk factors. *Arch Pediatr Adolesc Med* 151: 462-465, 1997.

57. Hager, R.L., L.A. Tucker, and G.T. Seljaas. Aerobic fitness, blood lipids, and body fat in children. *Am J Pub Health* 85: 1702-1706, 1995.

58. Hoffman, A., and H.J. Walter. The association between physical fitness and cardiovascular disease risk factors in children in a five year followup study. *Int J Epidemiol* 18: 830-835, 1989.

59. Horswill, C.A., W.B. Zipf, L. Klein, and E.B. Kahle. Insulin's contribution to growth in children and the potential for exercise to mediate insulin's action. *Pediatr Exerc Sci* 9: 18-32, 1997.

60. Hui, S.L., C.C. Johnston, and R.B. Mazess. Bone mass in normal children and young adults. *Growth* 49: 34-43, 1985.

61. Hull, S.S., A.R. Evans, E. Vanoli, P.B. Adamson, M. Badiale, D.E. Albert, R.D. Foreman, and P.J. Schwartz. Heart rate variability before and after myocardial infarction in conscious dogs at high and low risk of sudden death. *J Am Coll Cardiol* 16: 978-985, 1990.

62. Isasi, C.R., T.J. Starc, R.P. Tracy, R. Deckelbaum, L. Berglund, and S. Shea. Inverse association of physical fitness with plasma fibrinogen level in children. *Am J Epidemiol* 152: 212-218, 2000.

63. Janz, K.F., J.D. Dawson, and L.T. Mahoney. Changes in physical fitness and physical activity during puberty do not predict lipoprotein profile changes: the Muscatine Study. *Pediatr Exerc Sci* 12: 232-243, 2000.

64. Kahle, E.B., W.B. Zipfe, D.R. Lamb, C.A. Horswill, and K.M. Ward. Association between mild, routine exercise and improved insulin dynamics and glucose control in obese adolescents. *Int J Sports Med* 17: 1-6, 1996.

65. Kemper, H.C.G. Skeletal development during childhood and adolescence and the effects of physical activity. *Pediatr Exerc Sci* 12: 198-216, 2000.

66. Kirby, B., and R.M. Kirby. Heart rate variability in 11- to 16-year olds. In Armstrong, N., B. Kirby, and J. Welsman, eds., *Children and exercise XIX.* London: Spon, 1997, 434-439.

67. Kramsch, D.M., A.J. Aspen, B.M. Abrahamowitz, T. Kreimendahl, and W.B. Hood. Reduction of coronary atherosclerosis by moderate conditioning exercise in monkeys on an atherogenic diet. *N Eng J Med* 305: 1483-1489, 1981.

68. Lauer, R.M., W.R. Clarke, L.T. Mahoney, and J. Witt. Childhood predictors for high adult blood pressure. *Pediatr Clin N Amer* 40: 23-39, 1993.

69. Leibel, R.L., M. Rosenbaum, and J. Hirsch. Changes in energy expenditure resulting from added body weight. *N Eng J Med* 332: 621-628, 1995.

70. Linder, C.W., R.H. DuRant, R.G. Gray, and J.W. Harkness. The effect of exercise on serum lipids in children [abstract]. *Clin Res* 27: 297, 1979.

71. Linder, C.W., R.H. DuRant, and O.M. Mahoney. The effect of physical conditioning on serum lipids and lipoproteins in white male adolescents. *Med Sci Sports Exerc* 15: 232-236, 1983.

72. Mahon, A.D., C.C. Cheatham, K.Q. Kelsey, and J.D. Brown. Plasma fibrinogen, physical activity, and aerobic fitness in children. In Armstrong, N., B. Kirby, and J. Welsman, eds., *Children and exercise XIX.* London: Spon, 1997, 117-123.

73. Malina, R.M. Tracking of physical activity and physical fitness across the lifespan. *Res Q Exerc Sport* 67 (Suppl 3): S48-S57, 1996.

74. Manuck, S.B., J.R. Kaplan, and T.B. Clarkson. Behaviorally induced heart rate reactivity and atherosclerosis in cynomolgus monkeys. *Psychosom Med* 45: 95-108, 1983.

75. Margulies, J.Y., A. Simkin, I. Leichter, A. Bivas, R. Steinberg, M. Giladi, M. Stein, H. Kashtan, and C. Milgrom. Effect of intensive physical activity on the bone mineral content in power limbs of young adults. *J Bone Joint Surg* 68A: 1090-1093, 1986.

76. McMurray, R.G., J.S. Harrell, S.I. Bangdiwala, S. Dery, and C.B. Bradley. Effects of exercise and weight on lipid profiles of middle-school aged youth: the CHIC study [abstract]. *Med Sci Sports Exerc* 32 (Suppl): S339, 2000.

77. McMurray, R.G., J.S. Harrell, A.A. Levine, and S.A. Gansky. Childhood obesity elevates blood pressure and total cholesterol independent of physical activity. *Int J Obesity* 19: 881-886, 1995.

78. Melanson, E.L. Heart rate variability: relationship to physical activity level, response to training, and effect of maturation. 1998. Doctoral thesis. University of Massachusetts.

79. Moen, S.M., C.F. Sanborn, N.M. Dimarco, B. Gench, S.L. Bonnick, H.A. Keizer, and P.C.C.A. Menheere. Lumbar bone mineral density in adolescent female runners. *J Sports Med Phys Fitness* 38: 234-239, 1998.

80. Morris, F.L., G.A. Naughton, J.L. Gibbs, J.S. Carlson, and J.D. Wark. Prospective ten-month exercise intervention in premenarcheal girls: positive effects on bone and lean mass. *J Bone Miner Res* 12: 1453-1462, 1997.

81. Oalmann, M.C., J.P. Strong, R.E. Tracy, and G.T. Malcom. Atherosclerosis in youth: are hypertension and other coronary heart disease risk factors already at work? *Pediatr Nephrol* 1: 99-107, 1997.

82. Pate, R., S. Trost, and C. Williams. Critique of existing guidelines for physical activity in young people. In Biddle, S., J. Sallis, and N. Cavill, eds., *Young and active? Young people and health-enhancing physical activity—evidence and implications.* London: Health Education Authority, 1998, 162-176.

83. Pinhas-Hamiel, O., L.M. Dolan, S.R. Daniels, D. Standiford, P.R. Khoury, and P. Zeitler. Increasing incidence of non-insulin-dependent diabetes mellitus among adolescents. *J Pediatr* 128: 608-615, 1996.

84. Powell, K.E., M. Smith, I.M. Dick, R.I. Price, P.G. Webb, and N.K. Henderson. Physical activity and the incidence of coronary heart disease. *Ann Rev Pub Health* 8: 253-287, 1987.

85. Rabbia, F., F. Veglio, G. Pinna, S. Oliva, V. Surgo, B. Rolando, A. Bessone, et al. Cardiovascular risk factors in adolescents: prevalence and familial aggregation. *Prev Med* 23: 809-815, 1994.

86. Raitakari, O.T., K.V.K. Parkka, S. Taimela, R. Telama, L. Rasanen, and J.S.A. Viikari. Effects of persistent physical activity and inactivity on coronary risk factors in children and young adults. *Am J Epidemiol* 140: 195-205, 1994.

87. Riddoch, C. Relationships between physical activity and health in young people. In Biddle, S., J. Sallis, and N. Cavill, eds., *Young and active? Young people and health-enhancing physical activity—evidence and implications.* London: Health Education Authority, 1998, 17-48.

88. Riddoch, C., and C. Boreham. Physical activity, physical fitness, and children's health: current concepts. In Armstrong, N., and W. van Mechelen, eds., *Paediatric exercise science and medicine.* Oxford: Oxford University Press, 2000, 243-252.

89. Riddoch, C., J.M. Savage, N. Murphy, G.W. Cran, and C. Boreham. Long term implications of fitness and physical activity patterns. *Arch Dis Child* 66: 1426-1433, 1991.

90. Rimmer, J.H., and M.A. Looney. Effects of an aerobic activity program on the cholesterol levels of adolescents. *Res Q Exerc Sport* 68: 74-79, 1997.

91. Rippe, J.M. *Lifestyle medicine.* Malden, MA: Blackwell Scientific, 1999.

92. Ross, J.G., R.R. Pate, T.G. Lohman, and G.M. Christenson. Changes in body composition of children. *JOPERD* 58: 74-77, 1987.

93. Roth, D.M., F.L. White, and M.L. Nichols. Effect of long-term exercise on regional myocardial function and coronary collateral development after gradual coronary artery occlusion in pigs. *Circulation* 82: 1778-1779, 1990.

94. Rowland, T.W. Effects of obesity on aerobic fitness in adolescent females. *Am J Dis Child* 145: 764-768, 1991.

95. Rowland, T.W. Physical activity, fitness, and health in children: a close look. *Pediatrics* 93: 669-671, 1994.

96. Rowland, T.W. Preventive cardiology in children: faith, hope, and strategic policy-making. *Med Exerc Nutr Health* 3: 178-179, 1997.

97. Rowland, T.W., L. Martel, P. Vanderburgh, T. Manos, and N. Charkoudian. The influence of short-term aerobic training on blood lipids in healthy 10-12 year old children. *Int J Sports Med* 17: 487-492, 1996.

98. Ryan, A.J. Exercise and health: lessons from the past. In Eckert, H.M. and H.J. Montoye, eds., *Exercise and health.* Champaign, IL: Human Kinetics, 1984, 3-13.

99. Sallis, J.F. Age-related decline in physical activity: a synthesis of human and animal studies. *Med Sci Sports Exerc* 32: 1598-1600, 2000.

100. Sallis, J.F., and K. Patrick. Physical activity guidelines for adolescents: consensus statement. *Pediatr Exerc Sci* 6: 302-314, 1994.

101. Sasaki, J., M. Shindo, H. Tanaka, M. Ando, and K. Arakawa. A long-term aerobic exercise program decreases the obesity index and increases the high density lipoprotein cholesterol concentration in obese children. *Int J Obesity* 11: 339-345, 1987.

102. Savage, M.P., M.M. Petratis, W.H. Thompson, K. Berg, J. Smith, and S.P. Sady. Exercise training effects on serum lipids of prepubescent boys and adult men. *Med Sci Sports Exerc* 18: 197-204, 1986.

103. Shephard, R.J., and F. Trudeau. The legacy of physical education: influences on adult lifestyle. *Pediatr Exerc Sci* 12: 34-50, 2000.

104. Smith, B.W., W.P. Metheny, and W.D. Van Huss. Serum lipids and lipoprotein profiles in elite age-group endurance runners [abstract]. *Circulation* 68: 191, 1983.

105. Snow-Harter, C., M. Bouxsein, B. Lewis, D. Carter, and R. Marcus. Effects of resistance and endurance exercise on bone mineral status of young women: a randomized exercise intervention trial. *J Bone Miner Res* 7: 761-769, 1992.

106. Stark, O., E. Atkins, O. Wolff, and J. Douglas. Longitudinal study of obesity in the National Survey of Health and Development. *Br Med J* 283: 13-17, 1981.

107. Steinberger, J., and A.P. Rocchini. Is insulin resistance responsible for the lipid abnormalities seen in obesity. *Circulation* 84 (Suppl II): II-5, 1991.

108. Stergioulas, A., A. Tripolitsioti, D. Messinis, A. Bouloukos, and C. Nounopoulos. The effects of endurance training on selected coronary risk factors in children. *Acta Paediatr* 87: 401-404, 1998.

109. Stoedefalke, K., N. Armstrong, B.J. Kirby, and J.R. Welsman. Effect of training on peak oxygen uptake and blood lipids in 13 to 14-year old girls. *Acta Paediatr* 89: 1290-1294, 2000.

110. Strong, W.B., R.J. Deckelbaum, S.S. Gidding, R.W. Kavey, R. Washington, J.H. Wilmore, and C.L. Perry. Integrated cardiovascular health promotion in childhood. *Circulation* 85: 1638-1650, 1992.

111. Tolfrey, K., I.G. Campbell, and A.M. Batterham. Exercise training induced alterations in prepubertal children's lipid-lipoprotein profile. *Med Sci Sports Exerc* 30: 1684-1692, 1998.

112. Tolfrey, K., I.G. Campbell, and A.M. Jones. Predictor variables and the lipid-lipoprotein profile in prepubertal girls and boys. *Med Sci Sports Exerc* 31: 1550-1557, 1999.

113. Tolfrey, K., A.M. Jones, and I.G. Campbell. The effect of aerobic exercise training on the lipid-lipoprotein profile of children and adolescents. *Sports Med* 29: 99-112, 2000.

114. Tran, Z.V., A. Weltman, G.V. Glass, and D.P. Mood. The effects of exercise on blood lipids and lipoproteins: a meta-analysis of studies. *Med Sci Sports Exerc* 15: 393-402, 1983.

115. Troiano, R.P., K.M. Flegal, and R.J. Kuczmarksi. Overweight prevalence and trends for children and adolescents. *Arch Pediatr Adolesc Med* 149: 1085-1091, 1995.

116. Twisk, J.W.R. Physical activity, physical fitness, and cardiovascular health. In Armstrong, N. and W. van Mechelen, eds., *Paediatric exercise science and medicine.* Oxford: Oxford University Press, 2000, 251-263.

117. Twisk, J.W.R., H.C.G. Kemper, W. van Mechelen, G.B. Post, and F.J. van Lenthe. Body fatness: longitudinal relationship of body mass index and the sum of four skinfolds with other risk factors for coronary heart disease. *Int J Obesity* 22: 915-922, 1998.

118. Twisk, J.W.R., W. van Mechelen, H.C.G. Kemper, and G.B. Post. The relation between 'long-term exposure' to life style during youth and young adulthood and risk factors for cardiovascular disease. *J Adolesc Health* 20: 309-319, 1997.

119. U.S. Department of Health and Human Services. *Physical activity and health: a report of the Surgeon General.* Atlanta: U.S. Department of Health and Human Services, Centers for Disease Control and Prevention, 1996.

120. van Mechelen, W., and H.C.G. Kemper. Habitual physical activity in longitudinal perspective. In Kemper, H.C.G., ed., *The Amsterdam Growth Study: a longitudinal analysis of health, fitness, and lifestyle.* Champaign, IL: Human Kinetics, 1995, 135-158.

121. Ward, A. Exercise in the prevention of coronary artery disease. In Rippe, J.M., ed., *Lifestyle medicine.* Malden, MA: Blackwell Scientific, 1999, 90-97.

122. Ward, D.S., and O. Bar-Or. Role of the physician and physical education teacher in the treatment of obesity at school. *Pediatrician* 13: 44-51, 1986.

123. Ward, D.S., and R. Evans. Physical activity, aerobic fitness, and obesity in children. *Med Exerc Nutr Health* 4: 3-16, 1995.

124. Welten, D.C., H.C.G. Kemper, G.B. Post, W. van Mechelen, J. Twisk, P. Lips, and G.J. Teule. Weight-bearing activity during youth is a more important factor for peak bone mass than calcium intake. *J Bone Miner Res* 9: 1089-1096, 1994.

125. Weltman, A., C. Janney, C.B. Rains, K. Strand, and F.I. Katch. The effects of hydraulic-resistance strength training on serum lipid levels in prepubertal boys. *Am J Dis Child* 141: 777-780, 1987.

126. Widhalm, K., E. Maxa, and H. Zyman. Effect of diet and exercise upon the cholesterol and triglyceride content of plasma lipoproteins in overweight children. *Eur J Pediatr* 127: 121-126, 1978.

127. Williams, R.S., E.E. Logue, and J.L. Lewis. Physical conditioning augments the fibrinolytic response to venous occlusion in healthy adults. *N Eng J Med* 302: 987-991, 1980.

128. Williford, H.N., D.L. Blessing, and W.J. Duey. Exercise training in black adolescents: changes in blood lipids and VO$_2$max. *Ethn Dis* 6: 279-285, 1996.

Chapter 6

1. Agudo, A., S. Bardagi, P.V. Romero, and C.A. Gonzalez. Exercise-induced airways narrowing and exposure to environmental tobacco smoke in schoolchildren. *Am J Epidemiol* 140: 409-417, 1994.

2. Alsuwaidan, S., A. Li Wan Po, G. Morrison, A. Redmond, J.A. Dodge, J. McElnay, E. Stewart, and C.F. Stanford. Effect of exercise on the nasal transmucosal potential difference in patients with cystic fibrosis and normal subjects. *Thorax* 49: 1249-1250, 1994.

3. American Academy of Pediatrics. The asthmatic child's participation in sports and physical education. *Pediatrics* 74: 155-156, 1984.

4. American Academy of Pediatrics Committee on Sports Medicine and Fitness. Metered dose inhalers for young athletes with exercise-induced asthma. *Pediatrics* 94: 129-130, 1994.

5. Amin, N., and A.J. Dozor. Effects of administration of aerosolized recombinant human deoxyribonuclease on resting energy expenditure in patients with cystic fibrosis. *Pediatr Pulmonol* 18: 150-154, 1994.

6. Amirav, I., V. Panz, B.I. Joffe, R. Dowdswell, M. Plit, and H.C. Seftel. Effects of inspired air conditions on catecholamine response to exercise in asthma. *Pediatr Pulmonol* 18: 99-103, 1994.

7. Anderson, S., J.P. Seale, L. Ferris, R. Schoeffel, and D.A. Lindsay. An evaluation of pharmacotherapy for exercise-induced asthma. *J Allergy Clin Immunol* 64: 612-614, 1979.

8. Anderson, S.D. Physiological aspects of exercise-induced bronchoconstriction. 1972. Dissertation, University of London.

9. Anderson, S.D. Exercise-induced asthma: current views. *Patient Management* 6: 43-55, 1982.

10. Anderson, S.D. Exercise-induced asthma. The state of the art. *Chest* 87S: 191S-195S, 1985.

11. Anderson, S.D., N.M. Connolly, and S. Godfrey. Comparison of bronchoconstriction induced by cycling and running. *Thorax* 26: 396-401, 1971.

12. Anderson, S.D., S. Lambert, J.D. Brannan, R.J. Wood, H. Koskela, A.R. Morton, and K.D. Fitch. Laboratory protocol for exercise asthma to evaluate salbutamol given by two devices. *Med Sci Sports Exerc* 33: 893-900, 2001.

13. Anderson, S.D., R. Pojer, I.D. Smith, and D. Temple. Exercise-related changes in plasma levels of 15-keto-13,14-dihydro-prostaglandin F2 and noradrenaline in asthmatic and normal subjects. *Scand J Respir Dis* 57: 41-48, 1976.

14. Anderson, S.D., P.J. Rozea, R. Dolton, and D.A. Lindsay. Inhaled and oral bronchodilator therapy in exercise-induced asthma. *Austr NZ J Med* 5: 544-550, 1975.

15. Anderson, S.D., R.E. Schoeffel, R. Follet, et al. Sensitivity to heat and water loss at rest and during exercise in asthmatic patients. *Eur J Respir Dis* 63: 93-105, 1982.

16. Andreasson, B., B. Jonson, R. Kornfalt, E. Nordmark, and S. Sandstrom. Long-term effects of physical exercise on working capacity and pulmonary function in cystic fibrosis. *Acta Paediatr Scand* 76: 70-75, 1987.

17. Andreasson, B., M. Lindroth, W. Mortensson, N.W. Svenningsen, and B. Jonson. Lung function eight years after neonatal ventilation. *Arch Dis Child* 64: 108-113, 1989.

18. Araujo, C.J.G., and O. Bar-Or. Asthma, exercise-induced asthma, and aquatic physical activities. *J Back Muscoskel Rehab* 4: 309-314, 1994.

19. Argyros, G.J., Y.Y. Phillips, D.B. Rayburn, R.R. Rosenthal, and J.J. Jaeger. Water loss without heat flux in exercise-induced bronchospasm. *Am Rev Respir Dis* 147: 1419-1424, 1993.

20. Asher, M.I., R.L. Pardy, A.L. Coates, E. Thomas, and P.T. Macklem. The effects of inspiratory muscle training in patients with cystic fibrosis. *Am Rev Respir Dis* 126: 855-859, 1982.

21. Astin, R.W., and R.W.B. Penman. Airway obstruction due to hypoxemia in patients with chronic lung disease. *Am Rev Respir Dis* 95: 567-575, 1967.

22. Backman, A. Physiological and psychological aspects of the training of asthmatic children. In Oseid, S. and A.M. Edwards, eds., *The asthmatic child in play and sports.* London: Pitman, 1983, 277-282.

23. Bader, D., A.D. Ramos, C.D. Lew, A.C. Platzker, M.W. Stabile, and T.G. Keens. Childhood sequelae of infant lung disease: exercise and pulmonary function abnormalities after bronchopulmonary dysplasia. *J Pediatr* 110: 693-699, 1987.

24. Badiei, B., J. Faciane, and R.M. Sly. Effect of theophylline, ephedrine and their combination upon exercise-induced airway obstruction. *Ann Allergy* 35: 32-35, 1975.

25. Bar-Or, O. Climate and the exercising child—a review. *Int J Sports Med* 1: 53-65, 1980.

26. Bar-Or, O. Physical conditioning in children with cardiorespiratory disease. *Exerc Sport Sci Rev* 13: 305-334, 1985.

27. Bar-Or, O. Noncardiopulmonary pediatric exercise tests. In Rowland, T.W., ed., *Pediatric laboratory exercise testing: clinical guidelines.* Champaign, IL: Human Kinetics, 1993, 165-185.

28. Bar-Or, O. Home-based exercise programs in cystic fibrosis: are they worth it? *J Pediatr* 136: 279-280, 2000.

29. Bar-Or, O., C.J.R. Blimkie, J.A. Hay, J.D. Mac-Dougall, D.S. Ward, and W.M. Wilson. Voluntary dehydration and heat intolerance in cystic fibrosis. *Lancet* 339: 696-699, 1992.

30. Bar-Or, O., and O. Inbar. Swimming and asthma: benefits and deleterious effects. *Sports Med* 14: 397-405, 1992.

31. Bar-Or, O., I. Neuman, and R. Dotan. Effects of dry and humid climates on exercise-induced asthma in children and preadolescents. *J Allergy Clin Immunol* 69: 163-168, 1977.

32. Bar-Yishay, E., I. Gur, O. Inbar, et al. Differences between swimming and running as stimuli for exercise-induced asthma. *Eur J Appl Physiol* 48: 387-397, 1982.

33. Bardagi, S., A. Agudo, C.A. Gonzalez, et al. Prevalence of exercise-induced airways narrowing in schoolchildren from a Mediterranean town. *Am Rev Respir Dis* 147: 112-115, 1993.

34. Ben Dov, I., E. Bar-Yishay, and S. Godfrey. Refractory period after exercise-induced asthma unexplained by respiratory heat loss. *Am Rev Respir Dis* 125: 530-534, 1982.

35. Benckhuijsen, J., J.W. Van den Bos, E. Van Velzen, R. De Bruijn, and R. Aalbers. Differences in the effect of allergen avoidance on bronchial hyperresponsiveness as measured by methacholine, adenosine 5'-monophosphate, and exercise in asthmatic children. *Pediatr Pulmonol* 22: 147-153, 1996.

36. Benson, L.N., C.J.L. Newth, M. Desouza, R. Lobraico, W. Karthodihardjo, C. Corkey, and P.M. Olley. Radionuclide assessment of right and left ventricular function during bicycle exercise in young patients with cystic fibrosis. *Am Rev Respir Dis* 130: 987-992, 1984.

37. Berkowitz, R., E. Schwartz, D. Bukstein, M. Grunstein, and H. Chai. Albuterol protects against exercise-induced asthma longer than metaproterenol sulfate. *Pediatrics* 77: 173-178, 1986.

38. Bevegård, S., B.O. Eriksson, and V. Graff-Lonnevig. Respiratory function, cardiovascular dimensions and work capacity in boys with bronchial asthma. *Acta Paediatr Scand* 65: 289-296, 1976.

39. Bevegård, S., B.O. Eriksson, V. Graff-Lonnevig, S. Kraepelien, and B. Saltin. Circulatory and respiratory dimensions and functional capacity in boys aged 8 to 13 years with bronchial asthma. *Acta Paediatr Scand Suppl* 217: 86-89, 1971.

40. Bierman, C.W., and W.E. Pierson. Summary—symposium on exercise and asthma. *Pediatrics* 56: 950-952, 1975.

41. Bierman, C.W., G.G. Shapiro, W.E. Pierson, and Y.W. Cho. Exercise-induced bronchospasm in asthmatic children as a dose-response model for theophylline. *Int J Clin Pharmacol Biopharmacol* 16: 245-248, 1978.

42. Bierman, C.W., G.G. Shapiro, W.E. Pierson, and C.S. Dorsett. Acute and chronic theophylline therapy in exercise-induced bronchospasm. *Pediatrics* 60: 845-849, 1977.

43. Bierman, C.W., S.G. Spiro, and I. Petheram. Characterization of the late response in exercise induced asthma. *J Allergy Clin Immunol* 74: 701-706, 1984.

44. Bittleman, D.B., R.J.H. Smith, and J.M. Weiler. Abnormal movement of the arytenoid region during exercise presenting as exercise-induced asthma in an adolescent athlete. *Chest* 106: 615-616, 1994.

45. Blackhall, M.I. Ventilatory function in subjects with childhood asthma who have become symptom free. *Arch Dis Child* 45: 363-366, 1970.

46. Blomquist, M., U. Freyschuss, L.G. Wiman, and B. Strandvik. Physical activity and self-treatment in patients with cystic fibrosis. *Arch Dis Child* 61: 362-367, 1986.

47. Boas, S.R. Exercise recommendations for individuals with cystic fibrosis. *Sports Med* 24: 17-37, 1997.

48. Boas, S.R., M.J. Danduran, and S.A. McColley. Energy metabolism during anaerobic exercise in children with cystic fibrosis and asthma. *Med Sci Sports Exerc* 31: 1242-1249, 1999.

49. Boas, S.R., M.J. Danduran, S.A. McColley, K. Beaman, and M.R.G. O'Gorman. Immune modulation following aerobic exercise in children with cystic fibrosis. *Int J Sports Med* 21: 294-301, 2000.

50. Boas, S.R., M.J. Danduran, and S.K. Saini. Anaerobic exercise testing in children with asthma. *J Asthma* 35: 481-487, 1998.

51. Boas, S.R., M.L. Joswiak, A.P.A. Nixon, J.A. Fulton, and D.M. Orenstein. Factors limiting anaerobic performance in adolescent males with cystic fibrosis. *Med Sci Sports Exerc* 28: 291-298, 1996.

52. Boner, A.L., E. Spezia, P. Piovesan, E. Chiocca, and G. Maiocchi. Inhaled formeterol in the prevention of exercise-induced bronchoconstriction in asthmatic children. *Am J Respir Crit Care Med* 149: 935-939, 1994.

53. Boner, A.L., G. Vallone, M. Chiesa, E. Spezia, L. Fambri, and L. Sette. Reproducibility of late pulmonary response to exercise and its relationship to bronchial hyperreactivity in children with chronic asthma. *Pediatr Pulmonol* 14: 156-159, 1992.

54. Bouchard, C., A. Tremblay, C. Lebanc, G. Lortie, R. Savard, and G.A. Therialt. A method to assess energy expenditure in children and adults. *Am J Clin Nutr* 37: 461-467, 1983.

55. Boulet, L.-P., H. Turcotte, and S. Tennina. Comparative efficacy of salbutamol, ipratropium, and cromoglycate in the prevention of bronchospasm induced by exercise and hyperosmolar challenges. *J Allergy Clin Immunol* 83: 882-887, 1989.

56. Boulet, L.P., C. Legris, H. Turcotte, and J. Hebert. Prevalence and characteristics of late asthmatic responses to exercise. *J Allergy Clin Immunol* 80: 655-662, 1987.

57. Braback, L., and L. Kalvesten. Asthma in schoolchildren. Factors influencing morbidity in a Swedish survey. *Acta Paediatr Scand* 77: 826-830, 1988.

58. Braun, N.M.T., N. Arora, and D.F. Rochester. Force-length relationship of the normal human diaphragm. *J Appl Physiol* 53: 405-412, 1982.

59. Brenner, A., P.C. Weiser, K. Krogh, and M. Loren. Effectiveness of a portable face mask in attenuating exercise-induced asthma. *JAMA* 244: 2196-2198, 1980.

60. Bronstein, M.N., P.S.W. Davies, K.M. Hambidge, and F.J. Accurso. Normal energy expenditure in the infant with presymptomatic cystic fibrosis. *J Pediatr* 126: 28-33, 1995.

61. Brook, U., D. Stein, and Y. Alkalay. The attitude of asthmatic and nonasthmatic adolescents toward gymnastic lessons at school. *J Asthma* 31: 171-175, 1994.

62. Brudno, D.S., J.M. Wagner, and N.T. Rupp. Length of postexercise assessment in the determination of exercise-induced bronchospasm. *Ann Allergy* 73: 227-231, 1994.

63. Buchdahl, R.M., M. Cox, C. Fulleylove, J.L. Marchant, A.M. Tomkins, M.J. Brueton, and J.O. Warner. Increased resting energy expenditure in cystic fibrosis. *J Appl Physiol* 64: 1810-1816, 1988.

64. Bundgaard, A. Exercise and the asthmatic. *Sports Med* 2: 254-266, 1985.

65. Bundgaard, A. Physical training in bronchial asthma. *Scand J Sports Sci* 10: 97-105, 1989.

66. Burr, M.L., B.A. Eldridge, and L.K. Borysiewicz. Peak expiratory flow rates before and after exercise in schoolchildren. *Arch Dis Child* 49: 923-926, 1974.

67. Burr, M.L., E.S. Limb, S. Andrae, D.M. Barry, and F. Nagel. Childhood asthma in four countries: a comparative survey. *Int J Epidemiol* 23: 341-347, 1994.

68. Bury, J.D. Climate and chest disorders [letter]. *Br Med J* 4: 613, 1972.

69. Buttifant, D.C., J.S. Carlson, and G.A. Naughton. Anaerobic characteristics and performance of prepubertal asthmatic and nonasthmatic males. *Pediatr Exerc Sci* 8: 268-275, 1996.

70. Cabrera, M.E., M.D. Lough, C.F. Doerschuk, and A.E. Salvator. An expanded scoring system including an index on nutritional status for patients with cystic fibrosis. *Pediatr Pulmonol* 18: 199-205, 1994.

71. Cabrera, M.E., M.D. Lough, C.F. Doershuk, and G.A. DeRivera. Anaerobic performance—assessed by the Wingate test in patients with cystic fibrosis. *Pediatr Exerc Sci* 5: 78-87, 1993.

72. Carlsen, K.H., S. Oseid, H. Odden, and E. Mellbye. The response of children with and without bronchial asthma to heavy swimming exercise. In Oseid, S. and K.H. Carlsen, eds., *Children and exercise XIII.* Champaign, IL: Human Kinetics, 1989, 351-367.

73. Cerny, F., and L.M. Armitage. Exercise and cystic fibrosis: a review. *Pediatr Exerc Sci* 1: 116-126, 1989.

74. Cerny, F.J. Ventilatory control during exercise in children with cystic fibrosis. *Am Rev Respir Dis* 123: 195-202, 1981.

75. Cerny, F.J., G.J.A. Cropp, and M.R. Bye. Hospital therapy improves exercise tolerance and lung function in cystic fibrosis. *Am J Dis Child,* 138: 261-265, 1984.

76. Cerny, F.J., T. Pullano, and G.J.A. Cropp. Adaptation to exercise in children with cystic fibrosis. In Nagle, F.J. and H.J. Montoye, eds., *Exercise in health*

and disease. Springfield, IL: Charles C Thomas, 1982, 36-42.

77. Cerny, F.J., T.P. Pullano, and G.J.A. Cropp. Cardiorespiratory adaptations to exercise in cystic fibrosis. *Am Rev Respir Dis* 126: 217-220, 1982.

78. Chang-Yeung, M.M.W., M.N. Vyas, and S. Grzybowski. Exercise-induced asthma. *Am Rev Respir Dis* 104: 915-923, 1971.

79. Chen, W.Y., and D.J. Horton. Heat and water loss from the airways and exercise-induced asthma. *Respiration* 34: 305-313, 1977.

80. Chen, W.Y., and D.J. Horton. Airways obstruction in asthmatics induced by body cooling. *Scand J Respir Dis* 59: 13-20, 1978.

81. Chen, Y., R. Dales, and D. Krewski. Leisure-time energy expenditure in asthmatics and non-asthmatics. *Respir Med* 95: 13-18, 2001.

82. Chhabra, S.K., and U.C. Ojha. Late asthmatic response in exercise-induced asthma. *Ann Allergy Asthma Immunol* 80: 323-327, 1998.

83. Chipps, B.E., P.O. Alderson, J.M.A. Roland, A.V. Yang, C.R. Martinez, and B.J. Rosenstein. Noninvasive evaluation of ventricular function in cystic fibrosis. *J Pediatr* 95: 379-384, 1979.

84. Coates, A.L. Oxygen therapy, exercise and cystic fibrosis [editorial comment]. *Chest* 101: 2-4, 1992.

85. Coates, A.L., P. Boyce, D.G. Shaw, S. Godfrey, and M. Mearns. The role of nutritional status, airway obstruction, hypoxia, and abnormalities in serum lipid composition in limiting exercise tolerance in children with cystic fibrosis. *Acta Paediatr Scand* 69: 353-358, 1980.

86. Coates, A.L., G. Canny, R. Zinman, R. Grisdale, K. Desmond, D. Roumeliotis, and H. Levison. The effects of chronic airflow limitation, increased dead space, and the pattern of ventilation on gas exchange during maximal exercise in advanced cystic fibrosis. *Am Rev Respir Dis* 138: 1524-1531, 1988.

87. Coates, A.L., K. Desmond, M.I. Asher, J. Hortop, and P.H. Beaudry. The effect of digoxin in exercise capacity and exercising cardiac function in cystic fibrosis. *Chest* 82: 543-547, 1982.

88. Cole, P., R. Forsyth, and J.S.J. Haight. Effects of cold air and exercise on nasal patency. *Ann Otol Rhinol Laryngol* 92: 196-198, 1983.

89. Collins, C.E., E.V. O'Loughlin, and R. Henry. Discrepancies between males and females with cystic fibrosis in dietary intake and pancreatic enzyme use. *J Pediatr Gastro Nutr* 26: 258-262, 1998.

90. Corey, M., L. Edwards, H. Levison, and M. Knowles. Longitudinal analysis of pulmonary function decline in patients with cystic fibrosis. *J Pediatr* 131: 809-14, 1997.

91. Counil, F.P., C. Karila, A. Varray, S. Guillaumont, M. Voisin, and C. Préfaut. Anaerobic fitness in children

with asthma: adaptation to maximal intermittent short exercise. *Pediatr Pulmonol* 31: 198-204, 2001.

92. Counil, F.P., A. Varray, C. Karila, M. Hayot, M. Voisin, and C. Préfaut. Wingate test performance in children with asthma: aerobic or anaerobic limitation? *Med Sci Sports Exerc* 29: 430-435, 1997.

93. Cropp, G.J.A. The exercise bronchoprovocation test: standardization of procedures and evaluation of response. *J Allergy Clin Immunol* 64: 627-633, 1979.

94. Cropp, G.J.A., T.P. Pullano, F.J. Cerny, and I.T. Nathanson. Exercise tolerance and cardiorespiratory adjustments at peak work capacity in cystic fibrosis. *Am Rev Respir Dis* 126: 211-216, 1982.

95. Cropp, G.J.A., L.M. Silbiger, G.A. Pankow, A.K. Ford, and J.A. Hirsch. Effects of oxygen breathing on exercise tolerance in cystic fibrosis (CF). *Am Rev Respir Dis* 135: 286, 1987.

96. Custovic, A., N. Arifhodzic, A. Robinson, and A. Woodcock. Exercise testing revisited. The response to exercise in normal and atopic children. *Chest* 105: 1127-1132, 1994.

97. Dahl, R., and J.M. Henriksen. Effect of oral and inhaled sodium cromoglycate in exercise-induced asthma. *Allergy* 35: 363-365, 1980.

98. Day, G., and M.B. Mearns. Bronchial lability in cystic fibrosis. *Arch Dis Child* 48: 355-359, 1973.

99. de Benedictis, F.M., G. Tuteri, A. Bertotto, L. Bruni, and R. Vaccaro. Comparison of the protective effects of cromolyn sodium and nedocromil sodium in the treatment of exercise-induced asthma in children. *J Allergy Clin Immunol* 94: 684-688, 1994.

100. De Meer, K., V.A.M. Gulmans, and J. van der Laag. Peripheral muscle weakness and exercise capacity in children with cystic fibrosis. *Am J Respir Crit Care Med* 159: 748-754, 1999.

101. De Meer, K., J.A.L. Jeneson, V.A.M. Gulmans, J. van der Laag, and R. Berger. Efficiency of oxidative work performance of skeletal muscle in patients with cystic fibrosis. *Thorax* 50: 980-983, 1995.

102. De Meer, K., K.R. Westerterp, R.H.J. Houwen, H.A.A. Brouwers, R. Berger, and A. Okken. Total energy expenditure in infants with bronchopulmonary dysplasia is associated with respiratory status. *Eur J Pediatr* 156: 299-304, 1997.

103. Deal, E.C. Jr., E.R. McFadden Jr., R.H. Ingram Jr., and J.J. Jaeger. Effects of atropine on the potentiation of exercise-induced bronchospasm by cold air. *J Appl Physiol: Respir Environ Exerc Physiol* 45: 238-243, 1978.

104. Deal, E.C. Jr., E.R. McFadden Jr., R.H. Ingram Jr., and J.J. Jaeger. Esophageal temperature during exercise in asthmatic and nonasthmatic subjects. *J Appl Physiol: Respir Environ Exerc Physiol* 46: 484-490, 1979.

105. Deal, E.C. Jr., E.R. McFadden Jr., R.H. Ingram Jr., and J.J. Jaeger. Hyperpnea and heat flux: initial reaction

sequence in exercise-induced asthma. *J Appl Physiol: Respir Environ Exerc Physiol* 46: 476-483, 1979.

106. Deal, E.C. Jr., E.R. McFadden Jr., R.H. Ingram Jr., R.H. Strauss, and J.J. Jaeger. Role of respiratory heat exchange in production of exercise-induced asthma. *J Appl Physiol: Respir Environ Exerc Physiol* 46: 467-475, 1979.

107. Derrick, E.H. The seasonal variation of asthma in Brisbane: its relation to temperature and humidity. *Int J Biometeorol* 9: 239-251, 1965.

108. Dinh Xuan, A.T., C. Lebequ, R. Roche, A. Ferriére, and M. Chaussain. Inhaled terbutaline administered via a spacer fully prevents exercise-induced asthma in young asthmatic subjects: a double-blind, randomized, placebo-controlled study. *J Int Med Res* 17: 506-513, 1989.

109. Edelman, J.M., J.A. Turpin, E.A. Bronsky, J. Grossman, J.P. Kemp, A.F. Ghannam, P.T. DeLucca, G.J. Gormley, and D.S. Pearlman. Oral montelukast compared with inhaled salmeterol to prevent exercise-induced bronchoconstriction. A randomized, double-blind trial. Exercise Study Group. *Ann Intern Med* 132: 97-104, 2000.

110. Edlund, L.D., R.W. French, J.J. Herbst, H.D. Ruttenberg, R.O. Ruhling, and T.D. Adans. Effects of a swimming program on children with cystic fibrosis. *Am J Dis Child* 140: 80-83, 1986.

111. Edmunds, A.T., M. Tooley, and S. Godfrey. The refractory period after EIA: its duration and relation to the severity of exercise. *Am Rev Respir Dis* 117: 247-254, 1978.

112. Eggleston, P.A. Exercise-induced asthma in children with intrinsic and extrinsic asthma. *Pediatrics* 56: 856-859, 1975.

113. Eggleston, P.A. Laboratory evaluation of exercise-induced asthma: methodologic considerations. *J Allergy Clin Immunol* 64: 64-608, 1979.

114. Eggleston, P.A., and J.L. Geurrant. A standardized method of evaluating exercise-induced asthma. *J Allergy Clin Immunol* 58: 414-425, 1976.

115. Eggleston, P.A., R.R. Rosenthal, S.A. Anderson, R. Anderton, C. Bierman, E.R. Bleecker, C. Hyman, G.J.A. Cropp, J.D. Johnson, P. Konig, J. Morse, L.J. Smith, R.J. Summers, and J.J. Trautlein. Guidelines for the methodology of exercise challenge testing of asthmatics (Study Group on Exercise Challenge, Broncho-Provocation Committee, American Academy on Allergy). *J Allergy Clin Immunol* 64: 642-645, 1979.

116. Engstrom, I., K. Fallstrom, E. Karlberg, G. Sten, and J.B. Jure. Psychological and respiratory physiological effects of a physical exercise programme on boys with severe asthma. *Acta Paediatr Scand* 80: 1058-1065, 1991.

117. Engstrom, I., P. Karlberg, S. Kraepelien, and G. Wengler. Respiratory studies in children VIII. Respiratory adaptation during exercise tolerance test with special reference to mechanical properties of the lungs in asthmatic and healthy children. *Acta Paediatr Scand* 49: 850-858, 1960.

118. Falk, B., R. Dotan, D.G. Liebermann, R. Regev, S.T. Dolphin, and O. Bar-Or. Birth weight and physical ability in 5-8 year old healthy children born prematurely. *Med Sci Sports Exerc* 29: 1124-1130, 1997.

119. Fanta, C.H., E.R.J. McFadden, and R.H. Ingram Jr. Effects of cromolyn sodium on the response to respiratory heat loss in normal subjects. *Am Rev Respir Dis* 123: 161-164, 1981.

120. Feinstein, R.A., J. La Russa, A. Wang-Dohlman, and A.A. Bartolucci. Screening adolescent athletes for exercise-induced asthma. *Clin J Sport Med* 6: 119-123, 1996.

121. Feisal, K.A., and F.J.D. Fuleihan. Pulmonary gas exchange during exercise in young asthmatic patients. *Thorax* 34: 393-396, 1979.

122. Fernandes, A.L.G., N. Molfino, P.A. McClean, F. Silverman, S. Tarlo, M. Raizenne, A.S. Slutsky, and N. Zamel. The effect of pre-exposure to 0.12 ppm of ozone on exercise-induced asthma. *Chest* 106: 1077-1082, 1994.

123. Fink, G., C. Kaye, H. Blau, and S.A. Spitzer. Assessment of exercise capacity in asthmatic children with various degrees of activity. *Pediatr Pulmonol* 15: 41-43, 1993.

124. Finnerty, J.P. Role of leukotrienes in exercise-induced asthma. Inhibitory effect of ICI 204219, a potent leukotriene D4 receptor antagonist. *Am Rev Respir Dis* 145: 746-749, 1992.

125. Fisher, H.K., P. Hatton, R. St. J. Buxton, and J.A. Nudel. Resistance to breathing during exercise-induced asthma attacks. *Am Rev Respir Dis* 101: 885-896, 1970.

126. Fitch, K. Swimming medicine and asthma. In Eriksson, B. and B. Furberg, eds., *Swimming medicine IV.* Baltimore: University Park Press, 1978, 16-31.

127. Fitch, K.D. Exercise-induced asthma and competitive athletics. *Pediatrics* 56: 942-943, 1975.

128. Fitch, K.D. The use of anti-asthmatic drugs. Do they affect sports performance? *Sports Med* 3: 136-150, 1986.

129. Fitch, K.D., and S. Godfrey. Asthma and athletic performance. *JAMA* 236: 152-157, 1976.

130. Fitch, K.D., and A.R. Morton. Specificity of exercise in exercise-induced asthma. *Br Med J* 4: 577-581, 1971.

131. Fitch, K.D., A.R. Morton, and B.A. Blanksby. Effects of swimming training on children with asthma. *Arch Dis Child* 51: 190-194, 1976.

132. Fontana, V.J., A. Fost, and I. Rappaport. Effects of rapid change in humidity on pulmonary function studies in normal and asthmatic children in a controlled environment. *J Allergy* 43: 16-21, 1969.

133. Foresi, A., S. Mattoli, G.M. Corbo, A. Verga, A. Sommaruga, and G. Ciappi. Late bronchial response and increase in methacholine hyperresponsiveness after exercise and distilled water challenge in atopic subjects with asthma with dual asthmatic response to allergen inhalation. *J Allergy Clin Immunol* 78: 1130-1139, 1986.

134. Forsyth, R.D., P. Cole, and R.J. Shephard. Exercise and nasal patency. *J Appl Physiol: Respir Environ Exerc Physiol* 55(3): 860-865, 1983.

135. Freed, A.N. Models and mechanisms of exercise-induced asthma. *Eur Respir J* 8: 1770-1785, 1995.

136. Fried, M.D., P.R. Durie, L.-C. Tsui, M. Corey, H. Levison, and P.B. Pencharz. The cystic fibrosis gene and resting energy expenditure. *J Pediatr* 119: 913-916, 1991.

137. Friedman, M., K.L. Kovitz, S.D. Miller, M. Marks, and M.A. Sackner. Hemodynamics in teenagers and asthmatic children during exercise. *J Appl Physiol: Respir Environ Exerc Physiol* 46: 293-297, 1979.

138. Frischer, T., J. Kuehr, R. Meinert, W. Kermaus, R. Barth, E. Hermann-Keeuz, and R. Urbanek. Maternal smoking in early childhood: a risk factor for bronchial responsiveness to exercise in primary-school children. *J Pediatr* 121: 17-22, 1992.

139. Frischer, T., R. Meinert, W. Karmaus, R. Urbanek, and J. Kuehr. Relationship between atopy and frequent bronchial response to exercise in school children. *Pediatr Pulmonol* 17: 320-325, 1994.

140. Gergen, P., D. Mullay, and R. Evans. National survey of prevalence of asthma among children in the United States, 1977-1980. *Pediatrics* 81: 1-7, 1988.

141. Gerhard, H., and E.N. Schachter. Exercise-induced asthma. *Postgrad Med* 67: 91-102, 1980.

142. Germann Henke, K., J.A. Regnis, and P.T. Bye. Benefits of continuous positive airway pressure during exercise in cystic fibrosis and relationship to disease severity. *Am Rev Respir Dis* 148: 1272-1276, 1993.

143. Geubelle, F., J. Dechange, I. Louis, and M. Beyer. Respiratory function, energetic metabolism and work capacity in boys with asthma syndrome. *Acta Paediatr Belg* 31: 79-86, 1978.

144. Geubelle, F., C. Ernould, and M. Jovanovic. Working capacity and physical training in asthmatic children, at 1800m altitude. *Acta Paediatr Scand Suppl* 217: 93-98, 1971.

145. Giacoia, G.P., P.S. Venkataraman, K.I. West-Wilson, and M.J. Faulkner. Follow-up of school-age children with bronchopulmonary dysplasia. *J Pediatr* 130: 400-408, 1997.

146. Gilbert, I.A., J.M. Fouke, and E.R. McFadden Jr. Heat and water flux in the intrathoracic airways and exercise-induced asthma. *J Appl Physiol* 63: 1681-1691, 1987.

147. Gilbert, I.A., J.M. Fouke, and E.R. McFadden Jr. The effect of repetitive exercise on airway temperatures. *Am Rev Respir Dis* 142: 826-831, 1990.

148. Gilbert, I.A., and E.R. McFadden Jr. Airway cooling and rewarming. The second reaction sequence in exercise-induced asthma. *J Clin Invest* 90: 699-704, 1992.

149. Godfrey, S. *Exercise testing in children: applications in health and disease.* Philadelphia: Saunders, 1974.

150. Godfrey, S. Exercise-induced bronchial lability in wheezy children and their families. *Pediatrics* 56: 851-855, 1975.

151. Godfrey, S. Exercise-induced asthma: review article. *Allergy* 33: 229-237, 1978.

152. Godfrey, S., and P. Konig. Inhibition of exercise-induced asthma by different pharmacological pathways. *Thorax* 31: 137-143, 1976.

153. Godfrey, S., and M. Mearns. Pulmonary function and response to exercise in cystic fibrosis. *Arch Dis Child* 46: 144-151, 1971.

154. Godfrey, S., and M. Silverman. Demonstration by placebo response in asthma by means of exercise testing. *J Psychosom Res* 17: 293-297, 1973.

155. Godfrey, S., M. Silverman, and S. Anderson. The use of treadmill for assessing EIA and the effect of varying the severity and duration of exercise. *Pediatrics* 56: 893-899, 1975.

156. Godfrey, S., M. Silverman, and S.D. Anderson. Problems of interpreting exercise-induced asthma. *J Allergy Clin Immunol* 52(4): 199-209, 1973.

157. Graff-Lonnevig, V. Cardio-respiratory function, aerobic capacity and effect of physical activity in asthmatic boys. 1978. MD thesis. Karolinska Institute, Stockholm.

158. Graff-Lonnevig, V., S. Bevegård, and B.O. Eriksson. Cardiac output and blood pressure at rest and during exercise in boys with bronchial asthma. *Scand J Respir Dis* 60: 36-43, 1979.

159. Graff-Lonnevig, V., S. Bevegård, B.O. Eriksson, S. Kraepelien, and B. Saltin. Two years' follow-up of asthmatic boys participating in a physical activity programme. *Acta Paediatr Scand* 69: 347-352, 1980.

160. Green, C.P., and J.F. Price. Prevention of exercise induced asthma by inhaled salmeterol xinafoate. *Arch Dis Child* 67: 1014-1017, 1992.

161. Greenberg, L., F. Field, J.I. Reed, and C.L. Erhardt. Asthma and temperature change. *Arch Environ Health* 8: 642-647, 1964.

162. Griffiths, J., F.Y. Leung, S. Grzybowski, and M.M.W. Chan-Yeung. Sequential estimation of plasma catecholamines in exercise-induced asthma. *Chest* 62: 527-533, 1972.

163. Gruber, W., and D. Kiosz. Physical training and well-being in patients with cystic fibrosis [in German]. *Prav-Rehab Jahrgang* 8: S64-S69, 1996.

164. Grunow, J.E., M.P. Azcue, G. Berall, and P.B. Pencharz. Energy expenditure in cystic fibrosis during activities of daily living. *J Pediatr* 122: 243-246, 1993.

165. Gulmans, V.A.M., K. De Meer, H.J.L. Brackel, J.A.J. Faber, R. Berger, and P.J.M. Helders. Outpatient exercise training in children with cystic fibrosis: physiological effects, perceived competence, and acceptability. *Pediatr Pulmonol* 28: 39-46, 1999.

166. Gulmans, V.A.M., K. De Meer, H.J.L. Brackel, and P.J.M. Helders. Maximal work capacity in relation to nutritional status in children with cystic fibrosis. *Eur Respir J* 10: 2014-2017, 1997.

167. Haas, F., H. Pineda, K. Axen, D. Gaudino, and A. Haas. Effects of physical fitness on expiratory airflow in exercising asthmatic people. *Med Sci Sports Exerc* 17: 585-592, 1985.

168. Haby, M.M., S.D. Anderson, J.K. Peat, C.M. Mellis, B.G. Toelle, and A.J. Woolcock. An exercise challenge protocol for epidemiological studies of asthma in children: comparison with histamine challenge. *Eur Respir J* 7: 43-49, 1994.

169. Hack, M., and A.A. Fanaroff. Outcomes of children of extremely low birthweight and gestational age in the 1990's. *Early Hum Dev* 53: 193-218, 1999.

170. Hahn, A., S.D. Anderson, A.R. Morton, J.L. Black, and K.D. Fitch. A reinterpretation of the effect of temperature and water content of the inspired air in exercise-induced asthma. *Am Rev Respir Dis* 130: 575-579, 1984.

171. Hahn, A.G., S.G. Nogrady, G.R. Burton, and A.R. Morton. Absence of refractoriness in asthmatic subjects after exercise with warm, humid inspirate. *Thorax* 40: 418-421, 1985.

172. Hahn, A.G., S.G. Nogrady, D.M. Tumilty, S.R. Lawrence, and A.R. Morton. Histamine reactivity during the refractory period after exercise induced asthma. *Thorax* 39: 919-923, 1984.

173. Hamielec, C.M., P.J. Manning, and P.M. O'Byrne. Exercise refractoriness after histamine inhalation in asthmatic subjects. *Am Rev Respir Dis* 138: 794-798, 1988.

174. Hanning, R.M., C.J.R. Blimkie, O. Bar-Or, L.C. Lands, L.A. Moss, and W.M. Wilson. Relationships among nutritional status, skeletal and respiratory muscle function in cystic fibrosis: does early dietary supplementation make a difference? *Am J Clin Nutr* 57: 580-587, 1993.

175. Hansen-Flaschen, J., and H. Schotland. New treatments for exercise-induced asthma. *N Eng J Med* 339: 192-193, 1998.

176. Hebestreit, A., U. Kersting, B. Basler, R. Jeschke, and H. Hebestreit. Exercise inhibits epithelial sodium channels in patients with cystic fibrosis. *Am J Respir Crit Care Med* 164: 443-446, 2001.

177. Heijerman, H.G., W. Bakker, P.J. Sterk, and J.H. Dijkman. Oxygen-assisted exercise training in adult cystic fibrosis patients with pulmonary limitation to exercise. *Int J Rehab Research* 14: 101-115, 1991.

178. Heijerman, H.G.M., W. Bakker, P.J. Sterk, and J.H. Dijkman. Long-term effects of exercise training and hyperalimentation in adult cystic fibrosis patients with severe pulmonary dysfunction. *Int J Rehab Research* 15: 252-257, 1992.

179. Heldt, G.P., M.B. McIlroy, T.N. Hansen, and W.H. Tooley. Exercise performance of the survivors of hyaline membrane disease. *J Pediatr* 96: 995-999, 1980.

180. Henderson, R.C., and C.D. Madsen. Bone mineral content and body composition in children and young adults with cystic fibrosis. *Pediatr Pulmonol* 27: 80-84, 1999.

181. Henke, K.G., and D.M. Orenstein. Oxygen saturation during exercise in cystic fibrosis. *Am Rev Respir Dis* 129: 708-711, 1984.

182. Henriksen, J.M. Effect of inhalation of corticosteroids on exercise induced asthma: randomised double blind crossover study of budesonide in asthmatic children. *Br Med J* 291: 248-249, 1985.

183. Henriksen, J.M. Protective effect and duration of action of inhaled formoterol and salbutamol on exercise-induced asthma in children. *J Allergy Clin Immunol* 89(6): 1176-1182, 1992.

184. Henriksen, J.M., and R. Dahl. Effects of inhaled budesonide alone and in combination with low-dose terbutaline in children with exercise-induced asthma. *Am Rev Respir Dis* 128: 993-997, 1983.

185. Henriksen, J.M., R. Dahl, and G.R. Lundquist. Influence of relative humidity and repeated exercise on exercise-induced bronchoconstriction. *Alleryg* 36: 463-470, 1981.

186. Henriksen, J.M., and T. Nielsen. Effect of physical training on exercise-induced bronchoconstriction. *Acta Paediatr Scand* 72: 31-36, 1983.

187. Henriksen, J.M., and T.T. Nielsen. Effects of physical training on exercise-induced bronchoconstriction. In Oseid, S., ed., *The asthmatic child in play and sports*. London: Pitman, 1983.

188. Herxheimer, H. Hyperventilation asthma. *Lancet* 1: 83-87, 1946.

189. Hjeltnes, N., J.K. Stanghelle, and D. Skyberg. Pulmonary function and oxygen uptake during exercise in 16 year old boys with cystic fibrosis. *Acta Paediatr Scand* 73: 548-553, 1984.

190. Holzer, F.J., R. Schnall, and L.I. Landau. The effect of a home exercise programme in children with cystic fibrosis and asthma. *Austr Paediatr J* 20: 297-302, 1984.

191. Horton, D.J., and W.Y. Chen. Effects of breathing warm humidified air on bronchoconstriction induced by body cooling and inhalation of metacholine. *Chest* 75: 24-28, 1979.

192. Huang, S., R. Veiga, U. Sila, E. Reed, and S. Hines. The effect of swimming in asthmatic children-participants in swimming program in the city of Baltimore. *J Asthma* 26: 117-121, 1989.

193. Hyde, J.S., and C.L. Swarts. Effect of an exercise program on the perennially asthmatic child. *Am J Dis Child* 116: 383-396, 1968.

194. Inbar, O., D.X. Alvarez, and H.A. Lyons. Exercise-induced asthma—a comparison between two models of exercise stress. *Eur J Respir Dis* 62: 160-167, 1981.

195. Inbar, O., R.A. Dlin, A. Sheinberg, and M. Scheinowitz. Response to progressive exercise in patients with cystic fibrosis and asthma. *Med Exerc Nutr Health* 2: 55-61, 1993.

196. Inbar, O., R. Dotan, R.A. Dlin, and I. Neuman. Breathing dry or humid air and exercise-induced asthma during swimming. *Eur J Appl Physiol* 44: 43-50, 1980.

197. Inbar, O., S. Naiss, E. Neuman, and J. Daskalovich. The effect of body posture on exercise and hyperventilation-induced asthma. *Chest* 100: 1229-1234, 1991.

198. Inbar, O., Y. Weinstein, Y. Daskalovic, R. Levi, and I. Neuman. The effect of prone immersion on bronchial responsiveness in children with asthma. *Am C Sports Med* 9: 1098-1102, 1993.

199. Ionescu, A.A., K. Chatham, C.A. Davies, L.S. Nixon, S. Enright, and D.J. Shale. Inspiratory muscle function and body composition in cystic fibrosis. *Am J Respir Crit Care Med* 158: 1271-1276, 2000.

200. Jacob, S.V., L.C. Lands, A.L. Coates, G.M. Davis, C.F. MacNeish, L. Hornby, S.P. Riley, and E.W. Outerbridge. Exercise ability in survivors of severe bronchopulmonary dysplasia. *Am J Respir Crit Care Med* 155: 1925-1929, 1997.

201. James, L., J. Faciane, and R.M. Sly. Effect of treadmill exercise on asthmatic children. *J Allergy Clin Immunol* 57: 408-416, 1976.

202. Johnson, J.D. Statistical considerations in studies of exercise-induced bronchospasm. *J Allergy Clin Immunol* 64: 634-641, 1979.

203. Jones, A., and M. Bowen. Screening for childhood asthma using an exercise test. *Br J Gen Pract* 44: 127-131, 1994.

204. Jones, R.H.T., and R.S. Jones. Ventilatory capacity in young adults with a history of asthma in childhood. *Br Med J* 2: 976-978, 1966.

205. Jones, R.S. Assessment of respiratory function in the asthmatic child. *Br Med J* 2: 972-975, 1966.

206. Jones, R.S., M.J. Wharton, and M.H. Buston. The place of physical exercise and bronchodilator drugs in the assessment of the asthmatic child. *Arch Dis Child* 38: 539-545, 1963.

207. Joos, G.F., R.A. Pauwels, and M.E. Van Der Straeten. The effect of nedocromil sodium on the bronchoconstrictor effect of neurokinin A in subjects with asthma. *J Allergy Clin Immunol* 83: 663-668, 1989.

208. Josenhans, W.T., G.N. Melville, and W.T. Ulmer. The effect of facial cold stimulation on airway conductance in man. *Can J Physiol Pharmacol* 47: 453-457, 1969.

209. Juto, J.E., and C. Lindberg. Nasal mucosa reaction, catecholamines and lactate during physical exercise. *Acta Otolaryngol* 98: 533-542, 1984.

210. Kaplan, T.A., G. Moccia-Loos, M. Rabin, and R.M. McKey. Lack of effect of delta F508 mutation on aerobic capacity in patients with cystic fibrosis. *Clin J Sports Med* 6: 226-231, 1996.

211. Karila, C., J. de Blic, S. Waernessyckle, M.R. Benoist, and P. Scheinmann. Cardiopulmonary exercise testing in children: an individualized protocol for workload increase. *Chest* 120: 81-87, 2001.

212. Karjalainen, J., A. Lindqvist, and L.A. Laitinen. Seasonal variability of exercise-induced asthma especially outdoors. Effect of birch pollen allergy. *Clin Exp Allergy* 19: 273-278, 1989.

213. Katsardis, C.V., K.J. Desmond, and A.L. Coates. Measuring the oxygen cost of breathing in normal adults and patients with cystic fibrosis. *Respir Physiol* 65: 257-266, 1986.

214. Kattan, M., T.G. Keens, C.M. Mellis, and H. Levison. The response to exercise in normal and asthmatic children. *J Pediatr* 92: 718-721, 1978.

215. Katz, R.M., S.C. Siegel, and G.S. Rachelefsky. Blood gas in exercise-induced bronchospasm: a review. *Pediatrics* 56: 880-882, 1975.

216. Kawabori, I., W.E. Pierson, L.L. Conquest, and D.W. Bierman. Incidence of exercise-induced asthma in children. *J Allergy Clin Immunol* 58: 447-455, 1976.

217. Keens, T.G. Exercise training programs for pediatric patients with chronic lung disease. *Pediatr Clin N Am* 26: 517-524, 1979.

218. Keens, T.G., I.R.B. Krastins, E.M. Wannamaker, H. Levison, O.N. Crozier, and C. Bryan. Ventilatory muscle endurance training in normal subjects and patients with cystic fibrosis. *Am Rev Respir Dis* 116: 853-860, 1977.

219. Keller, H., B.V. Ayub, S. Saigal, and O. Bar-Or. Neuromotor ability in 5- to 7-year-old children with very low or extremely low birthweight. *Dev Med Child Neurol* 40: 661-666, 1998.

220. Kemp, J.P., R.J. Dockhorn, W.W. Busse, E.R. Bleecker, and A. van As. Prolonged effect of inhaled salmeterol against exercise-induced bronchospasm. *Am J Respir Crit Care Med* 150: 1612-1615, 1994.

221. Kemp, J.P., R.J. Dockhorn, G.G. Shapiro, H.H. Nguyen, T.F. Reiss, B.C. Seidenberg, and B. Knorr. Montelukast once daily inhibits exercise-induced bronchoconstriction in 6- to 14-year-old children with asthma. *J Pediatr* 133: 424-428, 1998.

222. Kilham, H., M. Tooley, and M. Silverman. Running, walking and hyperventilation causing asthma in children. *Thorax* 34: 582-586, 1979.

223. Kirkpatrick, M.B., D. Sheppard, J.A. Nadel, and H.A. Boushey. Effect of the oronasal breathing route on sulfur dioxide-induced bronchoconstric-

tion in exercising asthmatic subjects. *Am Rev Respir Dis* 125: 627-631, 1982.

224. Kivity, S., H. Bibi, Y. Schwarz, Y. Greif, M. Topilsky, and E. Tabachnick. Variable vocal cord dysfunction presenting as wheezing and exercise-induced asthma. *J Asthma* 23: 241-244, 1986.

225. Klesges, R.C., T.J. Coates, L.M. Klesges, B. Holzer, J. Gustavson, and J. Barnes. The FATS: an observational system for assessing physical activity in children and associated parent behavior. *Behav Assess* 6: 333-345, 1984.

226. Knöpfli, B.H., and O. Bar-Or. Vagal activity and airway response to ipratropium bromide before and after exercise in ambient and cold conditions in healthy cross-country runners. *Clin J Sport Med* 9: 170-176, 1999.

227. Koh, Y.Y., H.S. Lim, and K.U. Min. Airway responsiveness to allergen is increased 24 hours after exercise challenge. *J Allergy Clin Immunol* 94: 507-516, 1994.

228. Konig, P. Clinical implications of bronchial lability in relation to asthma. Dissertation, University of London, 1974.

229. Konig, P., and S. Godfrey. Exercise-induced bronchial lability and atopic status of families of infants with wheezy bronchitis. *Arch Dis Child* 48: 942-946, 1973.

230. Kornfalt, R., B. Andreasson, E. Nordmark, and B. Jonson. Chest physiotherapy and physical exercise in cystic fibrosis. *Laekartidningen* 3: 851-853, 1986.

231. Kriemler, S., B. Wilk, W. Schurer, W.M. Wilson, and O. Bar-Or. Preventing dehydration in children with cystic fibrosis who exercise in the heat. *Med Sci Sports Exerc* 31: 774-779, 1999.

232. Kruhlak, R.T., R.L. Jones, and N.E. Brown. Regional air trapping before and after exercise in young adults with cystic fibrosis. *Western J Med* 145: 196-199, 1986.

233. Lands, L., K.J. Desmond, D. Demizio, A. Pavilanis, and A.L. Coates. The effects of nutritional status and hyperinflation on respiratory muscle strength in children and young adults. *Am Rev Respir Dis* 141: 1506-1509, 1990.

234. Lands, L.C., G.J.F. Heigenhauser, and N.L. Jones. Analysis of factors limiting maximal exercise performance in cystic fibrosis. *Clin Sci* 83: 391-397, 1992.

235. Lands, L.C., G.J.F. Heigenhauser, and N.L. Jones. Maximal short-term exercise performance and ion regulation in cystic fibrosis. *Can J Physiol Pharmacol* 71: 12-16, 1993.

236. Lands, L.C., G.J.F. Heigenhauser, and N.L. Jones. Respiratory and peripheral muscle function in cystic fibrosis. *Am Rev Respir Dis* 147: 865-869, 1993.

237. Lebecque, P., J.G. Lapierre, A. Lamerre, and A.L. Coates. Diffusion capacity and oxygen desaturation effects on exercise in patients with cystic fibrosis. *Chest* 91: 693-697, 1987.

238. Lebowitz, M.D., P. Bendheim, G. Ceistea, D. Markovitz, J. Misiaszek, M. Staniec, and D. van Wyck. The effect of air pollution and weather on lung function in exercising children and adolescents. *Am Rev Respir Dis* 109: 262-273, 1979.

239. Lee, T.H., T. Nagakura, N. Papageorgiou, Y. Iikura, and A.B. Kay. Exercise-induced late asthmatic reactions with neutrophil chemotactic activity. *N Eng J Med* 308: 1502-1505, 1983.

240. Leff, J.A., W.W. Busse, D. Pearlman, E.A. Bronsky, J. Kemp, L. Hendeles, R. Dockhorn, S. Kundu, J. Zhang, B.C. Seidenberg, and T.F. Reiss. Montelukast, a leukotriene-receptor antagonist, for the treatment of mild asthma and exercise-induced bronchoconstriction. *N Eng J Med* 339: 147-152, 1998.

241. Leisti, S., M.-J. Finnila, and E. Kiura. Effects of physical training on hormonal responses to exercise in asthmatic children. *Arch Dis Child* 54: 524-528, 1979.

242. Lin, C.C., W. Jen-Liang, W.C. Huang, and C.Y. Lin. A bronchial response comparison of exercise and metacholine in asthmatic subjects. *J Asthma* 28: 31-40, 1991.

243. Ludwick, S.K., J.W. Jones, T.K. Jones, J.T. Fukuhara, and R.C. Strunk. Normalization of cardiopulmonary endurance in severely asthmatic children after bicycle ergometry therapy. *J Pediatr* 109: 446-451, 1986.

244. Makker, H.K., A.F. Walls, D. Goulding, S. Montefort, J.J. Varley, M. Carroll, P.H. Howarth, and S.T. Holgate. Airway effects of local challenge with hypertonic saline in exercise-induced asthma. *Am J Respir Crit Care Med* 149: 1012-1019, 1994.

245. Malo, J.L., S. Filiatrault, and R.R. Martin. Combined effects of exercise and exposure to outside cold air on lung functions of asthmatics. *Bull Eur Physiopath Respir* 16: 623-635, 1980.

246. Mangla, P.K., and M.P.S. Menon. Effect of nasal and oral breathing on exercise-induced asthma. *Clin Allergy* 11: 433-439, 1981.

247. Manning, P.J., R.M. Watson, J. Margolskee, V.C. Williams, J.I. Schwartz, and P. O'Byrne. Inhibition of exercise-induced bronchoconstriction by MK-571, a potent leukotriene D4 receptor antagonist. *N Eng J Med* 323(25): 1736-1739, 1990.

248. Mansfield, L., J. McDonnell, W. Morgan, and J.F. Souhrada. Airway response in asthmatic children during and after exercise. *Respiration* 38: 135-143, 1979.

249. Marcotte, J.E., G.J. Canny, R. Grisdale, K. Desmond, M. Corey, R. Zinman, H. Levison, and A.L. Coates.

Effect of nutritional status on exercise performance in advanced cystic fibrosis. *Chest* 90: 375-379, 1986.

250. Marcotte, J.E., R.K. Grisdale, H. Levison, A.L. Coates, and G.J. Canny. Multiple factors limit exercise capacity in cystic fibrosis. *Pediatr Pulmonol* 2: 274-281, 1986.

251. Marcus, C.L., D. Bader, M.W. Stabile, C.-I. Wang, A.B. Osher, and T.G. Keens. Supplemental oxygen and exercise performance in patients with cystic fibrosis with severe pulmonary disease. *Chest* 101: 52-57, 1992.

252. Matsumoto, I., H. Araki, K. Tsuda, H. Odajima, S. Nishima, Y. Higaki, H. Tanaka, M. Tanaka, and M. Shindo. Effects of swimming training on aerobic capacity and exercise induced bronchoconstriction in children with bronchial asthma. *Thorax* 54: 196-201, 1999.

253. McAlpine, L.G., and N.C. Thomson. Prophylaxis of exercise-induced asthma with inhaled formeterol, a long-acting beta-adrenergic agonist. *Respir Med* 84: 293-295, 1990.

254. McCarthy, P. Wheezing and breezing through exercise-induced asthma. *Phys Sportsmed* 17(7): 125-130, 1989.

255. McFadden, E.R. Jr., J.A. Nelson, M.E. Skowronski, and K.A. Lenner. Thermally induced asthma and airway drying. *Am J Respir Crit Care Med* 160: 221-226, 1999.

256. McFadden, E.R. Jr., and I.A. Gilbert. Exercise-induced asthma [review article]. *N Eng J Med* 330: 1362-1367, 1994.

257. McFadden, E.R. Jr., R.H. Ingram Jr., R.L. Haynes, and J.J. Wellman. Predominant site of flow limitation and mechanisms of postexertional asthma. *J Appl Physiol: Respir Environ Exerc Physiol* 42: 746-752, 1977.

258. McFadden, E.R. Jr., K.A. Lenner, and K.P. Strohl. Postexertional airway rewarming and thermally induced asthma. New insights into pathophysiology and possible pathogenesis. *J Clin Invest* 78: 18-25, 1986.

259. McKenzie, D.C., S.L. McLuckie, and D.R. Stirling. The protective effects of continuous and interval exercise in athletes with exercise-induced asthma. *Med Sci Sports Exerc* 26: 951-956, 1994.

260. McLoughlin, P., D. McKeogh, P. Byrne, G. Finlay, J. Hayes, and M.X. Fitzgerald. Assessment of fitness in patients with cystic fibrosis and mild lung disease. *Thorax* 52: 425-430, 1997.

261. McNally, J.F., P. Enright, and J.F. Souhrada. The role of the oropharynx in exercise-induced bronchoconstriction. *Am Rev Respir Dis* 117: 372, 1978.

262. McNeill, R.S., J.R. Nairn, J.S. Millar, and C.G. Ingram. Exercise-induced asthma. *Q J Med* 35: 55-67, 1966.

263. Meeuwisse, W., D.C. McKenzie, S. Hopkins, and J. Road. The effect of salbutamol on performance in elite nonasthmatic athletes. *Med Sci Sports Exerc* 24: 1161-1166, 1992.

264. Mellis, C.M., and H. Levison. Bronchial reactivity in cystic fibrosis. *Pediatrics* 61: 446-450, 1978.

265. Millar, J.S., J.R. Nairn, R.D. Unkles, and R.S. McNeill. Cold air and ventilatory function. *Br J Dis Chest* 59: 23-27, 1965.

266. Miller, G.J., B.H. Davies, T.J. Cole, and A. Seaton. Comparison of the bronchial response to running and cycling in asthma using an improved definition of the response to work. *Thorax* 30: 306-311, 1975.

267. Mitsubayashi, T. Effect of physical training on exercise-induced bronchospasm of institutionalized asthmatic children. *Arerugi* 33: 318-327, 1984.

268. Moorcroft, A.J., M.E. Dodd, and A.K. Webb. Exercise testing and prognosis in adult cystic fibrosis. *Thorax* 52: 291-293, 1997.

269. Morton, A.R., and K.D. Fitch. Asthmatic drugs and competitive sport. An update. *Sports Med* 14(4): 228-242, 1992.

270. Morton, A.R., K.D. Fitch, and T. Davis. The effect of "warm-up" on exercise-induced asthma. *Ann Allergy* 42: 257-260, 1979.

271. Morton, A.R., S.M. Papalia, and K.D. Fitch. Is salbutamol ergogenic? The effects of salbutamol on physical performance in high-performance nonasthmatic athletes. *Clin J Sport Med* 2: 93-97, 1992.

272. Morton, A.R., K.J. Turner, and K.D. Fitch. Protection from exercise-induced asthma by pre-exercise cromolyn sodium and its relationship to serum IgE levels. *Ann Allergy* 31: 265-271, 1973.

273. Moser, C., P. Tirakitsoontorn, E. Nussbaum, R. Newcomb, and D.M. Cooper. Muscle size and cardiorespiratory response to exercise in cystic fibrosis. *Am J Respir Crit Care Med* 162: 1823-1827, 2000.

274. Mukhtar, M.R., and J.M. Patrick. Bronchoconstriction: a component of the "diving response" in man. *Eur J Appl Physiol* 53: 153-158, 1984.

275. Mussaffi, H., C. Springer, and S. Godfrey. Increased bronchial responsiveness to exercise and histamine after allergen challenge in children with asthma. *J Allergy Clin Immunol* 77: 48-52, 1986.

276. Nelson, J.A., L. Strauss, M. Skowronski, R. Ciufo, R. Novak, and E.R. McFadden Jr. Effect of long-term salmeterol treatment on exercise-induced asthma. *N Eng J Med* 339: 141-146, 1998.

277. Nickerson, B.G., D.B. Bautista, M.A. Namey, W. Richards, and T.G. Keens. Distance running improves fitness in asthmatic children without pulmonary complications or changes in exercise-induced bronchospasm. *Pediatrics* 71: 147-152, 1983.

278. Ninan, R.K., and G. Russell. Is exercise testing useful in a community based asthma survey? *Thorax* 48: 1218-1221, 1993.

279. Nixon, P.A., D.M. Orenstein, S.E. Curtis, and E.A. Ross. Oxygen supplementation during exercise in cystic fibrosis. *Am Rev Respir Dis* 142: 807-811, 1990.

280. Nixon, P.A., D.M. Orenstein, S.F. Kelsey, and C.F. Doershuk. The prognostic value of exercise testing in patients with cystic fibrosis. *N Eng J Med* 327: 1785-1788, 1992.

281. Nordemar, R., U. Berg, B. Ekblom, and L. Edström. Changes in muscle fibre size and physical performance in patients with rheumatoid arthritis after 7 months of physical training. *Scand J Rheumatol* 5: 233-238, 1976.

282. Novembre, E., G. Grongia, E. Lombardi, G. Veneruso, and A. Vierucci. Respiratory pathophysiologic responses. The preventive effect of nedocromil or furosemide alone or in combination on exercise-induced asthma in children. *J Allergy Clin Immunol* 94: 201-206, 1994.

283. Nystad, W. The physical activity level in children with asthma based on a survey among 7-16 year old school children. *Scand J Med Sci Sports* 7: 331-335, 1997.

284. Nystad, W., J. Harris, and J.S. Borgen. Asthma and wheezing among Norwegian elite athletes. *Med Sci Sports Exerc* 32: 266-270, 2000.

285. Nystad, W., P. Magnus, A. Gulsvik, I.J. Skarpaas, and K.H. Carlsen. Changing prevalence of asthma in school children: evidence for diagnostic changes in asthma in two surveys 13 yrs apart. *Eur Respir J* 10: 1046-1051, 1997.

286. O'Byrne, P.M., and G.L. Jones. The effect of indomethacin on exercise-induced bronchoconstriction and refractoriness after exercise. *Am Rev Respir Dis* 134: 69-72, 1986.

287. Obata, T., and Y. Iikura. Comparison of bronchial reactivity to ultrasonically nebulized distilled water, exercise and methacholine challenge test in asthmatic children. *Ann Allergy* 72: 167-172, 1994.

288. Oldenburg, F.A., M.B. Dolovich, J.M. Montgomery, and M.T. Newhouse. Effects of postural drainage, exercise and cough on mucus clearance in chronic bronchitis. *Am Rev Respir Dis* 120: 739-745, 1979.

289. Olson, L.G., and K.P. Strohl. The response of the nasal airway to exercise. *Am Rev Respir Dis* 135: 356-359, 1987.

290. Onda, T., T. Nagakura, and Y. Ikura. Asthmatic children and swimming training. A comparison of NCF and $FEV_{1.0}$ changes. *Areragi* 34: 1015-1020, 1985.

291. Orenstein, D.M. Cystic fibrosis. In Goldberg, B., ed., *Sports and exercise for children with chronic health conditions.* Champaign, IL: Human Kinetics, 1995, 167-186.

292. Orenstein, D.M. Asthma and sports. In Bar-Or, O., ed., *The child and adolescent athlete.* Oxford: Blackwell Scientific, 1996, 433-454.

293. Orenstein, D.M., B.A. Franklin, C.F. Doerchuk, H.K. Hellerstein, K.J. Germann, J.G. Horowitz, and R.C. Stern. Exercise conditioning and cardiopulmonary fitness in cystic fibrosis. The effects of a three-month supervised running program. *Chest* 80: 392-398, 1981.

294. Orenstein, D.M., K.G. Henke, D.L. Costill, C.F. Doershuk, P.J. Lemon, and R.C. Stern. Exercise and heat stress in cystic fibrosis patients. *Pediatr Res* 17: 267-269, 1983.

295. Orenstein, D.M., K.G. Henks, and F.J. Cerny. Exercise and cystic fibrosis. *Phys Sportsmed* 11: 57-63, 1983.

296. Orenstein, D.M., and A.P.A. Nixon. Patients with cystic fibrosis. In Franklin, B.A., S. Gordon, and G.C. Timmins, eds., *Exercise in modern medicine: testing and prescription in health and disease.* Baltimore: Williams & Wilkins, 1989, 204-214.

297. Orenstein, D.M., and P. Nixon. Exercise performance and breathing patterns in cystic fibrosis: male-female differences and influence of resting pulmonary function. *Pediatr Pulmonol* 10: 101-105, 1991.

298. Orenstein, D.M., M.E. Reed, F.T. Grogan, and L.V. Crawford. Exercise conditioning in children with asthma. *J Pediatr* 105: 556-560, 1985.

299. Oseid, S. Exercise-induced asthma: a review. In Berg, K. and B.O. Eriksson, eds., *Children and exercise IX.* Baltimore: University Park Press, 1980, 277-288.

300. Oseid, S., and K. Haaland. Exercise studies on asthmatic children before and after regular physical training. In Eriksson, B. and B. Furberg, eds., *Swimming medicine IV.* Baltimore: University Park Press, 1978, 32-41.

301. Oseid, S., M. Kendall, R.B. Larsen, and R. Selbekk. Physical activity programs for children with exercise-induced asthma. In Eriksson, B. and B. Furberg, eds., *Swimming medicine IV.* Baltimore: University Park Press, 1978, 42-51.

302. Paul, D.W., J.M. Bogaard, and W.C. Hop. The bronchoconstrictor effect of strenuous exercise at low temperatures in normal athletes. *Int J Sports Med* 14: 433-436, 1993.

303. Pavord, I., H. Lazarowicz, D. Inchley, D. Baldwin, A. Knox, and A. Tattersfield. Cross refractoriness between sodium metabisulphite and exercise induced asthma. *Thorax* 49: 245-249, 1994.

304. Pavord, I.D., A. Wisniewski, and A.E. Tattersfield. Inhaled furosemide and exercise-induced asthma. Evidence for a role for inhibitory prostanoids. *Am Rev Respir Dis* 143: 210-215, 1991.

305. Pearson, R.B. The effect of exercise in asthma. *Acta Allergol (Kbh)* 5: 310-311, 1952.

306. Penny, P.T. Swimming pool wheezing. *Br Med J* 287(6390): 461-462, 1983.

307. Perrault, H., M. Coughlan, J.-E. Marcotte, S.P. Drblik, and A. Lamarre. Comparison of cardiac

output determinants in response to upright and supine exercise in patients with cystic fibrosis. *Chest* 101: 42-51, 1992.

308. Petersen, K.H., and T.R. McElhenney. Effects of a physical fitness program upon asthmatic boys. *Pediatrics* 35: 295-299, 1965.

309. Pianosi, P.T., and M. Fisk. Cardiopulmonary exercise performance in prematurely born children. *Pediatr Res* 47: 653-658, 2000.

310. Pierson, W.E., and C.W. Bierman. Free running test for exercise-induced bronchospasm. *Pediatrics* 56: 890-892, 1975.

311. Pierson, W.E., C.W. Bierman, and S.J. Stamm. Cycloergometer-induced bronchospasm. *J Allergy* 43: 136-144, 1969.

312. Pierson, W.E., and R.O. Voy. Exercise-induced bronchospasm in the XXIII summer Olympic games. *N Eng Reg Allergy Proc* 9: 209-213, 1988.

313. Prasad, S.A., and F.J. Cerny. Factors that influence adherence to exercise and their effectiveness: application to cystic fibrosis. *Pediatr Pulmonol* 34: 66-72, 2002.

314. Preece, M.A. Prepubertal and pubertal endocrinology. In Falkner, F. and J.M. Tanner, eds., *Human growth: a comprehensive treatise.* Vol. 2. New York: Plenum Press, 1985, 211-223.

315. Price, J.F., P.H. Weller, and D.J. Matthew. Response to bronchial provocation and exercise in children with cystic fibrosis. *Clin Allergy* 9: 563-570, 1979.

316. Proctor, D.F. The upper airway. I. Nasal physiology and defense of the lungs. *Am Rev Respir Dis* 115: 97-129, 1977.

317. Proctor, D.F., I. Andersen, and G.R. Lundqvist. Human nasal mucosal function at controlled temperatures. *Respir Physiol* 30: 109-124, 1977.

318. Ramonatxo, M., F.A. Amsalem, J.G. Mercier, R. Jean, and C. Prefaut. Ventilatory control during exercise in children with mild or moderate asthma. *Med Sci Sports Exerc* 21: 11-17, 1989.

319. Ramsey, B.W., P.M. Farrell, P.B. Pencharz, and the Consensus Committee. Nutritional assessment and management in cystic fibrosis: a consensus report. *Am J Clin Nutr* 55: 108-116, 1992.

320. Reggiani, E., L. Marugo, A. Delpino, G. Paistra, G. Chiodini, and G. Odaglia. A comparison of various exercise challenge tests on airway reactivity in atopical swimmers. *J Sports Med Phys Fitness* 28: 394-401, 1988.

321. Reiff, D.B., N.B. Choudry, N.B. Pride, and P.W. Ind. The effect of prolonged submaximal warm-up exercise on exercise-induced asthma. *Am Rev Respir Dis* 139: 479-484, 1989.

322. Reisman, J.J., B. Rivington-Law, M. Corey, J. Marcotte, E. Wannamaker, D. Harcourt, and H. Levison. Role of conventional physiotherapy in cystic fibrosis. *J Pediatr* 113: 632-636, 1988.

323. Richerson, H.B., and P.M. Seebohm. Nasal airway response to exercise. *J Allergy* 41: 269-284, 1968.

324. Roberts, W.O. Certifying wrestlers' minimum weight. *Phys Sports Med* 26: 79-81, 1998.

325. Robertson, C.F., E. Heycock, J. Bishop, T. Nolan, A. Olinsky, and P.D. Phelan. Prevalence of asthma in Melbourne schoolchildren: changes over 26 years. *Br Med J* 302: 1116-1118, 1991.

326. Robinson, D.M., D.M. Egglestone, P.M. Hill, H.H. Rea, G.N. Richards, and S.M. Robinson. Effects of a physical conditioning programme on asthmatic patients. *NZ Med J* 105: 253-256, 1992.

327. Robuschi, M., M. Scuri, S. Spagnotto, G. Gambaro, E. Lodola, R. Pisati, and S. Bianco. The protective effect of transdermal broxaterol on exercise-induced bronchoconstriction. *Eur J Clin Pharmacol* 47: 465-466, 1995.

328. Rosenfeld, M., R. Davis, S. FitzSimmons, M. Pepe, and B. Ramsey. Gender gap in cystic fibrosis mortality. *Am J Epidemiol* 145: 794-803, 1997.

329. Rubinstein, I., H. Levison, A.S. Slutsky, H. Hak, J. Wells, N. Zamel, and A.S. Rebuck. Immediate and delayed bronchoconstriction after exercise in patients with asthma. *N Eng J Med* 317: 482-485, 1987.

330. Rupp, N.T. Diagnosis and management of exercise-induced asthma. *Phys Sportsmed* 24: 77-87, 1996.

331. Rupp, N.T., D.S. Brudno, and M.F. Guill. The value of screening for risk of exercise-induced asthma in high school athletes. *Ann Allergy* 70: 339-342, 1993.

332. Rupp, N.T., M.F. Guill, and D.S. Brudno. Unrecognized exercise-induced bronchospasm in adolescent athletes. *Am J Dis Child* 146: 941-944, 1992.

333. Russell, G. Childhood asthma and growth—a review of the literature. *Respir Med* 88: 31-37, 1994.

334. Salcedo, P.A., M.R. Giron, and B.B. Beltran. Complementary therapies in cystic fibrosis: evidence of therapeutic benefits and treatment recommendations. *Ann Pediatr (Barc)* 58: 39-44, 2003.

335. Salh, W., D. Bilton, M. Dodd, and A.K. Webb. Effect of exercise and physiotherapy in aiding sputum expectoration in adults with cystic fibrosis. *Thorax* 44: 1006-1008, 1989.

336. Santuz, P., E. Baraldi, P. Zarmella, M. Filippone, and F. Zacchello. Factors limiting exercise performance in long-term survivors of bronchopulmonary dysplasia. *Am J Respir Crit Care Med* 152: 1284-1289, 1995.

337. Sawyer, E.H., and T.L. Clanton. Improved pulmonary function and exercise tolerance with inspiratory muscle conditioning in children with cystic fibrosis. *Chest* 104: 1490-1497, 1993.

338. Schachter, E.N., E. Lach, and M. Lee. The protective effect of a cold weather mask on exercise-induced asthma. *Ann Allergy* 46: 12-16, 1981.

339. Scherr, M.S., and L. Frankel. Physical conditioning program for asthmatic children. *JAMA* 168: 1996-2000, 1958.

340. Schnall, R., P. Ford, I. Gillam, and L. Landau. Swimming and dry land exercises in children with asthma. *Austr Paediatr J* 18: 23-27, 1982.

341. Schnall, R.P., and L.I. Landau. Protective effects of repeated short sprints in exercise-induced asthma. *Thorax* 35: 828-832, 1980.

342. Schnall, R.P., L.I. Landau, and P.D. Phelan. The use of short periods of exercise in the prevention and reversal of exercise-induced asthma. In Oseid, S. and A.M. Edwards, eds., *The asthmatic child in play and sport*. London: Pitman, 1983, 107-115.

343. Schneiderman-Walker, J., S.L. Pollock, M. Corey, D.D. Wilkes, G.J. Canny, L. Pedder, and J.J. Reisman. A randomized controlled trial of a 3-year home exercise program in cystic fibrosis. *J Pediatr* 136: 304-310, 2000.

344. Schoeffel, R.E., S.D. Anderson, and J.P. Seale. The protective effect and duration of action of metaproterenol aerosol on exercise-induced asthma. *Ann Allergy* 46: 273-275, 1981.

345. Schurer, G.W. Changes in body composition, anaerobic muscle power, and pulmonary function with age in children with cystic fibrosis. 1997. MSc thesis. McMaster University, 1-142.

346. Sears, M., D. Taylor, C. Print, D.C. Lake, Q.Q. Li, E.M. Flannery, D.M. Yates, M.K. Lucas, and G.P. Herbison. Regular inhaled beta-agonist treatment in bronchial asthma. *Lancet* 336: 1391-1396, 1990.

347. Selvadurai, H.C., C.J. Blimkie, N. Meyers, C.M. Mellis, P.J. Cooper, and P.P. Van Asperen. Randomized controlled study of in-hospital exercise training programs in children with cystic fibrosis. *Pediatr Pulmonol* 33: 194-200, 2002.

348. Selvadurai, H.C., K.O. McKay, C.J. Blimkie, P.J. Cooper, C.M. Mellis, and P.P. Van Asperen. The relationship between genotype and exercise tolerance in children with cystic fibrosis. *Am J Respir Crit Care Med* 165: 762-765, 2002.

349. Serra-Batlles, J., J.M. Montserrat, J. Mullol, E. Ballester, A. Xaubet, and C. Picado. Response of the nose to exercise in healthy subjects and in patients with rhinitis and asthma. *Thorax* 49: 128-132, 1994.

350. Seto-Poon, M., T.C. Amis, J.P. Kirkness, and J.R. Wheatley. Nasal dilator strips delay the onset of oral route breathing during exercise. *Can J Appl Physiol* 24: 538-547, 1999.

351. Shah, A.R., D. Gozal, and T.G. Keens. Determinants of aerobic and anaerobic exercise performance in cystic fibrosis. *Am J Respir Crit Care Med* 157: 1145-1150, 2000.

352. Shapiro, G.G., W.E. Pierson, C.T. Furukawa, and C.W. Bierman. A comparison of the effectiveness of free-running and treadmill exercise for assessing exercise-induced bronchospasm in clinical practice. *J Allergy Clin Immunol* 64: 609-611, 1979.

353. Shephard, R.J. Exercise-induced bronchospasm—a review. *Med Sci Sports* 9: 1-10, 1977.

354. Shephard, R.J. *Physical activity and growth.* Chicago: Year Book Medical, 1982.

355. Shepherd, R.W., T.L. Holt, L. Vasques-Velasquez, W.A. Coward, A. Prentice, and A. Lucas. Increased energy expenditure in young children with cystic fibrosis. *Lancet* 1: 1300-1303, 1988.

356. Sheppard, D., A. Saisho, J. Nadel, and H. Boushey. Exercise increases sulfur dioxide-induced bronchoconstriction in asthmatic subjects. *Am Rev Respir Dis* 123: 486-491, 1981.

357. Shturman-Ellstein, R., R.J. Zeballos, J.M. Buckley, and J.F. Souhrada. The beneficial effect of nasal breathing on exercise-induced bronchoconstriction. *Am Rev Respir Dis* 118: 65-73, 1978.

358. Signorile, J.F., T.A. Kaplan, B. Applegate, and A.C. Perry. Effects of acute inhalation of the bronchodilator, albuterol, on power output. *Med Sci Sports Exerc* 24: 638-642, 1992.

359. Silverman, M., and S.D. Anderson. Standardization of exercise tests in asthmatic children. *Arch Dis Child* 47: 882-889, 1972.

360. Silverman, M., and T. Andrea. Time course of effect of disodium cromoglycate on exercise-induced asthma. *Arch Dis Child* 47: 419-422, 1972.

361. Silverman, M., F.D. Hobbs, I.R. Gordon, and F. Carswell. Cystic fibrosis, atopy, and airways lability. *Arch Dis Child* 53: 873-877, 1978.

362. Simons, F.E.R., T.V. Gerstner, and M.S. Cheang. Tolerance to the bronchoprotective effect of salmeterol in adolescents with exercise-induced asthma using concurrent inhaled glucocorticoid treatment. *Pediatrics* 99: 655-659, 1997.

363. Simonsson, B.G., F.M. Jacobs, and J.A. Nadel. Role of autonomic nervous system and the cough reflex in the increased responsiveness of airways in patients with obstructive airway disease. *J Clin Invest* 46: 1812-1818, 1967.

364. Sims, D.G., M.A.P.S. Downham, P.S. Gardner, J.K.G. Webb, and D. Weightman. Study of 8-year-old children with a history of respiratory syncytial virus bronchiolitis in infancy. *Br Med J* 1: 11-14, 1978.

365. Skorecki, K., H. Levison, and D.N. Crozier. Bronchial lability in cystic fibrosis. *Acta Paediatr Scand* 65: 39-42, 1976.

366. Sly, R.M. Exercise related changes in airway obstruction: frequency and clinical correlates in asthmatic children. *Ann Allergy* 28: 1-16, 1970.

367. Sly, R.M. Effect of cromolyn sodium on exercise-induced airway obstruction in asthmatic children. *Ann Allergy* 29: 362-366, 1971.

368. Sly, R.M. Effect of beta-adrenoreceptor stimulants on exercise-induced asthma. *Pediatrics* 56: 910-915, 1975.

369. Sly, R.M. Mortality from asthma. *J Allergy Clin Immunol* 84: 421-434, 1989.

370. Smith, C.M., and S.D. Anderson. Hyperosmolarity as the stimulus to asthma induced by hyperventilation. *J Allergy Clin Immunol* 77: 729-736, 1986.

371. Spann, C., and M.E. Winter. Effect of clenbuterol on athletic performance. *Ann Pharmacother* 29: 75-77, 1995.

372. Speelberg, B., E.A. Panis, D. Bijl, C.L. van Herwaarden, and P.L. Bruynzeel. Late asthmatic responses after exercise challenge are reproducible. *J Allergy Clin Immunol* 87: 1128-1137, 1991.

373. Speelberg, B., N.J. van den Berg, C.H. Oosthoek, N.P. Verhoeff, and W.T. van den Brink. Immediate and late asthmatic responses induced by exercise in patients with reversible airflow limitation. *Eur Respir J* 2: 402-408, 1989.

374. Spicher, V., M. Roulet, and Y. Schutz. Assessment of total energy expenditure in free-living patients with cystic fibrosis. *J Pediatr* 118: 865-872, 1991.

375. Spitzer, W., S. Saissa, P. Ernst, R.I. Horwitz, B. Habbick, D. Cockcroft, J.F. Boivin, M. McNutt, A.S. Buist, and A.S. Rebuck. The use of beta-agonists and the risk of death and near death from asthma. *New Eng J Med* 326: 501-506, 1992.

376. Sprenkle, A.C., P.P. Van Arsdel, and C.W. Bierman. New drug evaluation using exercise-induced bronchospasm. *Pediatrics* 56: 937-939, 1975.

377. Stalcup, S.A., and R.B. Mellins. Mechanical forces producing pulmonary edema in acute asthma. *N Eng J Med* 297: 592-595, 1977.

378. Stallings, V.A., E.H. Archibald, and P.B. Pencharz. Potassium, magnesium, and calcium balance in obese adolescents on a protein-sparing modified fast. *Am J Clin Nutr* 47: 220-224, 1988.

379. Stanghelle, J.K. Physical exercise for patients with cystic fibrosis: a review. *Int J Sports Med* 9: 6-18, 1988.

380. Stanghelle, J.K., N. Hjeltnes, H.J. Bangstad, and H. Michalsen. Effect of daily short bouts of trampoline exercise during 8 weeks on the pulmonary function and the maximal oxygen uptake of children with cystic fibrosis. *Int J Sports Med* 9: 32-36, 1988.

381. Stanghelle, J.K., N. Hjeltnes, H. Michalsen, H.J. Bengstad, and D. Skyberg. Pulmonary function and oxygen uptake during exercise in 11-year-old patients with cystic fibrosis. *Acta Paediatr Scand* 75: 657-661, 1986.

382. Stanghelle, J.K., S. Maehlum, D. Skyberg, S. Landaas, H. Oftebro, A. Bardon, O. Ceder, and H. Kollberg. Biochemical changes and endocrine responses in cystic fibrosis in relation to incremental maximal exhaustive exercise. *Int J Sports Med* 9: 41-44, 1988.

383. Stanghelle, J.K., H. Michalsen, and D. Skyberg. Five-year follow-up of pulmonary function and peak oxygen uptake on 16-year-old boys with cystic fibrosis with special regard to the influence of regular physical exercise. *Int J Sports Med* 9: 19-24, 1988.

384. Stanghelle, J.K., and D. Skyberg. Cystic fibrosis patients running a marathon race. *Int J Sports Med* 9: 37-40, 1988.

385. Stanghelle, J.K., M. Winnem, K. Roaldsen, S. De Wit, J.H. Notgewitch, and B.R. Nilsen. Young patients with cystic fibrosis: attitude toward physical activity and influence on physical fitness and spirometric values of a 2-week training course. *Int J Sports Med* 9: 25-31, 1988.

386. Steer, R.G. Asthma and the weather [letter]. *Med J Austr* 7: 38, 1976.

387. Stern, R.C., G. Borkat, S.S. Hirschfeld, T.F. Boat, L.W. Matthews, J. Liebman, et al. Heart failure in cystic fibrosis. *Am J Dis Child* 114: 775-794, 1980.

388. Strauss, G., A. Osher, C.-I. Wang, et al. Variable weight training in cystic fibrosis. *Chest* 92: 273-276, 1987.

389. Strauss, R.H., R.L. Haynes, R.H. Ingram Jr., and E.R. McFadden Jr. Comparison of arm vs. leg work in induction of acute episodes of asthma. *J Appl Physiol* 42: 565-570, 1977.

390. Strauss, R.H., E.R. McFadden Jr., R.H. Ingram Jr., E. Chandler Deal Jr., and J.J. Jaeger. Influence of heat and humidity on the airway obstruction induced by exercise in asthma. *J Clin Invest* 61: 433-440, 1978.

391. Strauss, R.H., E.R. McFadden Jr., R.H. Ingram Jr., and J.J. Jaeger. Enhancement of exercise-induced asthma by cold air. *N Eng J Med* 297: 743-747, 1977.

392. Strohl, K.P., M.J. Decker, L.G. Olson, T.A. Falk, and P.L. Hoekje. The nasal response to exercise and exercise-induced bronchoconstriction in normal and asthmatic subjects. *Thorax* 43: 890-895, 1988.

393. Strunk, R.C., L.J. Kelly, and D.L. Rubin. Recreation therapy in the rehabilitation of asthmatic children. Making the most of leisure. In Oseid, S. and A.M. Edwards, eds., *The asthmatic child in play and sports.* London: Pitman, 1983, 283-292.

394. Strunk, R.C., D.A. Mrazek, J.T. Fukuhara, J. Masterson, S.K. Ludwick, and J.F. LaBrecque. Cardiovascular fitness in children with asthma correlates with psychologic functioning of the child. *Pediatrics* 84: 460-464, 1989.

395. Sturani, C., A. Sturani, and I. Tosi. Parasympathetic activity assessment by diving reflex and by airway response to methacholine in bronchial asthma. *Respiration* 48: 321-328, 1985.

396. Sue, D.Y., J.E. Hansen, and K. Wasserman. The value of exercise in testing beta blockade and airway reactivity in asthmatic patients. *Am Heart J* 104: 442-445, 1982.

397. Svenonius, E., M. Arborelius Jr., R. Wilberg, and P. Ekberg. Prevention of exercise-induced asthma by drugs inhaled from metered aerosols. *Allergy* 43: 252-257, 1988.

398. Svenonius, E., R. Kautto, and M. Arborelius Jr. Improvement after training of children with exercise-induced asthma. *Acta Paediatr Scand* 72: 23-30, 1983.

399. Swann, I.L., and C.A. Hanson. Double-blind prospective study of the effect of physical training on childhood asthma. In Oseid, S. and A.M. Edwards, eds., *The asthmatic child in play and sports*. Bath: Pitman, 1983, 318-322.

400. Szentágothal, K., I. Gyene, M. Szócska, and P. Osvàth. Physical exercise program for children with bronchial asthma. *Pediatr Pulmonol* 3: 166-172, 1987.

401. Tabka, Z., A.B. Jebria, J. Vergeret, and H. Guenard. Effect of dry warm air on respiratory water loss in children with exercise-induced asthma. *Chest* 94: 81-86, 1988.

402. Tanizaki, Y., H. Komagoe, M. Sudo, and H. Morinaga. Swimming training in a hot spring pool as therapy for steroid dependent asthma. *Arerugi* 33: 389-395, 1984.

403. Taylor, W.R., and P.W. Newacheck. Impact of childhood asthma on health. *Pediatrics* 90: 657-662, 1992.

404. Terblanche, E., and R.I. Stewart. The influence of exercise-induced bronchoconstriction on participation in organized sport. *S Afr Med J* 78: 741-743, 1990.

405. Thomson, M.A., R.W. Wilmott, C. Wainwright, B. Masters, P.J. Francis, and R.W. Shepherd. Resting energy expenditure, pulmonary inflammation, and genotype in the early course of cystic fibrosis. *J Pediatr* 129: 367-373, 1996.

406. Tomezsko, J.L., V.A. Stallings, A. Kawchak, J.E. Goin, G. Diamond, and T.F. Scanlin. Energy expenditure and genotype of children with cystic fibrosis. *Pediatr Res* 35: 451-460, 1994.

407. Tower, J. Office testing for exercise-induced asthma. *Alaska Med* 20: 70-72, 1978.

408. Tromp, S.W., and J. Bouma. Effect of weather on asthmatic children in the eastern part of the Netherlands. *Int J Biometeorol* 9: 233-238, 1965.

409. Unnithan, V.B., K.J. Thomson, T.C. Aitchison, and J.Y. Paton. β_2-agonists and running economy in prepubertal boys. *Pediatr Pulmonol* 17: 378-382, 1994.

410. Vaisman, N., P.B. Pencharz, M. Corey, G.I. Canny, and E. Hahn. Energy expenditure of patients with cystic fibrosis. *J Pediatr* 111: 496-500, 1987.

411. Varray, A., and C. Préfaut. Importance of physical exercise training in asthmatics. *J Asthma* 29(4): 229-236, 1992.

412. Varray, A.L., J.G. Mercier, C.M. Terral, and C.G. Préfaut. Individualized aerobic and high intensity training for asthmatic children in an exercise readaptation program. Is training always helpful for better adaptation to exercise? *Chest* 99(3): 579-586, 1991.

413. Vassallo, C.L., J.B.L. Gee, and B.M. Domm. Exercise-induced asthma. Observations regarding hypocapnia and acidosis. *Am Rev Respir Dis* 105: 42-49, 1972.

414. Vathenen, A.S., A.J. Knox, A. Wisniewski, and A.E. Tattersfield. Effect of inhaled budesonide on bronchial reactivity to histamine, exercise, and eucapnic dry air hyperventilation in patients with asthma. *Thorax* 46: 811-816, 1991.

415. Vávra, J., M. Măcek, B. Mrzena, and V. Spicák. Intensive physical training in children with bronchial asthma. *Acta Paediatr Scand* 217: 90-92, 1971.

416. Vávra, J., M. Măcek, and V. Spicák. Working capacity of asthmatic children [in French]. *Rev Pediatr* 5: 3-7, 1969.

417. Verma, S., and J.S. Hyde. Physical education programs and exercise-induced asthma. *Clin Pediatr* 15: 697-699, 1976.

418. Waalkens, H.J., E.E. Essen-Zandvliet, J. Gerritsen, E.J. Duiverman, K.F. Kerrebijn, and K. Knol. The effect of an inhaled corticosteroid (budesonide) on exercise-induced asthma in children. Dutch CNSLD Study Group. *Eur Respir J* 6: 652-656, 1993.

419. Weiler-Ravell, D., and S. Godfrey. Do exercise and antigen-induced asthma utilize the same pathway? *J Allergy Clin Immunol* 67: 391-397, 1981.

420. Weinstein, R.E., J.A. Anderson, P. Kvale, and L.C. Sweet. Effects of humidification on exercise-induced asthma (EIA). *J Allergy Clin Immunol* 57: 250-251, 1976.

421. West, J.V., C.F. Robertson, R. Roberts, and A. Olinsky. Evaluation of bronchial responsiveness to exercise in children as an objective measure of asthma in epidemiological surveys. *Thorax* 51: 590-595, 1996.

422. Wiens, L., R. Sabath, L. Ewing, R. Gowdamarajan, J. Portnoy, and D. Scagliotti. Chest pain in otherwise healthy children and adolescents is frequently caused by exercise-induced asthma. *Pediatrics* 90: 350-353, 1992.

423. Wilbourn, K. The long distance runner. *Runner's World* 13: 62-65, 1978.

424. Wilson, B.A., O. Bar-Or, and P.M. O'Byrne. The effects of indomethacin on refractoriness following exercise both with and without a bronchoconstrictor response. *Eur Respir J* 7: 2174-2178, 1994.

425. Wilson, B.A., O. Bar-Or, and L.G. Seed. Effects of humid air breathing during arm or treadmill exercise on exercise-induced bronchoconstriction and refractoriness. *Am Rev Respir Dis* 142: 349-352, 1990.

426. Wilson, B.A., and J.N. Evans. Standardization of work intensity for evaluation of exercise-induced bronchoconstriction. *Eur J Appl Physiol* 47: 289-294, 1981.

427. Wolff, R.K., M.B. Dolovich, G. Obminski, and M.T. Newhouse. Effects of exercise and eucapnic hyperventilation on bronchial clearance in man. *J Appl Physiol: Respir Environ Exerc Physiol* 43: 46-50, 1977.

428. Wolkove, N., H. Kreisman, H. Frank, and M. Gent. The effect of ipratropium on exercise-induced bronchoconstriction. *Ann Allergy* 47: 311-315, 1981.

429. Woolley, M., S.D. Anderson, and B.M. Quigley. Duration of protective effect of terbutaline sulfate and cromolyn sodium alone and in combination on exercise-induced asthma. *Chest* 97: 39-45, 1990.

430. Yeung, R., G.M. Nolan, and H. Levison. Comparison of the effects of inhaled SCH 1000 and fenoterol on exercise-induced bronchospasm in children. *Pediatrics* 66: 109-114, 1980.

431. Zach, M., B. Oberwaldner, and F. Hauslen. Cystic fibrosis: physical exercise vs. chest physiotherapy. *Arch Dis Child* 57: 587-589, 1982.

432. Zach, M., B. Purrer, and B. Oberwalder. Effect of swimming on forced expiration and sputum clearance in cystic fibrosis. *Lancet* 2: 1201-1203, 1981.

433. Zambie, M.F., S. Gupta, R.J. Lemen, B. Hilman, W.W. Wawring, and M. Sly. Relationships between response to exercise and allergy in patients with cystic fibrosis. *Ann Allergy* 42: 290-294, 1979.

434. Zawadski, D.K., K.A. Lenner, and E.R. McFadden Jr. Re-examination of the late asthmatic response to exercise. *Am Rev Respir Dis* 137: 837-841, 1988.

435. Zeballos, R.J., R. Shturman-Ellstein, J.F. McNally Jr., J.E. Hirsch, and J.F. Souhrada. The role of hyperventilation in exercise-induced bronchoconstriction. *Am Rev Respir Dis* 118: 877-884, 1978.

436. Zelkowitz, P.S., and S.T. Giammona. Effects of gravity and exercise on the pulmonary diffusing capacity in children with cystic fibrosis. *J Pediatr* 74: 393-398, 1969.

437. Zinman, R., M. Corey, A.L. Coates, G.J. Canny, J. Connolly, H. Levison, and P.H. Beaudry. Nocturnal home oxygen in the treatment of hypoxemic cystic fibrosis patients. *J Pediatr* 114: 368-377, 1989.

Chapter 7

1. Abrunzo, T.J. Commotio cordis. *Am J Dis Child* 145: 1279-1282, 1991.

2. Allen, S.W., E.M. Shaffer, L.A. Harrigan, R.R. Wolfe, M.P. Glode, and J.W. Wiggins. Maximal voluntary work and cardiorespiratory fitness in patients who have had Kawasaki syndrome. *J Pediatr* 121: 221-225, 1992.

3. Alpert, B.S., H.H. Bain, J.W. Balfe, B.S.L. Kidd, and P.M. Olley. Role of the renin-angiotensin-aldosterone system in hypertensive children with coarctation of the aorta. *Am J Cardiol* 43: 828-832, 1979.

4. Alpert, B.S., N.L. Flood, W.B. Strong, E.V. Dover, R.H. DuRant, A.M. Martin, and D.L. Booker. Responses to ergometer exercise in a healthy biracial population of children. *J Pediatr* 101: 538-545, 1982.

5. Alpert, B.S., and M.E. Fox. Hypertension. In Goldberg, B., ed., *Sports and exercise for children with chronic health conditions*. Champaign, IL: Human Kinetics, 1995, 197-205.

6. Alpert, B.S., W. Kartodihardjo, and P. Harp. Exercise blood pressure response—a predictor of severity of aortic stenosis in children. *J Pediatr* 98: 763-765, 1981.

7. Arensman, F.W., J. Christiansen, and W.B. Strong. The young athlete with hypertension. In Smith, N.J., ed., *Common problems in pediatric sports medicine*. Chicago: Year Book Medical, 1989, 105-111.

8. Atherton, J.J., A.C. Tweddel, and M.P. Frenneaux. Mechanisms for exercise limitation in chronic heart failure and the role of rehabilitation. *Q J Med* 90: 731-734, 1997.

9. Balderston, S.M., E. Daberkow, D.R. Clarke, and R.R. Wolfe. Maximal voluntary exercise variables in children with postoperative coarctation of the aorta. *J Am Coll Cardiol* 19: 154-158, 1992.

10. Balfour, I.C., A.M. Drimmer, S. Nouri, D.G. Pennington, C.L. Hemkens, and L.L. Harvey. Pediatric cardiac rehabilitation. *Am J Dis Child* 145: 627-630, 1992.

11. Barber, G. Training and the pediatric patient: a cardiologist's perspective. In Blimkie, J.R. and O. Bar-Or, eds., *New horizons in pediatric exercise science*. Champaign, IL: Human Kinetics, 1995, 137-146.

12. Basso, C., B.J. Maron, D. Corrado, and G. Thiene. Clinical profile of congenital coronary artery anomalies with origin from the wrong aortic sinus leading to sudden death in young competitive athletes. *J Am Coll Cardiol* 35: 1493-1501, 2000.

13. Bay, G., A.M. Abrahamsen, and C. Muller. Left-to-right shunt in atrial septal defect at rest and during exercise. *Acta Med Scand* 190: 205-209, 1971.

14. Beekman, R.H., B.P. Katz, C. Moorehead-Steffens, and A.P. Rocchini. Altered baroreceptor function in children with systolic hypertension after coarctation repair. *Am J Cardiol* 52: 112-117, 1983.

15. Benbasset, J., and P.F. Froom. Blood pressure response to exercise as a predictor of hypertension. *Arch Int Med* 146: 2053-2035, 1986.

16. Bergman, A.B., and S.J. Stamm. The morbidity of cardiac nondisease in schoolchildren. *N Eng J Med* 276: 1008-13, 1967.

17. Bernstein, D., V.A. Starnes, and D. Baum. Pediatric heart transplantation. *Adv Pediatr* 37: 413-440, 1990.

18. Bhatia, N.G., C. Heise, and G. Barber. Exercise and QT interval corrections [abstract]. *Pediatr Exerc Sci* 8: 94, 1996.

19. Bielen, E.C., R.H. Fagard, and A.K. Amery. Inheritance of acute cardiac changes during bicycle exercise: an echocardiographic study in twins. *Med Sci Sports Exerc* 23: 1254-1259, 1991.

20. Bisset, G.S., D.C. Schwartz, R.A. Meyer, F.W. James, and S. Kaplan. Clinical spectrum and long-term followup of isolated mitral valve prolapse in 119 children. *Circulation* 62: 423-429, 1980.

21. Bjarke, B. Spirometric data, pulmonary ventilation and gas exchange at rest and during exercise in adult patients with tetralogy of Fallot. *Scand J Respir Dis* 55: 47-61, 1974.

22. Bradley, L.M., F.M. Galioto, P. Vaccaro, D.A. Hansen, and J. Vaccaro. Effect of intense aerobic training on exercise performance in children after surgical repair of tetralogy of Fallot or complete transposition of the great arteries. *Am J Cardiol* 56: 816-818, 1985.

23. Braverman, A.C. Exercise and the Marfan syndrome. *Med Sci Sports Exerc* 30 (Suppl): S387-S395, 1998.

24. Bricker, J.T., A. Garson, S.M. Paridon, and T.A. Vargo. Exercise correlates of electrophysiological assessment of sinus node function in young individuals. *Pediatr Exerc Sci* 2: 163-168, 1990.

25. Calzolari, A., A. Turchetta, G. Biondi, F. Drago, C. DeRanieri, G. Gagliardi, I. Giambini, S. Giannico, and F. Perrotta. Rehabilitation of children after total correction of tetralogy of Fallot. *Int J Cardiol* 28: 151-158, 1990.

26. Carpenter, M.A., J.F. Dammann, D.D. Watson, R. Jedeikin, D.G. Tompkins, and G.A. Beller. Left ventricular hyperkinesia at rest and during exercise in normotensive patients 2 to 27 years after coarctation repair. *J Am Coll Cardiol* 6: 879-886, 1985.

27. Chandramouli, B., D.A. Ehmke, and R.M. Lauer. Exercise-induced electrocardiographic changes in children with congenital aortic stenosis. *J Pediatr* 87: 725-730, 1975.

28. Chiang, C.-E., and D.M. Roden. The long QT syndromes: genetic basis and clinical implications. *J Am Coll Cardiol* 36: 1-12, 2000.

29. Christos, S.C., V. Katch, D.C. Crowley, B.L. Eakin, A.L. Lindauer, and R.H. Beekman. Hemodynamic responses to upright exercise of adolescent cardiac transplant recipients. *J Pediatr* 121: 312-316, 1992.

30. Coats, A.J.S. Origins of symptoms of heart failure. *Cardiovasc Drugs Ther* 11: 265-272, 1997.

31. Colan, S. Mechanics of left ventricular systolic and diastolic function in physiologic hypertrophy of the athlete's heart. *Cardiol Clin* 15: 355-372, 1997.

32. Committee on Rheumatic Fever, Endocarditis, and Kawasaki Disease. Guidelines for long-term management of patients with Kawasaki disease. *Circulation* 89: 916-922, 1994.

33. Conner, T.M. Evaluation of persistent coarctation of aorta after surgery with blood pressure measurement and exercise testing. *Am J Cardiol* 43: 74-78, 1979.

34. Cortes, R.G.S., G. Satomi, M. Yoshigi, and K. Momma. Maximal hemodynamic response after the Fontan procedure: Doppler evaluation during the treadmill test. *Pediatr Cardiol* 15: 170-177, 1994.

35. Crawford, D.W., E. Simpson, and M.B. McIllroy. Cardiopulmonary function in Fallot's tetralogy after palliative shunting operations. *Am Heart J* 74: 463-472, 1967.

36. Cueto, L., and J.H. Moller. Haemodynamics of exercise in children with isolated aortic valvular disease. *Br Heart J* 35: 93-98, 1973.

37. Cumming, G.R. Maximal exercise capacity of children with heart defects. *Am J Cardiol* 42: 613-619, 1978.

38. Cumming, G.R. Maximal supine exercise haemodynamics after open heart surgery for Fallot's tetralogy. *Br Heart J* 41: 683-691, 1979.

39. Cyran, S.E., M. Grzeszczak, K. Kaufman, H.S. Weber, J.L. Myers, M.M. Gleason, and B.G. Baylen. Aortic 'recoarctation' at rest versus at exercise in children evaluated by stress Doppler echocardiography after a 'good' operative result. *Am J Cardiol* 71: 963-70, 1993.

40. Cyran, S.E., F.W. James, S. Daniels, W. Mays, R. Shukla, and S. Kaplan. Comparison of the cardiac output and stroke volume response to upright exercise in children with valvular and subvalvular aortic stenosis. *J Am Coll Cardiol* 11: 651-658, 1988.

41. Danforth, J.S., K.D. Allen, J.M. Fitterling, J.A. Danforth, D. Farrar, M. Brown, and R.S. Drabman. Exercise as a treatment for hypertension in low socioeconomic status black children. *J Consult Clin Psychol* 58: 237-239, 1990.

42. Del Torso, S., M.J. Kelly, V. Kalff, and A.W. Venables. Radionuclide assessment of ventricular contraction at rest and during exercise following the Fontan procedure for either tricuspid atresia or single ventricle. *Am J Cardiol* 55: 1127-1132, 1985.

43. Diano, R., C. Bouchard, J. Demesnil, C. Leblanc, and J.L. Laurenceau. Parent-child resemblance in left ventricular echocardiographic measurements [abstract]. *Can J Appl Sport Sci* 5: 4, 1980.

44. Dlin, R.A., N. Hanne, D. Silverberg, and O. Bar-Or. Followup of normotensive men with exaggerated blood pressure response to exercise. *Am Heart J* 106: 316-320, 1083.

45. Donovan, E.F., P.A. Mathews, P.A. Nixon, R.J. Stephenson, R.J. Robertson, F. Dean, F.J. Fricker, L.B. Beerman, and D.R. Fischer. An exercise program

for pediatric patients with congenital heart disease: psychosocial aspects. *J Cardiac Rehab* 3: 476-480, 1983.

46. Doyle, E.F., P. Arumugham, E. Lara, M.R. Rutkowski, and B. Kiely. Sudden death in young patients with congenital aortic stenosis. *Pediatrics* 53: 481-489, 1974.

47. Driscoll, D.J. Exercise rehabilitation programs for children with congenital heart disease: a note of caution. *Pediatr Exerc Sci* 2: 191-196, 1990.

48. Driscoll, D.J., G.K. Danielson, F.J. Puga, H.V. Schaff, C.T. Heise, and B.A. Staats. Exercise tolerance and cardiorespiratory response to exercise after the Fontan operation for tricuspid atresia or functional single ventricle. *J Am Coll Cardiol* 7: 1087-94, 1986.

49. Driscoll, D.J., and K. Durongpisitkul. Exercise testing after the Fontan operation. *Pediatr Cardiol* 20: 57-59, 1999.

50. Duffie, E.R., and F.H. Adams. The use of the working capacity test in the evaluation of children with congenital heart disease. *Pediatrics* 32 (Suppl): 757-768, 1963.

51. Duncan, J.J., J.E. Farr, J. Upton, R.D. Hagan, M.F. Oglesby, and S.N. Blair. The effects of aerobic exercise on plasma catecholamines and blood pressure in patients with mild essential hypertension. *JAMA* 254: 2609-2613, 1985.

52. Durongpisitkul, K., D.J. Driscoll, D.W. Mahoney, P.C. Wollan, C.D. Mottram, F.J. Puga, and G.K. Danielson. Cardiorespiratory response to exercise after modified Fontan operation: determinants of performance. *J Am Coll Cardiol* 29: 785-790, 1997.

53. Ehsani, A.A., J.M. Hagberg, and R.C. Hickson. Rapid changes in left ventricular dimensions and mass in response to physical conditioning and deconditioning. *Am J Cardiol* 42: 52-56, 1978.

54. Ensing, G.J., C.T. Heise, and D.J. Driscoll. Cardiovascular response to exercise after the Mustard operation for simple and complex transposition of the great arteries. *Am J Cardiol* 62: 617-622, 1988.

55. Fagard, R.H. Impact of different sports and training on cardiac structure and function. *Cardiol Clin* 15: 397-412, 1997.

56. Feinstein, R.A., and W.A. Daniel. Chronic chest pain in children and adolescents. *Pediatr Ann* 15: 685-686, 1986.

57. Fixler, D.E., W.P. Laird, R. Browne, V. Fitzgerald, S. Wilson, and R. Vance. Response of hypertensive adolescents to dynamic and isometric exercise stress. *Pediatrics* 64: 579-583, 1979.

58. Fixler, D.E., W.P. Laird, and K. Dana. Usefulness of exercise stress testing for prediction of blood pressure trends. *Pediatrics* 75: 1071-1075, 1985.

59. Franciosa, J.A., M. Park, and T.B. Levine. Lack of correlation between exercise capacity and indexes of resting left ventricular performance in heart failure. *Am J Cardiol* 47: 33-39, 1981.

60. Frick, M., S. Punsar, and T. Somer. The spectrum of cardiac capacity in patients with nonobstructive congenital heart disease. *Am J Cardiol* 17: 20-26, 1966.

61. Galanti, G., L. Toncelli, and M. Comeglio. Morphological and functional effects on athletes' heart after a period of absolute inactivity. *J Sports Card* 4: 102-106, 1987.

62. Galioto, F.M. Exercise rehabilitation programs for children with congenital heart disease: a note of enthusiasm. *Pediatr Exerc Sci* 2: 197-200, 1990.

63. Gardiner, H.M., D.S. Celermajer, K.E. Sorensen, D. Georgakopoulos, J. Robinson, O. Thomas, and J.E. Deanfield. Arterial reactivity is significantly impaired in normotensive young adults after successful repair of aortic coarctation in childhood. *Circulation* 89: 1745-50, 1994.

64. Garson, A. Ventricular arrhythmias. In Gillette, P.C. and A. Garson, eds., *Pediatric arrhythmias: electrophysiology and pacing*. Philadelphia: Saunders, 1990, 427-500.

65. George, K.P., L.A. Wolfe, and G.W. Burggraf. The 'athletic heart syndrome.' A critical review. *Sports Med* 11: 300-331, 1991.

66. Gewillig, M.H., U.R. Lundstrom, C. Bull, R.K.H. Wyse, and J.E. Deanfield. Exercise responses in patients with congenital heart disease after Fontan repair: patterns and determinants of performance. *J Am Coll Cardiol* 15: 1424-32, 1990.

67. Gillette, P.C., J.S. Heinle, and V.L. Zeigler. Cardiac pacing. In Gillette, P.C. and A. Garson, eds., *Clinical pediatric arrhythmias*. Philadelphia: Saunders, 1999, 190-220.

68. Gilljam, T., B.O. Eriksson, and R. Sixt. Cardiac output and pulmonary gas exchange at maximal exercise after atrial redirection for complete transposition. *Eur Heart J* 19: 1856-1864, 1998.

69. Goldberg, B., R.R. Fripp, G. Lister, J. Loke, J.A. Nicholas, and N.S. Talner. Effect of physical training on exercise performance of children following surgical repair of congenital heart disease. *Pediatrics* 68: 691-699, 1981.

70. Graham, T.P., J.T. Bricker, F.W. James, and W.B. Strong. Task force 1. Congenital heart disease. *Med Sci Sports Exerc* 26 (Suppl): S246-S253, 1994.

71. Grubb, B.P., P.N. Temsey-Armos, D. Samoil, D.A. Wolfe, H. Hahn, and L. Elliott. Tilt table testing in the evaluation and management of athletes with recurrent exercise-induced syncope. *Med Sci Sports Exerc* 25: 24-28, 1993.

72. Guenthard, J., and F. Wyler. Exercise-induced hypertension in the arms due to impaired arterial reactivity after successful coarctation resection. *Am J Cardiol* 75: 814-817, 1995.

73. Guenthard, J., U. Zumsteg, and F. Wyler. Arm-leg pressure gradients on late followup after coarctation repair. Possible causes and implications. *Eur Heart J* 17: 1572-1575, 1996.

74. Hagberg, J.M. Exercise, fitness, and hypertension. In Bouchard, C., R.J. Shephard, T. Stephens, J.R. Sutton, and B.D. McPherson, eds., *Exercise, fitness, and health.* Champaign, IL: Human Kinetics, 1990, 455-466.

75. Hagberg, J.M., A.A. Ehsani, D. Goldring, A. Hernandez, D.R. Sincore, and J.O. Holloszy. Effect of weight training on blood pressure and hemodynamics in hypertensive adolescents. *J Pediatr* 104: 147-151, 1984.

76. Hagberg, J.M., D. Goldring, A.A. Ehsani, G.W. Heath, A. Hernandez, K. Schiechtman, and J.O. Holloszy. Effect of exercise training on the blood pressure and hemodynamic features of hypertensive adolescents. *Am J Cardiol* 52: 763-768, 1983.

77. Halloran, K.H. The telemetered exercise electrocardiogram in congenital aortic stenosis. *Pediatrics* 47: 31-39, 1971.

78. Hanisch, D.G., M.E. Cabrera, R. Murtaugh, M.L. Spector, and J. Liebman. Comparison of exercise performance in children paced in VVI versus VVIR modes [abstract]. *PACE* 13: 1191, 1990.

79. Hansen, H.S., K. Froberg, N. Hyldebrandt, and J.R. Nielsen. A controlled study of eight months of physical training and reduction of blood pressure in children: the Odense Schoolchild Study. *Br Med J* 303: 682-685, 1991.

80. Harrison, D.A., P. Liu, J.E. Walters, J.M. Goodman, S.C. Siu, G.D. Webb, W.W. Williams, and P.R. McLaughlin. Cardiopulmonary function in adult patients late after Fontan repair. *J Am Coll Cardiol* 26: 1016-1021, 1995.

81. Helber, U., R. Baumann, H. Seboldt, U. Reinhard, and H.M. Hoffmeister. Atrial septal defect in adults: cardiopulmonary exercise capacity before and 4 months and 10 years after defect closure. *J Am Coll Cardiol* 29: 1345-1350, 1997.

82. Henein, M.Y., S. Dinarevic, C.A. O'Sullivan, D.G. Gibson, and E.A. Shinebourne. Exercise echocardiography in children with Kawasaki disease: ventricular long axis is selectively abnormal. *Am J Cardiol* 81: 1356-1359, 1998.

83. Hijazi, Z.M., J.E. Udelson, H. Snapper, J. Rhodes, G.R. Marx, S.L. Schwartz, and D.R. Fulton. Physiologic significance of chronic coronary aneurysms in patients with Kawasaki disease. *J Am Coll Cardiol* 24: 1633-1638, 1994.

84. Hsu, D.T., R.P. Garafano, J.D. Douglas, R.E. Michler, J.M. Quaegebeur, W.M. Gersony, and L.J. Addonizio. Exercise performance after pediatric heart transplantation. *Circulation* 88: 238-242, 1993.

85. Hugenholtz, P.G., and A. Nadas. Exercise studies in patients with congenital heart disease. *Pediatrics* 32: 769-775, 1963.

86. Ikkos, D., and J.S. Hanson. Response to exercise in congenital complete atrioventricular block. *Circulation* 22: 583-590, 1960.

87. Iserin, L., T.P. Chua, J. Chambers, A.J.S. Coats, and J. Somerville. Dyspnoea and exercise intolerance during cardiopulmonary exercise testing in patients with univentricular heart. *Eur Heart J* 18: 1350-1356, 1997.

88. Jacobsen, J.R., A. Garson, P.C. Gillette, and D.G. McNamara. Premature ventricular contractions in normal children. *J Pediatr* 92: 36-38, 1978.

89. James, F.W. Exercise testing in children and young adults: an overview. *Cardiovasc Clin* 9: 187-203, 1978.

90. James, F.W., S. Kaplan, and D.C. Schwartz. Response to exercise in patients after total surgical correction of tetralogy of Fallot. *Circulation* 54: 671-679, 1976.

91. Jeresaty, R.M. Mitral valve prolapse: definition and implications in athletes. *J Am Coll Cardiol* 7: 231-236, 1986.

92. Johnson, D., P. Bonnin, H. Perrault, T. Marchand, S.J. Vobecky, A. Fournier, and A. Davignon. Peripheral blood flow responses to exercise after successful correction of coarctation of the aorta. *J Am Coll Cardiol* 26: 1719-1724, 1995.

93. Johnson, D., H. Perrault, A. Fournier, J.-M. Leclerc, J.-L. Bigras, and A. Davignon. Cardiovascular responses to dynamic submaximal exercise in children previously treated with anthracycline. *Am Heart J* 133: 169-173, 1997.

94. Kaplan, N.M., R.D. Deveraux, and H.S. Miller. Systemic hypertension. *Med Sci Sports Exerc* 26 (Suppl): S268-S270, 1994.

95. Karpawich, P.P., B.L. Perry, Z.Q. Farooki, S.K. Clapp, W.L. Jackson, C.A. Cicalese, and E.W. Green. Pacing in children and young adults with nonsurgical atrioventricular block: comparison of single-rate ventricular and dual-chamber modes. *Am Heart J* 113: 316-321, 1987.

96. Katagiri-Kawade, M., T. Ohe, Y. Arakaki, T. Kurita, W. Shimizu, T. Kamiya, and T. Orii. Abnormal response to exercise, face immersion, and isoproterenol in children with the long QT syndrome. *PACE* 18: 2128-2134, 1995.

97. Kato, H., E. Ichinose, and F. Yoshioka. Fate of coronary aneurysms in Kawasaki disease: serial coronary angiography and long-term followup study. *Am J Cardiol* 49: 1758-66, 1982.

98. Kavey, R.W., M.S. Blackman, H.M. Sondheimer, and C.J. Brum. Ventricular arrhythmias and mitral valve prolapse in childhood. *J Pediatr* 105: 885-890, 1984.

99. Kelley, G., and Z.V. Tran. Aerobic exercise and normotensive adults: a meta-analysis. *Med Sci Sports Exerc* 27: 1371-1377, 1995.

100. Kimball, T.R., J.M. Reynolds, W.A. Mays, P. Khoury, R.P. Claytor, and S.R. Daniels. Persistent hyperdynamic cardiovascular state at rest and during exercise in children after successful repair of coarctation of the aorta. *J Am Coll Cardiol* 24: 194-200, 1994.

101. Klein, A.A., W.W. McCrory, M.A. Engle, R. Rosenthal, and K.H. Ehlers. Sympathetic nervous system and exercise tolerance response in normotensive and hypertensive adolescents. *J Am Coll Cardiol* 3: 381-386, 1984.

102. Klissouras, V., F. Pirnay, and J.M. Petit. Adaptation to maximal effort: genetics and age. *J Appl Physiol* 35: 288-293, 1973.

103. Kocis, K.C. Chest pain in pediatrics. *Ped Clin N Amer* 46: 189-204, 1999.

104. Lambert, J., R.J. Ferguson, A. Gervais, and G. Gilbert. Exercise capacity, residual abnormalities and activity habits following total correction of tetralogy of Fallot. *Cardiology* 66: 120-132, 1980.

105. Landry, F., C. Bouchard, and J. Dumesnil. Cardiac dimension changes with endurance training. *JAMA* 254: 77-80, 1985.

106. Lauer, R.M., W.E. Connor, and P.E. Leaverton. Coronary heart disease risk factors in school children: the Muscatine study. *J Pediatr* 86: 697-703, 1975.

107. Leandro, J., J.F. Smallhorn, L. Benson, N. Musewe, J.W. Balfe, J.D. Dyck, L. West, and R. Freedom. Ambulatory blood pressure monitoring and left ventricular mass and function after successful surgical repair of coarctation of the aorta. *J Am Coll Cardiol* 20: 197-204, 1992.

108. Lewis, A.B., M.A. Heymann, P. Stanger, J.I.E. Hoffman, and A.M. Rudolph. Evaluation of subendocardial ischemia in valvular aortic stenosis in children. *Circulation* 49: 978-984, 1974.

109. Link, M.S., B.J. Maron, P.J. Wang, and N.A.M. Estes. Sudden death and other cardiovascular manifestations of chest wall trauma in sports. In Thompson, P.J., ed., *Exercise and sports cardiology.* New York: McGraw-Hill, 2001, 249-263.

110. Link, M.S., P.J. Wang, N.G. Pandian, S. Bharati, J.E. Udelson, M.-Y. Lee, M.A. Vecchiotti, et al. An experimental model of sudden death due to low-energy chest-wall impact (commotio cordis). *N Eng J Med* 338: 1805-11, 1998.

111. Londe, S., J.J. Bourgoignie, and A.M. Robson. Hypertension in apparently normal children. *J Pediatr* 78: 569-575, 1971.

112. MacDougall, J.D., D. Tuxen, D.G. Sale, J.R. Moroz, and J.R. Sutton. Arterial blood pressure response to heavy resistance exercise. *J Appl Physiol* 58: 785-790, 1985.

113. Markel, H., A.P. Rocchini, and R.H. Beekman. Exercise-induced hypertension after repair of coarctation of the aorta: arm versus leg exercise. *J Am Coll Cardiol* 8: 165-171, 1986.

114. Maron, B.J., J.O. Humphries, R.D. Rowe, and E.D. Mellitis. Prognosis of surgically corrected coarctation of the aorta: a 20-year post operative appraisal. *Circulation* 47: 119-126, 1973.

115. Maron, B.J., A. Pelliccia, and P. Spirito. Cardiac disease in young trained athletes. *Circulation* 91: 1596-1601, 1995.

116. Maron, B.J., L.C. Poliac, J.A. Kaplan, and F.O. Mueller. Blunt impact to the chest leading to sudden death from cardiac arrest during sports activities. *N Eng J Med* 333: 337-42, 1995.

117. Maron, B.J., W.C. Roberts, and S.E. Epstein. Sudden death in hypertrophic cardiomyopathy: a profile of 78 patients. *Circulation* 65: 1388-1394, 1982.

118. Maron, B.J., J. Shirani, L.C. Poliac, R. Mathenge, W.O. Roberts, and F.O. Mueller. Sudden death in young competitive athletes. *JAMA* 276: 199-204, 1996.

119. Maron, B.J., P.D. Thompson, J.C. Puffer, C.A. McGrew, W.B. Strong, P.S. Douglas, L.T. Clark, M.J. Mitten, M.D. Crawford, D.L. Atkins, D.J. Driscoll, and A.E. Epstein. Cardiovascular preparticipation screening of competitive athletes. *Circulation* 94: 850-856, 1996.

120. Massin, M., H. Hovels-Gurich, S. Dabritz, B. Messmer, and G. von Bernuth. Results of the Bruce treadmill test in children after arterial switch operation for simple transposition of the great arteries. *Am J Cardiol* 81: 56-60, 1998.

121. Matitiau, A., A. Perez-Atayde, S.P. Sanders, T. Sluysmans, I.A. Parness, P.J. Spevak, and S.D. Colan. Infantile dilated cardiomyopathy. *Circulation* 90: 1310-1318, 1994.

122. Matthews, R.A., F.J. Fricker, and L.B. Beerman. Exercise studies after the Mustard operation in transposition of the great arteries. *Am J Cardiol* 51: 1526-1529, 1983.

123. Matthys, D. Pre- and postoperative exercise testing of the child with atrial septal defect. *Pediatr Cardiol* 20: 22-25, 1999.

124. McCrory, W.W., A.A. Klein, and F. Fallo. Predictors of blood pressure: humoral factors. In Loggie, J.M.H., M.J. Horan, A.B. Gruskin, A.R. Hohn, J.B. Dunbar, and R.J. Havlik, eds., *NHLBI workshop on juvenile hypertension.* New York: Biomedical Information Corporation, 1984, 181-204.

125. Meijboom, F., J. Hess, A. Szatmari, E. Utens, J. McGhie, J.W. Deckers, J. Roelandt, and E. Bos. Long-term follow-up (9 to 20 years) after surgical closure of atrial septal defect at a young age. *Am J Cardiol* 72: 1431-1434, 1993.

126. Mertens, L., T. Reybrouck, M. Dumoulin, M. Weymans, W. Daenen, and M. Gewillig. Cardiopulmo-

nary exercise testing late after surgical closure of a large ventricular septal defect [abstract]. *Pediatr Exerc Sci* 8: 89, 1996.

127. Michaelsson, M., and M.A. Engle. Congenital complete heart block: an international study of the natural history. *Cardiovasc Clin* 4: 85-98, 1972.

128. Miller, J.D., M.-L. Young, D.L. Atkins, and G.S. Wolff. Rate-responsive ventricular pacing in pediatric patients. *Am J Cardiol* 64: 1052-53, 1989.

129. Mirkin, G. Abdominal pains. In *Runner's World magazine. The complete runner*. New York: Avon, 1974, 111-112.

130. Molthan, M.E., R.A. Miller, and A.R. Hastreiter. Congenital heart block with fatal Adams-Stokes attacks in childhood. *Pediatrics* 30: 32-41, 1962.

131. Morganroth, J., B.J. Maron, W.L. Henry, and S.E. Epstein. Comparative left ventricular dimensions in trained athletes. *Ann Intern Med* 82: 521-524, 1975.

132. Moses, F.M. The effect of exercise on the gastrointestinal tract. *Sports Med* 9: 159-172, 1990.

133. Moss, A.J., J.L. Robinson, L. Gessman, R. Gillespie, W. Zareba, P.J. Schwartz, G.M. Vincent, J. Benhorin, E.L. Heilbron, J.A. Touwbin, S.G. Priori, C. Napolitano, L. Zhang, A. Medina, M.L. Andrews, and K. Timothy. Comparison of clinical and genetic variables of cardiac events associated with loud noise versus swimming among subjects with the long QT syndrome. *Am J Cardiol* 84: 876-879, 1999.

134. Murphy, A.M., M. Blades, S. Daniels, and F.W. James. Blood pressure and cardiac output during exercise: a longitudinal study of children undergoing repair of coarctation. *Am Heart J* 117: 1327-1332, 1989.

135. Myers, J.L., M.M. Gleason, S.E. Cyran, and B.G. Baylen. Surgical management of coronary insufficiency in a child with Kawasaki disease: use of bilateral internal mammary arteries. *Ann Thorac Surg* 46: 459-61, 1988.

136. Nagaoka, H., N. Isobe, S. Kubota, T. Iizuka, S. Imai, and T. Suzuki. Myocardial contractile reserve as prognostic determinant in patients with idiopathic dilated cardiomyopathy without overt heart failure. *Chest* 111: 344-50, 1997.

137. Nau, K.L., V.L. Katch, R.H. Beekman, and M. Dick. Acute intraarterial blood pressure response to bench press weight lifting in children. *Pediatr Exerc Sci* 2: 37-45, 1990.

138. Nielsen, J.S., and J. Fabricius. The effect of exercise on the size of the shunt in patients with atrial septal defects. *Acta Med Scand* 183: 91-95, 1968.

139. Nir, A., D.J. Driscoll, C.D. Mottram, K.P. Offord, F.J. Puga, H.V. Schaff, and G.K. Danielson. Cardiorespiratory response to exercise after the Fontan operation: a serial study. *J Am Coll Cardiol* 22: 216-220, 1993.

140. Nixon, P., J. Fricker, B.E. Noyes, S.A. Webber, D.M. Orenstein, and J.M. Armitage. Exercise testing in pediatric heart, heart-lung, and lung transplant recipients. *Chest* 107: 1328-1335, 1995.

141. Nudel, D.B., I. Hassett, and A. Gurain. Young long distance runners: physiologic characteristics. *Clin Pediatr* 28: 500-505, 1989.

142. O'Connor, F.G., R.G. Oriscello, and B.D. Levine. Exercise-related syncope in the young athlete: reassurance, restriction or referral? *Am Fam Physician* 60: 2001-8, 1999.

143. Oelberg, D.A., F. Marcotte, H. Kreisman, N. Wolkove, D. Langleben, and D. Small. Evaluation of right ventricular systolic pressure during incremental exercise by Doppler echocardiography in adults with atrial septal defect. *Chest* 113: 1459-65, 1998.

144. Orbach, P., and D.T. Lowenthal. Evaluation and treatment of hypertension in active individuals. *Med Sci Sports Exerc* 30 (Suppl): S354-S366, 1998.

145. Page, E., H. Perrault, and P. Flore. Cardiac output response to dynamic exercise after atrial switch repair for transposition of the great arteries. *Am J Cardiol* 77: 892-895, 1996.

146. Pahl, E., R. Sehgal, D. Chrystof, W.H. Neches, C.L. Webb, E. Duffy, S.T. Shulman, and F.A. Chaudhry. Feasibility of exercise stress echocardiography for the follow-up of children with coronary involvement secondary to Kawasaki disease. *Circulation* 91: 122-128, 1995.

147. Paridon, S.M., F.M. Galioto, J.A. Vincent, T.L. Tomassoni, N.M. Sullivan, and J.T. Bricker. Exercise capacity and incidence of myocardial perfusion defects after Kawasaki disease in children and adolescents. *J Am Coll Cardiol* 25: 1420-1424, 1995.

148. Paridon, S.M., R.A. Humes, and W.W. Pinsky. The role of chronotropic impairment during exercise after the Mustard operation. *J Am Coll Cardiol* 17: 729-732, 1991.

149. Paridon, S.M., R.D. Ross, L.R. Kuhns, and W.W. Pinsky. Myocardial performance and perfusion during exercise in patients with coronary artery disease caused by Kawasaki disease. *J Pediatr* 116: 52-56, 1990.

150. Paul, M.H., and H.U. Wessel. Exercise studies in patients with transposition of the great arteries after atrial repair operations (Mustard/Senning): a review. *Pediatr Cardiol* 20: 49-55, 1999.

151. Pelech, A.N., W. Kartodihardjo, J.A. Balfe, J.W. Balfe, P.M. Olley, and F.H.H. Leenen. Exercise in children before and after coarctectomy: hemodynamic, echocardiographic, and biochemical assessment. *Am Heart J* 112: 1263-1270, 1986.

152. Perrault, H. Benefits of exercise training after surgical repair of congenital heart disease: a theoretical perspective. In Blimkie, C.J.R. and O. Bar-Or, eds.,

New horizons in pediatric exercise science. Champaign, IL: Human Kinetics, 1995, 123-136.

153. Perrault, H., A. Davignon, G. Grief, A. Fournier, and C. Chartrand. Disturbance of heart rate during exercise following intracardiac repair: contribution of cardiopulmonary bypass. *Pediatr Exerc Sci* 4: 270-280, 1992.

154. Perrault, H.M., and R.A. Turcotte. Do athletes have 'the athlete heart'? *Progr Pediatr Cardiol* 2: 40-50, 1993.

155. Petersson, P.O. Atrial septal defect of secundum type. A clinical study before and after operation with special reference to hemodynamic function. *Acta Paediatr Scand* 174 (Suppl): 71-85, 1967.

156. Pyeritz, R.E., and V.A. McKusick. The Marfan syndrome: diagnosis and management. *N Eng J Med* 300: 772-777, 1979.

157. Ragonese, P., P. Guccione, F. Drago, A. Turchetta, A. Calzolari, and R. Formigari. Efficacy and safety of ventricular rate responsive pacing in children with complete atrioventricular block. *PACE* 17: 603-610, 1994.

158. Reid, M.J., N.E. Coleman, and J.G. Stevenson. Management of congenital aortic stenosis. *Arch Dis Child* 45: 201-205, 1970.

159. Reybrouck, T., A. Bisschop, M. Dumoulin, and L.G. van der Hauwaert. Cardiorespiratory exercise capacity after surgical closure of atrial septal defect is influenced by the age at surgery. *Am Heart J* 122: 1073-1077, 1991.

160. Rhodes, J., R.L. Geggel, G.R. Marx, L. Bevilacqua, Y.B. Dambach, and Z.M. Hijazi. Excessive anaerobic metabolism during exercise after repair of aortic coarctation. *J Pediatr* 131: 210-214, 1997.

161. Riopel, D.A., A.B. Tayler, and A.R. Hohn. Blood pressure, heart rate, pressure-rate product, and electrocardiographic changes in healthy children during treadmill exercise. *Am J Cardiol* 44: 697-703, 1979.

162. Rosenthal, M., A. Redington, and A. Bush. Cardiopulmonary physiology after surgical closure of asymptomatic secundum atrial septal defects in childhood. *Eur Heart J* 18: 1816-1822, 1997.

163. Ross, B.A. Congenital complete atrioventricular block. *Pediatr Clin N Amer* 37: 69-78, 1990.

164. Rost, R., and W. Hollman. Athlete's heart—a review of its historical assessment and new aspects. *Int J Sports Med* 4: 147-165, 1983.

165. Rowland, T.W. Cardiac characteristics of the child endurance athlete. Malina, R.M., and Clark, M.A. (eds.) Youth Sports. Perspectives for a New Century. Montery, CA: *Coaches Choice*, 53-68: 2003.

166. Rowland, T.W., B.C. Delaney, and S.F. Siconolfi. 'Athlete's heart' in prepubertal children. *Pediatrics* 79: 800-804, 1987.

167. Rowland, T.W., R.C. McFaul, and D.A. Burton. Syncope during athletic participation: a diagnostic case study. *Pediatr Exerc Sci* 3: 263-270, 1991.

168. Rowland, T., J. Potts, T. Potts, J. Son-Hing, G. Harbison, and G. Sandor. Cardiovascular responses to exercise in children and adolescents with myocardial dysfunction. *Am Heart J* 137: 126-133, 1999.

169. Rowland, T.W., and M.M. Richards. The natural history of idiopathic chest pain in children. A follow-up study. *Clin Pediatr* 25: 612-614, 1986.

170. Rowland, T.W., V.B. Unnithan, N.G. MacFarlane, N.G. Gibson, and J.Y. Paton. Clinical manifestations of the 'athlete's heart' in prepubertal male runners. *Int J Sports Med* 15: 515-519, 1994.

171. Rowland, T., M. Wehnert, and K. Miller. Cardiac responses to exercise in competitive child cyclists. *Med Sci Sports Exerc* 32: 747-752, 2000.

172. Ruttenberg, H.D. Pre- and postoperative exercise testing of the child with coarctation of the aorta. *Pediatr Cardiol* 20: 33-37, 1999.

173. Sabath, R., L. Ewing, M. Hubbell, R. Gowdamarajan, C. Ong, and D. Scagliotti. Effects of a 12 week program of cardiac rehabilitation on selected hemodynamic parameters in patients with congenital heart defects [abstract]. *Pediatr Exerc Sci* 5: 464, 1993.

174. Sarubbi, B., G. Pacileo, C. Pisacane, V. Ducceschi, C. Iacono, M.G. Russo, A. Iacono, and R. Calabro. Exercise capacity in young patients after total repair of tetralogy of Fallot. *Pediatr Cardiol* 21: 211-215, 2000.

175. Schieken, R.M., W.R. Clarke, and R.M. Lauer. The cardiovascular response to exercise in children across the blood pressure distribution. *Hypertension* 5: 71-78, 1983.

176. Schulze-Neick, I., T. Nieber, S. Hempel, M. Loebe, S. Schuler, M. Hummel, R. Hetzer, and P.E. Lange. Heart rate response to exercise in pediatric heart transplant recipients [abstract]. *Pediatr Exerc Sci* 8: 181, 1996.

177. Schwartz, P.J., S.G. Priori, C. Spazzolini, A.J. Moss, G.M. Vincent, C. Napolitano, I. Denjoy, P. Guicheney, G. Breithardt, M.T. Keating, J.A. Towbin, et al. Genotype-phenotype correlation in the long-QT syndrome. *Circulation* 103: 89-95, 2001.

178. Schwartz, P.J., A. Zaza, E. Locati, and A.J. Moss. Stress and sudden death. The case of the long QT syndrome. *Circulation* 83 (Suppl II): II-71-II-80, 1991.

179. Selzer, A. Changing aspects of the natural history of valvular aortic stenosis. *N Eng J Med* 317: 91-98, 1987.

180. Shen, W.F., G.S. Roubin, K. Hirasawn, C.Y. Choong, B.F. Hutton, and P.J. Harris. Left ventricular volume and ejection fraction response to exercise in chronic congestive heart failure: difference between dilated cardiomyopathy and previous myocardial infarction. *Am J Cardiol* 55: 1027-1031, 1985.

181. Shephard, R.J. Responses of the cardiac transplant patient to exercise and training. *Exerc Sport Sci Rev* 20: 297-320, 1992.

182. Shimizu, W., T. Ohe, T. Kurita, and K. Shimomura. Differential response of QTU interval to exercise, isoproterenol, and atrial pacing in patients with congenital long QT syndrome. *PACE* 14 (Part II): 1966-1970, 1991.

183. Sholler, G.F., and E.P. Walsh. Congenital complete heart block in patients without anatomic cardiac defects. *Am Heart J* 118: 1193-1198, 1989.

184. Sigurdardottir, L.Y., and H. Helgason. Exercise-induced hypertension after corrective surgery for coarctation of the aorta. *Pediatr Cardiol* 17: 301-307, 1996.

185. Simsolo, R., B. Grunfeld, M. Gimenez, M. Lopez, G. Berri, L. Becu, and M. Bartolini. Long-term systemic hypertension in children after successful repair of coarctation of the aorta. *Am Heart J* 115: 1268-1273, 1988.

186. Smith, R.T. Pacemakers for children. In Gillette, P.C. and A. Garson, eds., *Pediatric arrhythmias: electrophysiology and pacing*. Philadelphia: Saunders, 1990, 532-558.

187. Soffer, E.E., J. Wilson, G. Duethman, J. Launspach, and T.E. Adrian. Effect of graded exercise on esophageal motility and gastroesophageal reflux in nontrained subjects. *Dig Dis Sci* 39: 193-198, 1994.

188. Strong, W.B., R.E. Botti, D.R. Siebert, and J. Liebman. Peripheral and renal vein plasma renin activity in coarctation of the aorta. *Pediatrics* 45: 254-259, 1970.

189. Sullivan, M.J. Role of exercise conditioning in patients with severe systolic left ventricular dysfunction. In Fletcher, G.F., ed., *Cardiovascular response to exercise*. Mount Kisco, NY: Futura, 1994, 359-376.

190. Swan, H., L. Toivonen, and M. Viitasalo. Rate adaptation of QT intervals during and after exercise in children with congenital long QT syndrome. *Eur Heart J* 19: 508-513, 1998.

191. Taylor, M.R.H., and S. Godfrey. Exercise studies in congenital heart block. *Br Heart J* 34: 930-935, 1972.

192. Thilenius, O.G., J.A. Quinones, T.S. Husayni, and J. Novak. Tilt test for diagnosis of unexplained syncope in pediatric patients. *Pediatrics* 87: 334-338, 1991.

193. Thompson, P.D. Exercise for patients with coronary artery and/or coronary heart disease. In Thompson, P.D., ed., *Exercise and sports cardiology*. New York: McGraw-Hill, 2001, 354-370.

194. Thorén, C., P. Herin, and J. Vávra. Studies of submaximal and maximal exercise in congenital complete heart block. *Acta Paediatr Belg* 28 (Suppl): 132-143, 1974.

195. Tomassoni, T.L., V. Raczkowski, and F.M. Galioto. Cardiorespiratory fitness of children several years following recovery from Kawasaki disease [abstract]. *Pediatr Exerc Sci* 3: 82, 1991.

196. Tsuji, A., M. Nagashima, S. Hasegawa, N. Nagai, K. Nishibata, M. Goto, and M. Matsushima. Long-term followup of idiopathic ventricular arrhythmias in otherwise normal children. *Jpn Circ J* 59: 654-62, 1995.

197. Update on the 1987 task force report on high blood pressure in children and adolescents: a working group report from the National High Blood Pressure Education Program. *Pediatrics* 98: 649-658, 1996.

198. Vaccaro, P., F.M. Galioto, L.M. Bradley, and J. Vaccaro. Effect of physical training on exercise tolerance of children following surgical repair of D-transposition of the great arteries. *J Sports Med Phys Fitness* 27: 443-448, 1987.

199. Van Camp, S.P., C.M. Bloor, F.O. Mueller, R.C. Cantu, and H.G. Olson. Nontraumatic sports death in high school and college athletes. *Med Sci Sports Exerc* 27: 641-647, 1995.

200. Vincent, G.M. Long QT syndrome. *Cardiol Clin* 18: 309-325, 2000.

201. Vincent, G.M., D. Jaiswal, and K.W. Timothy. Effects of exercise on heart rate, QT, QTc, and QT/QS2 in the Romano-Ward inherited long QT syndrome. *Am J Cardiol* 68: 498-503, 1991.

202. Wagner, H.R., R.C. Ellison, J.F. Keane, J.O. Humphries, and A.S. Nadas. Clinical course in aortic stenosis. *Circulation* 56 (Suppl I): 47-56, 1977.

203. Webber, S.A. 15 years of pediatric transplantation at the University of Pittsburgh: lessons learned and future prospects. *Pediatr Transplantation* 1: 8-21, 1997.

204. Weber, H.S., S.E. Cyran, M. Grzeszczak, J.L. Myers, M.M. Gleason, and B.G. Baylen. Discrepancies in aortic growth explain aortic arch gradients during exercise. *J Am Coll Cardiol* 21: 1002-1007, 1993.

205. Weesner, K.M., M. Bledsoe, A. Chauvenet, and M. Wofford. Exercise echocardiography in the detection of anthracycline cardiotoxicity. *Cancer* 68: 435-438, 1991.

206. Weindling, S.N., G. Wernovsky, S.D. Colan, J.A. Parker, C. Boutin, S.M. Mone, J. Costello, A.R. Casteneda, and S.T. Treves. Myocardial perfusion, function and exercise tolerance after the arterial switch operation. *J Am Coll Cardiol* 23: 424-433, 1994.

207. Weintraub, R.G., R.M. Gow, and J.L. Wilkinson. The congenital long QT syndromes in childhood. *J Am Coll Cardiol* 16: 674-680, 1990.

208. Wessel, H.U., W.J. Cunningham, and M.H. Paul. Exercise performance in tetralogy of Fallot after intracardiac repair. *J Thorac Cardiovasc Surg* 80: 582-593, 1980.

209. Wessel, H.U., and M.H. Paul. Exercise studies in tetralogy of Fallot: a review. *Pediatr Cardiol* 20: 39-47, 1999.

210. Winkler, R.B., M.D. Freed, and A.S. Nadas. Exercise-induced ventricular ectopy in children and young adults with complete heart block. *Am Heart J* 99: 87-92, 1980.

211. Zellers, T.M., and D.J. Driscoll. Utility of exercise testing to assess aortic recoarctation. *Pediatr Exerc Sci* 1: 163-170, 1989.

212. Zellers, T.M., D.J. Driscoll, C.D. Mottram, F.J. Puga, H.V. Schaff, and G.K. Danielson. Exercise tolerance and cardiorespiratory response to exercise before and after the Fontan operation. *Mayo Clin Proc* 64: 1489-1497, 1989.

213. Zipes, D.G., and A. Garson. Arrhythmias. *Med Sci Sports Exerc* 26 (Suppl): S276-S283, 1994.

Chapter 8

1. Akerblom, H.K., R. Koivukangas, and J. Ilkka. Experiences from a winter camp for teenage diabetics. *Acta Paediatr Scand Suppl* 283: 50-56, 1980.

2. Akerblom, H.K., J.R.O.T. Viikari, and M. Uhari. Cardiovascular risk in young Finns study: general outline and recent developments. *Ann Med* 31: 45-54, 1999.

3. Allen, F.M. Note concerning exercise in the treatment of severe diabetes. *Boston Med Surg J* 173: 734-744, 1915.

4. Aman, J., I. Karlsson, and L. Wranne. Symptomatic hypoglycaemia in childhood diabetes: a population-based questionnaire study. *Diabet Med* 6: 257-261, 1989.

5. Annuzzi, G., G. Riccardi, B. Capaldo, and L. Kaijser. Increased insulin-stimulated glucose uptake by exercised human muscles one day after prolonged physical exercise. *Eur J Clin Invest* 21: 6-12, 1991.

6. Arslanian, S., P.A. Nixon, D. Becker, and A.L. Drasch. Impact of physical fitness and glycemic control on in vivo insulin action in adolescents with IDDM. *Diabetes Care* 13: 9-15, 1990.

7. Baevre, H., O. Sovik, A. Wisnes, and E. Heiervang. Metabolic responses to physical training in young insulin-dependent diabetics. *Scand J Clin Lab Invest* 45: 109-114, 1985.

8. Bailey, R.C., J. Olson, S.L. Pepper, J. Porszasz, T.J. Barstow, and D.M. Cooper. The level and tempo of children's physical activities: an observational study. *Med Sci Sports Exerc* 27: 1033-1041, 1995.

9. Bar-Or, O. Rating of perceived exertion in children and adolescents: clinical aspects. In Ljunggren, G. and S. Dornic, eds., *Psychophysics in action.* Heidelberg: Springer-Verlag, 1989, 105-113.

10. Bar-Or, O. Exertional perception in children and adolescents with a disease or a physical disability: assessment and interpretation. *Int J Sport Psychol* 32: 127-136, 2001.

11. Baraldi, E., C. Monciotti, M. Filippone, G. Magagnin, S. Zanconato, and F. Zacchello. Gas exchange during exercise in diabetic children. *Pediatr Pulmonol Suppl* 13: 155-160, 1992.

12. Baran, D., and H. Dorchy. Aptitude physique de l'adolescent diabetique. *Bull Eur Physiopath Respir* 18: 51-58, 1982.

13. Barkai, L., P. Kempler, I. Vamosi, K. Lukacs, A. Marton, and K. Keresztes. Peripheral sensory nerve dysfunction in children and adolescents with type 1 diabetes mellitus. *Diabet Med* 15: 228-233, 1998.

14. Barkai, L., M. Peja, and I. Vamosi. Physical work capacity in diabetic children and adolescents with and without cardiovascular autonomic dysfunction. *Diabet Med* 13: 254-258, 1996.

15. Becker, D. Individualized insulin therapy in children and adolescents with type 1 diabetes. *Acta Paediatr Suppl* 425: 20-24, 1998.

16. Berenson, G.S., B. Radhakrishnamurthy, B. Weihang, and S.R. Srinivasan. Does adult-onset diabetes mellitus begin in childhood? The Bogalusa Heart Study. *Am J Med Sci* 310 (Suppl 1): S77-S82, 1995.

17. Berger, M., P. Berchtold, H.J. Cuppers, and H. Drost, H.K. Kley, W.A. Müller, W. Wiegelmann, H. Zimmerman-Telschow, F.A. Gries, H.L. Kruskemper, and H. Zimmermann. Metabolic and hormonal effects of muscular exercise in juvenile type diabetics. *Diabetologia* 13: 355-365, 1977.

18. Berger, M., P. Berchtold, F.A. Gries, and H. Zimmermann. Die Bedeutung von Muskelarbeit und training fur die Therapie des Diabetes Mellitus. *Deutsch Med Wochensr* 103: 439-443, 1978.

19. Bergström, E., O. Hernell, L.A. Persson, and B. Vessby. Insulin resistance syndrome in adolescents. *Metabolism* 45: 908-914, 1996.

20. Borg, G. Perceived exertion as an indicator of somatic stress. *Scand J Rehab Med* 2: 92-98, 1970.

21. Buckler, J.M. The relationship between changes in plasma growth hormone levels and body temperature occurring with exercise in man. *Biomedicine* 193-197, 1973.

22. Buckler, J.M.H. Exercise as a screening test for growth hormone release. *Acta Endocrinol* 69: 219-229, 1972.

23. Buckler, J.M.H. Plasma growth hormone response to exercise as a diagnostic aid. *Arch Dis Child* 48: 565-567, 1973.

24. Buckler, J.M.H. Exercise as a physiological stimulus to growth hormone release. *Arch Dis Child* 50: 830, 1975.

25. Burstein, R. Changes in insulin resistance in trained athletes upon cessation of training. 1982. Master's thesis. McMaster University.

26. Campaigne, B.N., T.B. Gilliam, M.L. Spencer, R.M. Lampman, and M.A. Schork. Effects of a physical activity program on metabolic control and cardiovascular fitness in children with insulin-dependent diabetes mellitus. *Diabetes Care* 7: 57-62, 1984.

27. Campaigne, B.N., and R. Gunnarsson. The effects of physical training in people with insulin-dependent diabetes. *Diabet Med* 5: 429-433, 1988.

28. Chase, H.P. Elevation of resting and exercise blood pressures in subjects with type I diabetes and relation to albuminuria. *J Diabetes Complications* 6: 138-142, 1992.

29. Costill, D.L., P. Cleary, W.J. Fink, C. Foster, J.L. Ivy, and F. Witzmann. Training adaptations in skeletal muscle of juvenile diabetics. *Diabetes* 28: 818-822, 1979.

30. Dahl-Jørgensen, K., H.D. Meen, Kr., F. Hanssen, and O. Aagenaes. The effect of exercise on diabetic control and hemoglobin A1 (HbA1) in children. *Acta Paediatr Scand* 283: 53-56, 1980.

31. De Mondenard, J.P. Principes alimentaires d'un sportif diabétique—à propos du cyclisme. *Vie Méd Can Francais* 8: 643-648, 1979.

32. de San, L.C., J.M. Parkin, and S.J. Turner. Treadmill exercise test in short children. *Arch Dis Child* 59: 1179-1182, 1984.

33. Després, J.-P., C. Couillard, J. Bergeron, and B. Lamarche. Regional body fat distribution, the insulin-resistance-dyslipidemic syndrome, and the risk of type 2 diabetes and coronary heart disease. In Ruderman, N., J.T. Devlin, S.H. Schneider, and A. Kriska, eds., *Handbook of exercise in diabetes.* Alexandria, VA: American Diabetes Association, 2001, 197-234.

34. deVries, J.H., R.J.P. Noorda, G.A. Voetberg, and E.A. Vander Veen. Growth hormone release after sequential use of growth hormone releasing factor and exercise. *Horm Metab Res* 23: 397-398, 1991.

35. Donaubauer, J., J. Kratzsch, C. Fritzsch, B. Stach, W. Kiess, and E. Keller. The treadmill exhausting test is not suitable for screening of growth hormone deficiency! *Horm Res* 55: 137-140, 2001.

36. Dorchy, H., F. Ego, D. Baran, and H. Loeb. Effect of exercise on glucose uptake in diabetic adolescents. *Acta Paediatr Belg* 29: 83-85, 1976.

37. Dorchy, H., D. Haumont, H. Loeb, M. Jennes, G. Niset, and J. Poortmans. Decline of the blood glucose concentration after muscular effort in diabetic children. *Acta Paediatr Belg* 33: 105-109, 1980.

38. Dorchy, H., G. Niset, H. Ooms, J. Poortmans, D. Baran, and H. Loeb. Study of the coefficient of glucose assimilation during muscular exercise in diabetic adolescents deprived of insulin. *Diab. Metab* 3: 31-34, 1977.

39. Dorchy, H., and J. Poortmans. Sport and the diabetic child. *Sports Med* 7: 248-262, 1989.

40. Dorchy, H., and J.R. Poortmans. Juvenile diabetes and sports. In Bar-Or, O., ed., *The child and adolescent athlete.* Oxford: Blackwell Scientific, 1996, 455-479.

41. Draminsky-Petersen, H., B. Korsgaard, T. Deckert, and E. Nielsen. Growth, body weight and insulin requirement in diabetic children. *Acta Pediatr* 67: 453-457, 1978.

42. Elo, O., B. Hirvonen, T. Peltonen, and I. Välimäki. Physical working capacity of normal and diabetic children. *Ann Paediatr Fenn* 11: 25-31, 1965.

43. Engerbretson, D.L. The effect of physical conditioning upon the regulation of diabetes mellitus. 1970. PhD dissertation. Pennsylvania State University.

44. Eriksson, B.O., B. Persson, and J.I. Thorell. The effects of repeated prolonged exercise on plasma growth hormone, insulin, glucose, free fatty acids, glycerol, lactate and hydroxybutyric acid in 13 year-old boys and in adults. *Acta Pediatr* 217: 142-146, 1971.

45. Etkind, E.L., and L. Cunningham. Physical abilities in diabetic boys. *Isr J Med Sci* 8: 848-849, 1972.

46. Felig, P., and J. Wharen. Fuel homeostasis in exercise. *N Eng J Med* 293: 1078-1084, 1975.

47. Ferguson, M.A., B. Gutin, N.-A. Le, W. Karp, M. Litaker, M. Humphries, T. Okuyama, S. Riggs, and S. Owens. Effects of exercise training and its cessation on components of the insulin resistance syndrome in obese children. *Int J Obesity Relat Metab Disord* 23: 889-895, 1999.

48. Frasier, S.D. A review of growth hormone stimulation tests in children. *Pediatrics* 53: 929-937, 1974.

49. Frewin, D.B., A.G. Frantz, and J.A. Downey. The effect of ambient temperature on the growth hormone and prolactin response to exercise. *Austr J Exp Biol Med Sci* 54: 97-101, 1976.

50. Frid, A., J. Ostman, and B. Linde. Hypoglycemia risk during exercise after intramuscular injection of insulin in thigh in IDDM. *Diabetes Care* 13: 473-477, 1990.

51. Garlaschi, C., M.J. del Guercio, B. diNatale, A. Caccamo, and G. Chiumello. Muscular exertion: a test of pituitary function in children. *Acta Paediatr Scand* 64: 752-754, 1975.

52. Garlaschi, C., B. diNatale, and M.J. del Guercio. Effect of physical exercise on secretion of growth hormone, glucagon and cortisol in obese and diabetic children. *Diabetes* 24: 758-761, 1975.

53. Ghigo, E., J. Bellone, G. Aimaretti, S. Bellone, S. Loche, M. Cappa, F. Bartolotta, F. Dammacco, and F. Camanni. Reliability of provocative tests to assess growth hormone secretory status. Study in

472 normally growing children. *J Clin Endocrinol Metab* 81: 3323-3327, 2001.

54. Gil-Ad, I., N. Leibowitch, Z. Josefsberg, M. Wasserman, and Z. Laron. Effect of oral clonidine, insulin-induced hypoglycemia and exercise on plasma GHRH levels in short-stature children. *Acta Endocrinol* 122: 89-95, 1990.

55. Goodyear, L.J., M.F. Hirshman, P.M. Valyou, and E.S. Horton. Glucose transporter number, function, and subcellular distribution in rat skeletal muscle after exercise training. *Diabetes* 41: 1091-1099, 1992.

56. Greene, S.A., T. Torresani, and A. Prader. Growth hormone response to a standardised exercise test in relation to puberty and stature. *Arch Dis Child* 62: 53-56, 1987.

57. Guthrie, D.W. Exercise, diets and insulin for children with diabetes. *Nursing (Jenkintown)* 7: 48-54, 1977.

58. Gutin, B., and M. Humphries. Exercise, body composition, and health in children. In Lamb, D.R. and R. Murray, eds., *Exercise, nutrition, and weight control.* Carmel, IN: Cooper, 1998, 295-347.

59. Hagan, R.D., J.F. Marks, and P.A. Warren. Physiologic responses of juvenile-onset diabetic boys to muscular work. *Diabetes* 28: 1114-1119, 1979.

60. Hartley, L.H., J.W. Mason, R.P. Hogan, L.G. Jones, T.A. Kotchen, E.H. Mougey, F.E. Wherry, L.L. Pennington, and P.T. Ricketts. Multiple hormonal responses to graded exercise in relation to physical training. *J Appl Physiol* 33: 602-606, 1972.

61. Hebbelinck, M., H. Loeb, and H. Meersseman. Physical development and performance capacity in a group of diabetic children and adolescents. *Acta Paediatr Belg* 28 (Suppl): 151-161, 1974.

62. Hermansson, G., and J. Ludvigsson. Renal function and blood-pressure reaction during exercise in diabetic and non-diabetic children and adolescents. A pilot study. *Acta Pediatr* 283: 86-94, 1980.

63. Hough, D.O. Diabetes mellitus in sports. *Med Clin North Am* 78: 423-437, 1994.

64. Huttunen, N.P., M.L. Kaar, M. Knip, A. Mustonen, R. Puukka, and H.K. Akerblom. Physical fitness of children and adolescents with insulin-dependent diabetes mellitus. *Ann Clin Res* 16: 1-5, 1984.

65. Huttunen, N.P., S.L. Lankela, M. Knip, P. Lautala, M.L. Karr, K. Lassonen, and R. Puukka. Effect of once-a-week training program on physical fitness and metabolic control in children with IDDM. *Diabetes Care* 12: 737-740, 1989.

66. Jackson, R.L., and H.G. Kelly. A study of physical activity in juvenile diabetic patients. *J Pediatr* 33: 155-166, 1948.

67. Jacober, B., R.M. Schmulling, and M. Engstein. Carbohydrate and lipid metabolism in type 1 diabetics during exhaustive exercise. *Int J Sports Med* 4: 104-108, 1983.

68. Johanson, A.J., and G.L. Morris. A single growth hormone determination to rule out growth hormone deficiency. *Pediatrics* 59: 467-468, 1977.

69. Johansson, C. The diabetic's own view on physical exercise as a part of life. *Acta Paediatr Scand* 283: 117-119, 1980.

70. Johnsonbaugh, R.E., D.E. Bybee, and L.P. Georges. Exercise tolerance test. Single-sample screening technique to rule out growth-hormone deficiency. *JAMA* 240: 664-666, 1978.

71. Joslin, E.P., H.F. Root, P. White, and A. Marble. The treatment of diabetes mellitus. In Joslin, E.P., H.F. Root, P. White, and A. Marble, eds., *The treatment of diabetes mellitus.* Philadelphia: Lea & Febiger, 1959, 243-300.

72. Kaplan, S.L., C.A.L. Abrams, J.J. Bell, F.A. Conte, and M.M. Grumbach. Growth and growth hormone. I. Changes in serum level of growth hormone following hypoglycemia in 134 children with growth retardation. *Pediatr Res* 2: 43-63, 1968.

73. Kaufman, F.R., M. Halvorson, D. Miller, M. MacKenzie, L.K. Fisher, and P. Pitukcheewanont. Insulin pump therapy in type 1 pediatric patients: now and into the year 2000. *Diabetes Metab Res Rev* 15: 338-352, 1999.

74. Keenan, B.S., L.B. Killmer, and J. Sode. Growth hormone response to exercise. A test of pituitary function in children. *Pediatrics* 50: 760-764, 1972.

75. Kiess, W., A. Bottner, K. Raile, T. Kapellen, G. Muller, A. Galler, R. Paschke, and M. Wabitsch. Type 2 diabetes mellitus in children and adolescents: a review from a European perspective. *Horm Res* 59 (Suppl 1): 77-84, 2003.

76. Kitagawa, T., M. Owada, and T. Urakami. Increased incidence of non-insulin dependent diabetes mellitus among Japanese school children correlates with an increased intake of protein and fat. *Clin Pediatr* 37: 111-115, 1998.

77. Kleinbaum, J., and H. Shamoon. Impaired counterregulation of hypoglycemia in insulin-dependent diabetes mellitus. *Diabetes* 32: 493-498, 1983.

78. Koivisto, V.A., and P. Felig. Effects of leg exercise on insulin absorption in diabetic patients. *N Eng J Med* 298: 79-83, 1978.

79. Koivisto, V.A., V.R. Soman, R. DeFronzo, and P. Felig. Effects of acute exercise and training on insulin binding to monocytes and insulin sensitivity in vivo. *Acta Pediatr* 283: 70-78, 1980.

80. Lacey, K.A., A. Hewison, and J.M. Parkin. Exercise as a screening test for growth hormone deficiency in children. *Arch Dis Child* 48: 508-512, 1973.

81. Landt, K.W., B.N. Campaigne, F.W. James, and M.A. Sperling. Effects of exercise training on insulin sensitivity in adolescents with type I diabetes. *Diabetes Care* 8: 461-465, 1985.

82. Larsson, Y., B. Persson, G. Sterky, and C. Thorén. Functional adaptation to vigorous training and exercise in diabetic and non-diabetic adolescents. *J Appl Physiol* 19: 629-635, 1964.

83. Larsson, Y., G. Sterky, K. Ekengren, and T. Moller. Physical fitness and the influence of training in diabetic adolescent girls. *Diabetes* 11: 109-117, 1962.

84. Lawrence, R.D. The effect of exercise on insulin action in diabetes. *Br Med J* 1: 648-650, 1926.

85. Liberman, B., F.P. Cesar, and B.L. Wajchenberg. Human growth hormone (HGH) stimulation tests: the sequential exercise and L-dopa procedure. *Clin Endocrinol* 10: 649-654, 1979.

86. Lin, T., and J.R. Tucci. Provocative tests of growth-hormone release. A comparison of results with seven stimuli. *Ann Int Med* 80: 464-469, 1974.

87. Lohmann, D., F. Liebold, W. Heilmann, H. Senger, and A. Pobl. Diminished insulin response in highly trained athletes. *Metabolism* 27: 521-524, 1978.

88. Ludvigsson, J. Physical exercise in relation to degree of metabolic control in juvenile diabetes. *Acta Paediatr Scand* 283 (Suppl): 45-49, 1980.

89. Ludvigsson, J., Y. Larsson, and P.G. Svensson. Attitudes towards physical exercise in juvenile diabetics. *Acta Paediatr Scand* 283: 106-111, 1980.

90. MacDonald, M.J. Postexercise late-onset hypoglycemia in insulin-dependent diabetic patients. *Diabetes Care* 10: 584-588, 1987.

91. Maclaren, N.K., G.E. Taylor, and S. Raiti. Propranolol-augmented, exercise-induced human growth hormone release. *Pediatrics* 56: 804-807, 1975.

92. Maehlum, S., A.T. Hostmark, and L. Hermansen. Synthesis of muscle glycogen during recovery after prolonged severe exercise in diabetic subjects. Effect of insulin deprivation. *Scand J Clin Lab Invest* 38: 35-39, 1977.

93. Marble, A., and R.M. Smith. Exercise in diabetes mellitus. *Arch Int Med* 58: 577-588, 1936.

94. McNiven-Temple, M.Y., O. Bar-Or, and M.C. Riddell. The reliability and repeatability of the blood glucose response to prolonged exercise in adolescent boys with IDDM. *Diabetes Care* 18: 326-332, 1995.

95. Miller, J.D., G.S. Tannenbaum, E. Colle, and H.J. Guyda. Daytime pulsatile growth hormone secretion during childhood and adolescence. *J Clin Endocrinol Metab* 55: 989-994, 1982.

96. Mogensen, C.E., C.K. Christensen, and E.V. Vittinghus. The stages in diabetic renal disease with emphasis on the state of incipient diabetes nephropathy. *Diabetes* 32 (Suppl 2): 28-33, 1983.

97. Mogensen, C.E., and E.V. Vittinghus. Urinary albumin excretion during exercise in juvenile diabetes. A provocation test for early abnormalities. *Scand J Clin Lab Invest* 35: 295-300, 1975.

98. Mokan, M., A. Mitrakou, T. Veneman, C. Ryan, M. Korytkowski, P. Cryer, and J. Grich. Hypoglycemia unawareness in IDDM. *Diabetes Care* 17: 1397-1403, 1994.

99. Mosher, P.E., M.S. Nash, A.C. Perry, A.R. LaPerriere, and R.B. Goldberg. Aerobic circuit exercise training: effect on adolescents with well-controlled insulin-dependent diabetes mellitus. *Arch Phys Med Rehab* 79: 652-657, 1998.

100. Nicoll, A.G., P.J. Smail, and C.C. Forsyth. Exercise test for growth hormone deficiency. *Arch Dis Child* 59: 1177-1190, 1984.

101. Nordgren, H., U. Freyschuss, and B. Persson. Blood pressure response to physical exercise in healthy adolescents and adolescents with insulin-dependent diabetes mellitus. *Clin Sci* 86: 425-432, 1994.

102. Nurick, M.A., and S. Bennett Johnson. Enhancing blood glucose awareness in adolescents and young adults with IDDM. *Diabetes Care* 14: 1-7, 1991.

103. Okada, Y., T. Hikita, K. Ishitobi, M. Wada, Y. Santo, and Y. Harada. Human growth hormone secretion after exercise and oral glucose administration in patients with short stature. *J Clin Endocrinol Metab* 34: 1055-1058, 1972.

104. Okada, Y., T. Matsuoka, and Y. Kumahara. Human growth hormone secretion during exposure to hot air in normal adult male subjects. *J Clin Endocrinol Metab* 34: 759-763, 1972.

105. Okada, Y., K. Watanabe, T. Takenchi, T. Onishi, K. Tanaka, M. Tsuji, S. Morimoto, and Y. Kumahara. Re-evaluation of exercise as a screening test for ruling out human growth hormone deficiency. *Endocrinol Japan* 25: 437-442, 1978.

106. Orsetti, A., J.F. Brun, C. Fedou, and A.M. Puech-Cathala. Microalbuminuria during an exercise test. Relations to metabolic control and insulin antibodies in type I diabetics. *Diab Metab (Paris)* 14: 214-215, 1988.

107. Owens, S., B. Gutin, M. Ferguson, J. Allison, W. Karp, and N.-A. Le. Visceral adipose tissue and cardiovascular risk factors in obese children. *J Pediatr* 133: 41-45, 1998.

108. Palayew, K., P. Crock, P. Pianosi, A. Coates, G. Weitzner, and A. Schiffrin. Growth hormone response in very short children. *Clin Invest Med* 14: 331-337, 1991.

109. Pedersen, O., H. Beck-Nielsen, and L. Heding. Increased insulin receptors after exercise in patients with insulin-dependent diabetes mellitus. *N Eng J Med* 302: 886-892, 1980.

110. Persson, B., and C. Thorén. Prolonged exercise in adolescent boys with juvenile diabetes mellitus: circulatory and metabolic responses in relation to perceived exertion. *Acta Paediatr Scand Suppl* 283: 62-69, 1980.

111. Pinhas-Hamiel, O., L.M. Dolan, S.R. Daniels, D. Stanford, P.R. Khoury, and P. Zeitler. Increased incidence of non-insulin-dependent diabetes mellitus among adolescents. *J Pediatr* 128: 608-615, 1996.

112. Pombo, M., J.M. Martinon, F. Tato, and J. Peña. Propranolol and exercise test for growth hormone assays. *Pediatrics* 60: 778, 1977.

113. Poortmans, J., A. Dewancker, and H. Dorchy. Urinary excretion of total protein, albumin and B₂-microglobulin during exercise in adolescent diabetics. *Biomedicine* 25: 273-274, 1976.

114. Poortmans, J., H. Dorchy, and D. Toussaint. Urinary excretion of total proteins, albumin, and B2-microglobulin during rest and exercise in diabetic adolescents with and without retinopathy. *Diabetes Care* 5: 617-623, 1982.

115. Poortmans, J., B. Waterlot, and H. Dorchy. Training effect on postexercise microproteinuria in type I diabetic adolescents. *Pediatr Adolesc Endocrinol* 17: 166-172, 1988.

116. Poortmans, J.R., P. Saerens, R. Edelman, F. Vertongen, and H. Dorchy. Influence of the degree of metabolic control on physical fitness in type 1 diabetic adolescents. *Int J Sports Med* 7: 232-235, 1986.

117. Poortmans, J.R., and J. Vanderstraeten. Kidney function during exercise in healthy and diseased humans. An update. *Sports Med* 18: 419-437, 1994.

118. Pruett, E.D.R., and S. Maehlum. Muscular exercise and metabolism in male juvenile diabetics. I. Energy metabolism during exercise. *Scand J Clin Lab Invest* 32: 139-147, 1973.

119. Richardson, R., and A.L. Case. Factors determining the effect of exercise on blood sugar in the diabetic. *J Clin Invest* 13: 949-961, 1934.

120. Riddell, M.C., and O. Bar-Or. Children and adolescents. In Ruderman, N., J.T. Devlin, S.H. Schneider, and A. Kriska, eds., *Handbook of exercise in diabetes.* Alexandria, VA: American Diabetes Association, 2001, 547-566.

121. Riddell, M.C., O. Bar-Or, B.V. Ayub, R.E. Calvert, and G.J.F. Heigenhauser. Glucose ingestion matched with total carbohydrate utilization attenuates hypoglycemia during exercise in adolescents with IDDM. *Int J Sports Nutr* 9: 24-34, 1999.

122. Riddell, M.C., O. Bar-Or, H.C. Gerstein, and G.J.F. Heigenhauser. Perceived exertion with glucose ingestion in adolescent males with IDDM. *Med Sci Sports Exerc* 32: 167-173, 2000.

123. Riddell, M.C., O. Bar-Or, M. Hollidge-Horvat, H.P. Schwarcz, and G.J.F. Heigenhauser. Glucose ingestion and substrate utilization during exercise in boys with IDDM. *J Appl Physiol* 88: 1239-1246, 2000.

124. Ronnemaa, T., and V.A. Koivisto. Combined effect of exercise and ambient temperature on insulin absorption and postprandial glycemia in type I patients. *Diabetes Care* 11: 769-773, 1988.

125. Rose, S.R., J.L. Ross, M. Uriarte, K.M. Barnes, F.G. Cassorla, and G.B. Cutler Jr. The advantage of measuring stimulated as compared with spontaneous growth hormone levels in the diagnosis of growth hormone deficiency. *N Eng J Med* 319: 201-207, 1988.

126. Rosenbloom, A.L., R.S. Young, J.R. Joe, and W.E. Winter. Emerging epidemic of type 2 diabetes in youth. *Diabetes Care* 22: 345-354, 1999.

127. Rosenfeld, R.G., K. Albertsson-Wikland, F. Cassorla, S.D. Frasier, Y. Hasegawa, R.L. Hintz, S. Lafranchi, B. Lippe, L. Loriaux, and S. Melmed. Diagnostic controversy: the diagnosis of childhood growth hormone deficiency revisited. *J Clin Endocrinol Metab* 80: 1532-1540, 1995.

128. Roth, J., S.M. Glick, R.S. Yalow, and S.A. Berson. Secretion of human growth hormone: physiologic and experimental modification. *Metabolism* 12: 577-579, 1963.

129. Round table. Diabetes and exercise. *Phys Sportsmed* 7: 47-64, 1979.

130. Rowland, T.W., P.M. Martha Jr., E.O. Reiter, and L.N. Cunningham. The influence of diabetes mellitus on cardiovascular function in children and adolescents. *Int J Sports Med* 13: 431-435, 1992.

131. Rowland, T.W., L.A. Swadba, D.E. Biggs, E.J. Burke, and E.O. Reiter. Glycemic control with physical training in insulin-dependent diabetes mellitus. *Am J Dis Child* 139: 307-310, 1985.

132. Ruderman, N., J.T. Devlin, S.H. Schneider, and A. Kriska, eds. *Handbook of exercise in diabetes.* Alexandria, VA: American Diabetes Association, 2001, 1-699.

133. Rutenfranz, J., R. Mocellin, J. Bauer, and W. Herzig. Investigations of the physical working capacity of healthy and sick adolescents. II. The physical working capacity of diabetic children and adolescents [in German]. *Z Kinderheilkd* 103: 133-156, 1968.

134. Saggese, G., and C.B.E. Meossi. Physiological assessment of growth hormone secretion in the diagnosis of children with short stature. *Pediatrician* 14: 121-137, 1987.

135. Saltin, B., M. Houston, E. Nygaard, T. Graham, and J. Wahren. Muscle fiber characteristics in healthy men and patients with juvenile diabetes. *Diabetes* 28(1): 93-99, 1979.

136. Sanders, C.A., G.E. Levinson, W.H. Abelman, and N. Freinkel. Effect of exercise on peripheral utilization of glucose in man. *N Eng J Med* 271: 220-225, 1964.

137. Schiffrin, A., and S. Parikh. Accommodating planned exercise in type I diabetic patients on intensive treatment. *Diabetes Care* 8: 337-342, 1985.

138. Schneider, S.H., A. Vitug, R. Ananthakrishnan, and A.K. Khachadurian. Impaired adrenergic response to prolonged exercise in type I diabetes. *Metabolism* 40: 1219-1225, 1991.

139. Seip, R.L., A. Weltman, D. Goodman, and A.D. Rogol. Clinical utility of cycle exercise for the physiologic assessment of growth hormone release in children. *Am J Dis Child* 144: 998-1000, 1990.

140. Shanis, B.S., and T. Moshang. Propranolol and exercise as a screening test for growth hormone deficiency. *Pediatrics* 57: 712-714, 1976.

141. Sinha, R., G. Fisch, B. Teague, W.V. Tamborlane, B. Banyas, K. Allen, M. Savoye, V. Rieger, S. Taksali, G. Barbetta, R.S. Sherwin, and S. Caprio. Prevalence of impaired glucose tolerance among children and adolescents with marked obesity. *N Eng J Med* 346: 802-810, 2002.

142. Soman, V.R., V.A. Koivisto, D. Deibert, P. Felie, and R.A. DeFronzo. Increased insulin sensitivity and insulin binding to monocytes after physical training. *N Eng J Med* 301: 1200-1204, 1979.

143. Sonnenberg, G.E., F.W. Kemmer, and M. Berger. Exercise in type I (insulin-dependent) diabetic patients treated with continuous subcutaneous insulin infusion. *Diabetologia* 33: 696-703, 1990.

144. Sperling, M.A. Diabetes mellitus in children. In Behrman, R.E., R.M. Kliegman, and H.B. Jenson, eds., *Nelson's textbook of pediatrics*. Philadelphia: Saunders, 2000, 1783.

145. Sprague, R.G. Physical activity and the control of diabetes mellitus. In Hamwi, G.J. and T.S. Danowski, eds., *Diabetes mellitus: diagnosis and treatment*. New York: American Diabetes Association, 1967, 117-119.

146. Sterky, G. Physical work capacity in diabetic school children. *Acta Pediatr* 52: 1-10, 1963.

147. Stratton, R., D.P. Wilson, R.K. Endres, and D.E. Goldstein. Improved glycemic control after supervised 8-week exercise program in insulin-dependent diabetic adolescents. *Diabetes Care* 10: 589-593, 1987.

148. Sutton, J., and L. Lazarus. Effect of adrenergic blocking agents on growth hormone responses to physical exercise. *Horm Metab Res* 6: 427-430, 1974.

149. Sutton, J., and L. Lazarus. Growth hormone in exercise: comparison of physiological and pharmacological stimuli. *J Appl Physiol* 41: 523-527, 1976.

150. Sutton, J.R., and L. Lazarus. Growth hormone in exercise: comparison of physiological and pharmacological stimuli. *J Appl Physiol* 41: 523-527, 1976.

151. Sutton, J.R., J.D. Young, L. Lazarus, J.B. Hickie, and J. Maksvytis. The hormonal response to physical exercise. *Austr Ann Med* 18: 84-90, 1969.

152. Tamborlane, W.V., R.S. Sherwin, V. Koivisto, R. Hendler, M. Genel, and P. Felig. Normalization of the growth hormone and catecholamine response to exercise in juvenile onset diabetic subjects treated with a portable insulin infusion pump. *Diabetes* 28: 785-788, 1979.

153. Thompson, D.L., J.Y. Weltman, A.D. Rogol, D.L. Metzger, J.D. Veldhuis, and A. Weltman. Cholinergic and opioid involvement in release of growth hormone during exercise and recovery. *J Appl Physiol* 75: 870-878, 1993.

154. Treuth, M.S., G.R. Hunter, R. Figueroa-Colon, and M.I. Goran. Effects of strength training on intra-abdominal adipose tissue in obese prepubertal girls. *Med Sci Sports Exerc* 30: 1738-1743, 1998.

155. Trovati, M., Q. Carta, F. Cavalot, S. Vitali, G. Passarino, G. Rocca, G. Emanuelli, and G. Lenti. Continuous subcutaneous insulin infusion and postprandial exercise in tightly controlled type I (insulin-dependent) diabetic patients. *Diabetes Care* 7: 327-330, 1984.

156. Turnheim, E., E. Ogris, and W. Swoboda. Results of the exercise (ergometer) test as a diagnostic screening procedure in children with short stature. *Wien Klin Wochenschr* 87: 608-611, 1975.

157. Wahren, J., L. Hagenfeldt, and P. Felig. Splanchnic and leg exchange of glucose, amino acids, and free fatty acids during exercise in diabetes mellitus. *J Clin Invest* 55: 1303-1314, 1975.

158. Weltman, A., J.Y. Weltman, C.J. Womack, S.E. Davis, J.L. Blumer, G.A. Gaesser, and M.L. Hartman. Exercise training decreases the growth hormone (GH) response to acute constant-load exercise. *Med Sci Sports Exerc* 29: 669-676, 1997.

159. Wilkinson, P.W., and J. Parkin. Growth hormone response to exercise in obese children. *Lancet* 2: 55, 1974.

160. Wilson, T.A., I.L. Solomon, and E.J. Schoen. Exercise screening of short children for growth hormone deficiency in a family practice setting. *J Family Pract* 11: 547-549, 1980.

161. Winter, J.S.D. The metabolic response to exercise and exhaustion in normal and growth-hormone-deficient children. *Can J Physiol Pharmacol* 52: 575-582, 1974.

162. Wise, P.H., R.B. Burnet, T.D. Geary, and H. Berriman. Selective impairment of growth hormone response to physiological stimuli. *Arch Dis Child* 50: 210-214, 1975.

163. Wise, P.H., R.D. Burnet, R.D. Geary, and H. Berriman. Exercise as a physiological stimulus to growth hormone release. *Arch Dis Child* 50: 830, 1975.

164. Yasar, S.A., T. Tulassay, L. Madacsy, A. Korner, L. Szucs, I. Nagy, A. Szabo, and M. Miltenyi. Sympathetic-adrenergic activity and acid-base regulation under acute physical stress in type I (insulin-dependent) diabetic children. *Horm Res* 42: 110-115, 1994.

165. Yki-Jarvinen, H., R.A. DeFronzo, and V.A. Koivisto. Normalization of insulin sensitivity in type I diabetic subjects by physical training during insulin pump therapy. *Diabetes Care* 7: 520-527, 1984.

166. Zinman, B. Insulin pump therapy with continuous subcutaneous insulin infusion and exercise in patients with type 1 diabetes. In Ruderman, N., J.T. Devlin, S.H. Schneider, and A. Kriska, eds., *Handbook of exercise in diabetes*. Alexandria, VA: American Diabetes Association, 2001, 377-381.

167. Zinman, B., F.T. Murray, M. Vranic, A.M. Albisser, B.S. Leibel, P.A. McClean, and E.B. Marliss. Glucoregulation during moderate exercise in insulin treated diabetics. *J Clin Endocrinol Metab* 45: 641-652, 1977.

Chapter 9

1. Allon, N. Self-perceptions of the stigma of overweight in relationship to weight-losing pattern. *Am J Clin Nutr* 32: 470-480, 1979.

2. Amador, M., L.T. Ramos, M. Morono, and M.P. Hermelo. Growth rate reduction during energy restriction in obese adolescents. *Exp Clin Endocrinol* 96: 73-82, 1990.

3. American Psychiatric Association. *Diagnostic and statistical manual of mental disorders; DSM-IV*. Washington, D.C.: American Psychiatric Association, 1994.

4. Anton-Kuchly, B., P. Roger, and P. Varene. Determinants of increased energy cost of submaximal exercise in obese subjects. *J Appl Physiol: Respir Environ Exerc Physiol* 56: 18-23, 1984.

5. Areskog, N.H., R. Selinus, and B. Vahlquist. Physical work capacity and nutritional status in Ethiopian male children and young adults. *Am J Clin Nutr* 22: 471-479, 1969.

6. Åstrand, I., P.O. Åstrand, and A. Stunkard. Oxygen intake of obese individuals during work on a bicycle ergometer. *Acta Physiol Scand* 50: 294-299, 1960.

7. Ayub, B.V., and O. Bar-Or. Energy cost of walking in boys who differ in adiposity but are matched for body mass. *Med Sci Sports Exerc* 35:669-674, 2003.

8. Backman, L., U. Freyschuss, D. Hallberg, and A. Melcher. Cardiovascular function in extreme obesity. *Acta Med Scand* 193: 437-446, 1973.

9. Baile, C.A., W. Zinn, and C. McLaughlin. Exercise, blood lactate and food intake. *Experientia* 26: 1227-1235, 1970.

10. Bandini, L.G., D.A. Schoeller, and W.H. Dietz. Energy expenditure in obese and nonobese adolescents. *Pediatr Res* 27: 198-203, 1990.

11. Bandini, L.G., D.A. Schoeller, R.D.H. Cyr, V.R. Young, and W.H. Dietz. A validation of energy intake and energy expenditure in obese and nonobese adolescents. *Int J Obesity* 111: 437A, 1987.

12. Bar-Or, O. Obesity. In Goldberg, B., ed., *Sports and exercise for children with chronic health conditions*. Champaign, IL: Human Kinetics, 1995, 335-353.

13. Bar-Or, O., and T. Baranowski. Physical activity, adiposity, and obesity among adolescents. *Pediatr Exerc Sci* 6: 348-360, 1994.

14. Bar-Or, O., J. Foreyt, C. Bouchard, K.D. Brownell, W.H. Dietz, E. Ravussin, A.D. Salbe, S. Schwenger, S. St. Jeor, and B. Torun. Physical activity, genetic, and nutritional considerations in childhood weight management. *Med Sci Sports Exerc* 30: 2-10, 1998.

15. Bar-Or, O., and V. Unnithan. Nutritional requirements of young soccer players. *J Sport Sci* 12: S39-S42, 1994.

16. Barac-Nieto, M., G.B. Spurr, M.G. Maksud, and H. Lotero. Aerobic work capacity in chronically undernourished adult males. *J Appl Physiol* 44: 209-215, 1978.

17. Barry, A.J., and T.K. Cureton. Factorial analysis of physique and performance in prepubescent boys. *Res Q* 32: 283-300, 1961.

18. Barta, L., L. Szoke, and V. Vandor-Szobotka. Working capacity of obese children. *Acta Paediatr Acad Sci Hung* 9: 17-21, 1968.

19. Bassey, E.J., J.C. Bryant, E. Clark, et al. Factors affecting cardiac frequency during self paced walking: body composition, age, sex and habitual activity. *J Physiol (London)* 291: 46P, 1979.

20. Battle, E.K., and K.D. Brownell. Confronting a single tide of eating disorders and obesity: treatment vs prevention and policy. *Addict Behav* 21: 755-765, 1997.

21. Beaune, B., S. Blonc, N. Fellmann, M. Bedu, and J. Coudert. Serum insulin-like growth factor I and physical performance in prepubertal Bolivian girls of a high and low socio-economic status. *Eur J Appl Physiol* 76: 98-102, 1997.

22. Becque, M.D., V.L. Katch, A.P. Socchini, R.M. Charles, and C. Moorhead. Coronary risk incidence of obese adolescents: reduction by exercise plus diet interventions. *Pediatrics* 81: 605-612, 1988.

23. Benefice, E., T. Fouere, and R.M. Malina. Early nutritional history of motor performance of Senegalese children, 4-6 years of age. *Ann Hum Biol* 26: 443-455, 1999.

24. Berg, K. Body composition and nutrition of adolescent boys training for bicycle racing. *Nutr Metab* 14: 172-180, 1972.

25. Berndt, I., H.-J. Rehs, and J. Rutenfranz. Sportpadagogische Gesichtspunkte Zur Prophylaxe der Adipositas im Kindesalter. *Off Gesundh-Wesen* 37: 1-9, 1975.

26. Beunen, G., R.M. Malina, M. Ostyn, R. Rensen, J. Simons, and D. Van Gerven. Fatness, growth and motor fitness of Belgian boys 12 through 20 years of age. *Hum Biol* 55: 599-613, 1983.

27. Beunen, G.P., R.M. Malina, R. Renson, J. Simons, and J. Lefevre. Physical activity and growth, matu-

ration and performance: a longitudinal study. *Med Sci Sports Exerc* 24: 576-585, 1992.

28. Björntorp, P. Exercise in the treatment of obesity. *Clin Endocrinol Metab* 5: 431-453, 1976.

29. Björntorp, P. Obesity and physical exercise in relation to glucose tolerance and plasma lipids. In Carlson, L.A. and B. Pernow, eds., *Metabolic risk factors in ischemic cardiovascular disease.* New York: Raven Press, 1982.

30. Björntorp, P., K. de Jounge, L. Sjöström, and L. Sullivan. The effect of physical training on insulin production in obesity. *Metabolism* 19: 632-638, 1970.

31. Blaak, E.E., O. Bar-Or, K.R. Westerterp, and W.H.M. Saris. Effect of VLCD on daily energy expenditure and body composition in obese boys. *Int J Obesity* 14 (Suppl 2): 86, 1990.

32. Blaak, E.E., K.R. Westerterp, O. Bar-Or, L.J.M. Wouters, and W.H.M. Saris. Total energy expenditure and spontaneous activity in relation to training in obese boys. *Am J Clin Nutr* 55: 777-782, 1992.

33. Blimkie, C.J.R., B. Ebbesen, D. MacDougall, O. Bar-Or, and D. Sale. Voluntary and electrically evoked strength characteristics of obese and nonobese preadolescent boys. *Hum Biol* 61: 515-532, 1989.

34. Blimkie, C.J.R., D.G. Sale, and O. Bar-Or. Voluntary strength, evoked twitch contractile properties and motor unit activation of knee extensors in obese and non-obese adolescent males. *Eur J Appl Physiol* 61: 313-318, 1990.

35. Blomquist, B., M. Borjeson, V. Larsson, B. Persson, and G. Sterky. The effect of physical activity on the body measurements and work capacity of overweight boys. *Acta Paediatr Scand* 54: 566-572, 1965.

36. Blonc, S., N. Fellmann, M. Bedu, G. Falgairette, R. De Jonge, P. Obert, B. Beaune, H. Spielvogel, W. Tellez, A. Quintela, S. San Miguel, and J. Coudert. Effect of altitude and socioeconomic status on $\dot{V}O_2$max and anaerobic power in prepubertal Bolivian girls. *J Appl Physiol* 80: 2002-2008, 1996.

37. Boileau, R.A., E.R. Buskirk, D.H. Horstman, et al. Body composition changes in obese and lean men during physical conditioning. *Med Sci Sports* 3: 183-189, 1971.

38. Borg, G. *Physical performance and perceived exertion.* Lund: Gleerup, 1962.

39. Botvin, G.J., A. Cantlon, B.J. Carter, and C.L. Williams. Reducing adolescent obesity through a school health program. *J Pediatr* 15: 1060-1062, 1979.

40. Bouten, C.V., W.D. Van Marekn Lichtenbelt, and K.R. Westerterp. Body mass index and daily physical activity in anorexia nervosa. *Med Sci Sports Exerc* 28: 967-973, 1996.

41. Bradfield, R.B., J. Paulos, and L. Grossman. Energy expenditure and heart rate of obese high school girls. *Am J Clin Nutr* 24: 1482-1488, 1971.

42. Brook, C.G.D., J.K. Lloyd, and O.H. Wolf. Rapid weight loss in children. *Br Med J* 3: 44-45, 1974.

43. Brownell, K.D., and F.S. Kaye. A school-based behavior modification, nutrition education, and physical activity program for obese children. *Am J Clin Nutr* 35: 277-283, 1982.

44. Brownell, K.D., J.H. Kelman, and A.J. Stunkard. Treatment of obese children with and without their mothers. Changes in weight and blood pressure. *Pediatrics* 71: 515-523, 1983.

45. Bruch, H. Obesity in childhood. II. Basal metabolism and serum cholesterol of obese children. *Am J Dis Child* 58: 1001-1022, 1939.

46. Bruch, H. Obesity in childhood. IV. Energy expenditure of obese children. *Am J Dis Child* 60: 1082-1109, 1940.

47. Bruch, H. *Eating disorders: obesity, anorexia nervosa and the person within.* New York: Basic Books, 1973.

48. Bullen, B.A., L.F. Monello, H. Cohen, and J. Mayer. Attitudes towards physical activity, food and family in obese and nonobese adolescent girls. *Am J Clin Nutr* 12: 1-11, 1963.

49. Bullen, B.A., R.B. Reed, and J. Mayer. Physical activity of obese and nonobese adolescent girls appraised by motion picture sampling. *Am J Clin Nutr* 14: 211-223, 1964.

50. Bundred, P., D. Kitchiner, and I. Buchan. Prevalence of overweight and obese children between 1989 and 1998: population based series of cross sectional studies. *Br Med J* 322: 326-329, 2001.

51. Buskirk, E.R. Increasing energy expenditure: the role of exercise. In Wilson, N.L., ed., *Obesity.* Philadelphia: Davis, 1969, 163-176.

52. Caprio, S., L.D. Hyman, S. McCarthy, R. Lange, M. Bronson, and W.V. Tamborlane. Fat distribution and cardiovascular risk factors in obese adolescent girls: importance of the intraabdominal fat depot. *Am J Clin Nutr* 64: 12-17, 1996.

53. Casper, R.C., D.A. Schoeller, R. Kushner, J. Hnilicka, and S.T. Gold. Total daily energy expenditure and activity level in anorexia nervosa. *Am J Clin Nutr* 53: 1143-1150, 1991.

54. Chalmers, J., J. Catalan, A. Day, and C. Fairburn. Anorexia nervosa presenting as morbid exercising. *Lancet* 1: 1985.

55. Chaussain, M., B. Gamain, A.M. Latorre, P. Vaida, and J. deLattre. Respiratory function at rest in obese children [in French]. *Bull Eur Physiopath Respir* 13: 599-609, 1977.

56. Chavez, A., and C. Martinez. Behavioral measurements of activity in children and their relation to food intake in a poor community. In Pollitt, E. and P. Amante, eds., *Energy intake and activity.* New York: Liss, 1984, 303-321.

57. Christakis, G., S. Sajecki, R.W. Hillman, E. Miller, S. Blumenthal, and M. Archer. Effect of a combined nutrition education and physical fitness program on the weight status of obese high school boys. *Fed Proc* 25: 15-19, 1966.

58. Clancey-Hepburn, K., A.A. Hickey, and G. Nevill. Children's behavior responses to TV food advertisements. *J Nutr Educ* 6: 93-96, 1974.

59. Clark, D.G., and S.N. Blair. Physical activity and prevention of obesity in childhood. In Krasnegor, N.A., G.D. Grave, and N. Kretchmer, eds., *Childhood obesity: a biobehavioral perspective.* Caldwell, NJ: Telford Press, 1988, 121-142.

60. Coates, T.J., and C.E. Thoresen. Treating obesity in children and adolescents: a review. *Am J Pub Health* 68: 143-151, 1978.

61. Cohen, C.J. Physical activity and dietary patterns of lean versus obese middle-school children. *Pediatr Exerc Sci* 4: 187-188, 1992.

62. Cohen, C.J., C.S. McMillan, and D.R. Samuelson. Long-term effects of a lifestyle modification program on the fitness of sedentary, obese children. *J Sports Med Phys Fitness* 31: 183-188, 1991.

63. Cole, T.J., M.C. Bellizzi, K.M. Flegal, and W.H. Dietz. Establishing a standard definition for child overweight and obesity worldwide: international survey. *Br Med J* 320: 1240-1243, 2000.

64. Cooper, D.M., J. Poage, T.J. Barstow, and C. Springer. Are obese children truly unfit? Minimizing the confounding effect of body size on the exercise response. *J Pediatr* 116: 223-230, 1990.

65. Corbin, C.B., and P. Pletcher. Diet and physical activity patterns of obese and nonobese elementary school children. *Res Q* 39: 922-928, 1968.

66. Davies, C.T.M. Physiological responses to exercise in East African children. II. The effects of schistosomiasis, anaemia and malnutrition. *Environ Child Health* 19: 115-119, 1973.

67. Davies, C.T.M., L. Fohlin, and C. Thorén. Perception of exertion in anorexia nervosa patients. In Berg, K. and B.O. Eriksson, eds., *Children and exercise IX.* Baltimore: University Park Press, 1980, 327-332.

68. Davies, C.T.M., S. Godfrey, M. Light, A.J. Sargeant, and E. Zeidifard. Cardiopulmonary responses to exercise in obese girls and young women. *J Appl Physiol* 38: 373-376, 1975.

69. Davies, C.T.M., W. Von Döbeln, L. Fohlin, G. Freyschuss, and C. Thorén. Total body potassium, fat free weight and maximal aerobic power in children with anorexia nervosa. *Acta Paediatr Scand* 67: 229-234, 1978.

70. Davies, P.S.W. Diet composition and body mass index in pre-school children. *Eur J Clin Nutr* 51: 443-448, 1997.

71. Davies, P.S.W., J. Gregory, and A. White. Physical activity and body fatness in pre-school children. *Int J Obesity* 19: 6-10, 1995.

72. Davies, P.S.W., J.M. Wells, C.A. Fieldhouse, J.M.E. Day, and A. Lucas. Parental body composition and infant energy expenditure. *Am J Clin Nutr* 61: 1026-1029, 1995.

73. Davis, C., D.K. Katzman, S. Kaptein, C. Kirsh, H. Brewer, K. Kalmbach, M.P. Olmsted, D.B. Woodside, and A.S. Kaplan. The prevalence of high-level exercise in the eating disorders: etiological implications. *Compr Psychiatry* 38: 321-326, 1997.

74. Davis, C., S.H. Kennedy, E. Ravelski, and M. Dionne. The role of physical activity in the development and maintenance of eating disorders. *Psychol Med* 24: 957-967, 1994.

75. de Onis, M., and M. Blossner. *WHO global database on child growth and malnutrition.* WHO/NUT/77.4. Geneva: World Health Organization, 1997.

76. Delany, J.P., D.W. Harsha, J.C. Kime, J. Kumler, L. Melancon, and G.A. Bray. Energy expenditure in lean and obese prepubertal children. *Obesity Res* 1: 67-72, 1995.

77. Demeersman, R., D.C. Schaefer, and W.W. Miller. Effect of diet upon ventilation and effort perception during a physical stressor. *Physiol Behav* 35: 555-558, 1985.

78. Dempsey, J.A., J.A. Reddan, B. Balke, and J. Rankin. Work capacity determinants and physiologic cost of weight-supported work in obesity. *J Appl Physiol* 21: 1815-1820, 1966.

79. Després, J.-P., and B. Lamarche. Physical activity and the metabolic complications of obesity. In Bouchard, C., ed., *Physical activity and obesity.* Champaign, IL: Human Kinetics, 2000, 331-354.

80. Dietz, W.H. Is reduced metabolic rate associated with obesity? *J Pediatr* 129: 621-623, 1996.

81. Dietz, W.H., and S.L. Gortmaker. Do we fatten our children at the television set? Obesity and television viewing in children and adolescents. *Pediatrics* 75: 807-812, 1985.

82. Dietz, W.H. Jr., and R. Hartung. Changes in height velocity of obese preadolescents during weight reduction. *Am J Dis Child* 139: 705-707, 1985.

83. Donnelly, J.E., D.J. Jacobsen, J.E. Whatley, J.O. Hill, L.L. Swift, A. Cherrington, B. Polk, Z.V. Tran, and G. Reed. Nutrition and physical activity program to attenuate obesity and promote physical and metabolic fitness in elementary school children. *Obesity Res* 4: 229-243, 1996.

84. DuRant, R.H., T. Baranowski, M. Johnson, and W.O. Thompson. The relationship among television watching, physical activity, and body composition of young children. *Pediatrics* 94: 449-455, 1994.

85. Epstein, L.H. Methodological issues and ten-year outcomes for obese children. *Ann NY Acad Sci* 699: 237-249, 1993.

86. Epstein, L.H., K.J. Coleman, and M.D. Myers. Exercise in treating obesity in children and adolescents. *Med Sci Sports Exerc* 28: 428-435, 1996.

87. Epstein, L.H., and G. Goldfield. Physical activity in the treatment of childhood overweight and obesity: current evidence and research issues. *Med Sci Sports Exerc* 31: 553-559, 1999.

88. Epstein, L.H., M.D. Myers, H.A. Raynor, and B.E. Saelens. Treatment of childhood obesity. *Pediatrics* 101: 554-570, 1998.

89. Epstein, L.H., A. Valoski, R.R. Wing, and J. McCurley. Ten-year follow-up of behavioral, family-based treatment for obese children. *JAMA* 264: 2519-2523, 1990.

90. Epstein, L.H., A. Valoski, R.R. Wing, and J. McMurley. Ten-year outcomes of behavioral family-based treatment for childhood obesity. *Health Psych* 13: 373-383, 1994.

91. Epstein, L.H., A.M. Valoski, L.S. Vara, J. McCurley, L. Wisniewski, M.A. Kalarchian, K.R. Klein, and L.R. Shrager. Effects of decreasing sedentary behavior and increasing activity on weight change in obese children. *Health Psych* 14: 109-115, 1995.

92. Epstein, L.H., R.R. Wing, R. Koeske, and A. Valoski. A comparison of lifestyle exercise, aerobic exercise and calisthenics on weight loss in obese children. *Behav Therapy* 16: 345-356, 1985.

93. Ferguson, M.A., B. Gutin, N.-A. Le, W. Karp, M. Litaker, M. Humphries, T. Okuyama, S. Riggs, and S. Owens. Effects of exercise training and its cessation on components of the insulin resistance syndrome in obese children. *Int J Obesity Relat Metab Disord* 23: 889-895, 1999.

94. Ferro-Luzzi, A., A. D'Amicis, A.H. Ferrini, and G. Maiale. Nutrition, environment and physical performance of preschool children in Italy. *Bibl Nutr Dieta* 27: 85-106, 1979.

95. Fohlin, L. Body composition, cardiovascular and renal function in adolescent patients with anorexia nervosa. *Acta Paediatr Scand Suppl* 268: 5-20, 1977.

96. Fohlin, L. Exercise performance and body dimensions in anorexia nervosa before and after rehabilitation. *Acta Med Scand* 204: 61-65, 1978.

97. Fohlin, L., C.T.M. Davies, V. Freyschuss, B. Bjarke, and C. Thorén. Body dimensions and exercise performance in anorexia nervosa patients. In Borms, J. and M. Hebbelinck, eds., *Pediatric work physiology.* Karger: Basel, 1978, 102-107.

98. Fohlin, L., U. Freyschuss, B. Bjarke, C.T.M. Davies, and C. Thorén. Function and dimensions of the circulatory system in anorexia nervosa. *Acta Paediatr Scand Suppl* 67: 11-16, 1978.

99. Fohlin, L.P.M. The effects of growth, body composition, and circulatory function of anorexia nervosa in adolescent patients. In Berg, K. and B.O. Eriksson, *Children and exercise IX.* Baltimore: University Park Press, 1980, 317-326.

100. Fontvieille, A.M., A. Kriska, and E. Ravussin. Decreased physical activity in Pima Indian compared with Caucasian children. *Int J Obesity* 17: 445-452, 1993.

101. Friedman, M.A., and K.D. Brownell. Psychological correlates of obesity; moving to the next research generation. *Psychol Bull* 117: 3-20, 1995.

102. Gagnon, G., M. Brault-Dubuc, M. Nadeau, and A. Demirjian. Subcutaneous fat and nutrition in the first year of life. *Nutr Rep Int* 19: 541-551, 1979.

103. Gilliam, T.B., and M.B. Burke. Effects of exercise on serum lipids and lipoproteins in girls, ages 8 to 10 years. *Artery* 4: 203-213, 1978.

104. Goran, M.I., W.H. Carpenter, A. McGloin, R. Johnson, J.M. Hardin, and R.L. Weinsier. Energy expenditure in children of lean and obese parents. *Am J Physiol* 268: E917-E924, 1995.

105. Goran, M.I., G. Hunter, T.R. Nagy, and R. Johnson. Physical activity related energy expenditure and fat mass in young children. *Int J Obesity* 21: 171-178, 1997.

106. Gortmaker, S., W.H. Dietz Jr., and L.W.Y. Cheung. Inactivity, diet and the fattening of America. *J Am Diet Assoc* 90: 1247-1255, 1990.

107. Gortmaker, S.L., W.H. Dietz, A.M. Sobol, and C.A. Wehler. Increasing pediatric obesity in the United States. *Am J Dis Child* 141: 535-540, 1987.

108. Gortmaker, S.L., A. Must, J.M. Perrin, A.M. Sobot, and W.H. Dietz. Social and economic consequences of overweight in adolescence and young adulthood. *N Eng J Med* 329: 1008-1012, 1993.

109. Gortmaker, S.L., A. Must, A.M. Sobol, K. Peterson, G.A. Colditz, and W.H. Dietz. Television viewing as a cause of increasing obesity among children in the United States, 1986-1990. *Arch Pediatr Adolesc Med* 105: 356-362, 1996.

110. Gower, B.A., T.R. Nagy, C.A. Trowbridge, C. Dezenberg, and M.I. Goran. Fat distribution and insulin response in prepubertal African American and white children. *Am J Clin Nutr* 67: 821-827, 1998.

111. Gracey, M., N.E. Hitchcock, K.L. Wearne, P. Garcia-Webb, and R. Lewis. The 1977 Busselton Children's Survey. *Med J Austr* 2: 265-267, 1979.

112. Griffiths, M., and P.R. Payne. Energy expenditure in small children of obese and non-obese parents. *Nature* 260: 698-700, 1976.

113. Gutin, B., and P. Barbeau. Physical activity and body composition in children and adolescents. In Bouchard, C., ed., *Physical activity and obesity.* Champaign, IL: Human Kinetics, 2000, 213-245.

114. Gutin, B., N. Cucuzzo, S. Islam, C. Smith, R. Moffatt, and D. Pargman. Physical training improves body composition of black obese 7- to 11-year-old girls. *Obesity Res* 3: 305-312, 1995.

115. Gutin, B., N. Cucuzzo, S. Islam, C. Smith, and M.E. Stachura. Physical training, lifestyle education and

coronary risk factors in obese girls. *Med Sci Sports Exerc* 28: 19-23, 1996.

116. Gutin, B., and M. Humphries. Exercise, body composition, and health in children. In Lamb, D.R. and R. Murray, eds., *Exercise, nutrition, and weight control.* Carmel, IN: Cooper, 1998, 295-347.

117. Gutin, B., and S. Owens. Role of exercise intervention in improving body fat distribution and risk profile in children. *Am J Hum Biol* 11: 237-247, 1999.

118. Gutin, B., L. Ramsey, P. Barbeau, W. Cannady, M. Ferguson, M. Litaker, and S. Owens. Plasma leptin concentrations in obese children: changes during 4-mo periods with and without physical training. *Am J Clin Nutr* 69: 388-394, 1999.

119. Hampton, M.C., R.L. Hueneman, L.R. Shapiro, and B.W. Mitchell. *J Am Diet Assoc* 50: 385-396, 1967.

120. Hanning, R.M., C.J.R. Blimkie, O. Bar-Or, L.C. Lands, L.A. Moss, and W.M. Wilson. Relationships among nutritional status, skeletal and respiratory muscle function in cystic fibrosis: does early dietary supplementation make a difference? *Am J Clin Nutr* 57: 580-587, 1993.

121. Hansen-Smith, F.M., H.G. Maksud, and D.L. Van Horn. Influence of chronic undernutrition on oxygen consumption of rats during exercise. *Growth* 41: 115-121, 1977.

122. Harris, M.B., and E.S. Hallbauer. Self-directed weight control through eating and exercise. *Behav Respir Ther* 11: 523-529, 1973.

123. Hensley, L.D., W.B. East, and J.L. Stillwell. Body fatness and motor performance during preadolescence. *Res Q Exerc Sport* 53: 133-140, 1982.

124. Heywood, P.F., B. Rur, and M.C. Latham. Use of the SAMI heart-rate integrator in malnourished children. *Am J Clin Nutr* 24: 1446-1450, 1971.

125. Hills, A.P., and A.W. Parker. Gait asymmetry in obese children. *Neuro-orthopedics* 12: 29-33, 1991.

126. Hills, A.P., and A.W. Parker. Locomotor characteristics of obese children. *Child: Care, Health and Development* 18: 29-34, 1992.

127. Hills, A.P., and A.W. Parker. Electromyography of walking in obese children. *Electromyogr Clin Neurophysiol* 33: 225-233, 1993.

128. Holloszy, J.O. The effects of endurance exercise on body composition. In Vague, J. and J. Boyer, eds., *The regulation of the adipose tissue mass.* Excerpta Medica International Congress Series, 1974, 254-258.

129. Hueneman, R.L. Environmental factors associated with preschool obesity. *J Am Diet Assoc* 64: 480-487, 1974.

130. Inselman, L.S., A. Milanese, and A. Deurloo. Effect of obesity on pulmonary function in children. *Pediatr Pulmonol* 16: 130-137, 1993.

131. Ishii, K., and K. Amano. Study of maximal oxygen uptake as related to growth of Japanese boys (7-18 years): via per unit of body weight and muscle plus bone volume of the leg. *Lab Physiol Biomech Per* 12: 25-32, 1984.

132. Johnson, M.L., B.S. Burke, and J. Mayer. Relative importance of inactivity and overeating in the energy balance of obese high school girls. *Am J Clin Nutr* 4: 37-44, 1956.

133. Jokl, E. Physical activity and body composition: fitness and fatness. *Ann NY Acad Sci* 110: 778-794, 1963.

134. Joyce, J.M., D.L. Warren, L.L. Humphrie, A.J. Smith, and J.S. Coon. Osteoporosis in women with eating disorders: comparison of physical parameters, exercise, and menstrual status with SPA and DPA evaluation. *J Nucl Med* 31: 325-331, 1986.

135. Katch, V., M.D. Becque, C. Marks, C. Moorehead, and A. Rocchini. Oxygen uptake and energy output during walking of obese male and female adolescents. *Am J Clin Nutr* 47: 26-32, 1988.

136. Keys, A., J. Brozek, A. Henschel, O. Michelsen, and H.L. Taylor. *The biology of human starvation.* Minneapolis: University of Minnesota Press, 1950.

137. Kirschenbaum, D.S., E.S. Harris, and A.J. Tomarken. Effects of parental involvement in behavioral weight loss therapy for preadolescents. *Behav Ther* 15: 485-500, 1984.

138. Kitagawa, K., and M. Miyashita. Muscle strengths in relation to fat storage in young men. *Eur J Appl Physiol* 38: 189-196, 1978.

139. Klesges, R.C., L.H. Eck, C.L. Hanson, C.K. Haddock, and L.M. Klesges. Effects of obesity, social interactions, and physical environment on physical activity in preschoolers. *Health Psych* 9: 435-449, 1990.

140. Klesges, R.C., M.L. Shelton, and L.M. Klesges. Effects of television on metabolic rate: potential implications for childhood obesity. *Pediatrics* 91: 281-286, 1993.

141. Kooh, S.W., E. Noriega, K. Leslie, C. Muller, and J.E. Harrison. Bone mass and soft tissue composition in adolescents with anorexia nervosa. *Bone* 19: 181-188, 1996.

142. Kraemer, W.J., J.S. Volek, K.L. Clark, S.E. Gordon, S.M. Puhl, L.P. Koziris, J.M. McBride, N.T. Triplett-McBride, M. Putukian, R.U. Newton, K. Hakinnen, J.A. Bush, and W.J. Sebastianelli. Influence of exercise training on physiological and performance changes with weight loss in men. *Med Sci Sports Exerc* 31: 1320-1329, 1999.

143. Kriemler, S., H. Hebestreit, S. Mikami, T. Bar-Or, B.V. Ayub, and O. Bar-Or. Impact of a single exercise bout on energy expenditure and spontaneous physical activity of obese boys. *Pediatr Res* 46: 40-44, 1999.

144. Krotkiewski, M., K. Mandroukas, L. Sjoström, L. Sullivan, H. Wettergvist, and P. Björntorp. Effects of long-term physical training on body fat, metabo-

lism and blood pressure in obesity. *Metabolism* 28: 650-658, 1979.

145. Ku, L.C., L.R. Shapiro, P.B. Crawford, and R.L. Huenemann. Body composition and physical activity in 8 year old children. *Am J Clin Nutr* 34: 2770-2775, 1981.

146. Kuczmarski, R.J., C.L. Ogden, L.M. Grummer-Strawn, K.M. Flegal, S. Guo, R. Wei, Z. Mei, L.R. Curtin, A.F. Roche, and C.L. Johnson. CDC growth charts: United States. Advance data from vital and health statistics. Hyattsville, MD: National Center for Health Statistics, 2000, 314.

147. Kumar, A., O.P. Ghai, and N. Singh. Delayed nerve conduction velocities in children with protein-calorie malnutrition. *J Pediatr* 90: 149-153, 1977.

148. Lamb, K.L. Children's ratings of effort during cycle ergometry: an examination of the validity of two effort rating scales. *Pediatr Exerc Sci* 7: 407-421, 1995.

149. Lands, L., A. Pavilanis, A. Charge, and A.L. Coates. Cardiopulmonary response to exercise in anorexia nervosa. *Pediatr Pulmonol* 13: 101-107, 1992.

150. Leon, A.S., J. Conrad, D.B. Hunningnake, and R. Serfass. Effects of a vigorous walking program on body composition, and carbohydrate and lipid metabolism of obese young men. *Am J Clin Nutr* 32: 1776-1787, 1979.

151. Li, R., I. O'Connor, D. Buckley, and B. Specker. Relation of activity level to body fat in infants 6 to 9 months of age. *J Pediatr* 126: 353-357, 1995.

152. Lichtman, S.W., K. Pisarska, E.R. Berman, M. Pestone, H. Dowling, E. Offenbacher, H. Weisel, S. Heshka, D.E. Mathews, and S.B. Heymsfield. Discrepancy between self-reported and actual caloric intake and exercise in obese subjects. *N Eng J Med* 327: 1893-1898, 1992.

153. Linder, P. Techniques of management for the inactive obese child. In Collipp, P.J., *Childhood obesity*. Littleton, MA: PSG, 1980, 179-205.

154. Luepker, R.V., C.L. Perry, S.M. McKinlay, P.R. Nader, G.S. Parcel, E.J. Stone, L.S. Webber, J.P. Elder, H.A. Feldman, C.C. Johnson, S.H. Kelder, and M. Wu. Outcomes of a field trial to improve children's dietary patterns and physical activity. *JAMA* 275: 768-776, 1996.

155. Mack, R.W., and M.E. Kleinhenz. Growth, caloric intake and activity levels in early infancy: a preliminary report. *Hum Biol* 46: 345-354, 1974.

156. Maffeis, C., F. Schena, M. Zaffanello, L. Zoccante, Y. Schutz, and L. Pinelli. Maximal aerobic power during running and cycling in obese and non-obese children. *Acta Paediatr* 83: 113-116, 1994.

157. Maffeis, C., Y. Schutz, M. Zaffanello, R. Piccoli, and L. Pinelli. Elevated energy expenditure and reduced energy intake in obese prepubertal children: paradox of poor dietary reliability in obesity? *J Pediatr* 124: 348-354, 1994.

158. Maffeis, C., Y.Y. Schutz, F. Schena, M. Zaffanello, and L. Pinelli. Energy expenditure during walking and running in obese and nonobese prepubertal children. *J Pediatr* 123: 193-199, 1993.

159. Malina, R.M. Physical activity and motor development/performance in populations nutritionally at risk. In Pollitt, E. and P. Amante, eds., *Energy intake and activity.* New York: Liss, 1984, 285-302.

160. Malina, R.M., G. Beunen, A. Claessens, J. Lefevre, B. Vanden Eynde, R. Renson, B. Vanreusel, and J. Simons. Fatness and fitness of girls to 17 years. *Obesity Res* 3: 221-231, 1995.

161. Markuske, M. Obesity and school sports. *Sportarzt und Sportmedizin* 10: 404-406, 1969.

162. Marti, B., and E. Vartiainen. Relation between leisure time exercise and cardiovascular risk factors among 15-year-olds in Eastern Finland. *J Epidemiol Comm Health* 43: 228-233, 1989.

163. Mayer, J. Obesity during childhood. In Winnick, M., ed., *Childhood obesity.* New York: Wiley, 1975, 73-80.

164. Mayer, J., N.B. Marshall, J.J. Vitale, J.H. Christensen, M.B. Mashayeekhi, and F.J. Starve. Exercise, food intake and body weight in normal rats and genetically obese mice. *Am J Physiol* 177: 544-548, 1954.

165. Mayer, J., P. Roy, and K.P. Mitra. Relations between caloric intake, body weight and physical work: studies in industrial male population in West Bengal. *Am J Clin Nutr* 4: 169-175, 1956.

165a. Meeks-Gardner, J., M. Grantham-McGregor, S.M. Chang, and C.A. Powell. Dietary intake and observed activity of stunted and non-stunted children in Kingston, Jamacia. Part II: observed activity. *Eur J. Clin Nutr* 44: 585-593, 1990.

166. Messier, S.P., A.B. Davies, D.T. Moore, S.E. Davis, R.J. Pack, and S.C. Kazmar. Severe obesity: effect on foot mechanics during walking. *Foot and Ankle* 15: 29-34, 1994.

167. Mocellin, R., and J. Rutenfranz. Investigations on the physical working capacity of healthy and sick adolescents. *Z Kinderheilkd* 104: 179-196, 1968.

168. Mocellin, R., and J. Rutenfranz. Investigations of the physical working capacity of obese children. *Acta Paediatr Belg* (Suppl) 217: 77-79, 1971.

169. Monello, L.F., and J. Mayer. Obese adolescent girls: an unrecognized 'minority' group? *Am J Clin Nutr* 13: 35-39, 1963.

170. Montoye, H.J., W.M. Mikkelsen, W.D. Block, and R. Gayle. Relationship of oxygen uptake capacity, serum uric acid and glucose tolerance in males and females age 10-69. *Am J Epidemiol* 108: 274-282, 1978.

171. Moody, D.L., J.H. Wilmore, R.N. Girandola, and J.P. Royce. The effects of a jogging program on the body composition of normal and obese high school girls. *Med Sci Sports* 4: 210-213, 1972.

172. Moore, L.L., U.-S.D.T. Neguyen, K.J. Rothman, L.A. Cupples, and R.C. Ellison. Preschool physical

activity level and change in body fatness in young children. *Am J Epidemiol* 142: 982-988, 1995.

173. Must, A., G.E. Dallal, and W.H. Dietz. Reference data for obesity: 85th and 95th percentiles of body mass index (wt/ht²) and triceps skinfold thickness. *Am J Clin Nutr* 53: 839-846, 2001.

174. Must, A., P.F. Jacques, G.E. Dallal, C.J. Bajema, and W.H. Dietz. Long-term morbidity and mortality of overweight adolescents. A follow up of the Harvard Growth Study of 1922-1935. *New Eng J Med* 327: 1350-1355, 1992.

175. Nudel, D.B., N. Gootman, M.P. Nussbaum, and J.R. Shenker. Altered exercise performance and abnormal sympathetic responses to exercise in patients with anorexia nervosa. *J Pediatr* 105: 34-37, 1984.

176. O'Connell, J.K., J.H. Price, S.M. Roberts, S.G. Jurs, and R. McKinley. Utilizing the health belief model to predict dieting and exercising behavior of obese and nonobese adolescents. *Health Educ Q* 12: 343-351, 1985.

177. O'Hara, W.J., C. Allen, and R.J. Shephard. Loss of body fat during an arctic winter expedition. *Can J Physiol Pharmacol* 55: 1235-1241, 1977.

178. Oggiano, N., A. Kantar, E. Fabbrizi, and P.L. Giorgi. Pulmonary function in obese children. In Giorgi, P.L., R.M. Suskind, and C. Catassi, eds., *The obese child. Pediatric adolescent medicine.* Basel: Karger, 1992, 73-80.

179. Okamoto, E., L.L. Davidson, and D.R. Conner. High prevalence of overweight in inner-city schoolchildren. *Am J Dis Child* 147: 155-159, 1993.

180. Oscai, L.B., S.P. Babirak, J.A. McGarr, and C.N. Spirakis. Effect of exercise on adipose tissue cellularity. *Fed Proc* 33: 1956-1958, 1974.

181. Oscai, L.B., S.P. Babriak, F.B. Dubach, J.A. McGarr, and C.N. Spirakis. Exercise or food restriction: effect on adipose tissue cellularity. *Am J Physiol* 227: 901-904, 1974.

182. Owens, S., B. Gutin, J. Allison, S. Riggs, M. Ferguson, M. Litaker, and W. Thompson. Effect of physical training on total and visceral fat in obese children. *Med Sci Sports Exerc* 31: 143-148, 1999.

183. Palla, B., and I.F. Litt. Medical complications of eating disorders in adolescents. *Pediatrics* 81: 613-623, 1988.

184. Parcel, G.S., L.W. Green, and B.A. Bettes. School-based programs to prevent or reduce obesity. In Krasnegor, N.A., G.D. Grave, and N. Kretchmer, eds., *Childhood obesity: a biobehavioral perspective.* Caldwell, NJ: Telford Press, 1988, 143-155.

185. Pařizková, J., L. Stanková, S. Sprynarová, and M. Vamberová. Influence de l'exercice physique sur certains index metaboliques sanguins chez les garçons obèses apres l'effort. *Nutr Dieta* 7: 21-27, 1965.

186. Pařizková, J., and M. Vamberová. Body composition as a criterion of the suitability of reducing regimens

in obese children. *Dev Med Child Neurol* 9: 202-211, 1967.

187. Pařizková, J., M. Vanecková, and M. Vamberová. A study of changes in some functional indicators following reduction of excessive fat in obese children. *Physiol Bohemoslov* 11: 351-357, 1962.

188. Passmore, R. A note on the relation of appetite to exercise. *Lancet* 1: 29, 1958.

189. Pate, R., and J.G. Ross. The National Children and Youth Fitness Study II: factors associated with health-related fitness. *JOPERD* 58: 93-95, 1987.

190. Peckos, P.S. Caloric intake in relation to physique in children. *Science* 117: 631-633, 1953.

191. Peckos, P.S., J.A. Spargo, and F.P. Heald. Program and results of a camp for obese adolescent girls. *Postgrad Med* 27: 527-533, 1960.

192. Pinhas-Hamiel, O., L.M. Dolan, S.R. Daniels, D. Stanford, P.R. Khoury, and P. Zeitler. Increased incidence of non-insulin-dependent diabetes mellitus among adolescents. *J Pediatr* 128: 608-615, 1996.

193. Pirke, K.M., M. Eckert, B. Ofers, G. Goebl, B. Spyra, U. Schweiger, R.J. Tuschl, and M.M. Fichter. Plasma norepinephrine response to exercise in bulimia, anorexia nervosa and controls. *Biol Psychiatry* 25: 795-799, 1989.

194. Poehlman, E.T. Review: exercise and its influence on resting energy metabolism in man. *Med Sci Sports Exerc* 21: 515-525, 1989.

195. Raju, N.V. Effect of early malnutrition on muscle function and metabolism in rats. *Life Sci* 15: 949-960, 1974.

196. Raju, N.V. Effect of exercise during rehabilitation on swimming performance, metabolism and function of muscle in rats. *Br J Nutr* 38: 157-165, 1977.

197. Rehs, H.J., I. Berndt, and J. Rutenfranz. Study on the physical performance capacity of obese subjects with special reference to physical education [in German]. *Z Kinderheilkd* 115: 23-39, 1973.

198. Rehs, H.J., I. Berndt, and J. Rutenfranz. Untersuchungen zur frage der leistungsfahigkeir adiposer unter besonderer berucksichtigung des sportunterrichtes. *Z Kinderheilkd* 115: 23-29, 1973.

199. Resnicow, K. School-based interventions: population vs high-risk approach. *Ann NY Acad Sci* 699: 154-166, 1993.

200. Reybrouck, T., L. Mertens, and D. Schepers. Assessment of cardiorespiratory exercise function in obese children and adolescents by body mass-independent parameters. *Eur J Appl Physiol* 75: 478-483, 1997.

201. Reybrouck, T., M. Weymans, J. Vinckx, H. Stijns, and M. Vanderschueren-Lodeweyckx. Cardiorespiratory function during exercise in obese children. *Acta Paediatr Scand* 76: 342-348, 1987.

202. Roberts, S.B., J. Savage, W.A. Coward, B. Chew, and L. Lucas. Energy expenditure and intake in infants

born to lean and overweight mothers. *N Eng J Med* 318: 461-466, 1988.

203. Robinson, T.N. Reducing children's television viewing to prevent obesity. A randomized controlled trial. *JAMA* 282: 1561-1567, 1999.

204. Robinson, T.N., L.D. Hammer, J.D. Killen, H.C. Kraemer, D.M. Wilson, C. Hayward, and C.B. Taylor. Does television viewing increase obesity and reduce physical activity? Cross-sectional and longitudinal analysis among adolescent girls. *Pediatrics* 91: 273-280, 1993.

205. Rocchini, A.P., V. Katch, J. Anderson, J. Hinderliter, D. Becque, M. Martin, and C. Marks. Blood pressure in obese adolescents: effect of weight loss. *Pediatrics* 82: 16-23, 1988.

206. Rose, H.E., and J. Mayer. Activity, calorie intake, fat storage and the energy balance of infants. *Pediatrics* 41: 18-29, 1968.

207. Rowland, T.W. Effects of obesity on aerobic fitness in adolescent females. *Am J Dis Child* 145: 764-768, 1991.

208. Russell, D., P.J. Prendergast, P.L. Darby, P.E. Garfinkel, J. Whitwell, and K.N. Jeejeebhoy. A comparison between muscle function and body composition in anorexia nervosa: the effect of refeeding. *Am J Clin Nutr* 38: 229-237, 1983.

209. Sallis, J.F., and T.L. McKenzie. Physical education's role in public health. *Res Q Exerc Sport* 62: 124-137, 1991.

210. Sallis, J.F., R.L. Patterson, M.J. Buono, and P.R. Nader. Relation of cardiovascular fitness and physical activity to cardiovascular disease risk factors in children and adults. *Am J Epidemiol* 127: 933-941, 1988.

211. Sallis, J.F., T.L. Patterson, T.L. McKenzie, and P.R. Nader. Family variables and physical activity in preschool children. *J Dev Behav Pediatr* 9: 57-61, 1988.

212. Saris, W.H. Habitual physical activity in children: methodology and findings in health and disease. *Med Sci Sports Exerc* 18: 253-263, 1986.

213. Saris, W.H.M. Aerobic power and daily physical activity in children with special reference to methods and cardiovascular risk indicators. 1982. Dissertation. Catholic University, Krips Repro Meppal, Nijmegen.

214. Satyanarayana, K., and A. Nadamuni Naidu. Nutrition and menarche in rural Hyderabad. *Ann Hum Biol* 6: 163-165, 1979.

215. Savage, M.P., M.M. Petratis, W.H. Thomson, K. Berg, J.L. Smith, and S.P. Sady. Exercise training effects on serum lipids of prepubescent boys and adult men. *Med Sci Sports Exerc* 18: 197-204, 1986.

216. Schoeller, D.A., L.G. Bandini, L.L. Levitsky, and W.H. Dietz. Energy requirements of obese children and young adults. *Proc Nutr Soc* 47: 241-246, 1988.

217. Schrub, J.C., L.M. Wolf, H. Courtois, and F. Javet. Fasting with muscular exercise changes in weight and nitrogen balance [in French]. *Nouv Presse Méd* 4: 875-878, 1975.

218. Seltzer, C.C., and J. Mayer. An effective weight control program in a public school system. *Am J Pub Health* 60: 679-689, 1970.

219. Shephard, R.J. Metabolic adaptations to exercise in the cold. *Sports Med* 16: 266-289, 1993.

220. Shephard, R.J., W.J. O'Hara, C. Allen, and G. Allen. A controlled study of fat loss in the cold. *Med Sci Sports* 11: 98, 1979.

221. Slaughter, M.H., T. Lohman, and J.E. Misner. Relationship of somatotype and body composition to physical performance in 7- to 12-year old boys. *Res Q* 48: 159-168, 1977.

222. Slaughter, M.H., T.G. Lohman, and J.E. Misner. Association of somatotype and body composition to physical performance in 7-12 year-old girls. *J Sports Med Phys Fitness* 20: 189-198, 1980.

223. Soman, V.R., V.A. Koivisto, D. Deibert, P. Felie, and R.A. DeFronzo. Increased insulin sensitivity and insulin binding to monocytes after physical training. *N Eng J Med* 301: 1200-1204, 1979.

224. Sonka, J., I. Gregorová, Z. Tomosová, A. Pavlová, A. Zbirková, R. Rath, J. Urbanek, and M. Josifko. Plasma androsterone, dehydroepiandrosterone and 11-hydroxycorticoids in obesity. Effects of diet and physical activity. *Steroids Lipids Res* 3: 65-74, 1972.

225. Sprynarová, S., and J. Parizková. Changes in the aerobic capacity and body composition in obese boys after reduction. *J Appl Physiol* 20: 934-937, 1965.

226. Spurr, G.B., M. Barac-Nieto, and M.G. Maksud. Childhood undernutrition: implications for adult work capacity and productivity. In Folinsbee, L.J., ed., *Environmental stress. Individual human adaptations.* New York: Academic Press, 1978.

227. Spurr, G.B., and J.C. Reina. Influence of dietary intervention on artificially increased activity in marginally undernourished Colombian boys. *Eur J Clin Nutr* 42: 835-846, 1988.

228. Spurr, G.B., and J.C. Reina. Patterns of daily energy expenditure in normal and marginally undernourished school-aged Colombian children. *Eur J Clin Nutr* 42: 819-834, 1988.

229. Spurr, G.B., and J.C. Reina. Maximum oxygen consumption in marginally malnourished Colombian boys and girls 6-16 years of age. *Am J Hum Biol* 1: 11-19, 1989.

230. Stalonas, P.M. Jr., W.G. Johnson, and M. Christ. Behavior modification for obesity: the evaluation of exercise, contingency management, and program adherence. *J Consult Clin Psychol* 46: 463-469, 1978.

231. Stanley, E.J., H.H. Glaser, D.G. Levin, P.A. Adams, and I.L. Coley. Overcoming obesity in adolescents.

A description of a promising endeavor to improve management. *Clin Pediatr* 9: 29-36, 1970.

232. Stefanik, P.A., F.P. Heald, and J. Mayer. Caloric intake in relation to energy output of obese and non-obese adolescent boys. *Am J Clin Nutr* 7: 55-62, 1959.

233. Sterky, G. Clinical and metabolic aspects of obesity in childhood. In Pernow, B. and B. Saltin, eds., *Muscle metabolism during exercise.* New York: Plenum Press, 1971.

234. Stern, J.S. Is obesity a disease of inactivity? In Stunkard, A.J. and E. Stellar, *Eating and its disorders.* New York: Raven Press, 1984.

235. Stevenson, J.A.F., B.M. Box, V. Feleki, and J.R. Beaton. Bouts of exercise and food intake in the rat. *J Appl Physiol* 21: 118-122, 1966.

236. Story, M., and P. Faulkner. The prime-time diet: a content analysis of eating behavior in television program content and commercials. *Am J Pub Health* 80: 738-740, 1990.

237. Stunkard, A., and Y. Pestka. The physical activity of obese girls. *Am J Dis Child* 103: 116-121, 1962.

238. Sunnegardh, J., L.E. Bratteby, U. Hagman, G. Samuelson, and S. Sjolin. Physical activity in relation to energy intake and body fat in 8 and 13-year-old children in Sweden. *Acta Paediatr Scand* 75: 955-963, 1986.

239. Suskind, R.M., M.S. Sothern, R.P. Farris, T.K. von Almen, H. Schumacher, L. Carlisle, A. Vargas, O. Escobar, M. Loftin, G. Fuchs, R. Brown, and J.N. Udall Jr. Recent advances in the treatment of childhood obesity. *Ann NY Acad Sci* 699: 181-199, 1993.

240. Talbot, N.B., and J. Worcester. The basal metabolism of obese children. *J Pediatr* 16: 146-150, 1940.

241. Taras, H.L., J.F. Sallis, T.L. Patterson, P.R. Nader, and J.A. Nelson. Television's influence on children's diet and physical activity. *Dev Behav Ped* 10: 176-180, 1989.

243. Taylor, W., and T. Baranowski. Physical activity, cardiovascular fitness, and adiposity in children. *Res Q Exerc Sport* 62: 157-163, 1991.

244. Thorén, C. Physical training of handicapped school children. *Scand J Rehab Med* 3: 26-30, 1971.

245. Thorén, C. Working capacity in anorexia nervosa. In Borms, J. and M. Hebbelink, eds., *Pediatric work physiology.* Basel: Karger, 1978, 89-95.

246. Thorén, C., V. Seliger, M. Máček, J. Vavra, and J. Rutenfranz. The influence of training on physical fitness in healthy children and children with chronic diseases. In Linneweh, F., ed., *Current aspects of perinatology and physiology of children.* Berlin, Springer, 1973.

247. Tremblay, M.S., and J.D. Willms. Secular trends in the body mass index of Canadian children. *Can Med Assoc J* 163: 1429-1433, 2000.

248. Troiano, R.P., K.M. Flegal, R.J. Kuczmarski, S.M. Campbell, and C.L. Johnson. Overweight prevalence and trends for children and adolescents. The National Health and Nutrition Examination Surveys, 1963 to 1991. *Arch Pediatr Adolesc Med* 149: 1085-1091, 1995.

249. Tucker, L.A. The relationship of television viewing to physical fitness and obesity. *Adolescence* 21: 797-806, 1986.

250. Twisk, J.W.R., H.C.G. Kemper, W. van Mechelen, and G.B. Post. Tracking of risk factors for coronary heart disease over a 14-year period: a comparison between lifestyle and biologic risk factors with data from the Amsterdam Growth and Health Study. *Am J Epidemiol* 145: 888-898, 1997.

251. United Nations-ACC/SCN. Highlight of the world nutrition situation. *SCN News,* No. 8, 1-3, 1992.

252. Vaisman, N., M. Corey, M.F. Rossi, E. Goldberg, and P. Pencharz. Changes in body composition during refeeding of patients with anorexia nervosa. *J Pediatr* 113: 925-929, 1988.

253. Van Dale, D., W.H.M. Saris, P.F.M. Schoffelen, and F. Ten Hoor. Does exercise give an additional effect in weight reduction regimens? *Int J Obesity* 11: 367-375, 1987.

254. Van Lenthe, F.J., H.C.G. Kemper, W. van Mechelen, and J.W.R. Twisk. Longitudinal development and tracking of central patterns of subcutaneous fat in adolescents and adults: the Amsterdam Growth and Health Study. *Int J Epidemiol* 25: 1162-1171, 1996.

255. Vara, L., and S. Agras. Caloric intake and activity levels are related in young children. *Int J Obesity* 13: 613-617, 1989.

256. Von Noorden, C. *Obesity, metabolism and practical medicine.* Vol. 3. Chicago: Kenner, 1907, 696.

257. Waller, E.G., A.J. Wade, J. Treasure, A. Ward, T. Leonard, and J. Powell-Tuck. Physical measures of recovery from anorexia nervosa during hospitalised re-feeding. *Eur J Clin Nutr* 50: 165-170, 1996.

258. Ward, D.S., and O. Bar-Or. Role of the physician and the physical education teacher in the treatment of obesity at school. *Pediatrician* 13: 44-51, 1986.

259. Ward, D.S., and O. Bar-Or. Use of the Borg scale in exercise prescription for overweight youth. *Can J Sport Sci* 15: 120-125, 1990.

260. Ward, D.S., C.J.R. Blimkie, and O. Bar-Or. Rating of perceived exertion in obese adolescents. *Med Sci Sports Exerc* 18: S72, 1986.

261. Watson, A.W., and D.J. O'Donovan. Influence of level of habitual activity on physical working capacity and body composition of post-pubertal school boys. *Q J Exp Physiol* 62: 325-332, 1977.

262. Watson, A.W.S. Quantification of the influence of body fat content on selected physical performance variables in adolescent boys. *Ir J Med Sci* 157: 383-384, 1988.

263. Waxman, M., and A.J. Stunkard. Caloric intake and expenditure of obese boys. *J Pediatr* 96: 187-193, 1980.

264. Wells, J.C.K., T.J. Cole, and P.S.W. Davies. Total energy expenditure and body composition in early infancy. *Arch Dis Child* 75: 423-426, 1996.

265. Wells, J.C.K., M. Stanley, A.S. Laidlaw, J.M.E. Day, and P.S.W. Davies. The relationship between components of infant energy expenditure and childhood body fatness. *Int J Obesity* 20: 848-853, 1996.

266. Whipp, B.J., and J.A. Davis. The ventilatory stress of exercise in obesity. *Am Rev Respir Dis* 129 (Suppl): S90-S92, 1984.

267. Whipp, B.J., and W.K. Ruff. The effect of caloric restriction and physical training on the responses of obese adolescents to graded exercise. *J Sports Med Phys Fitness* 11: 146-153, 1971.

268. Whitaker, R.C., J.A. Wright, M.S. Pepe, K.D. Seidel, and W.H. Dietz. Predicting obesity in young adulthood from childhood and parental obesity. *New Eng J Med* 337: 869-873, 1997.

269. Widhalm, K., E. Maxa, and H. Zyman. Effect of diet and exercise upon the cholesterol and triglyceride content of plasma lipoproteins in overweight children. *Eur J Pediatr* 127: 121-126, 1978.

270. Wilcken, D.E.L., D.F. Lynch, M.D. Marshall, R.L. Scott, and X.L. Wang. Relevance of body weight to apolipoprotein levels in Australian children. *Med J Austr* 164: 22-25, 1996.

271. Wilkinson, P.W., J.M. Parkin, G. Pearlson, et al. Energy intake and physical activity in obese children. *Br Med J* 1: 756, 1977.

272. Wilmore, J.H. The 1983 C.H. McCloy Research Lecture: appetite and body composition consequent to physical activity. *Res Q* 54: 415-425, 1983.

273. Wilmore, J.H. Eating disorders in the young athlete. In Bar-Or, O., ed., *The child and adolescent athlete*. Oxford: Blackwell Scientific, 1996, 287-303.

274. Wolf, A.M., S.L. Gortmaker, L. Cheung, H.M. Gray, D.B. Herzog, and G.A. Colditz. Activity, inactivity, and obesity: differences related to race, ethnicity, and age among girls. *Am J Pub Health* 83: 1625-1627, 1993.

275. World Health Organization. *Obesity. Preventing and managing the global epidemic.* Report of a WHO consultation on obesity. Geneva: World Health Organization, 1997, 1-276.

276. Wurmser, H., R. Laessle, K. Jacob, S. Langhard, H. Uhl, A. Angst, A. Muller, and K.M. Pirke. Resting metabolic rate in preadolescent girls at high risk of obesity. *Int J Obesity* 22: 799, 1998.

277. Ylitalo, V. Treatment of obese schoolchildren with special reference to the mode of therapy, cardiorespiratory performance and the carbohydrate and lipid metabolism. *Acta Paediatr Scand Suppl* 290: 1-108, 1981.

278. Zahorska-Markiewicz, B. Effects of timing on energy expenditure during rest and exercise in obese women. *Nutr Metab* 24: 238-243, 1980.

279. Zakus, G., M.L. Chin, H. Cooper, et al. Treating adolescent obesity: a pilot project in school. *J School Health* 51: 663-666, 1981.

280. Zanconato, S., E. Baraldi, P. Santuz, F. Rigon, L. Vido, L. Da Dalt, and F. Zacchello. Gas exchange during exercise in obese children. *Eur J Pediatr* 148: 614-617, 1989.

281. Zuti, W.B., and L.A. Golding. Comparing diet and exercise as weight reduction tools. *Phys Sportsmed* 4: 49-53, 1976.

Chapter 10

1. Abramson, A.S., and J. Rogoff. Physical treatment in muscular dystrophy [abstract]. *Proceedings of the 2nd Medical Conference of the Muscular Dystrophy Association*, 1952, 123-124.

2. Adams, R., M. Andrews, and M. Sussman. Physical activity for patients wearing spinal orthoses. *Phys Sportsmed* 11: 75-83, 1983.

3. Agre, J.C., T.W. Findley, M.C. McNally, R. Habeck, A.S. Leon, L. Stradel, R. Birkebak, and R. Schmalz. Physical activity capacity in children with myelomeningocele. *Arch Phys Med Rehab* 68: 372-377, 1987.

4. Aird, R.B. The importance of seizure-inducing factors in the control of refractory forms of epilepsy. *Epilepsia* 24: 567-583, 1983.

5. American Academy of Pediatrics. Committee report: the epileptic child and competitive school athletics. *Pediatrics* 42: 700-703, 1968.

6. American Academy of Pediatrics. Medical conditions affecting sports participation. *Pediatrics* 94: 757-760, 1994.

7. Andrade, C.K., J. Kramer, M. Garber, and P. Longmuir. Changes in self-concept, cardiovascular endurance and muscular strength of children with spina bifida aged 8 to 13 years in response to a 10-week physical-activity programme: a pilot study. *Child Care Health Dev* 17: 183-196, 1991.

8. Angel, R.W., and W.W. Hofmann. The H reflex in normal, spastic and rigid subjects. *Arch Neurol* 8: 591-596, 1963.

9. Arida, R.M., A. Scorza, N.F. dos Santos, C.A. Peres, and E.A. Cavalheiro. Effect of physical exercise on seizure occurrence in a model of temporal lobe epilepsy in rats. *Epilepsy Res* 37: 45-52, 1999.

10. Ashworth, B. Preliminary trial of carisoprodol in multiple sclerosis. *Practitioner* 192: 540-542, 1964.

11. Athanasopoulos, S., T. Paxinos, K. Zachariou, and S. Chatziconstantinou. The effect of anaerobic training in girls with idiopathic scoliosis. *Scand J Med Sci Sports* 9: 36-40, 1999.

12. Axen, K., M. Bishop, and F. Haas. Respiratory load compensation in neuromuscular disorders. *J Appl Physiol* 64: 2659-2666, 1988.

13. Bader, D., A.D. Ramos, C.D. Lew, A.C. Platzker, M.W. Stabile, and T.G. Keens. Childhood sequelae of infant lung disease: exercise and pulmonary function abnormalities after bronchopulmonary dysplasia. *J Pediatr* 110: 693-699, 1987.

14. Bandini, L.G., D.A. Schoeller, N.K. Fukugawa, L.J. Wykes, and W.H. Dietz. Body composition and energy expenditure in adolescents with cerebral palsy or myelodysplasia. *Pediatr Res* 29: 70-77, 1991.

15. Bar-Or, O. The Wingate anaerobic test. An update on methodology, reliability and validity. *Sports Med* 4: 381-394, 1987.

16. Bar-Or, O. Noncardiopulmonary pediatric exercise tests. In Rowland, T.W., ed., *Pediatric laboratory exercise testing: clinical guidelines.* Champaign, IL: Human Kinetics, 1993, 165-185.

17. Bar-Or, O. Role of exercise in the assessment and management of neuromuscular disease in children. *Med Sci Sports Exerc* 28: 421-427, 1996.

18. Bar-Or, O., R. Dotan, O. Inbar, A. Rothstein, J. Karlsson, and P. Tesch. Anaerobic capacity and muscle fiber type distribution in man. *Int J Sports Med* 1: 89-92, 1980.

19. Bar-Or, O., O. Inbar, and R. Dotan. Proficiency, speed and endurance test for the wheelchair bound. Unpublished. 1977.

20. Bar-Or, O., O. Inbar, and R. Spira. Physiological effects of a sports rehabilitation program on cerebral palsied and post-poliomyelitic adolescents. *Med Sci Sports* 8: 157-161, 1976.

21. Bar-Or, O., O. Inbar, and R. Spira. Physiological effects of a sports rehabilitation program on cerebral palsied and post-poliomyelitic adolescents. *Med Sci Sports* 8: 157-161, 1976.

22. Baraldi, E., S. Zanconato, C. Zorzi, P. Santuz, F. Benini, and F. Zacchello. Exercise performance in very low birth weight children at the age of 7-12 years. *Eur J Pediatr* 150: 713-716, 1991.

23. Barnes, P.R., D.J. Taylor, G.J. Kemp, and G.K. Radda. Skeletal muscle bioenergetics in the chronic fatigue syndrome. *J Neurol Neurosurg Psychiatry* 56: 679-683, 1993.

24. Berbrayer, D., and P. Ashby. Reciprocal inhibition in cerebral palsy. *Neurology* 40: 653-656, 1990.

25. Berg, K. Effect of physical training of school children with cerebral palsy. *Acta Paediatr Scand Suppl* 270: 27-33, 1970.

26. Berg, K., and J. Bjure. Methods for evaluation of the physical working capacity of school children with cerebral palsy. *Acta Paediatr Scand Suppl* 204: 15-26, 1970.

27. Berg, K., and T. Olsson. Energy requirements of school children with cerebral palsy as determined from indirect calorimetry. *Acta Paediatr Scand Suppl* 204: 71-80, 1970.

28. Bergofsky, E.H., G.M. Turino, and A.P. Fishman. Cardiorespiratory failure in kyphoscoliosis. *Medicine* 38: 263-317, 1959.

29. Bhambhani, Y.N., L.J. Holland, and R.D. Steadward. Maximal aerobic power in cerebral palsied wheelchair athletes: validity and reliability. *Arch Phys Med Rehab* 73: 246-252, 1992.

30. Bjorholt, P.G., K.O. Nakken, K. Rohme, and H. Hansen. Leisure time habits and physical fitness in adults with epilepsy. *Epilepsia* 31: 83-87, 1990.

31. Bjure, J., G. Grimby, and A. Nachemson. The effect of physical training in girls with idiopathic scoliosis. *Acta Orthop Scand* 40: 325-333, 1969.

32. Bonsett, C.A. Pseudohypertrophic muscular dystrophy. Distribution of degenerative features as revealed by an anatomical study. *Neurology* 13: 728-738, 1963.

33. Borg, G. Perceived exertion as an indicator of somatic stress. *Scand J Rehab Med* 2: 92-98, 1970.

34. Boucharlat, J., A. Maitre, and J. Ledru. Sport et épilepsie de l'enfant. *Ann Méd Psychol (Paris)* 1: 392-401, 1973.

35. Bower, B.D. Epilepsy and school athletics. *Dev Med Child Neurol* 11: 244-245, 1969.

36. Bowker, J.H., and P.J. Halpin. Factors determining success in reambulation of the child with progressive muscular dystrophy. *Orthop Clin North Am* 9: 431-436, 1978.

37. Brussock, C.M., S.M. Haley, T.L. Munsat, and D.B. Bernhardt. Measurement of isometric force in children with and without Duchenne muscular dystrophy. *Phys Ther* 72: 105-114, 1992.

38. Butler, P., M. Engelbrecht, R.E. Major, J.H. Tait, J. Stallard, and J.H. Patrick. Physiological cost index of walking for normal children and its use as an indicator of physical handicap. *Dev Med Child Neurol* 26: 607-612, 1984.

39. Campbell, J., and J. Ball. Energetics of walking in cerebral palsy. *Orthop Clin North Am* 9: 374-377, 1978.

40. Carroll, J.E., J.M. Hagberg, G.H. Brooke, and J.B. Shumate. Bicycle ergometry and gas exchange measurements in neuromuscular diseases. *Arch Neurol* 36: 457-461, 1979.

41. Carter, G.T., M.A. Wineinger, S.A. Walsh, S.J. Horasek, R.T. Abresch, and W.M. Fowler Jr. Effect of voluntary wheel-running exercise on muscles of the mdx mouse. *Neuromuscul Disord* 5: 323-332, 1995.

42. Castle, M.E., T.A. Reyman, and M.E. Schneider. Pathology of spastic muscle in cerebral palsy. *Clin Orthop* 142: 223-233, 1979.

43. Chase, D. With epilepsy they take the medicine and play. *Phys Sportsmed* 2: 61, 1974.

44. Chong, K.C., R.M. Letts, and G.R. Cumming. Influence of spinal curvature on exercise capacity. *J Pediatr Orthop* 1: 251-254, 1981.

45. Chong, P.K.K., R.T. Jung, C.M. Scrimgeour, M.J. Rennie, and C.R. Paterson. Energy expenditure and body composition in growth hormone deficient adults on exogenous growth hormone. *Clin Endocrinol* 40: 103-110, 1994.

46. Coakley, J.H., A.J. Wagenmakers, and R.H. Edwards. Relationship between ammonia, heart rate, and exertion in McArdle's disease. *Am J Physiol* 262: E167-E172, 1992.

47. Committee on Children with Handicaps and Committee on Sports Medicine. American Academy of Pediatrics position statement on sports and the child with epilepsy. *Pediatrics* 72: 884-885, 1983.

48. Cooper, D.M., J.V. Rojas, R.B. Mellins, H.A. Keim, and A.L. Mansell. Respiratory mechanics in adolescents with idiopathic scoliosis. *Am Rev Respir Dis* 130: 16-22, 1984.

49. Cowart, V.S. Should epileptics exercise? *Phys Sportsmed* 14: 183-191, 1986.

50. Cragg, S. Patty's magnificent marathon. *Reader's Digest,* April, 75-78, 1978.

51. Cratty, B.J. *Adapted physical education for handicapped children and youth.* Denver: Love, 1980.

52. Damiano, D.L., L.E. Kelly, and C.L. Vaughn. Effects of quadriceps femoris muscle strengthening on crouch gait in children with spastic diplegia. *Phys Ther* 75: 658-667, 1995.

53. Damiano, D.L., L.J. Kelly, and C.L. Vaughan. Effects of quadriceps femoris muscle strengthening on crouch gait in children with spastic diplegia. *Phys Ther* 75: 658-667, 1995.

54. Damiano, D.L., C.L. Vaughan, and M.F. Abel. Muscle response to heavy resistance exercise in children with spastic cerebral palsy. *Dev Med Child Neurol* 37: 731-739, 1995.

55. Darrah, J., J.S.W. Fan, L.C. Chen, J. Nunweiler, and B. Watkins. Review of the effects of progressive resisted muscle strengthening in children with cerebral palsy: a clinical consensus exercise. *Pediatr Phys Ther* 9: 12-17, 1997.

56. de Lateur, B.J., and R.M. Giaconi. Effect on maximal strength of submaximal exercise in Duchenne muscular dystrophy. *Am J Phys Med* 58: 26-36, 1979.

57. Denio, L.S., M.E. Drake Jr., and A. Pakalnis. The effect of exercise on seizure frequency. *J Med* 20: 171-176, 1989.

58. DiRocco, P.J., A.L. Breed, J.I. Carlin, and W.G. Reddan. Physical work capacity in adolescent patients with mild idiopathic scoliosis. *Arch Phys Med Rehab* 64: 476-478, 1983.

59. DiRocco, P.J., and P. Vaccaro. Cardiopulmonary functioning in adolescent patients with mild idiopathic scoliosis. *Arch Phys Med Rehab* 69: 198-201, 1988.

60. Dowben, R.M. Treatment of muscular dystrophy with steroids. A preliminary report. *N Eng J Med* 268: 912-916, 1963.

61. Dresen, M.H.W., G. de Groot, J.R. Mesa Menor, and L.N. Bouman. Physical working capacity and daily physical activities of handicapped and non-handicapped children. *Eur J Appl Physiol* 48: 241-251, 1982.

62. Dresen, M.H.W., G. deGroot, J.R. Meso Menor, and L.N. Bouman. Aerobic energy expenditure of handicapped children after training. *Arch Phys Med Rehab* 6: 302-306, 1985.

63. Duffy, C.M., A.E. Hill, A.P. Cosgrove, I.S. Corry, and H.K. Graham. Energy consumption in children with spina bifida and cerebral palsy: a comparative study. *Dev Med Child Neurol* 38: 238-243, 1996.

64. Dupont-Versteegden, E.E., R.J. McCarter, and M.S. Katz. Voluntary exercise decreases progression of muscular dystrophy in diaphragm of mdx mice. *J Appl Physiol* 77: 1736-1741, 1994.

65. Edwards, R.H., H. Gibson, J.E. Clague, and T. Helliwell. Muscle histopathology and physiology in chronic fatigue syndrome. *Ciba Found Symp* 173: 102-117, 1993.

66. Edwards, R.H.T., J.M. Round, M.J. Jackson, R.D. Griffiths, and M.F. Lilburn. Weight reduction in boys with muscular dystrophy. *Dev Med Child Neurol* 26: 384-390, 1984.

67. Eickelberg, W.W.B., and M. Less. The effects of passive exercise of skeletal muscles on cardiac cost, respiratory function and associative learning in severe myopathic children. *J Hum Ergol* 3: 157-162, 1975.

68. Ekblom, B., and A. Lundberg. Effects of physical training on adolescents with severe motor handicaps. *Acta Pediatr* 57: 17-23, 1968.

69. Emons, H.J.G., D.C. Groenboom, Y.I. Burggraaff, T.L.E. Janssen, and M.A. van Baak. Wingate anaerobic test in children with cerebral palsy. In Coudert, J. and E. Van Praagh, eds., *Children and exercise XVI.* Paris: Masson, 1992, 187-189.

70. Engsberg, J.R., K.S. Olree, S.A. Ross, and T.S. Park. Quantitative clinical measure of spasticity in children with cerebral palsy. *Arch Phys Med Rehab* 77: 594-599, 1996.

71. Ericson, A., and B. Kallen. Very low birthweight boys at the age of 19. *Arch Dis Child Fetal Neonatal Ed* 78: F171-F174, 1998.

72. Esquivel, E., M. Chaussain, P. Plouin, G. Ponsot, and M. Arthuis. Physical exercise and voluntary hyperventilation in childhood absence epilepsy. *Electroenceph Clin Neurophysiol* 79: 127-132, 1991.

73. Falk, B., A. Eliakim, R. Dotan, D.G. Liebermann, R. Regev, and O. Bar-Or. Birth weight and physical ability. *Med Sci Sports Exerc* 29: 1124-1130, 1997.

74. Fallen, E.L., W.C. Elliott, and R. Gorlin. Mechanisms of angina in aortic stenosis. *Circulation* 36: 480-488, 1967.

75. Fan, J.S., J. Wessel, and J. Ellsworth. The relationship between strength and function in females with juvenile rheumatoid arthritis. *J Rheumatol* 25: 1399-1405, 1998.

76. Fernandez, J.E., and K.H. Pitetti. Training of ambulatory individuals with cerebral palsy. *Arch Phys Med Rehab* 74: 468-472, 1993.

77. Ferrari, K., P. Goti, A. Sanna, G. Misuri, F. Gigliotti, R. Duranti, I. Iandelli, S. Ceppatelli, and G. Scano. Short-term effects of bracing on exercise performance in mild idiopathic thoracic scoliosis. *Lung* 175: 299-310, 1997.

78. Findley, T.W., and J.C. Agre. Ambulation in adolescents with spina bifida. II. Oxygen cost of mobility. *Arch Phys Med Rehab* 69: 855-861, 1988.

79. Findley, T.W., J.C. Agre, R.V. Habeck, R. Schmalz, R.R. Birkebak, and M.C. McNally. Ambulation in the adolescent with myelomeningocele. I: Early childhood predictors. *Arch Phys Med Rehab* 68: 518-522, 1987.

80. Florence, J.M., and J.M. Hagberg. Effect of training on the exercise responses of neuromuscular disease patients. *Med Sci Sports Exerc* 16: 460-465, 1984.

81. Florence, J.M., S. Pandya, W.M. King, J.D. Robison, J. Baty, J.P. Miller, J. Schierbecker, and L.C. Signore. Intrarater reliability of manual muscle test (Medical Research Council Scale) grades in Duchenne's muscular dystrophy. *Phys Ther* 72: 115-122, 1992.

82. Florence, J.M., S. Pandya, W.M. King, J.D. Robison, L.C. Signore, M. Wentzell, and M.A. Province. Clinical trials in Duchenne dystrophy: standardization and reliability of evaluation procedures. *Phys Ther* 64: 41-45, 1984.

83. Ford, G.W., W.H. Kitchen, and L.W. Doyle. Muscular strength at 5 years of children with a birthweight under 1500 g. *Austr Paediatr J* 24: 295-296, 1988.

84. Fowler, W.M., R.T. Abresch, D.B. Larson, R.B. Sharman, and R.K. Entrikin. High-repetitive submaximal treadmill exercise training: effect on normal and dystropic mice. *Arch Phys Med Rehab* 71: 552-557, 1990.

85. Fowler, W.M. Jr., and G.W. Gardner. Quantitative strength measurements in muscular dystrophy. *Arch Phys Med Rehab* 48: 629-644, 1967.

86. Fowler, W.M. Jr., C.M. Pearson, G.H. Egstrom, and G.W. Gardner. Ineffective treatment of muscular dystrophy with an anabolic steroid and other measures. *N Eng J Med* 272: 875-882, 1965.

87. Fulcher, K.Y., and P.D. White. Randomised controlled trial of graded exercise in patients with the chronic fatigue syndrome. *BMJ* 314: 1647-1652, 1997.

88. Gailani, S., T.S. Danowski, and D.S. Fisher. Muscular dystrophy. Catheterization studies indicating latent congestive heart failure. *Circulation* 17: 585-588, 1958.

89. Giannini, M.J., and E.J. Protas. Aerobic capacity in juvenile rheumatoid arthritis patients and healthy children. *Arthritis Care Res* 4: 131-135, 1991.

90. Giannini, M.J., and E.J. Protas. Comparison of peak isometric knee extensor torque in children with and without juvenile rheumatoid arthritis. *Arthritis Care Res* 6: 82-88, 1993.

91. Gibson, H., N. Carroll, J.E. Clague, and R.H. Edwards. Exercise performance and fatiguability in patients with chronic fatigue syndrome. *J Neurol Neurosurg Psychiatry* 56: 993-998, 1993.

92. Goldberg, M.S., N.E. Mayo, B. Poitras, S. Scott, and J. Hanley. The Ste-Justine Adolescent Idiopathic Scoliosis Cohort Study. Part II: Perception of health, self and body image, and participation in physical activities. *Spine* 19: 1562-1572, 1994.

93. Gotze, W., S. Kubicki, M. Hunter, and J. Teichmann. Effect of physical exercise on seizure threshold. *Dis Nerv Syst* 28: 664-667, 1967.

94. Gozal, D. Pulmonary manifestations of neuromuscular disease with special reference to Duchenne muscular dystrophy and spinal muscular atrophy. *Pediatr Pulmonol* 29: 141-150, 2000.

95. Gundel, L. Proposal for exemption of epileptic children from school sports. *Deutsch Med Wochensr* 100: 491-494, 1975.

96. Hack, M., and A.A. Fanaroff. Outcomes of children of extremely low birthweight and gestational age in the 1990's. *Early Hum Dev* 53: 193-218, 1999.

97. Haller, R.G., and S.F. Lewis. Abnormal ventilation during exercise in McArdle's syndrome: modulation by substrate availability. *Neurology* 36: 716-719, 1986.

98. Hankard, R., F. Gottrand, D. Turck, A. Carpentier, M. Romon, and J.P. Farriaux. Resting energy expenditure and energy substrate utilization in children with Duchenne muscular dystrophy. *Pediatr Res* 40: 29-33, 1996.

99. Hauser, C.R., and D.M. Johnson. Breathing exercises for children with pseudohypertrophic muscular dystrophy. *Phys Ther* 51: 751-759, 1971.

100. Hayes, A., and D.A. Williams. Beneficial effects of voluntary wheel running on the properties of dystrophic mouse muscle. *J Appl Physiol* 80: 670-679, 1996.

101. Hebestreit, H., and O. Bar-Or. Exercise and the child born prematurely. *Sports Med* 31: 591-599, 2001.

102. Hebestreit, H., S. Dietz, and A. Hiermer. Body coordination and mechanical efficiency in children born prematurely [abstract]. *Pediatr Exerc Sci* 11: 304, 1999.

103. Hebestreit, H., J. Muller-Scholden, and H.I. Huppertz. Aerobic fitness and physical activity in patients with HLA-B27 positive juvenile spondyloarthropathy that is inactive or in remission. *J Rheumatol* 25: 1626-1633, 1998.

104. Henderson, C.J., D.J. Lovell, B.L. Specker, and B.N. Campaigne. Physical activity in children with juvenile rheumatoid arthritis: quantification and evaluation. *Arthritis Care Res* 8: 114-119, 1995.

105. Hoofwijk, M., V.B. Unnithan, and O. Bar-Or. Maximal treadmill walking test for children with cerebral palsy. *Pediatr Exerc Sci* 7: 305-313, 1995.

106. Horyd, W., J. Gryziak, K. Niedzielska, and J.J. Zielinski. Effect of physical exertion on seizure discharges in the EEG of epilepsy patients [in Polish]. *Neurol Neurochir Pol* 15: 545-552, 1981.

107. Hosking, G.P., U.S. Bhat, V. Dubowitz, and R.H.T. Edwards. Measurements of muscle strength and performance in children with normal and diseased muscle. *Arch Dis Child* 51: 957-963, 1976.

108. Huberman, G. Organized sports activities with cerebral palsied adolescents. *Rehab Lit* 37: 103-107, 1976.

109. Inbar, O., R. Dlin, A. Rothstein, and B.J. Whipp. Physiological responses to incremental exercise in patients with chronic fatigue syndrome. *Med Sci Sports Exerc* 33: 1463-1470, 2001.

110. Inkley, S.R., F.C. Oldenburg, and P.J. Vignos. Pulmonary function in Duchenne muscular dystrophy related to stages of disease. *Am J Med* 56: 297-306, 1974.

111. Jackson, M.J., J.M. Round, D.J. Newham, and R.H. Edwards. An examination of some factors influencing creatine kinase in the blood of patients with muscular dystrophy. *Muscle Nerve* 10: 15-21, 1987.

112. Jalava, M., and M. Sillanpaa. Physical activity, health-related fitness, and health experience in adults with childhood-onset epilepsy: a controlled study. *Epilepsia* 38: 424-429, 1997.

113. Jeng, S.-F., K.G. Holt, L. Fetters, and C. Certo. Self-optimization of walking in nondisabled children and children with spastic hemiplegic cerebral palsy. *J Motor Behav* 28: 15-27, 1996.

114. Johnson, E.W., and R. Braddom. Over-work weakness in facioscapulohumeral muscular dystrophy. *Arch Phys Med Rehab* 52: 333-336, 1971.

115. Jungst, B.K., M. Spranger, M. Rochel, D. Schranzu, and H. Stopfkuchen. Kontinuirliche EEG—Ableitung unter submaximaler fahrradergometrischer. Belastung epilepsiekranker Kinder. In Freanz, I.W., H. Mellerowitz, and U.W. Noack, eds., *Training und Sport zur Pravention und Rehabilitation in der technisierten Umwelt*. Berlin: Springer-Verlag, 1985.

116. Kaplan, S.L. Cycling patterns in children with and without cerebral palsy. *Dev Med Child Neurol* 37: 620-630, 1995.

117. Kawazoe, Y., M. Kobayashi, T. Tasaka, and M. Tamamoto. Effects of therapeutic exercise on masticatory function in patients with progressive muscular dystrophy. *J Neurol Neurosurg Psychiatry* 45: 343-347, 1982.

118. Kearon, C., G.R. Viviani, and K.J. Killian. Factors influencing work capacity in adolescent idiopathic thoracic scoliosis. *Am Rev Respir Dis* 148: 295-303, 1993.

119. Keller, H., B.V. Ayub, S. Saigal, and O. Bar-Or. Neuromotor ability. *Dev Med Child Neurol* 40: 661-666, 1998.

120. Keller, H., B.V. Ayub, S. Saigal, and O. Bar-Or. Neuromotor ability in 5- to 7-year-old children with very low or extremely low birthweight. *Dev Med Child Neurol* 40: 661-666, 1998.

121. Keller, H., O. Bar-Or, S. Kriemler, B.V. Ayub, and S. Saigal. Anaerobic performance in 5- to 7-yr-old children of low birthweight. *Med Sci Sports Exerc* 32: 278-283, 2000.

122. Kesten, S., S.K. Garfinkel, T. Wright, and A.S. Rebuck. Impaired exercise capacity in adults with moderate scoliosis. *Chest* 99: 663-666, 1991.

123. Kirchheimer, J.C., A. Wanivenhaus, and A. Engel. Does sport negatively influence joint scores in patients with juvenile rheumatoid arthritis. An 8-year prospective study. *Rheumatol Int* 12: 239-242, 1993.

124. Kirsch, R., and E. Wirrell. Do cognitively normal children with epilepsy have a higher rate of injury than their nonepileptic peers? *J Child Neurol* 16: 100-104, 2001.

125. Klepper, S.E. Effects of an eight-week physical conditioning program on disease signs and symptoms in children with chronic arthritis. *Arthritis Care Res* 12: 52-60, 1999.

126. Klepper, S.E., J. Darbee, S.K. Effgen, and B.H. Singsen. Physical fitness levels in children with polyarticular juvenile rheumatoid arthritis. *Arthritis Care Res* 5: 93-100, 1992.

127. Klepper, S.E., and M.J. Giannini. Physical conditioning in children with arthritis: assessment and guidelines for exercise prescription. *Arthritis Care Res* 7: 226-236, 1994.

128. Koessler, W., T. Wanke, G. Winkler, A. Nader, K. Toifl, H. Kurz, and H. Zwick. 2 years' experience with inspiratory muscle training in patients with neuromuscular disorders. *Chest* 120: 765-769, 2001.

129. Korczyn, A.D. Participation of epileptic patients in sports. *J Sports Med Phys Fitness* 19: 195-198, 1979.

130. Kramer, J.F., and H.E.A. Macphail. Relationships among measures of walking efficiency, gross motor ability, and isokinetic strength in adolescents with cerebral palsy. *Pediatr Phys Ther* 6: 3-8, 1994.

131. Kriemler, S., H. Keller, S. Saigal, and O. Bar-Or. Aerobic and respiratory performance in 5- to 7-year-old children born prematurely, with and without bronchopulmonary dysplasia. McMaster University, Ontario. Unpublished research data, 2002.

132. Kuijer, A. Epilepsy and exercise, electroencephalographical and biochemical studies. In Wada, J.A. and J.K. Penry, eds., *Advances in epileptology: the Xth Epilepsy International Symposium*. New York: Raven Press, 1980, 543.

133. Kuijer, A., and R. Van Wilsum. The influence of physical exercise on the occurrence of epileptic seizures [abstract]. *Electroenceph Clin Neurophysiol* 35: 105, 1973.

134. Kwong, K.L., S.N. Wong, and K.T. So. Parental perception, worries and needs in children with epilepsy. *Acta Paediatr* 89: 593-596, 2000.

135. Landin, S., L. Hagenfeldt, B. Saltin, and J. Wahren. Muscle metabolism during exercise in hemiparetic patients. *Clin Sci Mol Med* 53: 257-269, 1977.

136. Lane, R.J.M., D. Woodrow, and L.C. Archard. Lactate responses to exercise in chronic fatigue syndrome [letter, abstract]. *J Neurol Neurosurg Psychiatry* 57: 662-663, 1994.

137. Le Rumeur, E., N. Le Tallec, C.J. Lewa, X. Ravalec, and J.D. de Certaines. In vivo evidence of abnormal mechanical and oxidative functions in the exercised muscle of dystrophic hamsters by 31P-NMR. *J Neurol Sci* 133: 16-23, 1995.

138. Leroy-Willig, A., T.N. Willig, M.C. Henry-Feugeas, V. Frouin, E. Marinier, A. Boulier, F. Barzic, E. Schouman-Claeys, and A. Syrota. Body composition determined with MR in patients with Duchenne muscular dystrophy, spinal muscular atrophy, and normal subjects. *Magn Reson Imaging* 15: 737-744, 1997.

139. Lewis, S.F., R.G. Haller, J.D. Cook, and C.G. Blomqvist. Metabolic control of cardiac output response to exercise in McArdle's disease. *J Appl Physiol* 57: 1749-1753, 1984.

140. Lewis, S.F., R.G. Haller, J.D. Cook, and R.L. Nunnally. Muscle fatigue in McArdle's disease studied by 31P-NMR: effect of glucose infusion. *J Appl Physiol* 59: 1991-1994, 1985.

141. Lindh, M. Energy expenditure during walking in patients with scoliosis. The effect of surgical correction. *Spine* 3: 122-134, 1978.

142. Livingston, S. Should physical activity of the epileptic child be restricted? *Clin Pediatr* 10: 694-696, 1971.

143. Livingston, S. Physical activity for the epileptic child. In Livingston, S., ed., *Comprehensive management of epilepsy in infancy, childhood and adolescence.* Springfield, IL: Charles C Thomas, 1972, 143-148.

144. Livingston, S. Should epileptics be athletes? *Sports Med* 3: 67-72, 1975.

145. Livingston, S., L.L. Pauli, and I. Prince. Epilepsy and sports [letter]. *JAMA* 239: 2, 1978.

146. Longmuir, P.E., and O. Bar-Or. Physical activity of children and adolescents with a disability: methodology and effects of age and gender. *Pediatr Exerc Sci* 6: 168-177, 1994.

147. Longmuir, P.E., and O. Bar-Or. Factors influencing the physical activity levels of youths with physical and sensory disabilities. *Adapted Phys Act Q* 17: 40-53, 2000.

148. Lough, L.K., and D.H. Nielsen. Ambulation of children with myelomeningocele: parapodium versus parapodium with Orlau swivel modification. *Dev Med Child Neurol* 28: 489-497, 1986.

149. Lundberg, A. Changes in the working pulse during the school year in adolescents with cerebral palsy. *Scand J Rehab Med* 5: 12-17, 1973.

150. Lundberg, A. Mechanical efficiency in bicycle ergometer work of young adults with cerebral palsy. *Dev Med Child Neurol* 17: 434-439, 1975.

151. Lundberg, A. Oxygen consumption in relation to work load in students with cerebral palsy. *J Appl Physiol* 40: 873-875, 1976.

152. Lundberg, A. Maximal aerobic capacity of young people with spastic cerebral palsy. *Dev Med Child Neurol* 20: 205-210, 1978.

153. Lundberg, A., C.O. Ovenfors, and B. Saltin. Effect of physical training on schoolchildren with cerebral palsy. *Acta Paediatr Scand* 56: 182-188, 1967.

154. Lundberg, A., and B. Pernow. The effect of physical training on oxygen utilization and lactate formation in the exercising muscle of adolescents with motor handicaps. *Scand J Clin Lab Invest* 26: 89-96, 1970.

155. MacGregor, J. The evaluation of patient performance using long-term ambulatory monitoring technique in the domiciliary environment. *Physiotherapy* 67: 30-33, 1981.

156. MacPhail, H.E., and J.F. Kramer. Effect of isokinetic strength-training on functional ability and walking efficiency in adolescents with cerebral palsy. *Dev Med Child Neurol* 37: 763-775, 1995.

157. Malleson, P.N., S.M. Bennett, M. MacKinnon, D.K. Jespersen, K.D. Coutts, S.P. Turner, and D.C. McKenzie. Physical fitness and its relationship to other indices of health status in children with chronic arthritis. *J Rheumatol* 23: 1059-1065, 1996.

158. Maltais, D., I. Kondo, and O. Bar-Or. Arm cranking economy in spastic cerebral palsy: effects of different speed and force combinations yielding the same mechanical power. *Pediatr Exerc Sci* 12: 258-269, 2000.

159. Maltais, D.B., O. Bar-Or, V. Galea, and M. Pierrynowski. Use of orthoses lowers the O_2 cost of walking in children with spastic cerebral palsy. *Med Sci Sports Exerc* 33: 320-325, 2001.

160. Marlow, N., L. Roberts, and R. Cooke. Outcome at 8 years for children with birth weights of 1250 g or less. *Arch Dis Child* 68: 286-290, 1993.

161. Massin, M., and N. Allington. Role of exercise testing in the functional assessment of cerebral palsy children after botulinum A toxin injection. *J Pediatr Orthop* 19: 362-365, 1999.

162. McArdle, B. Myopathy due to a defect in muscle glycogen breakdown. *Clin Sci* 10: 13-35, 1951.

163. McCartney, N., D. Moroz, S.H. Garner, and A.J. McComas. The effects of strength training in patients with selected neuromuscular disorders. *Med Sci Sports Exerc* 20: 362-368, 1988.

164. McClenaghan, B.S., S. Hill, R. Kohell, and C. Okazaki. Interface design for indirect calorimetry in children with cerebral palsy. *Arch Phys Med Rehab* 69: 548-551, 1988.

165. McDonald, C.M., R.T. Abresch, G.T. Carter, W.M. Fowler Jr., E.R. Johnson, D.D. Kilmer, and B.J. Sigford. Profiles of neuromuscular diseases. Duchenne muscular dystrophy. *Am J Phys Med Rehab* 74: S70-S92, 1995.

166. McDonald, C.M., K.M. Jaffe, V.S. Mosca, and D.B. Shurtleff. Ambulatory outcome of children with myelomeningocele: effect of lower-extremity muscle strength. *Dev Med Child Neurol* 33: 482-490, 1991.

167. Merlini, L., D. Dell'Accio, A. Holzl, and C. Granata. Isokinetic muscle testing (IMT) in neuromuscular diseases. Preliminary report. *Neuromuscul Disord* 2: 201-207, 1992.

168. Milner-Brown, H.S., and R.G. Miller. Muscle strengthening through electric stimulation combined with low-resistance weights in patients with neuromuscular disorders. *Arch Phys Med Rehab* 69: 20-24, 1988.

169. Milner-Brown, H.S., and R.G. Miller. Muscle strengthening through high-resistance weight training in patients with neuromuscular disorders. *Arch Phys Med Rehab* 69: 14-19, 1988.

170. Minor, M.A. Arthritis and exercise: the times they are a-changin'. *Arthritis Care Res* 9: 79-81, 1996.

171. Molbech, S. Energy cost in level walking in subjects with an abnormal gait. In Evang, K. and K.L. Andersen, eds., *Physical activity in health and disease.* Oslo: Universitets Forlaget, 1966, 146.

172. Moncur, C., R. Marcus, and S. Johnson. Pilot study of aerobic conditioning of subjects with juvenile arthritis [abstract]. *Arthritis Care Res* 3: S16, 1990.

173. Nakken, K.O. Physical exercise in outpatients with epilepsy. *Epilepsia* 40: 643-651, 1999.

174. Nakken, K.O., P.G. Bjorholt, S.I. Johanssen, T. Loyning, and E. Lind. Effect of physical training on aerobic capacity, seizure occurrence, and serum level of antiepileptic drugs in adults with epilepsy. *Epilepsia* 31: 88-94, 1990.

175. Nakken, K.O., A. Loyning, T. Loyning, G. Gloersen, and P.G. Larsson. Does physical exercise influence the occurrence of epileptiform EEG discharges in children? *Epilepsia* 38: 279-284, 1997.

176. Neilson, P.D., N.J. O'Dwyer, and J. Nash. Control of isometric muscle activity in cerebral palsy. *Dev Med Child Neurol* 32: 778-788, 1990.

177. Nesvadbá, Z., L. Hosková, and A. Rennerová. Rehabilitation of children with muscular dystro-phy at the State Spa of Jansko Lazne. In Walton, L., N. Canal, and G. Scarlato, eds., *Muscle diseases.* Amsterdam: Excerpta Medica, 1970, 555-557.

178. Nishikawa, M., T. Ichiyama, T. Hayashi, and S. Furukawa. Investigation of swimming classes for school children with epilepsy—a questionnaire survey of school nursing teachers [in Japanese]. *No To Hattatsu* 30: 15-19, 1998.

179. O'Connell, D.G., and R. Barnhart. Improvement in wheelchair propulsion in pediatric wheelchair users through resistance training: a pilot study. *Arch Phys Med Rehab* 76: 368-372, 1995.

180. O'Connell, D.G., R. Bernhart, and L. Parks. Muscular endurance and wheelchair propulsion in children with cerebral palsy or myelomeningocele. *Arch Phys Med Rehab* 73: 709-711, 1992.

181. Oberg, T., A. Karsznia, B.A. Gare, and A. Lagerstrand. Physical training of children with juvenile chronic arthritis. Effects on force, endurance and EMG response to localized muscle fatigue. *Scand J Rheumatol* 23: 92-95, 1994.

182. Ogunyemi, A.O., M.R. Gomez, and D.W. Klass. Seizures induced by exercise. *Neurology* 38: 633-634, 1988.

183. Parker, D.F., L. Carriere, H. Hebestreit, and O. Bar-Or. Anaerobic endurance and peak muscle power in children with spastic cerebral palsy. *Am J Dis Child* 146: 1069-1073, 1992.

184. Parker, D.F., L. Carriere, H. Hebestreit, and O. Bar-Or. Anaerobic endurance and peak muscle power in children with spastic cerebral palsy. *Am J Dis Child* 146: 1069-1073, 1992.

185. Parker, D.L., L. Carriere, H. Hebestreit, A. Salsberg, and O. Bar-Or. Muscle performance and gross motor function of children with spastic cerebral palsy. *Dev Med Child Neurol* 35: 17-23, 1993.

186. Pernow, B.B., R.J. Havel, and D.B. Hennings. The second wind phenomenon in McArdle's syndrome. *Acta Med Scand Suppl* 472: 294-307, 1967.

187. Petrof, B.J. The molecular basis of activity-induced muscle injury in Duchenne muscular dystrophy. *Mol Cell Biochem* 179: 111-123, 1998.

188. Pianosi, P.T., and M. Fisk. Cardiopulmonary exercise performance in prematurely born children. *Pediatr Res* 47: 653-658, 2000.

189. Powell, P., R.P. Bentall, F.J. Nye, and R.H. Edwards. Randomised controlled trial of patient education to encourage graded exercise in chronic fatigue syndrome. *BMJ* 322: 387-390, 2001.

190. Powls, A., N. Botting, R.W. Cooke, and N. Marlow. Motor impairment in children 12 to 13 years old with a birthweight of less than 1250 g. *Arch Dis Child Fetal Neonatal Ed* 73: F62-F66, 1995.

191. Refsum, H.E., C.F. Naess-Andresen, and J.E. Lange. Pulmonary function and gas exchange at rest and exercise in adolescent girls with mild idiopathic

scoliosis during treatment with Boston thoracic brace. *Spine* 15: 420-423, 1990.

192. Ricker, K., and G. Hertel. Influence of local cooling on the muscle contracture and paresis of McArdle's disease. *J Neurol* 215: 287-290, 1977.

193. Rideau, Y., L.W. Jankowski, and J. Grellet. Respiratory function in the muscular dystrophies. *Muscle Nerve* 4: 155-164, 1981.

194. Rieckart, H., L. Brohn, U. Schwalm, and W. Schnizer. Ein Ausdauertraining im Rahmen des Schulsports bei vor weigend spastisch gelahmten Kindern. *Med Welt* 28: 1694-1701, 1977.

195. Rieckert, H., L. Bruhn, U. Schwalm, and W. Schnizer. Endurance training within a program of physical education in children predominantly with cerebral palsy [in German]. *Med Welt* 28: 1694-1701, 1977.

196. Riley, M., D.P. Nicholls, A.M. Nugent, I.C. Steele, N. Bell, P.M. Davies, C.F. Stanford, and V.H. Patterson. Respiratory gas exchange and metabolic responses during exercise in McArdle's disease. *J Appl Physiol* 75: 745-754, 1993.

197. Rintala, P., H. Lyytinen, and J.M. Dunn. Influence of a physical activity program on children with cerebral palsy: a single subject design. *Pediatr Exerc Sci* 2: 46-56, 1990.

198. Rodillo, E., C.M. Noble-Jamieson, V. Aber, J.Z. Heckmatt, F. Muntoni, and V. Dubowitz. Respiratory muscle training in Duchenne muscular dystrophy. *Arch Dis Child* 64: 736-738, 1989.

199. Rose, J., J.G. Gamble, A. Burgos, J. Medeiros, and W.L. Haskell. Energy expenditure index of walking for normal children and for children with cerebral palsy. *Dev Med Child Neurol* 32: 333-340, 1990.

200. Rose, J., J.G. Gamble, J. Medeiros, A. Burgos, and W.L. Haskell. Energy cost of walking in normal children and in those with cerebral palsy: comparison of heart rate and oxygen uptake. *J Pediatr Orthop* 9: 276-279, 1989.

201. Rose, K.D. Should epileptics be barred from contact sports? AMA changes position. *Med World News* 62B-63B, 1974.

202. Rosenbaum, P.L., D.J. Russell, D.T. Cadman, C. Gowland, S. Jarvis, and S. Hardy. Issues in measuring change in motor function in children with cerebral palsy: a special communication. *Phys Ther* 70: 125-131, 1990.

203. Rotzinger, H., and H. Stoboy. Comparison between clinical judgement and electromyographic investigations of the effect of a special training program for CP children. *Acta Paediatr Belg* 28: 121-128, 1974.

204. Roubertie, A., K. Patte, F. Rivier, A.M. Pages, I. Maire, and B. Echenne. McArdle's disease in childhood: report of a new case. *Eur J Paediatr Neurol* 2: 269-273, 1998.

205. Russell, D.J., L.M. Avery, P.L. Rosenbaum, P.S. Raina, S.D. Walter, and R.J. Palisano. Improved scaling of the gross motor function measure for children with cerebral palsy: evidence of reliability and validity. *Phys Ther* 80: 873-885, 2000.

206. Russell, D.J., P.L. Rosenbaum, D.T. Cadman, C. Gowland, S. Hardy, and S. Jarvis. The Gross Motor Function Measure: a means to evaluate the effects of physical therapy. *Dev Med Child Neurol* 31: 341-352, 1989.

207. Ryan, C.A., and G. Dowling. Drowning deaths in people with epilepsy. *Canad Med Assoc J* 148: 781-784, 1993.

208. Sabath, R.J., H.W. Kilbride, and V.M. Valsh. Peak exercise capacity and pulmonary function of children born extremely premature [abstract]. *Pediatr Exerc Sci* 11: 255-256, 1999.

209. Saigal, S., P. Rosenbaum, B. Staskopf, and J.C. Sinclair. Outcome in infants 501 to 1000 grams birth weight delivered to residents of the McMaster health region. *J Pediatr* 105: 969-976, 1984.

210. Saigal, S., P. Szatmari, P. Rosenbaum, D. Campbell, and S. King. Congenitive abilities and school performance of extremely low birth weight children and matched term control children at age 8 years: a regional study. *J Pediatr* 118: 751-760, 1991.

211. Santuz, P., E. Baraldi, P. Zarmella, M. Filippone, and F. Zacchello. Factors limiting exercise performance in long-term survivors of bronchopulmonary dysplasia. *Am J Respir Crit Care Med* 152: 1284-1289, 1995.

212. Sargeant, A.J. Short-term muscle power in children and adolescents. In Bar-Or, O., ed., *Advances in pediatric sports sciences*. Champaign, IL: Human Kinetics, 1989, 42-65.

213. Sayers, S.P. The role of exercise as a therapy for children with Duchenne muscular dystrophy. *Pediatr Exerc Sci* 12: 23-33, 2000.

214. Schmitt, B., L. Thun-Hohenstein, H. Vontobel, and E. Boltshauser. Seizures induced by physical exercise: report of two cases. *Neuropediatrics* 25: 51-53, 1994.

215. Scott, O.M., C.M. Goddard, and A. Dubowitz. Quantitation of muscle function in children: a prospective study in Duchenne muscular dystrophy. *Muscle Nerve* 5: 291-301, 1982.

216. Scott, O.M., S.A. Hyde, C. Goddard, R. Jones, and V. Dubowitz. Effect of exercise in Duchenne muscular dystrophy. *Physiotherapy* 67: 174-176, 1981.

217. Scott, O.M., S.A. Hyde, G. Vrbova, and V. Dubowitz. Therapeutic possibilities of chronic low frequency electrical stimulation in children with Duchenne muscular dystrophy. *J Neurol Sci* 95: 171-182, 1990.

218. Scott, O.M., G. Vrbova, S.A. Hyde, and V. Dubowitz. Responses of muscles of patients with Duchenne muscular dystrophy to chronic electrical stimulation. *J Neurol Neurosurg Psychiatry* 49: 1427-1434, 1986.

219. Shephard, R.J. Chronic fatigue syndrome: an update. *Sports Med* 31: 167-194, 2001.

220. Shneerson, J.M. Pulmonary artery pressure in thoracic scoliosis during and after exercise while breathing air and pure oxygen. *Thorax* 33: 747-754, 1978.

221. Shneerson, J.M. The cardiorespiratory response to exercise in thoracic scoliosis. *Thorax* 33: 457-463, 1978.

222. Shneerson, J.M., and M.A. Edgar. Cardiac and respiratory function before and after spinal fusion in adolescent idiopathic scoliosis. *Thorax* 34: 658-661, 1979.

223. Shneerson, J.M., and R. Madgwick. The effect of physical training on exercise ability in adolescent idiopathic scoliosis. *Acta Orthop Scand* 50: 303-306, 1979.

224. Siegel, I.M. Muscular dystrophy: interdisciplinary approach and management. *Postgrad Med* 69: 125-133, 1981.

225. Sisto, S.A., W.N. Tapp, J.J. Lamanca, W. Ling, L.R. Korn, A.J. Nelson, and B.H. Natelson. Physical activity before and after exercise in women with chronic fatigue syndrome. *Q J Med* 91: 465-473, 1998.

226. Small, E., O. Bar-Or, E. Van Mil, and S. Saigal. Muscle function of 11- to 17-year-old children of extremely low birthweight. *Pediatr Exerc Sci* 10: 327-336, 1998.

227. Smyth, R.J., K.R. Chapman, T.A. Wright, J.S. Crawford, and A.S. Rebuck. Ventilatory patterns during hypoxia, hypercapnia, and exercise in adolescents with mild scoliosis. *Pediatrics* 77: 692-697, 1986.

228. Sockolov, R., B. Irwin, R.H. Dressenderfer, and E.M. Bernauer. Exercise performance in 6-11 year old boys with Duchene muscular dystrophy. *Arch Phys Med Rehab* 58: 195-201, 1977.

229. Sommer, M. Improvement of motor skills and adaptation of the circulatory system in wheelchair-bound children in cerebral palsy. In Simri, U., ed., *Sports as a means of rehabilitation.* Natanya: Wingate Institute, 1971, 11/1-11/11.

230. Southwood, T.R. Classifying childhood arthritis. *Ann Rheum Dis* 56: 79-81, 1997.

231. Spira, R., and O. Bar-Or. *An investigation of the ambulation problems associated with severe motor paralysis in adolescents. Influence of physical conditioning and adapted sport activities.* 19-P58065-F01. U.S. Dept. HEW, Social Rehabilitation Services, Washington D.C., 1975.

232. Steinhoff, B.J., K. Neususs, H. Thegeder, and C.D. Reimers. Leisure time activity and physical fitness in patients with epilepsy. *Epilepsia* 37: 1221-1227, 1996.

233. Stern, L.M., A.J. Martin, N. Jones, R. Garrett, and J. Yeates. Training inspiratory resistance in Duchenne dystrophy using adapted computer games. *Dev Med Child Neurol* 31: 494-500, 1989.

234. Stoboy, H. Pulmonary function and spiro ergometric criteria in scoliotic patients before and after Harrington rod surgery and physical exercise. In Borms, J. and M. Hebbelinck, eds., *Pediatric work physiology.* Basel: Karger, 1978, 72-81.

235. Stoboy, H., and B. Speierer. Lungenfuntionswerte und spiroergometrische Parameter wahiend der Rehabilitation von Patienten mit idiopathischer Skoliose (Fusionsoperatinder WS mach Harrington und Training). *Arch Orthop Unfall-Chir* 81: 247-254, 1975.

236. Stone, B., C. Beekman, V. Hall, V. Guess, and H.L. Brooks. The effects of an exercise program on change in curve in adolescents with minimal idiopathic scoliosis: a preliminary study. *Phys Ther* 59: 759-763, 1979.

237. Stuberg, W.A., and W.K. Metcalf. Reliability of quantitative muscle testing in healthy children and in children with Duchenne muscular dystrophy using a hand-held dynamometer. *Phys Ther* 68: 977-982, 1995.

238. Sünram, F., H.G. Gotze, and K. Scheele. Arterielle Blutgase und Saure-Basen-Verhaltnisse nach dosierter Ergometerbelastung bei 12-18-jahrigen Madchen mit idiopathischen Thorakalskoliosen vor und nach einem vierwochigen Training. *Sportarzt Sportmed* 25: 6-12,-3-38, 1974.

239. Takken, T., A. Hemel, J. van der Net, and P.J. Helders. Aerobic fitness in children with juvenile idiopathic arthritis: a systematic review. *J Rheumatol* 29: 2643-2647, 2002.

240. Takken, T., J. Van der Net, and P.J.M. Helders. Short term muscle power in juvenile idiopathic arthritis [abstract]. *Phys Ther* 81: A77, 2001.

241. Takken, T., J. van der Net, and P.J. Helders. Do juvenile idiopathic arthritis patients benefit from an exercise program? A pilot study. *Arthritis Rheum* 45: 81-85, 2001.

242. Tirosh, E., O. Bar-Or, and P. Rosenbaum. New muscle power test in neuromuscular disease. Feasibility and reliability. *Am J Dis Child* 144: 1083-1087, 1990.

243. Tirosh, E., O. Bar-Or, and P. Rosenbaum. New muscle power test in neuromuscular disease: feasibility and reliability. *Am J Dis Child* 144: 1083-1087, 1990.

244. Unnithan, V., J. Dowling, G. Frost, B. Volpe Ayub, and O. Bar-Or. Cocontraction and phasic activity during gait in children with cerebral palsy. *Electromyogr Clin Neurophysiol* 36: 487-494, 1996.

245. Unnithan, V.B., C. Clifford, and O. Bar-Or. Evaluation by exercise testing of the child with cerebral palsy. *Sports Med* 26: 239-251, 1998.

246. Unnithan, V.B., J.J. Dowling, G. Frost, and O. Bar-Or. Role of cocontraction in the O_2 cost of walking in

children with cerebral palsy. *Med Sci Sports Exerc* 28: 1498-1504, 1996.

247. Unnithan, V.B., J.J. Dowling, G. Frost, and O. Bar-Or. Role of mechanical power estimates in the O$_2$ cost of walking in children with cerebral palsy. *Med Sci Sports Exerc* 31: 1703-1708, 1999.

248. Van Den Berg-Emons, H.J.G. Physical training of school children with spastic cerebral palsy. 1996. PhD thesis. Maastricht, Krips Repo Meppel, 1-136.

249. Van Den Berg-Emons, H.J.G., W.H.M. Saris, D.C. de Barbanson, K.R. Westerterp, A. Huson, and M.A. van Baak. Daily physical activity of school children with spastic diplegia and of healthy controls. *J Pediatr* 127: 578-584, 1995.

250. Van Den Berg-Emons, R.J., M.A. van Baak, D.C. de Barbanson, L. Speth, and W.H. Saris. Reliability of tests to determine peak aerobic power, anaerobic power and isokinetic muscle strength in children with spastic cerebral palsy. *Dev Med Child Neurol* 38: 1117-1125, 1996.

251. Van Den Berg-Emons, R.J., M.A. van Baak, L. Speth, and W.H. Saris. Physical training of school children with spastic cerebral palsy: effects on daily activity, fat mass and fitness. *Int J Rehab Res* 21: 179-194, 1998.

252. Van Den Berg-Emons, R.J., M.A. van Baak, L. Speth, and W.H. Saris. Physical training of school children with spastic cerebral palsy: effects on daily activity, fat mass and fitness. *Int J Rehab Res* 21: 179-194, 1998.

253. Van Mil, E., N. Schoeber, R.E. Calvert, and O. Bar-Or. Optimization of force in the Wingate test for children with a neuromuscular disease. *Med Sci Sports Exerc* 28: 1087-1092, 1996.

254. Van Praagh, E., G. Falgairette, M. Bedu, N. Fellmann, and J. Coudert. Laboratory and field tests in 7-year-old boys. In Oseid, S. and H.K. Carlsen, eds., *Children and exercise XIII.* Champaign, IL: Human Kinetics, 1989, 11-17.

255. Vignos, P.J., and M.P. Watkins. The effect of exercise in muscular dystrophy. *JAMA* 197: 843-848, 1966.

256. Vignos, P.J. Jr., and K.C. Archibald. Maintenance of ambulation in childhood muscular dystrophy. *J Chron Dis* 12: 273-290, 1960.

257. Vorgerd, M., T. Grehl, M. Jager, K. Muller, G. Freitag, T. Patzold, N. Bruns, K. Fabian, M. Tegenthoff, W. Mortier, A. Luttmann, J. Zange, and J.P. Malin. Creatine therapy in myophosphorylase deficiency (McArdle disease): a placebo-controlled crossover trial. *Arch Neurol* 57: 956-963, 2000.

258. Vostrejs, M., and J.R. Hollister. Muscle atrophy and leg length discrepancies in pauciarticular juvenile rheumatoid arthritis. *Am J Dis Child* 142: 343-345, 1988.

259. Wanke, T., K. Toifl, M. Merkle, D. Formanek, H. Lahrmann, and H. Zwick. Inspiratory muscle training in patients with Duchenne muscular dystrophy. *Chest* 105: 475-482, 1994.

260. Webster, C., L. Silberstein, A.P. Hays, and H.M. Blau. Fast muscle fibers are preferentially affected in Duchenne muscular dystrophy. *Cell* 52: 503-513, 1988.

261. Weiss, H.R. The effect of an exercise program on vital capacity and rib mobility in patients with idiopathic scoliosis. *Spine* 16: 88-93, 1991.

262. Whipp, B.J., and J.A. Davis. The ventilatory stress of exercise in obesity. *Am Rev Respir Dis* 129 (Suppl): S90-S92, 1984.

263. Whiting, P., A.M. Bagnall, A.J. Sowden, J.E. Cornell, C.D. Mulrow, and G. Ramirez. Interventions for the treatment and management of chronic fatigue syndrome: a systematic review. *JAMA* 286: 1360-1368, 2001.

264. Willig, T.N., L. Carlier, M. Legrand, H. Rivière, and J. Navarro. Nutritional assessment in Duchenne muscular dystrophy. *Dev Med Child Neurol* 35: 1074-1082, 1993.

265. Wineinger, M.A., R.T. Abresch, S.A. Walsh, and G.T. Carter. Effects of aging and voluntary exercise on the function of dystrophic muscle from mdx mice. *Am J Phys Med Rehab* 77: 20-27, 1998.

266. Winnick, J.P. *Adapted physical education and sport.* Champaign, IL: Human Kinetics, 1990.

267. Wood, N.S., N. Marlow, K. Costeloe, A.T. Gibson, and A.R. Wilkinson. Neurologic and developmental disability after extremely preterm birth. EPICure Study Group. *N Eng J Med* 343: 378-384, 2000.

268. Working Group for the Royal Australian College of Physicians. Chronic fatigue syndrome. Draft clinical practice guidelines on the evaluation of prolonged fatigue and the diagnosis and management of chronic fatigue syndrome. *Med J Austr* 1: 1-8, 1997.

269. Wormgoor, M.E.A., and Y.H.J. Gierlings. Revalidatie van lichamelijk gehandicapten met dehulp van sprtactiviteiten in Noorwegen. *Bewegen Hulpverlening* 3: 219-238, 1989.

270. Wright, N.C., D.D. Kilmer, M.A. McCrory, S.G. Aitkens, B.J. Holcomb, and E.M. Bernauer. Aerobic walking in slowly progressive neuromuscular disease: effect of a 12-week program. *Arch Phys Med Rehab* 77: 64-69, 1996.

271. Zupan, A. Long-term electrical stimulation of muscles in children with Duchenne and Becker muscular dystrophy. *Muscle Nerve* 15: 362-367, 1992.

272. Zupan, A., M. Gregoric, V. Valencic, and S. Vandot. Effects of electrical stimulation on muscles of children with Duchenne and Becker muscular dystrophy. *Neuropediatrics* 24: 189-192, 1993.

Chapter 11

1. Acar, P., S. Sebahoun, L. de Pontual, and C. Maunoury. Myocardial perfusion in children with sickle cell anaemia. *Pediatr Radiol* 30: 352-354, 2000.

2. Alpert, B.S., P.A. Gilman, W.B. Strong, M.F. Ellison, M.D. Miller, J. McFarlane, and T. Hayasidera. Hemodynamic and ECG responses to exercise in children with sickle cell anemia. *Am J Dis Child* 135: 362-366, 1981.

3. American Academy of Pediatrics Committee on Sports Medicine and Fitness. Medical conditions affecting sports participation. *Pediatrics* 94: 757-760, 1994.

4. Åstrand, P.O. *Experimental studies of physical working capacity in relationship to sex and age.* Copenhagen: Munksgaard, 1952.

5. Åstrand, P.O., and K. Rodahl. *Textbook of work physiology.* 2nd ed. New York: McGraw-Hill, 1977.

6. Balfour, I.C., W. Covitz, H. Davis, P.S. Rao, W.B. Strong, and B.S. Alpert. Cardiac size and function in children with sickle cell anemia. *Am Heart J* 108: 345-350, 1984.

7. Baraldi, E., G. Montini, S. Zanconato, G. Zacchello, and F. Zacchello. Exercise tolerance after anaemia correction with recombinant human erythropoietin in end-stage renal disease. *Pediatr Nephrol* 4: 623-626, 1990.

8. Beardsley, D.S. Hemophilia. In Goldberg, B., ed., *Sports and exercise for children with chronic health conditions.* Champaign, IL: Human Kinetics, 1995, 301-309.

9. Bernstein, L., R.K. Ross, R.A. Lobo, R. Hanisch, M.D. Krailo, and B.E. Henderson. The effects of moderate physical activity on menstrual cycle patterns in adolescence: implications for breast cancer prevention. *Br J Cancer* 55: 681-685, 1987.

10. Bonzel, K.E., B. Wildi, M. Weiss, and K. Sharer. Spirometric performance of children and adolescents with chronic renal failure. *Pediatr Nephrol* 5: 22-28, 1991.

11. Brown, J., M. Geer, W. Covita, W. Hillenbrand, S. Leff, N. Talner, and L. Robinson. Electrocardiographic responses to exercise in sickle cell anemia. In Doyle, E.F., M.A. Engle, W.M. Gersony, W.J. Rashkind, and N. Talner, eds., *Pediatric cardiology.* New York: Springer-Verlag, 1986, 1111-1116.

12. Bruner, A.B., A. Joffe, A.K. Duggan, J.F. Casella, and J. Brandt. Randomised study of cognitive effect of iron supplementation in non-anemic iron-deficient adolescent girls. *Lancet* 348: 992-996, 1996.

13. Carpenter, C.L., R.K. Ross, A. Paganini-Hill, and L. Bernstein. Lifetime exercise activity and breast cancer risk among post-menopausal women. *Br J Cancer* 80: 1852-1858, 1999.

14. Cazin, B., N.C. Corin, J.P. LaPorte, B. Gallet, L. Douay, and M. Lopez. Cardiac complications after bone marrow transplantation. *Cancer* 57: 2061-2069, 1986.

15. Convertino, V.A. Blood volume response to training. In Fletcher, G.F., ed., *Cardiovascular responses to exercise.* Mount Kisco, NY: Futura, 1994, 207-221.

16. Courneya, K.S. Exercise interventions during cancer treatment: biopsychosocial outcomes. *Exerc Sport Sci Rev* 29: 60-64, 2001.

17. Covitz, W., C. Eubig, I.C. Balfour, R. Jerath, B.S. Alpert, W.B. Strong, and R.H. DuRant. Exercise-induced cardiac dysfunction in sickle cell anemia. A radio nuclide study. *Am J Cardiol* 51: 570-575, 1983.

18. Dallman, P.R., and M.A. Siimes. Percentile curves for hemoglobin and red cell volume in infancy and childhood. *J Pediatr* 94: 26-31, 1979.

19. Dancaster, C.P., and S.J. Whereat. Renal function in marathon runners. *S Afr Med J* 45: 547-551, 1971.

20. Denenberg, B.S., G. Criner, R. Jones, and J.F. Spann. Cardiac function in sickle cell anemia. *Am J Cardiol* 51: 1674-1678, 1983.

21. Eichner, E.R. Other medical considerations in prolonged exercise. In Lamb, D.R. and R. Murray, eds., *Perspectives in exercise science and sports medicine.* Vol. 1. *Prolonged exercise.* Indianapolis: Benchmark Press, 1988, 415-442.

22. Eichner, E.R. Sickle cell trait, exercise, and altitude. *Phys Sportsmed* 14: 144-157, 1986.

23. Eichner, E.R. Hematuria—a diagnostic challenge. *Phys Sportsmed* 18: 53-63, 1990.

24. Ekblom, B., A.N. Goldbarg, and B. Gullberg. Response to exercise after blood loss and reinfusion. *J Appl Physiol* 33: 175-180, 1972.

25. Eriksson, B.O. Physical training, oxygen supply, and muscle metabolism in 11-13 year old boys. *Acta Physiol Scand Suppl* 384: 1-48, 1972.

26. Falk, R.H., and W.B. Hood. The heart in sickle cell anemia. *Arch Intern Med* 142: 1680-1688, 1982.

27. Finch, C.A., L.R. Miller, A.R. Inamidar, R. Person, K. Seiler, and B. Mackler. Iron deficiency in the rat. Physiological and biochemical studies of muscle dysfunction. *J Clin Invest* 58: 447-453, 1976.

28. Fukazawa, R., S. Ogawa, and T. Hirayama. Early detection of anthracycline cardiotoxicity in children using exercise-based echocardiography and Doppler echocardiography. *Jpn Circ J* 58: 625-634, 1994.

29. Gardner, G.W., V.R. Edgerton, and B. Senewiratne. Physical work capacity and metabolic stress in subjects with iron deficiency anemia. *Am J Clin Nutr* 30: 910-917, 1977.

30. Gerry, J.L., B.H. Buckley, and G.M. Hutchins. Clinicopathologic analysis of cardiac dysfunction in 52 patients with sickle cell anemia. *Am J Cardiol* 42: 211-219, 1978.

31. Giordano, U., A. Calzolari, M.C. Matteucci, E. Pastore, A. Turchetta, and G. Rizzoni. Exercise tolerance and blood pressure response to exercise testing in children and adolescents after renal transplantation. *Pediatr Cardiol* 19: 471-473, 1998.

32. Gledhill, N. Blood doping and related issues: a brief review. *Med Sci Sports Exerc* 14: 183-189, 1982.

33. Gleim, G.W. Renal responses to exercise and training. In Garrett, W.E. and D.T. Kirkendall, eds., *Exercise and sports science.* Philadelphia: Lippincott Williams & Wilkins, 2000, 217-225.

34. Gordon, F.H., D.G. Nathan, S. Piomelli, and J.F. Cummins. The erythrocythemic effects of androgen. *Br J Haemat* 14: 611-621, 1968.

35. Grant, G.R., J.H. Graziano, C. Seaman, and A.L. Mansell. Cardiorespiratory response to exercise in patients with thalassemia major. *Am Rev Respir Dis* 136: 92-97, 1987.

36. Hamilton, W., A. Rosenthal, D. Berwick, and A.S. Nadas. Angina pectoris in a child with sickle cell anemia. *Pediatrics* 61: 911-914, 1978.

37. Hertenstein, B., M. Stefanic, T. Schmeiser, M. Scholz, V. Goller, and M. Clausen. Cardiac toxicity to bone marrow transplantation: predictive value of cardiologic evaluation before transplant. *J Clin Oncol* 12: 998-1004, 1994.

38. Hogarty, A.N., A. Leahy, H. Zhao, M.D. Hogarty, N. Bunin, A. Cnaan, and S.M. Paridon. Longitudinal evaluation of cardiopulmonary performance during exercise after bone marrow transplantation in children. *J Pediatr* 136: 311-317, 2000.

39. Iarussi, D., M. Galderisi, G. Ratti, M.A. Tedesco, P. Indolfi, F. Casale, M.T. DiTullo, O. DeVitus, and A. Iacono. Left ventricular systolic and diastolic function after anthracycline chemotherapy in childhood. *Clin Cardiol* 24: 663-669, 2001.

40. Jenney, M.E.M., E.B. Faragher, P.H. Morris Jones, and A. Woodcock. Lung function and exercise capacity in survivors of childhood leukemia. *Med Pediatr Oncol* 24: 222-230, 1995.

41. Johnson, D., H. Perrault, A. Fournier, J.-M. Leclerc, J.-L. Bigras, and A. Davignon. Cardiovascular responses to dynamic submaximal exercise in children previously treated with anthracycline. *Am Heart J* 133: 169-173, 1997.

42. Kanstrup, I., and B. Ekblom. Blood volume and hemoglobin concentration as determinants of maximal aerobic power. *Med Sci Sports Exerc* 16: 256-262, 1984.

43. Kark, J.A., D.M. Posey, H.R. Schumacher, and C.J. Ruehle. Sickle cell trait as a risk factor for sudden death in physical training. *N Eng J Med* 317: 781-787, 1987.

44. Kark, J.A., and F.T. Ward. Exercise and hemoglobin S. *Semin Hematol* 31: 181-225, 1994.

45. Kennedy, T.L., and N.J. Siegel. Chronic renal disease. In Goldberg, B., ed., *Sports and exercise for children with chronic health conditions.* Champaign, IL: Human Kinetics, 1995, 265-278.

46. Koudi, E., A. Iacovides, P. Iordanidis, S. Vassiliou, A. Deligiannis, C. Ierodiakonou, and A. Tourkantonis. Exercise renal rehabilitation program: pyschosocial effects. *Nephron* 77: 152-158, 1997.

47. Krull, F., I. Schulze-Neick, A. Hatopp, G. Offner, and J. Brodehl. Exercise capacity and blood pressure response in children and adolescents after renal transplantation. *Acta Paediatr* 83: 1296-1302, 1994.

48. Lamanca, J., and E. Haymes. Effects of dietary iron supplementation on endurance [abstract]. *Med Sci Sports Exerc* 21 (Suppl): S77, 1989.

49. Larsen, R.L., G. Barber, C.T. Heise, and C.S. August. Exercise assessment of cardiac function in children and young adults before and after bone marrow transplantation. *Pediatrics* 89: 722-729, 1992.

50. Leiken, S.L., and D. Gallagher. Mortality in children and adolescents with sickle cell disease. *Pediatrics* 84: 500-508, 1989.

51. Lester, L.A., P.C. Sodt, N. Hutcheon, and R.A. Arcilla. Cardiac abnormalities in children with sickle cell anemia. *Chest* 98: 1169-74, 1990.

52. Lipshultz, S.E., S.D. Colan, R.D. Gelber, A.R. Perez-Atayde, S.E. Sallam, and S.P. Sanders. Late cardiac effects of doxorubin therapy for acute lymphoblastic leukemia in childhood. *N Eng J Med* 324: 808-815, 1991.

53. Lonsdorfer, J., P. Bogui, A. Otayeck, E. Bursaux, and R. Cabannes. Cardiorespiratory adjustments in chronic sickle cell anemia. *Bull Eur Physiopathol Respir* 19: 339-344, 1983.

54. Marcus, P.M., B. Newman, P.G. Moorman, R.C. Millikan, D.D. Baird, B. Qaqish, and B. Sternfeld. Physical activity at age 12 and adult breast cancer risk (United States). *Cancer Causes Control* 10: 293-302, 1999.

55. Martin, T.W., I.M. Weisman, and R.J. Zeballos. Exercise and hypoxia increase sickling in venous blood from an exercising limb in individuals with sickle cell trait. *Am J Med* 87: 48-56, 1989.

56. Matteucci, M.C., A. Calzolari, E. Pompei, F. Principato, A. Turchetta, and G. Rizzoni. Abnormal response during exercise test in normotensive transplanted children and adolescents. *Nephron* 73: 201-206, 1996.

57. McConnell, M.E., S.R. Daniels, J. Lobel, F.W. James, and S. Kaplan. Hemodynamic response to exercise in patients with sickle cell anemia. *Pediatr Cardiol* 10: 141-144, 1989.

58. McDonagh, K.T., and A.W. Nienhus. The thalassemias. In Nathan, D.G. and F.A. Oski, eds., *Hematology of infancy and childhood.* 4th ed. Philadelphia: Saunders, 1993, 783-879.

59. McEnery, P.T., D.M. Stablein, G. Arbus, and A. Tejani. Renal transplantation in children. *N Eng J Med* 326: 1727-1732, 1992.

60. Miller, D.M., R.M. Winslow, and H.G. Klein. Improved performance after exchange transfu-

sion in subjects with sickle cell anemia. *Blood* 56: 1127-1131, 1980.

61. Miller, G.J., G.R. Sergeant, A. Sivapragasam, and M.C. Petch. Cardiopulmonary responses and gas exchange during exercise in adults with sickle cell disease (sickle cell anaemia). *Clin Sci* 44: 113-128, 1973.

62. Mittendorf, R., M.P. Longnecker, P.A. Newcomb, A.T. Dietz, E.R. Greenberg, and G.F. Bogdau. Strenuous physical activity in young adulthood and risk of breast cancer (United States). *Cancer Causes Control* 6: 347-353, 1995.

63. Murphy, J.R. Sickle cell hemoglobin (HbAS) in black football players. *JAMA* 225: 981-982, 1973.

64. National Hemophilia Foundation Medical and Scientific Advisory Committee. *The hemophilia handbook.* New York: National Hemophilia Foundation, 1992, 189.

65. Nelson, M. Anaemia in adolescent girls: effects on cognitive function and activity. *Proc Nutr Soc 55*: 359-367, 1996.

66. Oski, F.A. Differential diagnosis of anemia. In Nathan, D.G. and F.A. Oski, eds., *Hematology of infancy and childhood.* 4th ed. Philadelphia: Saunders, 1993, 346-353.

67. Painter, P.L. Exercise and end-stage renal disease. *Exerc Sport Sci Rev* 16: 305-340, 1988.

68. Pearson, H.A. Sickle cell trait and competitive athletics: is there a risk? *Pediatrics* 83: 613-614, 1989.

69. Pianosi, P., S.J.A. D'Sousa, T.D. Charge, M.J. Beland, D.W. Esselltine, and A.L. Coates. Cardiac output and oxygen delivery during exercise in sickle cell anemia. *Am Rev Respir Dis* 143: 231-235, 1991.

70. Pianosi, P., S.J.A. D'Sousa, D.W. Esselltine, T.D. Charge, and A.L. Coates. Ventilation and gas exchange during exercise in sickle cell anemia. *Am Rev Respir Dis* 143: 226-230, 1991.

71. Pietri, M.M., W.R. Frontera, I.S. Pratts, and E.L. Suarez. Skeletal muscle function in patients with hemophilia A and unilateral hemarthrosis of the knee. *Arch. Phys Med Rehab* 73: 22-28, 1992.

72. Pikhala, J., U.M. Saarinen, U. Lundstrom, M. Salmo, K. Virkola, and K. Virtanen. Effects of bone marrow transplantation on myocardial function in children. *Bone Marrow Transplant* 13: 149-155, 1994.

73. Poortmans, J.R. Postexercise proteinuria in humans. Facts and mechanisms. *JAMA* 253: 236-240, 1985.

74. Poortmans, J.R., M.F. Engels, and D. Labilloy. The influence of the type of activity on post-exercise proteinuria in man [abstract]. *Med Sci Sports Exerc* 14: 118, 1982.

75. Poortmans, J.R., D. Labilloy, and O. Niset. Relationship between post-exercise proteinuria and venous lactate [abstract]. *Med Sci Sports Exerc* 13: 84, 1981.

76. Rees, A.H., M.A. Stefadouros, W.B. Strong, M.D. Miller, P. Gilman, J.A. Rigby, and J. McFarlane.

Left ventricular performance in children with homozygous sickle cell anemia. *Br Heart J* 40: 690-698, 1978.

77. Rovelli, A., C. Pezzini, D. Silvestri, F. Tana, M.A. Galli, and C. Uderzo. Cardiac and respiratory function after bone marrow transplantation in children with leukemia. *Bone Marrow Transplant* 16: 571-576, 1995.

78. Rowland, T.W. Iron deficiency in the adolescent athlete. In Bar-Or, O., ed., *The child and adolescent athlete.* Oxford: Blackwell Scientific, 1996, 274-286.

79. Rowland, T., K. Miller, P. Vanderburgh, D. Goff, L. Martel, and L. Ferrone. Cardiovascular fitness in premenarcheal girls and young women. *Int J Sports Med* 21: 117-121, 2000.

80. Rowland, T., B. Popowski, and L. Ferrone. Cardiac responses to maximal upright cycle exercise in healthy boys and men. *Med Sci Sports Exerc* 29: 1146-1151, 1997.

81. Rowland, T., J. Potts, T. Potts, J. Son-Hing, G. Harbison, and G. Sandor. Cardiovascular responses to exercise in children and adolescents with myocardial dysfunction. *Am Heart J* 137: 126-133, 1999.

82. Rowland, T.W., L. Stagg, and J.F. Kelleher. Iron deficiency in adolescent girls. Are athletes at increased risk? *J Adolesc Health* 12: 22-25, 1991.

83. Sears, D.A. The morbidity of sickle cell trait: a review of the literature. *Am J Med* 64: 1021-1036, 1978.

84. Selsky, C., and H.A. Pearson. Neoplasms. In Goldberg, B., ed., *Sports and exercise for children with chronic health conditions.* Champaign, IL: Human Kinetics, 1995, 311-322.

85. Sharkey, A.M., A.B. Carey, C.T. Heise, and G. Barber. Cardiac rehabilitation after cancer therapy in children and young adults. *Am J Cardiol* 71: 1488-1490, 1993.

86. Shoff, S.M., P.A. Newcomb, A. Trentham-Dietz, P.L. Remington, R. Mittendorf, E.R. Greenberg, and W.C. Willett. Early-life physical activity and postmenopausal breast cancer: effect of body size and weight change. *Cancer Epidem Bio Prev* 9: 591-595, 2000.

87. Shore, S., and R.J. Shephard. Immune responses to exercise in children treated for cancer. *J Sports Med Phys Fitness* 39: 240-243, 1999.

88. Siegel, A.J., C.H. Hennekens, and H.S. Solomons. Exercise-related hematuria. Findings in a group of marathon runners. *JAMA* 241: 391-392, 1979.

89. Sigler, A.T., and W.H. Zinkham. Anemia. In Goldberg, B., ed., *Sports and exercise for children with chronic health conditions.* Champaign, IL: Human Kinetics, 1995, 279-300.

90. Smith, K.E., S. Gotleib, R.H. Gurwitch, and A.D. Blotcky. Impact of a summer camp experience on

daily activity and family interactions among children with cancer. *J Pediatr Psych* 12: 533-542, 1987.

91. Sproule, B.J., J.H. Mitchell, and W.F. Miller. Cardiopulmonary physiological responses to heavy exercise in patients with anemia. *J Clin Invest* 39: 378-388, 1960.

92. Streiff, M., and W.R. Bell. Exercise and hemostasis in humans. *Semin Hematol* 31: 155-165, 1994.

93. Sundberg, S., and R. Elovainio. Cardiorespiratory function in competitive endurance runners aged 12-16 years compared with ordinary boys. *Acta Paediatr Scand* 71: 987-992, 1982.

94. Ulmer, H.E., H. Greiner, H.W. Schuler, and K. Scharer. Cardiovascular impairment and physical working capacity in children with chronic renal failure. *Acta Paediatr Scand* 67: 43-48, 1978.

95. U.S. Department of Health and Human Services. *Physical activity and health: a report of the Surgeon General.* Atlanta: Centers for Disease Control and Prevention, 1996, 112-124.

96. Val-Mejias, J., W.K. Lee, A.B. Weisse, and T.J. Ragan. Left ventricular performance during and after sickle cell crisis. *Am Heart J* 97: 585-591, 1979.

97. Villa, M.P., P.L. Rotili, F. Santamaria, A. Vania, E. Bonci, G. Tancredi, and R. Ronchetti. Physical performance in patients with thalassemia before and after transfusion. *Pediatr Pulmonol* 21: 367-372, 1996.

98. Warner, J.T., W. Bell, D.K.H. Webb, and J.W. Gregory. Relationship between cardiopulmonary response to exercise and adiposity in survivors of childhood malignancy. *Arch Dis Child* 76: 298-303, 1997.

99. Weinstein, H.J., J.M. Rappaport, and J.L.M. Ferra. Bone marrow transplantation. In Nathan, D.G. and F.A. Oski, eds., *Hematology of infancy and childhood.* 4th ed. Philadelphia: Saunders, 1993, 317-344.

100. Willens, H.J., C. Lawrence, W.H. Frishman, and J.A. Strom. A noninvasive comparison of left ventricular performance in sickle cell anemia and chronic aortic regurgitation. *Clin Cardiol* 6: 542-548, 1983.

101. Woodson, R.D., R.E. Wills, and C. Lenfant. Effect of acute and established anemia on O_2 transport at rest, submaximal, and maximal work. *J Appl Physiol* 44: 36-43, 1978.

Chapter 12

1. Aine, D., and D. Lester. Exercise, depression, and self-esteem [comment]. *Percept Motor Skills* 81: 890, 1995.

2. Aronen, E.T., M.H. Teicher, D. Geenens, S. Curtin, C.A. Glod, and K. Pahlavan. Motor activity and severity of depression in hospitalized prepubertal children. *J Am Acad Child Adolesc Psych* 35: 752-763, 1996.

3. Åstrand, P.O. *Experimental studies of physical working capacity in relationship to sex and age.* Copenhagen: Munksgaard, 1952, 121.

4. Bahrke, M.S., and R.G. Smith. Alterations in anxiety of children after exercise and rest. *Am Corr Ther J* 39: 90-94, 1995.

5. Barchas, J., and D. Freedman. Brain amines: response to physiological stress. *Biochem Pharmacol* 12: 1232-1235, 1962.

6. Bar-Or, O., J.S. Skinner, and V. Bergsteinová. Maximal aerobic capacity of 6-15 year old girls and boys with subnormal intelligence quotients. *Acta Paediatr Scand* 217 (Suppl): 108-113, 1971.

7. Baumert, P.W., J.M. Henderson, and N.J. Thompson. Health risk behaviors of adolescent participants in organized sports. *J Adolesc Health* 22: 460-465, 1998.

8. Birmaher, B., N.D. Ryan, D.E. Williamson, D.A. Brent, J. Kaufman, R.E. Dahl, J. Perel, and B. Nelson. Childhood and adolescent depression: a review of the past 10 years. *J Am Acad Child Adolesc Psychiatry* 35: 1427-1439, 1996.

9. Bluechardt, M.H., J. Weiner, and R.J. Shephard. Exercise programmes in the treatment of children with learning disorders. *Sports Med* 19: 55-72, 1995.

10. Boileau, R.A., J.E. Ballad, and R.L. Sprague. Effect of methylphenidate on cardiorespiratory responses in hyperactive children. *Res Q Am Assoc Health Phys Educ* 47: 590-596, 1976.

11. Boyd, K., and D. Hrycaiko. The effect of physical activity intervention package on the self-esteem of pre-adolescent and adolescent females. *Adolescence* 32: 693-708, 1997.

12. Brent, D.A., J.A. Perper, and C.E. Goldstein. Risk factors for adolescent suicide: a comparison of adolescent suicide victims with suicidal inpatients. *J Am Acad Adolesc Psychiatry* 45: 581-588, 1988.

13. Broocks, A., B. Bandelow, and G. Pekrun. Comparison of aerobic exercise, clomipramine, and placebo in the treatment of panic disorder. *Am J Psychiatry* 155: 603-609, 1998.

14. Broocks, A., T.F. Meyer, B. Bandelow, A. George, U. Bartman, E. Ruther, and U. Hillmer-Vogel. Exercise avoidance and impaired endurance capacity in patients with panic disorder. *Neuropsychobiology* 36: 182-187, 1997.

15. Brown, B.S., and W.D. Van Huss. Exercise and rat brain catecholamines. *J Appl Physiol* 34: 664-669, 1973.

16. Brown, B.J. The effect of an isometric strength program on the intellectual and social development of trainable retarded males. *Am Corr Ther J* 31: 44-48, 1977.

17. Brown, D.R. Exercise, fitness, and mental health. In Bouchard, C., R.J. Shephard, T. Stephens, J.R. Sutton, and B.D. McPherson, eds., *Exercise, fitness, and health. A consensus of current knowledge.* Champaign, IL: Human Kinetics, 1990, 607-626.

18. Brown, J.D., and M. Lawton. Stress and well-being in adolescence: the moderating role of physical exercise. *J Hum Stress* 12: 125-131, 1986.

19. Brown, R.S. Exercise and mental health in the pediatric population. *Clin Sports Med* 1: 515-527, 1982.

20. Brown, S.W., M. Welsh, E.E. Labbe, W.F. Vitulli, and P. Kulkaami. Aerobic exercise in the psychological treatment of adolescents. *Percept Mot Skills* 74: 555-560, 1992.

21. Calfas, K.J. The relationship between physical activity and the psychological well-being of youth. In J.M. Rippe, ed., *Lifestyle medicine*. Malden, MA: Blackwell Scientific, 1999, 967-979.

22. Campbell, J. Improving the physical fitness of retarded boys. *Ment Retard* 12: 31-35, 1974.

23. Chasey, W.C., and W. Wyrick. Effect of a gross motor development program on form perception skills of educable mentally retarded children. *Res Q* 41: 345-352, 1970.

24. Corder, W.W. Effects of physical education on the intellectual and social development of educable mentally retarded boys. *Except Child* 32: 357-364, 1966.

25. Covey, L.A., and D.L. Feltz. Physical activity and adolescent female psychological development. *J Youth Adolesc* 20: 463-474, 1991.

26. Craft, L.L., and D.M. Landers. The effect of exercise on clinical depression: a meta-analysis [abstract]. *Med Sci Sports Exerc* 30 (Suppl): S117, 1998.

27. Davis, J.M. Carbohydrates, branched-chain amino acids, and endurance: the central fatigue hypothesis. *Int J Sports Nutr* 5: S29-S38, 1995.

28. De Wilde, E.J., C.W.M. Klenhorst, R.F.W. Diekstra, and W.H.G. Wolters. Social support, life events, and behavioral characteristics of psychologically distressed adolescents at high risk for attempting suicide. *Adolescence* 29: 49-60, 1994.

29. Dua, J., and L. Hargreaves. Effect of aerobic exercise on negative affect, positive affect, and depression. *Percept Mot Skills* 75: 335-361, 1992.

30. Dykens, E.M., B.A. Rosner, and G. Butterbaugh. Exercise and sports in children and adolescents with developmental disabilities. *Child Adolesc Psych Clin N Amer* 7: 757-771, 1998.

31. Farmer, M.E., B.Z. Locke, E.K. Moscicki, A.L. Dannenberg, D.B. Larson, and L.S. Radloff. Physical activity and depressive symptoms: the NHANES I Epidemiologic Followup Study. *Am J Epidemiol* 128: 1340-1351, 1988.

32. Farrell, P.A., W.K. Gates, M.G. Maksud, and W.P. Morgan. Increases in plasma beta-endorphin/beta-lipotropin immunoreactivity after treadmill running in humans. *J Appl Physiol* 52: 1245-1249, 1982.

33. Fernhall, B. Physical fitness and exercise training of individuals with mental retardation. *Med Sci Sports Exerc* 25: 442-450, 1992.

34. Fernhall, B., and K.H. Pitetti. Leg strength is related to endurance run performance in children and adolescents with mental retardation. *Pediatr Exerc Sci* 12: 324-333, 2000.

35. Fernhall, B., and K.H. Pitetti. Limitations to physical work capacity in individuals with mental retardation. *Clin Exerc Physiol* 3: 176-185, 2001.

36. Fernhall, B., K. Pitetti, T. Hensen, and M. Vukovich. Cross-validation of the 20-m shuttle run in children with mental retardation. *Adapted Phys Act Q* 17: 402-412, 2000.

37. Fernhall, B., K.H. Pitetti, and J.H. Rimmer. Cardio-respiratory capacity of individuals with mental retardation including Down syndrome. *Med Sci Sports Exerc* 28: 366-371, 1996.

38. Fernhall, B., K.H. Pitetti, N. Stubbs, and L. Standler. Validity and reliability of the 1/2 mile run-walk as an indicator of aerobic fitness in children with mental retardation. *Pediatr Exerc Sci* 8: 130-142, 1996.

39. Fernhall, B., K.H. Pitetti, and M.D. Vukovich. Validation of cardiovascular fitness field tests in children with mental retardation. *Am J Ment Retard* 102: 602-612.

40. Fernhall, B., and G. Tymeson. Graded exercise testing of mentally retarded adults. A study of feasibility. *Arch Phys Med Rehab* 68: 363-365, 1987.

41. Fernhall, B., G. Tymeson, A.L. Millar, and L.N. Burkett. Cardiovascular fitness testing and fitness levels of adolescents and adults with mental retardation including Down syndrome. *Educ Training Mental Retard* 24: 133-138, 1989.

42. Gauvin, L., J.C. Spence, and S. Anderson. Exercise and psychological well-being in the adult population: reality or wishful thinking? In J.M. Rippe, ed., *Lifestyle medicine*. Malden, MA: Blackwell Scientific, 1999, 957-966.

43. Glyshaw, K., L.H. Cohen, and L.C. Towbes. Coping strategies and psychological stress: prospective analyses of early and middle adolescents. *Am J Comm Psych* 17: 607-623, 1989.

44. Gruber, J.J. Physical activity and self-esteem development in children: a meta-analysis. In G.A. Stull and H.M. Eckert, eds., *Effects of physical activity on children*. Champaign, IL: Human Kinetics, 1986, 30-48.

45. Hastings, J.E., and R.A. Barkley. A review of psychophysiological research with hyperkinetic children. *J Abnorm Child Psych* 6: 413-447, 1978.

46. Hay, J.A., B. Faught, R. Hawes, R.J. McKinlay, and N. Klentrou. The relationships between self-efficacy, aerobic capacity, body composition, and perceived exertion among adolescent youth [abstract]. *Pediatr Exerc Sci* 13: 195, 2001.

47. Herman-Tofler, L.R., and B.W. Tuckman. The effects of aerobic training on children's creativity, self-perception, and aerobic power. *Child Adolesc Psych Clin N Amer* 7: 773-790, 1998.

48. Hinkle, J.S., B.W. Tuckman, and J.P. Sampson. The psychology, physiology, and creativity of middle school aerobic exercisers. *Elem School Guidance Counseling* 28: 133-145, 1993.

49. Holloway, J., A. Beuter, and J. Duda. Self-efficacy and training for strength in adolescent girls. *J Appl Soc Psych* 18: 699-719, 1988.

50. Kasch, F.W., and S.A. Zasueta. Physical capacities of mentally retarded children. *Acta Paediatr Scand Suppl* 217: 114-118, 1971.

51. Kirkendall, D.R. Effects of physical activity on intellectual development and academic performance. In G.A. Stull and H.M. Eckert, eds., *Effects of physical activity on children.* Champaign, IL: Human Kinetics, 1986, 49-63.

52. Klerman, G., and M.M. Weissman. Increasing rates of depression. *JAMA* 261: 2229-2235, 1989.

53. Koniak-Griffin, D. Aerobic exercise, psychological well-being, and physical discomforts during adolescent pregnancy. *Res Nurs Health* 17: 253-263, 1994.

54. Kovacs, M. Presentation and course of major depressive disorder during childhood and later years of the life span. *J Am Acad Child Adolesc Psychiatry* 35: 705-715, 1996.

55. Londerlee, B.R., and L.E. Johnson. Motor fitness of TMR vs. EMR and normal children. *Med Sci Sports* 6: 247-252, 1974.

56. MacMahon, J.R., and R.T. Gross. Physical and psychological effects of aerobic exercise in boys with learning disabilities. *J Dev Behav Pediatr* 8: 274-277, 1987.

57. Maksud, M.G., and L.H. Hamilton. Physiological responses of EMR children to strenuous exercise. *Am J Ment Defic* 79: 32-38, 1974.

58. Markoff, R.A., P. Ryan, and T. Young. Endorphins and mood changes in long-distance running. *Med Sci Sports Exerc* 14: 11-15, 1982.

59. McDonald, D.G., and J.A. Hodgdon. *Psychological effects of aerobic fitness training.* New York: Springer, 1991, 1-224.

60. Michaud-Tomson, L.M. Childhood depressive symptoms, physical activity, and health related fitness. 1995. Doctoral dissertation. Arizona State University, Columbia, AZ.

61. Millar, A.L., B. Fernhall, L.N. Burkett, and G. Tymeson. Effect of aerobic training in adolescents with Down syndrome. *Med Sci Sports Exerc* 25: 260-264, 1993.

62. Milligan, R.A.K., V. Burke, L.J. Beilin, J. Richards, D. Dunbar, M. Spencer, E. Balde, and M.P. Gracey. Health-related behaviors and psycho-social characteristics of 18-year old Australians. *Soc Sci Med* 45: 1549-62, 1997.

63. Morgan, W.P. Affective beneficence of vigorous physical activity. *Med Sci Sports Exerc* 17: 94-100, 1985.

64. Morgan, W.P. Psychogenic factors and exercise metabolism. A review. *Med Sci Sports Exerc* 17: 309-316, 1985.

65. Morgan, W.P., K. Hirota, G.A. Weintz, and B. Balke. Hypnotic perturbation of perceived exertion: ventilatory consequences. *Am J Clin Hypn* 18: 182-190, 1976.

66. Norris, R., D. Carroll, and R. Cochrane. The effects of physical activity and exercise training on psychological stress and well-being in an adolescent population. *J Psychosom Res* 36: 55-65, 1992.

67. North, T.C., P. McCullagh, and Z.V. Tran. Effect of exercise on clinical depression. *Exerc Sport Sci Rev* 18: 379-415, 1990.

68. Oler, M.J., A.G. Mainous, C.A. Martin, E. Richardson, A. Haney, D. Wilson, and T. Adams. Depression, suicide ideation, and substance abuse among adolescents. Are athletes at less risk? *Arch Fam Med* 3: 781-785, 1994.

69. Oliver, J.N. The effects of physical conditioning exercises and activities on the mental characteristics of educationally subnormal boys. *Br J Educ Psych* 28: 155-165, 1958.

70. Paluska, S.A., and T.L. Schwenk. Physical and mental health. Current concepts. *Sports Med* 29: 167-180, 2000.

71. Pert, C.B., and D.L. Bowie. Behavioral manipulation of rats causes alterations in opiate receptor occupancy. In Usdin, E., W.E. Bunney, and N.S. Kline, eds., *Endorphins and mental health.* New York: Oxford University Press, 1979, 93-104.

72. Petruzzello, S.J., and D.M. Landers. State anxiety reduction and exercise: does hemispheric activation reflect such changes? *Med Sci Sports Exerc* 26: 1028-1035, 1994.

73. Pittetti, K., and B. Fernhall. Aerobic capacity as related to leg strength in youths with mental retardation. *Pediatr Exerc Sci* 9: 223-236, 1997.

74. Pitetti, K.H., A.L. Millar, and B. Fernhall. Reliability of a peak performance treadmill test for children and adolescents with and without mental retardation. *Adapted Phys Act Q* 17: 322-332, 2000.

75. Rarick, G.L., J.H. Widdop, and G.D. Broadhead. The physical fitness and motor performance of educable mentally-retarded children. *Except Child* 36: 504-519, 1970.

76. Ransford, C.P. The role for amines in the antidepressant effect of exercise: a review. *Med Sci Sports Exerc* 14: 1-10, 1982.

77. Rowland, T.W. The biological basis of physical activity. *Med Sci Sports Exerc* 30: 392-399, 1998.

78. Satterfield, J.H., D.P. Cantwell, and B.T. Satterfield. Pathophysiology of the hyperactive child syndrome. *Arch Gen Psychiatry* 31: 839-844, 1974.

79. Shephard, R.J. Curricular physical activity and academic performance. *Pediatr Exerc Sci* 9: 113-126, 1997.

80. Sonstroem, R.J. Exercise and self-esteem. *Exerc Sport Sci Rev* 12: 123-156, 1984.

81. Solomon, A.H., and R. Pangle. *The effects of a structured physical education program on physical, intellectual, and self-concept development of educable retarded boys.* Monograph No. 4. : Institute on Mental Retardation and Intellectual Development Behavioral Science, 1966.

82. Spence, J.C., P. Poon, and P. Dyck. The effect of physical activity on self-concept: a meta-analysis. *J Sport Exerc Psychol* 19 (Suppl): S109, 1997.

83. Steptoe, A., and N. Butler. Sports participation and emotional well-being in adolescents. *Lancet* 347: 1789-1792, 1996.

84. Taylor, C.B., J.F. Sallis, and R. Needle. The relationship of physical activity and exercise to mental health. *Pub Health Reports* 100: 195-202, 1985.

85. Teo-Koh, S.M., and J.A. McCubbin. Relationship between peak $\dot{V}O_2$ and 1-mile walk test performance of adolescents with mental retardation. *Pediatr Exerc Sci* 11: 144-157, 1999.

86. Thorlindssen, T., R. Vilhjalmsson, and G. Valgeirsson. Sport participation and perceived health status: a study of adolescents. *Soc Sci Med* 31: 551-556, 1990.

87. Tortolero, S.R., W.C. Taylor, and N.G. Murray. Physical activity, physical fitness and social, psychological and emotional health. In Armstrong, N. and W. van Mechelen, eds., *Paediatric exercise science and medicine*. Oxford: Oxford University Press, 2000, 273-294.

88. Tuckman, B.W., and J.S. Hinkle. An experimental study of the physical and psychological effects of aerobic exercise on school children. *Health Psych* 5: 197-207, 1986.

89. Weiss, M.R. Self-esteem and achievement in children's sport and physical activity. In D. Gould and M.R. Weiss, eds., *Advances in pediatric sports sciences*. Champaign, IL: Human Kinetics, 1987, 87-120.

90. Yoshizawa, S., T. Tadatoshi, and H. Honda. Aerobic work capacity of mentally retarded boys and girls in junior high school. *J Hum Ergol* 4: 15-26, 1975.

Appendix I

1. Andersen, K.L., V. Seliger, J. Rutenfranz, and R. Mocellin. Physical performance capacity of children in Norway. Part I. Population parameters in a rural inland community with regard to maximal aerobic power. *Eur J Appl Physiol* 33: 177-195, 1974.

2. Cumming, G.R. Exercise studies in clinical pediatric cardiology. In: Frontiers of Activity and Child Health, edited by H. Lavalée and R.J. Shephard. Quebec: Pélican, 1977, p. 17-45.

3. Cumming, G.R., D. Everatt, and L. Hastman. Bruce treadmill test in children: normal values in a clinic population. *Am J Cardiol* 4: 69-75, 1978.

4. Roche, P.D. The development of norms for run-walk tests for children aged 7 to 17. *CAHPER Journal* 46: 6-13, 1980.

5. Rutenfranz, J., I. Berndt, H. Frost, R. Mocellin, R. Singer, and W. Sbresny. Physical performance capacity determined as W170 in youth. In: Pediatric Work Physiology., edited by Bar-Or O. Natanya: Wingate Institute, 1973, p. 245-249.

6. Seliger, V. and Z. Bartunek. New values of various indices of physical fitness in the investigation of Czecioslovak population aged 12-55 years. *CSTV Praha* 1976.

7. Van Praagh, E. Development of anaerobic function during childhood and adolescence. *Pediatr Exerc Sci* 12: 150-173, 2000.

8. Wirth, A., E. Trager, K. Scheele, D. Meyes, K. Diehm, K. Reischle, and H. Weicker. Cardiopulmonary adjustment and metabolic response to maximal and submaximal physical exercise of boys and girls at different stages of maturity. *Eur J Appl Physiol* 29: 229-240, 1978.

Appendix II

1. Adams, F.H., E. Bengtsson, and H. Berven. Determination by means of a bicycle ergometer, of the physical working capacity of children. *Acta Paediatr. Sc. and Suppl.* 118: 120-122, 1959.

2. Andersen, K.L., V. Seliger, J. Rutenfranz, and R. Mocellin. Physical performance capacity of children in Norway. Part I. Population parameters in a rural inland community with regard to maximal aerobic power. *Eur J Appl Physiol* 33: 177-195, 1974.

3. Armstrong, N., J. Williams, J. Balding, P. Gentle, and B. Kirby. The peak oxygen uptake of British children with reference to age, sex and sexual maturity. *Eur J Appl Physiol* 62: 369-375, 1991.

4. Asmussen, E. Growth in muscular strength and power. In Rarick, L., ed., *Physical activity, human growth and development*. New York: Academic Press. 1973, 60-79.

5. Åstrand, P.O. *Experimental studies of physical working capacity in relation to sex and age*. Copenhagen: Munksgaard. 1952.

6. Åstrand, P.O., and I. Rhyming. A nomogram for calculation of aerobic capacity (physical fitness) from pulse rate during submaximal work. *J Appl Physiol* 7: 218-221, 1954.

7. Bailey, D.A., and R.L. Mirwald. A children's test of fitness. In Borms, J. and M. Hebbelinck, eds., *Pediatric Work Physiology*. Basel: Karger, 1978, 56-64.

8. Bangsbo, J. Is the O_2 deficit an accurate quantitative measure of the anaerobic energy production during intense exercise? *J Appl Physiol* 73: 1207-1208, 1992.

9. Bar-Or, O. A comparison of responses to exercise and lung functions of Israeli Arabic and Jewish 12 to 17 year-old boys. In Bar-Or, O., ed., *Pediatric work physiology*. Natanya: Wingate Institute, 1973, 59-68.

10. Bar-Or, O. Age-related changes in exercise perception. In Borg, G., ed., *Physical Work and Effort*. Oxford and New York: Pergamon Press. 1977, 255-266.

11. Bar-Or, O. Pathophysiologic factors which limit the exercise capacity of the sick child. *Med Sci Sports Exerc* 18: 276-282, 1986.

12. Bar-Or, O. The Wingate Anaerobic Test. An update on methodology, reliability and validity. *Sports Med* 4: 381-394, 1987.

13. Bar-Or, O. Noncardiopulmonary pediatric exercise tests. In Rowland, T.W., ed., *Pediatric laboratory exercise testing: Clinical guidelines*. Champaign, IL, Human Kinetics, 1993, 165-185.

14. Bar-Or, O. Anaerobic performance. In Docherty, D., ed., *Measurement techniques in pediatric exercise science*. Champaign, IL: Human Kinetics, 1996, 161-182.

15. Bar-Or, O. Role of exercise in the assessment and management of neuromuscular disease in children. *Med Sci Sports Exerc* 28: 421-427, 1996.

16. Bar-Or, O. ed. The child and adolescent athlete. *Encyclopedia of sports medicine*. Oxford, England: Blackwell, 1996.

17. Bar-Or, O. Exertional perception in children and adolescents with a disease or a physical disability: assessment and interpretation. *Int J Sport Psychol* 32: 127-136, 2001.

18. Bar-Or, O., R. Dotan, O. Inbar, A. Rothstein, J. Karlsson, and P. Tesch. Anaerobic capacity and muscle fiber type distribution in man. *Int J Sports Med* 1: 89-92, 1980.

19. Bar-Or, T., O. Bar-Or, H. Waters, A. Hirji, and S. Russell. Validity and social acceptability of the Polar Vantage XL for measuring heart rate in preschoolers. *Pediatr Exerc Sci* 8: 115-121, 1996.

20. Barber, G. Cardiovascular function. In Armstrong, N. and W. van Mechelen, eds., *Pediatric exercise science and medicine*. Oxford, Oxford Press, 2000, 57-64.

21. Bedu, M., N. Fellmann, H. Spielvogel, G. Falgairette, E. Van Praagh, and J. Coudert. Force-velocity and 30-s Wingate tests in boys at high and low altitude. *J Appl Physiol* 70: 1031-1037, 1991.

22. Bengtsson, E. The working capacity in normal children, evaluated by submaximal exercise on the bicycle ergometer and compared with adults. *Acta Med Scand* 154: 91-109, 1956.

23. Blimkie, C.J.R. Age- and sex-associated variation in strength during childhood: anthropometric, morphological, neurologic, biomechanical, endocrinologic, genetic and physical activity correlates. In Gisolfi, C.V. and D.R. Lamb, eds., *Perspectives in exercise science and sports medicine: youth exercise and sport*. Indianapolis: Benchmark Press, 1989, 99-161.

24. Blimkie, C.J.R., and D. Macauley. Muscle strength. In Armstrong, N. and W. Van Mechelen, eds., *Paediatric exercise science and medicine*. Oxford: Oxford Press, 2000, 23-36.

25. Borg, G. *Physical performance and perceived exertion*. Lund: Gleerup, 1962.

26. Borg, G. Psychophysical bases of perceived exertion. *Med Sci Sports Exercise* 14: 377-381, 1982.

27. Bruce, R.A., and J. R. McDonough. Stress testing in screening for cardiovascular disease. *Bull NY Acad Med* 45: 1288-1305, 1969.

28. Burnie, J. and D.A. Brodie. Isokinetic measurement in preadolescent males. *Int J Sports Med* 7: 205-209, 1986.

29. Cafarelli, E. Peripheral contributions to the perception of effort. *Med Sci Sports Exerc* 14: 382-389, 1982.

30. Calvert, R.E., Bar-Or, O., McGillis, L., and Suei, K. The total work during an isokinetic and Wingate endurance tests in circumpubescent boys. *Pediatr Exerc Sci* 5, 398. 1993.

31. Carlson, J.S., and G.A. Naughton. An examination of the anaerobic capacity of children using maximal accumulated oxygen deficit. *Pediatr Exerc Sci* 5: 60-71, 1993.

32. Carlson, J.S., and G.A. Naughton. Assessing accumulated oxygen debt in children. In Van Praagh, E., ed., *Pediatric anaerobic performance*. Champaign, IL: Human Kinetics, 1998, 119-136.

33. Carron, A.V., and D.A. Bailey. Strength development in boys from 10 through 16 years. *Monog Soc Res Child Develop* 39: 1-37, 1974.

34. Chia, M., N. Armstrong, and D. Childs. The assessment of children's anaerobic performance using modifications of the Wingate test. *Pediatr Exerc Sci* 9: 80-89, 1997.

35. Cieslak, T.J., G. Frost, and P. Klentrou. Effects of physical activity, body fat and salivary cortisol on mucosal immunity in children. *J Appl Physiol In press: 2003*.

36. Counil, F.P., C. Karila, A. Varray, S. Guillaumont, M. Voisin, and C. Préfaut. Anaerobic fitness in children with asthma: adaptation to maximal intermittent short exercise. *Pediatr Pulmonol* 31: 198-204, 2001.

37. Cumming, G.R. Hemodynamics of supine bicycle exercise in "normal" children. *Am Heart J* 93: 617-622, 1977.

38. Cumming, G.R., D. Everatt, and L. Hastman. Bruce treadmill test in children: normal values in a clinic population. *Am J Cardiol* 4: 69-75, 1978.

39. Cumming, G.R., and A. Hnativk. Establishment of normal values for exercise capacity in a hospital clinic. In Berg, K. and B.O. Eriksson, eds., *Children and exercise IX.* Baltimore: University Park Press 79-93, 1980.

40. Davies, C.T.M., C. Barnes, and S. Godfrey. Body composition and maximal exercise performance in children. *Hum. Biol.* 44: 195-214, 1972.

41. di Prampero, P.E. and P. Cerretelli. Maximal muscular power (aerobic and anaerobic) in African natives. *Erg* 12: 51-59, 1969.

42. Docherty, D., ed. *Measurement in pediatric exercise science.* Champaign, IL: Human Kinetics, 1998.

43. Docherty, D., and C.A. Gaul. Relationship of body size, physique, and composition to physical performance in young boys and girls. *Int J Sports Med* 12: 525-532, 1991.

44. Dotan, R., and O. Bar-Or. Climatic heat stress and performance in the Wingate anaerobic test. *Eur J Appl Physiol* 44: 237-243, 1980.

45. Driscoll, D.J., B.A. Staats, and K.C. Beck. Measurement of cardiac output in children during exercise. *Pediatr Exerc Sci* 1: 102-115, 1989.

46. Dunstheimer, D., H. Hebestreit, B. Staschen, H.M. Strassburg, and R. Jeschke. Bilateral deficit during short-term, high-intensity cycle ergometry in girls and boys. *Eur J Appl Physiol* 84: 557-561, 2001.

47. Eston, R.G., K.L. Lamb, A. Bain, A.M. Williams, and J.G. Williams. Validity of a perceived exertion scale for children: a pilot study. *Perceptual and Motor Skills* 78: 691-697, 1994.

48. Falk, B., Y. Weinstein, R. Dotan, D.A. Abramson, D. Mann-Segal, and J. R. Hoffman. A treadmill test of sprint running. *Scand J Med Sci Sports* 6: 259-264, 1996.

49. Ford, G.W., W.H. Kitchen, and L.W. Doyle. Muscular strength at 5 years of children with a birthweight under 1500 g. *Aust Paediatr J* 24: 295-296, 1988.

50. Frost, G., O. Bar-Or, J. Dowling, and C. White. Habituation of children to treadmill walking and running: Metabolic and kinematic criteria. *Pediatr Exerc Sci* 7: 162-175, 1995.

51. Gaul, C.A. Muscular strength and endurance. In Docherty, D., ed., *Measurement in pediatric exercise science.* Champaign, IL: Human Kinetics. 1996, 225-258.

52. Godfrey, S. *Exercise testing in children: applications in health and disease.* Philadelphia: W.B. Saunders, 1974.

53. Hanne, N. A step-test for 6 to 12 year old girls and boys. *Wingate Institute* 1971.

54. Hebestreit, H., D. Dunstheimer, B. Staschen, and H.M. Strassburg. Single-leg Wingate test in children: reliability and optimal braking force. *Med Sci Sports Exerc* 31: 1218-1225, 1999.

55. Hebestreit, H., K. Mimura, and O. Bar-Or. Recovery of muscle power after high-intensity short-term exercise: comparing boys and men. *J Appl Physiol* 74: 2875-2880, 1993.

56. Hermansen, L,. and J.I. Medbo. The relative significance of aerobic and anaerobic processes during maximal exercise of short duration. In Marconnet, P., J. Poortmans, and L. Hermansen, eds., *Physiological chemistry of training and detraining.* Basel: Karger, 1984.

57. Hermansen, L. and S. Oseid. Direct and indirect estimation of maximal oxygen uptake in prepubertal boys. *Acta Paediatr Scand Suppl* 217: 18-23, 1971.

58. Hoofwijk, M., V.B. Unnithan, and O. Bar-Or. Maximal treadmill walking test for children with cerebral palsy. *Pediatr Exerc Sci* 7: 305-313, 1995.

59. Hosking, G.P., U.S. Bhat, V. Dubowitz, and R.H.T. Edwards. Measurements of muscle strength and performance in children with normal and diseased muscle. *Arch Dis Child* 51: 957-963, 1976.

60. Hyde, S.A., C.M. Goddard, and O.M. Scott. The myometer: the development of a clinical tool. *Physiotherapy* 69: 424-427, 1983.

61. Inbar, O., and O. Bar-Or. The effects of intermittent warm-up on 7 - 9 year-old boys. *Eur J Appl Physiol* 34: 81-89, 1975.

62. Inbar, O., and O. Bar-Or. Anaerobic characteristics in male children and adolescents. *Med Sci Sports Exerc* 18: 264-266, 1986.

63. Inbar, O., O. Bar-Or, and J.S. Skinner. *The Wingate anaerobic test: development and application.* Champaign, IL: Human Kinetics, 1996.

64. Jacobs, I. The effects of thermal dehydration on performance of the Wingate anaerobic test. *Int J Sports Med* 1: 21-24, 1980.

65. Jacobs, I., O. Bar-Or, J. Karlsson, R. Dotan, P. Tesch, P. Kaiser, and O. Inbar. Changes in muscle metabolites in females with 30-s exhaustive exercise. *Med Sci Sports Exerc* 14: 457-460, 1982.

66. Jacobs, I., P.A. Tesch, O. Bar-Or, J. Karlsson, and R. Dotan. Lactate in human skeletal muscle after 10 and 30 s of supramaximal exercise. *J Appl Physio: Resp Environ Ex Physiol* 55: 365-367, 1983.

67. James, F.W. Exercise testing in children and young adults: an overview. *Cardiovasc Clin* 9: 187-203, 1978.

68. Johnson, T.J., L. Brouha, and J.R. Gallagher. Dynamic physical fitness in adolescence: VI. The use of the step test in the evaluation of the fitness in adolescents. *Yale J Biol & Med* 15: 781-785, 1943.

69. Killian, K.J. Dyspnea and leg effort during incremental cycle ergometry. *Am Rev Respir Dis* 145: 1339-1345, 1992.

70. Klentrou, P., J. Hay, and M. Plyley. Habitual physical activity levels and health outcomes of Ontario youth. *Eur J Appl Physiol* 89: 460-465, 2003.

71. Klimt, F., and G.B. Voigt. Investigations on the standardization of ergometry in children. *Acta Pediatr* 217: 35-36, 1971.

72. Krahenbuhl, G.S., R.P. Pangrazi, G.W. Petersen, L.N. Burkett, and H.J. Schneider. Field testing of cardiorespiratory fitness in primary school children. *Med Sci Sports* 10: 208-213, 1978.

73. Lakomy, H.K.A., and S. Wootoon. Discrimination of rapid changes in pedal frequency. *J Physiol* 316: 316, 1984.

74. Lamb, K.L. Children's ratings of effort during cycle ergometry: an examination of the validity of two effort rating scales. *Pediatr Exerc Sci* 7: 407-421, 1995.

75. Lamb, K.L. Exercise regulation during cycle ergometry using the CERT and RPE scales. *Pediatr Exerc Sci* 8: 337-350, 1996.

76. Léger, L., and M. Thivierge. Heart rate monitors: validity, stability, and functionality. *Phys Sportsmed* 143-151, 1988.

77. Lock, J.E., S. Einzig, and J.H. Moller. Hemodynamic responses to exercise in normal children. *Am J Cardiol* 41: 1278-1284, 1978.

78. Lowry, A. The development of a pictorial scale to assess perceived exertion in school-children. Thesis/dissertation. 1995. Chester, England: University College.

79. MacDougall, J.D., P.D. Roche, O. Bar-Or, and J.R. Moroz. Maximal aerobic capacity of Canadian schoolchildren: prediction based on age-related oxygen cost of running. *Int J Sports Med* 4: 194-198, 1983.

80. Mahon, A.D. Assessment of perceived exertion during exercise in children. In Welsman, J., N. Armstrong, and B. Kirby, eds., *Children and exercise XIX volume II*. Exeter: Washington Singer Press, 1997, 25-39.

81. Mahon, A.D. and C.C. Cheatham. Ventilatory threshold: a review. *Pediatr Exerc Sci* 14: 16-29, 2002.

82. Mahon, A.D., Gay, J.A., and Kelsey, K.Q. Differential ratings of perceived exertion at ventilatory threshold in boys and men. *Med Sci Spots Exerc* 27, S115. 1995.

83. Malina, R.M., C. Bouchard, and O. Bar-Or. *Growth, maturation, and physical activity*. Champaign, IL: Human Kinetics, 2004.

84. Maltais, D., I. Kondo, and O. Bar-Or. Arm cranking economy in spastic cerebral palsy: effects of different speed and force combinations yielding the same mechanical power. *Pediatr Exerc Sci* 12: 258-269, 2000.

85. Margaria, R., P. Aghemo, and E. Rovelli. Measurement of muscular power (anaerobic) in man. *J Appl Physiol* 21: 1662-1664, 1966.

86. Medbo, J.I., A. Mohn, I. Tabata, R. Bahr, O. Vaage, and O.M. Sejersted. Anaerobic capacity determined by maximal accumulated O_2 deficit. *J Appl Physiol* 64: 50-60, 1988.

87. Mercier, B., J. Mercier, P. Granier, and D.L. Gallais. Maximal anaerobic power: relationship to anthropometric characteristics during growth. *Int J Sports Med* 13: 21-26, 1992.

88. Mocellin, R., H. Lindemann, J. Rutenfranz, and W. Sbrtesny. Determination of W_{170} and maximal oxygen uptake in children by different methods. *Acta Pediatr* 217: 13-17, 1971.

89. Mrzena, B., and M. Máček. Use of treadmill and working capacity assessment in pre-school children. In Borms, J. and H. Hebbelinck, eds., *Pediatric work physiology*. Basel: Karger. 1978, 29-31.

90. Noble, B., and R. Robertson. *Perceived exertion*. Champaign, IL, Human Kinetics. 1996.

91. Nystad, W., S. Oseid, and E.B. Mellbye. Physical education for asthmatic children: the relationship between changes in heart rate, perceived exertion, and motivation for participation. In Oseid, S. and K. Carlsen, eds., *Children and exercise XIII*. Champaign, IL: Human Kinetics, 1989, 359-377.

92. Ohuchi, H., T. Nakajima, M. Kawade, M. Matsuda, and T. Kamiya. Measurement and validity of the ventilatory threshold in patients with congenital heart disease. *Pediatr Cardiol* 17: 7-14, 1996.

93. Parker, D.F., L. Carriere, H. Hebestreit, and O. Bar-Or. Anaerobic endurance and peak muscle power in children with spastic cerebral palsy. *Am J Dis Child* 146: 1069-1073, 1992.

94. Petajoki, M.L., M. Arstila, and I. Välimäki. Pulse-conducted exercise test in children. *Acta Paediatr Belg* 40: 47, 1974.

95. Riopel, D.A., A.B. Taylor, and A.R. Hohn. Blood pressure, heart rate, pressure rate product and electrocardiographic changes in healthy children during treadmill exercise. *Am J Cardiol* 44: 697-704, 1979.

96. Rivera-Brown, A.M., M.A. Rivera, and W.R. Frontera. Reliability of VO_2 max in adolescent runners: a comparison between plateau achievers and non-achievers. *Pediatr Exerc Sci* 7: 203-210, 1995.

97. Robertson, R.J., F.L. Goss, J.A. Bell, C.B. Dixon, K.I. Gallagher, K.M. Lagally, J.M. Timmer, K.L. Abt, J.D. Gallagher, and T. Thompkins. Self-regulated cycling using the Children's OMNI Scale of Perceived Exertion. *Med Sci Sports Exerc* 34: 1168-1175, 2002.

98. Robertson, R.J., F.L. Goss, N. Boer, J.D. Gallagher, T. Thompkins, K. Bufalino, G. Balasekaran, C. Meckes, J. Pintar, and A. Williams. OMNI Scale of Perceived Exertion at ventilatory breakpoint in children: response normalized. *Med Sci Sports Exerc* 33: 1946-1952, 2001.

99. Robertson, R.J., F.L. Goss, N.F. Boer, J.A. Peoples, A.J. Foreman, I.M. Dabayebeh, N.B. Millich, G. Balasekaran, S.E. Riechman, J.D. Gallagher, and T. Thompkins. Children's OMNI scale of perceived exertion: mixed gender and race validation. *Med Sci Sports Exerc* 32: 452-458, 2000.

99a. Robertson, R.J., F.L. Goss, J. Rutkowski, B. Lenz, C. Dixon, J. Timmer, K. Franzee, J. Dube, and J. Andreacci. Concurrent validation of the OMNI perceived exertion scale for resistance exercise. *Med Sci Sports Exerc* 35: 333-341, 2003.

100. Robertson, R.J., and B.J. Noble. Perception of physical exertion: methods, mediators and applications. *Exerc Sports Sci Rev.* 25: 407-452, 1997.

101. Rowland, T., and P. Obert. Doppler echocardiography for the estimation of cardiac output with exercise. *Sports Med* 32: 973-986, 2002.

102. Rowland, T.W. *Developmental exercise physiology.* Champaign, IL: Human Kinetics, 1996, 1-268.

103. Rowland, T.W., J.A. Auchinachie, T.J. Keenan, and G.M. Green. Physiologic responses to treadmill running in adult and prepubertal males. *Int J Sports Med* 8: 292-297, 1987.

104. Rowland, T.W. ed. *Pediatric laboratory exercise testing: clinical guidelines.* Champaign, IL: Human Kinetics. 1993.

105. Rutenfranz, J. *Entwicklung und Beurteilung der korperlichen Leistungsfahigkeit bei Kindern und Jugendlichen.* Basel: Karger. 1964.

106. Rutenfranz, J. Exercise tests in children and adolescents. In Andersen, K.L., R.J. Shephard, H. Denolin, E. Vernauskas, and R. Masironi, eds., *Fundamentals of exercise testing.* Geneva: World Health Organization. 1971, 105-109.

107. Sale, D.G. Testing strength and power. In MacDougall, J.D., H.A. Wenger, and H.J. Green, eds., *Physiological testing of the high-performance athlete.* Champaign, IL: Human Kinetics, 1991, 21-106.

108. Sargeant, A.J., E. Hoinville, and A. Young. Maximum leg force and power output during short-term dynamic exercise. *J Appl Physiol: Respirat Environ Exercise Physiol* 51: 1175: 1182, 1981.

109. Scheet, T.P., P.J. Mills, M.G. Ziegler, J. Stoppani, and D.M. Cooper. Effect of exercise on cytokine and growth mediators in prepubertal boys. *Pediatr Res* 46: 429-434, 1999.

110. Shephard, R.J., C. Allen, O. Bar-Or, C. T. M. Davies, S. Degre, R. Hedman, K. Ishii, M. Kaneko, J.R. Lacour, P. E. di Prampero, and V. Seliger. The Working Capacity of Toronto Schoolchildren. Part I, II. *Can Med Assoc J* 100: 560-567, 705-714, 1969.

111. Skinner, J.S., O. Bar-Or, V. Bergsteinová, C.W. Bell, D. Royer, and E.R. Buskirk. Comparison of continuous and intermittent tests for determining maximal oxygen uptake in children. *Acta Paediatr Scand Suppl* 217: 24-28, 1971.

112. Stewart, K.J., and B. Guti. The prediction of maximal oxygen uptake before and after physical training in children. *J Human Ergol* 4: 153-162, 1975.

113. Sunnegardh, J., L.-E. Bratteby, L.-O. Nordesjo, and B. Nordgren. Isometric and isokinetic muscle strength, anthropometry and physical activity in 8 and 13 year old Swedish children. *Eur J Appl Physiol* 58: 291-297, 1988.

114. Sutton, N.C., D.J. Childs, O. Bar-Or, and N. Armstrong. A nonmotorized treadmill test to assess children's short-term power output. *Pediatr Exerc Sci* 12: 91-100, 2000.

115. Task Force on Blood Pressure Control in Children. Methodology and instrumentation for blood pressure measurement in infants and children. *Pediatrics* 59: 800-801, 1977.

116. Timmons, B.W. and Bar-Or, O. Immune response to strenuous exercise and carbohydrate intake in boys and men. *Pediatr Res* In press.

117. Tirosh, E., O. Bar-Or, and P. Rosenbaum. New muscle power test in neuromuscular disease. Feasibility and reliability. *Am J Dis Child* 144: 1083-1087, 1990.

118. Unnithan, V.B., C. Clifford, and O. Bar-Or. Evaluation by exercise testing of the child with cerebral palsy. *Sports Med* 26: 239-251, 1998.

119. Utter, A.C., R.J. Robertson, D.C. Nieman, and J. Kang. Children's OMNI Scale of Perceived Exertion: walking/running evaluation. *Med Sci Sports Exerc* 34: 139-144, 2002.

120. Välimäki, I., H.L. Petajoki, M. Arstila, P. Vihera, and H. Wendelin. Automatically controlled ergometer for pulse-conducted exercise test. In Borms, J. and M. Hebbelinck, eds., *Pediatric work physiology.* Basel: Karger. 1978, 47-57.

121. Van Mil, E., N. Schoeber, R.E. Calvert, and O. Bar-Or. Optimization of force in the Wingate test for children with a neuromuscular disease. *Med Sci Sports Exerc* 28: 1087-1092, 1996.

122. Van Praagh, E., G. Falgairette, M. Bedu, N. Fellmann, and J. Coudert. Laboratory and field tests in 7-year-old boys. In Oseid, S. and H.K. Carlsen, eds., *Children and exercise XIII.* Champaign, IL: Human Kinetics, 1989, 11-17.

123. Van Praagh, E. ed. *Pediatric anaerobic performance.* Champaign, IL: Human Kinetics. 1998, 1-375.

124. Vincent, S.D. and R.P. Pangrazi. Does reactivity exist in children when measuring activity level with pedometers? *Pediatr Exerc Sci* 14: 56-63, 2002.

125. Wahlund, H. Determination of the physical work capacity. *Acta Med Scand Suppl* 215: 1948.

126. Ward, D.S. and O. Bar-Or. Use of the Borg scale in exercise prescription for overweight youth. *Canad J Sport Sci* 15: 120-125, 1990.

127. Ward, D.S., O. Bar-Or, P. Longmuir, and K. Smith. Use of rating of perceived exertion (RPE) to prescribe exercise intensity for wheelchair-bound children and adults. *Pediatr Exerc Sci* 7: 94-102, 1995.

128. Ward, D.S., J.D. Jackman, and F.D. Galiano. Exercise intensity reproduction: children versus adults. *Pediatr Exerc Sci* 3: 209-218, 1991.

129. Washington, R.L. Anaerobic threshold. In Rowland, T.W., ed., *Pediatric laboratory exercise testing: clinical guidelines*. Champaign, IL: Human Kinetics, 1993, 115-129.

130. Washington, R.L., J.T. Bricker, B.S. Alpert, S.R. Daniels, R.J. Deckelbaum, E.A. Fisher, S.S. Gidding, J. Isabel-Jones, R.E.W. Kavey, G.R. Marx, W.B. Strong, D.W. Teske, J.H. Wilmore, and M. Winston. Guidelines for exercise testing in the pediatric age group. *Circulation* 90: 2166-2179, 1994.

131. Waynarowska, B. The validity of indirect estimation of maximal oxygen uptake in children 11-12 years of age. *Eur J Appl Physiol* 43: 19-23, 1980.

132. Wessel, J., C. Kaup, J. Fan, R. Ehalt, J. Ellsworth, C. Speer, P. Tenove, and A. Dombrosky. Isometric strength measurements in children with arthritis: reliability and relation to function. *Arthritis Care Res* 12: 238-246, 1999.

133. Williams, J.G., R.G. Eston, and C. Stretch. Use of rating of perceived exertion to control exercise intensity in children. *Pediatr Exerc Sci* 3: 21-27, 1991.

134. Williams, J.G., Furlong, B., MacKintosh, C., and Hockley, T. Rating and regulation of exercise intensity in young children. *Medi & Sci Sports Exer* 25, S8. 1993.

Appendix III

1. Allor KM and Pivarnik JM. Use of heart rate cutpoints to assess physical activity intensity in sixth-grade girls. *Pediatr Exerc Sci* 12: 284-292, 2000.

2. Armstrong N. Young people's physical activity patterns as assessed by heart rate monitoring. *J Sports Sci* in press: 1998.

3. Armstrong N, Balding J, Gentle P and Kirby B. Patterns of physical activity among 11 to 16 year old British children. *Br Med J* 301: 203-205, 1990.

4. Atkins S, Stratton G, Dugdill L and Reilly T. The free-living physical activity of schoolchildren: a longitudinal study. In: Children and Exercise XIX, edited by Armstrong N, Kirby B and Welsman J. London: E. & F.N.Spon, 1997, p. 145-150.

5. Bailey RC, Olson J, Pepper SL, Porszasz J, Barstow TJ and Cooper DM. The level and tempo of children's physical activities: an observational study. *Med Sci Sports Exerc* 27: 1033-1041, 1995.

6. Ballor DL, Burke LM, Knudson DV, Olson JR and Montoye HJ. Comparison of three methods of estimating energy expenditure: Caltrac, heart rate, and video analysis. *Res Quart* 60: 362-368, 1989.

7. Bar-Or T, Bar-Or O, Waters H, Hirji A and Russell S. Validity and social acceptability of the Polar Vantage XL for measuring heart rate in preschoolers. *Pediatr Exerc Sci* 8: 115-121, 1996.

8. Bitar A, Vermorel M, Fellmann N, Bedu M, Chamoux A and Coudert J. Heart rate recording method validated by whole body indirect calorimetry in 10-yr-old children. *Journal of Applied Physiology* 81: 1169-1173, 1996.

9. Blaak EE, Westerterp KR, Bar-Or O, Wouters LJ and Saris WH. Total energy expenditure and spontaneous activity in relation to training in obese boys. *Am J Clin Nutr* 55: 777-782, 1992.

10. Bouchard C, Tremblay A, Leblanc C, Lortie G, Savard R and Thériault GA. A method to assess energy expenditure in children and adults. *Am J Clin Nutr* 37: 461-467, 1983.

11. Bray MS, Morrow JR, Pivarnik JM and Bricker JT. Caltrac Validity for estimating caloric expenditure with children. *Pediatr Exerc Sci* 4: 166-179, 1992.

12. Bray MS, Wong WW, J.R. Morrow, Jr., Butte NF and Pivarnik JM. Caltrac versus calorimeter determination of 24-h energy expenditure in female children and adolescents. *Med Sci Sports Exerc* 26: 1524-1530, 1994.

13. Bullen BA, Reed RB and Mayer J. Physical activity of obese and nonobese adolescent girls appraised by motion picture sampling. *Am J Clin Nutr* 14: 211-223, 1964.

14. Craig SB, Bandini LG, Lichtenstein AH, Schaefer EJ and Dietz WH. The impact of physical activity on lipids, lipoproteins, and blood pressure in preadolescent girls. *Pediatrics* 98: 389-395, 1996.

15. Davies PSW, Day JME and Lucas A. Energy expenditure in early infancy and later body fatness. *Int J Obesity* 15: 727-731, 1991.

16. Eisenmann JC, Katzmarzyk PT, Thériault G, Song TMK, Malina RM and Bouchard C. Physical activity and pulmonary function in youth: The Quebec Family Study. *Pediatr Exerc Sci* 11: 208-217, 1999.

17. Ellis MJ and Scholtz GJL. *Activity and Play of Children*. Englewood Cliffs, N.J.: Prentice Hall Inc., 1978.

18. Emons HJG, Groenenboom DC, Westerterp KR and Saris WHM. Comparison of heart rate monitoring combined with indirect calorimetry and the doubly labeled water ($^2H_2{}^{18}O$) method for the measurement of energy expenditure in children. *Eur J Appl Physiol* 65: 99-103, 1992.

19. Epstein LH, McGowan C and Woodall K. A behavioral observation system for free play activity in young overweight female children. *Res Quart Exercise & Sport* 55: 180-183, 1984.

20. Eston RG, Rowlands AV and Ingledew DK. Validation of the Tritrac-R3D activity monitor during typical children's activities. In: Children and Exercise XIX, edited by Armstrong N, Kirby B and Welsman J. London: E. & F.N.Spon, 1997, p. 132-138.

21. Eston RG, Rowlands AV and Ingledew DK. Validity of heart rate, pedometry, and accelerometry for predicting the energy cost of children's activities. *J Appl Physiol* 84: 362-371, 1998.

22. Fairweather SC, Reilly JJ, Grant S, Whittaker A and Paton JY. Using the Computer Science and Applications (CSA) activity monitor in preschool children. *Pediatr Exerc Sci* 11: 413-420, 1999.

23. Freedson PS. Electronic motion sensors and heart rate as measures of physical activity in chidlren. *J Sch Health* 61: 220-223, 1991.

24. Freedson PS, Sirard J, Debold N, Pate R, Dowda M and Sallis J. Validity of two physical activity monitors in children and adolescents. In: Children and Exercise XIX, edited by Armstrong N, Kirby B and Welsman J. London: E. & F.N.Spon, 1997, p. 127-131.

25. Garcia AW, Pender NJ, Antonakos CL and Ronis DL. Changes in physical activity beliefs and behaviors of boys and girls across the transition to junior high school. *J Adolesc Health* 22: 394-402, 1998.

26. Gilbey H and Gilbey M. The Physical Activity of Singapore Primary School Children as Estimated by Heart Rate Monitoring. *Pediatr Exerc Sci* 7: 26-35, 1995.

27. Gilliam TB, Freedson PS, Geenen DL and Shahraray B. Physical activity patterns determined by heart rate monitoring in 6- to 7-year-old children. *Med Sci Sports Exerc* 13: 65-67, 1981.

28. Goran MI, Hunter G, Nagy TR and Johnson R. Physical activity related energy expenditure and fat mass in young children. *Int J Obesity* 21: 171-178, 1997.

29. Goran MI and Sun M. Total energy expenditure and physical activity in prepubertal children: recent advances based on the application of the doubly labeled water method. *Am J Clin Nutr* 68: 944S-949S, 1998.

30. Halton J, Atkinson S, Fraher L, Webber C, Cockshutt WP, Tam C and Barr R. Mineral homeostasis and bone mass at diagnosis in children with acute lymphoblastic leukemia. *J Pediatr* 126: 557-564, 1995.

31. Hay JA. Adequacy in and predilection for physical activity in children. *Clin J Sport Med* 2: 192-201, 1992.

32. Hay, J. A. Development and testing of the habitual physical activity estimation scale. Welsman, J., Armstrong, N., and Kirby, B. 125-129. 1997. Exeter, Washington Singer Press. Children and Exercise XIX, Volume II.

33. Hebestreit H and Bar-Or O. Influence of climate on heart rate in children: comparison between intermittent and continuous exercise. *Eur J Appl Physiol Occup Physiol* 78: 7-12, 1998.

34. Hebestreit H, Bar-Or O, McKinty C, Riddell M and Zehr P. Climate-related corrections for improved estimation of energy expenditure from heart rate in children. *J Appl Physiol* 79: 47-54, 1995.

35. Jakicic JM, Winters C, Lagally K, Ho J, Robertson RJ and Wing RR. The accuracy of the TriTrac-R3D accelerometer to estimate energy expenditure. *Med Sci Sports & Exerc* 31: 747-754, 1999.

36. Janz KF. Validation of the CSA accelerometer for assessing children's physical activity. *Med Sci Sports Exerc* 26: 369-375, 1994.

37. Janz KF, Golden JC, Hansen JR and Mahoney LT. Heart rate monitoring of physical activity in children and adolescents: The Muscatine Study. *Pediatr* 89: 256-261, 1992.

38. Johnson RK, Russ J and Goran MI. Physical activity related energy expenditure in children by doubly labeled water as compared with the Caltrac accelerometer. *Int J Obes Relat Metab Disord* 22: 1046-1052, 1998.

39. Jones PJ, Winthrop AL, Schoeller DA, Swyer PR, Smith J, Filler RM and Heim T. Validation of doubly labeled water for assessing energy expenditure in infants. *Pediatr Res* 21: 242-246, 1987.

40. Kilanowski CK, Consalvi AR and Epstein LH. Validation of an electronic pedometer for measurement of physical activity in children. *Pediatr Exerc Sci* 11: 63-68, 1999.

41. Klesges LM and Klesges RC. The assessment of children's physical activity: a comparison of methods. *Med Sci Sports Exerc* 19: 511-517, 1987.

42. Klesges RC, Coates TJ, Klesges LM, Holzer B, Gustavson J and Barnes J. The FATS: an observational system for assessing physical activity in children and associated parent behavior. *Behav Assess* 6: 333-345, 1984.

43. Klesges RC, Klesges LM, Swenson AM and Pheley AM. A validation of two motion sensors in the prediction of child and adult physical activity levels. *Am J Epidemiol* 122: 400-410, 1985.

44. Leonard WR, Katzmarzyk PT, Stephen MA and Ross AGP. Comparison of the heart rate-monitoring and factorial methods: assessment of energy expenditure in highland and coastal Ecuadoreans. *Am J Clin Nutr* 61: 1146-1152, 1995.

45. Livingstone MBE, Coward WA, Prentice AM, Davies PSW, Strain JA, McKenna PG, Mahoney CA, White JA, Stewart CM and Kerr MJ. Daily energy expenditure in free-living children: comparison of heart-rate monitoring with the doubly labeled water ($^2H_2^{18}O$) method 1-3. *Am J Clin Nutr* 56: 343-352, 1992.

46. Longmuir PE and Bar-Or O. Physical activity of children and adolescents with a disability: methodology and effects of age and gender. *Pediatr Exerc Sci* 6: 168-177, 1994.

47. Louie L, Eston RG, Rowlands AV, Tong KK, Ingledew DK and Fu FH. Validity of heart rate, pedometry, and accelerometry for estimating energy cost of activity in Hog Kong Chinese boys. *Pediatr Exerc Sci* 11: 229-239, 1999.

48. Manios Y, Kafatos A and Marakis G. Physical activity of 6-year-old children: validation of two proxy reports. *Pediatr Exerc Sci* 10: 176-188, 2002.

49. Matkin CC, Bachrach L, Wang MC and Kelsey J. Two measures of physical activity as predictors of bone mass in young cohort. *Clin J Sports Med* 8: 201-208, 1998.

50. McKenzie T, Sallis JF and Nader PR. SOFT: system for observing fitness instructions time. *J Teaching Phys Ed* 62: 195-205, 1992.

51. McKenzie TL, Sallis JF, Nader PR, Patterson TL, Elder JP, Berry CC, Rupp JW, Atkins CJ, Buono MJ and Nelson JA. BEACHES: an observational system for assessimg children's eating and physical activity behaviors associated events. *J Appl Behav Anal* 24: 141-151, 1991.

52. Metcalf BS, Curnow JS, Evans C, Voss LD and Wilkin TJ. Technical reliability of the CSA activity monitor: The EarlyBird Study. *Med Sci Sports Exerc* 34: 1533-1537, 2002.

53. Mimura K, Hebestreit H and Bar-Or O. Activity and heart rate in preschool children of low and high motor ability: 24-hour profiles. *Med Sci Sports Exerc* 23: S12, 1991.

54. Mukeshi.M., Gutin B, Anderson W, Zybert P and Basch C. Validation of the Caltrac movement sensor using direct observation in young children. *Pediatr Exerc Sci* 2: 249-254, 1990.

55. Nichols JF, Morgan CG, Sarkin JA, Sallis JF and Calfas KJ. Validity, reliability, and calibration of the Tritrac accelerometer as a measure of physical activity. *Med Sci Sports Exerc* 31: 908-912, 1999.

56. Noland M, Danner F, DeWalt K, McFadden M and Kotchen JM. The measurement of physical activity in young children. *Res Q Exerc Sport* 61: 146-153, 1990.

57. Ott AE, Pate RR, Trost SG, Ward DS and Saunders R. The use of uniaxial and triaxial accelerometers to measure children's free-play physical activity. *Pediatr Exerc Sci* 14: 2002.

58. Puhl J, Greaves K, Hoyt M and Baranowski T. Children's Activity Rating Scale (CARS): description and calibration. *Res Q Exerc Sport* 61: 26-36, 1990.

59. Rowe PJ, Schuldheisz JM and Van der Mars H. Validation of SOFT for measuring physical activity of first- to eighth-grade students. *Pediatr exer sci* 9: 136-149, 1997.

60. Sallis JF. Self-report measures of children's physical activity. *J Sch Health* 61: 215-219, 1991.

61. Sallis JF, Buono MJ and Freedson PS. Bias in estimating caloric expenditure from physical activity in children: Implications for epidemiological studies. *Sports Med* 11: 203-209, 1991.

62. Sallis JF, Buono MJ, Roby JJ, Carlson D and Nelson JA. The Caltrac accelerometer as a physical activity monitor for school-age children. *Med Sci Sports Exerc* 22: 698-703, 1990.

63. Sallis JF, Buono MJ, Roby JJ, Micale FG and Nelson JA. Seven-day recall and other physical activity self-reports in children and adolescents. *Med Sci Sports Exerc* 25: 99-108, 1993.

64. Saris WH. Habitual physical activity in children: methodology and findings in health and disease. *Med Sci Sports Exerc* 18: 253-263, 1986.

65. Simons-Morton, B. G. and Huang, I. W. Heart rate monitor and Caltrac assessment of moderate-to-vigorous physical activity among preadolescent children. Med Sci Sports Exerc 23, S60. 1997.

66. Sirard JR and Pate RR. Physical activity assessment in children and adolescents. *Sports Med* 31: 439-454, 2001.

67. Spurr GB, Prentice AM, Murgatroyd PR, Goldberg GR, Reina JC and Christman NT. Energy expenditure from minute-by-minute heart-rate recording: comparison with indirect calorimetry. *Am J Clin Nutr* 48: 552-559, 1988.

68. Spurr GB, Prentice AM, Murgatroyd PR, Goldberg GR, Reina JC and Christman NT. Energy expenditure from minute-by-minute heart-rate recording: comparison with indirect calorimetry. *Am J Clin Nutr* 48: 552-559, 1988.

69. Torun B. Physiological Measurements of Physical Activity Among Children Under Free-Living Conditions. In *Energy Intake and Activity,* edited by Pollitt E and Amante P. New York: Alan R. Liss Inc., 1984, p. 159-184.

70. Tremblay MS, Inman JW and Willms JD. Preliminary evaluation of video questionnaire to assess activity levels of children. *Med Sci Sports Exerc* 33: 2139-2144, 2001.

71. Treuth MS, Adolph AL and Butte NF. Energy expenditure in children predicted from heart rate and activity calibrated against respiration calorimetry. *Am J Physiol* 275: E12-E18, 1998.

72. Trost SG. Objective measurement of physical activity in youth: current issues, future directions. *Exerc Sport Sci Rev* 29: 32-36, 2001.

73. Troutman SR, Allor KM, Hartmann DC and Pivarnik JM. MINI-LOGGER--reliability and validity for estimating energy expenditure and heart rate in adolescents. *Res Q Exerc Sport* 70: 70-74, 1999.

74. Verschurr R and Kemper HCG. The pattern of daily physical actitivy. In: Growth, Health anf Fitness of Teenagers, edited by Kemper HCG. Basel: 1985, p. 169-186.

75. Vincent SD and Pangrazi RP. Does reactivity exist in children when measuring activity level with pedometers? *Pediatr Exerc Sci* 14: 56-63, 2002.

76. Welk GJ and Corbin CB. The validity of the Tritrac-R3D Activity Monitor for the assessment of physical activity in children. *Res Q Exerc Sport* 66: 202-209, 1995.

77. Welk GJ, Corbin CB and Dale D. Measurement issues in the assessment of physical activity in children. *Res Q Exerc Sport* 71: S59-S73, 2000.

Appendix V

1. Armstrong, N., B.J. Kirby, A.M. McManus, and J.R. Welsman. Aerobic fitness of prepubescent children. *Ann Hum Biol* 22: 427-441, 1995.

2. Armstrong, N., and J. Welsman. Assessment and interpretation of aerobic fitness in children and adolescents. *Exerc Sport Sci Rev* 22: 435-476, 1994.

3. Armstrong, N., and J. Welsman. *Young people and physical activity.* Oxford: Oxford University Press, 1997, 1-369.

4. Armstrong, N., J. Welsman, and B.J. Kirby. Peak oxygen uptake and maturation in 12-year-old boys. *Med Sci Sports Exerc* 30: 165-169, 1998.

5. Asmussen, E. Growth in muscular strength and power. In Rarick, L., ed., *Physical activity, human growth and development.* New York, Academic Press, 1973, 60-79.

6. Asmussen, E., and K.R. Heeboll-Nielsen. A dimensional analysis of physical performance and growth in boys. *J Appl Physiol* 7: 593-603, 1955.

7. Åstrand, P.O. *Experimental studies of physical working capacity in relation to sex and age.* Copenhagen: Munksgaard, 1952.

8. Bailey, D.A., W.D. Ross, R.L. Mirwald, and C. Wesse. Size dissociation of maximal aerobic power during growth in boys. In Borms, J. and M. Hebbelinck, eds., *Pediatric work physiology.* Basel: Karger, 1978, 140-151.

9. Bar-Or, O., L.D. Zwiren, and H. Ruskin. Anthropometric and developmental measurements of 11-to 12-year-old boys, as predictors of performance 2 years later. *Acta Paediatr Belg* 28: 214-220, 1974.

10. Blimkie, C.J.R. Age- and sex-associated variation in strength during childhood: anthropometric, morphological, neurologic, biomechanical, endocrinologic, genetic and physical activity correlates. In Gisolfi, C.V. and D.R. Lamb, eds., *Perspectives in exercise science and sports medicine.* Vol. 2. *Youth, exercise, and sport.* Indianapolis: Benchmark Press, 1989, 99-161.

11. Chia, M., N. Armstrong, and D. Childs. The assessment of children's anaerobic performance using modifications of the Wingate test. *Pediatr Exerc Sci* 9: 80-89, 1997.

12. Cooper, D.M., and T.J. Barstow. Magnetic resonance imaging and spectroscopy in studying exercise in children. *Exerc Sport Sci Rev* 24: 475-499, 1996.

13. Cooper, D.M., and N. Berman. Ratios and regressions in body size and function: a commentary. *J Appl Physiol* 77: 2015-2017, 1994.

14. Davies, C.T.M., C. Barnes, and S. Godfrey. Body composition and maximal exercise performance in children. *Hum Biol* 44: 195-214, 1972.

15. Eriksson, B.O., and C. Thorén. Training girls for swimming from medical and physiological points of view, with special reference to growth. In Eriksson, B. and B. Furberg, eds., *Swimming medicine IV.* Baltimore: University Park Press, 1978, 3-15.

16. Goldstein, H. Efficient statistical modelling of longitudinal data. *Ann Hum Biol* 13: 129-141, 1996.

17. Gray, B.F. On the surface law and basal metabolic rate. *J Theor Biol* 93: 757-767, 1981.

18. Holliday, M.A., D. Potter, A. Jarrah, and S. Bearg. The relation of metabolic rate to body weight and organ size. *Pediatr Res* 1: 185-195, 167.

19. Katch, V.L. Use of oxygen/body weight ratio in correlational analyses: spurious correlations and statistics? *Med Sci Sports Exerc* 5: 252-257, 1973.

20. Kleiber, M. Body size and metabolism. *Hilgardia* 1932.

21. McMiken, D.F. Maximum aerobic power and physical dimensions of children. *Annals Hum Biol* 3: 141-147, 1976.

22. Nevill, A.M. Adjusting (scaling) health related fitness variables for differences in body size and age. *J Sports Med* 4: 12-22, 1997.

23. Nevill, A.M., R. Ramsbottom, and C. Williams. Scaling physiological measurements for individuals of different body size. *Eur J Appl Physiol* 65: 110-117, 1992.

24. Ross, W.D., D.T. Drinkwater, N.O. Whittingham, and R.A. Faulkner. Anthropometric prototypes: age six to eighteen years. In Berg, K. and B.O. Eriksson, eds., *Children and exercise IX.* Baltimore: University Park Press, 1980, 3-12.

25. Rowland, T.W. The development of aerobic fitness in children. In Armstrong, N., B. Kirby, and J. Welsman, eds., *Children and exercise XIX.* London: Spon, 1997, 179-190.

26. Schmidt-Nielsen, K. *Scaling: why is animal size so important?* Cambridge: Cambridge University Press, 1984.

27. Shephard, R.J., H. Lavallée, R. Labarre, J.C. Jequier, M. Volle, and M. Rajic. On data standardization in prepubescent children. In Ostyn, M., G. Beunen, and J. Simons, eds., *Proc. 2nd Int. Seminars Kinanthropometry.* Basel: Karger, 1979.

28. Sjodin, B., and J. Svedenhag. Oxygen uptake during running as related to body mass in circumpubertal boys: a longitudinal study. *Eur J Appl Physiol Occup Physiol* 65: 150-157, 1992.

29. Sprynarová, S., and R. Reisenauer. Body dimensions and physiological indicators of physical fitness during adolescence. In Shephard, R.J. and H. Lavallée, eds., *Physical fitness assessment*. Springfield, IL: Charles C Thomas, 1978, 32-37.

30. Tanner, J.M. Fallacy of per-weight and per-surface area standards, and their relation to spurious correlation. *J Appl Physiol* 2: 1-15, 1949.

31. Taylor, C.R., N.C. Heglund, and G.M.O. Maloiy. Energetics and mechanics of terrestrial locomotion. I. Metabolic energy consumption as a function of speed and body size in birds and mammals. *J Exp Biol* 97: 1-21, 1982.

32. Von Döbeln, W., and B.O. Eriksson. Physical training, maximal oxygen uptake and dimensions of the oxygen transporting and metabolizing organs in boys 11 to 13 years of age. *Acta Paediatr Scand* 61: 653-660, 1972.

33. Welsman, J. Interpreting young people's exercise performance: sizing up the problem. In Armstrong, N., B. Kirby, and J. Welsman, eds., *Children and exercise XIX*. London: Spon, 1997, 191-203.

34. Welsman, J.R., and N. Armstrong. Interpreting exercise performance data in relation to body size. In Armstrong, N. and W. van Mechelen, eds., *Paediatric exercise science and medicine*. Oxford: Oxford University Press, 3-9, 2000.

35. Welsman, J.R., N. Armstrong, A.M. Nevill, E.M. Winter, and B.J. Kirby. Scaling peak $\dot{V}O_2$ for differences in body size. *Med Sci Sports Exerc* 28: 259-265, 1996.

36. Winter, E.M. Scaling: partitioning out differences in size. *Pediatr Exerc Sci* 4: 296-301, 1992.

37. Winter, E.M. Importance and principles of scaling for size differences. In Bar-Or, O., ed., *The child and adolescent athlete*. Oxford: Blackwell Scientific, 673-679, 1996.

INDEX

Note: The italicized *f* and *t* following page numbers refer to figures and tables, respectively.

ABOUT THE AUTHORS

Oded Bar-Or, MD, is a professor of pediatrics and founder and director of the Children's Exercise and Nutrition Centre at McMaster University in Hamilton, Canada. For more than 35 years he has conducted research focused on the effects of physical activity and inactivity on the health, well-being, and physical performance of healthy children and those with chronic diseases. He received his MD degree from the Hebrew University in Jerusalem, Israel.

Dr. Bar-Or served as president of the Canadian Association of Sports Sciences, president of the International Council for Physical Fitness Research, and vice president of the American College of Sports Medicine (ACSM). He chairs the Foundation for Active Healthy Kids.

A widely published author, he earned the ACSM's Citation Award in 1997 and the North American Society for Pediatric Medicine's Honor Award in 1998. In 2000, he received an honorary doctorate from the University of Blaise Pascal in France.

Thomas Rowland, MD, is director of pediatric cardiology at the Baystate Medical Center in Springfield, Massachusetts, where he established an exercise-testing laboratory. He is a pediatric cardiologist with extensive research experience in the exercise physiology of children.

Dr. Rowland is author of *Developmental Exercise Physiology* (1996) and *Pediatric Laboratory Exercise Testing: Clinical Guidelines* (1993) and editor of *Pediatric Exercise Science*. He is a former president of the North American Society for Pediatric Exercise Medicine (NASPEM) and a former member of the board of trustees of the American College of Sports Medicine (ACSM). He is a former president of the New England Chapter of the ACSM and received the Honor Award in 1993. Dr. Rowland received BS and MD degrees from the University of Michigan in 1965 and 1969.

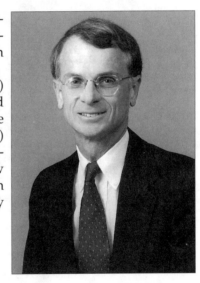